AF213438

PROTEIN KINASES

Edited by **Gabriela Da Silva Xavier**

Protein Kinases
Edited by Gabriela Da Silva Xavier

Published by InTech
Janeza Trdine 9, 51000 Rijeka, Croatia

Publishing Process Manager Romina Skomersic

Technical Editor Teodora Smiljanic

Cover Designer InTech Design Team

Additional hard copies can be obtained from orders@intechopen.com

Protein Kinases, Edited by Gabriela Da Silva Xavier
p. cm.
ISBN 978-953-51-0640-1

For EU product safety concerns:
IN TECH d.o.o., Prolaz Marije Krucifikse Kozulić 3, 51000 Rijeka, Croatia,
info@intechopen.com or visit our website at intechopen.com.

Meet the editor

My research interest is in the fuel sensing protein kinases - AMP-activated protein kinase (AMPK) and PAS-domain containing protein kinase (PASK)- and their roles in pancreatic endocrine cell function and development. I completed my PhD in Biochemistry at the University of Bristol in 2001. After two postdoctoral positions at the University of Bristol, I moved to Imperial College in 2006 where I was appointed lecturer in 2008 and have been involved in collaborative efforts to elucidate how the Type 2 Diabetes genes identified by Genome Wide Association Studies - TCF7L2, SLC30A8, WFS1, and HHEX - may affect beta-cell function and development. My work has been funded by the Juvenile Diabetes Research Foundation and European Foundation for the Study of Diabetes.

Contents

Preface

Proteins are the work horses of the cell. As regulators of protein function, protein kinases are involved in the control of cellular functions via intricate signalling pathways, allowing for fine tuning of physiological functions. This book is a collaborative effort, with contribution from experts in their respective fields, reflecting the spirit of collaboration - across disciplines and borders - that exists in modern science. Here, we review the existing literature and, on occasions, provide novel data on the function of protein kinases in various systems. We also discuss the implications of these findings in the context of disease, treatment, and drug development.

Gabriela Da Silva Xavier
Non-Clinical Lecturer in Cell Biology/Physiology, Department of Medicine,
Faculty of Medicine, Imperial College London,
UK

1

The Target of Rapamycin: Structure and Functions

Chang-Chih Wu, Po-Chien Chou and Estela Jacinto*
Department of Physiology and Biophysics, UMDNJ-Robert Wood Johnson Medical School, Piscataway, Piscataway, NJ USA

1. Introduction

The target of rapamycin (TOR, also called the mechanistic or mammalian target of rapamycin, mTOR) is an atypical protein kinase that is highly conserved in eukaryotes (Sarbassov et al. 2005; Wullschleger et al. 2006; Jacinto & Lorberg 2008). It modulates cell growth, metabolism, and cell survival in response to diverse extracellular and intracellular signals, such as growth factors, energy levels, and nutrient status (Reiling & Sabatini 2006; Wullschleger et al. 2006; Jacinto 2008). Inhibition of mTOR activity using rapamycin and more recently via mTOR active site inhibitors and disruption of mTOR complexes, has revealed important insights on how mTOR functions under physiological and pathological conditions (Sarbassov et al. 2005; Proud 2011; Zoncu et al. 2011).

TOR was first identified as the target of rapamycin, a potent antifungal macrolide originally purified from *Streptomyces hygroscopicus* in an Easter Island soil sample in 1975 (Sehgal et al. 1975; Vezina et al. 1975). This natural compound was later found to possess immunosuppressive and growth inhibitory properties on mammalian cells (Hall 1996; Thomas & Hall 1997; Young & Nickerson-Nutter 2005). A genetic screen in the budding yeast, *Saccharomyces cerevisiae*, identified three genes that conferred rapamycin resistance upon mutation. These genes include *TOR1, TOR2*, and *FPR1* (Heitman et al. 1991). Whereas TOR1 and TOR2 are relatively large proteins (around 300 kDa) and display homology to lipid kinases, FPR1 (also called FKBP12) is a small protein (about 12 kDa) that has cis-trans prolyl isomerase activity (Helliwell et al. 1994; Kunz et al. 2000). The activity of TOR/mTOR becomes inhibited by the complex formed by rapamycin and FKBP12. TOR orthologues were also discovered in mammalian cells (mTOR) and other higher eukaryotes (eg *C. elegans* TOR, CeTOR; *Drosophila* TOR, dTOR; *Arabidopsis thaliana* TOR, At TOR) (Brown et al. 1994; Oldham et al. 2000; Long et al. 2002; Menand et al. 2002). In this chapter, we will review the conserved structures of TOR, the regulation of the mTOR pathway, and summarize its conserved cellular functions. We also discuss the value of targeting mTOR function in therapeutic strategies.

2. Structure, conserved versus divergent sequences in TOR

TOR/mTOR is encoded by a single gene in most organisms although in some yeasts there are two *TOR* genes. The encoded proteins share about 40~60% identity in amino acid sequence among different species (Wullschleger et al. 2006). TOR belongs to the phosphatidylinositol-3 kinase-related kinase (PIKK) family, a subgroup of atypical protein kinases (Hanks & Hunter 1995; Manning et al. 2002; Miranda-Saavedra & Barton 2007). PIKKs are conserved from yeasts to mammals and have numerous functions in stress responses including DNA repair, transcription, and mRNA decay (Keith & Schreiber 1995). The PIKKs share some homology in the catalytic domain with lipid kinases including phosphatidylinositol-3 kinases (PI3Ks) (Keith & Schreiber 1995; Manning et al. 2002), but they possess serine/threonine kinase activity (Figure 1). The large size of PIKK family members (from 280 to 470 kDa) has been a major obstacle in studying the structure of these molecules (Knutson 2010). In general, these kinases are roughly defined by an α-helical N-terminal region and a catalytic C-terminal region (Choi et al. 1996; Lempiainen & Halazonetis 2009). From the structural prediction of amino acid sequences, PIKK family members contain HEAT (Huntington, elongation factor 3, alpha-regulatory subunit of protein phosphatase 2A and TOR1) repeats at the amino-terminus, FAT (FRAP-ATM-TRRAP) domain, kinase domain (KD), the PIKK-regulatory domain (PRD) and the FAT-C-

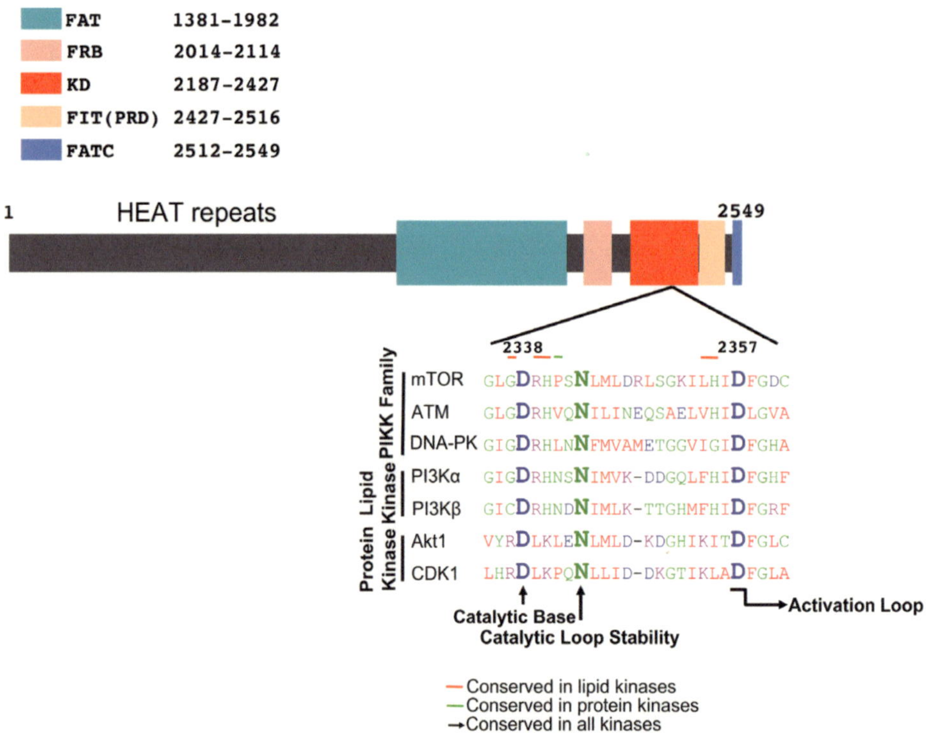

Fig. 1. Structural domains of mTOR and conserved amino acid sequences between lipid and protein kinases. Numbers indicate the residues in mTOR. FAT, FRAP-ATM-TRRAP domain; FRB, FKBP12-rapamycin-binding domain; KD, kinase domain; FIT, found in TOR domain; PRD, PIKK-regulatory domain; FATC, FAT-C-terminal domain; HEAT repeat, the protein domain found in **H**untingtin, **E**longation factor 3, protein phosphatase 2**A**, and **T**OR1; mTOR, mammalian target of rapamycin; ATM, Ataxia telangiectasia mutated; DNA-PK, DNA-dependent protein kinase; PI3K, Phosphatidylinositol 3-kinases; Akt, protein kinase B; CDK1, Cyclin-dependent kinase 1.

terminal (FATC) domain, at the carboxyl-terminus (Bosotti et al. 2000; Jacinto 2008; Hardt et al. 2011). The configuration of these protein motifs contributes to the catalytic activity and function of PIKKs, including TORs (Bosotti et al. 2000; Adami et al. 2007; Yip et al. 2010). Understanding the regulation of these motifs is key to unravelling the cellular function of TOR/mTOR and will provide insights on how we can manipulate the activity of this protein.

2.1 HEAT repeats

The amino-terminus of mTOR is characterized by HEAT repeats. This structural motif contains varied numbers of two anti-parallel α-helix repeats that are linked by inter-unit loops allowing flexibility in this structure (Perry & Kleckner 2003). A recent sequence analysis strategy revealed that there are 30 to 32 tandem HEAT repeats predicted in human, fly, plant, and yeast TORs (Knutson 2010). HEAT domains confer a curved-tubular shape, facilitating multiple protein-protein interactions in the N-terminal half of mTOR (Groves & Barford 1999; Adami et al. 2007). Accordingly, this region is shown to provide a platform for protein-protein interaction, where mTOR can bind with protein regulators or substrates. In the budding yeast, the TORC components, KOG1 (Kontroller Of Growth 1), AVO1 (Adheres VOraciously 1), and AVO3 (Adheres VOraciously 3), associate with TOR at the HEAT region of the N-terminus (Wullschleger et al. 2005; Adami et al. 2007; Yip et al. 2010). Similarly, raptor, the orthologue of KOG1 in mammalian cells, also interacts with the N-terminal region of mTOR and that the intact structure of the HEAT domain in mTOR is essential for this interaction (Kim et al. 2002; Adami et al. 2007; Yip et al. 2010).

2.2 FAT domain and FRB domain

FAT domain is a hallmark of the PIKK family (Bosotti et al. 2000). In mTOR and other PIKK members, the FAT domain is adjacent to the N-terminal portion of the catalytic region (Lempiainen & Halazonetis 2009; Hardt et al. 2011). Although the overall structure of FAT domains in mTOR is still unclear, it is suggested that this domain is composed entirely of α-helices according to sequence analysis and can be viewed as an extension of HEAT repeats. This would suggest that it serves as a platform for protein interaction as well (Perry & Kleckner 2003; Adami et al. 2007). Furthermore, the FAT domain may associate with another domain (FATC) to wedge the KD into a proper configuration and ensure the catalytic activity of mTOR (Bosotti et al. 2000).

The C-terminus of the FAT domain is where rapamycin, in complex with FKBP12, binds mTOR (Stan et al. 1994; Chen et al. 1995). This small structural motif, consisting of around 100 amino acids, has been termed as the FKBP12-rapamycin-binding (FRB) (Veverka et al. 2008). Structural and biochemical analyses have revealed that this motif can also bind to phosphatidic acid (PA), a lipid secondary messenger (Fang et al. 2001; Veverka et al. 2008). Upon mitogen stimulation, the level of PA is enhanced due to the activation of phospholipase D (English 1996). Although it is unclear whether PA can activate mTOR directly, it is suggested that PA may direct the membrane localization of mTOR (Veverka et al. 2008).

2.3 FATC domain

FATC, a protein motif containing around 30 amino acids, is conserved with high sequence similarity among the members of the PIKK family (Bosotti et al. 2000; Dames et al. 2005). In these kinases, the FATC domain is at the end of the C-terminal tail and exists in combination with the FAT domain to flank the kinase domain (Bosotti et al. 2000; Jacinto & Lorberg 2008) (Figure 1). Low-resolution structure of this domain in yeast TOR1, visualized by electron microscopy, suggests that this motif protrudes from the catalytic core domain (Adami et al. 2007; Lempiainen & Halazonetis 2009). However, it is also predicted that the attachment of FATC to the KD is required for the proper conformation and activation of the latter. Structural studies utilizing NMR spectroscopy uncover that the FATC domain contains an α-helix followed by a sharp turn, which is stabilized by a disulfide bond between two cysteine residues (Dames et al. 2005). The substitution of cysteine with serine increases the flexibility of FATC and leads to a lower expression level of TOR2 in budding yeast (Dames et al. 2005). Moreover, other mutagenesis assays have implied that the substitution and deletion of the hydrophobic residues in this domain abolish the autophosphorylation of mTOR and the mTOR-dependent phosphorylation of eukaryotic initiation factor 4E-binding protein (4E-BP) and p70 S6 kinase (S6K) (Peterson et al. 2000). Together, these studies indicate that the FATC domain modulates both kinase activity and stability of TOR.

2.4 PRD (FIT)

The PRD domain is a newly identified motif situated between the kinase and FATC domains in PIKK family members (Mordes et al. 2008). Unlike other C-terminal domains, this domain is not conserved in all PIKKs and its length varies between 16 to 82 amino acids (Mordes et al. 2008). In TOR, this region is also named the Found in TOR (FIT) domain (Sturgill & Hall 2009; Hardt et al. 2011). Its N-terminal half, which shows almost no sequence homology with other PIKKs, is defined as a suppressor of TOR activity. The deletion of residues 2430 to 2450 of rat mTOR enhances kinase activity of both mTOR and its downstream targets, *in vitro* and *in vivo* respectively (Brunn et al. 1997; Sekulic et al. 2000). The mitogen- and nutrient-induced post-translational phosphorylation on several residues in the FIT domain, *eg* Thr 2446, Ser 2448, and Ser 2481 is used as a marker of mTOR activation (Chiang & Abraham 2005; Holz & Blenis 2005; Copp et al. 2009). Thus, the phosphorylation in this region could relieve the suppressive action conferred by this domain.

2.5 Catalytic domain and kinase activity of mTOR

In the classification of eukaryotic protein kinases, mTOR belongs to the atypical group, a subset of protein kinases lacking sequence similarity to conventional protein kinases (Hanks & Hunter 1995) . In fact, the catalytic sequence of mTOR shares high homology with PI3K family, a lipid kinase family, but mTOR has been experimentally demonstrated to possess Serine/Threonine protein kinase activity (Alarcon et al. 1999). The segment from residues Lys^{2187} to Phe^{2421} comprises the catalytic region of mTOR (Hardt et al. 2011). Within this domain, two conserved structures are proposed to contribute to the kinase activity of mTOR: the catalytic loop, which contains the predicted catalytic base (Asp^{2338}) in the triplet DRH residues (Hardt et al. 2011), and the activation loop (also called T loop), consisting of twenty to thirty amino acids that connects the N- and C-lobes of the kinase domain (Lochhead 2009). By general definition, activation loop begins with the DFG sequence and ends with the PE sequence (Figure 1). In conventional protein kinases, there is one phosphorylatable Ser, Thr, or Tyr residue existing within the region and the phosphorylation at this residue is required for kinase activation. In most cases, either the kinase itself (autophosphorylation) or an upstream kinase mediates the phosphorylation of this site. However, in atypical protein kinases, such as mTOR, this phosphorylatable residue is substituted by an Asp or Glu, which mimics the phosphorylated state of the T loop. Hence, instead of the conventional phosphorylation-activation mechanism in the T loop, other cis- and trans-acting mechanisms modulate mTOR activity. As discussed above, the phosphorylation at the PRD could promote mTOR activity. mTOR could also be potentially regulated via FATC domain stability. The formation of disulfide bonds may stabilize the FATC structure, or even the whole mTOR protein (Sarbassov & Sabatini 2005). Due to this unique structural feature, it has been proposed that the FATC domain could act as a redox-sensor to regulate mTOR activity (Dames et al. 2005). In this model, the presence of nutrients could enhance mitochondria metabolism that alters the intracellular redox environment. Although it remains to be examined, this redox change could be sensed by the FATC domain and confer a conformation switch in mTOR thereby altering its activity in response to intracellular stimuli (Dames et al. 2005). Other regulatory mechanisms such as association with regulatory partners and subcellular localization are discussed in Section 4.

3. mTOR protein complexes

Forming protein complexes is a common and efficient way to acquire different functional modules in a spatial and temporal manner (Hartwell et al. 1999; Pereira-Leal et al. 2006). As described in the previous section, the multiple motif conformation and superhelical structure of mTOR enables this protein to associate with diverse cofactors. Early studies using gel filtration chromatography suggested that TOR could be part of multi-protein complexes (Yang & Guan 2007). This was supported by findings that TOR has a rapamycin-insensitive function in yeast, implying that two complexes could perform distinct functions (Zheng, 1995; Schmidt, 1996; Zheng, 1997). Indeed, in a number of organisms with perhaps the exception of plants and algae, there are two structurally and functionally distinct TOR complexes (Loewith et al. 2002; Wedaman et al. 2003; Matsuo et al. 2007; Diaz-Troya et al. 2008). In this section, both conserved and non-conserved components of TOR complexes will be discussed.

3.1 mTORC1

3.1.1 Raptor (KOG1)

KOG1 was first identified to co-purify with TOR1 in budding yeast (Loewith et al. 2002). In mammals, the KOG1 orthologue, raptor, was found to associate with mTOR (Hara et al. 2002; Kim et al. 2002). It is predicted that the C-terminal half of KOG1 and raptor consists of four HEAT repeats and seven WD40 repeats (Loewith et al. 2002; Wedaman et al. 2003; Adami et al. 2007; Yip et al. 2010). Due to their motif configuration, it is speculated that KOG1 and raptor function as scaffold proteins to facilitate the association between TORC1 (mTORC1) and downstream substrates (Loewith et al. 2002; Adami et al. 2007). Deletion of *kog1* in budding yeast and knockout of *raptor* in mice both led to lethality, implying that these genes are essential for normal development and cellular functions (Loewith et al. 2002; Murakami et al. 2004; Guertin et al. 2006).

In mammalian cells, raptor binds to mTOR in a rapamycin-sensitive or nutrient-responsive manner (Kim et al. 2002; Oshiro et al. 2004). This association promotes mTORC1 activity towards its substrates such as S6K and 4E-BP1 in response to insulin and nutrients (Kim et al. 2002). Presence of rapamycin/FKBP12 complex diminishes the association between raptor and mTOR, which could explain how rapamycin can inhibit mTORC1 function (Kim et al. 2002; Kim et al. 2003; Oshiro et al. 2004).

3.1.2 mLST8 (LST8)

LST8 is a 34 kDa protein composed of seven WD40 repeats (Chen & Kaiser 2003). Originally, LST8 was identified because of its function in the translocation of amino acid permease GAP1 from Golgi to cell surface (Roberg et al. 1997; Liu et al. 2001). Later, LST8 (Wat1 in fission yeast) was found in both TORC1 and TORC2 complexes (Chen & Kaiser 2003; Alvarez & Moreno 2006; Matsuo et al. 2007). Specifically, LST8 binds to the catalytic domain in the C-terminal region of TOR2 and regulates its kinase activity. In addition, it has been shown that LST8 is required for TORC2 complex integrity (Wullschleger et al. 2005). Similarly, mammalian LST8 (mLST8, also known as GβL) was first reported to interact with raptor and mTOR in a nutrient- and rapamycin- sensitive manner (Kim et al. 2003). mLST8 is only required for mTORC2 functions in the early development of mice (Guertin et al. 2006). Knockout of mLST8 in mice revealed that it is required for mTORC2 function but not for the mTORC1 function in S6K phosphorylation (Guertin et al. 2006).

3.2 mTORC2

3.2.1 Rictor (AVO3)

AVO3, a 164-kDa protein, is a conserved subunit of TORC2 in budding yeast (Loewith et al. 2002). It was first identified as a suppressor of sphingolipid biosynthesis mutants in a genetic screen (Dunn et al. 1998). The presence of AVO3 is required for the integrity of rapamycin-insensitive TOR complex but it is dispensable for the *in vitro* kinase activity of TOR2 (Wullschleger et al. 2005). Therefore, AVO3 is suggested to play a role in recruiting TORC2 substrates (Ho et al. 2008).

Rictor (rapamycin-insensitive companion of mTOR) is the mammalian orthologue of yeast AVO3 and is part of mTORC2 (Jacinto et al. 2004; Sarbassov et al. 2004). It lacks common or known structural motifs but its C-terminus is conserved among vertebrates. Knockdown or ablation of rictor in mammalian cells led to defective phosphorylation of several members of the AGC (protein kinase A, G, and C) kinase family, including Akt, SGK1 and PKC, decreased cell survival upon stress induction, and impaired reorganization of actin cytoskeleton (Jacinto et al. 2004; Sarbassov et al. 2004; Guertin et al. 2006; Shiota et al. 2006; Garcia-Martinez & Alessi 2008). In mouse models, rictor knockout is embryonic lethal and the rictor$^{-/-}$ MEFs (mouse embryonic fibroblasts) isolated from rictor null embryos display slower growth rate compared to wild type MEFs (Guertin et al. 2006; Shiota et al. 2006). Substitution of Gly^{934} in rictor prevented formation of rictor/SIN1 heterodimer and reduced mTORC2 activity (Aimbetov et al. 2011). Thus, the interaction between rictor and SIN1 is required to form an integral and active mTORC2.

3.2.2 SIN1 (AVO1)

AVO1 is another TORC2 component in budding yeast and binds to the N-terminus of TOR2. The depletion of AVO1 mimics the defective actin polarization phenotype observed in the *tor2* mutant strain (Loewith et al. 2002). Sin1, the orthologue of AVO1 in fission yeast, was first identified as a stress-responsive protein that interacts with Sty1/Spc1 mitogen-activated protein (MAP) kinase, a member of yeast stress-activated MAP kinase (SAPK) family (Wilkinson et al. 1999; Yang et al. 2006). Mammalian SIN1 is also implicated in the JNK (c-Jun N-terminal kinase) and MAPK (mitogen-activated protein kinase)/ERK (extracellular-regulated-protein kinase) pathways (Cheng et al. 2005; Schroder et al. 2005). It was later identified as a critical subunit of mTORC2 (Frias et al. 2006; Jacinto et al. 2006; Yang et al. 2006). To date, more than five alternatively spliced isoforms of mammalian SIN1 have been discovered (Schroder et al. 2004; Cheng et al. 2005). Three of these isoforms form distinct rapamycin-insensitive mTOR complexes with rictor and mTOR (Frias et al. 2006). SIN1 disruption affects both mTORC2 assembly and function. Loss of SIN1 is embryonic lethal, indicating an important role for this protein in development (Jacinto et al. 2006; Yang et al. 2006).

3.3 Other interactors

In addition to the main components of TOR complexes discussed above, there are many non-conserved proteins that associate with TOR/mTOR. Some of these mTORC interactors can affect mTOR activity. These mTORC-interacting molecules could also mediate crosstalk between the mTOR pathway and other signaling pathways (Woo et al. 2007).

In budding yeast, TCO89 (TOR complex one 89 kDa subunit) has been shown to associate with TORC1 (Reinke et al. 2004). Deletion of TOR1 and TCO89 results in rapamycin hypersensitivity and defective cell-wall integrity, respectively. AVO2 and BIT61 also associate with TORC2 but their roles in regulating TORC2 functions remain to be elucidated (Loewith et al. 2002; Reinke et al. 2004). BIT61 can associate with SLM1 and SLM2, which are also TORC2-associated proteins mediating actin cytoskeleton organization (Fadri et al. 2005).

PRAS40 (proline-rich Akt substrate of 40kDa) is a negative regulator of mTORC1 (Sancak et al. 2007; Wang et al. 2007). PRAS40 and mTORC1 substrates, such as 4E-BP-1 and S6K, share a similar raptor-binding motif, the TOR signaling (TOS) motif (Wang et al. 2007). Therefore, it is speculated that PRAS40 can directly bind to raptor and interfere with the ability of mTORC1 to interact with its substrates (Wang et al. 2007). This negative regulation of mTORC1 by PRAS40 is inhibited by the insulin signaling pathway, since activated Akt can phosphorylate PRAS40 and prevent its binding to raptor (Sancak et al. 2007).

PRR5 (proline-rich protein 5), also named Protor (protein observed with rictor), and PRR5L (PRR5-like) bind to rictor and non-essential subunits of mTORC2 (Pearce et al. 2007; Thedieck et al. 2007; Woo et al. 2007). Knockdown of these two proteins did not cause significant disruption of both complex integrity and kinase activity of mTORC2. Protor1 is required for the phosphorylation of SGK1, but not of Akt and PKCα, specifically in mouse kidney (Pearce et al. 2011). These findings suggest that this non-conserved interactor might regulate mTORC2 function in a tissue- and target- specific manner (Pearce et al. 2011). DEPTOR (DEPDC6, DEP domain-containing protein 6), is a negative regulator of mTOR that associates with both mTORC1 and mTORC2 (Peterson et al. 2009; Proud 2009). Loss of DEPTOR activates S6K1 and Akt, downstream substrates of mTORC1 and mTORC2, respectively. In most cancer cell lines DEPTOR expression is low, except for a subset of multiple myelomas harboring cyclin D1/D3 or c-MAF/MAFB translocations. The high DEPTOR levels in these cells are required for the activation of PI3K/Akt pathway and may suppress apoptosis (Peterson et al. 2009). Several studies have characterized how DEPTOR levels can be controlled (Duan et al. 2011; Gao et al. 2011; Zhao et al. 2011). DEPTOR is recognized and ubiquitinated by an F box protein, SCF (βTrCP) and degraded through the 26S-proteasome pathway (Zhao et al. 2011). Either expressing the dominant-negative mutant of βTrCP or interfering with the interaction between DEPTOR and βTrCP via mutagenesis causes the accumulation of DEPTOR and downregulation of mTOR activity (Zhao et al. 2011). Furthermore, mTORC1 and mTORC2 could directly phosphorylate DEPTOR (Gao et al. 2011). CK1α (casein kinase 1α) can generate a phosphodegron on the phosphorylated DEPTOR, which is bound by βTrCP to induce the degradation of DEPTOR (Duan et al. 2011; Gao et al. 2011). The degron mutant and βTrCP deletion can inhibit DEPTOR degradation and decrease mTOR activities (Gao et al. 2011). Together, these studies suggest that DEPTOR can regulate mTORC activity via a positive feedback loop involving mTOR itself and CK1.

4. Mode of regulation of mTOR complexes

mTOR serves to relay signals from growth cues to downstream events to consequently control cell growth and metabolism (Wullschleger et al. 2006; Zhou & Huang 2010). Below, we discuss how these growth cues alter mTOR activity via regulation of mTOR complex component modification, subcellular localization, and association with other regulatory molecules.

4.1 Phosphorylation

mTOR itself is regulated via phosphorylation. Ser^{1261} phosphorylation of mTOR is induced by insulin stimulation and is required for mTORC1 activity and mTOR autophosphorylation (Acosta-Jaquez et al. 2009). Furthermore, mTOR in the context of intact mTORC1 is predominantly phosphorylated at the Ser^{2448} residue (Copp et al. 2009). This site is phosphorylated by S6K in a mitogen- and nutrient-inducible manner (Chiang & Abraham 2005). The autophosphorylation site at Ser^{2481} is also growth-signal dependent (Peterson et al. 2000). A later report proposed that the Ser^{2481} phosphorylation event is an indicator of functional mTORC2 (Copp et al. 2009). Prolonged but not acute rapamycin treatment, which disrupts mTORC2 (Sarbassov et al. 2006), can abolish mTOR phosphorylation at this site (Copp et al. 2009). However, mTOR from raptor immunoprecipitates is also phosphorylated at Ser^{2481}. Furthermore, inhibition of mTORC1 by acute rapamycin treatment can reduce Ser^{2481} phosphorylation of mTOR that is associated with raptor (Soliman et al. 2010), implying that the phosphorylation of Ser^{2481} residue may also be involved in the regulation of mTORC1 functions. Thus, how Ser^{2481} phosphorylation affects the specific activity of mTORC1 vs mTORC2 would need to be clarified.

mTOR complex components are also phosphorylated at numerous sites. Phosphorylation of raptor at different residues may affect the kinase activity of mTOR. For example, AMPK mediates phosphorylation of raptor at $Ser^{722/792}$ upon nutrient depletion and inhibits mTORC1 function (Gwinn et al. 2008). In contrast, upon mitogen stimulation, p90 ribosomal S6 kinase (RSK) and mTORC1 mediate raptor phosphorylation at $Ser^{719/721/722}$ and Ser^{863}, respectively, which is essential for mTORC1 activation (Carriere et al. 2008). Rictor is predicted to be phosphorylated in at least 37 phosphorylation sites according to MS/MS analysis and phospho-proteome database (Dibble et al. 2009; Julien et al. 2010). These putative phosphorylation sites mainly localize in the C-terminal region of rictor, which is conserved only in vertebrates (Dibble et al. 2009; Julien et al. 2010). Thus, rictor could have acquired more diverse functions and complex regulation during evolution . Several studies have examined the function of rictor phosphorylation at Thr^{1135} residue located in the C-terminal region. This phosphorylation is mediated by S6K1 in an amino acid- and growth factor- dependent manner and is suggested to act as a feedback regulation of mTORC2 from mTORC1 signals (Dibble et al. 2009; Julien et al. 2010; Treins et al. 2010). Its effect on the mTORC2-mediated Akt activation is very minimal if any (Boulbes et al. 2010; Treins et al. 2010). Moreover, in $SIN1^{-/-}$ MEFs, in which mTORC2 complex integrity is disrupted, this phosphorylation is still detectable, suggesting that it might be involved in mTORC2-independent functions (Boulbes et al. 2010). Rictor is also phosphorylated at Ser^{1235} by GSK3β under ER stress conditions (McDonald et al. 2008). This phosphorylation event reduces the binding between mTORC2 and its substrate, Akt, hence negatively regulating mTORC2.

SIN1 can also be phosphorylated at multiple sites although the relevant sites remain to be identified. Hypophosphorylation of SIN1 interferes with its association with mTOR (Yang et al. 2006), but not with rictor (Rosner & Hengstschlager 2008). mTOR can phosphorylate SIN1 *in vitro*, which may prevent SIN1 degradation from lysosomal pathway *in vivo* (Chen & Sarbassov 2011). Other kinases that can phosphorylate SIN1 to regulate mTORC2 activity would need to be investigated.

4.2 Component stability and complex formation

The activity and specificity of mTOR can be modulated through complex assembly. Disruption of mTOR complexes via gene ablation or knockdown of a specific mTORC component has revealed the importance of an intact mTORC for phosphorylation of its downstream substrates. For instance, in the adipose-specific *raptor* knockout mice, which carry disrupted mTORC1, S6K phosphorylation in white adipose tissue was diminished (Boulbes et al. 2010). Similarly, the deletion of either SIN1 or rictor in MEFs and HeLa cells inhibited mTORC2 assembly and abolished Akt HM and TM phosphorylation (Guertin et al. 2006; Jacinto et al. 2006; Yang et al. 2006).

In mammalian cells, raptor binds to mTOR in a nutrient- responsive manner (Kim et al. 2002; Oshiro et al. 2004). Upon nutrient deprivation, raptor and mTOR form a stable interaction, which can inhibit mTORC1 activity (Kim et al. 2002). Under growth favorable conditions, the association between raptor and mTOR is less tight and presumably can promote mTORC1 activity towards its substrates such as S6K and 4E-BP (Kim et al. 2002). Instead of affecting the intrinsic kinase activity of mTOR, rapamycin/FKBP12 complex is proposed to attenuate the association between mTOR and raptor, thereby inhibiting mTORC1 (Chen et al. 1995; Choi et al. 1996; Kim et al. 2002). Supporting this model, mTOR purified from rapamycin-treated cells showed no defect of its autophosphorylation ability *in vivo* or kinase activity toward substrates *in vitro* (Peterson et al. 2000). In addition, recent studies revealed that not all mTORC1 functions can be inhibited by rapamycin, suggesting that it may only affect access to some substrates (Choo et al. 2008; Dowling et al. 2010).

While the FKBP12/rapamycin complex binds and inhibits mTORC1, it does not affect mTORC2 activity acutely perhaps because it does not bind to mTORC2 (Jacinto et al. 2004). However, chronic exposure to rapamycin could disrupt mTORC2 function in some cell lines presumably by blocking assembly of newly synthesized mTORC2 subunits (Sarbassov et al. 2006).

The integrity of mTORC2 is dependent on stability of the rictor/SIN1 heterodimer. These mTORC2 components interact tightly and deficiency in either one leads to destabilization of the other, suggesting they require each other for stability (Guertin et al. 2006; Jacinto et al. 2006; Yang et al. 2006). Other proteins associating with mTORC components have been identified that could affect mTORC activity or assembly. The folding chaperone Hsp70 interacts with rictor and its knockdown reduces rictor level as well as mTOR-rictor interaction, resulting in impaired mTORC2 formation and activity (Martin et al. 2008; Martin et al. 2008). The maturation and assembly of mTORCs was also shown to be dependent on Tel2 and Tti (Takai et al. 2007; Kaizuka et al. 2010 a; Kaizuka et al. 2010 b). Hsp90 was shown to mediate the formation of both TORCs, as well as other PIKKs (Horejsi et al. 2010; Takai et al. 2010). Whether mTORC signaling can be modulated by these interactors remains to be examined.

mTORC2 components have been found to associate with other proteins independently of mTOR. Rictor can form an E3 ligase complex with Cullin-1 and Rbx1 to promote the ubiquitination of SGK1 in an mTOR-independent manner (Gao et al. 2010). The interaction between rictor, Cullin-1, and Rbx1 is disrupted when rictor is phosphorylated at Thr^{1135} residue by multiple AGC kinases (Gao et al. 2010). Whether the phosphorylated rictor released from the E3 ligase complexes can affect mTORC2 assembly remains to be elucidated. SIN1 also interacts with other proteins independently of mTOR and rictor. The function of SIN1 when associated with these proteins remains unclear but these proteins are involved in stress responses including ras, MEKK2, JNK, p38, ATF2 and the stress-related cytokine receptors IFNAR2, TNFR1/2 (Schroder et al. 2005; Makino et al. 2006; Schroder et al. 2007; Ghosh et al. 2008). Unlike mTORC2 components, so far, the mTORC1 subunit raptor has not been found to associate with other proteins in an mTORC1-independent manner.

4.3 Localization of mTOR complexes

Compartmental localization enables a protein kinase to gain access to its regulators or effectors, thereby regulating its function. Supporting this concept, yeast TOR undergoes nuclear localization and binds to the 35S rDNA promoter to enhance 35S rRNA synthesis in a nutrient-sensitive fashion (Li et al. 2006). The HTH (helix-turn-helix) motif, a region in the HEAT domain of TOR, has been demonstrated to be essential for this association since *HTH* deletion interrupts the binding of TOR1 to 35S rDNA (Li et al. 2006). TOR1 and TOR2 also localize to membrane compartments that contain actin cytoskeleton and endocytosis regulators via their HEAT domain (Kunz et al. 2000; Aronova et al. 2007). Two endoplasmic reticulum (ER) and Golgi localization sequences were characterized in the HEAT domain, supporting that the TOR complexes may localize at the membrane periphery of these organelles (Liu & Zheng 2007). The best example for how mTORC1 can be regulated via localization is the recent finding that amino acid stimulation induces shuttling of mTORC1 to late endosomes and lysosomes (LELs) by interaction with Rag GTPases (Sancak et al.

2008). How the presence of amino acids would be sensed by mTORC1 interactors in a particular organelle such as the endosomes remain to be further elucidated.

Less is known on how mTORC2 can be compartmentalized and activated in response to growth signals. It co-localizes predominantly in the ER periphery (Boulbes et al. 2011). mTORC2 associates with actively translating ribosomes and specifically interacts with the proteins from the 80S large ribosomal subunit (Oh et al. 2010; Zinzalla et al. 2011). mTORC2 components can stably interact with ribosomal proteins that line the tunnel exit of the 80S and could function in this site by modifying emerging nascent polypeptides such as Akt (Oh et al. 2010). mTORC2 becomes activated upon association with ribosomes although the precise mechanism is currently unclear (Zinzalla et al. 2011). Since the protein synthesis machinery can physically associate with cell surface receptors (Tcherkezian et al. 2010), it can be speculated that mTORC2 could be activated upon nucleation of translation machinery with a signaling receptor in the membrane periphery.

5. Upstream regulators of mTOR complexes

5.1 Activation of mTORC1

mTORC1 integrates signals from extracellular and intracellular sources of nutrients with other growth, energy and mitogenic cues (Figure 2). mTORC1 activity is sensitive to intracellular amino acid concentration, particularly leucine depletion (Hay & Sonenberg 2004). Previously, intracellular amino acid levels were proposed to modulate mTORC1 activity via TSC1/2 (tuberous sclerosis complex 1/2) (Gao et al. 2002). However, in $TSC2^{-/-}$ cells, attenuated mTORC1 activity upon amino acid withdrawal was still detected (Smith et al. 2005), indicating the presence of additional pathways regulating mTORC1 activity in response to amino acid levels. Furthermore, hVPS34, a class III PI3K, can signal amino acid availability to mTORC1 bypassing the TSC1/2-Rheb axis (Byfield et al. 2005; Nobukuni et al. 2005). Recently, in both yeast and mammals, TORC1/mTORC1 activity has been shown to be regulated by Rag GTPases in response to amino acids. In yeast, Rag orthologues Gtr1 and Gtr2, as part of the EGO complex (consisting of EGO1, EGO3, Gtr1 and Gtr2) localize to the vacuolar/lysosomal membrane and could mediate amino acid signals to TORC1 (Dubouloz et al. 2005; Binda et al. 2009). Mammalian Rags function as heterodimers, wherein Rag A/B associates with Rag C/D (Sekiguchi et al. 2001). A protein complex termed the Ragulator consisting of MAPK scaffold protein 1, p14, p18 localizes Rag to these membrane compartments. In nutrient replete conditions, Rag complexes are fully activated as Rag A/B in GTP form and Rag C/D in GDP form (Sekiguchi et al. 2001). Activation of Rag heterodimers recruits mTORC1 to the membrane compartment where Rheb (Ras homolog enriched in the brain) is enriched, thus promoting mTORC1 activity (Rubio-Texeira & Kaiser 2006; Meijer & Codogno 2008; Sancak et al. 2008). p62, an adaptor protein that associates with mTORC1 and Rag complex was shown to be required for mTORC1 activation (Duran et al. 2011). It was proposed that p62 can promote the localization of mTORC1 and Rag complex to the lysosomes, which is a critical step of mTORC1 activation, in an amino-acid dependent manner (Duran et al. 2011). Rheb is a small GTPase belonging to the Ras family (Garami et al. 2003; Manning & Cantley 2003; Long et al. 2005). Upon binding to mTOR, GTP-loaded Rheb induces a conformational change in the KD of mTOR to promote its activation (Long et al. 2005). In the lysosomes, the amino acid transporter PAT1 (proton-assisted amino acid transporter 1; SLC36A1) is abundant and has been described as a lysosomal amino acid transporter (Russnak et al. 2001). This transporter could play a role in mTORC1 activation in an amino acid-dependent manner (Heublein et al. 2010). Knockdown of *PAT1* in MCF-7 cells led to decreased S6K and 4E-BP-1 phosphorylation (Heublein et al. 2010). Two membrane transporters, SLC7A5 and SLC3A2, were also shown to be required for mTORC1 activation (Nicklin et al. 2009). These two transporters function in the cellular uptake and the subsequent efflux of glutamine in the presence of essential amino acids (Nicklin et al. 2009). This process can increase intracellular concentration of leucine and can enhance mTORC1 activation via Rag complexes (Sancak et al. 2008). Thus, several regulatory molecules in the endosome/lysosomes and membrane compartments could regulate mTORC1 function in response to the presence of amino acids.

Numerous cellular inputs that convey growth or stress conditions are coupled to mTORC1 via the TSC1/2 tumor suppressor complex (Castro et al. 2003; Garami et al. 2003; Tee et al. 2003; Hay & Sonenberg 2004; Zoncu et al. 2011). Loss of *TSC1* or *TSC2* genes, encoding hamartin and tuberin respectively, is observed in tuberous sclerosis, a human genetic disorder characterized by benign tumors (Green et al. 1994; Onda et al. 1999). Through phosphorylation at different residues, the activity of the TSC 1/2 complex can be regulated by different upstream signaling pathways (Huang & Houghton 2003). In the presence of growth factors, such as insulin, the insulin receptor is activated and recruits the insulin receptor substrate (IRS). IRS serves to couple signals to downstream molecules including PI3K, which converts phosphatidylinositol-4,5-phosphate (PIP2) into phosphatidylinositol-3,4,5-phosphate (PIP3) in the plasma membrane. PIP3 can recruit both PDK1 and Akt to the membrane, where Akt is phosphorylated and activated by PDK1. The activated Akt can directly phosphorylate TSC2 and block the activity of TSC1/2 complex (Huang & Houghton 2003), which consequently results in the activation of mTORC1 and the phosphorylation of S6K and 4E-BP1, the two best-characterized effectors of mTORC1 (Valentinis & Baserga 2001). The regulation of mTORC1 activity via Akt can be counteracted by GADD34 (growth arrest and DNA damage protein 34) (Minami et al. 2007; Watanabe et al. 2007). By interacting with TSC1/2 complexes, GADD34 inhibits phosphorylation of TSC2 at the Akt phosphorylation site Thr^{1462}, thus negatively regulating mTORC1 (Minami et al. 2007; Watanabe et al. 2007).

The mitogen activated protein kinase (MAPK) pathway can also modulate mTORC1 functions via negative regulation of TSC1/2 in response to growth cues (Ma et al. 2005). In the presence of growth factors, receptor tyrosine kinases activate Ras-Erk1/2 signaling (Pearson et al. 2001). Activated Erk can phosphorylate TSC2 at Ser^{664} and induce dissociation of TSC1/2 complex, which suppresses TSC1/2 functions toward cell proliferation and transformation (Ma et al. 2005). Active Ras can also inactivate TSC via Rsk-mediated phosphorylation of TSC2 at Ser^{1798}, resulting in increased mTORC1 signaling (Roux et al. 2004).

AMP-activated protein kinase (AMPK) pathway is another well-established regulatory input of mTORC1 in response to energy conditions. Under low cellular energy level (high AMP/ATP ratio), LKB1, a tumor suppressor, activates AMPK to directly phosphorylate TSC2 at Thr^{1227} and Ser^{1345} residues (Inoki et al. 2003; Corradetti et al. 2004; Shaw et al. 2004). Unlike Akt and Erk, AMPK-mediated phosphorylation enhances the GAP activity of TSC1/2 complex and inhibits mTORC1 function. This would provide a mechanism for downregulating energy-consuming cellular processes under low ATP levels.

5.2 Activation of mTORC2

Early studies in yeast have elucidated how TORC2 can control actin cytoskeleton polarization but the signals that control TORC2 activity has been elusive (Schmidt et al. 1996; Schmidt et al. 1997; Loewith et al. 2002). Since this function of TORC2 is viewed to control growth spatially, it is reasonable to speculate that TORC2 is activated by growth signals. In mammals, mTORC2 activity is promoted by growth factors in a PI3K-dependent manner. After insulin stimulation, the phosphorylation of Akt at its hydrophobic motif (HM; Ser^{473}) mediated by mTORC2 is significantly increased *in vivo* (Hresko & Mueckler 2005; Sarbassov et al. 2005). Similar results were also observed in *in vitro* kinase assays (Sarbassov et al. 2005). Since mTORC2 also phosphorylates sites in Akt and PKC in a constitutive manner (Facchinetti et al. 2008; Ikenoue et al. 2008), its responsiveness to growth factors has been puzzling. However, further studies revealed that mTORC2 could phosphorylate the constitutive site in Akt during translation (Oh et al. 2010). This would suggest that conditions that enhance translation, such as the presence of growth factors, promote mTORC2 activity. Supporting this notion, mTORC2 was found to associate with translating ribosomes and this association is enhanced in cells with increased PI3K signaling (Oh et al. 2010; Zinzalla et al. 2011). Since nutrients are essential for promoting protein synthesis, this would suggest that mTORC2 would also be regulated by nutrients. However, it remains to be determined if this is the case and how mTORC2 becomes activated upon association with ribosomes.

6. mTOR functions

A number of diverse functions have been ascribed to mTOR but it is now emerging that it is a central hub to regulate growth and metabolism (Figure 2). In the whole organism, mTOR is required during early development. Rapamycin treatment inhibited amino acid stimulation of embryo outgrowth in mice at blastocyst stage (Martin & Sutherland 2001). Similarly, the embryonic development of *mTOR*$^{-/-}$ mice is aberrant and arrested at E5.5 (Gangloff et al. 2004). Conditional knockout strategies have shed light on tissue-specific mTOR functions. The inactivation of mTOR in T cells caused defects in differentiation of peripheral T lymphocytes (Delgoffe et al. 2009). The muscle-specific mTOR knockout mice displayed reduced levels of dystrophin and severe myopathy, which led to premature death (Risson et al. 2009). While there was no significant phenotype in mice with specific deletion of *mTOR* in the prostate, tumor initiation and progression in PTEN-/- mice, which possess hyperactivated mTORC1, was suppressed (Nardella et al. 2009). These findings support a critical role for mTOR itself in organism and organ development as well as disease progression.

The cellular functions of mTOR have surfaced from numerous studies over the years. The most well characterized function of mTOR is the rapamycin-sensitive control of protein synthesis by mTORC1. More recent studies have revealed that mTOR has rapamycin-insensitive functions both as part of mTORC1 and mTORC2. Most of the mTOR functions that have been elucidated are mediated by the AGC kinases S6K, Akt, SGK, and PKC. Another well characterized mTOR substrate is 4E-BP1, a translation regulator. With the recent phospho-proteomic studies that identified hundreds of direct and indirect targets of mTOR, we can begin to elucidate how the mTOR complexes could perform its wide array of cellular functions (Hsu et al. 2011; Yu et al. 2011).

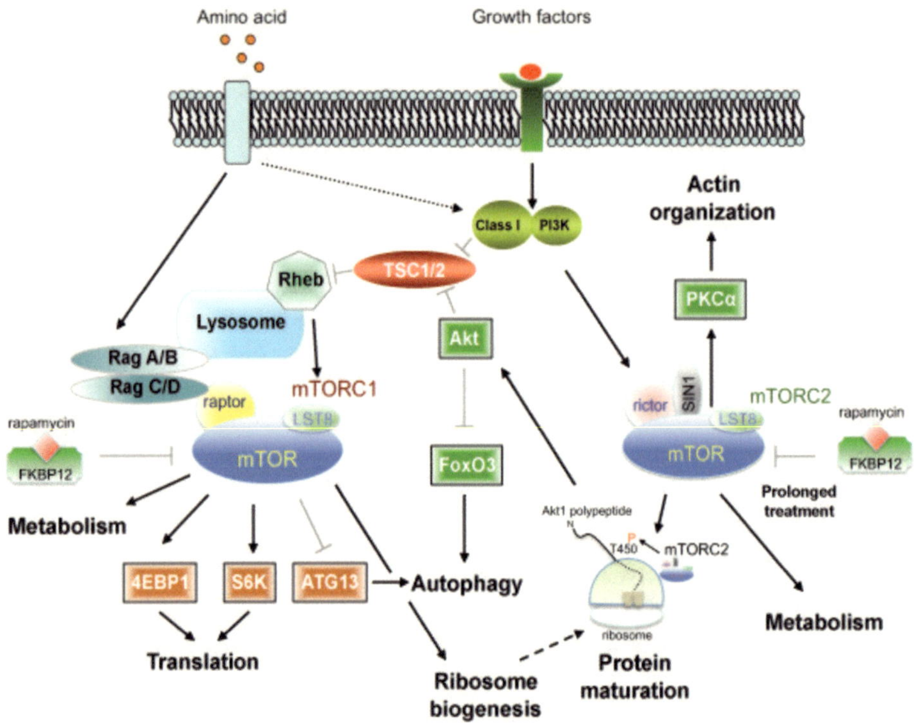

Fig. 2. Overview of mTOR complex signaling and functions. Black arrows and gray lines represent activating and inhibitory connections, respectively. Dotted lines indicate possible links.

6.1 Protein synthesis and maturation

Protein synthesis, ie translation, utilizes huge amounts of energy and machinery (ie. ribosomes), and consequently, the entire process is under tight control (Acker & Lorsch 2008; Malys & McCarthy 2011). mTOR, primarily mTORC1, is involved in different aspects of protein synthesis (Ma & Blenis 2009; Sonenberg & Hinnebusch 2009). Its most well characterized function is the phosphorylation of proteins involved in translation initiation, namely S6K1 and 4E-BP. 4E-BP is a negative regulator of translation initiation (Pause et al. 1994). When unphosphorylated, it interacts with eIF-4E (eukaryotic initiation factor 4E) cap-binding protein, preventing cap-dependent translation. mTORC1-mediated phosphorylation dissociates 4E-BP1 from eIF-4F and releases the inhibition of translation (Ma & Blenis 2009; Sonenberg & Hinnebusch 2009). However, there are four phosphorylation sites on 4E-BP1 that are regulated by mTORC1 but only two of them are sensitive to acute rapamycin treatment (Gingras et al. 1999; Gingras et al. 2001; Wang et al. 2003). In the presence of growth signals, translation of proteins specifically involved in cell proliferation is upregulated in a 4E-BP-dependent fashion (Dowling et al. 2010). This would suggest that mTORC1 can specifically regulate the translation of a subset of mRNA under a particular growth condition.

mTORC1 can directly phosphorylate and activate S6K1, which promotes mRNA translation by modulating multiple substrates involved in different stages of translation, from mRNA surveillance, initiation to translation elongation (Ma & Blenis 2009). Phosphorylated S6K1 binds to newly spliced mRNAs through its binding partner, SKAR (S6K1 aly/REF-like target) and potentially facilitate translation initiation and/or elongation (Ma et al. 2008). This protein complex interacts with EJC (exon junction complex), which monitors the quality of newly spliced mRNA (Ma et al. 2008). Active S6K1 also phosphorylates the 40S ribosomal protein S6 (rpS6) at several sites. Phosphorylated rpS6 is widely used as a readout of mTORC1 activity. Its phosphorylation appears to play a role in the control of cell size but is dispensable for translation of mRNA with 5′ terminal oligopyrimidine tract (TOP mRNAs) (Ruvinsky et al. 2005). S6K also phosphorylates PDCD4 (programmed cell death 4), an inhibitor of RNA helicase eIF4A. PDCD4 levels become downregulated upon phosphorylation (Dorrello et al. 2006). The degradation of PDCD4 greatly enhances eIF4A helicase activity and facilitates 40S ribosomal subunit scanning to the initiation codon. Moreover, S6K can also augment eIF-4A activity through increasing the levels of eIF-4A, eIF-4B, and eIF3 complex. The S6K-mediated phosphorylation of eIF4B can enhance its binding to eIF3 (Vornlocher et al. 1999). Thus, S6K can modulate a number of proteins involved in translation initiation. It could also regulate the elongation phase of translation. Phosphorylation of eEF2K (eukaryotic elongation factor 2 kinase) by S6K inhibits its activity towards eEF2 and consequently enhances elongation (Wang et al. 2001).

In addition to the regulation of the translation process, mTORC1 directly affects the biosynthesis of the translational machinery as well. The assembly of functional ribosomes is an energy-demanding process requiring a series of building components and assembly factors (Mayer & Grummt 2006). mTORC1 has been found to regulate ribosome biogenesis at different levels, including the production of ribosomal proteins (Cardenas et al. 1999; Hardwick et al. 1999), pre-rRNA processing (Powers & Walter 1999), and the rRNA synthesis (Hardwick et al. 1999; Hannan et al. 2003). In response to extracellular conditions, mTORC1 can coordinate all three nuclear RNA polymerases, Pol I, Pol II and Pol III, to control ribosome synthesis (Beck & Hall 1999; Miller et al. 2001; Yuan et al. 2002; Martin et al. 2004; White 2005). mTORC1 regulates the nuclear translocation of TIF1A, a transcription factor that is essential for Pol I-associated transcription initiation (Mayer et al. 2004). Furthermore, mTORC1 activity can enhance the tRNA levels via regulation of the transcription of Pol III (Shor et al. 2010). Maf1, a Pol III suppressor, directly associates with and inhibits Pol III apparatus (Reina et al. 2006). Under growth-favorable conditions, mTORC1 phosphorylates Maf1 to dissociate it from Pol III and promotes its cytoplasmic translocation (Shor et al. 2010). mTORC1 also interacts with rDNA (ribosomal DNA) promoters to promote Pol I and Pol III transcription in a growth factor-dependent and rapamycin-sensitive manner (Tsang et al. 2010).

mTORC2 is also emerging to play a role in translation. A more pronounced defect in protein synthesis and polysome assembly occurs upon mTOR inhibition with active site inhibitors, in contrast to rapamycin treatment (Yu et al. 2009; Carayol et al. 2010; Oh et al. 2010; Evangelisti et al. 2011). Although it can be argued that the exacerbated defects could be due to inhibition of rapamycin-insensitive mTORC1 functions, there is some evidence that mTORC2 inhibition could contribute to these defects. First, polysome recovery is somewhat defective in mTORC2-disrupted cells (Oh et al. 2010; Wu and Jacinto, unpublished results). Second, phosphorylation of eEF2 is aberrant in these cells. Most importantly, mTORC2 associates with actively translating ribosomes and SIN1 deficiency disengages mTOR or rictor from the ribosomes (Oh et al. 2010; Zinzalla et al. 2011). mTORC2 components can stably associate with ribosomal proteins that are present at the tunnel exit. This would be consistent with the finding that mTORC2 can cotranslationally phosphorylate the emerging nascent Akt polypeptide (Oh et al. 2010). Thus, mTORC2 could function in cotranslational maturation of newly synthesized proteins by phosphorylating relevant sites. The maturation of conventional PKC is also mediated by mTORC2 via phosphorylation (Facchinetti et al. 2008; Ikenoue et al. 2008).It would be interesting to see if mTORC2 could also cotranslationally phosphorylate PKC and whether it has additional cotranslational targets other than Akt.

6.2 Autophagy

Autophagy is a catabolic process that recycles intracellular components in the lysosomes to salvage substrates for energy production when nutrients become limiting (Noda, 1998; Janku, 2011). The control of this process by TOR was first discovered in yeast. In yeast, active TORC1 correlates with hyperphosphorylation of Atg13, a regulatory component of the Atg1 complex. Under this condition, assembly of the Atg1-Atg13 complex is inhibited thereby preventing autophagy (Yorimitsu et al. 2009; Kamada et al. 2010; Kijanska et al. 2010; Yeh et al. 2010). Increased autophagy is observed upon rapamycin treatment, supporting the role of TORC1 in regulating this cellular process (Kamada et al. 2000).

Under nutrient-replete conditions, mTOR negatively regulates autophagy by interacting with a protein complex composed of ULK1 (UNC51 like kinase), Atg13, and FIP200 (Ganley et al. 2009; Hosokawa et al. 2009; Jung et al. 2009). This complex is involved in the formation of the autophagosomes. The phosphorylation of ULK1 and Atg13 is inhibited by rapamycin treatment and leucine deprivation, implying the link between mTORC1 and autophagy (Hosokawa et al. 2009; Jung et al. 2009).

mTOR is also involved in regulating expression of genes that are involved in autophagosome or lysosome biogenesis. mTORC1 could mediate phosphorylation of the transcription factor EB (TFEB), thereby controlling its nuclear shuttling and activity. TFEB binds E box related DNA sequences and controls lysosomal gene transcription (Pena-Llopis et al. 2011; Settembre et al. 2011). mTORC2 could also be involved in the control of autophagy via Akt-FoxO signaling (Mammucari et al. 2007). Disruption of rictor enhanced autophagosome formation in skeletal muscle and that this effect was abrogated upon expression of constitutively active Akt in the absence or presence of rapamycin, indicating that in this cell type, autophagy is dependent on mTORC2 function, not mTORC1 (Mammucari et al. 2007). Indeed, attenuated Akt activity caused by downregulation of mTORC2 leads to decreased phosphorylation and increased nuclear translocation of FoxO3 (Shiojima & Walsh 2006). FoxO3 induces the expression of many autophagy-related genes, such as *Atg12l* and *Ulk2*, and increases autophagosome formation in isolated adult mouse muscle fibers (Zhao et al. 2007). Together, these results indicate that by directly interacting with autophagic proteins or modulating the transcription of autophagy-related genes, mTOR complexes regulate autophagy.

6.3 Metabolism

Early studies showing reduced fungal amino acid, nucleic acid, and lipid metabolism after rapamycin treatment have linked TOR to metabolic functions (Singh et al. 1979). Later transcription profiling screening of lymphoma cells treated with rapamycin revealed a tendency towards catabolism and that the levels of many mRNA involved in lipid, nucleotide, and protein synthesis were downregulated (Peng et al. 2002). In the whole organism, the function of mTOR in metabolism is underscored by findings on the effect of rapamycin treatment on insulin-responsive tissues. Inhibition of mTORC1 in mice via feeding with rapamycin induced diabetes due to smaller pancreatic islets and abolished insulin secretion (Bussiere et al. 2006). This effect could be mediated via S6K and S6 since these two mTORC1 pathway effectors are also required for the normal morphology and function of pancreatic islet cells and that removal of these two proteins led to a diabetic phenotype (Ruvinsky et al. 2005). Corollary to this, hyperactivation of mTORC1 occuring in *TSC* knockout resulted in larger islet size and higher number of β-cell (Rachdi et al. 2008). Together, these findings support a role for mTORC1 in maintaining metabolic homeostasis.

Recent studies, discussed below, that have employed mTORC component gene ablation or knockdown further demonstrate the central role of mTOR in cellular and systemic metabolism. The critical role of mTORC1 in cellular metabolism is illustrated by its involvement in the biogenesis of mitochondria. In muscle-specific *raptor* knockout mice, there are reduced levels of PGC1α (PPARγ coactivator 1), which is required for mitochondrial gene expression (Cunningham et al. 2007). In these mice, the skeletal muscle has lower mitochondria number, reduced oxidative capacity, and elevated glycogen storage, which led to muscle dystrophy (Bentzinger et al. 2008). Moreover, by genomic analysis, YY1 (yin-yang 1), a transcription factor that regulates mitochondrial gene expression and oxygen consumption, was identified as a downstream effector of mTORC1 (Cunningham et al. 2007). Both mTOR and raptor can bind to YY1 while rapamycin treatment inhibited YY1 activity by preventing its interaction with the coactivator, PGC1α (Cunningham et al. 2007). Rapamycin treatment of or knockdown of mTOR or raptor in muscle cells also reduced mitochondrial gene transcription and respiratory metabolism (Cunningham et al. 2007). Interestingly, this function of mTORC1 appears to be S6K1-independent. In line with these findings, enhanced muscular levels of PGC1α and mitochondria were found in *S6K1* knockout mice (Um et al. 2004). Thus, mTORC1, via YY1 and PGC1α, could regulate mitochondrial gene expression and thereby enhance mitochondrial oxidative functions.

In addition to skeletal muscle, raptor knockout in other insulin-responsive tissues further illustrate the role of mTORC1 in metabolism. In adipose-specific *raptor* knockout mice, adipose tissue was reduced and these mice were protected against diet-induced obesity and hypercholesterolemia. There was elevated expression of genes involved in mitochondrial respiration in white adipose tissue and the leanness of these mice could be explained by enhanced energy expenditure due to mitochondrial uncoupling (Polak et al. 2008). These mice also display higher glucose tolerance and insulin sensitivity. This could be due to defective S6K feedback regulation of IRS-1 activity, causing hyperactivated insulin receptor signaling (Polak et al. 2008). These findings underscore the role of adipose mTORC1 in whole body energy homeostasis.

In liver, inhibition of mTORC1 promotes hepatic ketogenesis in response to fasting (Sengupta et al. 2010). Active mTORC1 negatively regulates PPARα (peroxisome proliferator activated receptor α), the master transcriptional activator of ketogenic gene expression, through control of its corepressor, NCoR1 (nuclear receptor corepressor 1) (Sengupta et al. 2010).

Thus, mice with hyperactivation of mTORC1 upon loss of *TSC1* in the liver manifest defects in ketone body production and enlarged liver size during fasting (Sengupta et al. 2010).

Other effectors of mTORC1 in the control of metabolic processes have emerged in recent reports. mTORC1 can modulate sterol and lipid biosynthesis through SREBP-1 (sterol regulatory element binding protein-1), a transcription factor that controls lipo- and sterolgenic gene transcription (Porstmann et al. 2008; Duvel et al. 2010). The mTORC1 target S6K1 can partially promote the activity of SREBP-1 via posttranslational modification. More recently, one critical regulator of mTORC1-SREBP-1 pathway has been identified. Lipin 1, which is directly phosphorylated and sequestered in the cytoplasm by mTORC1, induces the translocation of SREBP-1 into cytoplasm and negatively regulates its activity as a transcription factor (Peterson et al. 2011). Under high-fat and –cholesterol diet, mTORC1 activity is required for SREBP-1 function to promote fat accumulation and hypercholesterolemia in mice (Peterson et al. 2011). mTORC1 can also modulate the expression level of Hif1α (hypoxia-inducible factor α), which activates numerous hypoxia-induced genes involved in cellular metabolic processes including those involved in glycolysis and glucose uptake (Goldberg et al. 1988; Brugarolas et al. 2003; Duvel et al. 2010).

There is also accumulating evidence that mTORC2 is required in metabolic processes. Knockdown of *rictor* in MEFs diminished metabolic activity (Shiota et al. 2006). Furthermore, deficiency of rictor in Jurkat cells, a leukemic T cell line, increased oxygen consumption (Schieke et al. 2006). However, mTORC2 could play a more complex function in mitochondrial metabolism since a PTEN-deficient cell line that is mTORC2 addicted/IL3-independent was shown to require a number of genes involved in mitochondrial functions (Colombi et al. 2011). Adipose-specific knockout of rictor in mice revealed that mTORC2 can function to control whole body growth (Cybulski et al. 2009). In these mice, there was increased size of non-adipose organs, such as heart, kidney, spleen, and pancreas (Cybulski et al. 2009). In addition, these mice also displayed hyperinsulinemia and elevated levels of IGF (insuline-like growth factor) and IGFBP3 (IGF binding protein 3) (Cybulski et al. 2009). Conversely, the deletion of rictor in pancreatic β-cells decreased their proliferation and mass, which led to reduced insulin secretion, hyperglycemia, and glucose intolerance in mice (Gu et al. 2011). Specific effectors of mTORC2 function in metabolism remain to be characterized. Future investigation should reveal how mTORC1 and mTORC2 signaling pathways impinge on metabolic pathways. This would be important in light of understanding how defects in cellular metabolism that occurs in cancer and other pathological conditions are linked to aberrant mTOR signaling.

6.4 Actin cytoskeleton reorganization

The regulation of actin cytoskeleton reorganization is a conserved function of mTORC2. In *S. cerevisiae, tor2* mutations or depletion of TORC2 components depolarizes the actin cytoskeleton (Schmidt et al. 1996; Loewith et al. 2002). Normal polarization of actin towards the growing bud controls spatial growth in yeast. In mammals, mTORC2 can control actin polymerization and cell spreading via Rho and Rac, members of the Rho family of GTPases that regulates F-actin assembly (Jacinto et al. 2004). Rex1, a Rac GEF, links mTOR signaling to Rac activation and regulates cell migration (Hernandez-Negrete et al. 2007). PKCα, which is phosphorylated by mTORC2, is also linked to actin cytoskeleton reorganization in mammalian cells (Sarbassov et al. 2004). However, it remains unclear how the mTORC2-

dependent phosphorylation of PKCα can promote the actin reorganization function of this AGC kinase.

Studies using rapamycin have also linked actin cytoskeleton reorganization to the mTORC1 pathway. Rapamycin inhibits the reorganization of F-actin and the phosphorylation of focal adhesion proteins through S6K1 (Berven et al. 2004; Liu et al. 2008). S6K1 is localized to actin stress fibers in fibroblasts (Crouch 1997) and S6K1, Akt, PDK1, and PI3K colocalized with the actin arc, a caveolin-enriched cytoskeletal structure located at the leading edge of migrating Swiss 3T3 cells (Berven et al. 2004). The mTORC1 pathway is also linked to cell motility and migration. Rapamycin inhibits cell motility in several cell types, such as neutrophils (Gomez-Cambronero 2003), vascular smooth muscle cells (Poon et al. 1996), and T-lymphocytes (Finlay & Cantrell 2010). The mTORC1 target, S6K, mediates cell migration via its regulation of focal adhesion formation (Liu et al. 2008), reorganization of F-actin (Berven et al. 2004; Liu et al. 2008), as well as the upregulation of the matrix metalloproteinase 9 (MM9) (Zhou & Wong 2006), and the activity and expression of RhoA (Liu et al. 2010). How mTORC1 or mTORC2 can more directly regulate its effectors in actin cytoskeleton reorganization remains to be examined. Future studies on how the mTORC-mediated function in actin cytoskeleton reorganization is coupled to the other growth-regulatory functions of mTORCs would need to be addressed.

7. mTOR inhibitors and therapeutic significance

Due to the central role of mTOR in cell survival, growth and proliferation, deregulation of the mTOR signaling pathway is implicated in many human diseases including benign and malignant tumors, neurological and metabolic disorders and cardiovascular diseases (Pei & Hugon 2008; Krymskaya & Goncharova 2009; Hwang & Kim 2011; Ibraghimov-Beskrovnaya & Natoli 2011). Moreover, its role in organismal aging is highlighted by findings that inhibition of the TOR/mTOR pathway can prolong lifespan in several organisms (Vellai et al. 2003; Kapahi et al. 2004; Medvedik et al. 2007; Harrison et al. 2009). Drawing lessons from rapamycin, numerous mTOR inhibitors have been developed and are currently being refined to achieve more specific inhibition. We discuss some recent findings on the use of these inhibitors at the bench and in the clinic.

Rapamycin and its analogs (rapalogs) are allosteric mTOR inhibitors and form a complex with FKBP12 and mTOR. By binding at the FRB domain of mTOR, the interaction between mTOR and raptor is diminished and can uncouple mTOR from its substrates (Oshiro et al. 2004). Rapamycin and derivatives such as CCI-779 (temsirolimus, Torisel), RAD001 (everolimus), and AP23573 (ridaforolimus) act as cytostatic agents that slow down or arrest the growth of cells derived from several cancer types such as rhabdomyosarcoma (Hosoi et al. 1999), prostate cancer (van der Poel et al. 2003), breast cancer (Pang & Faber 2001), and B-cell lymphoma (Muthukkumar et al. 1995). Early results from clinical trials reveal that they have antiproliferative activity in a subset of cancer, such as endometrial cancer (Oza et al. 2011), pancreatic neuroendocrine tumors (Goldstein & Meyer 2011), gastric cancer (Doi et al. 2010), and malignant glioma (Reardon et al. 2011). However, only a subset of mTORC1 functions is sensitive to rapamycin treatment, hence the antiproliferative properties of this drug can be limited (Wang et al. 2005; Choo et al. 2008; Choo & Blenis 2009). In a number of cell types, the inhibition of mTORC1 results in the upregulation of the PI3K/Akt pathway. Normally, mTORC1 activates S6K1 and the active S6K1 negatively regulates the insulin receptor substrate-1 (IRS-1) by phosphorylation at serine residues (Zhande et al. 2002; Shah & Hunter 2006; Tzatsos & Kandror 2006). The inhibition of mTORC1 by rapalogs disrupts this feedback loop and results in increased IRS1 signaling and Akt activity that may compromise the anti-tumor activity of mTOR inhibitors (Harrington et al. 2004; Shah et al. 2004; Sun et al. 2005; O'Reilly et al. 2006). Since mTORC2 is a positive regulator of Akt, several new mTOR inhibitors that can block both mTORC1 and mTORC2 have been developed to more effectively inhibit mTOR signaling and cell proliferation.

The pyrazolopyrimidine analogs PP242 and PP30 are ATP-competitive inhibitors of mTOR that bind to its ATP-binding site and as a result, block the kinase activities of both mTORC1 and mTORC2 (Feldman et al. 2009). PP242 and PP30 both inhibit the mTORC2-induced phosphorylation of Akt at Ser473, indicating that these inhibitors can indeed interfere with mTORC2 functions. Furthermore, these two mTOR kinase domain inhibitors attenuate protein synthesis and proliferation of mouse embryonic fibroblasts (MEFs) (Feldman et al. 2009). PP242 has been shown to induce cytoreduction and apoptosis in multiple myeloma cells (Hoang et al. 2010) and cause death of mouse and human leukemia cells and delay leukemia onset *in vivo* (Janes et al. 2010), highlighting the potential therapeutic application of this compound. Another ATP-competitive mTOR inhibitor, Torin1, induces cell cycle arrest and inhibits cell growth and proliferation more efficiently than rapamycin (Thoreen et al. 2009). Preclinical studies substantiate the therapeutic value of Torin1. For example, Torin1 treatment prevented the anti-inflammatory potency of glucocorticoids both in human monocytes and myeloid dendritic cells (Weichhart et al. 2011). Moreover, Torin1 significantly inhibited the translation of viral proteins during human cytomegalovirus infection (Clippinger et al. 2011). Recently, Torin2, a novel mTOR inhibitor with improved pharmacokinetic properties and synthetic route, has been described (Liu et al. 2011). Torin2 inhibits mTOR complexes with IC_{50} of 0.25 nM, compared to Torin1 at IC_{50} of 2 to 10 nM. Therefore, Torin2 is suggested to be a more potent and stable mTORC inhibitor than Torin1.

Since mTORC1 inhibition leads to upregulation of IRS1 and subsequently PI3K, which in turn activates Akt, simultaneously blocking the activities of mTOR complexes and PI3K may inhibit cell proliferation and growth more effectively. As a result, several mTOR/PI3K dual inhibitors have been developed. NVP-BEZ235 inhibits PI3K and mTOR kinase activity by interacting with their ATP-binding domains (Maira et al. 2008). It has been implicated in the treatment of non-small cell lung cancer (Konstantinidou et al. 2009), melanoma (Marone et al. 2009), pancreatic cancer (Cao et al. 2009), and acute myeloid leukemia (Chapuis et al. 2010). Given the compensatory mechanisms that cells employ to adapt to growth-inhibitory conditions, it would be important to identify signaling pathways that impinge on the mTOR/PI3K pathway in order to develop combinatorial therapy for preventing malignancy.

8. Conclusion

Two decades after the discovery of TOR/mTOR, the function of this protein in orchestrating cellular processes in response to growth signals particularly nutrients has been established. Some key discoveries in the last few years that allowed more extensive analysis of mTOR function and regulation include identification of the mTOR complexes and regulatory proteins that link the mTOR pathway to nutrient and energy responses and the development of mTOR inhibitors. Studies to dissect the systemic function of mTOR complexes are also gaining momentum with the use of tissue-specific mTORC component knockout mice. There are still numerous outstanding questions that need to be addressed such as the precise regulation of mTOR complexes by nutrients, how it links signals from nutrients to cellular metabolism and other processes, and distinct functions and regulation of mTORCs in different cellular compartments. The recent identification of the myriad possible direct and indirect targets of the mTORCs should provide clues on the mechanisms involved in mTOR functions. Animal models would also provide insights on the role of mTOR in physiological and pathological conditions. Finally, development of specific mTORC1 and mTORC2 inhibitors would not only be useful to determine the distinct functions of these complexes but also would have numerous clinical applications.

9. Acknowledgements

We thank Won Jun Oh for comments on this manuscript and acknowledge support from the NIH (GM079176), American Cancer Society (RSG0721601TBE), Cancer Research Institute, and SU2C-AACR-IRG0311 (E.J.).

10. References

Acker, M. G.&Lorsch, J. R. (2008). "Mechanism of ribosomal subunit joining during eukaryotic translation initiation." *Biochemical Society transactions* 36(Pt 4): 653-657.

Acosta-Jaquez, H. A.; Keller, J. A.; Foster, K. G.; Ekim, B.; Soliman, G. A.; Feener, E. P.; Ballif, B. A.&Fingar, D. C. (2009). "Site-specific mTOR phosphorylation promotes mTORC1-mediated signaling and cell growth." *Molecular and cellular biology* 29(15): 4308-4324.

Adami, A.; Garcia-Alvarez, B.; Arias-Palomo, E.; Barford, D.&Llorca, O. (2007). "Structure of TOR and its complex with KOG1." *Molecular cell* 27(3): 509-516.

Aimbetov, R.; Chen, C. H.; Bulgakova, O.; Abetov, D.; Bissenbaev, A. K.; Bersimbaev, R. I.&Sarbassov, D. D. (2011). "Integrity of mTORC2 is dependent on the rictor Gly-934 site." *Oncogene.*

Alarcon, C. M.; Heitman, J.&Cardenas, M. E. (1999). "Protein kinase activity and identification of a toxic effector domain of the target of rapamycin TOR proteins in yeast." *Molecular biology of the cell* 10(8): 2531-2546.

Alvarez, B.&Moreno, S. (2006). "Fission yeast Tor2 promotes cell growth and represses cell differentiation." *Journal of cell science* 119(Pt 21): 4475-4485.

Aronova, S.; Wedaman, K.; Anderson, S.; Yates, J., 3rd&Powers, T. (2007). "Probing the membrane environment of the TOR kinases reveals functional interactions between TORC1, actin, and membrane trafficking in Saccharomyces cerevisiae." *Molecular biology of the cell* 18(8): 2779-2794.

Beck, T.&Hall, M. N. (1999). "The TOR signalling pathway controls nuclear localization of nutrient-regulated transcription factors." *Nature* 402(6762): 689-692.

Bentzinger, C. F.; Romanino, K.; Cloetta, D.; Lin, S.; Mascarenhas, J. B.; Oliveri, F.; Xia, J.; Casanova, E.; Costa, C. F.; Brink, M.; Zorzato, F.; Hall, M. N.&Ruegg, M. A. (2008). "Skeletal muscle-specific ablation of raptor, but not of rictor, causes metabolic changes and results in muscle dystrophy." *Cell metabolism* 8(5): 411-424.

Berven, L. A.; Willard, F. S.&Crouch, M. F. (2004). "Role of the p70(S6K) pathway in regulating the actin cytoskeleton and cell migration." *Exp Cell Res* 296(2): 183-195.

Binda, M.; Peli-Gulli, M. P.; Bonfils, G.; Panchaud, N.; Urban, J.; Sturgill, T. W.; Loewith, R.&De Virgilio, C. (2009). "The Vam6 GEF controls TORC1 by activating the EGO complex." *Molecular cell* 35(5): 563-573.

Bosotti, R.; Isacchi, A.&Sonnhammer, E. L. (2000). "FAT: a novel domain in PIK-related kinases." *Trends in biochemical sciences* 25(5): 225-227.

Boulbes, D.; Chen, C. H.; Shaikenov, T.; Agarwal, N. K.; Peterson, T. R.; Addona, T. A.; Keshishian, H.; Carr, S. A.; Magnuson, M. A.; Sabatini, D. M.&Sarbassov dos, D. (2010). "Rictor phosphorylation on the Thr-1135 site does not require mammalian target of rapamycin complex 2." *Molecular cancer research : MCR* 8(6): 896-906.

Boulbes, D. R.; Shaiken, T.&Sarbassov dos, D. (2011). "Endoplasmic reticulum is a main localization site of mTORC2." *Biochemical and biophysical research communications* 413(1): 46-52.

Brown, E. J.; Albers, M. W.; Shin, T. B.; Ichikawa, K.; Keith, C. T.; Lane, W. S.&Schreiber, S. L. (1994). "A mammalian protein targeted by G1-arresting rapamycin-receptor complex." *Nature* 369(6483): 756-758.

Brugarolas, J. B.; Vazquez, F.; Reddy, A.; Sellers, W. R.&Kaelin, W. G., Jr. (2003). "TSC2 regulates VEGF through mTOR-dependent and -independent pathways." *Cancer cell* 4(2): 147-158.

Brunn, G. J.; Fadden, P.; Haystead, T. A.&Lawrence, J. C., Jr. (1997). "The mammalian target of rapamycin phosphorylates sites having a (Ser/Thr)-Pro motif and is activated by antibodies to a region near its COOH terminus." *The Journal of biological chemistry* 272(51): 32547-32550.

Bussiere, C. T.; Lakey, J. R.; Shapiro, A. M.&Korbutt, G. S. (2006). "The impact of the mTOR inhibitor sirolimus on the proliferation and function of pancreatic islets and ductal cells." *Diabetologia* 49(10): 2341-2349.

Byfield, M. P.; Murray, J. T.&Backer, J. M. (2005). "hVps34 is a nutrient-regulated lipid kinase required for activation of p70 S6 kinase." *The Journal of biological chemistry* 280(38): 33076-33082.

Cao, P.; Maira, S. M.; Garcia-Echeverria, C.&Hedley, D. W. (2009). "Activity of a novel, dual PI3-kinase/mTor inhibitor NVP-BEZ235 against primary human pancreatic cancers grown as orthotopic xenografts." *Br J Cancer* 100(8): 1267-1276.

Carayol, N.; Vakana, E.; Sassano, A.; Kaur, S.; Goussetis, D. J.; Glaser, H.; Druker, B. J.; Donato, N. J.; Altman, J. K.; Barr, S.&Platanias, L. C. (2010). "Critical roles for mTORC2- and rapamycin-insensitive mTORC1-complexes in growth and survival of BCR-ABL-expressing leukemic cells." *Proceedings of the National Academy of Sciences of the United States of America* 107(28): 12469-12474.

Cardenas, M. E.; Cutler, N. S.; Lorenz, M. C.; Di Como, C. J.&Heitman, J. (1999). "The TOR signaling cascade regulates gene expression in response to nutrients." *Genes & development* 13(24): 3271-3279.

Carriere, A.; Cargnello, M.; Julien, L. A.; Gao, H.; Bonneil, E.; Thibault, P.&Roux, P. P. (2008). "Oncogenic MAPK signaling stimulates mTORC1 activity by promoting RSK-mediated raptor phosphorylation." *Current biology : CB* 18(17): 1269-1277.

Castro, A. F.; Rebhun, J. F.; Clark, G. J.&Quilliam, L. A. (2003). "Rheb binds tuberous sclerosis complex 2 (TSC2) and promotes S6 kinase activation in a rapamycin- and farnesylation-dependent manner." *The Journal of biological chemistry* 278(35): 32493-32496.

Chapuis, N.; Tamburini, J.; Green, A. S.; Vignon, C.; Bardet, V.; Neyret, A.; Pannetier, M.; Willems, L.; Park, S.; Macone, A.; Maira, S. M.; Ifrah, N.; Dreyfus, F.; Herault, O.; Lacombe, C.; Mayeux, P.&Bouscary, D. (2010). "Dual inhibition of PI3K and mTORC1/2 signaling by NVP-BEZ235 as a new therapeutic strategy for acute myeloid leukemia." *Clin Cancer Res* 16(22): 5424-5435.

Chen, C. H.&Sarbassov, D. D. (2011). "Integrity of mTORC2 is dependent on phosphorylation of SIN1 by mTOR." *The Journal of biological chemistry*.

Chen, E. J.&Kaiser, C. A. (2003). "LST8 negatively regulates amino acid biosynthesis as a component of the TOR pathway." *The Journal of cell biology* 161(2): 333-347.

Chen, J.; Zheng, X. F.; Brown, E. J.&Schreiber, S. L. (1995). "Identification of an 11-kDa FKBP12-rapamycin-binding domain within the 289-kDa FKBP12-rapamycin-associated protein and characterization of a critical serine residue." *Proceedings of the National Academy of Sciences of the United States of America* 92(11): 4947-4951.

Cheng, J.; Zhang, D.; Kim, K.; Zhao, Y.&Su, B. (2005). "Mip1, an MEKK2-interacting protein, controls MEKK2 dimerization and activation." *Molecular and cellular biology* 25(14): 5955-5964.

Chiang, G. G.&Abraham, R. T. (2005). "Phosphorylation of mammalian target of rapamycin (mTOR) at Ser-2448 is mediated by p70S6 kinase." *The Journal of biological chemistry* 280(27): 25485-25490.

Choi, J.; Chen, J.; Schreiber, S. L.&Clardy, J. (1996). "Structure of the FKBP12-rapamycin complex interacting with the binding domain of human FRAP." *Science* 273(5272): 239-242.

Choo, A. Y.&Blenis, J. (2009). "Not all substrates are treated equally: implications for mTOR, rapamycin-resistance and cancer therapy." *Cell Cycle* 8(4): 567-572.

Choo, A. Y.; Yoon, S. O.; Kim, S. G.; Roux, P. P.&Blenis, J. (2008). "Rapamycin differentially inhibits S6Ks and 4E-BP1 to mediate cell-type-specific repression of mRNA translation." *Proceedings of the National Academy of Sciences of the United States of America* 105(45): 17414-17419.

Clippinger, A. J.; Maguire, T. G.&Alwine, J. C. (2011). "The changing role of mTOR kinase in the maintenance of protein synthesis during human cytomegalovirus infection." *Journal of virology* 85(8): 3930-3939.

Colombi, M.; Molle, K. D.; Benjamin, D.; Rattenbacher-Kiser, K.; Schaefer, C.; Betz, C.; Thiemeyer, A.; Regenass, U.; Hall, M. N.&Moroni, C. (2011). "Genome-wide shRNA screen reveals increased mitochondrial dependence upon mTORC2 addiction." *Oncogene* 30(13): 1551-1565.

Copp, J.; Manning, G.&Hunter, T. (2009). "TORC-specific phosphorylation of mammalian target of rapamycin (mTOR): phospho-Ser2481 is a marker for intact mTOR signaling complex 2." *Cancer research* 69(5): 1821-1827.

Corradetti, M. N.; Inoki, K.; Bardeesy, N.; DePinho, R. A.&Guan, K. L. (2004). "Regulation of the TSC pathway by LKB1: evidence of a molecular link between tuberous sclerosis complex and Peutz-Jeghers syndrome." *Genes & development* 18(13): 1533-1538.

Crouch, M. F. (1997). "Regulation of thrombin-induced stress fibre formation in Swiss 3T3 cells by the 70-kDa S6 kinase." *Biochem Biophys Res Commun* 233(1): 193-199.

Cunningham, J. T.; Rodgers, J. T.; Arlow, D. H.; Vazquez, F.; Mootha, V. K.&Puigserver, P. (2007). "mTOR controls mitochondrial oxidative function through a YY1-PGC-1alpha transcriptional complex." *Nature* 450(7170): 736-740.

Cybulski, N.; Polak, P.; Auwerx, J.; Ruegg, M. A.&Hall, M. N. (2009). "mTOR complex 2 in adipose tissue negatively controls whole-body growth." *Proceedings of the National Academy of Sciences of the United States of America* 106(24): 9902-9907.

Dames, S. A.; Mulet, J. M.; Rathgeb-Szabo, K.; Hall, M. N.&Grzesiek, S. (2005). "The solution structure of the FATC domain of the protein kinase target of rapamycin suggests a role for redox-dependent structural and cellular stability." *The Journal of biological chemistry* 280(21): 20558-20564.

Delgoffe, G. M.; Kole, T. P.; Zheng, Y.; Zarek, P. E.; Matthews, K. L.; Xiao, B.; Worley, P. F.; Kozma, S. C.&Powell, J. D. (2009). "The mTOR kinase differentially regulates effector and regulatory T cell lineage commitment." *Immunity* 30(6): 832-844.

Diaz-Troya, S.; Perez-Perez, M. E.; Florencio, F. J.&Crespo, J. L. (2008). "The role of TOR in autophagy regulation from yeast to plants and mammals." *Autophagy* 4(7): 851-865.

Dibble, C. C.; Asara, J. M. & Manning, B. D. (2009). "Characterization of Rictor phosphorylation sites reveals direct regulation of mTOR complex 2 by S6K1." *Molecular and cellular biology* 29(21): 5657-5670.

Doi, T.; Muro, K.; Boku, N.; Yamada, Y.; Nishina, T.; Takiuchi, H.; Komatsu, Y.; Hamamoto, Y.; Ohno, N.; Fujita, Y.; Robson, M.&Ohtsu, A. (2010). "Multicenter phase II study of everolimus in patients with previously treated metastatic gastric cancer." *J Clin Oncol* 28(11): 1904-1910.

Dorrello, N. V.; Peschiaroli, A.; Guardavaccaro, D.; Colburn, N. H.; Sherman, N. E.&Pagano, M. (2006). "S6K1- and betaTRCP-mediated degradation of PDCD4 promotes protein translation and cell growth." *Science* 314(5798): 467-471.

Dowling, R. J.; Topisirovic, I.; Alain, T.; Bidinosti, M.; Fonseca, B. D.; Petroulakis, E.; Wang, X.; Larsson, O.; Selvaraj, A.; Liu, Y.; Kozma, S. C.; Thomas, G.&Sonenberg, N. (2010). "mTORC1-mediated cell proliferation, but not cell growth, controlled by the 4E-BPs." *Science* 328(5982): 1172-1176.

Duan, S.; Skaar, J. R.; Kuchay, S.; Toschi, A.; Kanarek, N.; Ben-Neriah, Y.&Pagano, M. (2011). "mTOR Generates an Auto-Amplification Loop by Triggering the betaTrCP- and CK1alpha-Dependent Degradation of DEPTOR." *Molecular cell* 44(2): 317-324.

Dubouloz, F.; Deloche, O.; Wanke, V.; Cameroni, E.&De Virgilio, C. (2005). "The TOR and EGO protein complexes orchestrate microautophagy in yeast." *Molecular cell* 19(1): 15-26.

Dunn, T. M.; Haak, D.; Monaghan, E.&Beeler, T. J. (1998). "Synthesis of monohydroxylated inositolphosphorylceramide (IPC-C) in Saccharomyces cerevisiae requires Scs7p, a protein with both a cytochrome b5-like domain and a hydroxylase/desaturase domain." *Yeast* 14(4): 311-321.

Duran, A.; Amanchy, R.; Linares, J. F.; Joshi, J.; Abu-Baker, S.; Porollo, A.; Hansen, M.; Moscat, J.&Diaz-Meco, M. T. (2011). "p62 Is a Key Regulator of Nutrient Sensing in the mTORC1 Pathway." *Molecular cell* 44(1): 134-146.

Duvel, K.; Yecies, J. L.; Menon, S.; Raman, P.; Lipovsky, A. I.; Souza, A. L.; Triantafellow, E.; Ma, Q.; Gorski, R.; Cleaver, S.; Vander Heiden, M. G.; MacKeigan, J. P.; Finan, P. M.; Clish, C. B.; Murphy, L. O.&Manning, B. D. (2010). "Activation of a metabolic gene regulatory network downstream of mTOR complex 1." *Molecular cell* 39(2): 171-183.

English, D. (1996). "Phosphatidic acid: a lipid messenger involved in intracellular and extracellular signalling." *Cellular signalling* 8(5): 341-347.

Evangelisti, C.; Ricci, F.; Tazzari, P.; Tabellini, G.; Battistelli, M.; Falcieri, E.; Chiarini, F.; Bortul, R.; Melchionda, F.; Pagliaro, P.; Pession, A.; McCubrey, J. A.&Martelli, A. M. (2011). "Targeted inhibition of mTORC1 and mTORC2 by active-site mTOR inhibitors has cytotoxic effects in T-cell acute lymphoblastic leukemia." *Leukemia : official journal of the Leukemia Society of America, Leukemia Research Fund, U.K* 25(5): 781-791.

Facchinetti, V.; Ouyang, W.; Wei, H.; Soto, N.; Lazorchak, A.; Gould, C.; Lowry, C.; Newton, A. C.; Mao, Y.; Miao, R. Q.; Sessa, W. C.; Qin, J.; Zhang, P.; Su, B.&Jacinto, E. (2008). "The mammalian target of rapamycin complex 2 controls folding and stability of Akt and protein kinase C." *The EMBO journal* 27(14): 1932-1943.

Fadri, M.; Daquinag, A.; Wang, S.; Xue, T.&Kunz, J. (2005). "The pleckstrin homology domain proteins Slm1 and Slm2 are required for actin cytoskeleton organization in yeast and bind phosphatidylinositol-4,5-bisphosphate and TORC2." *Molecular biology of the cell* 16(4): 1883-1900.

Fang, Y.; Vilella-Bach, M.; Bachmann, R.; Flanigan, A.&Chen, J. (2001). "Phosphatidic acid-mediated mitogenic activation of mTOR signaling." *Science* 294(5548): 1942-1945.

Feldman, M. E.; Apsel, B.; Uotila, A.; Loewith, R.; Knight, Z. A.; Ruggero, D.&Shokat, K. M. (2009). "Active-site inhibitors of mTOR target rapamycin-resistant outputs of mTORC1 and mTORC2." *PLoS biology* 7(2): e38.

Finlay, D.&Cantrell, D. (2010). "Phosphoinositide 3-kinase and the mammalian target of rapamycin pathways control T cell migration." *Ann N Y Acad Sci* 1183: 149-157.

Frias, M. A.; Thoreen, C. C.; Jaffe, J. D.; Schroder, W.; Sculley, T.; Carr, S. A.&Sabatini, D. M. (2006). "mSin1 is necessary for Akt/PKB phosphorylation, and its isoforms define three distinct mTORC2s." *Current biology : CB* 16(18): 1865-1870.

Gangloff, Y. G.; Mueller, M.; Dann, S. G.; Svoboda, P.; Sticker, M.; Spetz, J. F.; Um, S. H.; Brown, E. J.; Cereghini, S.; Thomas, G.&Kozma, S. C. (2004). "Disruption of the mouse mTOR gene leads to early postimplantation lethality and prohibits embryonic stem cell development." *Molecular and cellular biology* 24(21): 9508-9516.

Ganley, I. G.; Lam du, H.; Wang, J.; Ding, X.; Chen, S.&Jiang, X. (2009). "ULK1.ATG13.FIP200 complex mediates mTOR signaling and is essential for autophagy." *The Journal of biological chemistry* 284(18): 12297-12305.

Gao, D.; Inuzuka, H.; Tan, M. K.; Fukushima, H.; Locasale, J. W.; Liu, P.; Wan, L.; Zhai, B.; Chin, Y. R.; Shaik, S.; Lyssiotis, C. A.; Gygi, S. P.; Toker, A.; Cantley, L. C.; Asara, J. M.; Harper, J. W.&Wei, W. (2011). "mTOR Drives Its Own Activation via SCF(betaTrCP)-Dependent Degradation of the mTOR Inhibitor DEPTOR." *Molecular cell* 44(2): 290-303.

Gao, D.; Wan, L.; Inuzuka, H.; Berg, A. H.; Tseng, A.; Zhai, B.; Shaik, S.; Bennett, E.; Tron, A. E.; Gasser, J. A.; Lau, A.; Gygi, S. P.; Harper, J. W.; DeCaprio, J. A.; Toker, A.&Wei, W. (2010). "Rictor forms a complex with Cullin-1 to promote SGK1 ubiquitination and destruction." *Molecular cell* 39(5): 797-808.

Gao, X.; Zhang, Y.; Arrazola, P.; Hino, O.; Kobayashi, T.; Yeung, R. S.; Ru, B.&Pan, D. (2002). "Tsc tumour suppressor proteins antagonize amino-acid-TOR signalling." *Nature cell biology* 4(9): 699-704.

Garami, A.; Zwartkruis, F. J.; Nobukuni, T.; Joaquin, M.; Roccio, M.; Stocker, H.; Kozma, S. C.; Hafen, E.; Bos, J. L.&Thomas, G. (2003). "Insulin activation of Rheb, a mediator of mTOR/S6K/4E-BP signaling, is inhibited by TSC1 and 2." *Molecular cell* 11(6): 1457-1466.

Garcia-Martinez, J. M.&Alessi, D. R. (2008). "mTOR complex 2 (mTORC2) controls hydrophobic motif phosphorylation and activation of serum- and glucocorticoid-induced protein kinase 1 (SGK1)." *The Biochemical journal* 416(3): 375-385.

Ghosh, D.; Srivastava, G. P.; Xu, D.; Schulz, L. C.&Roberts, R. M. (2008). "A link between SIN1 (MAPKAP1) and poly(rC) binding protein 2 (PCBP2) in counteracting environmental stress." *Proceedings of the National Academy of Sciences of the United States of America* 105(33): 11673-11678.

Gingras, A. C.; Gygi, S. P.; Raught, B.; Polakiewicz, R. D.; Abraham, R. T.; Hoekstra, M. F.; Aebersold, R.&Sonenberg, N. (1999). "Regulation of 4E-BP1 phosphorylation: a novel two-step mechanism." *Genes & development* 13(11): 1422-1437.

Gingras, A. C.; Raught, B.; Gygi, S. P.; Niedzwiecka, A.; Miron, M.; Burley, S. K.; Polakiewicz, R. D.; Wyslouch-Cieszynska, A.; Aebersold, R.&Sonenberg, N. (2001). "Hierarchical phosphorylation of the translation inhibitor 4E-BP1." *Genes & development* 15(21): 2852-2864.

Goldberg, M. A.; Dunning, S. P.&Bunn, H. F. (1988). "Regulation of the erythropoietin gene: evidence that the oxygen sensor is a heme protein." *Science* 242(4884): 1412-1415.

Goldstein, R.&Meyer, T. (2011). "Role of everolimus in pancreatic neuroendocrine tumors." *Expert Rev Anticancer Ther* 11(11): 1653-1665.

Gomez-Cambronero, J. (2003). "Rapamycin inhibits GM-CSF-induced neutrophil migration." *FEBS Lett* 550(1-3): 94-100.

Green, A. J.; Smith, M.&Yates, J. R. (1994). "Loss of heterozygosity on chromosome 16p13.3 in hamartomas from tuberous sclerosis patients." *Nature genetics* 6(2): 193-196.

Groves, M. R.&Barford, D. (1999). "Topological characteristics of helical repeat proteins." *Current opinion in structural biology* 9(3): 383-389.

Gu, Y.; Lindner, J.; Kumar, A.; Yuan, W.&Magnuson, M. A. (2011). "Rictor/mTORC2 is essential for maintaining a balance between beta-cell proliferation and cell size." *Diabetes* 60(3): 827-837.

Guertin, D. A.; Stevens, D. M.; Thoreen, C. C.; Burds, A. A.; Kalaany, N. Y.; Moffat, J.; Brown, M.; Fitzgerald, K. J.&Sabatini, D. M. (2006). "Ablation in mice of the mTORC components raptor, rictor, or mLST8 reveals that mTORC2 is required for signaling to Akt-FOXO and PKCalpha, but not S6K1." *Developmental cell* 11(6): 859-871.

Gwinn, D. M.; Shackelford, D. B.; Egan, D. F.; Mihaylova, M. M.; Mery, A.; Vasquez, D. S.; Turk, B. E.&Shaw, R. J. (2008). "AMPK phosphorylation of raptor mediates a metabolic checkpoint." *Molecular cell* 30(2): 214-226.

Hall, M. N. (1996). "The TOR signalling pathway and growth control in yeast." *Biochemical Society transactions* 24(1): 234-239.

Hanks, S. K.&Hunter, T. (1995). "Protein kinases 6. The eukaryotic protein kinase superfamily: kinase (catalytic) domain structure and classification." *The FASEB journal : official publication of the Federation of American Societies for Experimental Biology* 9(8): 576-596.

Hannan, K. M.; Brandenburger, Y.; Jenkins, A.; Sharkey, K.; Cavanaugh, A.; Rothblum, L.; Moss, T.; Poortinga, G.; McArthur, G. A.; Pearson, R. B.&Hannan, R. D. (2003). "mTOR-dependent regulation of ribosomal gene transcription requires S6K1 and is mediated by phosphorylation of the carboxy-terminal activation domain of the nucleolar transcription factor UBF." *Molecular and cellular biology* 23(23): 8862-8877.

Hara, K.; Maruki, Y.; Long, X.; Yoshino, K.; Oshiro, N.; Hidayat, S.; Tokunaga, C.; Avruch, J.&Yonezawa, K. (2002). "Raptor, a binding partner of target of rapamycin (TOR), mediates TOR action." *Cell* 110(2): 177-189.

Hardt, M.; Chantaravisoot, N.&Tamanoi, F. (2011). "Activating mutations of TOR (target of rapamycin)." *Genes to cells : devoted to molecular & cellular mechanisms* 16(2): 141-151.

Hardwick, J. S.; Kuruvilla, F. G.; Tong, J. K.; Shamji, A. F.&Schreiber, S. L. (1999). "Rapamycin-modulated transcription defines the subset of nutrient-sensitive signaling pathways directly controlled by the Tor proteins." *Proceedings of the National Academy of Sciences of the United States of America* 96(26): 14866-14870.

Harrington, L. S.; Findlay, G. M.; Gray, A.; Tolkacheva, T.; Wigfield, S.; Rebholz, H.; Barnett, J.; Leslie, N. R.; Cheng, S.; Shepherd, P. R.; Gout, I.; Downes, C. P.&Lamb, R. F. (2004). "The TSC1-2 tumor suppressor controls insulin-PI3K signaling via regulation of IRS proteins." *The Journal of cell biology* 166(2): 213-223.

Harrison, D. E.; Strong, R.; Sharp, Z. D.; Nelson, J. F.; Astle, C. M.; Flurkey, K.; Nadon, N. L.; Wilkinson, J. E.; Frenkel, K.; Carter, C. S.; Pahor, M.; Javors, M. A.; Fernandez, E.&Miller, R. A. (2009). "Rapamycin fed late in life extends lifespan in genetically heterogeneous mice." *Nature* 460(7253): 392-395.

Hartwell, L. H.; Hopfield, J. J.; Leibler, S.&Murray, A. W. (1999). "From molecular to modular cell biology." *Nature* 402(6761 Suppl): C47-52.

Hay, N.&Sonenberg, N. (2004). "Upstream and downstream of mTOR." *Genes & development* 18(16): 1926-1945.

Heitman, J.; Movva, N. R.&Hall, M. N. (1991). "Targets for cell cycle arrest by the immunosuppressant rapamycin in yeast." *Science* 253(5022): 905-909.

Helliwell, S. B.; Wagner, P.; Kunz, J.; Deuter-Reinhard, M.; Henriquez, R.&Hall, M. N. (1994). "TOR1 and TOR2 are structurally and functionally similar but not identical phosphatidylinositol kinase homologues in yeast." *Molecular biology of the cell* 5(1): 105-118.

Hernandez-Negrete, I.; Carretero-Ortega, J.; Rosenfeldt, H.; Hernandez-Garcia, R.; Calderon-Salinas, J. V.; Reyes-Cruz, G.; Gutkind, J. S.&Vazquez-Prado, J. (2007). "P-Rex1 links mammalian target of rapamycin signaling to Rac activation and cell migration." *J Biol Chem* 282(32): 23708-23715.

Heublein, S.; Kazi, S.; Ogmundsdottir, M. H.; Attwood, E. V.; Kala, S.; Boyd, C. A.; Wilson, C.&Goberdhan, D. C. (2010). "Proton-assisted amino-acid transporters are conserved regulators of proliferation and amino-acid-dependent mTORC1 activation." *Oncogene* 29(28): 4068-4079.

Ho, H. L.; Lee, H. Y.; Liao, H. C.&Chen, M. Y. (2008). "Involvement of Saccharomyces cerevisiae Avo3p/Tsc11p in maintaining TOR complex 2 integrity and coupling to downstream signaling." *Eukaryotic cell* 7(8): 1328-1343.

Hoang, B.; Frost, P.; Shi, Y.; Belanger, E.; Benavides, A.; Pezeshkpour, G.; Cappia, S.; Guglielmelli, T.; Gera, J.&Lichtenstein, A. (2010). "Targeting TORC2 in multiple myeloma with a new mTOR kinase inhibitor." *Blood* 116(22): 4560-4568.

Holz, M. K.&Blenis, J. (2005). "Identification of S6 kinase 1 as a novel mammalian target of rapamycin (mTOR)-phosphorylating kinase." *The Journal of biological chemistry* 280(28): 26089-26093.

Horejsi, Z.; Takai, H.; Adelman, C. A.; Collis, S. J.; Flynn, H.; Maslen, S.; Skehel, J. M.; de Lange, T.&Boulton, S. J. (2010). "CK2 phospho-dependent binding of R2TP complex to TEL2 is essential for mTOR and SMG1 stability." *Molecular cell* 39(6): 839-850.

Hosoi, H.; Dilling, M. B.; Shikata, T.; Liu, L. N.; Shu, L.; Ashmun, R. A.; Germain, G. S.; Abraham, R. T.&Houghton, P. J. (1999). "Rapamycin causes poorly reversible inhibition of mTOR and induces p53-independent apoptosis in human rhabdomyosarcoma cells." *Cancer Res* 59(4): 886-894.

Hosokawa, N.; Hara, T.; Kaizuka, T.; Kishi, C.; Takamura, A.; Miura, Y.; Iemura, S.; Natsume, T.; Takehana, K.; Yamada, N.; Guan, J. L.; Oshiro, N.&Mizushima, N. (2009). "Nutrient-dependent mTORC1 association with the ULK1-Atg13-FIP200 complex required for autophagy." *Molecular biology of the cell* 20(7): 1981-1991.

Hresko, R. C.&Mueckler, M. (2005). "mTOR.RICTOR is the Ser473 kinase for Akt/protein kinase B in 3T3-L1 adipocytes." *The Journal of biological chemistry* 280(49): 40406-40416.

Hsu, P. P.; Kang, S. A.; Rameseder, J.; Zhang, Y.; Ottina, K. A.; Lim, D.; Peterson, T. R.; Choi, Y.; Gray, N. S.; Yaffe, M. B.; Marto, J. A.&Sabatini, D. M. (2011). "The mTOR-regulated phosphoproteome reveals a mechanism of mTORC1-mediated inhibition of growth factor signaling." *Science* 332(6035): 1317-1322.

Huang, S.&Houghton, P. J. (2003). "Targeting mTOR signaling for cancer therapy." *Current opinion in pharmacology* 3(4): 371-377.

Hwang, S. K.&Kim, H. H. (2011). "The functions of mTOR in ischemic diseases." *BMB Rep* 44(8): 506-511.

Ibraghimov-Beskrovnaya, O.&Natoli, T. A. (2011). "mTOR signaling in polycystic kidney disease." *Trends Mol Med.*

Ikenoue, T.; Inoki, K.; Yang, Q.; Zhou, X.&Guan, K. L. (2008). "Essential function of TORC2 in PKC and Akt turn motif phosphorylation, maturation and signalling." *The EMBO journal* 27(14): 1919-1931.

Inoki, K.; Zhu, T.&Guan, K. L. (2003). "TSC2 mediates cellular energy response to control cell growth and survival." *Cell* 115(5): 577-590.

Jacinto, E. (2008). "What controls TOR?" *IUBMB life* 60(8): 483-496.

Jacinto, E.; Facchinetti, V.; Liu, D.; Soto, N.; Wei, S.; Jung, S. Y.; Huang, Q.; Qin, J.&Su, B. (2006). "SIN1/MIP1 maintains rictor-mTOR complex integrity and regulates Akt phosphorylation and substrate specificity." *Cell* 127(1): 125-137.

Jacinto, E.; Loewith, R.; Schmidt, A.; Lin, S.; Ruegg, M. A.; Hall, A.&Hall, M. N. (2004). "Mammalian TOR complex 2 controls the actin cytoskeleton and is rapamycin insensitive." *Nature cell biology* 6(11): 1122-1128.

Jacinto, E.&Lorberg, A. (2008). "TOR regulation of AGC kinases in yeast and mammals." *The Biochemical journal* 410(1): 19-37.

Janes, M. R.; Limon, J. J.; So, L.; Chen, J.; Lim, R. J.; Chavez, M. A.; Vu, C.; Lilly, M. B.; Mallya, S.; Ong, S. T.; Konopleva, M.; Martin, M. B.; Ren, P.; Liu, Y.; Rommel, C.&Fruman, D. A. (2010). "Effective and selective targeting of leukemia cells using a TORC1/2 kinase inhibitor." *Nat Med* 16(2): 205-213.

Janku, F.; McConkey, D. J.; Hong, D. S.&Kurzrock, R. (2011). "Autophagy as a target for anticancer therapy." *Nat Rev Clin Oncol* 8(9): 528-539.

Julien, L. A.; Carriere, A.; Moreau, J.&Roux, P. P. (2010). "mTORC1-activated S6K1 phosphorylates Rictor on threonine 1135 and regulates mTORC2 signaling." *Molecular and cellular biology* 30(4): 908-921.

Jung, C. H.; Jun, C. B.; Ro, S. H.; Kim, Y. M.; Otto, N. M.; Cao, J.; Kundu, M.&Kim, D. H. (2009). "ULK-Atg13-FIP200 complexes mediate mTOR signaling to the autophagy machinery." *Molecular biology of the cell* 20(7): 1992-2003.

Kaizuka, T.; Hara, T.; Oshiro, N.; Kikkawa, U.; Yonezawa, K.; Takehana, K.; Iemura, S.; Natsume, T.&Mizushima, N. (2010). "Tti1 and Tel2 are critical factors in mammalian target of rapamycin complex assembly." *J Biol Chem* 285(26): 20109-20116.

Kaizuka, T.; Hara, T.; Oshiro, N.; Kikkawa, U.; Yonezawa, K.; Takehana, K.; Iemura, S.; Natsume, T.&Mizushima, N. (2010). "Tti1 and Tel2 are critical factors in mammalian target of rapamycin complex assembly." *The Journal of biological chemistry* 285(26): 20109-20116.

Kamada, Y.; Funakoshi, T.; Shintani, T.; Nagano, K.; Ohsumi, M.&Ohsumi, Y. (2000). "Tor-mediated induction of autophagy via an Apg1 protein kinase complex." *The Journal of cell biology* 150(6): 1507-1513.

Kamada, Y.; Yoshino, K.; Kondo, C.; Kawamata, T.; Oshiro, N.; Yonezawa, K.&Ohsumi, Y. (2010). "Tor directly controls the Atg1 kinase complex to regulate autophagy." *Molecular and cellular biology* 30(4): 1049-1058.

Kapahi, P.; Zid, B. M.; Harper, T.; Koslover, D.; Sapin, V.&Benzer, S. (2004). "Regulation of lifespan in Drosophila by modulation of genes in the TOR signaling pathway." *Current biology : CB* 14(10): 885-890.

Keith, C. T.&Schreiber, S. L. (1995). "PIK-related kinases: DNA repair, recombination, and cell cycle checkpoints." *Science* 270(5233): 50-51.

Kijanska, M.; Dohnal, I.; Reiter, W.; Kaspar, S.; Stoffel, I.; Ammerer, G.; Kraft, C.&Peter, M. (2010). "Activation of Atg1 kinase in autophagy by regulated phosphorylation." *Autophagy* 6(8): 1168-1178.

Kim, D. H.; Sarbassov, D. D.; Ali, S. M.; King, J. E.; Latek, R. R.; Erdjument-Bromage, H.; Tempst, P.&Sabatini, D. M. (2002). "mTOR interacts with raptor to form a nutrient-sensitive complex that signals to the cell growth machinery." *Cell* 110(2): 163-175.

Kim, D. H.; Sarbassov, D. D.; Ali, S. M.; Latek, R. R.; Guntur, K. V.; Erdjument-Bromage, H.; Tempst, P.&Sabatini, D. M. (2003). "GbetaL, a positive regulator of the rapamycin-sensitive pathway required for the nutrient-sensitive interaction between raptor and mTOR." *Molecular cell* 11(4): 895-904.

Knutson, B. A. (2010). "Insights into the domain and repeat architecture of target of rapamycin." *Journal of structural biology* 170(2): 354-363.

Konstantinidou, G.; Bey, E. A.; Rabellino, A.; Schuster, K.; Maira, M. S.; Gazdar, A. F.; Amici, A.; Boothman, D. A.&Scaglioni, P. P. (2009). "Dual phosphoinositide 3-kinase/mammalian target of rapamycin blockade is an effective radiosensitizing strategy for the treatment of non-small cell lung cancer harboring K-RAS mutations." *Cancer Res* 69(19): 7644-7652.

Krymskaya, V. P.&Goncharova, E. A. (2009). "PI3K/mTORC1 activation in hamartoma syndromes: therapeutic prospects." *Cell Cycle* 8(3): 403-413.

Kunz, J.; Loeschmann, A.; Deuter-Reinhard, M.&Hall, M. N. (2000). "FAP1, a homologue of human transcription factor NF-X1, competes with rapamycin for binding to FKBP12 in yeast." *Molecular microbiology* 37(6): 1480-1493.

Kunz, J.; Schneider, U.; Howald, I.; Schmidt, A.&Hall, M. N. (2000). "HEAT repeats mediate plasma membrane localization of Tor2p in yeast." *The Journal of biological chemistry* 275(47): 37011-37020.

Lempiainen, H.&Halazonetis, T. D. (2009). "Emerging common themes in regulation of PIKKs and PI3Ks." *The EMBO journal* 28(20): 3067-3073.

Li, H.; Tsang, C. K.; Watkins, M.; Bertram, P. G.&Zheng, X. F. (2006). "Nutrient regulates Tor1 nuclear localization and association with rDNA promoter." *Nature* 442(7106): 1058-1061.

Liu, L.; Chen, L.; Chung, J.&Huang, S. (2008). "Rapamycin inhibits F-actin reorganization and phosphorylation of focal adhesion proteins." *Oncogene* 27(37): 4998-5010.

Liu, L.; Luo, Y.; Chen, L.; Shen, T.; Xu, B.; Chen, W.; Zhou, H.; Han, X.&Huang, S. (2010). "Rapamycin inhibits cytoskeleton reorganization and cell motility by suppressing RhoA expression and activity." *J Biol Chem* 285(49): 38362-38373.

Liu, Q.; Wang, J.; Kang, S. A.; Thoreen, C. C.; Hur, W.; Ahmed, T.; Sabatini, D. M.&Gray, N. S. (2011). "Discovery of 9-(6-aminopyridin-3-yl)-1-(3-(trifluoromethyl)phenyl)benzo[h][1,6]naphthyr idin-2(1H)-one (Torin2) as a potent, selective, and orally available mammalian target of rapamycin (mTOR) inhibitor for treatment of cancer." *J Med Chem* 54(5): 1473-1480.

Liu, X.&Zheng, X. F. (2007). "Endoplasmic reticulum and Golgi localization sequences for mammalian target of rapamycin." *Molecular biology of the cell* 18(3): 1073-1082.

Liu, Z.; Sekito, T.; Epstein, C. B.&Butow, R. A. (2001). "RTG-dependent mitochondria to nucleus signaling is negatively regulated by the seven WD-repeat protein Lst8p." *The EMBO journal* 20(24): 7209-7219.

Lochhead, P. A. (2009). "Protein kinase activation loop autophosphorylation in cis: overcoming a Catch-22 situation." *Science signaling* 2(54): pe4.

Loewith, R.; Jacinto, E.; Wullschleger, S.; Lorberg, A.; Crespo, J. L.; Bonenfant, D.; Oppliger, W.; Jenoe, P.&Hall, M. N. (2002). "Two TOR complexes, only one of which is rapamycin sensitive, have distinct roles in cell growth control." *Molecular cell* 10(3): 457-468.

Long, X.; Lin, Y.; Ortiz-Vega, S.; Yonezawa, K.&Avruch, J. (2005). "Rheb binds and regulates the mTOR kinase." *Current biology : CB* 15(8): 702-713.

Long, X.; Spycher, C.; Han, Z. S.; Rose, A. M.; Muller, F.&Avruch, J. (2002). "TOR deficiency in C. elegans causes developmental arrest and intestinal atrophy by inhibition of mRNA translation." *Current biology : CB* 12(17): 1448-1461.

Ma, L.; Chen, Z.; Erdjument-Bromage, H.; Tempst, P.&Pandolfi, P. P. (2005). "Phosphorylation and functional inactivation of TSC2 by Erk implications for tuberous sclerosis and cancer pathogenesis." *Cell* 121(2): 179-193.

Ma, X. M.&Blenis, J. (2009). "Molecular mechanisms of mTOR-mediated translational control." *Nature reviews. Molecular cell biology* 10(5): 307-318.

Ma, X. M.; Yoon, S. O.; Richardson, C. J.; Julich, K.&Blenis, J. (2008). "SKAR links pre-mRNA splicing to mTOR/S6K1-mediated enhanced translation efficiency of spliced mRNAs." *Cell* 133(2): 303-313.

Maira, S. M.; Stauffer, F.; Brueggen, J.; Furet, P.; Schnell, C.; Fritsch, C.; Brachmann, S.; Chene, P.; De Pover, A.; Schoemaker, K.; Fabbro, D.; Gabriel, D.; Simonen, M.; Murphy, L.; Finan, P.; Sellers, W.&Garcia-Echeverria, C. (2008). "Identification and characterization of NVP-BEZ235, a new orally available dual phosphatidylinositol 3-kinase/mammalian target of rapamycin inhibitor with potent in vivo antitumor activity." *Mol Cancer Ther* 7(7): 1851-1863.

Makino, C.; Sano, Y.; Shinagawa, T.; Millar, J. B.&Ishii, S. (2006). "Sin1 binds to both ATF-2 and p38 and enhances ATF-2-dependent transcription in an SAPK signaling pathway." *Genes to cells : devoted to molecular & cellular mechanisms* 11(11): 1239-1251.

Malys, N.&McCarthy, J. E. (2011). "Translation initiation: variations in the mechanism can be anticipated." *Cellular and molecular life sciences : CMLS* 68(6): 991-1003.

Mammucari, C.; Milan, G.; Romanello, V.; Masiero, E.; Rudolf, R.; Del Piccolo, P.; Burden, S. J.; Di Lisi, R.; Sandri, C.; Zhao, J.; Goldberg, A. L.; Schiaffino, S.&Sandri, M. (2007). "FoxO3 controls autophagy in skeletal muscle in vivo." *Cell Metab* 6(6): 458-471.

Manning, B. D.&Cantley, L. C. (2003). "Rheb fills a GAP between TSC and TOR." *Trends in biochemical sciences* 28(11): 573-576.

Manning, G.; Whyte, D. B.; Martinez, R.; Hunter, T.&Sudarsanam, S. (2002). "The protein kinase complement of the human genome." *Science* 298(5600): 1912-1934.

Marone, R.; Erhart, D.; Mertz, A. C.; Bohnacker, T.; Schnell, C.; Cmiljanovic, V.; Stauffer, F.; Garcia-Echeverria, C.; Giese, B.; Maira, S. M.&Wymann, M. P. (2009). "Targeting melanoma with dual phosphoinositide 3-kinase/mammalian target of rapamycin inhibitors." *Mol Cancer Res* 7(4): 601-613.

Martin, D. E.; Soulard, A.&Hall, M. N. (2004). "TOR regulates ribosomal protein gene expression via PKA and the Forkhead transcription factor FHL1." *Cell* 119(7): 969-979.

Martin, J.; Masri, J.; Bernath, A.; Nishimura, R. N.&Gera, J. (2008). "Hsp70 associates with Rictor and is required for mTORC2 formation and activity." *Biochem Biophys Res Commun* 372(4): 578-583.

Martin, J.; Masri, J.; Bernath, A.; Nishimura, R. N.&Gera, J. (2008). "Hsp70 associates with Rictor and is required for mTORC2 formation and activity." *Biochemical and biophysical research communications* 372(4): 578-583.

Martin, P. M.&Sutherland, A. E. (2001). "Exogenous amino acids regulate trophectoderm differentiation in the mouse blastocyst through an mTOR-dependent pathway." *Developmental biology* 240(1): 182-193.

Matsuo, T.; Otsubo, Y.; Urano, J.; Tamanoi, F.&Yamamoto, M. (2007). "Loss of the TOR kinase Tor2 mimics nitrogen starvation and activates the sexual development pathway in fission yeast." *Molecular and cellular biology* 27(8): 3154-3164.

Mayer, C.&Grummt, I. (2006). "Ribosome biogenesis and cell growth: mTOR coordinates transcription by all three classes of nuclear RNA polymerases." *Oncogene* 25(48): 6384-6391.

Mayer, C.; Zhao, J.; Yuan, X.&Grummt, I. (2004). "mTOR-dependent activation of the transcription factor TIF-IA links rRNA synthesis to nutrient availability." *Genes & development* 18(4): 423-434.

McDonald, P. C.; Oloumi, A.; Mills, J.; Dobreva, I.; Maidan, M.; Gray, V.; Wederell, E. D.; Bally, M. B.; Foster, L. J.&Dedhar, S. (2008). "Rictor and integrin-linked kinase interact and regulate Akt phosphorylation and cancer cell survival." *Cancer research* 68(6): 1618-1624.

Medvedik, O.; Lamming, D. W.; Kim, K. D.&Sinclair, D. A. (2007). "MSN2 and MSN4 link calorie restriction and TOR to sirtuin-mediated lifespan extension in Saccharomyces cerevisiae." *PLoS biology* 5(10): e261.

Meijer, A. J.&Codogno, P. (2008). "Nutrient sensing: TOR's Ragtime." *Nature cell biology* 10(8): 881-883.

Menand, B.; Desnos, T.; Nussaume, L.; Berger, F.; Bouchez, D.; Meyer, C.&Robaglia, C. (2002). "Expression and disruption of the Arabidopsis TOR (target of rapamycin) gene." *Proceedings of the National Academy of Sciences of the United States of America* 99(9): 6422-6427.

Miller, G.; Panov, K. I.; Friedrich, J. K.; Trinkle-Mulcahy, L.; Lamond, A. I.&Zomerdijk, J. C. (2001). "hRRN3 is essential in the SL1-mediated recruitment of RNA Polymerase I to rRNA gene promoters." *The EMBO journal* 20(6): 1373-1382.

Minami, K.; Tambe, Y.; Watanabe, R.; Isono, T.; Haneda, M.; Isobe, K.; Kobayashi, T.; Hino, O.; Okabe, H.; Chano, T.&Inoue, H. (2007). "Suppression of viral replication by stress-inducible GADD34 protein via the mammalian serine/threonine protein kinase mTOR pathway." *Journal of virology* 81(20): 11106-11115.

Miranda-Saavedra, D.&Barton, G. J. (2007). "Classification and functional annotation of eukaryotic protein kinases." *Proteins* 68(4): 893-914.

Mordes, D. A.; Glick, G. G.; Zhao, R.&Cortez, D. (2008). "TopBP1 activates ATR through ATRIP and a PIKK regulatory domain." *Genes & development* 22(11): 1478-1489.

Murakami, M.; Ichisaka, T.; Maeda, M.; Oshiro, N.; Hara, K.; Edenhofer, F.; Kiyama, H.; Yonezawa, K.&Yamanaka, S. (2004). "mTOR is essential for growth and proliferation in early mouse embryos and embryonic stem cells." *Molecular and cellular biology* 24(15): 6710-6718.

Muthukkumar, S.; Ramesh, T. M.&Bondada, S. (1995). "Rapamycin, a potent immunosuppressive drug, causes programmed cell death in B lymphoma cells." *Transplantation* 60(3): 264-270.

Nardella, C.; Carracedo, A.; Alimonti, A.; Hobbs, R. M.; Clohessy, J. G.; Chen, Z.; Egia, A.; Fornari, A.; Fiorentino, M.; Loda, M.; Kozma, S. C.; Thomas, G.; Cordon-Cardo, C.&Pandolfi, P. P. (2009). "Differential requirement of mTOR in postmitotic tissues and tumorigenesis." *Science signaling* 2(55): ra2.

Nicklin, P.; Bergman, P.; Zhang, B.; Triantafellow, E.; Wang, H.; Nyfeler, B.; Yang, H.; Hild, M.; Kung, C.; Wilson, C.; Myer, V. E.; MacKeigan, J. P.; Porter, J. A.; Wang, Y. K.; Cantley, L. C.; Finan, P. M.&Murphy, L. O. (2009). "Bidirectional transport of amino acids regulates mTOR and autophagy." *Cell* 136(3): 521-534.

Nobukuni, T.; Joaquin, M.; Roccio, M.; Dann, S. G.; Kim, S. Y.; Gulati, P.; Byfield, M. P.; Backer, J. M.; Natt, F.; Bos, J. L.; Zwartkruis, F. J.&Thomas, G. (2005). "Amino acids mediate mTOR/raptor signaling through activation of class 3 phosphatidylinositol 3OH-kinase." *Proceedings of the National Academy of Sciences of the United States of America* 102(40): 14238-14243.

Noda, T & Ohsumi, Y. (1998). "Tor, a phosphatidylinositol kinase homologue, controls autophagy in yeast." *The Journal of biological chemistry* 273(7): 3963-3966.

O'Reilly, K. E.; Rojo, F.; She, Q. B.; Solit, D.; Mills, G. B.; Smith, D.; Lane, H.; Hofmann, F.; Hicklin, D. J.; Ludwig, D. L.; Baselga, J.&Rosen, N. (2006). "mTOR inhibition induces upstream receptor tyrosine kinase signaling and activates Akt." *Cancer Res* 66(3): 1500-1508.

Oh, W. J.; Wu, C. C.; Kim, S. J.; Facchinetti, V.; Julien, L. A.; Finlan, M.; Roux, P. P.; Su, B.&Jacinto, E. (2010). "mTORC2 can associate with ribosomes to promote cotranslational phosphorylation and stability of nascent Akt polypeptide." *The EMBO journal* 29(23): 3939-3951.

Oldham, S.; Montagne, J.; Radimerski, T.; Thomas, G.&Hafen, E. (2000). "Genetic and biochemical characterization of dTOR, the Drosophila homolog of the target of rapamycin." *Genes & development* 14(21): 2689-2694.

Onda, H.; Lueck, A.; Marks, P. W.; Warren, H. B.&Kwiatkowski, D. J. (1999). "Tsc2(+/-) mice develop tumors in multiple sites that express gelsolin and are influenced by genetic background." *The Journal of clinical investigation* 104(6): 687-695.

Oshiro, N.; Yoshino, K.; Hidayat, S.; Tokunaga, C.; Hara, K.; Eguchi, S.; Avruch, J.&Yonezawa, K. (2004). "Dissociation of raptor from mTOR is a mechanism of rapamycin-induced inhibition of mTOR function." *Genes to cells : devoted to molecular & cellular mechanisms* 9(4): 359-366.

Oza, A. M.; Elit, L.; Tsao, M. S.; Kamel-Reid, S.; Biagi, J.; Provencher, D. M.; Gotlieb, W. H.; Hoskins, P. J.; Ghatage, P.; Tonkin, K. S.; Mackay, H. J.; Mazurka, J.; Sederias, J.; Ivy, P.; Dancey, J. E.&Eisenhauer, E. A. (2011). "Phase II study of temsirolimus in women with recurrent or metastatic endometrial cancer: a trial of the NCIC Clinical Trials Group." *J Clin Oncol* 29(24): 3278-3285.

Pang, H.&Faber, L. E. (2001). "Estrogen and rapamycin effects on cell cycle progression in T47D breast cancer cells." *Breast Cancer Res Treat* 70(1): 21-26.

Pause, A.; Methot, N.; Svitkin, Y.; Merrick, W. C.&Sonenberg, N. (1994). "Dominant negative mutants of mammalian translation initiation factor eIF-4A define a critical role for eIF-4F in cap-dependent and cap-independent initiation of translation." *The EMBO journal* 13(5): 1205-1215.

Pearce, L. R.; Huang, X.; Boudeau, J.; Pawlowski, R.; Wullschleger, S.; Deak, M.; Ibrahim, A. F.; Gourlay, R.; Magnuson, M. A.&Alessi, D. R. (2007). "Identification of Protor as a novel Rictor-binding component of mTOR complex-2." *The Biochemical journal* 405(3): 513-522.

Pearce, L. R.; Sommer, E. M.; Sakamoto, K.; Wullschleger, S.&Alessi, D. R. (2011). "Protor-1 is required for efficient mTORC2-mediated activation of SGK1 in the kidney." *The Biochemical journal* 436(1): 169-179.

Pearson, G.; Robinson, F.; Beers Gibson, T.; Xu, B. E.; Karandikar, M.; Berman, K.&Cobb, M. H. (2001). "Mitogen-activated protein (MAP) kinase pathways: regulation and physiological functions." *Endocrine reviews* 22(2): 153-183.

Pei, J. J.&Hugon, J. (2008). "mTOR-dependent signalling in Alzheimer's disease." *J Cell Mol Med* 12(6B): 2525-2532.

Pena-Llopis, S.; Vega-Rubin-de-Celis, S.; Schwartz, J. C.; Wolff, N. C.; Tran, T. A.; Zou, L.; Xie, X. J.; Corey, D. R.&Brugarolas, J. (2011). "Regulation of TFEB and V-ATPases by mTORC1." *The EMBO journal* 30(16): 3242-3258.

Peng, T.; Golub, T. R.&Sabatini, D. M. (2002). "The immunosuppressant rapamycin mimics a starvation-like signal distinct from amino acid and glucose deprivation." *Molecular and cellular biology* 22(15): 5575-5584.

Pereira-Leal, J. B.; Levy, E. D.&Teichmann, S. A. (2006). "The origins and evolution of functional modules: lessons from protein complexes." *Philosophical transactions of the Royal Society of London. Series B, Biological sciences* 361(1467): 507-517.

Perry, J.&Kleckner, N. (2003). "The ATRs, ATMs, and TORs are giant HEAT repeat proteins." *Cell* 112(2): 151-155.

Peterson, R. T.; Beal, P. A.; Comb, M. J.&Schreiber, S. L. (2000). "FKBP12-rapamycin-associated protein (FRAP) autophosphorylates at serine 2481 under translationally repressive conditions." *The Journal of biological chemistry* 275(10): 7416-7423.

Peterson, T. R.; Laplante, M.; Thoreen, C. C.; Sancak, Y.; Kang, S. A.; Kuehl, W. M.; Gray, N. S.&Sabatini, D. M. (2009). "DEPTOR is an mTOR inhibitor frequently overexpressed in multiple myeloma cells and required for their survival." *Cell* 137(5): 873-886.

Peterson, T. R.; Sengupta, S. S.; Harris, T. E.; Carmack, A. E.; Kang, S. A.; Balderas, E.; Guertin, D. A.; Madden, K. L.; Carpenter, A. E.; Finck, B. N.&Sabatini, D. M. (2011). "mTOR complex 1 regulates lipin 1 localization to control the SREBP pathway." *Cell* 146(3): 408-420.

Polak, P.; Cybulski, N.; Feige, J. N.; Auwerx, J.; Ruegg, M. A.&Hall, M. N. (2008). "Adipose-specific knockout of raptor results in lean mice with enhanced mitochondrial respiration." *Cell metabolism* 8(5): 399-410.

Poon, M.; Marx, S. O.; Gallo, R.; Badimon, J. J.; Taubman, M. B.&Marks, A. R. (1996). "Rapamycin inhibits vascular smooth muscle cell migration." *J Clin Invest* 98(10): 2277-2283.

Porstmann, T.; Santos, C. R.; Griffiths, B.; Cully, M.; Wu, M.; Leevers, S.; Griffiths, J. R.; Chung, Y. L.&Schulze, A. (2008). "SREBP activity is regulated by mTORC1 and contributes to Akt-dependent cell growth." *Cell metabolism* 8(3): 224-236.

Powers, T.&Walter, P. (1999). "Regulation of ribosome biogenesis by the rapamycin-sensitive TOR-signaling pathway in Saccharomyces cerevisiae." *Molecular biology of the cell* 10(4): 987-1000.

Proud, C. G. (2009). "Dynamic balancing: DEPTOR tips the scales." *Journal of molecular cell biology* 1(2): 61-63.

Proud, C. G. (2011). "mTOR Signalling in Health and Disease." *Biochemical Society transactions* 39(2): 431-436.

Rachdi, L.; Balcazar, N.; Osorio-Duque, F.; Elghazi, L.; Weiss, A.; Gould, A.; Chang-Chen, K. J.; Gambello, M. J.&Bernal-Mizrachi, E. (2008). "Disruption of Tsc2 in pancreatic beta cells induces beta cell mass expansion and improved glucose tolerance in a TORC1-dependent manner." *Proceedings of the National Academy of Sciences of the United States of America* 105(27): 9250-9255.

Reardon, D. A.; Wen, P. Y.; Alfred Yung, W. K.; Berk, L.; Narasimhan, N.; Turner, C. D.; Clackson, T.; Rivera, V. M.&Vogelbaum, M. A. (2011). "Ridaforolimus for patients with progressive or recurrent malignant glioma: a perisurgical, sequential, ascending-dose trial." *Cancer Chemother Pharmacol.*

Reiling, J. H.&Sabatini, D. M. (2006). "Stress and mTORture signaling." *Oncogene* 25(48): 6373-6383.

Reina, J. H.; Azzouz, T. N.&Hernandez, N. (2006). "Maf1, a new player in the regulation of human RNA polymerase III transcription." *PloS one* 1: e134.

Reinke, A.; Anderson, S.; McCaffery, J. M.; Yates, J., 3rd; Aronova, S.; Chu, S.; Fairclough, S.; Iverson, C.; Wedaman, K. P.&Powers, T. (2004). "TOR complex 1 includes a novel component, Tco89p (YPL180w), and cooperates with Ssd1p to maintain cellular integrity in Saccharomyces cerevisiae." *The Journal of biological chemistry* 279(15): 14752-14762.

Risson, V.; Mazelin, L.; Roceri, M.; Sanchez, H.; Moncollin, V.; Corneloup, C.; Richard-Bulteau, H.; Vignaud, A.; Baas, D.; Defour, A.; Freyssenet, D.; Tanti, J. F.; Le-Marchand-Brustel, Y.; Ferrier, B.; Conjard-Duplany, A.; Romanino, K.; Bauche, S.; Hantai, D.; Mueller, M.; Kozma, S. C.; Thomas, G.; Ruegg, M. A.; Ferry, A.; Pende, M.; Bigard, X.; Koulmann, N.; Schaeffer, L.&Gangloff, Y. G. (2009). "Muscle inactivation of mTOR causes metabolic and dystrophin defects leading to severe myopathy." *The Journal of cell biology* 187(6): 859-874.

Roberg, K. J.; Bickel, S.; Rowley, N.&Kaiser, C. A. (1997). "Control of amino acid permease sorting in the late secretory pathway of Saccharomyces cerevisiae by SEC13, LST4, LST7 and LST8." *Genetics* 147(4): 1569-1584.

Rosner, M.&Hengstschlager, M. (2008). "Cytoplasmic and nuclear distribution of the protein complexes mTORC1 and mTORC2: rapamycin triggers dephosphorylation and delocalization of the mTORC2 components rictor and sin1." *Human molecular genetics* 17(19): 2934-2948.

Roux, P. P.; Ballif, B. A.; Anjum, R.; Gygi, S. P.&Blenis, J. (2004). "Tumor-promoting phorbol esters and activated Ras inactivate the tuberous sclerosis tumor suppressor complex via p90 ribosomal S6 kinase." *Proceedings of the National Academy of Sciences of the United States of America* 101(37): 13489-13494.

Rubio-Texeira, M.&Kaiser, C. A. (2006). "Amino acids regulate retrieval of the yeast general amino acid permease from the vacuolar targeting pathway." *Molecular biology of the cell* 17(7): 3031-3050.

Russnak, R.; Konczal, D.&McIntire, S. L. (2001). "A family of yeast proteins mediating bidirectional vacuolar amino acid transport." *The Journal of biological chemistry* 276(26): 23849-23857.

Ruvinsky, I.; Sharon, N.; Lerer, T.; Cohen, H.; Stolovich-Rain, M.; Nir, T.; Dor, Y.; Zisman, P.&Meyuhas, O. (2005). "Ribosomal protein S6 phosphorylation is a determinant of cell size and glucose homeostasis." *Genes & development* 19(18): 2199-2211.

Sancak, Y.; Peterson, T. R.; Shaul, Y. D.; Lindquist, R. A.; Thoreen, C. C.; Bar-Peled, L.&Sabatini, D. M. (2008). "The Rag GTPases bind raptor and mediate amino acid signaling to mTORC1." *Science* 320(5882): 1496-1501.

Sancak, Y.; Thoreen, C. C.; Peterson, T. R.; Lindquist, R. A.; Kang, S. A.; Spooner, E.; Carr, S. A.&Sabatini, D. M. (2007). "PRAS40 is an insulin-regulated inhibitor of the mTORC1 protein kinase." *Molecular cell* 25(6): 903-915.

Sarbassov, D. D.; Ali, S. M.; Kim, D. H.; Guertin, D. A.; Latek, R. R.; Erdjument-Bromage, H.; Tempst, P.&Sabatini, D. M. (2004). "Rictor, a novel binding partner of mTOR, defines a rapamycin-insensitive and raptor-independent pathway that regulates the cytoskeleton." *Current biology : CB* 14(14): 1296-1302.

Sarbassov, D. D.; Ali, S. M.&Sabatini, D. M. (2005). "Growing roles for the mTOR pathway." *Current opinion in cell biology* 17(6): 596-603.

Sarbassov, D. D.; Ali, S. M.; Sengupta, S.; Sheen, J. H.; Hsu, P. P.; Bagley, A. F.; Markhard, A. L.&Sabatini, D. M. (2006). "Prolonged rapamycin treatment inhibits mTORC2 assembly and Akt/PKB." *Molecular cell* 22(2): 159-168.

Sarbassov, D. D.; Guertin, D. A.; Ali, S. M.&Sabatini, D. M. (2005). "Phosphorylation and regulation of Akt/PKB by the rictor-mTOR complex." *Science* 307(5712): 1098-1101.

Sarbassov, D. D.&Sabatini, D. M. (2005). "Redox regulation of the nutrient-sensitive raptor-mTOR pathway and complex." *The Journal of biological chemistry* 280(47): 39505-39509.

Schieke, S. M.; Phillips, D.; McCoy, J. P., Jr.; Aponte, A. M.; Shen, R. F.; Balaban, R. S.&Finkel, T. (2006). "The mammalian target of rapamycin (mTOR) pathway regulates mitochondrial oxygen consumption and oxidative capacity." *The Journal of biological chemistry* 281(37): 27643-27652.

Schmidt, A.; Bickle, M.; Beck, T.&Hall, M. N. (1997). "The yeast phosphatidylinositol kinase homolog TOR2 activates RHO1 and RHO2 via the exchange factor ROM2." *Cell* 88(4): 531-542.

Schmidt, A.; Kunz, J.&Hall, M. N. (1996). "TOR2 is required for organization of the actin cytoskeleton in yeast." *Proceedings of the National Academy of Sciences of the United States of America* 93(24): 13780-13785.

Schroder, W.; Bushell, G.&Sculley, T. (2005). "The human stress-activated protein kinase-interacting 1 gene encodes JNK-binding proteins." *Cellular signalling* 17(6): 761-767.

Schroder, W.; Cloonan, N.; Bushell, G.&Sculley, T. (2004). "Alternative polyadenylation and splicing of mRNAs transcribed from the human Sin1 gene." *Gene* 339: 17-23.

Schroder, W. A.; Buck, M.; Cloonan, N.; Hancock, J. F.; Suhrbier, A.; Sculley, T.&Bushell, G. (2007). "Human Sin1 contains Ras-binding and pleckstrin homology domains and suppresses Ras signalling." *Cellular signalling* 19(6): 1279-1289.

Sehgal, S. N.; Baker, H.&Vezina, C. (1975). "Rapamycin (AY-22,989), a new antifungal antibiotic. II. Fermentation, isolation and characterization." *The Journal of antibiotics* 28(10): 727-732.

Sekiguchi, T.; Hirose, E.; Nakashima, N.; Ii, M.&Nishimoto, T. (2001). "Novel G proteins, Rag C and Rag D, interact with GTP-binding proteins, Rag A and Rag B." *The Journal of biological chemistry* 276(10): 7246-7257.

Sekulic, A.; Hudson, C. C.; Homme, J. L.; Yin, P.; Otterness, D. M.; Karnitz, L. M.&Abraham, R. T. (2000). "A direct linkage between the phosphoinositide 3-kinase-AKT signaling pathway and the mammalian target of rapamycin in mitogen-stimulated and transformed cells." *Cancer research* 60(13): 3504-3513.

Sengupta, S.; Peterson, T. R.; Laplante, M.; Oh, S.&Sabatini, D. M. (2010). "mTORC1 controls fasting-induced ketogenesis and its modulation by ageing." *Nature* 468(7327): 1100-1104.

Settembre, C.; Di Malta, C.; Polito, V. A.; Garcia Arencibia, M.; Vetrini, F.; Erdin, S.; Erdin, S. U.; Huynh, T.; Medina, D.; Colella, P.; Sardiello, M.; Rubinsztein, D. C.&Ballabio, A. (2011). "TFEB links autophagy to lysosomal biogenesis." *Science* 332(6036): 1429-1433.

Shah, O. J.&Hunter, T. (2006). "Turnover of the active fraction of IRS1 involves raptor-mTOR- and S6K1-dependent serine phosphorylation in cell culture models of tuberous sclerosis." *Molecular and cellular biology* 26(17): 6425-6434.

Shah, O. J.; Wang, Z.&Hunter, T. (2004). "Inappropriate activation of the TSC/Rheb/mTOR/S6K cassette induces IRS1/2 depletion, insulin resistance, and cell survival deficiencies." *Current biology : CB* 14(18): 1650-1656.

Shaw, R. J.; Bardeesy, N.; Manning, B. D.; Lopez, L.; Kosmatka, M.; DePinho, R. A.&Cantley, L. C. (2004). "The LKB1 tumor suppressor negatively regulates mTOR signaling." *Cancer cell* 6(1): 91-99.

Shiojima, I.&Walsh, K. (2006). "Regulation of cardiac growth and coronary angiogenesis by the Akt/PKB signaling pathway." *Genes Dev* 20(24): 3347-3365.

Shiota, C.; Woo, J. T.; Lindner, J.; Shelton, K. D.&Magnuson, M. A. (2006). "Multiallelic disruption of the rictor gene in mice reveals that mTOR complex 2 is essential for fetal growth and viability." *Developmental cell* 11(4): 583-589.

Shor, B.; Wu, J.; Shakey, Q.; Toral-Barza, L.; Shi, C.; Follettie, M.&Yu, K. (2010). "Requirement of the mTOR kinase for the regulation of Maf1 phosphorylation and control of RNA polymerase III-dependent transcription in cancer cells." *The Journal of biological chemistry* 285(20): 15380-15392.

Singh, K.; Sun, S.&Vezina, C. (1979). "Rapamycin (AY-22,989), a new antifungal antibiotic. IV. Mechanism of action." *The Journal of antibiotics* 32(6): 630-645.

Smith, E. M.; Finn, S. G.; Tee, A. R.; Browne, G. J.&Proud, C. G. (2005). "The tuberous sclerosis protein TSC2 is not required for the regulation of the mammalian target of rapamycin by amino acids and certain cellular stresses." *The Journal of biological chemistry* 280(19): 18717-18727.

Soliman, G. A.; Acosta-Jaquez, H. A.; Dunlop, E. A.; Ekim, B.; Maj, N. E.; Tee, A. R.&Fingar, D. C. (2010). "mTOR Ser-2481 autophosphorylation monitors mTORC-specific catalytic activity and clarifies rapamycin mechanism of action." *The Journal of biological chemistry* 285(11): 7866-7879.

Sonenberg, N.&Hinnebusch, A. G. (2009). "Regulation of translation initiation in eukaryotes: mechanisms and biological targets." *Cell* 136(4): 731-745.

Stan, R.; McLaughlin, M. M.; Cafferkey, R.; Johnson, R. K.; Rosenberg, M.&Livi, G. P. (1994). "Interaction between FKBP12-rapamycin and TOR involves a conserved serine residue." *The Journal of biological chemistry* 269(51): 32027-32030.

Sturgill, T. W.&Hall, M. N. (2009). "Activating mutations in TOR are in similar structures as oncogenic mutations in PI3KCalpha." *ACS chemical biology* 4(12): 999-1015.

Sun, S. Y.; Rosenberg, L. M.; Wang, X.; Zhou, Z.; Yue, P.; Fu, H.&Khuri, F. R. (2005). "Activation of Akt and eIF4E survival pathways by rapamycin-mediated mammalian target of rapamycin inhibition." *Cancer research* 65(16): 7052-7058.

Takai, H.; Wang, R. C.; Takai, K. K.; Yang, H.&de Lange, T. (2007). "Tel2 regulates the stability of PI3K-related protein kinases." *Cell* 131(7): 1248-1259.

Takai, H.; Xie, Y.; de Lange, T.&Pavletich, N. P. (2010). "Tel2 structure and function in the Hsp90-dependent maturation of mTOR and ATR complexes." *Genes & development* 24(18): 2019-2030.

Tcherkezian, J.; Brittis, P. A.; Thomas, F.; Roux, P. P.&Flanagan, J. G. (2010). "Transmembrane receptor DCC associates with protein synthesis machinery and regulates translation." *Cell* 141(4): 632-644.

Tee, A. R.; Manning, B. D.; Roux, P. P.; Cantley, L. C.&Blenis, J. (2003). "Tuberous sclerosis complex gene products, Tuberin and Hamartin, control mTOR signaling by acting as a GTPase-activating protein complex toward Rheb." *Current biology : CB* 13(15): 1259-1268.

Thedieck, K.; Polak, P.; Kim, M. L.; Molle, K. D.; Cohen, A.; Jeno, P.; Arrieumerlou, C.&Hall, M. N. (2007). "PRAS40 and PRR5-like protein are new mTOR interactors that regulate apoptosis." *PloS one* 2(11): e1217.

Thomas, G.&Hall, M. N. (1997). "TOR signalling and control of cell growth." *Current opinion in cell biology* 9(6): 782-787.

Thoreen, C. C.; Kang, S. A.; Chang, J. W.; Liu, Q.; Zhang, J.; Gao, Y.; Reichling, L. J.; Sim, T.; Sabatini, D. M.&Gray, N. S. (2009). "An ATP-competitive mammalian target of rapamycin inhibitor reveals rapamycin-resistant functions of mTORC1." *The Journal of biological chemistry* 284(12): 8023-8032.

Treins, C.; Warne, P. H.; Magnuson, M. A.; Pende, M.&Downward, J. (2010). "Rictor is a novel target of p70 S6 kinase-1." *Oncogene* 29(7): 1003-1016.

Tsang, C. K.; Liu, H.&Zheng, X. F. (2010). "mTOR binds to the promoters of RNA polymerase I- and III-transcribed genes." *Cell Cycle* 9(5): 953-957.

Tzatsos, A.&Kandror, K. V. (2006). "Nutrients suppress phosphatidylinositol 3-kinase/Akt signaling via raptor-dependent mTOR-mediated insulin receptor substrate 1 phosphorylation." *Molecular and cellular biology* 26(1): 63-76.

Um, S. H.; Frigerio, F.; Watanabe, M.; Picard, F.; Joaquin, M.; Sticker, M.; Fumagalli, S.; Allegrini, P. R.; Kozma, S. C.; Auwerx, J.&Thomas, G. (2004). "Absence of S6K1 protects against age- and diet-induced obesity while enhancing insulin sensitivity." *Nature* 431(7005): 200-205.

Valentinis, B.&Baserga, R. (2001). "IGF-I receptor signalling in transformation and differentiation." *Molecular pathology : MP* 54(3): 133-137.

van der Poel, H. G.; Hanrahan, C.; Zhong, H.&Simons, J. W. (2003). "Rapamycin induces Smad activity in prostate cancer cell lines." *Urol Res* 30(6): 380-386.

Vellai, T.; Takacs-Vellai, K.; Zhang, Y.; Kovacs, A. L.; Orosz, L.&Muller, F. (2003). "Genetics: influence of TOR kinase on lifespan in C. elegans." *Nature* 426(6967): 620.

Veverka, V.; Crabbe, T.; Bird, I.; Lennie, G.; Muskett, F. W.; Taylor, R. J.&Carr, M. D. (2008). "Structural characterization of the interaction of mTOR with phosphatidic acid and a novel class of inhibitor: compelling evidence for a central role of the FRB domain in small molecule-mediated regulation of mTOR." *Oncogene* 27(5): 585-595.

Vezina, C.; Kudelski, A.&Sehgal, S. N. (1975). "Rapamycin (AY-22,989), a new antifungal antibiotic. I. Taxonomy of the producing streptomycete and isolation of the active principle." *The Journal of antibiotics* 28(10): 721-726.

Vornlocher, H. P.; Hanachi, P.; Ribeiro, S.&Hershey, J. W. (1999). "A 110-kilodalton subunit of translation initiation factor eIF3 and an associated 135-kilodalton protein are encoded by the Saccharomyces cerevisiae TIF32 and TIF31 genes." *The Journal of biological chemistry* 274(24): 16802-16812.

Wang, L.; Harris, T. E.; Roth, R. A.&Lawrence, J. C., Jr. (2007). "PRAS40 regulates mTORC1 kinase activity by functioning as a direct inhibitor of substrate binding." *The Journal of biological chemistry* 282(27): 20036-20044.

Wang, X.; Beugnet, A.; Murakami, M.; Yamanaka, S.&Proud, C. G. (2005). "Distinct signaling events downstream of mTOR cooperate to mediate the effects of amino acids and insulin on initiation factor 4E-binding proteins." *Molecular and cellular biology* 25(7): 2558-2572.

Wang, X.; Li, W.; Parra, J. L.; Beugnet, A.&Proud, C. G. (2003). "The C terminus of initiation factor 4E-binding protein 1 contains multiple regulatory features that influence its function and phosphorylation." *Molecular and cellular biology* 23(5): 1546-1557.

Wang, X.; Li, W.; Williams, M.; Terada, N.; Alessi, D. R.&Proud, C. G. (2001). "Regulation of elongation factor 2 kinase by p90(RSK1) and p70 S6 kinase." *The EMBO journal* 20(16): 4370-4379.

Watanabe, R.; Tambe, Y.; Inoue, H.; Isono, T.; Haneda, M.; Isobe, K.; Kobayashi, T.; Hino, O.; Okabe, H.&Chano, T. (2007). "GADD34 inhibits mammalian target of rapamycin signaling via tuberous sclerosis complex and controls cell survival under bioenergetic stress." *International journal of molecular medicine* 19(3): 475-483.

Wedaman, K. P.; Reinke, A.; Anderson, S.; Yates, J., 3rd; McCaffery, J. M.&Powers, T. (2003). "Tor kinases are in distinct membrane-associated protein complexes in Saccharomyces cerevisiae." *Molecular biology of the cell* 14(3): 1204-1220.

Weichhart, T.; Haidinger, M.; Katholnig, K.; Kopecky, C.; Poglitsch, M.; Lassnig, C.; Rosner, M.; Zlabinger, G. J.; Hengstschlager, M.; Muller, M.; Horl, W. H.&Saemann, M. D. (2011). "Inhibition of mTOR blocks the anti-inflammatory effects of glucocorticoids in myeloid immune cells." *Blood* 117(16): 4273-4283.

White, R. J. (2005). "RNA polymerases I and III, growth control and cancer." *Nature reviews. Molecular cell biology* 6(1): 69-78.

Wilkinson, M. G.; Pino, T. S.; Tournier, S.; Buck, V.; Martin, H.; Christiansen, J.; Wilkinson, D. G.&Millar, J. B. (1999). "Sin1: an evolutionarily conserved component of the eukaryotic SAPK pathway." *The EMBO journal* 18(15): 4210-4221.

Woo, S. Y.; Kim, D. H.; Jun, C. B.; Kim, Y. M.; Haar, E. V.; Lee, S. I.; Hegg, J. W.; Bandhakavi, S.&Griffin, T. J. (2007). "PRR5, a novel component of mTOR complex 2, regulates platelet-derived growth factor receptor beta expression and signaling." *The Journal of biological chemistry* 282(35): 25604-25612.

Wullschleger, S.; Loewith, R.&Hall, M. N. (2006). "TOR signaling in growth and metabolism." *Cell* 124(3): 471-484.

Wullschleger, S.; Loewith, R.; Oppliger, W.&Hall, M. N. (2005). "Molecular organization of target of rapamycin complex 2." *The Journal of biological chemistry* 280(35): 30697-30704.

Yang, Q.&Guan, K. L. (2007). "Expanding mTOR signaling." *Cell research* 17(8): 666-681.

Yang, Q.; Inoki, K.; Ikenoue, T.&Guan, K. L. (2006). "Identification of Sin1 as an essential TORC2 component required for complex formation and kinase activity." *Genes & development* 20(20): 2820-2832.

Yeh, Y. Y.; Wrasman, K.&Herman, P. K. (2010). "Autophosphorylation within the Atg1 activation loop is required for both kinase activity and the induction of autophagy in Saccharomyces cerevisiae." *Genetics* 185(3): 871-882.

Yip, C. K.; Murata, K.; Walz, T.; Sabatini, D. M.&Kang, S. A. (2010). "Structure of the human mTOR complex I and its implications for rapamycin inhibition." *Molecular cell* 38(5): 768-774.

Yorimitsu, T.; He, C.; Wang, K.&Klionsky, D. J. (2009). "Tap42-associated protein phosphatase type 2A negatively regulates induction of autophagy." *Autophagy* 5(5): 616-624.

Young, D. A.&Nickerson-Nutter, C. L. (2005). "mTOR--beyond transplantation." *Current opinion in pharmacology* 5(4): 418-423.

Yu, K.; Toral-Barza, L.; Shi, C.; Zhang, W. G.; Lucas, J.; Shor, B.; Kim, J.; Verheijen, J.; Curran, K.; Malwitz, D. J.; Cole, D. C.; Ellingboe, J.; Ayral-Kaloustian, S.; Mansour, T. S.; Gibbons, J. J.; Abraham, R. T.; Nowak, P.&Zask, A. (2009). "Biochemical, cellular, and in vivo activity of novel ATP-competitive and selective inhibitors of the mammalian target of rapamycin." *Cancer research* 69(15): 6232-6240.

Yu, Y.; Yoon, S. O.; Poulogiannis, G.; Yang, Q.; Ma, X. M.; Villen, J.; Kubica, N.; Hoffman, G. R.; Cantley, L. C.; Gygi, S. P.&Blenis, J. (2011). "Phosphoproteomic analysis identifies Grb10 as an mTORC1 substrate that negatively regulates insulin signaling." *Science* 332(6035): 1322-1326.

Yuan, X.; Zhao, J.; Zentgraf, H.; Hoffmann-Rohrer, U.&Grummt, I. (2002). "Multiple interactions between RNA polymerase I, TIF-IA and TAF(I) subunits regulate preinitiation complex assembly at the ribosomal gene promoter." *EMBO reports* 3(11): 1082-1087.

Zhande, R.; Mitchell, J. J.; Wu, J.&Sun, X. J. (2002). "Molecular mechanism of insulin-induced degradation of insulin receptor substrate 1." *Molecular and cellular biology* 22(4): 1016-1026.

Zhao, J.; Brault, J. J.; Schild, A.; Cao, P.; Sandri, M.; Schiaffino, S.; Lecker, S. H.&Goldberg, A. L. (2007). "FoxO3 coordinately activates protein degradation by the autophagic/lysosomal and proteasomal pathways in atrophying muscle cells." *Cell Metab* 6(6): 472-483.

Zhao, Y.; Xiong, X.&Sun, Y. (2011). "DEPTOR, an mTOR Inhibitor, Is a Physiological Substrate of SCF(betaTrCP) E3 Ubiquitin Ligase and Regulates Survival and Autophagy." *Molecular cell* 44(2): 304-316.

Zheng, X. F.; Florentino, D.; Chen, J.; Crabtree, G. R.&Schreiber, S. L. (1995). "TOR kinase domains are required for two distinct functions, only one of which is inhibited by rapamycin." *Cell* 82(1): 121-130.

Zheng, X. F.&Schreiber, S. L. (1997). "Target of rapamycin proteins and their kinase activities are required for meiosis." *Proceedings of the National Academy of Sciences of the United States of America* 94(7): 3070-3075.

Zhou, H.&Huang, S. (2010). "The complexes of mammalian target of rapamycin." *Current protein & peptide science* 11(6): 409-424.

Zhou, H. Y.&Wong, A. S. (2006). "Activation of p70S6K induces expression of matrix metalloproteinase 9 associated with hepatocyte growth factor-mediated invasion in human ovarian cancer cells." *Endocrinology* 147(5): 2557-2566.

Zinzalla, V.; Stracka, D.; Oppliger, W.&Hall, M. N. (2011). "Activation of mTORC2 by association with the ribosome." *Cell* 144(5): 757-768.

Zoncu, R.; Efeyan, A.&Sabatini, D. M. (2011). "mTOR: from growth signal integration to cancer, diabetes and ageing." *Nature reviews. Molecular cell biology* 12(1): 21-35.

2

SNF1/AMP-Activated Protein Kinases: Genes, Expression and Biological Role

Dmytro O. Minchenko[1,2] and Oleksandr H. Minchenko[1]
[1]*Department of Molecular Biology, Palladin Institute of Biochemistry, National Academy of Sciences of Ukraine,*
[2]*National Bogomolets Medical University, Kyiv, Ukraine*

1. Introduction

Most of the physiological and metabolic processes in any organism are controlled by a regulatory factor network, which includes a lot of protein kinases, protein phosphatases and transcription factors. Protein kinases and phosphatases are key regulators of the majority of transcription factors which control metabolism both in normal and in different pathological conditions; it is a circadian type of regulation [1–5]. AMPK-related kinases SNARK and NUAK1 as well as many others AMPK-related kinases represent molecular components of signalling cascades that control metabolism, gene expression and perhaps cell proliferation in response to cellular, metabolic and environmental stresses [6–8].

The sucrose-non-fermenting protein kinase (SNF1) from *Saccharomyces cerevisiae* and its mammalian counterpart, AMP-activated protein kinase (AMPK), form a family of serine/threonine kinases that acts as a master sensor and regulator of the energy balance at the cellular level as well as the stress response systems, has been critical to our understanding of the whole body energy homeostasis [9]. This family of protein kinases is highly conserved between animals, fungi and plants and is commonly activated in response to cellular and environmental stresses such as nutrient deprivation. Yeast SNF1 responds to glucose deprivation by derepressing genes implicated in carbon source utilization and by modulating the transcription of glucose-regulated genes involved in gluconeogenesis, respiration, sporulation, thermotolerance, peroxisome biogenesis and cell cycle regulation. Activated by environmental stresses AMPK switches off anabolic pathways (e.g. fatty acid and cholesterol synthesis) and induces ATP generating catabolic pathways [9]. Twelve protein kinases (NUAK1, NUAK2, BRSK1, BRSK2, SIK, QIK, QSK, MARK1, MARK2, MARK3, MARK4 and MELK) in the human kinome are closely related to AMPKα_1 and AMPKα_2, thus forming a 14 kinase phylogenetic tree known as "AMPK-related kinases" which represent components of signalling cascades that control metabolism, gene expression and perhaps cell proliferation in response to cellular, metabolic and environmental stresses [9].

The AMP-activated protein kinase system acts as a sensor of cellular energy status that is conserved in all eukaryotic cells. It is activated by a large variety of cellular stresses that increase cellular AMP and decrease ATP levels and also by physiological stimuli, such as muscle contraction, or by hormones such as leptin and cellular adiponectin as well as by metabolic stresses that either interferes with ATP production or that accelerate ATP consumption [10]. AMPK modulates multiple metabolic pathways. Activation in response to an increase in AMP involves phosphorylation by an upstream kinase, the tumour suppressor LKB1. Once activated, AMPK switches on catabolic pathways that generate ATP, while switching off ATP-consuming processes such as biosynthesis and cell growth and proliferation. Thus, it is a key player in the development of new treatments for obesity, the metabolic syndrome, type 2 diabetes or even cancer. In fact, it has been recently reported that drugs used in the treatment of diabetes, such as metformin and thiazolidinediones, exert their beneficial effects through the activation of AMPK.

The sucrose-non-fermenting protein kinase (SNF1)/AMP-activated protein kinase-related kinase (SNF1/AMP-activated protein kinase; SNARK) is a member of AMPK kinases (NUAK family SNF1-like kinase 2) which is related to serine/threonine protein kinases [7, 9]. SNARK activity is regulated by glucose- or glutamine-deprivation, induction of endoplasmic reticulum stress by dithyothreitol or homocysteine, elevation of cellular AMP and/or depletion of ATP, hyperosmotic stress, salt stress and oxidative stress caused by hydrogen peroxide. However, the regulation of SNARK activity in response to cellular stresses depends greatly upon cell type. It was also shown that SNARK is also regulated by metabolic stress and diabetes [11]. Nuclear localization of SNARK has shown its impact on gene expression [12].

Tsuchihara et al. [13] demonstrated that SNARK(+/-) mice exhibit mature-onset obesity and related metabolic disorders. Moreover, the incidence of both adenomas and aberrant crypt foci were significantly higher in SNARK(+/-) mice than in their wild-type counterparts, suggesting that SNARK deficiency contributed to the early phase of tumourigenesis via obesity-dependent and obesity-independent mechanisms [14]. Recently, Namiki et al. [14] have shown that AMP kinase-related kinase SNARK affects tumour growth, migration, and clinical outcome of human melanoma, further supporting the importance of this protein kinase in cancer development and tumour progression, while AMPK has antioncogenic properties. We have also shown that the SNF1/AMP-activated protein kinase-related kinase is a sensitive marker for the action of ecotoxicant methyl tert-butyl ether (MTBE) as well as silver nanoparticles [15, 16]. These observations support a role for SNARK as a molecular component of the cellular stress response.

SNF1-like kinase 1 (NUAK1) is an AMP-activated protein kinase family member 5, ARK5, which regulates ploidy and senescence, tumour cell survival, malignancy and invasion downstream of Akt signaling, acts as an ATM kinase under the conditions of nutrient starvation [17-20]. Moreover, NUAK1 suppresses the apoptosis, induced by nutrient starvation, and death receptors via inhibition of caspase-8 and caspase-6 activation [21, 22]. Importantly, AMPK-related kinase NUAK1 as well as many others AMPK kinases (including MARK/PAR-1) is regulated by protein kinase LKB1 and USP9X [23, 24].

2. NUAK family SNF1-like kinase 2 (NUAK2), Sucrose nonfermenting AMPK-related kinase (SNARK)

Human NUAK family SNF1-like kinase 2 (NUAK2; EC_number "2.7.11.1") also known as sucrose nonfermenting AMPK-related kinase or skeletal muscle sucrose, nonfermenting 1/adenosine monophosphate activated protein kinase-related kinase (SNARK) is an AMP-activated protein kinase family member 4, which was identified 10 years ago as an potential mediator of cellular response to metabolic stress [9].

2.1 NUAK2 gene, transcripts and encoded proteins

The human NUAK2 (SNARK) gene (geneID: 81788) is localized on chromosome 1 (1q32,1). The SNARK gene encodes mRNA (GenBank accession number NM_030952) of seven exons. Northern blotting demonstrated that mRNA transcripts (at least two variants) for the SNARK were widely expressed in rodent tissues, but most abundant in rat kidney. Reverse-transcriptase-mediated PCR detected two SNARK cDNA products in RNA from rat heart, skin, spleen, lung, uterus, liver and a neonatal rat keratinocyte cell line, NRKC. The two different SNARK PCR products were cloned, sequenced and found to encode either authentic SNARK (1437 bp) or an internally deleted SNARK transcript (1247 bp) [9]. Whereas rat kidney contained predominantly the intact SNARK transcript and testes expressed only the 1247 bp SNARK transcript, both intact and internally deleted SNARK transcripts were detected in other tested tissues. The ORF encodes a putative protein of 630 amino acid residues with a predicted molecular mass of 70 kDa and a theoretical pI of 9.35. Translation of the SNARK-deleted transcript is predicted to give rise to a prematurely terminated protein of approximately 415 amino acid residues [9].

Although no autophosphorylated products were detected in samples of immunoprecipitated endogenous SNARK from wild type BHK cells, one major phosphorylated band, possibly a protein doublet, was detected in the immunoprecipitates from SNARK-transfected BHK cells [9]. The size of the phosphorylated band(s) corresponds to the size of SNARK detected in these cell lines by Western blot analysis. Thus, these results demonstrate that SNARK is a protein kinase capable of autophosphorylation *in vitro*. Besides that, immunoprecipitated SNARK protein exhibits phosphotransferase activity with the synthetic peptide substrate HMRSAMSGLHLVKRR as a kinase substrate [9]. SNARK was translated in vitro to yield a single protein band of approximately 76 kDa, possibly a protein doublet, however, Western analysis of transfected BHK (baby hamster kidney) cells detected two SNARK-immunoreactive bands of approximately 76 - 80 kDa.

The NUAK family SNF1-like kinase 2 (NUAK2 or SNARK) is a member 4 of AMPK kinases which are related to serine/threonine protein kinases and contains all 11 catalytic subdomains conserved in these protein kinases. Analysis of the catalytic domain of SNARK with the Prosite program revealed a protein kinase ATP-binding region signature (residues 63 - 89) and a serine/threonine protein kinase active-site signature (residues 175 - 187). The sequences at the C-terminus of SNARK were distinct and not well conserved with C-terminal sequences of other SNF1/AMPK family members. The instability index is computed to be 58.40 with the Protparam Tool program, classifying protein kinase SNARK as an unstable protein [9].

Comparison of the SNARK catalytic subdomains I - XI to other SNF-1/AMPK family members demonstrates that protein kinase SNARK originated very early in eukaryotic evolution, diverging before the divergence of yeast and humans [9]. On the basis of the phylogeny of the catalytic subdomains, SNARK is no more closely related to SNF1 than it is to AMPK and represents a new branch of the SNF1/AMPK family of protein kinases.

2.2 Protein kinase SNARK, its activity and regulation

The NUAK family SNF1-like kinase 2 is a member of AMPK kinases, it is commonly activated in response to cellular and environmental stresses, and it is a molecular component of the cellular stress response, but its precise mechanisms remain unclear [7, 9, 11, 25]. Its activity is regulated by glucose- or glutamine-deprivation, induction of endoplasmic reticulum stress by homocysteine or dithiothreitol, hyperosmotic stress, salt stress, elevation of cellular AMP and/or depletion of ATP, ultraviolet B radiation and oxidative stress caused by hydrogen peroxide. However, the regulation of protein kinase SNARK activity in response to cellular stresses depends greatly upon cell type. Several aspects of SNARK activation and regulation are broadly similar to AMPK [8]. For example, SNARK and AMPK are both AMP-responsive and activated by treatments known to increase the AMP:ATP ratio, including glucose deprivation and chemical ATP production [9, 12]. Nevertheless, the metabolic role of SNARK at the cellular level, particularly in humans, especially in skeletal muscle, is incompletely resolved.

Kuga et al. [12] identified the subcellular localization of SNARK protein. Unlike cytoplasmic localizing AMPKα, SNARK was predominantly localized in the nucleus. This protein kinase is constitutively distributed in the nucleus; even when SNARK is activated by metabolic stimuli such as the AMP-mimetic agent, 5-aminoimidazole-4-carboxamide riboside (AICAR) or glucose-deprivation. Conserved nuclear localization signal was identified at the N-terminal portion (^{68}KKAR^{71}) of protein kinase SNARK. Deletion and point mutation of this part resulted in the cytoplasmic translocation of mutant proteins. Furthermore, GFP fused with the SNARK fragment containing ^{68}KKAR^{71} translocated to the nucleus.

A microarray analysis revealed that nuclear localized SNARK alteres transcriptome profiles and a considerable part of these alterations were canceled by the mutation of nuclear localization signal (first two core lysine residues of ^{68}KKAR^{71} were altered to alanine (^{68}AAAR^{71})), suggesting the ability of SNARK to modulate gene expression is dependent on its nuclear localization. It has been shown that overexpression of protein kinase SNARK in human liver hepatoma cells results in the upregulation (more than 2.0-fold) of 76 mRNA targets and in the downregulation (more than 2.0-fold) of 32 mRNA targets, suggesting that this protein kinase can work as a stress-responsive transcriptional modulator in the nucleus [12].

Moreover, transcriptome profiles of wild-type and nuclear localized signal-mutant SNARK expressing cells were compared to identify the impact of the nuclear localization of SNARK on the regulation of mRNA levels of potential downstream genes. Among the 76 up-regulated probe sets by overexpressed SNARK, only eight probe sets increased more than 2.0-fold in ^{68}AAAR^{71}-overexpressing cells compared with vector-transfected cells. On the other hand, among the 32 down-regulated probe sets by overexpressed SNARK, only 13 probe sets decreased more than 2.0-fold in ^{68}AAAR^{71}-overexpressing cells compared with vector-transfected cells. Thus, overexpressed SNARK altered the gene expression profiles more than nuclear localization signal-mutant SNARK. This result implied that protein kinase SNARK in the nucleus, but not the cytoplasm, has a remarkable impact on gene expression and can work in the nucleus as a transcriptional modulator in response to stress. This data may become a platform to elucidate the molecular mechanism and the physiological signification of protein kinase SNARK.

AMPK and AMPK-related kinases are believed to be activated by increased AMP:ATP ratio through a direct activation mechanism of the allosteric effect and/or indirectly activated by phosphorylation at threonine residue in the activation loop by upstream kinases, LKB1 (serine/threonine protein kinase 11, STK11), CaMKK (calcium/calmodulin-dependent protein kinase kinase 1, alpha), and TAK1 (mitogen-activated protein kinase kinase kinase 7; MAP3K7) [25]. CaMKK and TAK1 are localized in the cytoplasm, but LKB1 is localized in both nucleus and cytoplasm. Therefore SNARK might be phosphorylated in the nucleus by protein kinase LKB1 [12].

Rune et al. [11] showed that skeletal muscle SNARK expression is also regulated by metabolic stress and increases in human obesity, and in response to metabolic stressors. This increase in SNARK mRNA expression may occur as a consequence of systemic factors associated with metabolic impairments in obesity, since exposure of myotubes to elevated levels of TNF-α or palmitate acutely increased SNARK mRNA expression. siRNA against SNARK failed to rescue TNFα- or palmitate-induced insulin resistance, indicating that changes in SNARK expression occur as a consequence, rather than a cause of insulin resistance. Based on this data in human skeletal muscle, in the insulin-resistant and obesity phenotype in whole-body SNARK-haploinsufficient mice [13], SNARK expression in metabolically active tissues beyond skeletal muscle may play a role in whole body energy and glucose homeostasis.

Interestingly, SNARK has anti-apoptotic properties, acting through a TNF-α-sensitive nuclear NF-κB-mediated mechanism. Thus, the SNF1/AMP kinase-related kinase 2, which is induced in response to various forms of metabolic stress, was identified as an NF-κB-regulated anti-apoptotic kinase that contributes to the tumour-promoting activity of death receptor CD95 (APO-1/Fas) in apoptosis-resistant tumour cells [26]. The death receptor CD95 induces apoptosis in many tissues. However, in apoptosis-resistant tumour cells, stimulation of CD95 induces up-regulation of a defined number of mostly anti-apoptotic genes, resulting in increased motility and invasiveness of tumour cells. The majority of these genes are known NF-κB target genes. One of the CD95-regulated genes is the serine/threonine kinase (SNARK). It was shown that up-regulation of SNARK in response to CD95 ligand and tumour necrosis factor α depends on activation of NF-κB. Overexpression of SNARK rendered tumour cells more resistant, whereas a kinase-inactive mutant of SNARK sensitizes cells to CD95-mediated apoptosis. Furthermore, small interfering RNA-mediated knockdown of SNARK increases the sensitivity of tumour cells to death receptor CD95 ligand- and TRAIL-induced apoptosis. Importantly, cells with reduced expression of SNARK also showed reduced motility and invasiveness in response to CD95 engagement. SNARK therefore represents an NF-κB-regulated anti-apoptotic gene that contributes to the tumour-promoting activity of CD95 in apoptosis-resistant tumour cells [26].

Kim et al. [27] have investigated the effect of Epstein-Barr virus (EBV) latent membrane protein 1 (LMP1) on human cancer cells for identification of potential target genes. It was found that LMP1 upregulated the expression of protein kinase SNARK compared with the empty vector transfected control cells. Moreover, SNARK expression increased drug resistance in response to doxorubicin, whereas knockdown of SNARK by siRNA effectively inhibited LMP-1-mediated increase of cell survival. SNARK stimulates the expression of anti-apoptotic genes BCL6 and BIRC2; knockdown of these genes decreased the SNARK-mediated increase of cell survival. These results suggest that SNARK is a downstream cellular target of LMP1 in malignant cells [27].

2.3 Protein kinase SNARK and tumourigenesis

Members of the AMP kinase family play an important role in tumourigenesis [6]. This activity is believed to be due to their activation by various forms of metabolic stress such as glucose deprivation, a condition to be expected within solid tumours [28]. Recently, Namiki et al. [14] showed that AMP kinase-related kinase NUAK2 affects tumour growth, migration, and clinical outcome of human melanoma. This study further supports the importance of NUAK2 in cancer development and tumour progression, while AMPK has antioncogenic properties.

Although several *in vitro* studies have suggested that metabolic stress as well as genotoxic or osmotic stresses induce SNARK activation, the physiological roles of protein kinase SNARK remain uncertain. Using SNARK-deficient mice helps to clarify the *in vivo* function of this kinase. Interestingly, SNARK(+/-) mice exhibited mature-onset obesity and related metabolic disorders [13]. Thus, an increased bodyweight in these mice is accompanied by fat deposition, fatty changes of the liver, and increased serum triglyceride concentration. These mice also exhibited hyperinsulinemia, hyperglycemia, and glucose intolerance, symptoms which are similar to those of human type II diabetes mellitus accompanied with obesity. Obesity is regarded as a risk factor for colorectal cancer. To investigate whether SNARK deficiency is involved in tumorigenesis in the large intestine, obese SNARK(+/-) mice were treated with a chemical carcinogen, azoxymethane, a chemical carcinogen that induces aberrant crypt foci, colorectal adenoma, and adenocarcinoma. The incidences of both adenomas and aberrant crypt foci were significantly higher in SNARK(+/-) mice than in their wild-type counterparts, suggesting that SNARK deficiency contributed to the early phase of tumourigenesis via obesity-dependent and -independent mechanisms [13].

2.4 Activation of protein kinase SNARK during muscle contraction

Sakamoto et al. [29] have shown that deficiency of LKB1 in skeletal muscle prevents AMPK activation and glucose uptake during contraction. In LKB1-lacking muscle, the basal activity of the AMPKalpha2 isoform was greatly reduced and was not increased by the AMP-mimetic agent, 5-aminoimidazole-4-carboxamide riboside (AICAR), by the antidiabetic drug phenformin, or by muscle contraction. Moreover, phosphorylation of acetyl CoA carboxylase-2, a downstream target of AMPK, was profoundly reduced. Glucose uptake stimulated by AICAR or muscle contraction, but not by insulin, was inhibited in the absence of LKB1. Contraction increased the AMP:ATP ratio to a greater extent in LKB1-deficient muscles than in LKB1-expressing muscles. These studies establish the importance of LKB1 in regulating AMPK activity and cellular energy levels in response to contraction and phenformin.

Recently, Koh et al. [30] showed that muscle contraction also increases protein kinase SNARK activity and that this effect blunted in the muscle-specific LKB1 knockout mice. It is known that the signaling mechanisms that mediate the important effects of contraction, to increase glucose transport in skeletal muscle, occur through an insulin-independent mechanism. Moreover, muscle-specific knockout of protein kinase LKB1, an upstream kinase for AMPK and AMPK-related protein kinases, significantly inhibited contraction-stimulated glucose transport, suggests that one or more AMPK-related protein kinases are important for this process. It has been shown that expression of a mutant SNARK in mouse tibialis anterior muscle impaired contraction-stimulated, but not insulin-stimulated, glucose transport. The impaired contraction-stimulated glucose transport was also observed in skeletal muscle of whole-body SNARK heterozygotic knockout mice [30]. Thus, SNARK, the fourth member of the AMP-activated protein kinase catalytic subunit family, is activated by muscle contraction and is a unique mediator of contraction-stimulated glucose transport in skeletal muscle.

There is data that NUAK2 is a TNFalpha-induced kinase which regulates myosin phosphatase target subunit 1 (MYPT1) activity by phosphorylation at a site other than known Rho-kinase phosphorylation sites (Thr696 or Thr853) responsible for inhibition of myosin phosphatase activity [31]. Moreover, Suzuki et al. [32] observed the induction of cell-cell detachment during glucose starvation through F-actin conversion by protein kinase SNARK. Recently, Vallenius et al. [33] have shown that an association between AMP kinase-related kinase SNARK and myosin phosphatase Rho-interacting protein (MRIP) reveals a novel mechanism for regulation of actin stress fibers via activation of MLCP (myosin light chain phosphatase). Moreover, new roles for the LKB1-NUAK pathway in controlling myosin phosphatase complexes and cell adhesion have been shown [34].

2.5 Protein kinase SNARK as a regulator of whole-body metabolism

Ichinoseki-Sekine et al. [35] provide evidence for a robust effect on whole body metabolism by hemiallelic SNARK deficiency, suggesting that this AMPK-related kinase is a previously unrecognized important regulator of whole-body metabolism. They have investigated the *in vivo* effects of altering expression of SNARK by using hemiallelic loss of SNARK on whole body metabolic homeostasis and physical activity behaviour. Homozygous SNARK-deficient mice have a high incidence of embryonic lethality, whereas the heterozygous SNARK-deficient mice have an obvious metabolic phenotype with mature-onset obesity and increased white adipose tissue mass evident after the age of 4 month.

Activation of SNARK by upstream kinase LKB1 occurs by phosphorylation of Thr^{208}, a conserved threonine residue equivalent in position to Thr^{172} within the activation loop of $AMPK\alpha_2$ [36]. LKB1 is attractive as a regulator of SNARK activity by virtue of its nuclear localization, which is coincident with the predominant nuclear localization of SNARK [12], but SNARK may also directly mediate some physiological effects of LKB1. Several aspects of SNARK regulation and activity are broadly similar to those of AMPK, which can be summarized as follows. First, SNARK possesses AMPK-like phosphotransferase activity; second, activation of SNARK is AMP responsive; third, SNARK activity is increased by AICAR, albeit in a cell-specific manner; and fourth, SNARK is activated by treatments known to increase AMP/ATP ratio or disrupt ATP production, including glucose deprivation and chemical ATP depletion among others [36].

Possible AMPK activation of SNARK, secondary to the activation of AMPK by these treatments, has not been investigated but raises the possibility that one or more activities previously attributed to the AMPK-signaling cascade may be attributable, in part, to SNARK activation. In addition, similarities between AMPK and SNARK regulation do not necessarily infer that SNARK activity directly mirrors AMPK activity in the context of cellular metabolism. Cell-specific differences are reported between SNARK and AMPK activity and pharmacological activation as well as in the relative rates of phosphorylation and peptide substrates phosphorylated [36].

Examining metabolic and anthropometric effects of SNARK deficiency, the core finding in Ichinoseki-Sekine et al. [35] investigation is that the provision of voluntary exercise opportunities to SNARK(+/-) mice results in habitually increased daily physical activity (~2-fold) compared with SNARK(+/+) mice, commensurate with the prevention of mature-onset obesity to which these animals are genetically predisposed. Physical activity resulted in a reduction in total body mass, liver mass, and white and brown adipose tissue mass in both exercise groups compared with sedentary controls. At termination of the study, body mass was similar between genotypes in the physically active mice. The prevention of weight gain in the active SNARK(+/-) mice occurred despite a 10% increase in food intake. Differences in physical activity were not attributable to sex, age, or disrupted circadian rhythm, nor were they attributable to any intrinsic deficit in forced exercise capacity/muscle energetic associated with SNARK deficiency. Direct SNARK-dependent modulation of whole body metabolism, similar to AMPK effects in the context of carbohydrate and lipid metabolism has not been demonstrated yet. However, SNARK is predominantly and constitutively localized in the nucleus, where it is likely to be regulated by protein kinase LKB1 or other unidentified kinases [12].

Interestingly, the SNARK gene expression and kinase activity is tissue specific, and its activity profile differs significantly from the AMPKα_2 activity profile. Therefore, targeting SNARK could potentially affect whole body metabolic homeostasis, and a more thorough examination of the physiological role of SNARK is warranted. Thus, protein kinase SNARK is a novel regulator of whole body metabolic homeostasis and highlights yet another protein kinase as an exciting new addition to the already extensive paradigm of homeostatic regulation by cellular energy sensors.

2.6 Protein kinase SNARK and PFKFB-3 alternative splicing

The new aspect of the biological role of protein kinase SNARK was demonstrated in SNARK-deficient mice by investigation of the expression of 6-phosphofructo-2-kinase/fructose-2,6-bisphosphatase-3 (PFKFB-3) mRNA and its alternative splice variants in the liver, lungs, testes, heart, and skeletal muscle [37]. Bifunctional enzyme PFKFB is a key regulatory enzyme of glycolysis which also participates in glucose phosphorylation [38, 39]. The PFKFB-3 expression significantly increased due to hypoxia in different normal and cancer cell lines via hypoxia inducible transcription factor (HIF)-dependent mechanism [40, 41]. Hypoxia also induces expression of PFKFB-3 in different mouse organs in vivo, except skeletal muscle [42]. High expression level and phosphorylation status of PFKFB-3 as an important glycolytic regulator was determined in different malignant tumours [43–47]. Because SNARK deficiency contributed to the early phase of tumourigenesis and is important in cancer development and tumour progression [13, 14], investigation of the expression of PFKFB-3 and its alternative splice variants, which have different proliferative properties [48], is necessary for understanding the role of SNARK deficiency in tumourigenesis.

As shown in Fig. 1, the expression levels of PFKFB-3 mRNA significantly increases in the liver and lung of SNARK(-/-) knockout mice as compared to corresponding tissues of control C57BL/6 mice [37]. At the same time, PFKFB-3 mRNA expression level in skeletal muscle of SNARK knockout mice decreases without significant changes in the heart as compared to control animals. Thus, SNARK deficiency leads to variable changes of PFKFB-3 mRNA expression in different mouse tissues.

Reverse-transcriptase-mediated PCR of the carboxyl-terminus of PFKFB-3 mRNA detected three - four cDNA products in RNA from the liver, lung, testis, heart, and skeletal muscle of control C57BL/6 and SNARK knockout mice (Fig. 2). This heterogeneity is a result of alternative splicing of the PFKFB-3 mRNA in tissue specific manner. Alternative splice variants of PFKFB-3 mRNA were identified by sequence analysis of cloned fragments. The major difference among the members of these bifunctional enzyme alternative splice variants is the length and composition of the carboxyl-terminal region (Fig. 3 and 4), supporting the idea that this terminus of the various enzyme isoforms serve to adapt the kinetic properties of the catalytic core to metabolic exigencies of a particular tissue. Alternative splice variants of PFKFB-3 also have different amounts and sequence positions of serine residues which are very important in the regulation of isozyme activity via phosphorylation [39, 46]. It was shown that the pattern of alternative splice variants of the PFKFB-3 mRNA differs in different mouse organs (Fig. 2).

Results of this study strongly support the SNARK dependent regulation of PFKFB alternative splicing. Thus, the level of smallest alternative splice variant increases in the heart and liver of SNARK knockout mice compared with control mice. However, the level of longest alternative splice variant decreases in the skeletal muscle of SNARK knockout mice compared with control mice (Fig, 2). Therefore, investigation of different alternative splice variants of PFBFB-3 isozymes is important for comprehension of tissue-specific regulation mechanisms of glycolysis. The precise molecular mechanisms, whereby SNARK participates in the splicing of PFKFB as well as a role of different isoenzymes in the regulation of glycolysis, await further study.

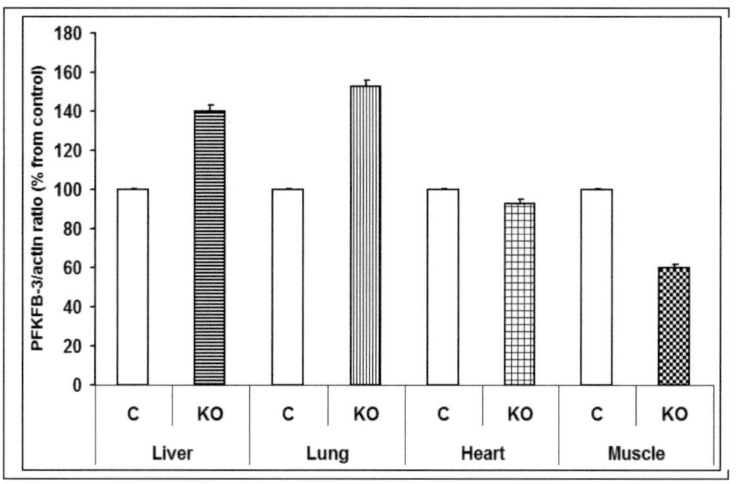

Fig. 1. Real time PCR analysis of PFKFB-3 mRNA expression in liver, lung, heart and skeletal muscle of control C57BL/6 male mice (C) and SNARK(-/-) knockout mice (KO). Amplification of PFKFB-3 mRNA was carried out using M4 forward and M5 reverse primers. Intensities of PFKFB-3 mRNA expression were normalized to β-actin mRNA [37].

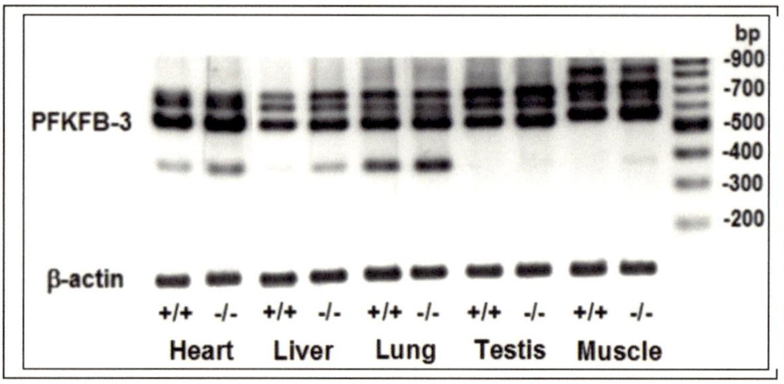

Fig. 2. RT-PCR analysis of PFKFB-3 mRNA expression in the heart, liver, lung, testis and skeletal muscle of control C57BL/6 (+/+) and SNARK knockout mice (-/-). Amplification of PFKFB-3 mRNA was carried out using M3 forward and M6 reverse primers. The amplified PCR products were run on an agarose gel. Intensities of PFKFB-3 mRNA bands were normalized to β-actin mRNA [37].

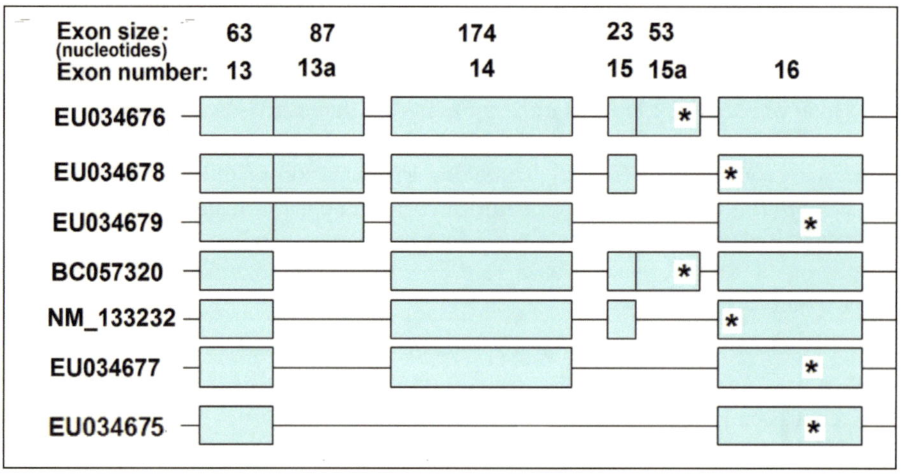

Fig. 3. Schematic representation of exon structure of mouse PFKFB-3 mRNA alternative splice variants. Most of these splice variants do not have exon 13th. Some of splice variants have 15th and 15tha exons which alters the reading frame, amino acid sequence and length of C-terminus. One splice variant is shortest because it does not have exon 14th. Position of three possible stop codons for the different alternative splice variants of PFKFB-3 mRNA are shown by asterisk. The GenBank accession number of alternative splice variants of mouse PFKFB-3 is noted on the left [38].

2.7 SNARK kinase as a stress sensor and sensitive marker of silver nanoparticles and methyl tert-butyl ether action

Recently we have shown that the expression of SNF1/AMP-activated protein kinase (SNARK) increases in different organs of male Wistar rats intratracheally instilled by 30% silver nanoparticles (28-30 nm) in sodium chloride matrix aerosol in dose 50 μg/kg (or 0.05 mg/kg) body weight (=15 μg of silver) [15]. Silver nanoparticles were prepared in the

Laboratory No. 84 of the Paton Electric Welding Institute of The National Academy of Sciences of Ukraine. The expression levels of the SNARK mRNA were analyzed in the lung, liver, brain, heart, kidney and testis using quantitative polymerase chain reaction on the 1st, 3rd or 14th day after one-time intratracheally treated rats with silver nanoparticles.

```
                    450       460       470       480       490       500
EU034676 - THRERSEAVKIQHFASVVRPSSYTELDFQSVESAKQDAKKGPNPLMRRNSVTPLASPEPT
EU034678 - ************************************************************
EU034679 - ************************************************************
BC057320 - *******-----------------------------************************
NM_133232- *******-----------------------------************************
EU034677 - *******-----------------------------************************
EU034675 - *******-----------------------------------------------------

                    510       520       530       540       550
EU034676 - KKPRINSFEERVASTSAALPSCLPPEVPTQLPGQPLLGKACLRSVCHIFSKFSPY
EU034678 - **********************************NMRSPRSGAESSQKH
EU034679 - **********************************PLLGKACLT
BC057320 - **********************************PLLGKACLRSVCHIFSKFSPY
NM_133232- **********************************NMRSPRSGAESSQKH
EU034677 - **********************************PLLGKACLT
EU034675 - ----------------------------------NMRSPRSGAESSQKH
```

Fig. 4. Amino acid sequence of alternative splice variants of mouse PFKFB-3. Differences in amino acid sequences and length of C-terminus of different alternative splice variants of mouse PFKFB-3 are shown. Serine residues are underlined. The GenBank accession number of alternative splice variants of mouse PFKFB-3 is noted on the left [38].

It was shown that the expression of protein kinase SNARK mRNA increases in the liver, lung and brain on the 1st, 3rd and 14th day after one-time treatment of rats with silver nanoparticles, being more intense (more than 2 fold) on the 3rd and 14th day in the brain and liver and on the 1st day in the lung (Fig. 5 and 6). Results of Fig. 6 and 7 also indicated that SNARK mRNA expression does not change significantly in the heart and testes on the 1st

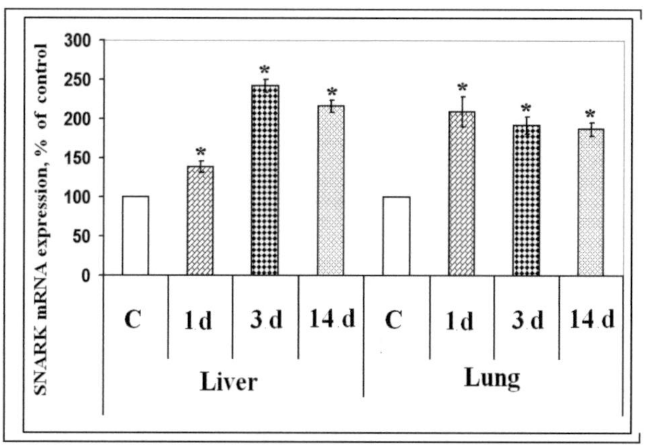

Fig. 5. The effect of silver nanoparticles on the expression of protein kinase SNARK mRNA in the liver and lung in 1, 3 or 14 days (d) after treatment. Values of SNARK mRNA expressions were normalized to β-actin mRNA expression; C – control; $n = 3$; * - $P < 0.05$ [15].

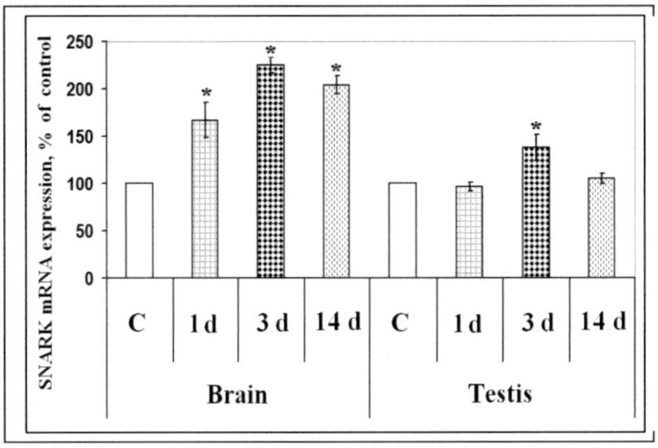

Fig. 6. The effect of silver nanoparticles on the expression of protein kinase SNARK mRNA in the brain and testis in 1, 3 or 14 days (d) after treatment. Values of SNARK mRNA expressions were normalized to β-actin mRNA expression; C – control; $n = 3$; * - $P < 0.05$ [15].

and 14th days after treatment of rats with silver nanoparticles, but there is a clear increase of SNARK expression on the 3rd day as compared to control animals. Results of Fig. 7 show that SNARK mRNA expression also increases in the kidney on the 1st, 3rd and 14th day after treatment of rats with silver nanoparticles, being more intense on the 3rd and 14th days; however, this induction is significantly less when compared to the liver, lung or brain.

Thus, one-time intratracheally instilled silver nanoparticles change the expression of the protein kinase SNARK in different rat organs, not only in the lung tissue. Moreover, this effect of silver nanoparticles on the expression of the protein kinase SNARK strongly depends on time after the treatment of rats with these nanoparticles in a tissue-specific manner. These results correlate to data from Lefebvre and Rosen [7] whom have shown that the regulation of protein kinase SNARK activity in response to cellular stresses greatly depends upon cell type.

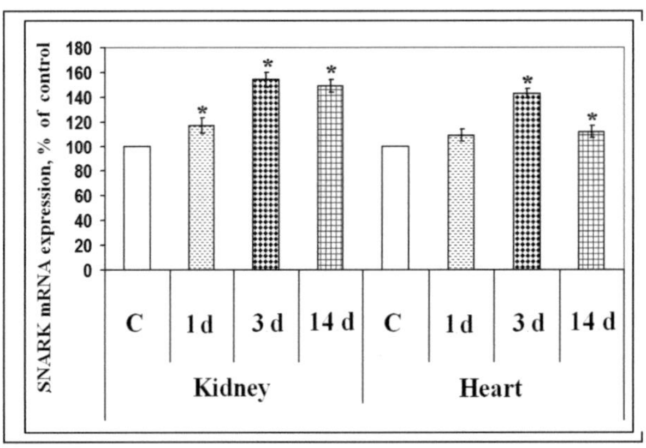

Fig. 7. The effect of silver nanoparticles on the expression of protein kinase SNARK mRNA in the kidney and heart in 1, 3 or 14 days (d) after treatment. Values of SNARK mRNA expressions were normalized to β-actin mRNA expression; C – control; $n = 3$; * - $P < 0.05$ [15].

Shimada et al. [49] demonstrated that the intratracheally instilled ultrafine nanoparticles are able to translocate from the mouse lung into systemic circulation. Precise mechanisms of the anatomical translocation (crossing the air-blood barrier) of inhaled nanoparticles at the alveolar wall are not fully understood. Silver nanoparticles are widely used in the field of biomedicine, but a comprehensive understanding of how silver nanoparticles distribute in the body and the induced toxicity remains largely unknown. Tang et al. [50] investigated the distribution and accumulation of silver nanoparticles in rats with subcutaneous injection. Rats were injected with either silver nanoparticles SNPs or silver microparticles (SMPs) at 62.8 mg/kg, and then sacrificed at predetermined time points. Silver content analysis by Inductively coupled plasma mass spectrometry was used for determination of silver content in different organs. Results indicated that silver nanoparticles translocated into the blood circulation and distributed throughout the main organs, especially in the kidney, liver, spleen, brain and lung in the form of particles. Ultrastructural observations indicate that those silver nanoparticles that had accumulated in organs could enter different kinds of cells. Moreover, silver nanoparticles also induced blood-brain barrier (BBB) destruction and astrocyte swelling, and caused neuronal degeneration [50].

There is data that silver nanoparticles are more toxic than silver microparticles or ions [51–53]. Powers et al. [53] have shown that silver nanoparticles have the potential to evoke developmental neurotoxicity even more potently than known neurotoxicants. Silver ions inhibited replication and increased cell death in undifferentiated cells, and selectively impaired neurite formation. Silver nanoparticles in D. melanogaster induce heat shock stress, oxidative stress, DNA damage and apoptosis [54]. Thus, silver nanoparticles up-regulate the expression of heat shock protein 70, the cell cycle checkpoint p53 and cell signaling protein p38 all of which are involved in the DNA damage repair pathway. Moreover, the activity of caspase-3 and caspase-9, markers of apoptosis was significantly higher in silver nanoparticles exposed organisms.

It is possible that silver nanoparticles create stress conditions which affect the expression of protein kinase SNARK mRNA via unfolded protein response signals through activation of inositol requiring enzyme-1 (endoplasmic reticulum–nuclei-1) and alternative splicing of XBP-1 [55 - 57]. Endoplasmic reticulum stress signalling activates inositol requiring enzyme-1 and XBP-1 which control diverse cell type- and condition-specific transcriptional regulatory networks. However, the cellular mechanism for survival under stress conditions is complex and further investigation of the mechanism by which silver nanoparticles affects protein kinase SNARK expression as well as biologic significance of silver nanoparticles induced alteration in the expression of these genes is needed. The stage is now set for the elucidation of the molecular mechanisms responsible for these important SNARK responses to silver nanoparticles action.

Results of investigations clearly demonstrate that silver nanoparticles have a significant effect on important regulatory mechanisms which control metabolic processes in different tissues via SNARK gene expression, which can be considered as a sensitive marker for silver nanoparticles action. These results suggest that more caution is needed in biomedical applications of silver nanoparticles as well as higher level of safety in the silver nanoparticles production industry.

We have also studied the effect of different doses of ecotoxicant methyl tertbutyl ether on the expression protein kinase SNARK in the liver, lung and heart [16]. Results of this investigation demonstrated that methyl tertbutyl ether affects the expression protein kinase SNARK in the liver, lung and heart in dose dependent and tissue specific manner and that very small dose induce the expression of protein kinase SNARK in all tested vital organs in rats (Fig. 8 and 9). There is data that methyl tertbutyl ether can initiate the variety of neurotoxic, allergic and respiratory illnesses, liver hypertrophy and leukaemia in humans as well as the following cancers in rats and mice: kidney, liver, testicular and lymph nodes, initiate development of leukemia [58, 59]. We have recently shown that methyl tertbutyl ether affects the expression of PFKFB-4 mRNA and its alternative splicing [60]. Thus, SNARK gene expression can be considered as sensitive markers for the methyl tertbutyl ether action.

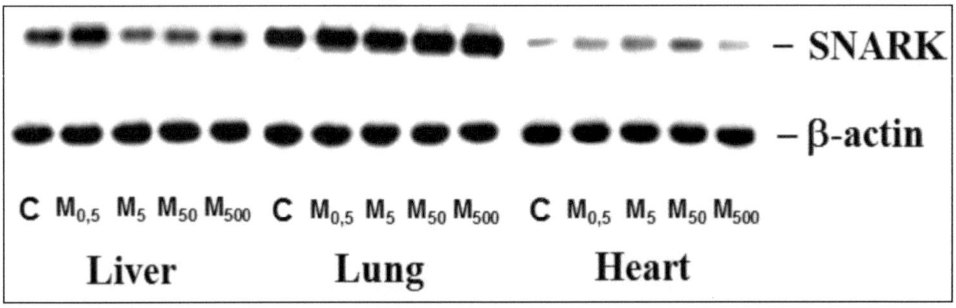

Fig. 8. Effect of methyl tertbutyl ether [0.5 ($M_{0,5}$); 5 (M_5); 50 (M_{50}) та 500 (M_{500}) mg/kg body weight during two months] on SNARK mRNA expression in the liver, lung and heart by reverse-transcriptase-mediated PCR. C - control rats.

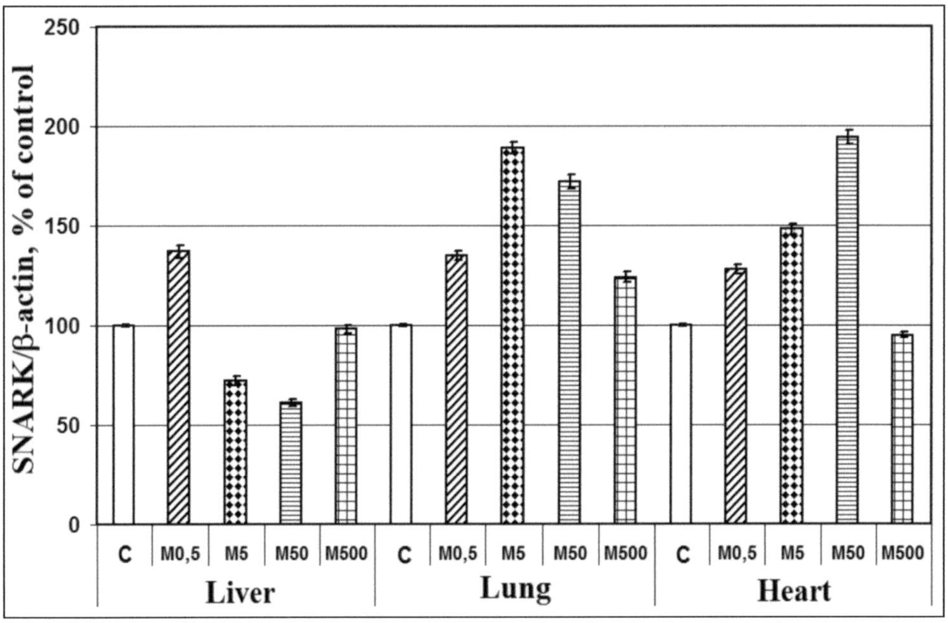

Fig. 9. Effect of methyl tertbutyl ether [0.5 ($M_{0,5}$); 5 (M_5); 50 (M_{50}) та 500 (M_{500}) mg/kg body weight during two months] on the expression levels of SNARK mRNA in the liver, lung and heart by quantitative PCR. C - control rats.

3. SNF1-like kinase, 1 (NUAK1) AMP-activated protein kinase family member 5, ARK5

Human NUAK family SNF1-like kinase 1 (NUAK1; EC_number "2.7.11.1") is an AMP-activated protein kinase family member, ARK5. AMP-activated protein kinases (AMPKs) are a class of serine/threonine protein kinases that are activated by an increase in intracellular AMP concentration. They are a sensitive indicator of cellular energy status and have been found to promote tumor cell survival during nutrient starvation. The human gene encoded protein kinase NUAK1 (ARK5) is located on chromosome 12 ("12q23.3"; GeneID: 9891). This gene encodes a protein of 661 amino acid residues. ARK5, which is the fifth member of the AMPK catalytic subunit family, is a tumor malignancy-associated factor at the downstream of Akt [19]. ARK5 is a tumour cell survival and invasion-associated factor. The activated ARK5 induces cell survival during nutrient starvation and death receptor activation, and tumor cell invasion and metastasis [18–21].

However, the precise mechanisms of how ARK5 activity inhibits caspase dependent cell death remains to be determined. Both cell death and cell survival are important for cellular homeostasis; therefore, an imbalance of their signalling causes several disease states, including tumourogenesis.

ARK5, as an AMP-activated protein kinase family member 5, is a tumour progression-associated factor that is directly phosphorylated by AKT at serine 600 in the regulatory domain, but phosphorylation at the conserved threonine residue on the active T loop has been found to be required for its full activation [18, 19]. Suzuki et al. [61] identified serine/threonine protein kinase NDR2 as a protein kinase that also phosphorylates and activates ARK5 during insulin-like growth factor-1 (IGF-1) signalling. Upon stimulation with IGF-1, protein kinase NDR2 was found to directly phosphorylate the conserved threonine 211 on the active T loop of protein kinase ARK5 and to promote cell survival and invasion of colorectal cancer cell lines through ARK5.

During IGF-1 signaling, phosphorylation at three residues (threonine 75, serine 282 and threonine 442) was also found to be required for NDR2 activation. Among these three residues, phosphorylation of serine 282 seemed to be most important for NDR2 activation (the same as for the mouse homologue) because its aspartic acid-converted mutant (NDR2/S282D) induced ARK5-mediated cell survival and invasion activities even in the absence of IGF-1. Threonine 75 in protein kinase NDR2 was required for interaction with protein S100B, and binding was in a calcium ion-dependent and phospholipase C-gamma-dependent manner [61]. Thus, protein kinase NDR2 is an upstream kinase of ARK5 that plays an essential role in tumour progression through an AMP-activated protein kinase ARK5.

NUAK1 acts as an ATM kinase under the conditions of nutrient starvation and plays a key role in tumour malignancy downstream of Akt signalling [19]. Matrigel invasion assays demonstrated that both overexpressed and endogenous ARK5 showed strong Akt dependent activity. In addition, ARK5 expression induced activation of matrix metalloproteinase 2 (MMP-2) and MMP-9. In nude mice, ARK5 expression was associated with a significant increase in tumour growth and significant suppression of necrosis in tumour tissue. Interestingly, only the ARK5-overexpressing PANC-1 cell line tumour showed invasion and metastasis in nude mice, although Akt was activated in tumours derived from both PANC-1 and ARK5-overexpressing PANC-1 cell lines.

Suzuki et al. [21] have investigated the mechanisms of induction of cell survival by protein kinase ARK5 and have shown that ARK5 suppresses the apoptosis induced by nutrient starvation and death receptors via inhibition of caspase 8 activation. Thus, human hepatoma HepG2 cells undergo necrotic cell death within 24 h after the start of glucose starvation, and the cell death signaling has been found to be mediated by death-receptor-independent activation of caspase 8. When HepG2 cells were transfected with ARK5 expression vector and subjected to several cell death stimuli, ARK5 was found to suppress cell death by glucose starvation and TNF-alpha, but not by camptothecin or doxorubicin. Western blotting analysis revealed that glucose starvation induced Bid cleavage and FLIP degradation following caspase 8 activation in a time-dependent manner, and ARK5 overexpression clearly delayed Bid cleavage, FLIP degradation, and caspase 8 activation. These results demonstrated that cell survival induced by ARK5 is, at least in part, due to inhibition of caspase 8 activation.

AMP-activated protein kinase family member 5 also negatively regulates procaspase-6 by phosphorylation at serine 257, leading to resistance to the FasL/Fas system the key regulator promoting cell death and cell survival [22]. Fas is a type I transmembrane protein mediating intracellular cell death signalling upon the stimulation of Fas ligand (FasL). When Fas is activated by the ligation of FasL, an intracellular interaction of Fas death domain (Fas-DD), FADD, and caspase-8 (death inducing signalling complex (DISC) recruitment) is initiated for the activation of executioner caspase; and cellular FLIP is well known as the inhibitor of DISC recruitment. The serine/threonine protein kinase Akt induces cell survival as a result of phosphorylation and several cell death-associated factors, such as Bad, caspase-9 and Forkhead, upon the stimulation of growth factor receptor and integrin-induced cell signaling. Although active caspase-6 overexpression induced cell death in SW480 and DLD-1 cell lines, SW480 cells, but not DLD-1 cells, exhibite strong resistance to procaspase-6 overexpression. Moreover, mutant caspase-6, in which the serine 257 was substituted by alanine (caspase-6/SA), induced cell death and FLIP degradation, even in SW480 cells. Active ARK5 was found to phosphorylate wild-type caspase-6 in vitro, but not caspase-6/SA, and the prevented activation of caspase-6 was promoted due to its phosphorylation by active ARK5 in vitro.

AMPK-related kinases NUAK1 and many others (including MARK/PAR-1) are regulated by protein kinase LKB1 and USP9X [23, 24]. Moreover, there is data that the LKB1-NUAK pathway plays important role in controlling myosin phosphatase complexes and cell adhesion [34]. NUAK1 regulates ploidy and senescence; cells that constitutively express NUAK1 suffer gross aneuploidies and show diminished expression of the genomic stability regulator LATS1, whereas depletion of NUAK1 with shRNA exerts opposite effects [17].

AMP-activated protein kinase-related kinase 5 (ARK5/NUAK1) is expressed in rat skeletal muscle and phosphorylated by electrically elicited contractions and 5-aminoimidazole-4-carboxamide-1-beta-d-ribofuranoside (AICAR). Increased phosphorylation of ARK5 by muscle contractions or exposure to AICAR, however, is insufficient to activate this protein kinase in skeletal muscle, suggesting that some other modification (e.g., phosphorylation on tyrosine or by Akt) may be necessary for its activity in muscle [62].

4. Conclusions

NUAK family SNF1-like kinase includes two kinases, NUAK1 and NUAK2; both are members of AMP-activated protein kinases which are related to serine/threonine protein kinases. The sucrose-non-fermenting protein kinase (SNF1)/AMP-activated protein kinase-related kinase (SNF1/AMP-activated protein kinase; SNARK) is a member 4 of AMPK kinases (NUAK family SNF1-like kinase 2; NUAK2). SNF1-like kinase 1 (NUAK1) is an AMP-activated protein kinase family member 5, ARK5. Protein kinase NUAK2 (SNARK) is a molecular component of the cellular stress response and an important regulator of whole-body metabolism. Protein kinase SNARK was consistently localized in the nuclei. It has been shown that the nuclear localizing SNARK alters transcriptome profiles. It therefore represents a NF-κB-regulated anti-apoptotic gene that contributes to the tumour-promoting activity of death receptor CD95 in apoptosis-resistant tumour cells and plays an important role in cancer development and tumour progression. SNARK affects tumour growth, migration, and clinical outcome of human melanoma. Protein kinase SNARK is also activated by muscle contraction and is a unique mediator of contraction-stimulated glucose transport in skeletal muscle. Moreover, association between AMP kinase-related kinase SNARK and myosin phosphatase Rho-interacting protein reveals a novel mechanism for regulation of actin stress fibers via activation of myosin light chain phosphatase. Protein kinase NUAK1 (ARK5) regulates ploidy and senescence, tumour cell survival, malignancy and invasion downstream of Akt signalling and suppresses apoptosis induced by nutrient starvation and death receptors via inhibition of caspase-8 and caspase-6 activation. It is interesting to note that the expression of SNARK is a sensitive marker of silver nanoparticles and methyl tert-butyl ether toxic effect.

5. References

[1] Mizoguchi, T., Putterill, J. and Ohkoshi, Y. (2006) Kinase and phosphatase: the cog and spring of the circadian clock. Int. Rev. Cytol., 250, 47–72.

[2] Walton, K.M., Fisher, K., Rubitski, D., Marconi, M., Meng, Q.J., Sládek, M., Adams, J., Bass, M., Chandrasekaran, R., Butler, T., Griffor, M., Rajamohan, F., Serpa, M., Chen, Y., Claffey, M., Hastings, M., Loudon, A., Maywood, E., Ohren, J., Doran, A. and Wager, T.T. (2009) Selective inhibition of casein kinase 1 epsilon minimally alters circadian clock period. J. Pharmacol. Ex. Ther., 330, 430–439.

[3] Meng, Q.J., Maywood, E.S., Bechtold, D.A., Lu, W.Q., Li, J., Gibbs, J.E., Dupré, S.M., Chesham, J.E., Rajamohan, F., Knafels, J., Sneed, B., Zawadzke, L.E., Ohren, J.F., Walton, K.M., Wager, T.T., Hastings, M.H. and Loudon, A.S. (2010) Entrainment of disrupted circadian behavior through inhibition of casein kinase 1 (CK1) enzymes. Proc. Natl. Acad. Sci. U.S.A., 107, 15240–15245.

[4] Huang, W., Ramsey, K.M. and Marcheva, B. (2011) Circadian rhythms, sleep, and metabolism. J. Clin. Invest., 121, 2133–2141.

[5] Kovac J., Husse J., Oster H. (2009) A time to fast, a time to feast: the crosstalk between metabolism and the circadian clock. *Mol. Cells*, 28, 75–80.

[6] Hardie, G. D., Carling, D. and Carlson, M. (1998) The AMP-activated/SNF1 protein kinase subfamily : metabolic sensors of the eukaryotic cell? *Annu. Rev. Biochem.* 67, 821-855.
[7] Lefebvre, D.L. and Rosen, C.F. (2005) Regulation of SNARK activity in response to cellular stresses. *Biochim. Biophys. Acta*, 1724, 71–85.
[8] Egan, B. and Zierath, J.R. (2009) Hunting for the SNARK in metabolic disease. Am. J. Physiol. Endocrinol. Metab., 296, E969–E972.
[9] Lefebvre, D.L., Bai, Y., Shahmolky, N., Sharma, M., Poon, R., Drucker, D.J. and Rosen, C.F. (2001) Identification and characterization of a novel sucrose-non-fermenting protein kinase/AMP-activated protein kinase-related protein kinase, SNARK. *Biochem. J.*, 355, 297–305.
[10] Sanz, P. (2008) AMP-activated protein kinase: structure and regulation. Curr. Protein Pept. Sci., 9, 478–492.
[11] Rune, A., Osler, M.E., Fritz, T. and Zierath, J.R. (2009) Regulation of skeletal muscle sucrose, non-fermenting 1/AMP-activated protein kinase-related kinase (SNARK) by metabolic stress and diabetes. *Diabetologia*, 52, 2182–2189.
[12] Kuga, W., Tsuchihara, K., Ogura, T., Kanehara, S., Saito, M., Suzuki, A. and Esumi, H. (2008) Nuclear localization of SNARK; its impact on gene expression. *Biochem. Biophys. Res. Commun.*, 377, 1062–1066.
[13] Tsuchihara, K., Ogura, T., Fujioka, R., Fujii, S., Kuga, W., Saito, M., Ochiya, T., Ochiai, A., Esumi, H. (2008) Susceptibility of Snark-deficient mice to azoxymethaneinduced colorectal tumorigenesis and the formation of aberrant crypt foci. *Cancer Sci.*, 99, 677–682.
[14] Namiki, T., Tanemura, A., Valencia, J.C., Coelho, S.G., Passeron, T., Kawaguchi, M., Vieira, W.D., Ishikawa, M., Nishijima, W., Izumo, T., Kaneko, Y., Katayama, I., Yamaguchi, Y., Yin, L., Polley, E.C., Liu, H., Kawakami, Y., Eishi, Y., Takahashi, E., Yokozeki, H. and Hearing, V.J. (2011) AMP kinase-related kinase NUAK2 affects tumor growth, migration, and clinical outcome of human melanoma. *Proc. Natl. Acad. Sci. U.S.A.*, 108, 6597–6602.
[15] Minchenko, D., Bozhko, I., Zinchenko, T., Yavorovsky, O., Minchenko, O. (2011) Expression of SNF1/AMP-activated protein kinase and casein kinase-1epsilon in different rat tissues is a sensitive marker of in vivo silver nanoparticles action. *Materialwissenschaft und Werkstofftechnik*, 42, N 2, 21–25.
[16] Minchenko, O.H., Minchenko, D.O., Yavorovsky, O.P., Zavhorodny, I.V., Paustovsky, Y.O., Tsuchihara, K., Esumi, H. (2008) Expression of casein kinase-1□ and SNARK in the liver, lung and heart as a marker of methyl tertbutyl ether action on the organism of laboratory animals. *Scientific Bulletin National Bogomolets Medical University*, 21, 21–27.
[17] Humbert, N., Navaratnam, N., Augert, A., Da Costa, M., Martien, S., Wang, J., Martinez, D., Abbadie, C., Carling, D., de Launoit, Y., Gil, J. and Bernard, D. (2010) Regulation of ploidy and senescence by the AMPK-related kinase NUAK1. *EMBO J.*, 29, 376–386.
[18] Suzuki, A., Kusakai, G., Kishimoto, A., Lu, J., Ogura, T., Lavin, M.F. and Esumi, H. (2003) Identification of a novel protein kinase mediating Akt survival signaling to the ATM protein. *J. Biol. Chem.*, 278, 48--53.

[19] Suzuki, A., Lu, J., Kusakai, G., Kishimoto, A., Ogura, T. and Esumi, H. (2004) ARK5 is a tumor invasion-associated factor downstream of Akt signaling. *Mol. Cell. Biol.*, 24, 3526–3535.

[20] Kusakai, G., Suzuki, A., Ogura, T., Kaminishi, M. and Esumi, H. (2004) Strong association of ARK5 with tumor invasion and metastasis. J. Exp. Clin. Cancer Res., 23, 263–268.

[21] Suzuki, A., Kusakai, G., Kishimoto, A., Lu, J., Ogura, T. and Esumi, H. (2003) ARK5 suppresses the cell death induced by nutrient starvation and death receptors via inhibition of caspase 8 activation, but not by chemotherapeutic agents or UV irradiation. *Oncogene*, 22, 6177–6182.

[22] Suzuki, A., Kusakai, G., Kishimoto, A., Shimojo, Y., Miyamoto, S., Ogura, T., Ochiai, A. and Esumi, H. (2004) Regulation of caspase-6 and FLIP by the AMPK family member ARK5. *Oncogene*, 23, 7067–7075.

[23] Al-Hakim, A.K., Zagorska, A., Chapman, L., Deak, M., Peggie, M. and Alessi, D.R. (2008) Control of AMPK-related kinases by USP9X and atypical Lys(29)/Lys(33)-linked polyubiquitin chains. *Biochem. J.*, 411, 249–260.

[24] Lizcano, J.M., Goransson, O., Toth, R., Deak, M., Morrice, N.A., Boudeau, J., Hawley, S.A., Udd, L., Makela, T.P., Hardie, D.G. and Alessi, D.R. (2004) LKB1 is a master kinase that activates 13 kinases of the AMPK subfamily, including MARK/PAR-1. *EMBO J.*, 23, 833–843.

[25] Williams, T. and Brenman, J.E. (2008) LKB1 and AMPK in cell polarity and division, *Trends Cell. Biol.*, 18, 193–198.

[26] Legembre, P., Schickel, R., Barnhart, B.C. and Peter, M.E. (2004) Identification of SNF1/AMP kinase-related kinase as an NF-kappaB-regulated anti-apoptotic kinase involved in CD95-induced motility and invasiveness. *J. Biol. Chem.*, 279, 46742–46747.

[27] Kim, J.H., Kim, W.S. and Park, C. (2008) SNARK, a novel downstream molecule of EBV latent membrane protein 1, is associated with resistance to cancer cell death. Leuk. Lymphoma, 49, 1392–1398.

[28] Denko, N.C. (2008) Hypoxia, HIF1 and glucose metabolism in the solid tumour. *Nature Reviews Cancer*, 8, 705–713.

[29] Sakamoto, K., McCarthy, A., Smith, D., Green, K.A., Grahame Hardie, D., Ashworth, A. and Alessi, D.R. (2005) Deficiency of LKB1 in skeletal muscle prevents AMPK activation and glucose uptake during contraction. EMBO J., 24, 1810–1820.

[30] Koh, H.J., Toyoda, T., Fujii, N., Jung, M.M., Rathod, A., Middelbeek, R.J., Lessard, S.J., Treebak, J.T., Tsuchihara, K., Esumi, H., Richter, E.A., Wojtaszewski, J.F., Hirshman, M.F. and Goodyear, L.J. (2010) Sucrose nonfermenting AMPK-related kinase (SNARK) mediates contraction-stimulated glucose transport in mouse skeletal muscle. *Proc. Natl. Acad. Sci. U.S.A.*, 107, 15541–15546.

[31] Yamamoto,H., Takashima,S., Shintani,Y., Yamazaki,S., Seguchi,O., Nakano,A., Higo,S., Kato,H., Liao, Y., Asano, Y., Minamino, T., Matsumura, Y., Takeda, H. and Kitakaze, M. (2008) Identification of a novel substrate for TNFalpha-induced kinase NUAK2. *Biochem. Biophys. Res. Commun.*, 365, 541–547.

[32] Suzuki, A., Kusakai, G., Kishimoto, A., Minegichi, Y., Ogura, T. and Esumi, H. (2003) Induction of cell-cell detachment during glucose starvation through F-actin

conversion by SNARK, the fourth member of the AMP-activated protein kinase catalytic subunit family. *Biochem. Biophys. Res. Commun.*, 311, 156-161.

[33] Vallenius, T., Vaahtomeri, K., Kovac, B., Osiceanu, A.M., Viljanen, M. and Makela, T.P. (2011) An association between NUAK2 and MRIP reveals a novel mechanism for regulation of actin stress fibers. *J. Cell. Sci.*, 124, 384–393.

[34] Zagorska, A., Deak, M., Campbell, D.G., Banerjee, S., Hirano, M. Aizawa, S., Prescott, A.R. and Alessi, D.R. (2010) New roles for the LKB1-NUAK pathway in controlling myosin phosphatase complexes and cell adhesion. *Sci. Signal*, 3, RA25.

[35] Ichinoseki-Sekine, N., Naito, H., Tsuchihara, K., Kobayashi, I., Ogura Y., Kakigi, R., Kurosaka, M., Fujioka, R. and Esumi, H. (2009) Provision of a voluntary exercise environment enhances running activity and prevents obesity in Snark-deficient mice. Am. J. Physiol. Endocrinol. Metab., 296, E1013–E1021.

[36] Kahn, B.B., Alquier, T., Carling, D. and Hardie, D.G. (2005) AMP-activated protein kinase: ancient energy gauge provides clues to modern understanding of metabolism. *Cell Metab.*, 1, 15–25.

[37] Minchenko, D.O., Tsuchihara, K., Komisarenko, S.V., Moenner, M., Bikfalvi, A., Esumi, H., Minchenko O.H. (2008) Unique alternative splice variants of mouse PFKFB-3 mRNA: tissue specific expression. *Scientific Bulletin National Bogomolets Medical University*, 20, 22–31.

[38] Minchenko, D.O., Bobarykina, A.Y., Kundieva, A.V., Lypova N.M., Bozhko, I.V., Ratushna, O.O. and Minchenko O.H. (2009) 6-phosphofructo-2-kinase/fructose-2,6-biphosphatase genes: structural organization, expression and regulation of the expression. *Studia Biologica*, 3(3), 123–140.

[39] Rider, M.H., Bertrand, L., Vertommen, D., Michels, P.A., Rousseau, G.G. and Hue, L. (2004) 6-phosphofructo-2-kinase/fructose-2,6-biphosphatase: head-head with a bifunctional enzyme that controls glycolysis. *Biochem. J.*, 381, 561–579.

[40] Minchenko, A.G., Leshchinsky, I., Opentanova, I., Sang, N., Srinivas, V., Armstead, V.E. and Caro, J. (2002) Hypoxia-inducible factor-1-mediated expression of the 6-phosphofructo-2-kinase/fructose-2,6-bisphosphatase-3 (PFKFB3) gene. Its possible role in the Warburg effect. *J. Biol. Chem.*, 277, 6183–6187.

[41] Bobarykina, A.Y., Minchenko, D.O., Opentanova, I.L., Moenner, M., Caro, J., Esumi, H. and Minchenko, O.H. (2006) Hypoxic regulation of PFKFB-3 and PFKFB-4 gene expression in gastric and pancreatic cancer cell lines and expression of PFKFB genes in gastric cancers. *Acta Biochim. Pol.*, 53, 789–799.

[42] Minchenko, O., Opentanova, I. and Caro, J. (2003) Hypoxic regulation of the 6-phosphofructo-2-kinase/fructose-2,6-bisphosphatase gene family (PFKFB-1-4) expression in vivo. *FEBS Lett.*, 554, 264–270.

[43] Minchenko, O.H., Ochiai, A., Opentanova, I.L., Ogura, T., Minchenko, D.O., Caro, J., Komisarenko, S.V. and Esumi, H. (2005) Overexpression of 6-phosphofructo-2-kinase/fructose-2,6-bisphosphatase-4 in the human breast and colon malignant tumors. *Biochimie*, 87, 1005–1010.

[44] Minchenko, D.O., Bobarykina, A.Y., Senchenko, T.Y., Hubenya, O.V., Tsuchihara, K., Ochiai, A., Moenner, M., Esumi, H. and Minchenko, O.H. (2009) Expression of the VEGF, Glut1 and 6-phosphofructo-2-kinase/fructose-2,6-bisphosphatase-3 and -4 in human cancers of the lung, colon and stomach. *Studia Biologica*, 3(1), 25–34.

[45] Atsumi, T., Chesney, J., Metz, C., Leng, L., Donnelly, S., Makita, Z., Mitchell, R. and Bucala, R. (2002) High expression of inducible 6-phosphofructo-2-kinase/fructose-2,6-bisphosphatase-3 (iPFK-2; PFKFB3) in human cancers. *Cancer Res.*, 62, 5881–5887.

[46] Bando, H., Atsumi, T., Nishio, T., Niwa, H., Mishima, S., Shimizu, C., Yoshioka, N., Bucala, R. and Koike, T. (2005) Phosphorylation of the 6-phosphofructo-2-kinase/fructose 2,6-bisphosphatase/PFKFB3 family of glycolytic regulators in human cancer. Clin. Cancer Res., 11, 5784–5792.

[47] Kessler, R., Bleichert, F., Warnke, J.P. and Eschrich, K. (2008) 6-Phosphofructo-2-kinase/fructose-2,6-bisphosphatase (PFKFB3) is up-regulated in high-grade astrocytomas. J. Neurooncol., 86, 257–264.

[48] Duran, J., Gómez, M., Navarro-Sabate, A., Riera-Sans, L., Obach, M., Manzano, A., Perales, J.C. and Bartrons, R. (2008) Characterization of a new liver- and kidney-specific pfkfb3 isozyme that is downregulated by cell proliferation and dedifferentiation. Biochem. Biophys. Res. Commun., 367, 748–754.

[49] Shimada, A., Kawamura, N., Okajima, M., Kaewamatawong, T., Inoue, H. and Morita, T. (2006) Translocation pathway of the intratracheally instilled ultrafine particles from the lung into the blood circulation in the mouse. *Toxicol. Pathol.*, 34, 949–957.

[50] Tang, J., Xiong, L., Wang, S., Wang, J., Liu, L., Li, J., Yuan, F. and Xi, T. (2009) Distribution, translocation and accumulation of silver nanoparticles in rats. J. Nanosci. Nanotechnol., 9, 4924–4932.

[51] Gaiser, B.K., Fernandes, T.F., Jepson, M., Lead, J.R., Tyler, C.R. and Stone, V. (2009) Assessing exposure, uptake and toxicity of silver and cerium dioxide nanoparticles from contaminated environments. Environ. Health., 8, Suppl. 1, S2.

[52] Lubick, N. (2008) Nanosilver toxicity: ions, nanoparticles or both? *Environ. Sci. Technol.*, 42, 8617.

[53] Powers, C.M., Wrench, N., Ryde, I.T., Smith, A.M., Seidler, F.J. and Slotkin, T.A. (2010) Silver impairs neurodevelopment: studies in PC12 cells. Environ Health Perspect., 118, 73–79.

[54] Ahamed, M., Posgai, R., Gorey, T.J., Nielsen, M., Hussain, S.M. and Rowe, J.J. (2010) Silver nanoparticles induced heat shock protein 70, oxidative stress and apoptosis in Drosophila melanogaster. Toxicol. Appl. Pharmacol., 242, 263–269.

[55] Lin, J.H., Li, H., Yasumura, D., Cohen, H.R., Zhang, C., Panning, B., Shokat, K.M., Lavail, M.M. and Walter, P. (2007) IRE1 signaling affects cell fate during the unfolded protein response. *Science*, 318, 944–949.

[56] Aragón, T., van Anken, E., Pincus, D., Serafimova, I.M., Korennykh, A.V., Rubio, C.A. and Walter, P. (2009) Messenger RNA targeting to endoplasmic reticulum stress signalling sites. Nature, 457, 736–740.

[57] Acosta-Alvear, D., Zhou, Y., Blais, A., Tsikitis, M., Lents, N.H., Arias, C., Lennon, C.J., Kluger, Y. and Dynlacht, D.D. (2007) XBP1 controls diverse cell type- and condition-specific transcriptional regulatory networks. *Mol. Cell*, 27, 53–66.

[58] McGregor, D. (2007) Ethyl tertiary-butyl ether: a toxicological review. Critical Review Toxicol., 2007; 37(4): 287–312.

[59] Hutcheon, D. E., Arnold, J. D., Hove, W. and Boyle, J. 3rd. (1996) Disposition, metabolism, and toxicity of methyl tertiary butyl ether, an oxygenate for reformulated gasoline. J. Toxicol. Environ. Health, 47, 453–464.

[60] Minchenko, D.O., Mykhalchenko, V.G., Tsuchihara, K., Kanehara, S., Yavorovsky, O.P., Zavgorodny, I.V., Paustovsky, Y.O., Komisarenko, S.V., Esumi, H. and Minchenko, O.H. (2008) Unique alternative splice variants of rat 6-phosphofructo-2-kinase/fructose-2,6-bisphosphatase-4 mRNA. *Ukr. Biochem. J.*, 80(4), 66–73.

[61] Suzuki, A., Ogura T. and Esumi, H. (2006) NDR2 acts as the upstream kinase of ARK5 during insulin-like growth factor-1 signaling. J. Biol. Chem., 281, 13915–13921.

[62] Fisher, J.S., Ju, J.S., Oppelt, P.J., Smith, J.L., Suzuki, A. and Esumi, H. (2005) Muscle contractions, AICAR, and insulin cause phosphorylation of an AMPK-related kinase. Am. J. Physiol. Endocrinol. Metab., 289, E986-E992.

3

Alternating Phosphorylation with O-GlcNAc Modification: Another Way to Control Protein Function

Victor V. Lima and Rita C. Tostes*
Department of Pharmacology, School of Medicine of Ribeirao Preto, University of Sao Paulo, Ribeirao Preto-SP, Brazil

1. Introduction

As widely known, reversible phosphorylation of proteins, or the addition of a phosphate ($PO4^{3-}$) molecule to a polar R group of an amino acid residue, is an important regulatory mechanism that switches many enzymes and receptors "on" or "off" and therefore controls a range of cellular functions. Regulatory roles of phosphorylation include biological thermodynamics of energy-requiring reactions, enzyme and receptors' activation or inhibition, protein-protein interaction via recognition domains, protein degradation.

Kinases and phosphatases are involved in this process and these enzymes induce phosphorylation and dephosphorylation, respectively, of target proteins. Phosphorylation usually occurs on serine, threonine, and tyrosine (*O*-linked), or histidine (*N*-linked) residues of proteins, although arginine and lysine residues can also be phosphorylated.

O-GlcNAcylation, or glycosylation with *O*-linked β-N-acetylglucosamine, is similar to protein phosphorylation in that both modifications occur on serine and threonine residues, both are dynamically added and removed in response to cellular signals, and both alter the function and associations of the modified protein. *O*-GlcNAcylation also modulates many cellular functions by mechanisms that include protein targeting to specific substrates, transient complex formation with other proteins, subcellular compartmentalization upon glycosylation of specific proteins and a complex interplay with protein *O*-phosphorylation, the main topic of this chapter. Accordingly, in this chapter we will discuss the biology of the *O*-GlcNAc modification, the interplay between *O*-GlcNAcylation and *O*-phosphorylation, signaling pathways modified by *O*-GlcNAcylation, and the physiological implications of alternating *O*-GlcNAcylation and *O*-phosphorylation.

2. The biology of the *O*-GlcNAc

Glycosylation is the site-specific enzymatic addition of saccharides [from the Greek word *sákkharon* (meaning sugar); also known in biochemistry as carbohydrates or hydrates of

* Corresponding Author

carbon due to the chemical empirical formula $C_m(H_2O)_n$] to proteins and lipids. Glycosylation has many functions in a cell: it allows correct folding of proteins (some proteins do not fold correctly unless they are glycosylated first); confers stability (some unglycosylated proteins are more rapidly degraded); allows cell-cell adhesion (e.g. surface glycoproteins are directly involved in the biological functions of lymphocytes); and modulates intracellular signaling pathways (glycosylation of proteins may enhance or inhibit enzymes' activities) (Spiro, 2002; Taylor & Drickamer, 2006; Varki et al., 2009).

There are many types of glycosylation: *N*-linked, where the carbohydrate is attached to a nitrogen of asparagine or arginine side-chains; *O*-linked, where glycans are attached to the hydroxy oxygen of serine, threonine, tyrosine, hydroxylysine, or hydroxyproline side-chains; phospho-linked, where the sugar is attached via the phosphate of a phospho-serine; *C*-linked, where the carbohydrate is added to a carbon on a tryptophan side-chain; the formation of a glycosylphosphatidylinositol (GPI) anchor (glypiation), where the sugar is linked to phosphoethanolamine, which in turn is attached to the terminal carboxyl group of the protein (Spiro, 2002; Taylor & Drickamer, 2006; Varki et al., 2009). However, great interest has been directed to *O*-GlcNAcylation, or glycosylation of proteins with *O*-linked β-N-acetylglucosamine.

Cellular glycoproteins were initially thought to be targeted, after their synthesis, only to luminal or extracellular compartments. However, in 1984, Torres and Hart, who were interested in characterizing the role of cell-surface saccharides in the development and functions of lymphocytes, described a novel carbohydrate (N-acetylglucosamine, GlcNAc)-peptide linkage, which was present on proteins localized in the cytosol and the cyto- and nucleoplasmic faces of membranous organelles (Torres & Hart, 1984). In 1989, Kelly and Hart described that Drosophila polytene chromosomes (i.e., polytene chromosome spreads prepared from the salivary glands of third instar stage *Drosophila melanogaster* larvae) contained a surprisingly large amount of terminal GlcNAc residues along their lengths. Nearly all of the chromatin-associated GlcNAc moieties existed as single monosaccharide residues attached to protein by an *O*-linkage (*O*-GlcNAc) (Kelly & Hart, 1989). Also in the late 80's, the glycosyltransferase responsible for the addition of GlcNAc to proteins was found to be oriented with its active site in the cytoplasm and the first proteins modified with *O*-GlcNAc were described (Hart et al., 1988, 1989; Hart, 1997). These initial observations, which indicated a functional or biological significance for the *O*-linkage of GlcNAc to proteins, led to the term *O*-GlcNAcylation. Accordingly, *O*-GlcNAcylation is currently defined as an unusual form of protein glycosylation, where a single-sugar [N-acetylglucosamine (*O*-GlcNAc)] is added (β-attachment) to the hydroxyl moiety of serine (Ser) and threonine (Thr) residues of nuclear and cytoplasmic proteins.

It is unusual in that it is found in nuclear and cytoplasmic proteins, representing the first reported example of glycosylated proteins found outside of the secretory channels. Unlike other peptide-linked monosaccharides, the β-linked GlcNAc-Ser/Thr does not become further substituted by other sugars, remaining a single monosaccharide modification of the protein to which it is attached. *O*-GlcNAcylation is widely dispersed among eukaryotes, from protozoa to higher mammals. The amino acid consensus sequence or glycosylation motifs for the formation of *O*-GlcNAc bonds have not yet been found. However, information relating to the polypeptide domains that favors *O*-GlcNAc attachment has been obtained and seems to involve PEST [proline (P), glutamic acid (E), serine (S), and threonine (T)] sequences (Haltiwanger et al., 1997; Rogers et al., 1986).

Results from recent proteomic studies, from different laboratories, suggest that more than 1500 proteins in the cell are modified by *O*-GlcNAc. These proteins belong to almost every functional class of proteins including transcription or translation factors, cytoskeletal proteins, nuclear pore proteins, RNA polymerase II, tumor suppressors, hormone receptors, phosphatases, and kinases (Khidekel et al., 2007; Nandi et al., 2006; Wang et al., 2008; Vosseller et al., 2006). A database of *O*-GlcNAcylated proteins and sites, dbOGAP, was recently created and is primarily based on literature published since *O*-GlcNAcylation was first described in 1984. The database currently contains ~800 proteins with experimental *O*-GlcNAcylation information. The *O*-GlcNAcylated proteins are primarily nucleocytoplasmic, and include membrane- and non-membrane bounded organelle-associated proteins (Wang et al., 2011). An *O*-GlcNAcylation site prediction system (*O*-GlcNAcScan) based on nearly 400 *O*-GlcNAcylation sites was also developed (Hu, 2010). Both the database and the prediction system are publicly available at *http://cbsb.lombardi.georgetown.edu/OGAP.html* and *http://cbsb.lombardi.georgetown.edu/filedown.php*, respectively.

The attachment of the single-sugar ß-N-acetylglucosamine via an *O*-linkage to Ser/Thr residues is controlled by two highly conserved enzymes, *O*-GlcNAc transferase (OGT or uridine diphospho-N-acetyl glucosamine; polypeptide β-N-acetylglucosaminyl transferase; UDP-NAc transferase) and β-N-acetylglucosaminidase (OGA or *O*-GlcNAcase). Whereas OGT catalyses the addition of *O*-GlcNAc to the hydroxyl group of Ser and Thr residues of a target protein using UDP-GlcNAc as the obligatory substrate, OGA catalyses the hydrolytic cleavage of *O*-GlcNAc from post-translationally-modified proteins (Hart et a;, 2007; Zachara & Hart, 2006) **(Figure 1)**.

A single OGT gene is located on the X chromosome in humans and mice (Kreppel et al., 1997; Nolte & Muller, 2002). In some tissues, such as skeletal muscle, kidney, and liver, three distinct isoforms of OGT have been identified, including two 110-kDa subunits and one 78-kDa subunit, which can assemble into multimers, and smaller mitochondrial isoforms (Kreppel & Hart, 1999, Lazarus et al., 2006; Lubas & Hanover, 2000). Each variant contains a C-terminal catalytic domain, but differs in the number of tetratricopeptide repeats (TPRs) within its N-terminal domain. The TPRs serve as protein-protein interaction modules that appear to target OGT to accessory proteins and potential substrates, such as the related *O*-GlcNAc transferase interacting protein (OIP106) and protein phosphatase-1 (PP1) (Wells et al., 2004). Phylogenetic analysis of eukaryotic OGTs indicate that plants have two distinct OGTs, SEC (secret agent)- and SPY (spindly)-like, that originated in prokaryotes and that are involved in diverse plant processes, including response to hormones and environmental signals, circadian rhythms, development, intercellular transport and virus infection (Olszewski et al., 2009; Swain et al, 2001). Animals and some fungi have a SEC-like enzyme while plants have both. Green algae and some members of the Apicomplexa and amoebozoa have the SPY-like enzyme (Olszewski et al., 2009).

The donor substrate for OGT activity, UDP-GlcNAc or uridine-diphosphate-N-acetylglucosamine, is a terminal product of the hexosamine biosynthesis pathway (HBP – **Figure 1**). Flux through the HBP and UDP-GlcNAc levels changes rapidly in response to many different nutrients, such as glucose, fatty acids, and amino acids (Hanover et al., 2009) altering the extent of *O*-GlcNAcylation of many proteins. It is estimated that 2–5% of total cellular glucose is funneled into the HBP, although the glucose flux is potentially different in various cell types (Hart et al., 2007, Hanover et al., 2009). Free fatty acids can increase HBP flux by inhibiting glycolysis, resulting in elevated fructose-6-phosphate levels. Acetyl-CoA, produced by fatty acid metabolism, serves as the donor for the acetylation of glucosamine in the formation of UDP-GlcNAc (Wang et al., 1998). Exogenously, small amounts of glucosamine can dramatically increase UDP-GlcNAc pools in cells (Zou et al., 2009).

The HBP shares its first two steps with glycolysis. First, hexokinase phosphorylates glucose to produce glucose 6-phosphate, which is then converted into fructose 6-phosphate. At this point the pathways diverge, fructose 6-phosphate is converted by the HBP rate-limiting enzyme glutamine fructose-6-phosphate transferase (GFAT) into glucosamine 6-phosphate (Slawson et al., 2010). Because OGT activity is exquisitely sensitive to UDP-GlcNAc concentrations (Haltiwanger et al., 1992) (**Figure 1**), *O*-GlcNAcylation may act as a sensor for the general metabolic state of the cell.

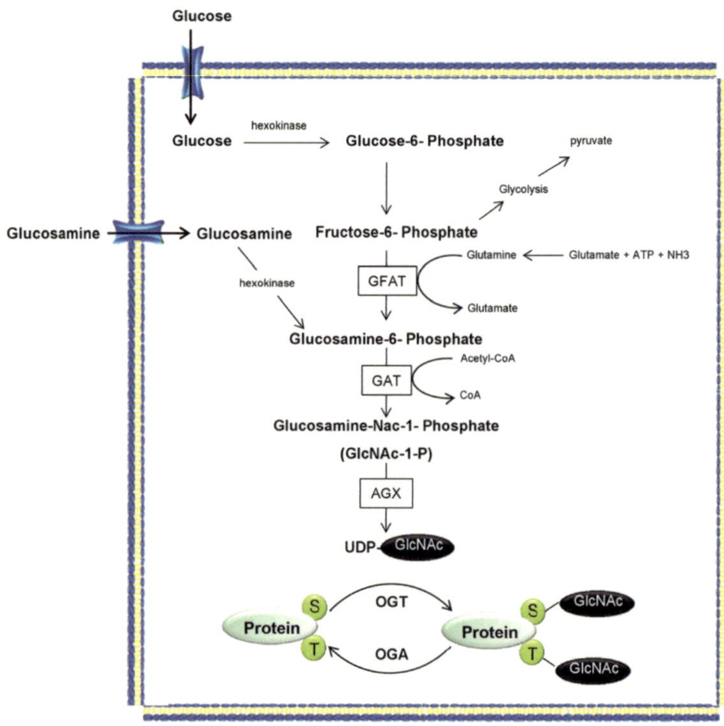

Fig. 1. The hexosamine biosynthesis pathway. After entering the cell via a glucose transporter and being converted to glucose-6-phosphate (glucose-6P) by a hexokinase and to fructose-6-phosphate (fructose-6P), glucose can either be used in the glycolytic or the hexosamine biosynthesis (HBP) pathways. The HBP uses fructose-6P to form glucosamine-6-phosphate (glucosamine-6P), with glutamine serving as the donor of the aminogroup. The reaction is catalyzed by the rate-limiting enzyme glutamine:fructose-6-phosphate transferase (GFAT). Glucosamine-6P is rapidly acetylated through the action of acetyl-CoA:d-glucosamine-6-phosphate N-acetyltransferase (GAT), and isomerized to N-Acetylglucosamine-1-phosphate (GlcNAc-1-P) and activated, via the action of UDP-GlcNAc pyrophosphorylase (AGX), to UDP-N-acetylglucosamine (UDP-GlcNAc) that serves as the donor of *O*-GlcNAc for OGT activity. Glucosamine can also enter the cell through the glucose transporter and is rapidly phosphorylated by hexokinase yielding glucosamine-6P, thereby bypassing the rate-limiting first step of the HBP. S, serine; T, threonine, OGT, *O*-GlcNAc transferase; OGA, *O*-GlcNAcase.

O-GlcNAcase or OGA was initially identified as hexosaminidase C. However, OGA activity is specific for N-acetyl-β-D-glucosaminides and, unlike hexosaminidase, has an optimum pH near neutral and mainly a cytosolic localization (Dong & Hart, 1994; Zachara & Hart, 2006). OGA appears to use substrate catalysis involving the 2-acetamido group and contains an N-terminal glycosidase domain and a putative C-terminal histone acetyltransferase domain (Macauley et al., 2005; Toleman et a., 2004). To date, two distinct isoforms of OGA have been described, a 130-kDa and a 75-kDa variant, which differ in their C terminus. Whereas the 130-kDa or "long OGA" contains a distinct N-terminal glycosidase domain and the C-terminal histone acetyltransferase domain, the 75-kDa or "short OGA" lacks the C-terminal domain. One important functional aspect in the existence of these two splices is their differential sensitivity to previously described potent OGA inhibitors. For example, the short OGA exhibits comparative resistance to PugNAc and NAG-thiazoline, but is very sensitive to alpha-GlcNAc thiolsulfonate (Zachara & Hart, 2006). Inhibition of OGT and OGA represents an area of great interest on *O*-GlcNAcylation research, which is evident from the increasing number of studies addressing the enzymes molecular mechanisms for the addition and removal of *O*-GlcNAc (Borodkin & van Aalten, 2010; Dorfmueller et al., 2010, 2011; Dorfmueller & van Aalten, 2010; Gloster et al., 2011; Gloster & Vocadlo, 2010; Lameira et al., 2011; Lazarus et al., 2011; Li et al., 2011; Macauley & Vocadlo, 2010; Martinez-Fleites et al., 2010).

3. The interplay between *O*-GlcNAcylation and Protein *O*-Phosphorylation

The dynamic addition of *O*-GlcNAc to proteins has been implicated in modulating protein behavior via one potential mechanism that includes a complex interplay between *O*-GlcNAcylation and phosphorylation. Many phosphorylation sites are also known glycosylation sites, and this reciprocal occupancy may produce different activities or alter stability in the target protein (Hu et al., 2010; Zeidan & Hart, 2010) (**Figure 2**). In support of this model, an earlier report has shown that activation of PKC and PKA reduced glycosylation in a detergent insoluble cytoskeletal and cytoskeleton-associated protein fraction. Conversely, inhibition of PKC and PKA increased *O*-GlcNAc protein modification in this fraction (Griffith & Schmitz, 1999). The competition between *O*-GlcNAcylation and phosphorylation for the same or neighboring residues has been termed the "yin-yang" hypothesis and has been reported in a variety of proteins (Hart et al., 1995).

However, it should be emphasized that the interplay between these two PTMs is not always reciprocal. For example, some proteins, such as p53 and vimentin, can be concomitantly phosphorylated and *O*-GlcNAcylated, and the adjacent phosphorylation or *O*-GlcNAcylation can regulate the addition of either moiety (Wang et al., 2007; Yang et al., 2006).

In addition to the reciprocal crosstalk at same or proximal sites of the proteins, crosstalk between *O*-GlcNAcylation and phosphorylation also exists among distantly located sites, such as on the C-terminal domain of RNA polymerase II and on cytokeratins (Chou et al., 1992; Comer & Hart, 2001). Furthermore, the crosstalk between phosphorylation and *O*-GlcNAcylation also influences each other by regulating the activities or localization of other cycling enzymes. For example, OGT is directly activated by tyrosine phosphorylation and is

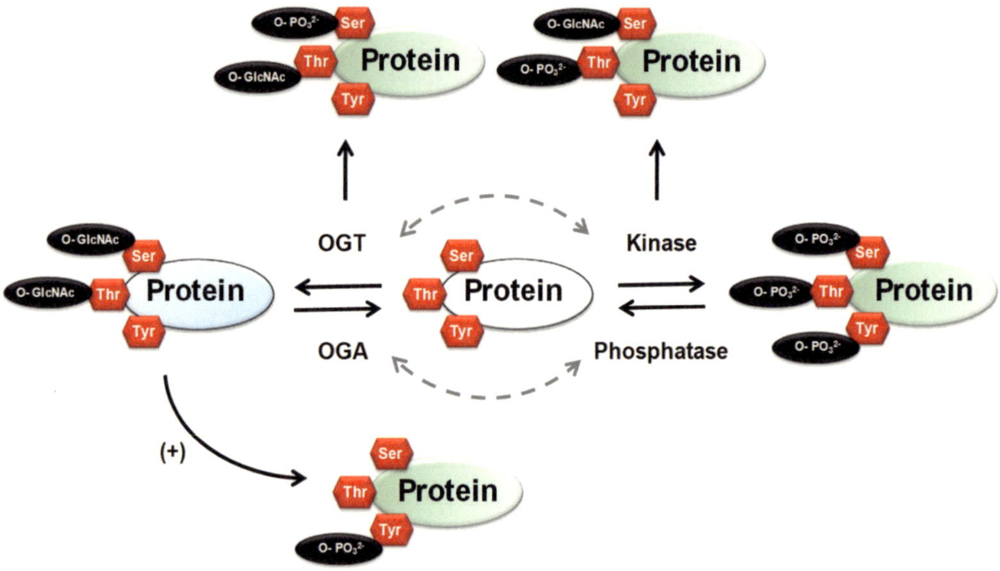

Fig. 2. The interplay between *O*-GlcNAcylation and *O*-phosphorylation of proteins. Both phosphorylation and *O*-GlcNAcylation occur on serine/threonine (Ser/Thr) residues of proteins. In specific proteins, there is a competitive relationship between *O*-GlcNAc and *O*-phosphate for the same Ser/Thr residues, although there can be adjacent or multiple occupancy for phosphorylation and *O*-GlcNAcylation on the same protein. The interplay between phosphorylation and *O*-GlcNAcylation creates molecular diversity by altering specific protein sites that regulate protein functions and signaling events. OGT, *O*-GlcNAc transferase; OGA, *O*-GlcNAcase; Tyr, tyrosine. Reproduced with permission, from Lima et al., 2012, *Clinical Science*, vol__, pp__-__. © the Biochemical Society.

itself *O*-GlcNAc-modified [49]. OGT also forms a stable and active complex with protein phosphatase-1 (PP1β and PP1γ) in rat brain [50]. The association between OGT and PP1 is particularly intriguing, as it may provide a direct mechanism to couple *O*-GlcNAc to dephosphorylation of specific substrates. As with OGT, OGA has been shown to interact with specific proteins, including protein phosphatase-2β (Wells et al., 2002).

A recent report showed that rat brain assembly protein AP180, which is involved in the assembly of clathrin-coated vesicles in synaptic vesicle endocytosis, contains a phosphorylated *O*-GlcNAc (*O*-GlcNAc-P) within a highly conserved sequence (*O*-GlcNAc or *O*-GlcNAc-P, but not phosphorylation alone, was found at Thr^{310}) (Graham et al., 2011). *O*-GlcNAcylation was thought to be a terminal modification, i.e. the *O*-GlcNAc was not found to be additionally modified. The existence of protein glycosyl phosphorylation (*O*-GlcNAc-P) adds further complexity to the phosphorylation-*O*-GlcNAcylation interplay.

Lastly, the interplay between *O*-GlcNAc modification and phosphorylation may not be limited to Ser/Thr phosphorylation, but may also include tyrosine (Tyr) phosphorylation. Based on the higher prevalence of Tyr phosphorylation among *O*-GlcNAc-modified proteins (~68% vs. ~2% in non-*O*-GlcNAc-modified proteins), Mishra and colleagues suggested that Tyr phosphorylation plays a role in the interplay between *O*-GlcNAc modification and Ser/Thr phosphorylation in proteins (Mishra et al., 2011).

This clearly shows that the interplay between *O*-GlcNAcylation and phosphorylation is both complex and very extensive. As with any PTM, mapping the attachment sites is a prerequisite toward understanding the biological functions of *O*-GlcNAcylation. With the development of sample enrichment methods and new mass spectrometry fragmentation methods, such as electron capture dissociation and electron transfer dissociation, now hundreds of *O*-GlcNAc sites have been mapped, and some cellular stimuli were shown to increase both modifications. For further information on the complex interplay between *O*-GlcNAcylation and phosphorylation, please refer to the following comprehensive and excellent reviews (Copeland et al., 2008; Hart et al., 2011; Hu et al., 2010; Wang et al., 2008; Zeidan & Hart 2010).

4. Signaling pathways modified by *O*-GlcNAcylation

Many proteins, mainly kinases, involved in signaling pathways that regulate cell growth, apoptosis, ion channel activities, and actin cytoskeleton are target for *O*-GlcNAc modification (Lima et al., 2009, 2011). In this section we will briefly comment general aspects of some of the signaling proteins that have been identified as targets for *O*-GlcNAcylation.

The protein kinase C (PKC) family constitutes a group of multifunctional Ser/Thr protein kinases that are classified into three groups: the classic PKCs [PKCalpha(α), PKCbeta(βI), PKCbeta(βII), PKCgamma(γ)], the novel PKCs [PKCdelta(δ), PKCepsilon(ε), PKCeta(η), PKCmu(μ), PKCtheta(θ)], and the atypical PKCs [PKCzeta(ζ), PKCiota/lambda(ι/λ)] (Salamanca & Khalil, 2005).

Functional studies have demonstrated that interaction of PKC with its protein substrate triggers activation of a cascade of kinases that ultimately stimulate many cellular functions, including contraction, hypertrophy, growth, proliferation and cell survival. As an example, PKC phosphorylates CPI-17, which in turn inhibits myosin light chain (MLC) phosphatase, increases MLC phosphorylation and enhances vascular smooth muscle contraction. PKC also phosphorylates the actin-binding protein calponin, and thereby reverses its inhibition of actin-activated myosin ATPase, allowing more actin to interact with myosin and increases vascular contraction (Budzyn et al., 2006; Salamanca & Khalil, 2005; Woodsome et al, 2001).

Initial studies indicated that activation of PKC or cAMP-dependent protein kinase significantly decreased overall *O*-GlcNAcylation in neuronal cytoskeletal proteins. Conversely, inhibition of PKC, cAMP-dependent protein kinase, cyclin-dependent protein kinases, or S6 kinase increased overall *O*-GlcNAc levels in fractions from these cells (Griffith et al., 1995). Stimulation of the transactivation of Sp1, which is *O*-GlcNAcylation–dependent, can be blocked by molecular and pharmacological inhibition of PKC (Fantus et al., 2006). In cerebellar neurons from early postnatal mice, activation of cAMP-dependent protein kinase or PKC results in reduced levels of *O*-GlcNAc specifically in the fraction of cytoskeletal and cytoskeleton-associated proteins, whereas inhibition of the same kinases results in increased levels of *O*-GlcNAc (Griffith & Schmitz, 1999).

In the reverse direction, all PKC isoforms expressed in rat hepatocytes are dynamically modified by *O*-GlcNAc. *O*-GlcNAcylation of PKC-α negatively correlates with enzyme activity (Robles-Flores et al., 2008). Increased *O*-GlcNAc modification in a human astroglial cell line, in response to glucosamine (which increases the production of glucosamine 6-phosphate and stimulates *O*-GlcNAc modification of proteins) or PUGNAc (which blocks *O*-

GlcNAcase activity, mimicking the enzyme-stabilized transition state), results in a decrease in membrane-associated PKC-ε and PKC-α, but not PKC-ι, indicating that increased levels of the *O*-GlcNAc modification regulates specific PKC isoforms (Matthews et al., 2005). Therefore, it is likely that *O*-GlcNAc modification of PKC isoforms, such as PKC-α, PKC-β, PKC-γ, PKC-ε, and PKC-ζ can interfere with cellular processes regulated by these enzymes.

Mitogen-activated protein kinases (MAPKs) are a family of serine/threonine kinases which are classically associated with cell contraction, migration, adhesion, collagen deposition, cell growth, differentiation, and survival (Pearson et al., 2001). Of the major MAPKs, extracellular signal-regulated kinases (ERK1/2), p38 MAPK, and stress-activated protein kinase/c-Jun N-terminal kinases (SAPK/JNK) are the best characterized. The complex signaling networks that underlie MAPK activation typically require phosphorylation by a MAPK kinase also known as MEK. The ERK1/2 phosphorylation cascade involves MEK1/2 (MAP/ERK kinase) whereas the signaling processes leading to SAPK/JNK and p38 MAPK activation involve MEK4/7 and MEK3/6, respectively (Pearson et al., 2001). Activation of MAPKs has been reported to be primarily dependent on the nonreceptor tyrosine kinase c-Src in different cell types. To date, at least 14 Src-related kinases have been identified, of which the 60 kDa c-Src is the most abundantly expressed isoform in vascular smooth muscle cells and rapidly activated by G protein-coupled receptors. Other proximal regulators of MEK include the Ras-Raf pathway, which may not necessarily involve c-Src (Kolch, 2005; Martin, 2001; Oda et al., 1999).

The MAPKs p38 and ERK1/2 have been reported to be phosphorylated in response to increased *O*-GlcNAc levels (Laczy et al., 2009). A positive correlation between phosphorylation of the MAPK cascade (ERK1/2 and p38) and nuclear *O*-GlcNAcylation was observed in fetal human cardiac myocytes exposed to high glucose (Gross et al., 2005). In isolated rat hearts, perfusion with 5 mM glucosamine increases *O*-GlcNAc levels and confers cardioprotection after ischemia-reperfusion (Zou et al., 2009). Interestingly, although glucosamine does not alter the response of either ERK1/2 or Akt (protein kinase B) to ischemia-reperfusion, it significantly attenuates the ischemia-induced increase in p38 phosphorylation, as well as the increased p38 phosphorylation at the end of reperfusion, suggesting that glucosamine-induced cardioprotection may be mediated via the p38 MAPK pathway (Jones et al., 2008).

Augmented *O*-GlcNAc levels in mouse hippocampal synapses increases phosphorylation of synapsin I/II at Ser^{9} (cAMP-dependent protein kinase substrate site), $Ser^{62/67}$ (ERK1/2 [MAPK 1/2] substrate site), and Ser^{603} (calmodulin kinase II site). Activation-specific phosphorylation events on ERK1/2 and calmodulin kinase II are also increased in response to elevation of *O*-GlcNAc levels (Rexach et al., 2008).

Advanced glycation end-products induce ROS accumulation, apoptosis, MAPK activation, and nuclear *O*-GlcNAcylation in human cardiac myocytes (Li et al., 2007). In addition, exposure of neutrophils to PUGNAc or glucosamine also stimulates the small GTPase Rac, which is an important upstream regulatory element in p38 and ERK1/2 MAPK signaling in neutrophils, and these MAPKs are implicated in chemotactic signal transduction.

Conversely, alterations in MAPK pathways can also have effects on the enzymes responsible for the regulation of *O*-GlcNAc (Laczy et al., 2009, Lima et al., 2011). In neuro-2a neuroblastoma cells, increased OGT expression on glucose deprivation occurs in an AMP-activated protein kinase-dependent manner, whereas OGT enzymatic activity is regulated in a p38 MAPK-dependent manner. OGT is not phosphorylated by p38, but rather it interacts directly with p38 through its C terminus. The interaction with p38 does not change the catalytic activity of OGT, but p38 regulates OGT activity within the cell by recruiting it to specific targets (Cheung & Hart, 2008).

Together, these data indicate that *O*-GlcNAcylation is an important signaling element and it modulates the activities of several critical signaling kinases (Kneass & Marchase, 2005). Thus, it is possible that signaling kinases, such as proteins from MAPK, PKC, and RhoA/Rho kinase pathways, are also regulated by *O*-GlcNAc modifications and that this post-translational modification not only modulates many cellular responses, but also may play a role in the abnormal function of kinases observed in various pathological conditions.

Ca^{+2} sensitization in smooth muscle cells is a well known process mediated by the small GTPase Rho and its downstream target Rho-kinase. The exchange of bound guanosine diphosphate (GDP) for guanosine triphosphate (GTP) activates Rho and stimulates its translocation from the cytosol to the plasma membrane. Rho-GTP phosphorylates Rho-kinase, which inhibits MLC phosphatase activity by phosphorylation of the MLC phosphatase target subunit (MYPT1). A decrease in MLC phosphatase activity increases phosphorylation of myosin and therefore contributes to smooth muscle contraction at low levels of intracellular Ca^{+2} (Somlyo & Somlyo, 2000). RhoA/Rho kinase signaling has been implicated in many cellular processes including contraction, reactive oxygen species generation, inflammation, and cell migration (Calo & Pessina, 2007).

Rho-kinase activation also suppresses eNOS activity/expression, and decreased sensitivity of contractile proteins to Ca^{2+} is considered a key mechanism in NO-induced relaxation of vascular smooth muscle cells. Accordingly, NO also induces vasodilation through the inhibition of the RhoA/Rho-kinase signaling pathway. Accordingly, NO-mediated increases in cGMP and activation of cGMP-dependent protein kinase (cGK) lead to inhibition of RhoA (Chitaley & Webb, 2002; Sauzeau et al., 2001; Sawada et al., 2001).

The small G-protein RhoA and its downstream target, Rho-kinase, play a direct role in the regulation of MLC phosphatase activity. In the active state, RhoA engages downstream effectors, such as Rho-kinase, which then phosphorylates the myosin binding subunit of MLC phosphatase (MYPT1 Thr^{853}), inhibiting its activity, and thus promoting the phosphorylated state of MLC (Chitaley et al., 2001). Data from our laboratory and others indicate that increased *O*-GlcNAcylation augments vascular reactivity to constrictor stimuli via changes in the RhoA/Rho-kinase pathway (Lima et al, 2011; Kim et al, 2011).

Since increased *O*-GlcNAcylation decreases eNOS/NO signaling (Musick et al., 2005) and NO inhibits RhoA/Rho-kinase signaling, increased RhoA/Rho-kinase activity observed in many pathological conditions may be associated with augmented *O*-GlcNAc levels.

5. Physiological implications of alternating *O*-GlcNAcylation and *O*-phosphorylation

The physiological significance of the crosstalk between *O*-GlcNAcylation and *O*-phosphorylation certainly warrants further investigation. However, data available so far indicate that the "on" or "off" state of many enzymes and receptors are not simply determined by the kinases- and phosphatases-driven phosphorylation of specific aminoacid residues. The complex interplay between *O*-GlcNAcylation and *O*-phosphorylation, within reciprocal or proximal sites, makes the activation/deactivation or the "on"/"off" switch of enzymes and receptors a much more elaborated process. Since both post-translational modifications modulate many cellular functions via protein targeting to specific substrates, transient complex formation with other proteins, subcellular compartmentalization of specific proteins, activation/inhibition of many signaling pathways, the interplay between *O*-GlcNAcylation and *O*-phosphorylation adds great complexity to our knowledge of protein activity regulation.

New techniques allowing the recognition of several *O*-GlcNAc sites will further clarify how different cellular stimuli interfere with these post-translational modifications. One big challenge in the field has been to map the sites where the attachments are simultaneously occurring. The development and improvement of some techniques such as electron capture dissociation and electron transfer dissociation has opened new possibilities to map *O*-GlcNAcylation and *O*-phosphorylation sites. Please, refer to the following comprehensive and excellent reviews for further information regarding *O*-GlcNAc enrichment methods (Macauley & Vocadlo, 2009; 120. Peter-Katalinic, 2005; Wang et al, 2010; Zachara, 2009).

6. Conclusions

Our understanding of the *O*-GlcNAcylation process (enzymatic regulation, cellular targets and sites for *O*-GlcNAc addition, modulation by other pathways) as well as of its functional importance and its contribution to (dys)regulation of many cellular processes is rapidly increasing. It is also evident that the direct interactions between *O*-GlcNAcylation and *O*-phosphorylation and the fact that both post-translational modifications can interfere with many signaling pathways and cellular processes, not only add great complexity to our knowledge of protein activity regulation, but warrant intense research in the field.

Future investigations focusing on the characterization of specific *O*-GlcNAcylated and *O*-phosphorylated sites/proteins, as well as studies addressing and identifying the factors involved in the regulation of OGT and OGA activity are needed. They will provide a greater understanding as to how *O*-GlcNAc modulates cellular function and potentially provide an avenue for targeted interventions and therapies.

7. Acknowledgments

Financial Support from FAPESP (Fundacao de Amparo a Pesquisa do Estado de Sao Paulo) and CNPq (Conselho Nacional de Desenvolvimento Cientifico e Tecnologico) - Brazil.

8. References

Borodkin, V.S. & van Aalten, D.M. (2010). An efficient and versatile synthesis of GlcNAcstatins-potent and selective O-GlcNAcase inhibitors built on the tetrahydroimidazo[1,2-a]pyridine scaffold. *Tetrahedron*. 66, pp.7838-7849. ISSN 0040-4020.

Budzyn, K.; Paull, M.; Marley, P.D. & Sobey, C.G. (2006). Segmental differences in the roles of rho-kinase and protein kinase C in mediating vasoconstriction. *Journal of Pharmacology and Experimental Therapeutics*. 317, pp.791-796. ISSN 0022-3565.

Calo, L.A. & Pessina, A.C. (2007). RhoA/Rho-kinase pathway: much more than just a modulation of vascular tone. Evidence from studies in humans. *Journal of Hypertension*. 25, pp.259-264. ISSN 0263-6352.

Cheung, W.D. & Hart, G.W. (2008). AMP-activated protein kinase and p38 MAPK activate O-GlcNAcylation of neuronal proteins during glucose deprivation. *Journal of Biological Chemistry*. 283, pp.13009-13020. ISSN 0021-9258.

Chitaley, K. & Webb, R.C. (2002). Nitric oxide induces dilation of rat aorta via inhibition of rho-kinase signaling. *Hypertension*. 39, pp.438-442. ISSN 0194-911X.

Chitaley, K.; Weber, D. & Webb, R.C. (2001). RhoA/Rho-kinase, vascular changes, and hypertension. *Current Hypertension Reports* 3, pp.139-144. ISSN 1522-6417.

Chou, C.F.; Smith, A.J. & Omary, M.B. (1992). Characterization and dynamics of O-linked glycosylation of human cytokeratin 8 and 18. *Journal of Biological Chemistry*. 267, pp.3901-3906. ISSN 0021-9258.

Comer, F.I. & Hart, G.W. (2001). Reciprocity between O-GlcNAc and O-phosphate on the carboxyl terminal domain of RNA polymerase II. *Biochemistry*. 40, pp.7845-7852. ISSN 0006-2960.

Copeland, R.J.; Bullen, J.W. & Hart, G.W. (2008). Cross-talk between GlcNAcylation and phosphorylation: roles in insulin resistance and glucose toxicity. *American Journal of Physiology (Endocrinology and Metabolism)*. 295, pp.E17-E28. ISSN 0193-1849.

Dong, D.L. & Hart, G.W. (1994). Purification and characterization of an O-GlcNAc selective N-acetyl-beta-D-glucosaminidase from rat spleen cytosol. *Journal of Biological Chemistry*. 269, pp.19321-19330. ISSN 0021-9258.

Dorfmueller, H.C.; Borodkin, V.S.; Blair, D.E.; Pathak, S.; Navratilova, I. & van Aalten, D.M. (2011). Substrate and product analogues as human O-GlcNAc transferase inhibitors. *Amino Acids*. 40, pp.781-792. ISSN 0939-4451.

Dorfmueller, H.C.; Borodkin, V.S.; Schimpl, M.; Zheng, X.; Kime, R.; Read, K.D. & van Aalten, D.M. (2010). Cell-penetrant, nanomolar O-GlcNAcase inhibitors selective against lysosomal hexosaminidases. *Chemistry and Biology*. 17, pp.1250-1255. ISSN 1074-552.

Dorfmueller, H.C. & van Aalten, D.M. (2010). Screening-based discovery of drug-like O-GlcNAcase inhibitor scaffolds. *FEBS Letters*. 584, pp.694-700. ISSN 0014-5793.

Fantus, G.I.; Goldberg, H.J.; Whiteside, C.I. & Topic, D. (2006). The Hexosamine biosynthesis pathway: Contribution to the pathogenesis of diabetic nephropathy. In: The diabetic kidney, edited by P. C, and C.E. Humana Press, p.120-133. ISBN 978-1-59745-153-6.

Gloster, T.M. & Vocadlo, D.J. (2010). Mechanism, structure, and inhibition of O-GlcNAc processing enzymes. *Current Signal Transduction Therapy*. 5, pp.74-91. ISSN 1574-3624.

Gloster, T.M.; Zandberg, W.F.; Heinonen, J.E.; Shen, D.L.; Deng, L. & Vocadlo, D.J. (2011). Hijacking a biosynthetic pathway yields a glycosyltransferase inhibitor within cells. *Nature Chemical Biology*. 7, pp.174-181. ISSN 1552-4450.

Graham, M.E.; Thaysen-Andersen, M.; Bache, N.; Craft, G.E.; Larsen, M.R.; Packer, N.H. & Robinson, P.J. (2011). A novel post-translational modification in nerve terminals: O-linked N-acetylglucosamine phosphorylation. *Journal of Proteome Research*. 10, pp.2725-2733. ISSN 1535-3893.

Griffith, L.S. & Schmitz, B. (1999). O-linked N-acetylglucosamine levels in cerebellar neurons respond reciprocally to pertubations of phosphorylation. *European Journal of Biochemistry*. 262, pp.824-831. ISSN 0014-2956.

Griffith, L.S.; Mathes, M. & Schmitz, B. (1995). Beta-amyloid precursor protein is modified with O-linked N-acetylglucosamine. *Journal of Neuroscience Research.* 41, pp.270-278. ISSN 0360-4012.

Gross, B.J.; Kraybill, B.C. & Walker, S. (2005). Discovery of O-GlcNAc transferase inhibitors. *Journal of the American Chemistry Society.* 127, pp.14588-14589. ISSN 0002-7863.

Haltiwanger, R.S.; Blomberg, M.A. & Hart, G.W. (1992). Glycosylation of nuclear and cytoplasmic proteins. Purification and characterization of a uridine diphospho-N-acetylglucosamine:polypeptide beta-N-acetylglucosaminyltransferase. *Journal of Biological Chemistry.* 267, pp.9005-9013. ISSN 0021-9258.

Haltiwanger, R.S.; Busby, S.; Grove, K.; Li, S.; Mason, D.; Medina, L.; Moloney, D.; Philipsberg, G. & Scartozzi, R. (1997). O-glycosylation of nuclear and cytoplasmic proteins: regulation analogous to phosphorylation? *Biochemistry and Biophysical Research Communication.* 231, pp.237-242. ISSN 0006-291X.

Hanover, J.A.; Krause, M.W. & Love, D.C. (2009). The hexosamine signaling pathway: O-GlcNAc cycling in feast or famine. *Biochimica et Biophysica Acta.* 1800, pp.80-95. ISSN 0006-3002.

Hart, G.W. (1997). Dynamic O-linked glycosylation of nuclear and cytoskeletal proteins. *Annual Review of Biochemistry.* 66, pp.315-335. ISSN 0066-4154.

Hart, G.W.; Greis, K.D.; Dong, L.Y.; Blomberg, M.A.; Chou, T.Y.; Jiang, M.S.; Roquemore, E.P.; Snow, D.M.; Kreppel, L.K.; Cole, R.N., et al. (1995). O-linked N-acetylglucosamine: the "yin-yang" of Ser/Thr phosphorylation? Nuclear and cytoplasmic glycosylation. *Advances in Experimental Medical Biology.* 376, pp.115-123. ISSN 0065-2598.

Hart, G.W.; Haltiwanger, R.S.; Holt, G.D. & Kelly, W.G. (1989). Glycosylation in the nucleus and cytoplasm. *Annual Review of Biochemistry.* 58, pp.841-874. ISSN 0066-4154.

Hart, G.W.; Holt, G.D. & Haltiwanger, R.S. (1988). Nuclear and cytoplasmic glycosylation: novel saccharide linkages in unexpected places. *Trends in Biochemical Sciences.* 13, pp.380-384. ISSN 0968-0004.

Hart, G.W.; Housley, M.P. & Slawson, C. (2007). Cycling of O-linked beta-N-acetylglucosamine on nucleocytoplasmic proteins. *Nature.* 446, pp.1017-1022. ISSN 0028-0836.

Hart, G.W.; Slawson, C.; Ramirez-Correa, G. & Lagerlof, O. (2011). Cross Talk Between O-GlcNAcylation and Phosphorylation: Roles in Signaling, Transcription, and Chronic Disease. *Annual Review of Biochemistry.* 80, pp.825-858. ISSN 0066-4154.

Hu, P.; Shimoji, S. & Hart, G.W. (2010). Site-specific interplay between O-GlcNAcylation and phosphorylation in cellular regulation. FEBS Letters. 584, pp.2526-2538. ISSN 0014-5793.

Hu, Z.Z. dbOGAP: Database of O-GlcNAcylated Proteins and Sites. (2010) [cited; Available from: http://cbsb.lombardi.georgetown.edu/OGAP.html.

Jones, S.P.; Zachara, N.E.; Ngoh, G.A.; Hill, B.G.; Teshima, Y.; Bhatnagar, A.; Hart, G.W. & Marban, E. (2008). Cardioprotection by N-acetylglucosamine linkage to cellular proteins. *Circulation.* 117, pp.1172-1182. ISSN 0009-7322.

Kelly, W.G. & Hart, G.W. (1989). Glycosylation of chromosomal proteins: localization of O-linked N-acetylglucosamine in Drosophila chromatin. *Cell.* 57, pp.243-251. ISSN 0092-8674.

Khidekel, N.; Ficarro, S.B.; Clark, P.M.; Bryan, M.C.; Swaney, D.L.; Rexach, J.E.; Sun, Y.E.; Coon, J.J.; Peters, E.C. & Hsieh-Wilson, L.C. (2007). Probing the dynamics of O-

GlcNAc glycosylation in the brain using quantitative proteomics. *Nature Chemical Biology*. 3, pp.339-348. ISSN 1552-4450.

Kim do, H.; Seok, Y.M.; Kim, I.K.; Lee, I.K.; Jeong, S.Y. & Jeoung, N.H. (2011). Glucosamine increases vascular contraction through activation of RhoA/Rho kinase pathway in isolated rat aorta. *BMB Reports*. 44, pp.415-420. ISSN 1976-6696.

Kneass, Z.T. & Marchase, R.B. (2005). Protein O-GlcNAc modulates motility-associated signaling intermediates in neutrophils. *Journal of Biological Chemistry*. 280, pp.14579-14585. ISSN 0021-9258.

Kolch W. (2005). Coordinating ERK/MAPK signalling through scaffolds and inhibitors. *Nature Reviews Molecular Cell Biology*. 6, pp.827-837. ISSN 1471-0072.

Kreppel, L.K.; Blomberg, M.A. & Hart, G.W. (1997). Dynamic glycosylation of nuclear and cytosolic proteins. Cloning and characterization of a unique O-GlcNAc transferase with multiple tetratricopeptide repeats. *Journal of Biological Chemistry*. 272, pp.9308-9315. ISSN 0021-9258.

Kreppel, L.K. & Hart, G.W. (1999). Regulation of a cytosolic and nuclear O-GlcNAc transferase. Role of the tetratricopeptide repeats. *Journal of Biological Chemistry*. 274, pp.32015-32022. ISSN 0021-9258.

Laczy, B.; Hill, B.G.; Wang, K.; Paterson, A.J.; White, C.R.; Xing, D.; Chen, Y.F.; Darley-Usmar, V.; Oparil, S. & Chatham, J.C. (2009). Protein O-GlcNAcylation: a new signaling paradigm for the cardiovascular system. *American Journal of Physiology (Heart and Circulatory Physiology)*. 296, pp.H13-H28. ISSN 0363-6135.

Lameira, J.; Alves, C.N.; Tunon, I.; Marti, S. & Moliner, V. (2011). Enzyme molecular mechanism as a starting point to design new inhibitors: a theoretical study of O-GlcNAcase. *The Journal of Physical Chemistry B*. 115, pp.6764-6775. ISSN 1520-6106.

Lazarus, B.D.; Love, D.C. & Hanover, J.A. (2006). Recombinant O-GlcNAc transferase isoforms: identification of O-GlcNAcase, yes tyrosine kinase, and tau as isoform-specific substrates. *Glycobiology*. 16, pp.415-421. ISSN 0959-6658.

Lazarus, M.B.; Nam, Y.; Jiang, J.; Sliz, P. & Walker, S. (2011). Structure of human O-GlcNAc transferase and its complex with a peptide substrate. *Nature*. 469, pp.564-567. ISSN 0028-0836.

Li, S.Y.; Sigmon, V.K.; Babcock, S.A. & Ren, J. (2007). Advanced glycation endproduct induces ROS accumulation, apoptosis, MAP kinase activation and nuclear O-GlcNAcylation in human cardiac myocytes. *Life Sciences*. 80, pp.1051-1056. ISSN 0024-3205.

Li, T.; Guo, L.; Zhang, Y.; Wang, J.; Li, Z.; Lin, L.; Zhang, Z.; Li, L.; Lin, J.; Zhao, W.; Li, J. & Wang, P.G. (2011). Design and synthesis of O-GlcNAcase inhibitors via 'click chemistry' and biological evaluations. *Carbohydrate Research*. 346, pp.1083-1092. ISSN 0008-6215.

Lima, V.V.; Rigsby, C.S.; Hardy, D.M.; Webb, R.C. & Tostes, R.C. (2009). O-GlcNAcylation: a novel post-translational mechanism to alter vascular cellular signaling in health and disease: focus on hypertension. *Journal of the American Society of Hypertension*. 3, pp.374-387. ISSN 1933-1711.

Lima, V.V.; Giachini, F.R.; Carneiro, F.S.; Carvalho, M.H.; Fortes, Z.B.; Webb, R.C. & Tostes, R.C. (2011). O-GlcNAcylation contributes to the vascular effects of ET-1 via activation of the RhoA/Rho-kinase pathway. *Cardiovascular Research*. 89, pp.614-622. ISSN 0008-6363.

Lima, V.V.; Giachini, F.R.; Hardy, D.M.; Webb, R.C. & Tostes, R.C. (2011). O-GlcNAcylation: a novel pathway contributing to the effects of endothelin in the vasculature.

American Jounral of Physiology (Regulatory and Integrative Comparative Physiology). 300, pp.R236-R250. ISSN 0363-6119.

Lubas, W.A. & Hanover, J.A. (2000). Functional expression of O-linked GlcNAc transferase. Domain structure and substrate specificity. *Journal of Biological Chemisty*. 275, pp.10983-10988. ISSN 0021-9258.

Macauley, M.S. & Vocadlo, D.J. (2010). Increasing O-GlcNAc levels: An overview of small-molecule inhibitors of O-GlcNAcase. *Biochimica et Biophysica Acta*. 1800, pp.107-121. ISSN 0006-3002.

Macauley, M.S.; Whitworth, G.E.; Debowski, A.W.; Chin, D. & Vocadlo, D.J. (2005). O-GlcNAcase uses substrate-assisted catalysis: kinetic analysis and development of highly selective mechanism-inspired inhibitors. *Journal of Biological Chemisty*. 280, pp.25313-25322. ISSN 0021-9258.

Martin GS. (2001). The hunting of the Src. *Nature Reviews Molecular Cell Biology*. 2, pp.467-475. ISSN 1471-0072.

Martinez-Fleites, C.; He, Y. & Davies, G.J. (2010). Structural analyses of enzymes involved in the O-GlcNAc modification. *Biochimica et Biophysica Acta*. 1800, pp.122-133. ISSN 0006-3002.

Matthews, J.A.; Acevedo-Duncan, M. & Potter, R.L. (2005). Selective decrease of membrane-associated PKC-alpha and PKC-epsilon in response to elevated intracellular O-GlcNAc levels in transformed human glial cells. *Biochimica et Biophysica Acta* 1743, pp.305-315. ISSN 0006-3002.

Mishra, S.; Ande, S.R. & Salter, N.W. (2011). O-GlcNAc modification: why so intimately associated with phosphorylation? Cell Communication & Signaling. 9, p.1. ISSN 1478811X.

Musicki, B.; Kramer, M.F.; Becker, R.E. & Burnett, A.L. (2005). Inactivation of phosphorylated endothelial nitric oxide synthase (Ser-1177) by O-GlcNAc in diabetes-associated erectile dysfunction. *Proceedings of the National Academy of Sciences USA* 102, pp.11870-11875. ISSN 1091-6490.

Nandi, A.; Sprung, R.; Barma, D.K.; Zhao, Y.; Kim, S.C. & Falck, J.R. (2006). Global identification of O-GlcNAc-modified proteins. *Analytical Chemistry*. 78, pp.452-458. ISSN 0003-2700.

Nolte, D. & Muller, U. (2002). Human O-GlcNAc transferase (OGT): genomic structure, analysis of splice variants, fine mapping in Xq13.1. *Mammalian Genome*. 13, pp.62-64. ISSN 0938-8990.

Oda, Y.; Renaux, B.; Bjorge, J.; Saifeddine, M.; Fujita, D.J. & Hollenberg, M.D. (1999). c-Src is a major cytosolic tyrosine kinase in vascular tissue. *Canadian Journal of Physiology and Pharmacology*. 77, pp.606-617. ISSN 0008-4212.

Olszewski, N.E.; West, C.M.; Sassi, S.O. & Hartweck, L.M. (2009). O-GlcNAc protein modification in plants: Evolution and function. *Biochimica et Biophysica Acta*. 1800, pp.49-56. ISSN 0006-3002.

Pearson, G.; Robinson, F.; Beers Gibson, T.; Xu, B.E.; Karandikar, M.; Berman, K. & Cobb, M.H. (2001). Mitogen-activated protein (MAP) kinase pathways: regulation and physiological functions. *Endocrine Reviews* 22, pp.153-183. ISSN 0163-769X.

Peter-Katalinic, J. (2005). Methods in enzymology: O-glycosylation of proteins. *Methods in Enzymology* 405, pp.139-171. ISSN 0076-6879.

Rexach, J.E.; Clark, P.M. & Hsieh-Wilson, L.C. (2008). Chemical approaches to understanding O-GlcNAc glycosylation in the brain. *Nature Chemical Biology*. 4, pp.97-106. ISSN 1552-4450.

Robles-Flores, M.; Melendez, L.; Garcia, W.; Mendoza-Hernandez, G.; Lam, T.T.; Castaneda-Patlan, C. & Gonzalez-Aguilar, H. (2008). Posttranslational modifications on protein kinase c isozymes. Effects of epinephrine and phorbol esters. *Biochimica et Biophysica Acta* 1783, pp.695-712. ISSN 0006-3002.

Rogers, S.; Wells, R. & Rechsteiner, M. (1986). Amino acid sequences common to rapidly degraded proteins: the PEST hypothesis. *Science*. 234, pp.364-368. ISSN 0036-8075.

Salamanca, D.A. & Khalil, R.A. (2005). Protein kinase C isoforms as specific targets for modulation of vascular smooth muscle function in hypertension. *Biochemical Pharmacology*. 70, pp.1537-1547. ISSN 0006-2952.

Sauzeau, V.; Le Jeune, H.; Cario-Toumaniantz, C.; Smolenski, A.; Lohmann, S.M.; Bertoglio, J.; Chardin, P.; Pacaud, P. & Loirand, G. (2000). Cyclic GMP-dependent protein kinase signaling pathway inhibits RhoA-induced Ca2+ sensitization of contraction in vascular smooth muscle. *Journal of Biological Chemistry*. 275, pp.21722-21729. ISSN 0021-9258.

Sawada, N.; Itoh, H.; Yamashita, J.; Doi, K.; Inoue, M.; Masatsugu, K.; Fukunaga, Y.; Sakaguchi, S.; Sone, M.; Yamahara, K.; Yurugi, T. & Nakao, K. (2001). cGMP-dependent protein kinase phosphorylates and inactivates RhoA. *Biochemical and Biophysical Research Communication*. 280, pp.798-805. ISSN 0006-291X.

Slawson, C.; Copeland, R.J. & Hart, G.W. (2010). O-GlcNAc signaling: a metabolic link between diabetes and cancer? *Trends in Biochemical Sciences*. 35, pp.547-555. ISSN 0968-0004.

Somlyo, A.P. & Somlyo, A.V. (2000). Signal transduction by G-proteins, rho-kinase and protein phosphatase to smooth muscle and non-muscle myosin II. *Journal of Physiology*. 522, Pt 2, pp.177-185. ISSN 0022-3751.

Spiro, R.G. (2002) Protein glycosylation: nature, distribution, enzymatic formation, and disease implications of glycopeptide bonds. *Glycobiology*. 12, pp.43R-56R. ISSN 0959-6658.

Swain, S.M.; Tseng, T.S. & Olszewski, N.E. (2001). Altered expression of SPINDLY affects gibberellin response and plant development. *Plant Physiology*. 126, pp.1174-1185. ISSN 0032-0889.

Taylor, M.E. & Drickamer, K. (2006). Introduction to glycobiology / Maureen E. Taylor, Kurt Drickamer. 2nd ed. Oxford, New York. Oxford University Press. xix, 255 p. : ill. ISBN-10: 0199258686.

Toleman, C.; Paterson, A.J.; Whisenhunt, T.R. & Kudlow, J.E. (2004). Characterization of the histone acetyltransferase (HAT) domain of a bifunctional protein with activable O-GlcNAcase and HAT activities. *Journal of Biological Chemistry*. 279, pp.53665-53673. ISSN 0021-9258.

Torres, C.R. & Hart, G.W. (1984). Topography and polypeptide distribution of terminal N-acetylglucosamine residues on the surfaces of intact lymphocytes. Evidence for O-linked GlcNAc. *Journal of Biological Chemistry*. 259, pp.3308-3317. ISSN 0021-9258.

Varki, A.; Cummings, R.D.; Esko, J.D.; Freeze, H.H.; Stanley, P.; Bertozzi, C.R.; Hart, G.W. & Etzler, M.E. (2009). Essentials of Glycobiology. 2nd edition ed, ed. T.C.o.G. Editors: Cold Spring Harbor (NY): Cold Spring Harbor Laboratory Press. ISBN-13: 9780879697709.

Vosseller, K.; Trinidad, J.C.; Chalkley, R.J.; Specht, C.G.; Thalhammer, A.; Lynn, A.J.; Snedecor, J.O.; Guan, S.; Medzihradszky, K.F.; Maltby, D.A.; Schoepfer, R. & Burlingame, A.L. (2006). O-linked N-acetylglucosamine proteomics of postsynaptic density preparations using lectin weak affinity chromatography and mass spectrometry. *Molecular and Cellular Proteomics*. 5, pp.923-934. ISSN 1535-9476.

Wang, Z.; Udeshi, N.D.; O'Malley, M.; Shabanowitz, J.; Hunt, D.F. & Hart, G.W. (2010). Enrichment and site mapping of O-linked N-acetylglucosamine by a combination of chemical/enzymatic tagging, photochemical cleavage, and electron transfer dissociation mass spectrometry. *Molecular and Cellular Proteomics*. 9, pp.153-160. ISSN 1535-9476.

Wang, J.; Liu, R.; Hawkins, M.; Barzilai, N. & Rossetti, L. (1998). A nutrient-sensing pathway regulates leptin gene expression in muscle and fat. *Nature*. 393, pp.684-688. ISSN 0028-0836.

Wang, J.; Torii, M.; Liu, H.; Hart, G.W. & Hu, Z.Z. (2011). dbOGAP - an integrated bioinformatics resource for protein O-GlcNAcylation. *BMC Bioinformatics*. 12, pp.91. ISSN 1471-2105.

Wang, Z.; Gucek, M. & Hart, G.W. (2008). Cross-talk between GlcNAcylation and phosphorylation: site-specific phosphorylation dynamics in response to globally elevated O-GlcNAc. *Proceedings of the National Academy of Sciences USA*. 105, pp.13793-13798. ISSN 1091-6490.

Wang, Z.; Pandey, A. & Hart, G.W. (2007). Dynamic interplay between O-linked N-acetylglucosaminylation and glycogen synthase kinase-3-dependent phosphorylation. *Molecular and Cellular Proteomics*. 6, pp.1365-1379. ISSN 1535-9476.

Wells, L.; Gao, Y.; Mahoney, J.A.; Vosseller, K.; Chen, C.; Rosen, A. & Hart, G.W. (2002). Dynamic O-glycosylation of nuclear and cytosolic proteins: further characterization of the nucleocytoplasmic beta-N-acetylglucosaminidase, O-GlcNAcase. *Journal of Biological Chemistry*. 277, pp.1755-1761. ISSN 0021-9258.

Wells, L.; Kreppel, L.K.; Comer, F.I.; Wadzinski, B.E. & Hart, G.W. (2004). O-GlcNAc transferase is in a functional complex with protein phosphatase 1 catalytic subunits. *Journal of Biological Chemistry*. 279, pp.38466-38470. ISSN 0021-9258.

Woodsome, T.P.; Eto, M.; Everett, A.; Brautigan, D.L. & Kitazawa, T. (2001). Expression of CPI-17 and myosin phosphatase correlates with Ca(2+) sensitivity of protein kinase C-induced contraction in rabbit smooth muscle. *Journal of Physiology*. 535, pp.553-564. ISSN 0022-3751.

Yang, W.H.; Kim, J.E.; Nam, H.W.; Ju, J.W.; Kim, H.S.; Kim, Y.S. & Cho, J.W. (2006). Modification of p53 with O-linked N-acetylglucosamine regulates p53 activity and stability. *Nature Cell Biology*. 8, pp.1074-1083. ISSN 1465-7392.

Zachara, N.E. (2009). Detecting the "O-GlcNAc-ome"; detection, purification, and analysis of O-GlcNAc modified proteins. *Methods in Molecular Biology*. 534, pp.251-279. ISSN 1064-3745.

Zachara, N.E. & Hart, G.W. (2006). Cell signaling, the essential role of O-GlcNAc! *Biochimica et Biophysica Acta*. 1761, pp.599-617. ISSN 0006-3002.

Zeidan, Q. & Hart, G.W. (2010). The intersections between O-GlcNAcylation and phosphorylation: implications for multiple signaling pathways. *Journal of Cell Science*. 123, pp.13-22. ISSN 0021-9533.

Zou, L.; Yang, S.; Champattanachai, V.; Hu, S.; Chaudry, I.H.; Marchase, R.B. & Chatham, J.C. (2009). Glucosamine improves cardiac function following trauma-hemorrhage by increased protein O-GlcNAcylation and attenuation of NF-{kappa}B signaling. American Journal of Physiology (Heart and Circulatory Physiology). 296, pp.H515-H523. ISSN 0363-6135.

4

Technologies for the Use of Protein Kinases into Medical Applications

Yoshiki Katayama
Kyushu University
Japan

1. Introduction

Living systems continuously monitor and respond to the surrounding environment. These processes are made possible by cellular signal transduction systems. When particular information reaches the cell (in many cases to the surface of cells), corresponding molecular networks are activated to process the information. These cascade-type reactions change many enzymes in the cell and ultimately the enzymatic reactions taking place determine the cellular response. Although the system includes a large number of enzymes, protein kinases are the most important group of enzymes and play key roles of signal transduction events. The human genome encodes nearly 500 types of protein kinases (Cohen, 2001). One-third of all cellular proteins act as substrates of protein kinases. Therefore, the monitoring of the activities of protein kinases is a crucial technology not only for understanding life processes, but also for the development of efficient diagnostics or effective drug discovery programs. Modification of certain protein kinase activities will also be an important medical technology for therapy against many diseases. In this context, this chapter will introduce recent technologies that have been developed to monitor protein kinases. In addition, if we use the activity of the protein kinases as a tool for medical engineering, we may be able to control cellular function when needed. In this category, new technologies that use protein kinase activities for controlling transgene regulation will also be introduced.

2. Overview of kinase assays

The activity of protein kinases is easily detected through the incorporation of a radioactive phosphate to a protein or peptide substrate using ^{32}P- or ^{33}P-ATP (Schutkowski et al., 2004; Panse et al., 2004; Diks et al., 2004). Although such assays are highly sensitive and quantitative, there are some important drawbacks such as the requirement of a special facility to handle radioactive materials, production of radioactive waste, the short half-lives (14 days) of the radioactive phosphate, and the potential risk to health. Thus, many types of non-radioactive protein kinase assays have been developed. In these assays, fluorescence-based approaches represent a promising way for high-throughput analysis of protein kinase activities. Alternatively, colorimetry may be easier to handle and cheaper when compared with fluorimetry; however, the sensitivity is generally lower. Mass spectrometry represents another way to monitor protein kinase activity.

When fluorescence techniques are used for the design of protein kinase assays, many useful properties of the fluorescence phenomena can be used such as fluorescence intensity, fluorescence polarization, fluorescence energy transfer and the fluorescence life-time.

What is important in assay design is the ability of the system to distinguish between phosphorylated and non-phosphorylated forms of the substrate. An anti-phospho antibody is a convenient way to recognize phosphorylation of a substrate. However, antibody with sufficient affinity for phorsporykated Ser/Thr is not commercially available. Therefore, other artificial molecules such as metal complexes, polycationic polymers or beads are used for the recognition of phosphor-serine or -threonine. Phos-Tag (Kinoshita et al., 2006; Inamori et al., 2005) or Pro-Q Diamond dye (Steinberg et al., 2003) are typical examples for this category (Fig. 1). Phos-Tag is a zinc complex that was designed by using the alkaline phosphatase structure, and this compound binds to phosphor-amino acids. The molecule also possesses a biotin moiety so that avidin derivatives can also bind tightly. Pro-Q Diamond dye is a gallium complex of a fluorescent molecule. Phosphorylation of a peptide or protein can be detected by agarose gel electrophoresis using this probe. Instead of binding other molecules to the phosphorylation site, phosphorylated amino acid residues may be derivatized to other chemical groups by attaching a marker molecule to detect the phosphorylated substrate (Oda et al., 2001).

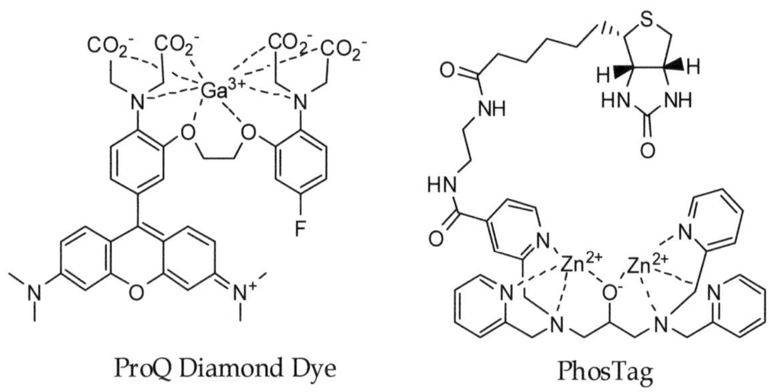

Fig. 1. Chemical structures of ProQ Diamond dye and Phos Tag

3. Protein kinase assay with measurement of fluorescence intensity

Probably, the simplest way of using fluorimetry in kinase assays is the design of a fluorescent substrate that changes its fluorescence intensity upon phosphorylation. However, it is not so easy to change the fluorescence intensity of a single fluorescent molecule through phosphorylation. The first example was an acrylodan-labeled peptide substrate of protein kinase C (PKC) (McIlroy et al., 1991). The molecule, Acrylodan-CKKKKRFSFKKSFKLSGFSFKKNKK-OH, decreased its fluorescence intensity by 20% upon phosphorylation by PKC. The time course of the fluorescence decrease was found to correlate well with that of [^{32}P]phosphate incorporation. The assay detected 0.02 nM of PKC. Although the assay cannot be applied to living cells, the PKC activity in a brain homogenate was easily detected. On the other hand, Higashi et al, 1996. reported a cell permeable acrylodan-labeled peptide (syntide 2) for the detection of calcium calmodulin dependent kinase II (CaMKII) activity (Higashi et al., 1996). The probe was used for imaging CaMKII activity in mice hippocampus slices. However, this type of assay using simply fluorescence-labeled substrates provides only a small change in the fluorescence intensity following phosphorylation. Consequently, the sensitivity of this approach is generally low.

Generally it is difficult to obtain large changes in fluorescence intensity if the fluorophore involves the simple labeling of a peptide. Kupcho et al, 2003. reported a unique fluorescent probe for the detection of protein kinase A (PKA) (Kupcho et al., 2003) (Fig. 2a). In this case,

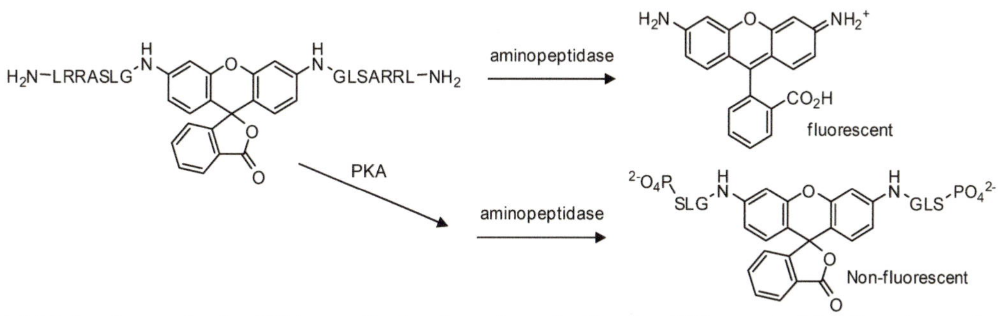

(a) Molecular probe for PKA monitoring based on rhodamine 110 reported by Kupcho et al.

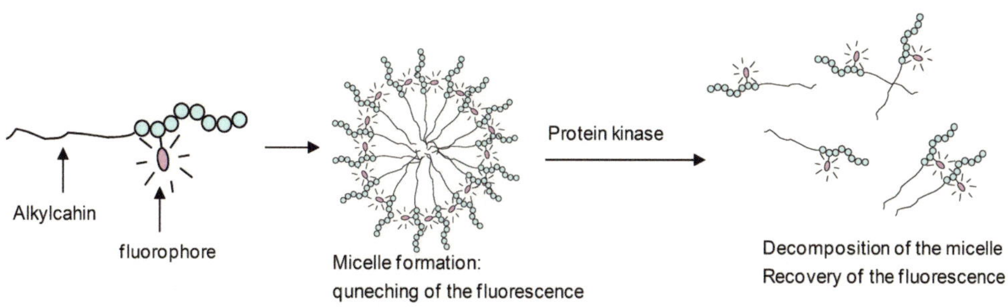

(b) Micelle-based protein kinase probe for fluorescence monitoring reported by Sun et al.

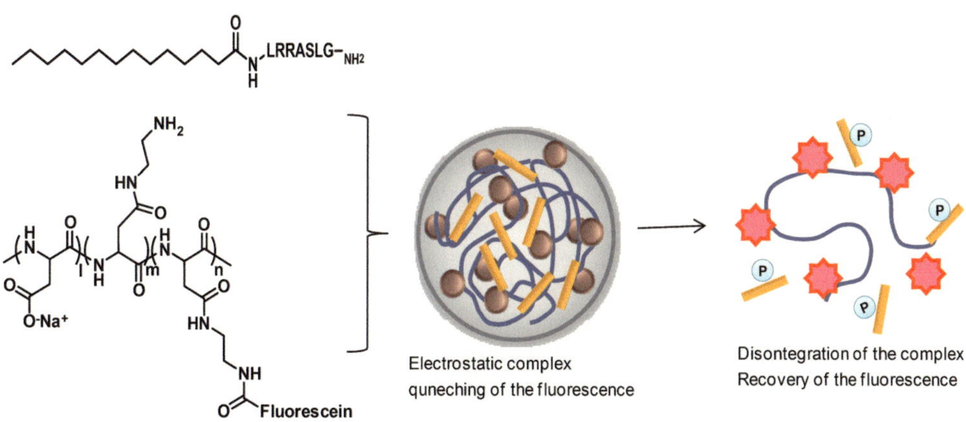

(c) Electrostatic complex consisting of lipid-type substrate and polyanion for kinase monitoring

Fig. 2. Protein kinase assay with measurement of fluorescence intensity (for the comment: This is an original figure, although it resembles to that in our pepare published in Bioconjugate chemistry, 22, 1526-(2011).)

the substrate peptide of PKA was introduced into the both sides of rhodamine 110 through amide bonds. The molecule is practically non-fluorescent due to the formation of a lactone ring in the rhodamine structure. Peptidase degrades the peptide moieties from their amino termini. When the peptides were completely digested, the fluorescence of the rhodamine 110 recovers. However, if the peptide is phosphorylated with PKA, digestion by peptidase is inhibited at the phosphor-serine such that the fluorescence never recovers. This method can clearly detect PKA activity with an 'on-off' strategy; however, the molecular design is not versatile and therefore not suitable for many kinases. In addition, the method cannot be applied to living cells and *in vivo*. It often requires the design of a complicated probe molecule to obtain large changes in the fluorescence intensity with phosphorylation when using small probe molecules. However, if we use a molecular assembly system such as micelles or polyionic complexes, it becomes easier to obtain larger changes in the fluorescence intensity when probing phosphorylation. Sun et al. 2005 reported a micelle system for the detection of protein kinase activity (Sun et al., 2005) (Fig. 2b). In this system, an aliphatic chain was connected to a peptide substrate of a target kinase that was labeled with a fluorescent molecule. If the length of the hydrocarbon chain is optimized, the material forms a micelle-like assembly. The fluorescence is then quenched due to the concentrating of the fluorophores. However, if the substrate is phosphorylated by a target kinase the fluorescence intensity increases several fold because of the decomposition of the micelle. Phosphorylation of the peptide moiety dramatically changes the hydrophilic-hydrophobic balance of the alkylated peptide substrate. We also developed a polyion complex consisting of an alkylated cationic peptide substrate and a fluorescein-labeled polyaspartic acid for monitoring protein kinase A or protein kinase Cα activity (Koga et al. 2011) (Fig. 2c). Such a polyion complex formed a nano-particle with a size of 100-200 nm. In this particle, the fluorescence is quenched because of the high concentration. Phosphorylation of the peptide moiety decreases the cationic net charges of the peptide so that an electrostatic interaction between the lipid-type peptide and the fluorescence-labeled polyanion decreases and this leads to the disintegration of the polyion complex. Such an event leads to an increase in the fluorescence intensity by several fold. In this system, the peptide substrate does not require the direct labeling of a fluorophore, which sometimes affects the ability of the molecule to act as a kinase substrate. This system was successfully applied to validate kinase inhibitors.

4. Protein kinase assay based on fluorescent polarization

Fluorescence polarization is a technique to detect the rotational property of a target molecule. When a fluorescent molecule is excited with polarized light, the extent of the remaining polarization of the emitted light depends on the rotation of the molecule. If the molecule rotates within the period of its excitation, the emitted light loses the polarization in the plane of excited light. Since the molecular rotation is dependent on the molecular weight of the protein, the technique can detect the binding of a large molecule to the target molecule. This strategy can be applied in the design of protein kinase assays. Seethala et al. 1997 reported the first kinase assay based on a fluorescence polarization experiment (Seethala, 1997). Once a fluorescently labeled peptide substrate is phosphorylated with a protein kinase, an anti-phospo-amino acid antibody binds to the substrate (Fig. 3a). Due to the dramatic increase of the molecular size, the fluorescence polarization signal was observed to increase because of a reduction in the rate of rotation. This direct monitoring of fluorescence polarization is simple, but it usually requires a relatively large amount of anti-phospho antibody. The method also needs a small substrate such as a peptide and the technique cannot be applied using a protein substrate. On the other hand, if the strategy involves a competition assay, the protein substrate can also be available (Seethala et al., 1998; Kristjansdottir et al., 2003) (Fig. 3b). In this system, after the protein substrate is phosphorylated by the protein kinase, it is added into the complex of the fluorescence-labeled phosphorpeptide and anti-phopho antibody to compete for binding to the antibody. Thereafter, we can evaluate the kinase activity by evaluating the decrease in the fluorescence polarization signal. The advantage of the fluorescence polarization assay is that this approach is independent of the concentration or fluorescence intensity of the fluorophore used. However, only an anti-phospho antibody for tyrosine is available. Therefore, another molecule that can bind to phosphoserine or threonine is needed if this type of assay is to be applied to monitor the activity of serine/threonine protein kinases. Polycationic peptides and trivalent cation-containing particles can be used for this purpose (Coffin et al., 2000). However, such compounds have poor specificity. Moreover, polycationic peptides are also limited to the use of neutral substrate peptides and trivalent cation-containing particles sometimes suffer from weak binding to phosphorylated sites. The Phos-Tag may be another practical possibility because of its relatively high specificity and binding constant. Recently, a fluorescent polarization assay was applied to a high-throughput assay for screening inhibitors of a protein kinase (Kumar et al., 2011).

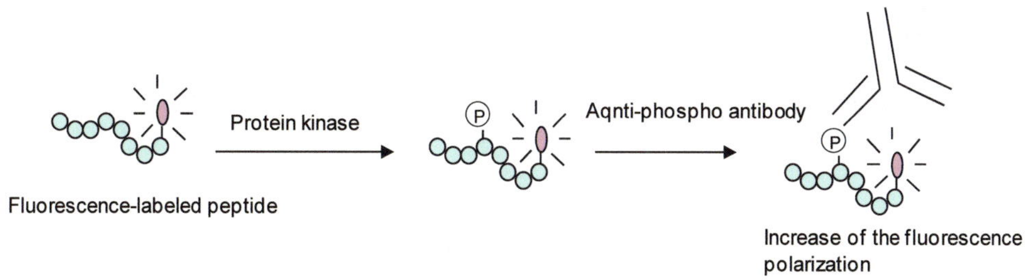

(a) Basic concept of protein kinase assay base on fluorescence polarization

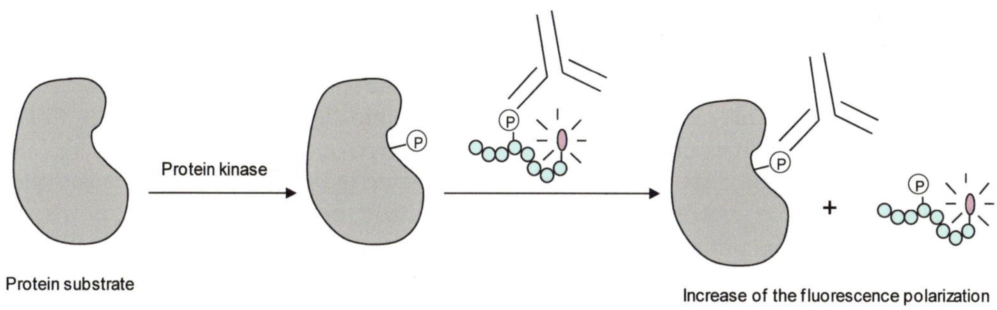

(b) Protein kinase assay base on fluorescence polarization by using protein substrate

Fig. 3. Protein kinase assay based on fluorescence polarization

5. Use of FRET for protein kinase assay

Fluorescence resonance energy transfer (FRET) is the non-radiation energy transfer between two different fluorophores. If the emission spectrum of the donor fluorophore and the excitation spectrum of the acceptor fluorophore overlap and these two fluorophores exist in close proximity, the excitation light for the donor produces an emitted light derived from the acceptor. This phenomenon can be observed when the two molecules exist within 10 nm. Consequently, FRET is highly sensitive to distances between donors and acceptors. This methodology can be applied to similar systems characterized by fluorescence polarization-based assays using peptide substrates and an anti-phospho antibody. If the peptide substrate and the antibody are labeled with acceptor and donor molecules, respectively, binding of the antibody to the phosphorylated substrate generates FRET between the donor and acceptor. Therefore, protein kinase activity can be detected by monitoring the ratio of the fluorescence intensity at two emission wavelengths for the donor and acceptor. FRET measurements are sometimes disturbed by background fluorescence derived from other biomaterials and are also affected by direct excitation of the acceptor with excitation light for the donor molecule. To avoid such disturbances, time-resolve FRET is often used. Riddle et al. 2006 reported a FRET system using a GFP-fused peptide substrate and an anti-phosphotyrosine antibody labeled with a terbium ion complex (Riddle et al. 2006 (fig.4a)). The time-resolved FRET technique can be applied because rare earth metal complexes, such as terbium or europium complexes, produce long life-time fluorescences. However, a GFP fusion sometimes disturbs the phosphorylation of the substrate because of its large size. To avoid this effect, small organic fluorophores such as Alexa dyes are also used as acceptors (Zhang et al., 2005). A microplate based-high throughput assay has also been reported using a FRET-based kinase assay (Gratz et al., 2010) (Fig. 4b). In this assay, the substrate peptide of casein kinase 2 was labeled with fluorophore (EDANS) and the quencher (DABSYL) at the C and N terminus, respectively. The fluorescence of EDANS was quenched due to the FRET with DABSYL. Phosphorylation of the peptide by CK2 prohibited the cleavage of the peptide with elastase. Thus, CK2 activity was evaluated by the decrease in fluorescence. This approach was applied to a microplate-based assay and CK2 inhibitors were screened.

Although these systems are applied only to solution samples, if a FRET system can be applied to living cells, it has an advantage of ratiometry, in which, the assay can be performed independent of the thickness of the sample. Phocus is a good example of such a system. In this probe, CFP and YFP are fused with a kinase substrate, linker and phosphor-recognition domain (Sato & Umezawa, 2004) (Fig. 4c). In the free form, FRET between CFP and YFP does not occur because of the long distance between the two molecules. On the other hand, phosphorylation of the substrate domain causes the binding of the phosphor-recognition domain. This moves the two fluorophores into close proximity to cause FRET. Using this probe, activities of c-Jun and Src were observed successfully in living cells. The advantage of this probe is that the probe can be expressed in living cell spontaneously after the transfection of the encoding genes. On the other hand, optimization of the construct is required to design the probe for each kinase, and the large fluorescence moieties, CFP and YFP, may disturb the access of the protein kinase to the substrate domain in some cases. The method is also inconvenient for HTS systems.

An additional monitoring system of protein kinases uses an alpha-screen assay (Pedro et al., 2010). Alpha-screening is not exactly a FRET system, but the excitation energy of a donor bead transfers to an acceptor bead indirectly via singlet oxygen. Excitation of the donor produces singlet oxygen with a photodynamic effect. Although the lifetime of singlet oxygen is very short, 4 μsec, if the acceptor exists within 200 nm from the donor, the singlet oxygen can reach the acceptor. The acceptor then produces an emission light with singlet oxygen. Pedro et al. 2010 reported the monitoring of a leucine-rich repeat kinase (LRRK2), which is sometimes active in Parkinson's disease, using the alpha-screen system. Moesin is a substrate of LRRK2 fused with a GST tag. Donor and acceptor beads were modified with GST and protein A, respectively. An antiphospho-antibody was then introduced onto the acceptor bead through protein A. After the phosphorylation of moecine with LRRK2, the donor and acceptor beads bound the moecine through GST and the phosphorylation site, respectively. In this case, excitation of the donor bead with 680 nm light produced an emission light at 520–600 nm from the acceptor bead through singlet oxygen. The system was applied to HTS analysis using a 384-well plate.

The QTL Light Speed Kinase Activity Assay™ is also a sensitive detection system for detecting protein kinase activity (Moon et al., 2007). This assay uses a highly fluorescent microsphere and quencher-labeled substrate peptide. The peptide can bind to the microsphere if the peptide is phosphorylated by the target kinase, because its surface is modified with a gallium complex. The fluorescence of the microsphere is then quenched with the quencher on the peptide. The system was used in a HTS method involving a microarray.

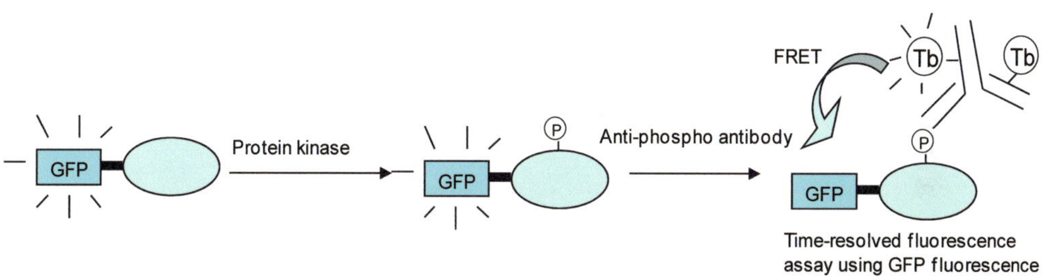

(a) Conceptual design of Phocus which is FRET based kinase probe for intracellular imaging

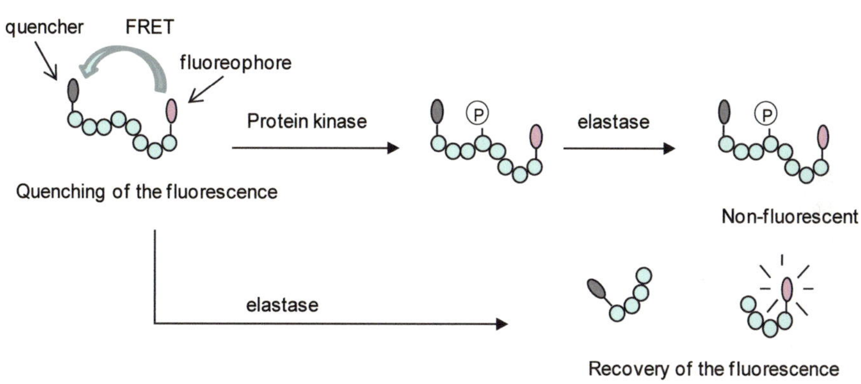

(b) FRET assay of protein kinase for high throughput inhibitor screening

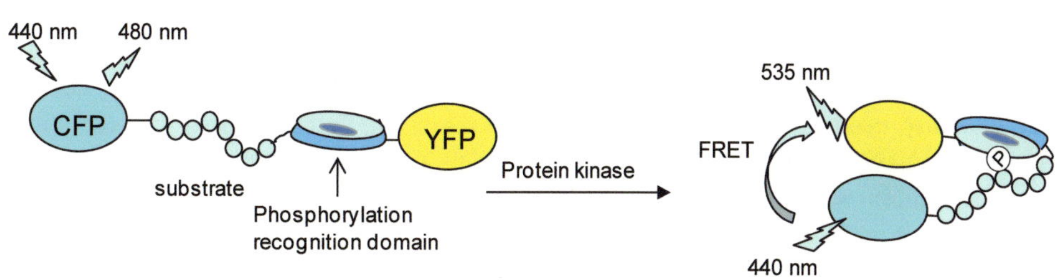

(c) Protein kinase assay based on time resolved FRET measurement

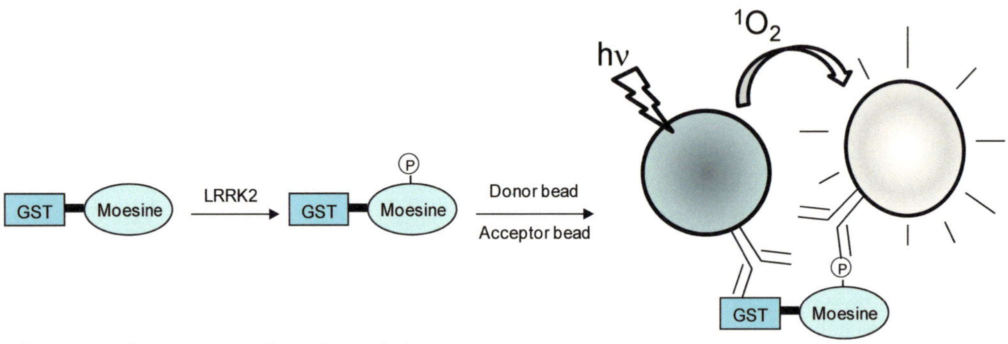

(d) Protein kinase assay based on alpha-screen assay

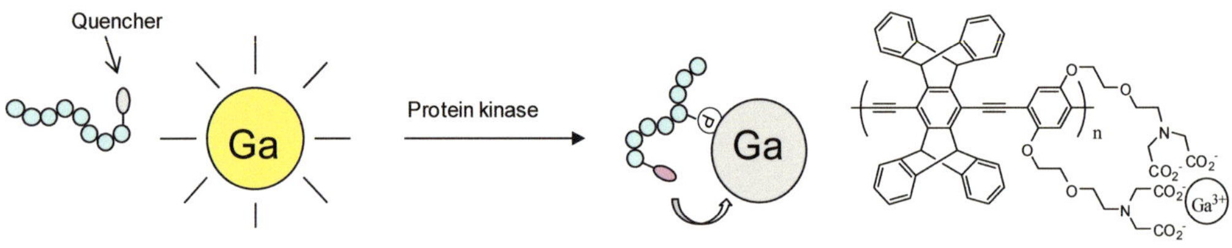

(e) Protein kinase assay based on QTL system

Fig. 4. FRET-based protein kinase assays

6. HTS assay using Gold Nano-Particle (GNP) with colorimetry

Although fluorimetry is sensitive and flexible, it can be affected by factors such as background substances, temperature, pH and the concentration of the fluorophore. Colorimetry, on the other hand, is simple and robust. We reported a label-free kinase assay using gold nano-particle (GNP) (Oishi et al., 2007; Oishi et al., 2008) (Fig. 5). Cationic peptides causes an aggregation of anionic GNP prepared by citrate reducing. This changes the color of the GNP dispersion from red to blue. This aggregation is highly sensitive to the peptide. Cationic peptide aggregates are 1000-fold more effective than inorganic cations that have same cationic charges. However, if the peptide is phosphorylated by the target kinase, the ability to aggregate is reduced dramatically so that the color of the dispersion remains red upon the addition of the peptide. Thus, phosphorylation of the peptide can easily be detected by monitoring the absorbance at 670 nm. The assay has been sufficiently sensitive to detect PKA, PKCα, MAPK, p38 and Src activity in solution, cell lysates and tissue extracts. Detection of PKCα activity in tumor and normal tissues from human patients suggests that the assay can be applied as a diagnostic of breast cancer (Kang et al., 2010). The assay was also used to screen for protein kinase inhibitors using a micro-titer plate format (Oishi et al., 2008; Asami et al., 2011). Using a chemical library containing 3000 chemicals, new PKA inhibitors, which have similar inhibitory activity to current PKA inhibitors, were actually identified using this assay. The aggregation of GNP is affected by the cationic net charges of the peptide and ionic strength. Therefore, the ionic concentration of the detecting solution has to be optimized for each peptide sequence. However, such conditions can be optimized easily, because the conditions of the phosphorylation and detection steps can be set independently. In addition, this assay does not require any labeling steps to the substrate peptide and is simple, rapid and widely applicable from solution to tissue samples. Since the assay depends on decreasing net charge of the substrate peptide, the original net charge has to be cationic. On the other hand, some protein kinases require anionic peptide sequences as their substrates. This issue can be overcome by the addition of some cationic amino acids at one end of the peptide through a flexible triethylene glycol linker. If gold nano-rods are covered with a cationic surfactant, Cetyltrimethylammonium bromide (CTAB) is used instead of GNP, and an anionic substrate can be used without any addition of cationic amino acids (Kitazaki et al., 2011).

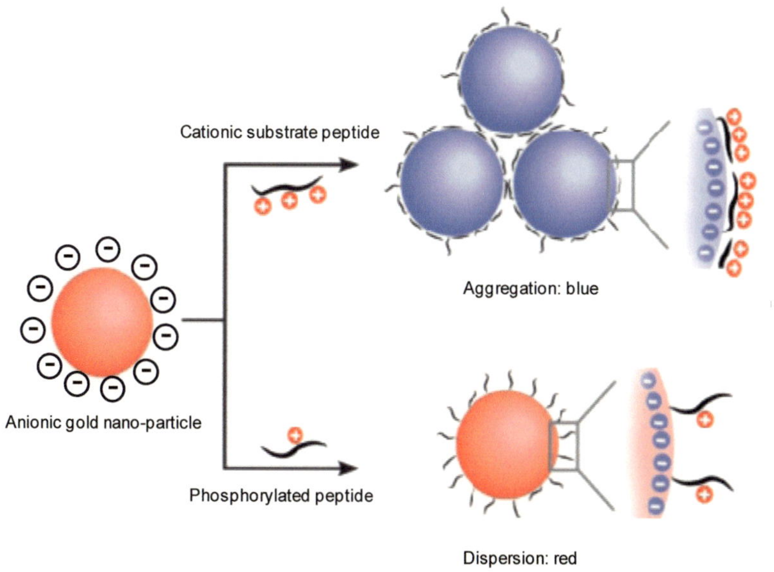

Fig. 5. Concept of colorimetrical assay of protein kinase using gold nano-particle

7. Peptide array

Genomic and post-genomic research has enabled us to understand life at the molecular level. Many aspects of molecular processes have been elucidated. As a result, evolutionary changes have been made in methodologies in drug discovery, diagnostics and other medical technologies. Developments of molecular targeted drugs that primarily target tyrosine kinases for cancer therapy represent a typical example. However, life was not constructed through a simple combination of such pieces, but involves many pieces that interact with each other to generate a complicated network system. Therefore, diseases are not simply treated by inhibiting a single target molecule. For example, cancer cells often acquire a tolerance against molecular targeted drugs during their applications, even though the drugs continue to inhibit the target protein kinases. Therefore, we have to clarify the condition of entire signal network to know the cellular condition exactly. From research using gene chip technology, it has been clarified that a major part of the transcriptome is necessary to maintain the basic functions of living cells, and only a part of the transcriptome relates to each cell-specific function. Tiny fluctuations of the transcriptome sometimes cause a significant change in the enzymatic network determining cell function (Irish et al., 2004).

This indicates that monitoring enzymatic activities in cellular signal transduction events must be efficient and effective for precisely evaluating cell conditions. Kinome, the entire profile of protein kinases in cells, is a new concept presented in this issue. The most practical format of the kinome analysis is peptide array. Peptide array involves chip technology, in which many peptide substrates of protein kinases are immobilized on a solid support. Some peptide array systems are now commercially available. The PepScan array is probably the first example in which kinome analysis was carried out using a peptide array (Diks et al., 2004). In this array, many peptide substrates are immobilized onto a glass slide and phosphorylation is detected with the incorporation of [^{33}P]-phosphate from RI-labeled ATP. Using the array immobilizing 192 peptides, changes in the activities in particular kinases after LPS stimulation was analyzed. Following this report, kinome in Barrett's esophagus, endothelial cells and also c-Met activity in colon cancer after the inhibition of cycloocygenase-2 were monitored using more than 1000 peptides (van Baal et al., 2011). Although the array seems to give reasonable evaluations in activation of particular kinases, the system assumes that one peptide is phosphorylated by a single kinase; however, such short peptides may be phosphorylated by plural protein kinases. The CelluSpot system is another similar peptide array format (Olaussen et al., 2009). In this array, peptide substrates that were synthesized on a nitrocellulose membrane are cleaved and set on a solid support. The effect of tyrosine kinase inhibitors on the kinome in a carcinoma cell line was evaluated using this system with 144 peptides. Phosphorylation of the peptides was detected using a fluorescence-labeled anti-phosphotyrosine antibody. Although this system is simple and does not use radio-active material, the size of each spot is large (1 mm) such that a relatively large amount of sample is required.

Enzymatic reactions are sometimes inconvenient to perform on solid surfaces. The Pamchip is an array to address this issue (Jinnin et al., 2008; Maat et al., 2009). In this system, peptides are immobilized in a well consisting of porous material. The porous material is 60 μm thick and has long branched interconnected capillaries with a diameter of 200 nm. It results in a 500-fold increase in the reactive surface and the reaction and washing steps can also be performed by pulsing back and forth through the porous material many times. The profile of tyrosine kinase activities in a pediatric brain tumor was evaluated and compared with kinomes from other various cancer cell lines in 144 peptides (Sikkema et al., 2009). A group of peptides that were phosphorylated by all cancer cell lines and by particular cell lines were identified. However, it is relatively difficult to identify each kinase from such profiles. Activation of vascular endothelial growth factor receptor type2 (VEGEFR2) and Src signaling were also confirmed in infantile hemangioma and melanoma cells, respectively, with this system.

The above mentioned peptide arrays are useful for kinome analysis. However, it is unclear whether they can be used for quantitative analysis. For monitoring kinase activity, it is important to know how much activity has changed. In many cases, cellular function will be influenced by the degree of activation of particular kinases. Recently we developed a peptide array in which peptide substrates are immobilized onto gold or glass support for surface Plasmon resonance (SPR) and fluorescence detection, respectively (Shigaki et al., 2007; Inamori et al. 2005) (Fig. 6a). By careful optimization of the surface chemistry, these arrays secured quantitative data describing the detection of the phosphorylation ratio in each peptide (Han et al., 2008; Shimomura et al., 2011). For example, the efficiency of peptide immobilization influenced the results from the methodology (Inamori et al., 2008) (Fig. 6b).

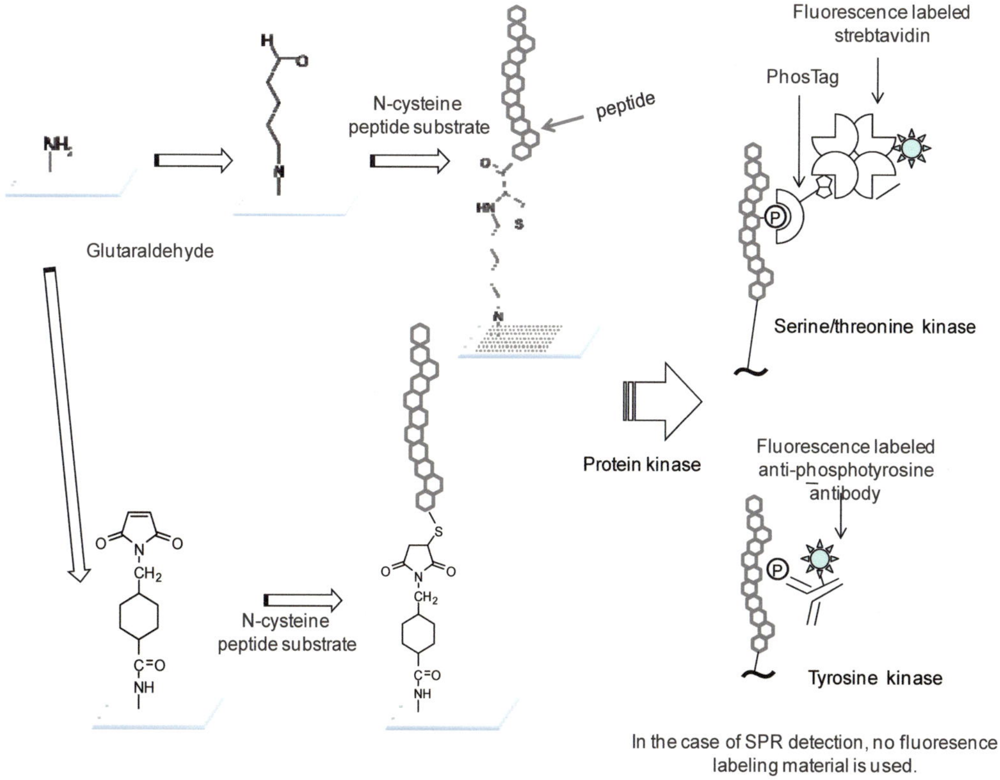

(a) Schematic illustration of peptide immobilization and detection in quantitative peptide array

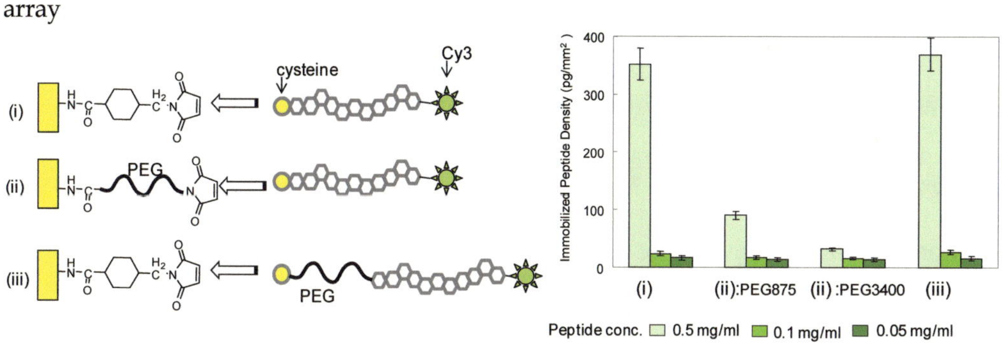

(b) Correlation between immobilized peptide density and immobilization protocol

Fig. 6. Quantitative peptide array: concept and immobilization method of peptide substrate

Figure 6b shows the efficiency of peptides immobilized onto a gold chip. Although the polyethylene glycol (PEG) moiety is effective in suppressing non-specific adsorption of bio-substances, peptide immobilization was also suppressed if the PEG moiety was modified on the chip in advance. On the other hand, when the PEG-linked peptide was reacted with a rigid group on the chip, the amount of the immobilized peptide increased dramatically. In our system, phosphorylation of the peptide was achieved by using an anti-phosphotyrosine antibody or PhosTag molecules followed by the addition of streptavidin. Changes in the activities of particular protein kinases were detected using the SPR chip in solution and a cell lysate after the cell was stimulated with NGF (Han et al., 2009). An advantage of the SPR detection system is that it does not require any labeling for detection. However, this advantage contains a risk that binding of any other substances gives rise to a detectable signal at the same time. In this case, fluorescence detection will be more practical. In the quantitative detection of the kinome on a peptide array, peptides that have cysteines at the amino terminus were immobilized through a formyl group or maleimide. In the former case, high-density amino-modified glass surfaces were treated with glutaraldehyde, and then the peptide was linked to the formyl group at the thiol group of the cysteine residue by forming a thiazoline ring (Mori et al., 2009). The surface of the chip was then blocked from unspecific adsorption with Blocking One-P, which is a commercial cocktail. We recently used a plastic plate that was modified, called the 'S-Bio' system, to avoid the adsorption of biomacromolecules. In this case, cysteine-containing peptides were immobilized on the chip using maleimide chemistry. Detection of phosphorylation was achieved using a Cy-3 labeled anti-phosphotyrosine antibody or a PhosTag and Alexa647-labeled streptavidin. After the optimization of the conditions of immobilization, the obtained chip provides quantitative phosphorylation ratios of the peptide. This quantitative analysis is sufficient for measuring peptide phosphorylation immobilized on a single chip. However, an internal standard is required for the inter-plate comparison. Alexa647 labeled peptide is used for this purpose. Using this array, Src activity in various cancer cell lines and mouse tissues was successfully monitored. Changes in the kinome profile with drug stimulation such as NGF or Iressa was also obtained (Han et al., 2010). Such arrays were also applied to screen kinase inhibitors (Inamori et al., 2009).

Kinome analysis using peptide array has not been a well-establish technology. Especially, bio-informatics technique which converts obtained phosphorylation profile into actual signal network of protein kinases. Reproducibility of peptide array has also to be improved.

8. New technology using protein kinase activity in artificial bio-regulation

As mentioned above, many technologies for monitoring protein kinase activity have been developed. On the other hand, any technology that uses protein kinase activity should also be useful for medicine, because protein kinases play key roles in determining cellular functions. Abnormal activation of particular kinases is often observed in many diseases. For example, hyper-activation of EGFR, c-MET, bcl-Abl, PKCα, or Src has been reported in many types of tumors. In myocardiac infarction, over expression of Rho kinase is also reported. Activation of I-κ-kinase is a key signal to initiate inflammation. Therefore, abnormal activity of such kinases can be markers to distinguish disease cells. In this context, if these signals can be converted to other information artificially, such signal conversion should identify disease cells specifically. Such artificial signal converters will offer a new strategy for cell-specific medicines. We recently reported some artificial gene regulators that activated transgene expression in response to target protein kinase activity (Oishi et al., 2006; Sonoda et al., 2005; Kawamura et al., 2005) (Fig.7). The regulators consist of polymer backbone and some cationic peptide side-chains. The peptide is also designed as a specific substrate of a target protein kinase. Since the polymer-peptide conjugates are polycationic, they can form electrostatic complexes with DNA such as an expression vector. In the complex, this type of conjugate suppresses the gene transcription much more efficiently

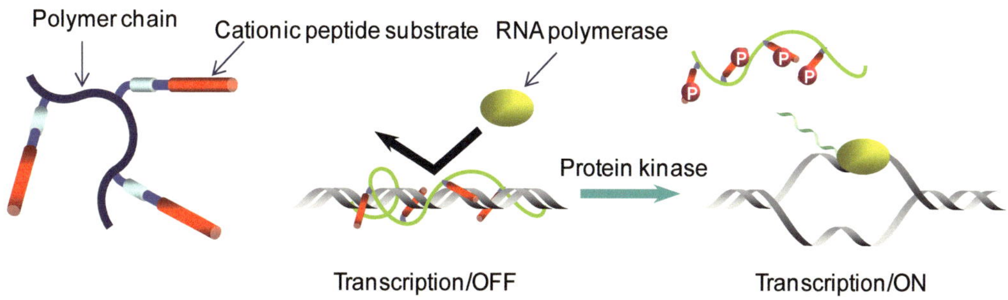

Fig. 7. Structure of artificial gene regulator (left) and concept of its gene regulation in response to protein kinase

than ordinary polycations such as polyethyleneimine, or poly-L-lysine. When the complex is taken up by target disease cells in which target protein kinases are hyper-activated, the peptide side-chains are phosphorylated. This introduction of anionic charges decreases the net cationic charges of the conjugate, and the electrostatic interaction between the conjugate and DNA is attenuated. As a result, the gene can be expressed due to the disintegration of the complex. This system is the first strategy for cell-specific gene therapy using kinase activity as a marker of cellular identification. Using this strategy, various gene regulators have been developed for Src, PKCα, I-κ-kinase, Rho kinase and PKA as target signals (Sato et al., 2010; Kang et al., 2010; Asai et al., 2009; Tsuchiya et al., 2011; Oishi et al., 2006). These materials realize highly cell-specific gene expression. Figure 8b indicates examples of such signal-responsive gene expression in I-κ-kinase and PKCα-responsive systems (Asai et al., 2009). The I- κ-kinase responsive system activated expression of a GFP encoding gene only following stimulation of NHI 3T3 cells with LPS or TNF-α, thereby initiating inflammation. However, if the serine residue in the peptide side-chain, which is a phosphorylation site, was replaced with alanine, such gene expression was not observed even following stimulation by LPS or TNF-α. PLCα is another important kinase for the proliferation of many types of cancer cells (Kang et al., 2009). Therefore, transfection of GFP encoding plasmid as a complex with the PKCα responsive regulator gave massive expression of GFP in various cancer cell lines (Asai et al., 2009). Conversely, a negative control-regulator, in which the serine residue was replaced with alanine, did not show any expression in such cell lines. In addition, no activation of GFP expression was observed when such cells were pre-treated with an inhibitor of PKC. These results clearly indicate that such systems regulated gene expression in response to target kinases (Toita et al., 2009). In particular, the PKCα responsive system worked also in tumor bearing mice (Kang et al., 2008; Toita et al., 2009; Kang et al., 2010) (fig. 8c). When a complex between a PKCα-responsive conjugate and the luciferase encoding gene was injected into a tumor directly, expression of luciferase was observed successfully. On the other hand, injection of the complex in normal subcutaneous tissue or injection of the complex using a negative control conjugate into a tumor did not show any expression of luciferase. The obtained image of luciferase indicates a proliferation activity, because the enzymatic activity regulates cancer proliferation directly and is closely related to the cancer malignancy. Thus, this system will be useful for cancer imaging, because this is the first functional imaging of cancer in contrast to ordinary imaging techniques of cancer that mainly visualize the existence of a tumor. Such functional images should provide much more sharpshooting information for prognosis than currently used imaging technologies. Such a system can also be applied to cancer cell-specific gene therapy (Tomiyama et al., 2010 and 2009). Using the caspase-8 encoding gene as a therapeutic gene, shrinkage of the tumor was also observed in HepG2 tumor bearing mice. HSV thymidine kinase encoding gene and the gancyclovir system also worked well in this system. Since this method is highly disease cell-specific, gene activity can be masked in other normal tissues or organs due to the absence of continuous target kinase activity (Kang et al., 2010). Therefore, many therapeutic genes, which were abandoned as clinical targets because of their side effects derived from undesired activation of such genes in non-target organs, should be revived using this methodology. These techniques are potentially useful for future medicines; although the gene complex has to be stabilized in blood flow. By covering the complex with sugar chains such as chondroitin sulfate or hyaluronic acid offers a promising way to access this issue (Tomiyama et al., 2011).

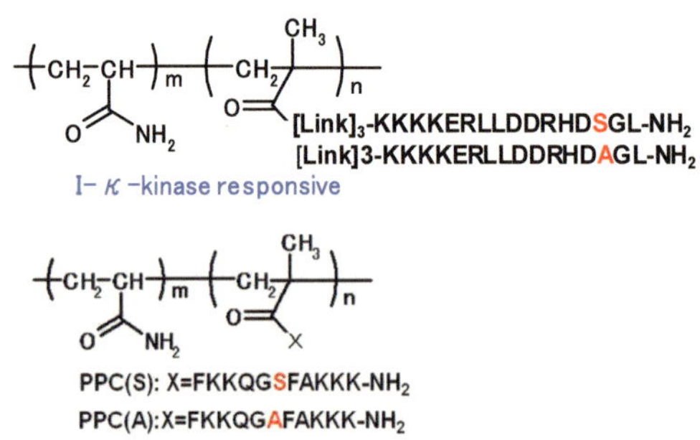

a) Chemical structure of I-κ-kinase and PKCα reponsive regulator

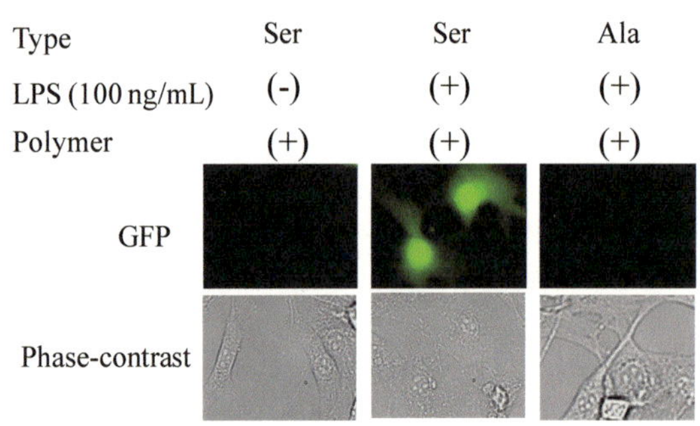

b) I-κ-kinase responsive expression of GFP by using artificial regulator

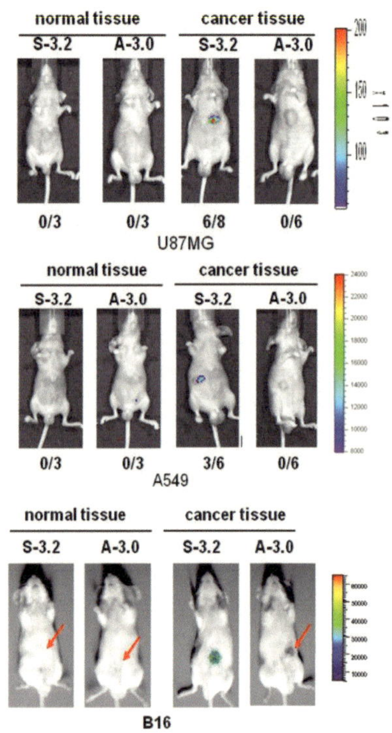

c) Tumor specific luciferase expression in PKC Responsive system. The complex was injected In tumor or normal skin directly.

Fig. 8. Intracellular protein kinase-responsive gene regulation system

9. Conclusion and future prospective

Recent technologies for the monitoring or handling of protein kinase activities have been described. Monitoring protein kinase activities using current technologies offers a way to understand basic biological processes of life. Such technologies are becoming significant in medical and medicinal fields, because protein kinases represent major drug targets and can also be used as diagnostic markers. However, such technologies have to be high-throughput to satisfy medical or pharmaceutical demands. Detecting dysfunctional activity of particular kinases and relating this to a disease condition will require the development of simple and rapid assays. Fluorescence polarization, bead techniques using fluorimetry and colorimetric assays offer a way to reach this goal. Mass spectrometry is also a promising approach (Kang et al., 2008; Kang et al., 2007; Shigaki et al., 2006). Peptide arrays are another promising technology for detail evaluation of cellular conditions. Since cellular function is determined by a network of signaling reactions governed by enzymes including protein kinases, the exact state of living cells in various diseases cannot be evaluated by a single protein kinase assay. In this context, kinome analysis will be crucial for providing a detailed diagnosis before medication, prognosis after medication and validation of drugs in pharmaceutical testing. However, problems with the current peptide array systems, including their low reproducibility of obtaining similar phosphorylation profiles on chips, are hampering progress towards fully accurate kinome analysis. Relatively low specificity of peptide substrates is another issue. It is difficult to convert the obtained phosphorylation profile into a profile that represents the actual protein kinase activity. Bioinformatics and mathematical technology should be combined with array technologies in the future.

Protein kinases are also attractive as a marker to distinguish between disease cells and normal cells. Our gene regulation system that responds to target protein kinase activity is the first artificial system that uses intracellular signaling as a trigger to output another biological signal. Such signal engineering represents a new cell-specific medicine approach. We need such a new technological field, termed "Cell Signalomics", to link basic biological findings to clinical approaches. Protein kinases will be one of the most important elements for such a technology.

10. References

Asai, D., Kang, J-H., Toita, R., Tsuchiya, A., Niidome, T., Nakashima, H. & Katayama, Y., (2009). Regulation of transgene expression in tumor cells by exploiting endogenous intracellular signals, *Nanoscale Res. Lett.* Vol. 4, 229-233.

Asai, D., Tsuchiya,A., Kang, J-H., Kawamura, K., Oishi, J., Mori, T., Niidome, T., Shoji, Y., Nakashima, H. & Katayama, Y., (2009). Inflammatory cell-specific gene regulation system responding to Ikappa-B kinase beta activation, *J. Gene Med.*, Vol. 11, 624-632.

Asami, Y., Oishi, J., Kitazaki, H., Kamimoto, J., Kang, J-H., Niidome, T., Mori, T. & Katayama, Y., (2011). A simple set-and-mix assay for screening of protein kinase inhibitors in cell lysates, *Anal. Biochem., Vol.* 418, 44-49.

Coffin, J., Latev, M., Bi, X. & Nikiforov, T. T. (2000). Detection of phosphopeptides by fluorescence polarization in the presence of cationic polyamino acids: Application to kinase assays, *Anal. Biochem.*, Vol. 278, 206-212.

Cohen, P. (2001). The role of protein phosphorylation in human health and disease, *Eur. J. Biochem.* Vol. 268: 5001–5010.

Diks, S. H., Kok, K., Toole, T. O., Hommes, D. W., van Dijken, P., Joore, J. & Peppelenbosch, M. P., (2004). Kinome profiling for studying lipopolysaccharide signal transduction in human peripheral blood mononuclear cells, *J. Biol. Chem.*, Vol. 279 (No. 47), 49206–49213.

Diks, S.H., Kok, K., Toole, T., Hommes, D.W., Dijken, P., Joore, J. & Peppelenbosch, M.P. (2004). Kinome profiling for studying lipopolysaccharide signal transduction in human peripheral blood mononuclear cells, *J. Biol. Chem.*, Vol.279 (No. 47), 49206-49213.

Gratz, A., Gotz, C. & Jose, J. (2004). A FRET-based microplate assay for human protein kinase CK2, a target in neoplastic disease, *J. Enzyme Inhibit.Med. Chem.*, Vol. 25(No. 2): 234–239.

Han, X., Shigaki, S., Yamaji, T., Yamanouchi, G., Mori, T., Niidome, T. & Katayama, Y., (2008). A quantitative peptide array for evaluation of protein kinase activity, *Anal. Biochem.*, Vol. 372, 106-115.

Han, X., Sonoda, T., Mori, T., Yamanouchi, G., Yamaji, T., Shigaki, S., Niidome, Y. & Katayama, Y., (2010). Protein kinase substrate profiling with a high-density peptide microarray, *Comb. Chem. High Throughput Screen. Vol.* 13, 777-789.

Han, X., Yamanouchi, G., Mori, T., Kang, J-H., Niidome, T. & Katayama, Y., (2009). Monitoring protein kinase activity in cell lysates using a high density peptide microarray, *J. Biomol. Screen. Vol.* 14, 256-262.

Higashi, H., Sato, K., Omori, A., Sekiguchi, M., Ohtake, A. & Kudo, Y. (1996) Imaging of Ca2+/calmodulin-dependent protein kinase II activity in hippocampus neurons, *Neuroreport* Vol. 7(No. 15-17), 2695-2700.

Inamori, K., Kyo, M., Matsukawa, K., Inoue, Y., Sonoda, T., Mori, T., Niidome, T. & Katayama, Y., (2009). Establishment of screening system toward discovery of kinase inhibitors using label-free on-chip phosphorylation assays, *BioSystems, Vol.* 97, 179-185.

Inamori, K., Kyo, M., Matsukawa, K., Inoue, Y., Sonoda, T., Tatematsu, K., Tanizawa, K., Mori, T. & Katayama, Y., (2008). Optimal surface chemistry for peptide immobilization in on-chip phosphorylation analysis, *Anal. Chem.*, Vol. 80, 643-650.

Inamori, K., Kyo, M., Nishiya, Y., Inoue, Y., Sonoda, T., Kinoshita, E., Koike, T. & Katayama, Y., (2005). Detection and quantification of on-chip phosphorylated peptides by surface plasmon resonance imaging techniques using a phosphate capture molecule, *Anal. Chem.*, Vol. 77, 3979-3985.

Inamori, K., Kyo. M., Nishiya, Y., Inoue, Y., Sonoda, T., Kinoshita, E., Koike, T. & Katayama, Y. (2005). Detection and quantification of on-chip phosphorylated peptides by surface Plasmon resonance imaging techniques using a ohosphate capture molecule, *Anal. Chem.*, Vol. 77, 3979-3985.

Irish, J. M., Hovland, R., Krutzik, P. O., Perez,O. D., Bruserud, Ø., Gjertsen, B. T. & Nolan, G. P., (2004). Single cell profiling of potentiated phospho-protein networks in cancer cells, *Cell*, Vol. 118, (No. 2) 217–228.

Jinnin, M., Medici, D., Park, L., Limaye, N., Liu,Y., Boscolo, E., Bischoff, J., Vikkula, M., Boye, E. & Olsen, B. R., (2008), Suppressed NFAT-dependent VEGFR1 expression and constitutive VEGFR2 signaling in infantile hemangioma, *Nat. Med.*, Vol. 14 (No. 11), 1236-1246.

Kang, J-H., Asai, D., Kim, J-H., Mori, T., Toita, R., Tomiyama, T., Asami, Y., Oishi, J. Sato, Y. T., Niidome, T., Jun, B., Nakashima, H. & Katayama, Y., (2008). Design of polymeric carriers for cancer-specific gene targeting: Utilization of abnormal protein kinase Cα activation in cancer cells, *J. Am. Chem. Soc.*, Vol. 130, 14906-14907.

Kang, J-H., Asai, D., Toita, R., Kitazaki, H. & Katayama, Y., (2009). Plasma protein kinase C (PKC)α as a biomarler for the diagnosis of cancers, *Carcinogenesis*, Vol. 30, 1927-1931.

Kang, J-H., Asami, Y., Murata, M., Kitazaki, H., Sadanaga, N., Tokunaga, E., Shiotani, S., Okada, S., Maehara, Y., Niidome, T., Hashizume, M., Mori, T. & Katayama, Y., (2010). Gold nanoparticle-based colorimetric assay for cancer diagnosis, *Biosens. Bioelectron.*, Vol. 25, 1869-1874.

Kang, J-H., Han, A., Shigaki, S., Oishi, S., Kawamura, K., Toita, R., Han, X., Mori, T., Niidome, T. & Katayama, Y., (2007). Mass-tag technology responding to intracellular signals as a novel assay system for the diagnosis of tumor, *J. Am. Soc. Mass Spectr. Vol. 18, 106-112.*

Kang, J-H., Kuramoto, M., Tsuchiya, A., Toita, R., Asai, D., Sato, Y. T., Mori, T., Niidome, T. & Katayama, Y., (2008). Correlation between phosphorylation ratios by MALDI-TOF MS analysis and enzyme kinetics, *Eur. J. Mass Spectrom.*, Vol. 14, 261-265.

Kang, J-H., Oishi, J., Kim, J-H., Ijuin, M., Toita, R., Jun, B., Asai, D., Mori, T., Niidome, T., Tanizawa, K., Kuroda, S. & Katayama, Y., (2010). Hepatoma-targeted gene delivery using a tumor cell-specific gene regulation system combined with a human liver cell-specific bionanocapsule, *Nanomedicine Vol. 6, 583-589.*

Kang, J-H., Toita, R. & Katayama, Y., (2010). Bio and nanotechnological strategies for tumor-targeted gene therapy, *Biotechnol. Adv.* Vol. 28, 757-763.

Kawamura, K., Oishi, J., Kang, J-H., Kodama, K., Sonoda, T., Murata, M., Niidome, T. & Katayama, Y., (2005). Intracellular Signal-Responsive Gene Carrier for Cell-Specific Gene Expression, *Biomacromolecules.* Vol. 6, 908-913.

Kinoshita, E., Kinoshita-Kikuta, E., Takiyama, K. & Koike T. (2006). Phosphate-binding tag: A new tool to visualize phosphorylated proteins, *Molecular & Cellular Proteomics*, Vol. 5, 749-757.

Kitazaki, H., Mori, T., Kang, J-H., Niidome, T., Murata, M., Hashizume, M. & Katayama, Y., (2011). A colorimetric assay of protein kinase activity based on peptide-induced coagulation of gold nanorods, *Coll. Surf. B,* in press.

Koga, H., Toita, R., Mori, T., Tomiyama, T., Kang, J-H., Niidome, T. & Katayama Y., (2011). Fluorescent nanoparticles consisting of lipopeptides and fluorescein-modified polyanions for monitoring of protein kinase activity, *Bioconjugate Chem.* Vol. 22, 1526–1534.

Kristjansdottir K. & Rudolph J. (2003). A fluorescence polarization assay for native protein substrates, *Anal. Biochem.*, Vol. 316, 41-49.

Kumar, E. A., Charvet, C. D., Lokesh, G.L. & Natarajan, A. (2011). High-throughput fluorescence polarization assay to identify inhibitors of Cbl(TKB)–protein tyrosine kinase interactions, *Anal. Biochem.*, Vol. 411, 254-260.

Kupcho, K., Somberg, R., Bullent, B., Goueli, S.A. (2003). A homogeneous, nonradioactive high-throughput fluorogenic protein kinase assay, *Anal. Biochem.* Vol. 317, 210-217.

Maat, W., el Filali, M., Dirks-Mulder, A., Luyten, G. P. M., Gruis, N. A., Desjardins, L., Boender, P., Jager, M. J. & van der Velden, P. A., (2009). Episodoc Src activation in uveal melanoma revealed by kinase activity profiling, *Br. J. Cancer*, Vol. 101, 312-319.

McIlroy, B. K., Walters, J. D. & Johnson, J. D. (1991). A Continuous Fluorescence Assay for Protein Kinase *Anal. Biochem.* Vol. 195,148-152.

Moon, J. H., MacLean, P., McDaniel W. & Hancock, L. F., (2007). Conjugated polymer nanoparticles for biochemical protein kinase assay, *Chem. Commun.*, 4910–4912

Mori, M., Yamanouchi, G., Han, X., Inoue, Y., Shigaki, S., Yamaji, T. Sonoda, T., Yasui, K., Hayashi, H., Niidome, T. & Katayama, Y., (2009). Signal-to-noise ratio improvement of peptide microarrays by using hyperbranched-polymer materials, *J. Appl. Phys.*, Vol. 105, 102020.

Oda, Y., Nagasu, T. & Chait, B. T. (2001). Enrichment analysis of phosphorylated proteins as a tool for probing the phosphoproteome Vol. 19 (No. 4) 379-382.

Oishi, J., Asami, Y., Mori, T., Kang, J-H., Tanabe, M., Niidome, T. & Katayama, Y., (2007). Measurement of homogeneous kinase activityfor cell lysates based on the aggregation of gold nanoparticles, *ChemBioChem*, Vol. 8, 875-879.

Oishi, J., Asami, Y., Mori, T., Kang, J-H., Tanabe, M., Niidome, T. & Katayama, Y., (2008). Colorimetric enzymatic activity assay based on "non-crosslinking aggregation" of gold nanoparticles induced by adsorption of substrate peptides, *Biomacromolecules, Vol.* 9, 2301-2308.

Oishi, J., Han, X., Kang, J-H., Asami, Y., Mori, T., Niidome, T. & Katayama, Y., (2008). High-throughput colorimetric detection of tyrosine kinase inhibitors based on the aggregation of gold nanoparticles, *Anal. Biochem., Vol.* 373, 161-163 .

Oishi, J., Ijuin, M., Sonoda, T., Kang, J-H., Kawamura, K., Mori, T., Niidome, T. & Katayama, Y., (2006). A protein kinase signal-responsive gene carrier modified RGD peptide, *Bioorg. Med. Chem. Lett.* Vol. 16, 5740-5743.

Oishi, J., Jung, J., Tsuchiya, A., Toita, R., Kang, J-H., Mori, T., Niidome, T., Tanizawa, K., Kuroda, S. & Katayama, Y., (2010). A gene delivery system specific for hepatoma cells and an intracellular kinase signal based on human liver-specific bionanocapsules and signal-responsive artificial polymer, *Int. J. Pharm., Vol. 396, 174-178.*

Oishi, J., Kawamura, K., Kang, J-H., Kodama, K., Sonoda, T., Murata, M., Niidome, T. & Katayama, Y., (2006). An intracellular kinase signal-responsible gene carrier for disordered cell-specific gene therapy, *J. Controlled Release Vol.* 110, 431-436.

Olaussen, K. A., Commo, F., Tailler, M., Lacrix, L., Vitale, I., Raza, S. Q., Richon, C., Dessen, P., Lazar, V., Sorita, J-C. & Kroemer, G., (2009). Synergistic proapoptotic effects of the two tyrosine kinase inhibitors pazopanib and lapatinib on multiple carcinoma cell lines, *Oncogene*, Vol. 28, 4249-4260.

Panse, S., Dong, L., Burian, A., Carus, R., Schutkowski, M., Reimer, U. & Schneider-Mergener, J. (2004). Profiling of generic antiphosphopeptide antibodies and kinases with peptide microarrays using radioactive and fluorescence-based assays, *Mol. Divers.*, Vol. *8*(No. 3), 291-299.

Pedro, R., Padros, J., Beaudet, L., Schubert, H-D., Gillardon, F. & Dahan, S., (2010). Development of a high-throughput AlphaScreen assay measuring full-length LRRK2(G2019S) kinase activity using moesin protein substrate, *Anal. Biochem.*, Vol. 404, 45-51.

Riddle, S. M., Vedvik, K. L., Hanson, G. T. & Vogel, K. W. (2006). Time-resolved fluorescence resonance energy transfer kinase assays using physiological protein substrates: Applications of terbium–fluorescein and terbium–green fluorescent protein fluorescence resonance energy transfer pairs, *Anal. Biochem.* Vol. 356, 108-116.

Sato, M. & Umezawa, Y., (2004). Imaging protein phosphorylation by fluorescence in single living cells, *Methods*, Vol. 32, 451-455.

Sato, Y., Kawamura, K., Niidome, T. & Katayama, Y., (2010). Characterization of gene expression regulation using D-RECS polymer by enzymatic reaction for an effective design of enzyme-responsive gene regulator, *J. Controlled Release,* Vol. 143, 344-349.

Schutkowski, M., Reimer, U., Panse, S., Dong, L., Lizacano, J.M., Alessi, D. R. & Schneider-Mergener, J. (2004). High-content peptide microarrays for deciphering kinase specificity and biology, *Anew. Chem. Int. Ed.*, Vol. *43*(No. 20), 2671-2674.

Seethala, R. & Menzel, R. (1997). A homogeneous,fluorescence polarization assay for Src-family tyrosine kinases, *Anal. Biochem.*, Vol. 253, 210-218.

Seethala, R. & Menzel, R. (1998). A Fluorescence polarization competition immunoassay, *Anal. Biochem.*, Vol. 255, 257-262.

Shigaki, S., Sonoda, T., Nagashima, T., Okitsu, O., Kita, Y., Niidome, T. & Katayama, Y., (2006). A new method for evaluation of intracellular protein kinase signals using mass spectrometry, *Science and Technology of Advanced Materials* Vol. 7, 699-704.

Shigaki, S., Yamaji, T., Han, X., Yamanouchi, G., Sonoda, T., Okitsu, O., Mori, T., Niidome, T. & Katayama, Y., (2007). A peptide microarray for the detection of protein kinase activity in cell lysate, *Anal. Sci.*, Vol. 23, 271-275.

Shimomura, T., Han, X., Hata, A., Niidome, T., Mori, T. & Katayama, Y., (2011). Optimization of peptide density on microarray surface for quantitative phosphoproteome, *Anal. Sci., Vol.* 27(No. 11), 13-17.

Sikkema, A. H., Diks, S. H., den Dunnen, F. A., ter Elst, A., Scherpen, F. J. G., Hoving, E. W., Ruijtenbeek, R., Boender, P. J., de Wijn, R., Kamps, W. A., Peppelenbosch, M. P. & de Bont, S. J. M., (2009). Kinome profiling in pediatric brain tumors as a new approach for target discovery, Cancer Res., Vol. 69 (No. 14), 5987-5995.

Sonoda, T., Niidome, T. & Katayama,Y., (2005). Controlled gene delivery responding to cell signals using peptide-polymer conjugates, *Recent. Res. Devl. Bioconjugate Chem.* Vol. 2, 145-158.

Steinberg, T. H., Agnew, B. J., Gee. K. R., Leung. W. Y., Goodman, T., Schulenberg, B., Hendrickson, J., Beechem, J. M., Haugland, R. P. & Patton, W. F. (2003). Global quantitative phosphoprotein analysis using multiplexed proteomics technology. *Proteomics* Vol. 3, 1128-1144.

Sun, H., Low, K.E., Woo, S., Noble, R.L., Graham, R.J., Connaughton, S.S., Gee, M.A. & Lee, L-G. (2005). Real-time protein kinase assay, *Anal. Chem.* Vol. 77 (No. 7), 2043-2049.

Toita,R., Kang, J-H., Kim, J-H., Tomiyama, T., Mori, T., Niidome, T., Jun, B. & Katayama, Y., (2009). Protein kinase Cα-specific peptide substrate graft-type copolymer for cancer cell-specific gene regulation systems, *J. Controlled Release,* Vol. 139, 133-139.

Tomiyama, T., Kang, J-H., Toita, R., Niidome, T. & Katayama, Y., (2009). Protein kinase Cα-responsive polymeric carrier: Its application for gene delivery into human cancers, *Cancer Sci.*, Vol. 100, 1532-1536.

Tomiyama, T., Toita, R., Kang, J-H., Asai, D., Shiosaki, S., Mori, T., Niidome, T. & Katayama, Y., (2010). Tumor therapy by gene regulation system responding to cellular signal, *J. Controlled Release*, Vol. 148, 101-105.

Tomiyama, T., Toita, R., Kang, J-H., Koga, H., Shiosaki, H., Mori, T., Niidome, T. & Katayama, Y., (2011). Effect of the introduction of condroitin sulfate into polymer-peptide conjugate responding to intracellular signals, *Nanoscale Res. Lett.*, Vol. 6, 532.

Tsuchiya, A., Kang, J-H., Asai, D., Mori, T., Niidome, T. & Katayama, Y., (2011). Transgene regulation system responding to Rho associated coiled-coil kinase (ROCK) activation, *J. Controlled Release*, Vol. 155, 40-46.

van Baal, J. W. P., Diks, S. H., Wanders, R. J. A., Rygiel, A. M., Milano, F., Joore, J., Bergman, J. G. H. M., Peppelembosch, M. P. & Krishnadath, K., (2011). Comparison of kinome profile of Barrett's esophagus with normal squamous esophagus and normal gastric cardia, *Cancer Res.*, Vol. 66 (No. 24), 11605-11612.

Zhang, W. X., Wang, R., Wisniewski, D., Marcy, A. I., LoGrasso, P., Lisnock, J-M., Cummings, R. T. & Thompson, J. E. (2005). Time-resolved Forster resonance energy transfer assays for the binding of nucleotide and protein substrates to p38_ protein kinase, *Anal. Biochem.* Vol. 343, 76-83.

5

Role of Kinases and Phosphatases in Host-Pathogen Interactions

Horacio Bach
Department of Medicine, Division of Infectious Diseases, University of British Columbia Vancouver, Canada

1. Introduction

Living organisms are constantly exposed to changing environmental stimuli and insults; dynamic adaptation is crucial for survival. The ability of cells to sense their surrounding environment and respond in an appropriate manner is essential for the normal functioning of every living organism, and although cells are constantly exposed to numerous stimuli, they are usually able to accurately identify them and respond accordingly. These correct responses are based on a multitude of intracellular signalling networks that are able to decode and translate the incoming stimuli.

Rapid adaptation to a changing environment requires a fast response from the organism. Thus, organisms have developed specific pathways based on cascades of chemical reactions, which culminate in gene transcription and a fast metabolic adaptation. These rapid changes are important especially when responses have to be orchestrated from different cellular compartments. Thus it is not surprising, given the importance of signalling in the normal functioning of the host cell, that pathogens exploit host cellular signalling networks in order to optimize their infectious cycles. The final goal of pathogens is to erode host-cell functions and therefore establish a permissive niche in which they can successfully survive and replicate. Although most microorganisms invading the human body are contained by an efficient immune response, some microbes have evolved to successfully establish infection by bypassing defensive hostile environments mounted by the host.

Professional phagocytes, such as macrophages, neutrophils and dendritic cells are uniquely qualified to engulf and destroy microorganisms. These cells initiate immune responses and respond to microorganisms based on signal transduction pathways which are largely dependant on phosphorylation/dephosphorylation processes mediated by kinases/phosphatases. These signalling pathways are versatile and sophisticated regulatory mechanisms that play a central role in forming an integrated, information-processing network capable of coordinating multiple cellular processes in response to a wide spectrum of internal and external signals. Thus, this ubiquitous mechanism is responsible for the adaptation of cells to changes in the environment and is based on a cascade of events involving protein kinases.

This chapter will focus on: (a) the effects of secreted bacterial kinases and phosphatases on the progress of bacterial infections within their hosts, and (b) the involvement of virulence factors used by bacterial pathogens to modulate signal transduction pathways associated with the immunological response of the host. Only in the interaction of macrophages with pathogens will be explored. Mechanisms of signal transduction activated within bacteria in response to infection will not be discussed.

2. Proteins and phospholipids involved in signalling

2.1 Protein kinases

Protein kinases are enzymes that phosphorylate a protein substrate by transferring a phosphate group from a high-energy donor, such as ATP or GTP onto specific serine, threonine, and tyrosine residues of a protein substrate. As a result, the phosphorylated substrate is activated to perform either a specific activity or to continue with the transfer of the phosphate group downstream to another protein substrate initiating a cascade of reactions. To suppress the activity of phosphorylated proteins, phosphatases catalyze the reverse reaction by dephosphorylating the phosphorylated substrate, turning the protein substrates to their initial inactivated state (not phosphorylated) preparing the system for the next signalling event. Thus, kinases and phosphatases function as ON/OFF switches modulating specific signal transduction pathways.

2.2 Phosphatidylinositol (PI) signalling

PIs are small lipids derived from inositol and are key components of cell membranes. They participate in essential roles in a wide range of cellular processes, such as membrane dynamics, actin cytoskeleton arrangements and vesicle trafficking (Di Paolo & De Camilli, 2006) (Table 1). The differential distribution of PIs in cell membranes is tightly regulated by localized PI kinases and phosphatases, which convert diverse PI species (Fig. 1). This dynamic diversity enables effective temporal and spatial regulation of membrane-associated signalling events.

PI	Distribution	Functions
PI(3P)	Endosomes	Endocytic membane traffic, phagosome maturation, autophagy
PI(4P)	Golgi	Golgi trafficking
PI(5P)	Nucleus	Apoptosis
PI(3,4)P2	Plasma membrane	Signalling, cytoskeleton dynamics
PI(3,5)P2	Endosomes	Signalling, vacuole homeostasis
PI(4,5)P2	Nucleus and plasma membrane	Endocytosis, cytoskeleton dynamics
PI(3,4,5)P3	Plasma membrane	Signalling, cytoskeleton dynamics

Table 1. Functions and distribution of PIs (adapted from Rusten & Stenmark, 2006)

Since PIs are involved in a wide range of cellular functions, their metabolism is often targeted by bacterial virulence factors that act as PI phosphatases or PI adaptor proteins. The signalling pathway of PIs is based on the well-established hydrolysis of phosphatidylinositol 4,5-bisphosphate PI-(4,5)P giving rise to the second messengers diacylglycerol and inositol 1,4,5-trisphosphate and the phosphorylation of PI(4,5)P2 yielding the novel lipid phosphatidylinositol 3,4,5-trisphosphate (PI(3,4,5)P3) (Fig. 1). PIs often act in concert with small GTPases to recruit cytosolic proteins to host membranes. This allows PIs and small GTPases to exert regulatory control on each other (Di Paolo & De Camilli, 2006). PIs can bind GTPase-activating proteins (GAPs) and guanine nucleotide exchange factors (GEFs), whereas GTPases control PI metabolism by regulating PI kinases and phosphatases (Di Paolo & De Camilli, 2006). Manipulation of this close functional interplay between PI metabolism and GTPase signalling can be observed in many bacterial infections.

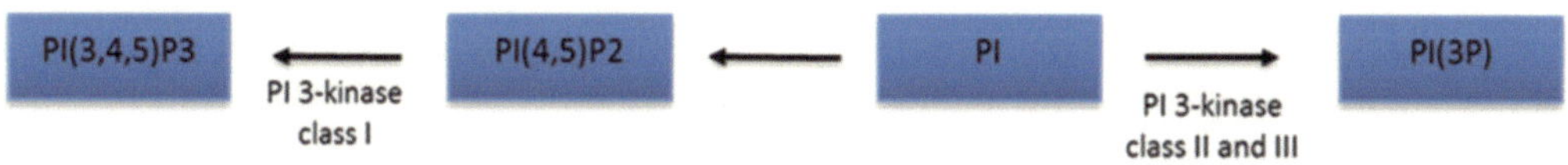

Fig. 1. Biochemical activities of PI 3-kinases.

2.3 Small GTPases

Proteins that hydrolyse GTP to GDP, called GTPases or G proteins, use this hydrolysis to serve a multitude of functions in the eukaryotic cells, such as actin dynamics, vesicle trafficking, phagocytosis, cell growth and cell differentiation. The Ras superfamily of small GTPases consists of several subfamilies, including the Rab, Rho, ADP-ribosylation factor (Arf), Ran, and Ras families (Sprang, 1997). Small GTPases function as molecular switches that cycle between an inactive guanosine dihosphate (GDP-bound state) and an active GTP-bound state. In the active GTP-bound conformation, each small GTPase binds to a subset of downstream effectors, which in turn activate downstream proteins to generate the appropriate outcome. This cycle is facilitated by two classes of regulatory proteins: GAPs and GEFs. GAPs turn the GTPase 'off' by accelerating the intrinsic rate of GTP hydrolysis, resulting in the formation of GDP and phosphate. By contrast, GEFs turn the switch 'on' by facilitating the dissociation of GDP and allowing the more abundant GTP to bind.

2.4 Src family

The regulation and activity of Src family kinases (SFKs) in response to external and internal cues is important during many cellular processes including cell adhesion, migration, polarity, and division (Bromann et al., 2004). SFKs are membrane-associated enzymes that can recognize and bind their specific substrates and transfer a phosphate group onto a target protein's tyrosine residues. SKFs are regulated themselves by tyrosine phosphorylation, which controls intramolecular interactions within the molecule that fix the kinase in an inactive closed conformation, or allow the kinase to adopt an active conformation.

SFKs also activate the cytoplasmic domain of tyrosine-based immunoreceptors (Fcγ receptors and complement receptor 3) once the extracellular domain binds opsonins, such as immunoglobulins G (IgG).

3. Microbial pattern recognition

Since macrophages need to recognize a plethora of foreign microbes rapidly, they express a diverse repertoire of receptors that bind conserved microbial molecular patterns. These receptors have evolved to recognize molecular patterns that have remained unchanged over the evolution of the microbes. Signalling that is initiated as a result of these pattern recognition receptors increases the macrophages antimicrobial abilities. To provide a fast response against microbe invasion, mammals have developed an early immune response defined as the innate response, which does not provide a long-lasting protection, but is an essential first line of defence against bacterial pathogens.

3.1 Pathogen-Associated Molecular Patterns (PAMPs) and receptors

Since unchangeable molecular patterns, such as the bacterial cell wall, is essentially conserved across Gram-positive and Gram-negative bacteria, eukaryotic organisms evolved specific receptors which recognize these molecular patterns. These specific receptors are encoded by germlines and termed pattern recognition receptors (PRRs) (Ishii et al., 2008). Recognition of Pathogen-Associated Molecular patterns (PAMPs) stimulates intracellular signalling leading to gene expression and ultimately the activation of antimicrobial and inflammatory activities. Therefore, the innate response exerts two functions: (a) a rapid line of defence against pathogens, and (b) the initiation of a signalling process leading to the development of adaptive immune responses and the establishment of an immunological memory.

To avoid detection by macrophages, some bacteria have evolved to modify their cellular surface and avoid the stimulation of receptors on phagocyte membranes. For example, many Gram-negative bacteria can alter their lipopolysaccharide structure during infection, to avoid recognition or to protect themselves from antibacterial products generated by the host, such as antibacterial peptides. In parallel, the innate recognition of microbes activates a cascade of kinase reactions, which in turn, will activate a cellular response capable of eliminating the invading microorganism. Since this innate immune response can be accompanied by tissue damage, tissue repair mechanisms are also activated (Medzhitov, 2008). In addition, the activation of transcription factors represents the culmination or endpoint of many signal transduction pathways activated in response to microbial recognition. These transcription factors can access the nucleus and bind to specific DNA sequences activating gene transcription upon binding to the respective promoters or, as in the case of the IFN-B promoter, activating different transcription factors such as nuclear factor kappa B (NF-kB), interferon (IFN)-regulatory factors (IRFs) and AP-1 (Honda & Taniguchi, 2006).

Most PRRs able to recognize bacterial patterns are Toll-like receptors (TLRs) and nucleotide-binding oligomerization domain containing proteins (NOD)-like receptors (NLRs) (Table 2). TLRs are transmembrane receptors able to recognize PAMPs in the extracellular space and in the cytoplasm or endosomes. In humans, ten TLRs have been identified (Takeuchi & Akira, 2010). Structurally, TLRs comprise of a single membrane-spanning domain separating the cytoplasmic domain involved in signalling from the recognizing receptor able to bind the ligand. The extracellular domain is involved in PAMP recognition, whereas the cytoplasmic domain is essential for downstream signalling (O'Neill & Bowie, 2007). TLRs are highly expressed by professional phagocytes such as macrophages and dendritic cells, but can also be expressed by other cell types, such as epithelial cells (Iwasaki & Medzhitov, 2004).

NLRs consist of an N′-terminal effector domain, responsible for downstream signalling and a C′-terminal region similar to TLRs, which are involved in the PAMP recognition (Inohara & Nunez, 2003). NOD1 and NOD2 are the best-characterized NLRs and are involved in the detection of intracellular bacteria (Chamaillard et al., 2003; Girardin et al., 2003). For instance, peptidoglycans (PGNs) are structural units of cell walls common to all bacteria (Gay & Gangloff, 2007). Degradation of PGNs leads to the release of several structural units including muramyl dipeptide, which is sensed in the cytosol by the NLR NOD2, which in turn activates NF-κB.

PAMP	PRR
TLR2	Lipopeptides
	Lipoteichoic acid
	Peptidoglycan
TLR4	Lipopolysaccharides
TLR5	Flagellin
TLR9	Unmethylated CpG DNA
Diaminopimelic acid	NOD1
Muramyl dipeptide	NOD2

Table 2. Receptors involved in bacterial pattern recognition

3.2 Major signalling pathways involved in host-pathogen interaction

Upon the perception of bacterial patterns, the immune response activates an intricate and complex network of kinases, which will ultimately result in the transcription of genes. The products of these genes will generate the immune response. Then, PRRs are able to activate a sequence of three major signalling pathways in mammals: mitogen-activated protein kinases (MAPKs), IRFs, and the nucler factor NF-kB, which will culminate in the transcription and release of proteins involved in the immune response.

The MAPKs are a group of protein serine/threonine kinases that are activated in mammalian cells in response to a variety of extracellular stimuli and mediate signal transduction from the cell surface to the nucleus where they can alter the phosphorylation status of specific transcription factors (Johnson & Lapadat, 2002). Three major types of MAPK pathways have been reported so far in mammalian cells. The extracellular signal-related kinases (ERKs 1 and 2) pathway is involved in cell proliferation and differentiation, whereas the c-Jun N-terminal kinases (JNKs 1, 2 and 3), and p38 MAPK (p38 α, β, γ and δ) pathways are involved in response to stress stimuli. These three factors -ERK, JNK and p38- dictate the fate of cells in concert (Johnson & Lapadat, 2002). As an illustration, TLR4 recognizes lipopolysaccharides (LPSs) of Gram-negative bacteria. Then, when TLR4 recognizes this pattern, an activation of the MAPKs' cascade is initiated. At the onset of this process, the cytoplasmic TIR domain of TLR4 mediates the activation of the cascade through the four adaptor proteins: (a) myeloid differentiation primary response protein 88 (MyD88), (b) TIR-domain-containing adaptor inducing IFN-B (TRIF), (3) TRIF related adaptor, and (4) MyD88-adaptor-like (Mal) (Fitzgerald et al., 2001, 2003; Yamamoto et al., 2003a; 2003b). Upon activation, these adaptor proteins communicate the signal via the kinases IL-1 receptor associated kinase (IRAK)-4, IRAK-1/2, and RIPI, which in collaboration with TNF receptor-associated factor (TRAF) 6, activate transforming growth factor B-activated kinase (TAK) 1 in association with TAB2/3, through a mechanism dependent on the E3 ubiquitin ligase activity of the TRAF molecules (Akira & Takeda 2004; Kawagoe et al., 2008, Sato et al., 2005). TAK1 activates IkB kinase (IKK)-alfa/beta to release NF-kB from the inhibitory subunit of IkB, as well as MAPKs (Sato et al., 2005).

4. Phagocytosis and intracellular survival

Upon infection, bacterial pathogens interact with host membranes through different mechanisms. The interaction between the bacterium and the host plasma membrane (and its embedded receptors) results in the activation of multiple host-signalling pathways that can alter actin cytoskeleton dynamics or vesicle trafficking. Three membrane-associated signalling events are targeted by bacterial pathogens: phosphoinositide (PI) metabolism, GTPase signalling and autophagy.

4.1 Avoiding phagocytosis

Some bacteria evolved to remain in the extracellular milieu to avoid being killed within the macrophage. This advantage also minimizes bacteria–macrophage interactions and as a consequence, the macrophage signalling required to activate an adaptive immune response is impaired. To avoid their engulfment, extracellular pathogens have to interfere with phagocytosis. One of the best study pathogens is *Yersinia*, which interferes with phagocytosis by a set of virulence proteins with an array of enzymatic activities that is delivered into macrophages. Some of the bacterial proteins interfere with the signal transduction of macrophages. For instance, YopH is a protein tyrosine phosphatase that targets host focal adhesion proteins, such as p130cas, paxillin, and focal adhesion kinase (FAK). Then, by dephosphorylating these substrates, YopH prevents uptake of bacteria by the host immune cells by destabilizing the focal adhesions involved in the internalization of bacteria by eukaryotic cells (Black et al., 1997), and allowing the pathogen to proliferate extracellularly. *Yersinia* also secretes YopE, a GTPase-activating protein that inactivates the small GTPases RhoA, Rac, and Cdc42 to prevent the actin polymerization that is required for phagocytosis (von Pawel-Rammingen, 2000). YopT is a papain-like cysteine protease that cleaves the lipid moiety of RhoA to depolymerize actin filaments, leading to their irreversible detachment from the plasma membrane and their inactivation (Shao et al., 2002). Thus, YopT contributes to the inhibition of bacterial phagocytosis by preventing rearrangements of the actin cytoskeleton. *Yersinia* also secretes the kinase YpkA into the host cytoplasm, where it phosphorylates specific proteins to prevent bacterial uptake and the killing by macrophages (Hakansson et al., 1996). Finally, YopO, a serine/threonine kinase activated by actin, contributes to the antiphagocytic activity in *Y. enterocolitica* by binding to Rho GTPases (Grosdent et al., 2002).

Other microorganisms such as enteropathogenic *Escherichia coli* (EPEC), target a different signalling pathway by secreting an unidentified bacterial protein into macrophages to inhibit the activity of phosphatidylinositol 3-kinase (PI3K) (Celi et al., 2001). Although pathogens that subvert macrophage phagocytic signalling remain outside the cell to avoid phagolysosomal degradation, they still have mechanisms to cope with extracellular defences, such as killing by complement or antimicrobial peptides (Wurzner, 1999).

4.2 Modulating the interacting membrane by disruption of PI signalling

PIs are key players in maintaining cell membrane structure by regulating the actin cytoskeleton underneath the plasma membrane and by tagging and targeting vesicles inside the cell. The disruption of PI homeostasis at the plasma membrane can destabilize actin dynamics changing membrane morphologies, and then intracellular pathogens can modulate the membrane integrity.

The inositol phosphate phosphatase IpgD is an effector from the facultative intracellular pathogen *Shigella flexneri* that is directly translocated into host cells through a type III secretion system (Niebuhr et al., 2000). IpgD hydrolyses PI(4,5)P2 to produce PI5P, at an early stage in the infection (Niebuhr et al., 2002). The removement of PI(4,5)P2 causes a rearrangement in the cytoskeleton by changing the extent of interaction of membrane visualized as a massive cell blebbing, facilitating the invasion of bacteria (Charras & Paluch, 2008).

Listeria invasion is mediated by interaction of the bacterial surface protein InlB with the host receptor Met receptor tyrosine kinase (Shen et al., 2000). InlB-Met interaction triggers activation, by tyrosine phosphorylation, of the Met receptor and subsequent rearrangements in the actin cytoskeleton of the mammalian cell (Mostowy & Cossart, 2009). Ultimately, these cytoskeletal changes remodel the host cell surface, resulting in the engulfment of adherent *Listeria*. The human GAP ARAP2 is required for InlB-mediated cytoskeletal changes and entry of the pathogen. ARAP2 is known to bind PI(3,4,5)P3, resulting in upregulation of a GAP domain that inactivates the mammalian GTPase Arf6 (Wong & Isberg, 2003). Then, one of the likely ways that PI3-kinase controls entry of *Listeria* is through regulation of ARAP2. In addition, cholesterol-rich lipid rafts at the plasma membrane are needed for InlB-mediated uptake of *Listeria* (Seveau et al., 2004).

Small GTPases Sar1, Rab1 and Arf1 are required for the *Legionella*-containing vesicles to acquire vesicle trafficking protein Sec22b (Kagan & Roy, 2002). *Legionella* secretes the effectors DrrA/SidM and LepB, which impaire the recruitment of Rab1 (Ingmundson et al., 2007). The association of these secreted effectors with the *Legionella*-containing vesicles surface is mediated by their affinity for the abundant lipid PI4P on the *Legionella*-containing vesicles surface (Brombacher et al., 2009; Ragaz et al., 2008).

SopB, a type III secretion system effector from *Salmonella typhimurium*, is a PI phosphatase that affects multiple processes during the course of infection, including bacterial invasion, *Salmonella*-containing vesicle formation and maturation (Hernandez et al., 2004). SopB hydrolyses PI(4,5)P2 both at the plasma membrane and on the *Salmonella*-containing vesicle membrane surface (Bakowski et al., 2010). Decreased levels of PI(4,5)P2 at the plasma membrane promote membrane fission by reorganizing the actin cytoskeleton during bacterial internalization (Mason et al., 2007).

SopB also mediates the production and maintenance of high levels of PI3P on the *Salmonella*-containing vesicle surface through an indirect effect of its phosphatase activity. SopB recruits Rab5 and its effector VPS34, a PI3-kinase (that generates PI3P), to the *Salmonella*-containing vesicle through a process that is dependent on the reduction of PI(4,5)P2 (Mallo et al., 2008). Then, by manipulating the lipid composition of the *Salmonella*-containing vesicle, SopB impairs the recruitment of Rabs avoiding lysosomal degradation.

4.3 Surviving and living within the host

The ability to survive intracellularly is crucial for several pathogenic bacteria after they invade their eukaryotic target cells. Following engulfment by macrophages, bacteria are internalized within a membrane-bound vacuole termed a phagosome. Phagosomes are pivotal organelles in the ability of mammalian cells, including professional and non-professional phagocytes, to restrict the establishment and spread of infectious diseases.

Rapidly after their formation, phagosomes modify their composition by recycling plasma membrane molecules, and by acquiring markers of the early endocytic pathway such as Rab5 and EEA1 (Steele-Mortimer et al., 1999). Phagosomes have been shown to fuse sequentially with endosomes of increasing age or of increasing maturation level (Jahraus et al., 1994). A variety of Rab proteins have been identified on phagosomes, including Rab5, Rab7, and Rab11 (Desjardins et al., 1994; Cox et al., 2000).

Under normal circumstances, the phagosome progressively acidifies and ultimately, in a tightly regulated process, will fuse with the lysosome, in an event known as phagosome-lysosome (phagolysosome) fusion.

The process of phagocytosis itself determines some of the characteristics of the first compartment in which pathogens are going to reside within the host cells. Newly formed phagosomes are immature organelles unable to kill and degrade microorganisms. In order to acquire and exert their microbicidal function, phagosomes must engage in a maturation process referred to as phagolysosome biogenesis. Then, to successfully invade and replicate intracellularly, pathogens must find ways to avoid the harsh environment of lysosomes, organelles containing an arsenal of potent microbicidal compounds. Therefore, the final goal of the majority of intracellular pathogens is to prevent their arrival to lysosomes, where their killing is dictated.

While the majority of bacteria grow outside of eukaryotic cells, some bacteria are facultative or even obligate intracellular pathogens; such is the case with *L. monocytogenes*, *Mycobacterium tuberculosis*, and *Chlamydia trachomatis*. Bacterial replication, therefore, takes place in the endosomal compartments, or in the case of *Listeria* in the cytoplasm due to a mechanism of escape from the phagosomes.

The life style of cytosolic bacteria can be divided into three main stages: (a) escape from the phagosome, (b) replication within the cytosol, and (c) manipulation of the innate immune responses triggered in the cytosol. The escape from the phagosome is a crucial step in the life cycle of cytosolic pathogens. This occurs rapidly following invasion, and most pathogens are detected free in the cytosol within 30 minutes of invasion. In order to evade the lysosome, a process lasting between 30-45 minutes post-engulfment (Yates et al., 2005), pathogens must escape before the fusion with lysosomes (Haas, 2007).

4.4 Disruption of MAPK signalling pathways

MAPK signalling is crucial for many responses to infection, representing a strategic target for bacterial subversion strategies. The extent of MAPK phosphorylation (kinase signalling kinetics) may influence the responses of macrophages. For instance, the duration of signalling through MAPK pathways determines whether a macrophage proliferates or activates in response to a stimulus (Velledor et al., 2000). Likewise, modification of MAPK pathways by bacteria may contribute to induction of host cell death, which is an important feature of bacterial pathogenesis, promoting bacterial tissue colonization.

In the case of *S. enterica* serovar Typhi, the tyrosine phosphatase SptP, a translocated protein from the pathogen within the host, inhibits the activation of the MAPK pathway by dephosphorylating Raf, an intermediate in this pathway (Lin et al., 2003).

A way to alter the MAPK pathway is the degradation of members involved in the response cascade. For example, *Bacillus anthracis* interrupts several MAPK signalling pathways by proteolytically degrading all MAPK kinases (MAPKKs) except MAPKK5. This interference is mediated by the delivery of a metalloproteinase to the cytosol, where it deactivates MEK1 by cleaving between its amino terminus and catalytic domain. Cleavage of the MAPKK that activates p38 MAPK, which is mediated by lethal factor, induces macrophage apoptosis, possibly by interfering with the p38-dependent expression of NF-κB target genes that are necessary for cell survival (Park et al., 2002).

Other pathogens interfere by blocking or inhibiting post-translational modifications, such as prevention of phosphorylation. Members of the genus *Yersinia* use an alternative mechanism to disrupt MAPK signalling and, as a result of this disruption, the downstream activation of NF-κB in macrophages is impaired. Specifically, *Y. pseudotuberculosis* delivers YopJ, a cysteine protease, which inhibits kinase activity by preventing phosphorylation (Orth et al., 1999). YopJ also interferes with the post-translational modification of proteins that are involved in MAPK signalling by disturbing the ubiquitin-like protein SUMO-1, and then inhibiting its conjugation to target proteins for degradation (Orth et al., 2000).

In conclusion, pathogens can modify the antibacterial response of macrophages not only towards a targeted kinase pathway, but also by the timing of the activation or inhibition.

An important downstream response of normal macrophage signalling is the production of cytokines. Cytokines are essential for modulation of inflammation, recruitment of other cells to the site of infection, and mediation of the link between innate and adaptive immune responses. As mentioned above, macrophages must control signalling that leads to inflammatory responses tightly, to avoid an inflammation dysregulation. One level of control is to regulate the intensity and duration of signalling, which often originates from TLRs. For example, the macrophage protein IRAK-M has a pivotal role in downregulating macrophage responses to LPS by inducing tolerance. Without IRAK-M, *Salmonella* infection causes increased tissue damage (Kobayashi et al., 2002). Another level of control is the balance between proinflammatory cytokines, such as TNF-α and IL-12, and predominantly anti-inflammatory cytokines, such as IL-10 and transforming growth factor-β (TGF-β), which are produced during infection. Bacterial pathogens target signalling that leads to the expression of cytokine genes or their post-translational modifications that perturb the balance of cytokines to their advantage. Macrophages and bacteria can therefore both control the extent of the immune response through cytokine production.

One effector protein secreted intracellularly by *Shigella* is OspF, which possesses phosphothreonine lyase activity. Once translocated into the nucleus, OspF irreversibly dephosphorylates host MAPKs, and therefore prevents the phosphorylation of histone H3 (Li et al., 2007; Arbibe et al., 2007). Interestingly, other bacterial virulence factors, such as SpvC from *S. typhimurium* possess the same phosphothreonine lyase activity as OspF and also target MAP kinases of their hosts (Mazurkiewicz et al., 2008). In addition to these factors, the *Yersinia* YopJ/P effector can inactivate host MAP kinases by catalyzing their acetylation (Mittal et al., 2006; Mukherjee et al., 2006). Finally, the anthrax lethal factor, a subunit of the anthrax toxin encoded by *B. anthracis*, cleaves host MAP kinases, leading to their irreversible inactivation (Turk, 2007).

Helicobacter pylori has been reported to activate MAPK3 enzymes (Asim et al., 2010). When invading the human-derived monocyte cell line THP-1, *H. pylori*-stimulates the extression of IL-18 that was reduced by either ERK or p38 inhibitors (Yamauchi et al., 2008). Inhibition of ERK and, to a greater degree, inhibition of p38 have been shown to reduce *H. pylori*-stimulated IL-8 expression in THP-1 cells (Bhattacharyya et al., 2002). Taken together, these studies suggested that at least MAPKs are involved in biological effects of *H. pylori* infection in macrophages.

4.5 Disruption of interferon signalling

Macrophages possess a robust tyrosine kinase signalling network that includes the Janus kinase (JAK) and the signal transducer and activator of transcription (STAT). Both pathways are activated as a result of IFN binding to their receptors on the cell surface. IFN-γ amplifies the antibactericidal activity of macrophages (Boehm et al., 1997) by activating various enzymes within the macrophage that increase the production of damaging reactive oxygen and nitrogen species, starve the bacteria of tryptophan within the phagolysosome, and increase lysosomal degradation of the bacteria. In addition, IFN-γ enhances the adaptive response of the organism by increasing the Major Histocompatibility Complex (MHC) class I and II antigen presentation and synthesis of cytokines such as IL-12 and TNF-α (Shtrichman & Samuel, 2001). In conclusion, the IFN-γ signalling network allows macrophages to respond more rapidly to bacterial infection. Bacterial impairment of IFN-γ signalling is best characterized in macrophages infected by *Mycobacteria* species. *M. avium* infection causes a decreased transcription of the IFN-γ receptor leading to impaired downstream STAT activation (Hussain et al., 1999). *M. tuberculosis* uses an uncharacterized mycobacterial surface component to affect a later step in IFN-γ signalling. Although STAT phosphorylation, dimerization, nuclear translocation and DNA binding is intact in *M. tuberculosis*-infected macrophages, there is still a decrease in the association of STAT with transcriptional co-activators, causing an impaired transcription of IFN-γ-responsive genes (Ting et al., 1999).

4.6 Disruption and amplification of NF-κB signalling

NF-κB signalling relies on the targeting of its inhibitor IκB. As a result of binding to IκB, NF-κB avoid translocation from the cytosol to the nucleus where it activates gene transcription. Inhibition of NF-κB signalling leads to the decreased release of proinflammatory cytokines, such as TNF-α, and increased apoptosis, both of which can protect pathogens from the immune response.Virulence proteins secreted by pathogens such as *Y. enterocolitica* bind to the IKK to prevent the phosphorylation of IκB, which is essential for its degradation, thereby trapping NF-κB in the cytosol and avoiding its gene target interactions (Schesser et al., 1998). *M. ulcerans* inhibits nuclear translocation of NF-κB independently of IκB, possibly by altering the phosphorylation of NF-κB or interfering with its DNA-binding ability (Pehleven et al., 1999).

On the other hand, pathogens can use an opposite strategy by actively increasing the NF-κB activity. In this way, the production of proinflammatory cytokines can recruit more host cells to the site of infection, facilitating the bacterial spread. For instance, listeriolysin O and InlB, two virulence proteins secreted by *L. monocytogenes*, activate NF-κB in a PI3K-dependent manner. As a result of an increase in the inflammatory response, pathogens spread by recruiting more monocyte to the site of infection (Kayaal et al., 2002). Another advantage for the pathogen is a protective environment because of the anti-apoptotic signalling activated by NF-κB.

L. monocytogenes secretes InlC intracellularly, which directly interacts with the IKKa protein to block the phosphorylation of IkBa (Gouin et al., 2010). Similarly, YopJ/P, an effector produced by pathogenic *Yersinia* species, mediates the acetylation of the IKKa and b proteins, which prevents their activation and subsequent IkBa phosphorylation (Mittal et al., 2006).

The effectors NleH1 and NleH2 of the enterohaemorrhagic *E. coli* (EHEC) are autophosphorylated serine/threonine kinases translocated by the pathogen. Both effectors bind directly to RPS3, a NF-κB non-Rel subunit. Although autophosphorylated, their binding to RPS3 is independent of kinase activity (Gao et al., 2009).

4.7 Disruption of small GTPase signalling

After initial attachment to the host cell membrane, many pathogens Gram-negative bacteria use a type III secretion system to inject virulence proteins into the host cell cytoplasm (Ghosh, 2004). A number of injected proteins bind directly to actin to modulate its dynamic leading to changes in the organization of the actin cytoskeleton (Patel & Galan, 2005) by regulating small GTPases. In a variety of pathogens, a family of conserved type III secreted proteins influences the actin cytoskeleton dynamic by mimicking the GTP-bound form of the Rho GTPases (Alto et al., 2006). These proteins, which share no obvious sequences homology with Rho GTPases, use a conserved WxxE motif to directly activate downstream effectors of Cdc42, Rac1, and Rho (Alto et al., 2006).

Several bacterial pathogens also use the type IV secretion systems to inject effector proteins into the cytoplasm of host cells (Cascales & Christie, 2003; Galan & Wolf-Watz, 2006). After translocation, these effectors target various components of eukaryotic signal transduction pathways, which subvert host cell functions for the benefit of the pathogen.

Rho-family GTPases, such as Rho, Rac1, and Cdc42, regulate actin dynamics by induction of actin, lamellipodia, and filopodia formation, respectively. Inactivation of these GTPases leads to a decrease in F-actin and increase in monomeric actin (G-actin), resulting in loss of cell shape, motility, and ability to phagocytose or endocytose pathogens.

S. typhimurium manipulates this members using the effectors SopE and SptP. SopE acts as a GEF for Cdc42 and Rac1, whereas SptP acts as a GAP for Cdc42 and Rac1 (Fu & Galan, 1999). SopE is translocated into the cell to induce actin rearrangement and membrane disruption to facilitate pathogen entry into the cell and formation of *Salmonella*-containing vesicles, while SptP disrupts these actin filaments to restore actin organization in the cell (Hardt et al., 1998). SptP possesses both a GAP and tyrosine phosphatase activities (Fu & Galan, 1999). It disrupts the actin cytoskeleton by binding to Rac1 and catalysing the hydrolysis of GTP to GDP. Although SopE and SptP are antagonists, they are coordinately regulated. While SopE acts early in the infection to facilitate the uptake of the pathogen, SptP disasssembles F-actin organization, allowing the pathogen to proliferate in the vesicle (Kubori & Galan, 2003).

IpgB1, a type III secretion system effector of *S. flexneri*, binds to the host cell engulfment and cell motility ELMO–DOCK180 complex activating Rac1 (Handa et al., 2007). As a result, IpgB1 increases infection efficiency. Another effector secreted by the same pathogen, IpgB2, induces membrane disruption by mimicking the Rho-GEF (Klink et al., 2010).

The effectors YopE and YopT secreted by *Yersinia* inhibit actin rearrangements by inactivating host Rho GTPases. YopE is known to act as a GAP (Black & Bliska, 2000) inhibiting RhoA, Rac-1 and Cdc42 by accelerating the conversion of the GTP-bound form of the Rho GTPase to the GDP-bound inactive form. The GAP activity of YopE is also needed to prevent the formation of pores generated by insertion of the translocation machinery in the host cell plasma membrane (Viboud & Bliska, 2001). YopT has been found previously to inhibit Rho GTPases by releasing them from the membrane (Zumbihl et al., 1999). YopT acts as a cysteine protease that cleaves the prenyl group of lipid-modified Rho GTPases (Shao et al., 2002).

The effector SidM from *L. pneumophila* targets Rab1 proteins involved in ER–Golgi transport. SidM is a bifunctional enzyme; the C′-terminus possesses a RAB/GEF activity, whereas the N′-terminus catalyses AMPylation. Then, SidM catalyses the exchange of GDP for GTP by changing the conformation of Rab1 residues that are important for nucleotide stabilization (Murata et al., 2006). AMPylation induces cell rounding and shrinkage, which contribute to the disruption of cell homeostasis and to cytotoxicity (Muller et al., 2010). SidM is localized to the membrane through its interaction with PI4P (see above), and recruits Rab1/GTP to the *Legionella*-containing vesicles, mimicking a Rab1/GEF and delaying GAP activity by AMPylation. SidM-mediated Rab1 activation and recruitment to the *Legionella*-containing vesicles promotes fusion of ER-derived vesicle with the *Legionella*-containing vesicles. Another *L. pneumophila* effector is LepB, that functions as a GAP for Rab1 (Ingmundson et al., 2007), inactivating the GTPase and releasing it from the *Legionella*-containing vesicles, promoting ist fusion with the ER. During the initial phase of infection, *L. pneumophila* resides in the ER-derived vesicle that interacts with the secretory pathway, whereas during the later stages of infection, when bacterial replication occurs, these vesicles acquire lysosomal markers (Sturgill-Koszycki & Swanson, 2000).

4.8 A unique infection model: *Mycobacterium tuberculosis*

The infection of a macrophage by *M. tuberculosis* is complex, and since a variety of pathways are orchestrated by the pathogen, a separated section is dedicated to analyze this pathogen.

M. tuberculosis is able to survive, reside, and multiply in macrophages as an intracellular parasite, circumventing all defence pathways of the host. The hallmarks of *Mycobacterium* infection are (a) the manipulation of phagolysosome maturation (Koul et al., 2004; Hestvik et al., 2005), (b) the prevention of antigen presentation (Moreno et al., 1988), (c) a decrease in stimulators of apoptosis (Balcewicz-Sablinska et al., 1998), (d) alteration of IFN-γ activity (Sibley et al., 1988), and (e) modulation of MAPK and JAK/STAT signalling pathways (Koul et al., 2004).

Upon internalization by macrophage phagocytosis, *M. tuberculosis* is able to arrest phagolysosome fusion (Pethe et al., 2004) and modulate other macrophage defences to promote its survival (Gan et al., 2008). Arrested *M. tuberculosis*-containing phagosomes are characterized by the presence of Rab5a, but the recruitment of its effectors, such as EEA1 and hVPS34, is impaired (Fratti et al., 2001). *M. tuberculosis* uses a range of protein and lipid effectors to alter the PI(3)P signalling (Vergne et al., 2005) and the concentration of cytosolic Ca^{2+}, both events essential for the proper phagosomal maturation (Jaconi et al., 1990). The mycobacterial mannosylated lipoarabinomanan (Man-LAM), a shed component of the cell wall, is distributed throughout the endocytic network (Beatty et al., 2000), preventing the increase in cytosolic [Ca^{2+}], a process necessary for phagocytosis upon activation of hVPS34 by calmodulin (Vergne et al., 2003). Inhibition of the PI3K pathway by Man-LAM also blocks the delivery of lysosomal proteins, such as hydrolases (e.g. cathepsin D) and the membrane-docking fusion protein syntaxin 6, from the trans-Golgi network to phagosomes (Fratti et al., 2003). In addition, the pathogen further impairs cytosolic Ca^{2+} flux by inhibiting sphingosine kinase, which converts sphingosine to sphingosine-1-phosphate, which in turn promotes Ca^{2+} efflux from the endoplasmic reticulum (Malik et al., 2003). *M. tuberculosis* also produces the phosphatase SapM, which has been shown to specifically inhibit hydrolysis of PI(3)P *in vitro* (Vergne et al., 2005). These findings indicate that Man-LAM blocks phagosome maturation by inhibiting a signalling cascade based on [Ca^{2+}], calmodulin, and PI3K. Mycobacterial phagosomes also recruit early phagosomal proteins such as coronin-1 (Ferrari et al., 1999), but avoid acidification as the bacteria specifically exclude the vesicular proton ATPase from the phagosomal membrane (Sturgill-Koezycki et al., 1994, Wong et al., 2011).

Macrophages infected with harmful bacteria activate their own apoptotic program when the infected cell cannot resolve its infection. However, many bacterial pathogens alter host apoptotic pathways (Spira et al., 2003). Mycobacteria-induced macrophage apoptosis is a complex mechanism that is modulated by mycobacterial virulence factors (Nigou et al., 2002). Ca^{2+} is thought to facilitate apoptosis by increasing the permeability of mitochondrial membranes, and then promoting the release of pro-apoptotic factors such as cytochrome c (Szalai et al., 1999). Interestingly, Man-LAM also stimulates the phosphorylation of the apoptotic protein Bad, preventing it from binding to the anti-apoptotic proteins Bcl-2 and Bcl-XL (Maiti et al., 2001).

Pro-inflammatory cytokines, such as IL-1, IL-6, TNF-α, and IFNs, are able to induce a cellular innate immune response when macrophages sense invading bacteria.

The activation of MAPK signalling in macrophages that are infected with non-pathogenic mycobacteria leads to the synthesis of various microbicidal molecules, including TNF-α, which mediate antibacterial and inflammatory immune responses (Roach & Schorey, 2002). These observations are supported by a study that demonstrated that the secretion of TNF-α by macrophages infected with *M. avium* is dependent on MEK1 and ERK1 and 2 activation (Reiling et al., 2001). A high level of TNF-α is a crucial factor for controlling primary infection, as it induces the expression of other pro-inflammatory cytokines, such as IL-1, and of several chemotactic cytokines, which attract immune cells to the site of infection.

Tyrosine phosphorylation of JAK and STAT has been shown to be essential for the antibacterial response of macrophages (Decker et al., 2002). Pathogenic mycobacteria have evolved mechanisms to suppress the IFN-γ and JAK/STAT signalling pathways (Hussain et al., 1999) by mechanisms not yet elucidated.

Surprisingly, two protein tyrosine phosphatases PtpA and PtpB are annotated in the genome sequence of *M. tuberculosis* (Cole et al., 1998). The presence of such proteins is interesting, since their partners, protein tyrosine kinases, are not predicted from the genome sequence, which suggests they play a role in the survival of the pathogen in host macrophages. The role of PtpA has been elucidated (Bach et al., 2008). This phosphatase is secreted within human macrophages upon infection and translocates into the cytosol to dephosphorylate VPS33B, an ubiquitously expressed protein essential for vesicle trafficking. Then, by dephosphorylating VPS33B, the pathogen prevents the maturation of the phagosome. Interestingly, a study reported the first protein tyrosine kinase, PtkA, in *M. tuberculosis* (Bach et al., 2009). This tyrosine kinase phosphorylates PtpA· although the role of this activity remains to be elucidated. The second protein tyrosine phosphatase annotated in the *M. tuberculosis* genome is PtpB. Interestingly, PtpB orthologs are restricted to pathogenic Mycobacteria. It has been reported that PtpB blocks the ERK*1* and 2 pathways in murine macrophages, but its mechanism has not yet been elucidated (Zhou et al., 2010).

Another virulence protein secreted by *M. tuberculosis* is the protein kinase G. Although this protein has been shown to participate in the inhibition of phagolysosome fusion in *M. bovis* strain BCG (Walburger et al., 2004), its mechanism has not yet been elucidated.

5. Conclusion

In this chapter, the mechanisms through which virulence proteins and conserved microbial structures can initiate macrophage signalling were discussed. Macrophages can use specific receptors and common signalling pathways to integrate this information to mount an immunological response, but they are still vulnerable to subversion by bacterial pathogens that can interfere with crucial kinase, trafficking or transcriptional networks. However, there are redundancies in macrophage signalling pathways, and the recent discovery of a cytosolic detection system in macrophages is a good example of how avoiding one component of a macrophage's arsenal makes pathogens vulnerable to another. It seems that the combination of mechanisms that a pathogen has to modify specific macrophage signalling cascades dictates their most successful niche.

Genome sequencing projects have identified an overwhelming number of host and bacterial genes that encode proteins with unknown functions. The characterization of the biological functions of these proteins will probably add to the ever-increasing number and diversity of strategies that are used by macrophages to detect and contain the invaders and by bacterial pathogens to subvert and evade host responses.

Finally, the development of new technologies, such as improvements in mass spectrometry techniques, will undoubtedly increase the currently known post-translational modification and facilitate the understanding of their roles in host-pathogen interactions. Identifying pathogen-encoded enzymes that catalyze specific post-translational modification critical for infection will provide valuable new targets for drug development. Indeed, the selective inhibition of these enzymes may constitute a promising strategy to contain and restrict the proliferation of pathogens. However, only few interaction partners have been identified so far. Systematic mapping of protein-protein interactions can provide valuable insights into biological systems. However, current large-scale screening methods fail to provide information about these interactions.

6. Acknowledgment

The author thanks Jeffrey Helm and Eviatar Bach for helpful discussions and technical support.

7. References

Akira, S. & Takeda, K. (2004). Toll-like receptor signalling. *Nature Reviews Immunology*, Vol.4, No.7, (July 2004), pp. 499-511

Alto, N.; Shao, F.; Lazar, C.; Brost, R.; Chua, G.; Mattoo, S.; McMahon, S.; Ghosh, P.; Hughes, T.; Boone, C. & Dixon, J. (2006). Identification of a bacterial type III effector family with G protein mimicry funcions. *Cell*, Vol.124, No.1, (January 2006), pp. 133-145

Arbibe, L.; Kim, D.; Batsche, E.; Pedron, T.; Mateescu, B.; Muchardt, C.; Parsot, C. & Sansonetti, P. (2007). An injected bacterial effector targets chromatin access for transcription factor NF-kappaB to alter transcription of host genes involved in immune responses. *Nature Immunology*, Vol.8, No.1, (January 2007), pp. 47-56

Asim, M.; Chaturvedi, R.; Hoge, S.; Lewis, N.; Singh, K.; Barry, D.; Algood, H.; de Sablet, T.; Gobert, A. & Wilson, K. (2010). *Helicobacter pylori* induces ERK-dependent formation of a phospho-c-Fos·c-Jun activator protein-1 complex that causes apoptosis in macrophages. *Journal of Biological Chemistry*, Vol.285, No.26, (June 2010), pp. 20343-20357

Bach, H.; Papavinasasundaram, K.; Wong, D.; Hmama, Z. & Av-Gay, Y. (2008). *Mycobacterium tuberculosis* virulence is mediated by PtpA dephosphorylation of human vacuolar protein sorting 33B. *Cell Host Microbe*, Vol.15, No.3, (May 2008), pp. 316-322

Bach, H.; Wong, D. & Av-Gay, Y. (2009). *Mycobacterium tuberculosis* PtkA is a novel protein tyrosine kinase whose substrate is PtpA. *Biochemical Journal*, Vol.420, No.2, (May 2009), pp. 155-160

Balcewicz-Sablinska, M.; Keane, J.; Kornfeld, H. & Remold, H. (1998). Pathogenic *Mycobacterium tuberculosis* evades apoptosis of host macrophages by release of TNF-R2, resulting in inactivation of TNF-alpha. *Journal of Immunology*, Vol.161, No. 5, (September 1998), pp. 2636-2641

Bakowsky, M.; Braun, V.; Lam, G.; Yeung, T.; Heo, W.; Meyer, T.; Finlay, B.; Grinstein, S. & Brumell, J. (2010). The phosphoinositide phosphatase SopB manipulates membrane surface charge and trafficking of the *Salmonella*-containing vacuole. *Cell Host Microbe*, Vol.7, No.6, (June 2010), pp. 453-462

Bhattacharyya, A.; Pathak, S.; Datta, S.; Chattopadhyay, S.; Basu, J. & Kundu, M. (2002). Mitogen-activated protein kinases and nuclear factor-kappaB regulate *Helicobacter pylori*-mediated interleukin-8 release from macrophages. *Biochemical Journal*, Vol.368, No.1, (November 2002), pp. 121-129

Beatty, W.; Rhoades, E.; Ullrich, H.; Chatterjee, D.; Heuser, J. & Russell, D. (2000). Trafficking and release of mycobacterial lipids from infected macrophages. *Traffic*, Vol.1, No.3, (March 2000), pp. 235-247

Black, D. & Bliska, J. (1997). Identification of p130Cas as a substrate of *Yersinia* YopH (Yop51), a bacterial protein tyrosine phosphatase that translocates into mammalian cells and targets focal adhesions. *EMBO Journal*, Vol.16, No.10, (May 1997), pp. 2730-2744

Boehm, U.; Klamp, T.; Groot, M. & Howard, J. (1997).Cellular response to interferon-gamma. *Annual Review of Immunology*, Vol. 15, pp. 749-795

Brombacher, E.; Urwyler, S.; Ragaz, C.; Weber, S.; Kami, K.; Overduin, M. & Hilbi, H. (Rab1 guanine nucleotide exchange factor SidM is a major phosphatidylinositol 4-phosphate-binding effector protein of *Legionella pneumophila*. *Journal of Biological Chemistry*, Vol. 284, No.8, (February 2009), pp. 4846-4856

Bromman, P.; Korkaya, S. & Courtneidge, S. (2004). The interplay between Src family kinases and receptor tyrosine kinases. *Oncogenes*, Vol.23, No.48 (October 2004), pp. 7957-7968

Cascales, E. & Christie, P. (2003). The versatile bacterial type IV secretion systems. *Nature Review Microbiology*, Vol.1, No.2, (November 2003), pp. 137-149

Celi, J.; Olivier, M. & Finlay, F. (2001). Enteropathogenic *Escherichia coli* mediates antiphagocytosis through the inhibition of PI 3-kinase-dependent pathways. *EMBO Journal*, Vol.20, No.6, (March 2001), pp. 1245-1258

Chamaillard, M.; Hashimoto, M.; Horie, Y.; Masumoto, J.; Qui, S.; Saab, L., Ogura, Y.; Kawasaki, A.; Fukase, K.; Kusumoto, S.; Valvano, M.; Foster, S.; Mak, T,; Nunez, G & Inohara, N. An essential role for NOD1 in host recognition of bacterial peptidoglycan containing diaminopimelic acid. *Nature Immunology*, Vol.4, No.7, (July 2003), pp. 702-707

Charras, G. & Paluch, E. (2008). Blebs lead the way: how to migrate without lamellipodia. *Nature Review Molecular Cell Biology*, Vol.9, No.9, (September 2008), pp. 730-736

Cole, S.; Brosch, R.; Parkhill, J.; Garnier, T.; Churcher, C.; Harris D., *et al.* (1998). Deciphering the biology of *Mycobacterium tuberculosis* from the complete genome sequence. *Nature*, Vol.393, No.6685, (June 1998), pp. 537-544

Cox, D.; Lee, D.; Dale, B.; Calafat, J. & Greenberg, S. (2000). A Rab11-containing rapidly recycling compartment in macrophages that promotes phagocytosis. *Proceedings of the National Academy of Science United States of America*, Vol.97, No.2, (January 2000), pp. 680-685

Decker, T.; Stockinger, S.; Karaghiosoff, M.; Muller, M. & Kovarik P. (2002). IFNs and STATs in innate immunity to microorganisms. *Journal of Clinical Investigation*, Vol.109, No.10, (May 2002), pp. 1271-1277

Desjardins, M.; Huber, L.; Parton, R. & Griffiths, G. (1994). Biogenesis of phagolysosomes proceeds through a sequential series of interactions with the endocytic apparatus. *Journal of Cell Biology*, Vol.124, No.5, (March 1994), pp. 677-688

Di Paolo, G. & De Camilli, P. (2006). Phosphoinositides in cell regulation and membrane dynamics. *Nature*, Vol.443, No.7112, (October 2006), pp. 651-657

Ferrari, G.; Langen, H.; Naito, M. & Pieters, J. (1999). A coat protein on phagosomes is involved in the intracellular survival of mycobaceria. *Cell*, Vol.97, No.4, (May 1999), pp. 435-447

Fitzgerald, K.; Palsson-McDermott, E.; Bowie, A.; Jefferies C.; Mansell, A.; Brady, G. et al., (2001). Mal (MyD88-adapter-like) is required for Toll-like receptor-4 signal transduction. *Nature*, Vol.413, N0.6851, (September 2001), pp. 78-83

Fitzgerald, K.; Rowe, D.; Barnes, B.; Caffrey, D.; Visintin, A.; Latz, E. et al. (2003). LPS-TLR4 signalling to IRF-3/7 and NF-kappaB involves the toll adapters TRAM and TRIF. *Journal of Experimental Medicine*, Vol.198, No.7, (October 2003), pp. 1043-1055

Fratti, R.; Becker, J.; Gruenberg, J.; Corvere, S. & Deretic, V. (2001). Role of phosphatidylinositol 3-kinase and Rab5 effectors in phagosomel biogenesis and mycobacterial phagosome maturation arrest. *Journal of Cell Biology*, Vol.154, No.3, (August 2001), pp. 631-644

Fu, Y. & Galan, J. (1998). The *Salmonella typhimurium* tyrosine phosphatase SptP is translocated into host cells and disrupts the actin cytoskeleton. *Molecular Microbiology*, Vol.27, No.2, (January 1998), pp. 359-368

Fu, Y. & Galan, J. (1999). A *Salmonella* protein antagonizes Rac-1 and Cdc42 to mediate host-cell recovery after bacterial invasion. *Nature*, Vol.401, No.6750, (September 1999), pp. 293-297

Galan, J. & Wolf-Watz, H. (2006). Protein delivery into eukaryotic cells by type III secretion machines. *Nature*, Vol.444, No.7119, (November 2006), pp. 567-573

Gan, H.; Lee, J.; Ren, F.; Chen, M.; Kornfeld, H. & Remold, H. (2008). *Mycobacterium tuberculosis* blocks crosslinking of annexin-1 and apoptotic envelope formation on infected macrophages to maintain virulence. *Nature Immunology*, Vol.9, No.10, (October 2008), pp. 1189-1197

Gao, X.; Wan, F.; Mateo, K.; Callegari, E.; Wang, D.; Deng, W.; Punte, J.; Li, F.; Chaussee, M.; Finlay, B.; Lenardo, M. & Hardwidge, P. (2009). Bacterial effector binding to ribosomal protein s3 subverts NF-kappaB function. *PLoS Pathogens*, Vol.5, No.12, (December 2009), pp. e1000708

Gay, N. & Gangloff, M. (2007). Structure and function of toll receptors and their ligands. *Annual Review of Biochemistry*, Vol.76, pp. 23.1-23.25

Ghosh, P. (2004). Process of protein transport by the type III secretion system. *Microbiology and Molecular Biology Reviews*, Vol.68, No.4, (December 2004), pp. 771-795

Girardin, S.; Boneca, I.; Viala, J.; Chamaillard, M.; Labigne, A.; Thomas, G.; Philpott, D. & Sansonetti, P. (2003). Nod2 is a general sensor of peptidoglycan through muramyl dipeptide (MDP) detection. *Journal of Biological Chemistry*, Vol.278, No.11, (March 2003), pp. 8869-8872

Gouin, E.; Adib-Conquy, M.; Balestrino, D.; Nahori, M.; Villiers, V.; Colland, F.; Dramsi, S.; Dussurget, O. & Cossart, P. (2010). The *Listeria monocytogenes* InlC protein interferes with innate immune responses by targeting the IkappaB kinase subunit IKKalpha. *Proceedings of the National Academy of Science United States of America*, Vol.107, No.40, (October 2010), pp. 17333-17338

Grosdent, N.; maridonneau-Parini, I.; Sory, M. & Cornelis, G. (2002). Role of Yops and adhesins in resistance of yersinia enterolitica to phagocytosis. *Infection & Immunity*, Vol.70, No.8, (August 2002), pp. 4165-4176

Haas, A. (2007). The phagosome: compartment with a license to kill. *Traffic*, Vol.8, No.4, (April 2007), pp. 311-330

Hakansson, S.; Galyov, E.; Rosqvist, R. & Wolf-Watz, H. (1996). The *Yersinia* YpkA Ser/Thr kinase is translocated and subsequently targeted to the inner surface of the HeLa cell plasma membrane. *Molecular Microbiology*, Vol.20, No.3, (May 1996), pp. 593-603

Handa, Y.; Suzuki, M.; Ohya, K.; Iwai, H.; Ishujima, N.; Koleske,A.; Fukui, Y. & Sasakawa, C. (2007). *Shigella* IpgB1 promotes bacterial entry through the ELMO-Dock180 machinery. *Nature Cell Biology*, Vol.9, No.1, (January 2007), pp. 121-128

Hardt, W.; Chen, L.; Schuebel, K.; Bustelo, X. & Alan, J. (1998). *S. typhimurium* encodes an activator of Rho GTPases that induces membrane ruffling and nuclear responses in host cells. *Cell*, Vol.93, No5., (May 1998), pp. 815-826

Hernandez, L.; Hueffer, K.; Wenk, M. & Galan, J. (2004). *Salmonella* modulates vesicular traffic by altering phosphoinositide metabolism. *Science*, Vol.304, No.5678, (June 2004), pp. 1805-1807.

Hestvik, A.; Hmama, Z. & Av-Gay Y. (2005). Mycobacterial manipulation of the host cell. *FEMS Microbiology Reviews*, Vol.29, No.5, (November 2005), pp. 1041-1050

Honda, K. & Taniguchi, T. (2006). IRFs: master regulators of signalling by Toll-receptors and cytosolic pattern-recognition receptors. *Nature Reviews Immunology*, Vol.6, No.9, (September 2006), pp. 644-658

Hussain, S.; Zwilling, B. & Lafuse, W. (1999). *Mycobacterium avium* infection of mouse macrophages inhibits IFN-gamma Janus kinase-STAT signalling and gene induction by downregulation of the IFN-gamma receptor. *Journal of Immunology*, Vol.163, No.4, (August 1999), pp. 2041-2048

Ingmundson, A.; Delprato, A.; Lambright, D. & Roy, C. (2007). *Legionella pneumophila* proteins that regulate Rab1 membrane cycling. *Nature*, Vol.450, No.7168, (November 2007), pp. 365-369

Inohara, N. & Nunez, G. (2003). NODs: intracellular proetins involved in inflammation and apoptosis. *Nature Reviews Immunology*, Vol.3, No.5, (May 2003), pp. 371-382

Ishii, K.; Koyama, S.; Nakagawa, A.; Coban, C. & Akira, S. (2008). Host innate immune receptors and beyond: making sense of microbial infections. *Cell Host Microbe*, Vol.3, No.8, (June 2008), pp. 352-363

Iwasaki, A. & Medzhitov, R. (2004). Toll-like receptor control of the adaptive immune responses. *Nature Immunology*, Vol.5, No.10, (October 2004), pp. 987-995

Jaconi, M.; Lew, D.; Carpentier, J.; Magnusson, K.; Sjogren, M. & Stendahl, O. (1990) Cytosolic free calcium elevation mediates the phagosome-lysosome fusion during phagocytosis in human neutrophils. *Journal of Cell Biology*, Vol.110, No.5, (May 1990), pp. 1555-1564

Jahraus, A.; Storrie, B.; Griffiths, G. & Desjardins, M. (1994). Evidence for retrograde traffic between terminal lysosomes and the prelysosomal/late endosome compartment. *Journal of Cell Biology*, Vol.107, No.1, (January 1994), pp. 145-157

Johnson, G. & Lapadat, R. (2002). Mitogen-activated protein kinase pathways mediated by ERK, JNK, and p38 protein kinases. *Science*, Vol.298, No.5600, (December 2002), pp. 1911-1912

Kagan, J. & Roy, C. (2002). *Legionella* phagosomes intercept vesicular traffic from endoplasmic reticulum exit sites. *Nature Cell Biology*, Vol.4, No.12, (December 2002), pp. 945-954

Kawagoe, T.; Sato, S.; Matsushita, K.; Kato, H.; Matsui, K.; Kumagai, Y. et al. (2008). Sequential control of toll-like receptor-dependent responses by IRAK1 and IRAK2. *Nature Immunology*, Vol.9, No.6, (June 2008), pp. 684-691

Kayal, S.; Lilienbaum, A.; Join-Lambert, O.; Li, X.; Israel, A. & Berche, P. (2002). Listeriolysin O secreted by *Listeria monocytogenes* induces NF-kappaB signalling by activating the IkappaB kinase complex. *Molecular Microbiology*, Vol. 44, No.5, (June 2002), pp. 1407-1419

Klink, B.; Barden, S.; Heidler, T.; Borchers, C.; Ladwein, M.; Stradal, T.; Rottner, K. & Heinz, D. (2010). Structure of *Shigella* IpgB2 in complex with human RhoA: implications for the mechanism of bacterial guanine nucleotide exchange factor mimicry. *Journal of Biological Chemistry*, Vol.285, No.22, (May 2010), pp. 17197-17208

Kobayashi, K.; Hernandez, L.; Galan, J.; Janeway, C.; Medzhitov, R. & Flavelli, R. (2002). IRAK-M is a negative regulator of Toll-like receptors signalling. *Cell*, Vol.110, No.2, (July 2002), pp. 191-202

Koul, A.; Herget, T.; Klebl, B. & Ullrich, A. (2004). Interplay between mycobacteria and host signalling pathways. *Nature Reviews Microbiology*, Vol.2, No.3 (March 2004), pp. 189-202

Kubori, T. & Galan, J. (2003). Temporal regulation of *Salmonella* virulence effector function by proteosome-dependent protein degradation. *Cell*, Vol.115, No.3, (October 2003), pp. 333-342

Li, H.; Xu, H.; Zhou, Y.; Zhang, J.; Lomg, C.; Li, S.; Chen, S.; Zhou, J. & Shao, F. (2007). The phosphothreonine lyase activity of a bacterial type III effector family. *Science*, Vol.315, No.5814, (February 2007), pp. 1000-1003

Lin, S.; Le, T. & Cowen, D. (2003). SptP, a *Salmonella typhimurium* type III-secreted protein, inhibits the mitogen-activated protein kinase pathway by inhibiting Raf activation. *Cellular Microbiology*, Vol.5, No.4, (April 2003), pp. 267-275

Maiti, D.; Bhattacharyya, A. & Basu, J. (2001). Lipoarabinomannan from *Mycobacterium tuberculosis* promotes macrophage survival by phosphorylating Bad through a phosphatidylinositol 3-kinase/Akt pathway. *Journal of Biological Chemistry*, Vol.276, No.1, (January 2001), pp. 329-333

Malik, Z.; Thompson, C.; Hashimi, S.; Porter, B.; Iyer, S. & Kusner, D. (2003). Cutting edge: *Mycobacterium tuberculosis* blocks Ca^{2+} signalling and phagosome maturation in human macrophages via specific inhibition of sphingosine kinase. *Journal of Immunology*, Vol.170, No.6, (March 2003), pp. 2811-2815

Mallo, G.; Espina, M.; Smith, A.; Terebiznik, M.; Aleman, A.; Finlay, B.; Rameh, L.; Grinstein, S. & Brummel, J. (2008). SopB promotes phosphatidylinositol 3-phosphate formation on *Salmonella* vacuoles by recruiting Rab5 and Vps34. *Journal of Cell Biology*, Vol.182, No.4, (August 2008), pp. 741-752

Mason, D.; mallo, G.; Terebiznik, M.; Payrastre, B.; Finlay, B.; Brumell, J.; Rameh, L. & Grinstein, S. (2007). Alteration of epithelial structure and function associated with PtdIns(4,5)P2 degradation by a bacterial phosphatase. *Journal of General Physiology*, Vol.129, No.4, (April 2007), pp. 267-283

Mazurkiewicz, P.; Thomas, J.; Thompson, J.; Liu, M.; Arbibe, L.; Sansonetti, P. & Holden, D. (2008). SpvC is a *Salmonella* effector with phosphothreonine lyase activity on host mitogen-activated protein kinases. *Molecular Microbiology*, Vol.67, No.6, (March 2008), pp. 1371-1383

Medzhitov, R. (2008). Origin and physiological roles of inflammation. *Nature*, Vol.454, No.7203, (July 2008), pp. 428-435

Mittal, R.; Peak-Chew, S. & McMahon, H. (2006). Acetylation of MEK2 and I kappa B kinase (IKK) activation loop residues by YopJ inhibits signalling. *Proceedings of the National Academy of Science United States of America*, Vol.103, No.49, (October 2010), pp. 18574-18579

Moreno, C.; Mehlert, A. & Lamb, J. (1988). The inhibitory effects of mycobacterial lipoarabinomannan and polysaccharides upon polyclonal and monoclonal human T cell proliferation. *Clinical and Experimental Immunology*, Vol.74, No.2, (November 1988), pp. 206-210

Mostowy, S. & Cossart, P. (2009). Cytoskeleton rearrangements during *Listeria* infection: clathrin and septins as new players in the game. *Cell Motility and the Cytoskeleton*, Vol.66, No.10, (October 2009), pp. 816-823

Mukherjee, S.; Keitany, G.; Li, Y.; Wang, Y.; Ball, H.; Goldsmith, E. & Orth, K. (2006). *Yersinia* YopJ acetylates and inhibits kinase activation by blocking phosphorylation. *Science*, Vol.312, No.5777, (May 2006), pp. 1211-1214

Muller, M.; Peters, H.; Blumer, J.; Blankenfeldt, W.; Goody, R. & Itzen, A. The *Legionella* effector protein DrrA AMPylates the membrane traffic regulator Rab1b. *Science*, Vol.329, No.5994, (August 2010), pp. 946-949

Murata, T.; Delprato, A.; Ingmundson, A.; Toomre, D.; Lambright, D. & Roy, C. (2006). The *Legionella pneumophila* effector protein DrrA is a Rab1 guanine-nucleotide-exchange factor. *Nature Cell Biology*, Vol.8, No.9, (September 2006), pp. 971-979

Niebuhr, K.; Jouihri, N.; Allaoui, A.; Gounon, P.; Sansonetti, P. & Parsot, C. (2000). IpgD, a protein secreted by the type III secretion machinery of *Shigella flexneri*, is chaperoned by IpgE and implicated in entry focus formation. *Molecular Microbiology*, Vol.38, No.1, (October 2000), pp. 8-19

Niebuhr, K.; Giuriato, S.; Pedron, T.; Philpott, D.; Gaits, F.; Sable, J.; Sheetz, M.; Parsot, C.; Sansonetti, P. &Payrastre, B. (2002). Conversion of PtdIns(4,5)P(2) into PtdIns(5)P by the *S. flexneri* effector IpgD reorganizes host cell morphology. *EMBO Journal*, Vol.21, No.19, (October 2002), pp. 5069-5078

Nigou, J.; Guilleron, M.; Rojas, M.; Garcia, L.; Thurnher, M. & Puzo, G. (2002). Mycobacterial lipoarabinomannans: modulators of dendritic cell function and the apoptotic response. *Microbes and Infection*, Vol.4, No.9, (July 2003), pp. 945-953

O'Neill, L. & Bowie, A. (2007). The family of five: TIR-domain containing adaptors in Toll-like receptor signalling. *Nature Reviews Immunology*, Vol.7, No.5, (May 2007), pp. 353-364

Orth, K.; Palmer, L.; Bao, Z.; Stewart, S.; Rudolph, A.; Bliska, J. & Dixon, J. (1999). Inhibition of the mitogen-activated protein kinase kinase superfamily by a *Yersinia* effector. *Science*, Vol.285, No.5435, (September 1999), pp. 1920-1923

Orth, K.; Xu, Z.; Mudgett, M.; Bao, Z.; Palmer, L.; Bliska, J.; Mangel, W.; Staskawicz, B. & Dixon, J. (2000). Disruption of signalling by *Yersinia* effector YopJ, a ubiquitin-like protein protease. *Science*, Vol.290, No.5496, (November 2000), pp. 1594-1597

Park, J.; Greten, F.; Li, Z. & Karin, M. (2002). Macrophage apoptosis by anthrax lethal factor through p38 MAP kinase inhibition. *Science*, Vol.297, No.5589, (September 2002), pp. 2048-2051

Patel, J. & Galan, J. (2005). Manipulation of the host actin cytoskeleton by *Salmonella*-all in the name of entry. *Current Opinion in Microbiology*, Vol.8, No.1, (February 2005), pp. 10-15

Pehleven, A.; Wrighy, D.; Andrews, C.; George, K.; Small, P. & Foxwell, B. (1999). The inhibitory action of *Mycobacterium ulcerans* soluble factor on monocyte/T cell cytokine production and NF-kappa B function. *Journal of Immunology*, Vol.163, No.7, (October 1999), pp. 3928-3935

Pethe, K.; Swenson, D.; Alonso, S.; Anderson, J.; Wang, C. & Russell, D. (2004). Isolation of *Mycobacterium tuberculosis* mutants defective in the arrest of phagosome maturation. *Proceedings of the National Academy of Science United States of America*, Vol.101, No.37, (September 2004), pp. 13642-13647

Ragaz, C.; Pietsch, H.; Urwyler, S.; Tiaden, A.; Weber, S. & Hilbi, H. (2008). The *Legionella pneumophila* phosphatidylinositol-4 phosphate-binding type IV substrate SidC recruits endoplasmic reticulum vesicles to a replication-permissive vacuole. *Cellular Microbiology*, Vol.10, No.12, (December 2008), pp. 2416-2433.

Reiling, N.; Blumenthal, A.; Flad, H.; Ernst, M. & Ehlers, S. (2001). Mycobacteria-induced TNF-alpha and IL-10 formation by human macrophages is differentially regulated at the level of mitogen-activated protein kinase activity. *Journal of Immunology*, Vol.167, No.6, (September 2001), pp. 3339-3345

Roach, S. & Schorey, J. (2002). Differential regulation of the mitogen-activated protein kinases by pathogenic and nonpathogenic mycobacteria. *Infection & Immunity*, Vol.70, No.6, (June 2002), pp. 3040-3052

Rusten E. & Stenmark, H. (2006). Analyzing phosphoinositides and their interacting proteins. *Nature Methods*, Vol.3, No.4, (April 2006), pp. 251-258

Sato, S.; Sanjo, H.; Takeda, K.; Ninomiya-Tsuji, J.; Yamamoto, M.; Kawai, T. et al. (2005). Essential function for the kinase TAK1 in innate and adaptive immune response. *Nature Immunology*, Vol.6, No.11, (November 2005), pp. 1087-1095

Schesser, K.; Spiik, A.; Dukuzumuremyi, J.; Neurath, M.; Petterson, S. & Wolf-Watz, H. (1998). The *yopJ* locus is required for *Yersinia*-mediated inhibition of NF-kappaB activation and cytokine expression: YopJ contains a eukaryotic SH2-like domain that is essential for its repressive activity. *Molecular Microbiology*, Vol.28, No.6, (June 1998), pp. 1067-1079

Seveau, S.; Bierne, H.; Giroux, S.; Prevost, M. & Cossart, P. (2004). Role of lipid rafts in E-cadherin- and HGF-R/Met-mediated entry of *Listeria monocytogenes* into host cells. *Journal of Cell Biology*, Vol.166, No.5, (August 2004), pp. 743-753

Shao, F.; Merritt, P.; Bao, Z.; Innes, R. & Dixon, J. (2002). A *Yersinia* effector and a *Pseudomonas* avirulence protein define a family of cysteine proteses functioning in bacterial pathogenesis. *Cell*, Vol.109, No.5, (May 2002), pp. 575-588

Shen, Y.; Naujokas, M.; Park, M. & Ireton, K. (2000). InlB-dependent internalization of *Listeria* is mediated by the Met receptor tyrosine kinase. *Cell*, Vol.103, No.3, (October 2000), pp. 501-510

Shtrichman, R.; & Samuel, C. (2001). The role of gamma interferon in antimicrobial immunity. *Current Opinion in Microbiology*, Vol. 4, No.3, (June 2001), pp. 251-259

Sibley, L.; Hunter, S.; Brennan, P. & Krahenbuhl, J. (1988). Mycobacterial lipoarabinomannan inhibits gamma interferon-mediated activation of macrophages. *Infection & Immunity*, Vol.56, No.5, (May 1988), pp. 1232-1236

Spira, A.; Carroll, J.; Liu, G.; Aziz, Z.; Shah, V.; Kornfeld, H. & Keane, J. (2003). Apoptosis genes in human alveolar macrophages infected with virulent or attenuated *M. tuberculosis*. *American Journal of Respiratory Cell Molecular Biology*, Vol.29, No.5, (November 2003), pp. 545-551

Sprang, S. (1997). G protein mechanisms: Insights from structural analysis. *Annual Reviews Biochemistry*, Vol.66, pp. 639-678

Steele-Mortimer, O.; Meresse, S.; Gorvel, J.; Toh, B. & Finlay, B. (1999). Biogenesis of *Salmonella typhimurium*-containing vacuoles in epithelial cells involves interactions with the early endocytic pathway. *Cellular Microbiology*, Vol.1, No.1, (July 1999), pp. 33-49

Sturgill-Koszycki, S.; Schlesinger, P.; Chakraborty, P.; Haddix, P.; Collins, H.; Fok, A.; Allen, R.; Gluck, S.; Heuser, J. & Russell, D. (1994). Lack of acidification in *Mycobacterium* phagosomes produced by exclusion of the vesicular proton-ATPase. *Science*, Vol.263, No.5147, (February 1994), pp. 678-681

Sturgill-Koszycki, S. & Swanson, M. (2000). *Legionella pneumophila* replication vacuoles mature into acidic, endocytic organelles. *Journal of Experimental Medicine*, Vol.192, No.9, (November 2000), pp. 1261-1272

Szalai, G.; Krishnamurthy, R. & Hajnoczky, G. (1999). Apoptosis driven by IP_3-linked mytochondrial calcium signals. *EMBO Journal*, Vol.18, No.22, (November 1999), pp. 6349-6361

Takeuchi, O. & Akira, S. (2010). Pattern recognition receptors and inflammation. *Cell*, Vol.140, No.6, (March 2010), pp. 805-820

Ting, L.; Kim, A.; Cattamanchi, A. & Ernst, J. (1999). *Mycobacterium tuberculosis* unhibits IFN-gamma transcriptional responses without inhibiting activation of STAT1. *Journal of Immunology*, Vol.163, No.7, (October 1999), pp. 3898-3906

Turk, B. (2007). Manipulation of host signalling pathways by anthrax toxins. *Biochemical Journal*, Vol.402, No.3, (March 2007), pp. 405-417

Velledor, A.; Comalada, M.; Xaus, J. & Celada, A. (2000). The differential time-course of extracellular-regulated kinase activity correlates with the macrophage response toward proliferation or activation. *Journal of Biological Chemistry*, Vol.275, No.10, (March 2010), pp. 7403-7409

Vergne, I.; Chua, J. & Deretic, V. (2003). Tuberculosis toxin blocking phagosome maturation inhibits a novel Ca^{2+}/calmodulin-PI3K hVPS34 cascade. *Journal of Experimental Medicine*, Vol.198, No.4, (August 2003), pp. 653-659

Vergne, I.; Chua, J.; Lee H.; Lucas, M.; Belisle, J. & Deretic, V. (2005). Mechanism of phagolysosome biogenesis block by viable *Mycobacterium tuberculosis*. *Proceedings of the National Academy of Science United States of America*, Vol.102, No.11, (March 2005), pp. 4033-4038

Viboud, G. & Bliska, J. (2001). A bacterial type III secretion system inhibits actin polymerization to prevent pore formation in host cell membranes. *EMBO Journal*, Vol.20, No.19, (October 2001), pp. 5373-5382

Von Pawel-Rammingen, U.; Telepnev, M.; Schmidt, G.; Aktories, K.; Wolf-Watz, H. & Rosqvist, R. (2000). GAP activity of the *Yersinia* YopE cytotoxin specifically targets the Rho pathway: a mechanism for disruption of actin microfilament structure. *Molecular Microbiology*, Vol.36, No.3, (May 2000), pp. 737-748

Walburger, A.; Koul, A.; Ferrari, G.; Nguyen, L.; Prescianotto-Baschong, C.; Huygen, K.; Klebl, B.; Thompson, C.; Bacher, G. & Pieters, J. (2004). Protein kinase G from pathogenic mycobacteria promotes survival within macrophages. *Science*, Vol.304, No.5678, (June 2004), pp. 1800-1804

Wong, K. & Isberg, R. (2003). Arf6 and phosphoinositol-4-phosphate-5-kinase activities permit bypass of the Rac1 requirement for beta1 integrin-mediated bacterial uptake. *Journal of Experimental Medicine*, Vol.198, No.4, (August 2003), pp. 603-614

Wong, D.; Bach, H.; Sun, J.; Hmama, Z. & Av-Gay, Y. (2011). *Mycobacterium tuberculosis* protein tyrosine phosphatase (PtpA) excludes host vacuolar-H^+-ATPase to inhibit phagosome acidification. *Proceedings of the National Academy of Science United States of America*, Vol.108, No.48, (November 2011), pp. 19371-19376

Wurzner, R. (1999). Evasion of pathogens by avoiding recogmition or eradication by complement, in part via molecular mimicry. *Molecular Immunology*, Vol.36, No.4-5, (March-April 1999), pp. 249-260

Yamamoto, M.; Sato, S.; Hemmi, H.; Hoshino, K.; Kaisho, T.; Sanjo, H. et al. (2003a). Role of adaptor TRIF in the MyD88-independent toll-like receptor signalling pathway. *Science*, Vol.301, No.5633, (August 2003), pp. 640-643

Yamamoto M.; Sato, S.; Hemmi, H.; Uematsu, S.; Hoshino, K.; Kaisho, T. et al. (2003b). TRAM is specifically involved in the Toll-like receptor 4-mediated MyD88-independent signalling pathway. *Nature Immunology*, Vol.4, No.11, (November 2003), pp. 1144-1150

Yamauchi, K.; Choi, I.; Lu, H.; Ogiwara, H.; Graham, D. & Yamaoka, Y. (2008). Regulation of IL-18 in *Helicobacter pylori* infection. *Journal of Immunology*, Vol.180, No.2, (January 2008), pp. 1207-1216

Yates, R.; Hermetter, A. & Russell, D. (2005). The kinetics of phagosome maturation as a function of phagosome/lysosome fusion and acquisition of hydrolytic activity. *Traffic*, Vol.6, No.5, (May 2005), pp. 413-420

Zhou, B.; He, Y.; Zhang, X.; Xu, J.; Luo, Y.; Wang, Y., Franzblau, S.; Yang, Z.; Chan, R.; Liu, Y.; Zheng, J. & Zhang, Z. (2010). Targeting *Mycobacterium* protein tyrosine phosphatase B for antituberculosis agents. *Proceedings of the National Academy of Science United States of America*, Vol.107, No.10, (March 2010), pp. 4573-4578

Zumbihl, R.; Aepfelbacher, M.; Andor, A.; Jacobi, C.; Ruckdeschel, K.; Rouot, B. & Heesemann, J. (1999). The cytotoxin YopT of *Yersinia enterolitica* induces modification and cellular redistribution of the small GTP-binding protein RhoA. *Journal of Biological Chemistry*, Vol.274, No.41, (October 1999), pp. 29289-29293

6

Pathogen Strategies to Evade Innate Immune Response: A Signaling Point of View

Bruno Miguel Neves[1,2], Maria Celeste Lopes[1] and Maria Teresa Cruz[1]
[1]*Faculty of Pharmacy and Centre for Neuroscience and Cell Biology, University of Coimbra,*
[2]*Department of Chemistry, Mass Spectrometry Center, QOPNA, University of Aveiro,*
Portugal

1. Introduction

An effective host defense against pathogens requires appropriate recognition of the invading microorganism by immune cells, conducing to an inflammatory process that involves recruitment of leukocytes to the site of infection, activation of antimicrobial effector mechanisms and induction of an adaptive immune response that ultimately will promote the clearance of infection. All these events require the coordination of multiple signaling pathways, initially triggered by the contact of the pathogen with innate immune cells. The "signal alarm" is normally triggered by ligation of microorganism, or microorganism's components, to pattern-recognition receptors, causing their phosphorylation and recruitment of adapter molecules, which in turn will activate second messengers within the cytosol of the cells, allowing the transduction of the signal. The second messengers are often protein kinases that in a cascade process ultimately activate the transcription factors responsible for the expression of effector molecules like, cytokines, chemokines and reactive oxygen species, crucial elements to mount an adequate immune response. The activity of such critical intracellular signaling pathways is a process extremely well controlled by a balance of positive and negative regulation, being the activation of a given protein kinase normally counterbalanced by the activation of its opposing phosphatase. However, as part of their pathogenic strategies, several microorganisms exploit host cell signaling mechanisms by distorting this balance between positive and negative signals. They hijack crucial immune-cell signaling pathways, subverting the immunogenic abilities of these cells and evading this way the host immune response. In the last few years a great effort has gone into understanding the molecular mechanisms behind this subversion, and various signaling cascades were identified as main targets of pathogens and virulence factors. Among these targets, assume particular importance the transcription factor nuclear factor-κB (NF-κB), a cornerstone of innate immunity and inflammatory responses, as well as the mitogen activated protein kinases (MAPKs), signaling cascades implicated in the regulation of crucial aspects of immunity. Overall in this chapter, we will provide an overview of the current understanding of how pathogens interact with host cells and how these microorganisms exploit host immune response in a signaling point of view.

2. Immune response to invading microorganisms

In mammals, immune system can be subdivided into two branches: innate and adaptive immunity. Following infection, innate immune cells like macrophages, dendritic cells (DCs) and neutrophils (collectively called phagocytes) engulf and destroy microorganisms, representing that way a rapid first defense barrier against infection. In turn, adaptive immunity is mediated via the generation of antigen-specific B and T lymphocytes, through a process of gene rearrangement resulting in the production and development of specific antibodies and killer T cell, respectively. Adaptive immunity is also behind immunological memory, allowing the host to rapidly respond when exposed again to the same pathogen. Contrarily to the originally thought, the innate immune response is not completely nonspecific, but rather is able to discriminate between self antigens and a variety of pathogens (Akira et al., 2006). Furthermore, much evidence has demonstrated that pathogen-specific innate immune recognition is a prerequisite to the induction of antigen-specific adaptive immune responses (Hoebe et al., 2004; Iwasaki & Medzhitov, 2010), being dendritic cells central players in this linking (Steinman, 2006). DCs are specialized antigen-presenting cells that function as sentinels, scanning changes in their local microenvironment and transferring the information to the cells of the adaptive immune system (Banchereau & Steinman, 1998; Banchereau et al., 2000). Upon activation by microorganisms or microorganism components, immature DCs suffer a complex process of morphological, phenotypical and functional modifications to become mature DCs that enter draining lymphatic vessels and migrate to the T-cell zones of draining lymph nodes where they present antigens to T lymphocytes. Depending on their maturation/activation profile, DCs

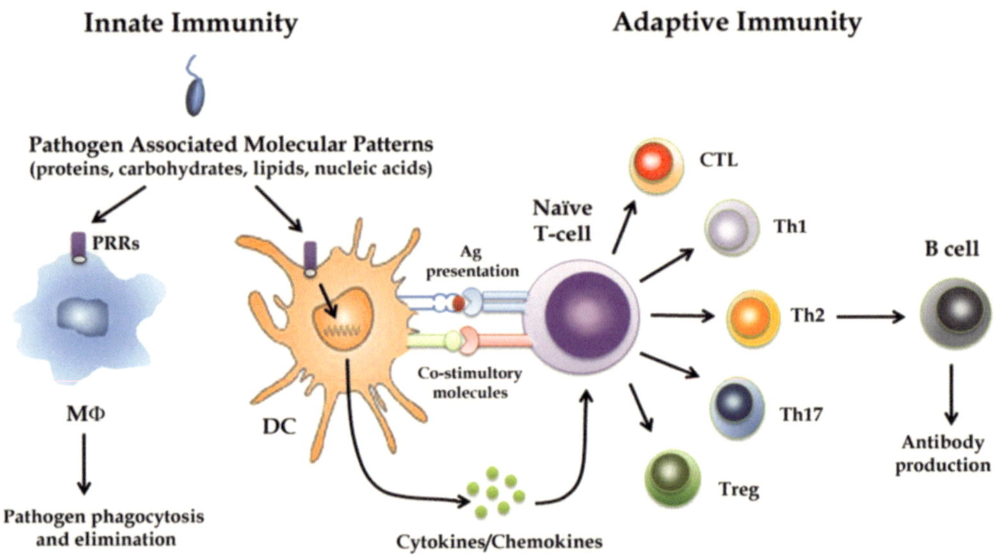

Fig. 1. Dendritic cells link innate to adaptive immunity. Once in contact with microbial antigens, DCs mature and migrate to draining lymph nodes where they present antigens to naïve T lymphocytes. Different pathogens trigger disticnt DCs maturation profiles, leading to the polarization of different T-cell subsets. The adaptive immune response is therefore modulated, in some extent, to match the nature of the pathogen. Ag: antigen; CTL: cytotoxic T cell; DC: dendritic cell; Mφ: macrophage.

will polarize and expand distinct T-cell subsets (T-helper cells [Th1, Th2, and Th17], regulatory T cells, and cytotoxic T cells) (Sporri & Reis e Sousa, 2005; Diebold, 2009) and given that the recognition of different microorganisms lead to distinct DC maturation/activation profiles, the adaptive immune response is, therefore, modulated to match the nature of the pathogen (Figure 1)

2.1 Recognition of microorganisms by innate immune cells

To a rational understanding of molecular mechanisms by which pathogens escape the immune system, we need first to know how our immune cells sense microorganisms and spread the "alarm".

Innate immune cells, like macrophages and DCs, recognize microorganisms through sensing conserved microbial components, globally designated as pathogen associated molecular patterns (PAMPs) (Kawai & Akira, 2010; Takeuchi & Akira, 2010; Medzhitov, 2007). These molecular patterns are normally essential components of microbial metabolism, including proteins, lipids, carbohydrates and nucleic acids, not subjected to antigenic variability. Another important feature of PAMPs is that they are markedly distinct from self-antigens, allowing the innate immune system to discriminate between self and non-self.

The recognition of PAMPs is mediated by constitutively expressed host´s germline-encoded pattern-recognition receptors (PRRs), such as Toll-like receptors (TLRs), C-type lectin receptors (CLRs), retinoic acid-inducible gene-1(RIG-1)-like receptors and nucleotide-oligomerization domain (NOD)-like receptors. The beauty of this evolutionary sensor mechanism is that different PRRs react with specific PAMPs, triggering a signaling pathway profile that ultimately lead to distinct anti-pathogen responses (Akira, 2009). Therefore, innate immunity is a key element in the infection-induced non specific inflammatory response as well as in the conditioning of the specific adaptive immunity to the invading pathogens (Akira et al., 2001; Iwasaki & Medzhitov, 2004).

2.1.1 Toll-like receptors

Among PPRs, TLRs are by far the most intensively studied and the more expressive group, being considered the primary sensors of pathogen components. TLRs are type I membrane glycoproteins formed by extracellular leucine rich repeats involved in PAMP recognition, and a cytoplasmic signaling domain homologous to that of the interleukin 1 receptor (IL-1R), know as Toll/IL-1R homology (TIR) domain. These receptors were originally identified in *Drosophila* as essential elements for the establishment of the dorso-ventral pattern in developing embryos (Hashimoto et al., 1988). However, in 1996, Hoffmann and colleagues would initiate a novel era in our understanding of innate immunity, demonstrating that Toll-mutant flies were highly susceptible to fungal infection, showing that way that TLRs were involved in the defense against invading microorganisms (Lemaitre et al., 1996). Afterward, mammalian homologues of Toll receptor were progressively identified, and actually most mammalian species are believed to have between ten and thirteen types of TLRs. In human, ten functional receptors (TLR1-10) have been identified so far and an

TLR family	Cellular location	Microbial components	Pathogens
TLR1/2	Cell surface	Tri-acyl lipopeptides Soluble factors	Bacteria, mycobacteria *Neisseria meningitides*
TLR2	Cell surface	Diacyl lipopeptides Triacyl lipopeptides Peptidoglycan Lipoteichoic acid Porins Lipoarabinomannan Phenol-soluble modulin tGPI-mutin Glycolipids Hemagglutinin protein Zymosan Phospholipomannan Glucuronoxylomannan	Mycoplasma Bacteria and mycobacteria Gram-positive bacteria Gram-positive bacteria *Neisseria* Mycobacteria *Staphylococcus epidermidis* *Trypanosoma Cruzi* *Treponema maltophilum* Measles virus Fungi *Candida albicans* *Cryptococcus neoformans*
TLR3	Endolysosome	Viral double-stranded RNA	Vesicular stomatitis virus, lymphocytic choriomeningitis virus reovirus
TLR4	Cell surface	LPS Fusion protein Envelope proteins HSP60 Manan Glycoinositolphospholipids	Gram-negative bacteria Respiratory syncytial vírus Mouse mammary tumor virus *Chlamydia pneumoniae* *Candida albicans* *Trypanosoma*
TLR5	Cell surface	Flagellin	Flagellated bacteria
TLR6/2	Cell surface	Diacyl lipopeptides Lipoteichoic acid Zymosan	Mycoplasma Group B Streptococcus *Saccharomyces cerevisiae*
TLR7	Endolysosome	Viral single-stranded RNA RNA	Several virus Bacteria from group B *Streptococcus*

TLR family	Cellular location	Microbial components	Pathogens
TLR8 (only human)	Endolysosome	Viral single-stranded RNA	Several virus
TLR9	Endolysosome	CpG-DNA dsDNA viruses Hemozoin	Bacteria and mycobacteria Herpes simplex virus and murine Cytomegalovirus Plasmodium
TLR10	Cell surface	Unknown	Unknown
TLR11 (only mouse)	Endosome	Profilin-like molecule	*Toxoplasma gondii* Uropathogenic *E. coli*
TLR12 (only mouse)	Cell surface	ND	Unknown
TLR13 (only mouse)	Cell surface	ND	Virus

Table 1. Toll-like receptors cellular location and microbial ligands. ND: not determined.

eleventh has been found to be encoded at gene level but, as it contains several stop codons, protein is not expressed (Zhang et al., 2004). TLRs are involved in sensing a wide panel of microbial products (Kawai & Akira, 2010), including lipids, peptidoglycans, proteins, and nucleic acids (Table 1). Regarding their cellular location, these receptors are either found at cell surface membrane or within intracellular compartments. A growing body of data suggests that TLRs involved in sensing bacterial chemical structures (TLR1, TLR2, TLR4 and TLR5) are located on the cell surface, while nucleic acid-recognizing TLRs (TLR3, TLR7, TLR8 and TLR9) are uniquely positioned intracellularly (McGettrick & O'Neill, ; Barton & Kagan, 2009).

2.1.1.1 Signaling through TLRs

Recognition of microbial components by TLRs leads to the activation of an intricate network of intracellular signaling pathways that ultimately result in the induction of molecules crucial to the resolution of infection such, proinflammatory cytokines, type I interferon (IFN), chemokines, and co-stimulatory molecules (Takeuchi & Akira, 2010 ; Kumar et al., 2010). These signaling cascades originate from cytoplasmic TIR domains and are meditated via the recruitment of different TIR domain-containing adaptor molecules, such as myeloid differentiation primary response gene 88 (MyD88), TIR-containing adaptor protein/ MyD88-adaptor-like (TIRAP/MAL), TIR-containing adaptor inducing interferon-β (IFN-β)/TIR-domain-containing adaptor molecule 1 (TRIF/TICAM1) and TIR-domain-containing adaptor molecule/TRIF-related adaptor molecule 2 (TRAM/TICAM2) (Fitzgerald et al.,

2001; Horng et al., 2001; Yamamoto et al., 2002; Takeda & Akira, 2004; Yamamoto et al., 2004).

In the signaling pathways downstream of the TIR domain, the TIR domain-containing adaptor MyD88 assumes a crucial role. With exception for TLR3, all TLRs recruit MyD88 and initiate MyD88-dependent signaling cascades to activate NF-κB and MAP kinases. MyD88 is used as the sole adapter in TLR5, TLR7 and TLR9 signaling, while TLR1, TLR2, and TLR6, additionally recruit the adaptor TIRAP. TLR4 uses the four adaptors, including MyD88, TIRAP, TRIF and TRAM (Yamamoto et al., 2002; Yamamoto et al., 2003)

In a general point of view, TLR signaling could be divided into two major pathways: MyD88-dependent and TRIF-dependent pathways.

MyD88-dependent pathway

Following stimulation, MyD88 recruits IL-1 receptor-associated kinase proteins (IRAK) to TLRs, resulting in IRAK phosphorylation and subsequent association and activation of tumor necrosis factor receptor (TNFR)-associated factor 6 (TRAF6) (Swantek et al., 2000; Suzuki et al., 2002). The IRAK-1/TRAF6 complex dissociates from the TLR receptor and associates with TGF-β-activated kinase 1 (TAK1) and TAK1 binding proteins, TAB1 and TAB2. From this new formed complex, IRAK-1 is degraded, whereas the remaining complex of TRAF6, TAK1, TAB1, and TAB2 is transported across the cytosol where it forms large complexes with E2 protein ligases such as the Ubc13 and Uev1A. As result, TRAF6 is polyubiquitinated and thereby induces TAK1 activation (Deng et al., 2000) which, in turn, activates the IκB kinases complex (IKK). The active IKK complex promotes the phosphorylation and subsequent ubiquitination of the NF-κB inhibitory protein IκB-α, leading to its proteosomal degradation. This allows the NF-κB subunits to be translocated to the nucleus, where they initiate the transcription of genes involved in inflammatory response (Wang et al., 2001). Additionally to NF-κB activation, MyD88-dependent signaling cascade also culminates into the activation of the three MAPK pathways (extracellular signal–regulated kinase (ERK), Jun N-terminal kinase (JNK) and p38), regulating both, the transcription of inflammatory genes and the mRNA stability of those transcripts (Figure 2).

TRIF-dependent pathway

Besides this MyD88-dependent pathway, NF-κB could also be activated follow TLR3 and TLR4 engagement in a TRIF-dependent manner. In TLR3 signaling, TRIF interacts directly with the TIR domain of the receptor, whereas for TLR4 another TIR domain containing adaptor, TRAM/TICAM-2, acts as a bridging between TLR4 and TRIF (Oshiumi et al., 2003; Oshiumi et al., 2003). In this pathway, TRIF recruits TRAF-6 and RIP1, molecules that cooperate in TAK1 activation, and lead to robust NF-κB activation.

TRIF-dependent signaling cascade also assumes a crucial role in the expression of type I IFN and IFN-inducible genes (ISGs). These genes are mainly potent antiviral molecules and their expression, follow TLR3 sensing of viral double stranded RNA, is of critical importance for the control of viral infections (review by Taniguchi et al., 2001). In this pathway, TRIF associates with TBK1 and IKKi, which in turn phosphorylate IRF3 and IRF7, leading to their nuclear translocation and induction of type I IFN genes and co-stimulatory molecules (Figure 2).

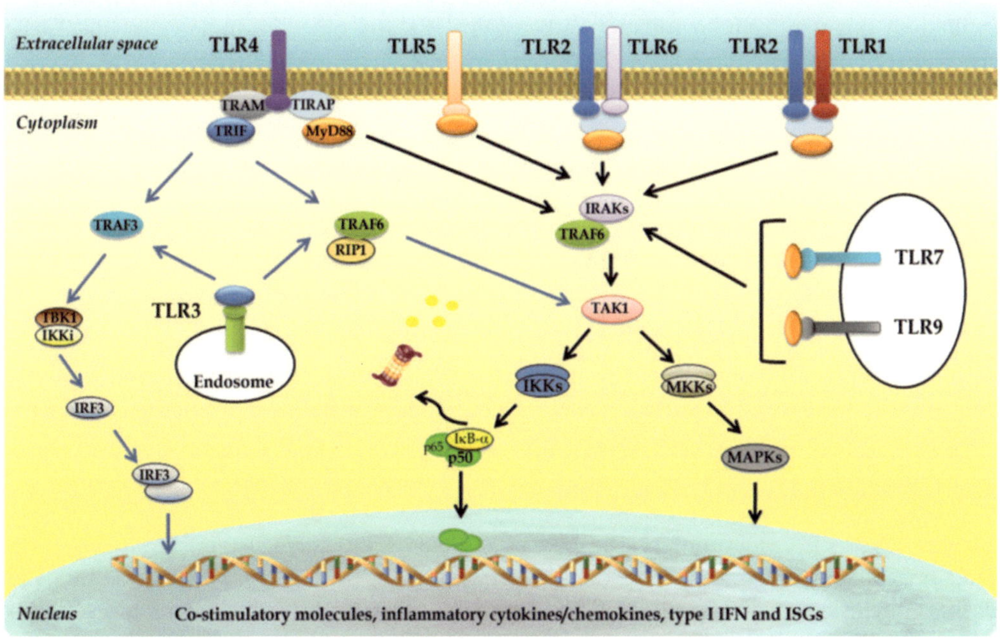

Fig. 2. Schematic representation of TLRs-mediated signaling. TLR signaling pathways were triggered by recognition of PAMPs by plasma membrane-localized TLRs, such as TLR4, TLR5, and TLR2 (TLR1 and TLR6 form heterodimers with TLR2 becoming functional receptor complexes) and endosomal-localized TLRs, such as TLR3, TLR7, and TLR9. Depending on the adaptor molecules involved, two major pathways could be established: the MyD88-dependent pathway (black arrows) and the TRIF-dependent pathway (blue arrows). MyD88-dependent signaling is initiated through the recruitment and activation of IRAK that associates and activates TRAF6. The IRAK-1/TRAF6 complex subsequently activates the TAK1 kinase, which in turn activates the IKK complex. The active IKK complex activates NF-κB subunits leading to their translocation to nucleus where they initiate the transcription of inflammatory cytokines/chemokines genes. In the TRIF-dependent signaling pathway, TRIF recruits TRAF-6 and RIP1, molecules that cooperate in TAK1 activation, leading to NF-κB activation. Besides, TRIF also recruits TBK1 and IKKi, leading to phosphorylation and nuclear translocation of IRF3 and IRF7, which results in transcription of type I IFN genes and co-stimulatory molecules

2.1.2 C-Type lectin receptors

C-type lectin receptors are a large superfamily of proteins characterized by the presence of one or more C-type lectin-like domains (CTLDs) that were originally described as Ca^{2+}-dependent, carbohydrate binding proteins (Weis et al., 1998). Over the past decade more than 60 CLRs have been identified in human immune cells (van Vliet et al., 2008). In recent years, some of these CLRs have emerged as PRRs with important roles in the induction of immune responses against numerous pathogens. Although the TLRs have a well defined role in alerting innate immune cells to the presence of pathogens, CLRs are mainly involved in the recognition and subsequent endocytosis or phagocytosis of microorganisms. These receptors have also crucial functions in recognizing glycan structures expressed by the host, facilitating this way cellular interaction between DCs and other immune cells, like T-cells and neutrophils (Geijtenbeek et al., 2000; van Gisbergen et al., 2005; Bogoevska et al., 2006).

Based on their structural features, C-type lectin receptors are sorted into two major groups: type I and type II receptors. Type I receptors are transmembrane proteins with multiple carbohydrate recognition domains (CRDs), being members of this group the mannose receptor (MR), DEC-205 (CD205), and Endo 180 (CD280), among others. Type II receptors are also transmembrane proteins, but in contrast, they have just a single CRD. DC-specific intercellular adhesion molecule (ICAM)-3 grabbing nonintegrin (DC-SIGN), Langerin, DC-associated C-type lectin-1 (Dectin 1), Dectin 2, DC-immunoreceptor (DCIR) and macrophage-inducible C-type lectin (Mincle) are examples of type II CLRs.

Originally, CLRs were thought to be predominantly involved in antifungal immunity, but are currently recognized to participate in immune responses induced by a wide spectrum of other pathogens, including bacteria, viruses and nematodes (Table 2).

2.1.2.1- Signaling through C-Type lectin receptors

Besides its roles in recognition and uptake of antigens, CLRs have also important signaling functions, shaping the immune responses to innumerous pathogens. Whereas some CLRs possess intrinsic signaling properties and are thus capable of directly activate transcription factors leading to cytokines expression, others predominantly act as modulators of responses to other PRRs, such as TLRs. This crosstalk between groups of PRRs is actually seen as a crucial event by which immune responses are balanced through collaborative induction of positive or negative feedback mechanisms. While TLRs engagement triggers

CLR Group	CLR	Microbial components	Pathogens
Type I	Mannose receptor (CD206)	High-mannose oligosaccharides, Fucose, Sulphated sugars and N-Acetylgalactosamine	*M. tuberculosis* *M. kansasii* *Francisella tularensis,* *Klebsiella pneumoniae,* HIV-1 and Dengue vírus *Candida albicans* *Cryptococcus neoformans* *Pneumocystis carinii* *Leishmania spp.*
	DEC205 (CD205)	ND	ND
Type II	DC-SIGN (CD209)	High-mannose oligosaccharides and fucose	*M. tuberculosis,* *M.leprae* BCG, *Lactobacilli* spp. *Helycobacter pylori* HIV-1 and Dengue vírus *Schistosoma mansoni*

			Leishmania spp. *Candida albicans* *Ixodes scapularis* Salp15 protein
	Langerin (CD207)	High-mannose oligosaccharides, Fucose and N-Acetylgalactosamine	HIV-1 *M.leprae*
	CLEC5A	ND	Dengue virus
	MGL (CD301)	Terminal N-Acetylgalactosamine	*Schistosoma mansoni* Filoviruses
	Dectin 1 (CLEC7A)	β-1,3 glucans	*Pneumocystis carinii* *Candida albicans* *M. tuberculosis* *Aspergillus fumigatos* *Histoplasma capsulatum*
	CLEC2 (CLEC1B)	ND	HIV-1
	MICL (CLEC12A)	ND	ND
	CLEC12B	ND	ND
	DNGR1 (CLEC9A)	ND	ND
	Dectin 2 (CLEC6A)	High-mannose oligosaccharides	*Aspergillus fumigatos* *M. tuberculosis* *Candida albicans* *Trichophyton rubrum* *Paracoccoides brasiliensis* Soluble components of *Schistosoma mansoni* eggs
	Mincle (CLEC4E)	α-mannose Trehalose-6,6-dimycolate	*Malassezia spp* Mycobacteria.
	BDCA2 (CD303)	ND	ND
	DCIR (CLEC4A)	ND	HIV-1

Table 2. Major C-type lectin receptors involved in pathogen recognition. BCG: Bacillus Calmette-Guérin; HIV-1: Human immunodeficiency virus type 1; ND: not determined.

intracellular signaling cascades that result in macrophage activation, DC maturation and ultimately T cell activation, binding of ligands to CLRs normally results in tolerogenic signals. Therefore the cross talk between TLRs and CLRs may fine-tune the balance between immune activation and tolerance. In terms of immunity this represents a paradox: if crucial to maintain tolerance to self-antigens, CLRs could be used by pathogens to escape immune system. Several pathogens exploit this "security breach", taking part of their capacity to activate C-type lectin receptors to promote an unresponsive state against their antigens recognized by other PPRs and increasing, this way, their chances of survival in host.

2.1.2.1.1 Mannose receptor

The mannose receptor is a type I transmembrane protein expressed on the surface of macrophages and immature dendritic cells. This receptor is primarily involved in recognition, phagocytosis and processing of glycans structures containing mannose, fucose and N-acetylglucosamine, molecules commonly found on the cell walls of pathogenic micro-organisms, such as mycobacteria, fungus, parasites and yeast (East & Isacke, 2002).

While MR has been shown to be involved in the expression of several pro and anti-inflammatory cytokines, the lack of an intracellular signaling motif on its cytoplasmic tail indicates that it requires an interaction with other PPRs in order to trigger any signaling cascade (Gazi & Martinez-Pomares, 2009). In fact, it was recently showed an intriguing interplay between the mannose receptor and another main CLR, Dectin-1. The recognition of fungi species, like *Candida albicans, Aspergillus fumigates* and *Pneumocystis carinii* by Dectin-1 enhances MR shedding in a serine/threonine protein kinase Raf-1 and phosphatidylinositol 3-kinase (PI3K)-dependent pathways. As these cleaved MR-cysteine-rich domains are capable of binding fungi particles and are recognized by tissues lacking mannose receptors this could represent a system delivery of MR-ligands to organs that do not possess MR receptors.

2.1.2.1.2 DC-SIGN

DC-SIGN is one of the most extensively studied type II CLRs. This receptor is primarily expressed in myeloid DCs being involved in numerous functions, like egress of DC-precursors from blood to tissues, DC-T-cell interactions and antigen recognition (Geijtenbeek et al., 2000; van Kooyk & Geijtenbeek, 2003). The receptor is involved in recognition of carbohydrate antigens of viruses, bacteria and protozoa, modulating the TLR signaling triggered by these pathogens.

Binding of several pathogens, including *M. tuberculosis, C. albicans* and HIV-1, to DC-SIGN triggers three routes that converge to activate Raf-1: the activation of the small GTPase Ras protein leads to its association with Raf-1 allowing Raf-1 phosphorylation at residues Ser338, Tyr340 and Tyr341, by p21-activated kinases (PAKs) and Src kinases, respectively. Raf-1 activation leads in turn to the modulation of TLR-induced NF-κB activation. After TLR-induced nuclear translocation of NF-κB, activated Raf-1 mediates the phosphorylation of NF-κB subunit p65 at the Ser276, which in turn leads to p65 acetylation. Acetylated p65 prolongs and increases the transcription of *IL-10* gene resulting in an augmented production of the immunosuppressive cytokine IL-10 (Gringhuis et al., 2007) (Figure 3a).

Recently, a different mechanism of TLR modulation by DC-SIGN was described after the interaction of the receptor with Salp15, an immunosuppressive protein of tick saliva (Hovius et al., 2008). Binding of Salp15, from the tick *Ixodes scapularis,* to DC-SIGN activates RAF1 that, with another not yet defined receptor, leads to MAPK/ERK kinase (MEK) activation. MEK-dependent signaling attenuates, in turn, the TLR-induced proinflammatory cytokine production at two distinct levels: enhancing the decay of *Il6* and *Tnf* (tumor necrosis factor) mRNA and decreasing nucleosome remodeling at the *IL-12p35* promoter, resulting in impaired IL-12p70 cytokine production (Figure 3b).

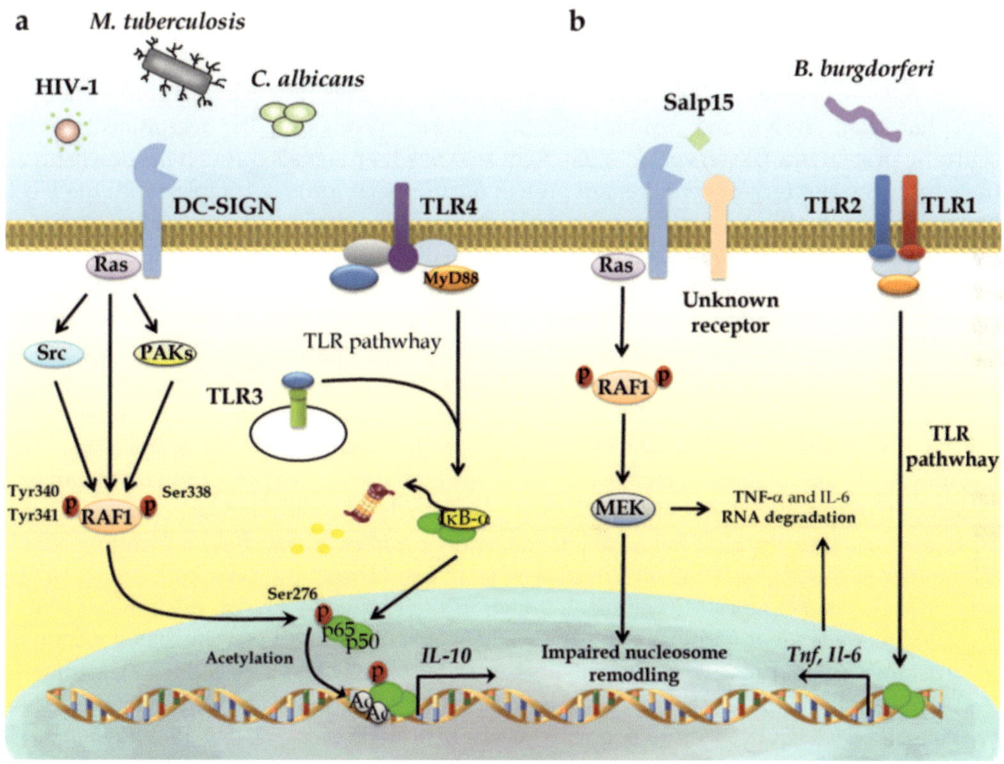

Fig. 3. Signaling through DC-SIGN. a) Carbohydrate antigens of HIV-1, *Mycobacterium tuberculosis* and *Candida albicans* are recognized by DC-SIGN, leading to activation of the small GTPase Ras proteins which associate with the serine/threonine protein kinase RAF1. RAF1 is then phosphorylated at residues Ser338, and Tyr340 and Tyr341 by PAKs and Src kinases, respectively. RAF1 activation leads to modulation of TLR-induced NF-κB activation by inducing the phosphorylation of p65 at Ser276 and its subsequent acetylation (Ac). Acetylated p65 exhibits enhanced transcriptional activity, particularly for *Il-10* gene, thereby increasing the production of IL-10. b) Binding of the salivary protein Salp15 from the tick *Ixodes scapularis* to DC-SIGN activates RAF1, and by a yet unknown co-receptor, changes downstream effectors of RAF1, leading to MEK activation. MEK-dependent signaling modulates *B. burgdorferi*-induced TLR1–TLR2-dependent pro-inflammatory cytokine production by enhancing the decay of *Il-6* and *Tnf* mRNA.

2.1.2.1.3 Dectin 1

In humans, Dectin-1 is mainly expressed in myeloid cells, such as macrophages, neutrophils and dendritic cells (Taylor et al., 2002), although it was also been found in other cell types, like B-cells, eosinophiles and mast cells (Ahren et al., 2003; Olynych et al., 2006). Unlike many other CLRs, Dectin-1 recognizes β-glucans in a Ca^{2+}-independent fashion (Brown & Gordon, 2001) and it lacks the conserved residues within its CRD that are typically necessary for binding carbohydrate ligands (Weis et al., 1998). The receptor contains a single CRD in the extracellular region and an immunoreceptor tyrosine-based activation (ITAM)-like motif within its intracellular tail.

It was the first non-TLR PPR shown to possess intrinsic signaling properties, being able to signal through both, spleen tyrosine kinase (Syk)-dependent and Syk-independent pathways (Brown, 2006) (Figure 4).

In the Syk-dependent pathway, and upon binding to Dectin-1, the ITAM-like motif is phosphorylated in tyrosine residues via Src kinases, promoting the recruitment of the signaling protein Syk (Rogers et al., 2005). Activated Syk then signals through the downstream transducer caspase recruitment domain protein (Card)9, that forms a complex with the B cell lymphoma 10 (Bcl10) and the mucosa associated lymphoid tissue translocation protein 1 (Malt1) (Gross et al., 2006). This activated CARD9–BCL10–MALT1 (CBM) complex controls NF-κB activation and subsequent expression of cytokines/chemokines, like TNF-α, IL-1β, IL-10, IL-6, IL-23, CCL2 and CCL3 (LeibundGut-Landmann et al., 2007). Dendritic cells, trough this Dectin-1-Syk-Card9 axis and by orchestration of the cytokines IL-1β, IL-6 and IL-23, promote the differentiation of Th17 helper cells, establishing this way a crucial host response against extracellular bacteria and fungi (Osorio et al., 2008). Moreover, there are evidences of a collaborative Dectin-1/TLR2 pathway for the induction of a specific *Candida albicans*-Th17 response, by the induction of prostaglandin E2, which in turn up-regulates the Th17 polarizing cytokines IL-6 and IL-23 (Smeekens et al., 2010). Besides the canonical NF-kB activation, Detin-1 can also activate, through Syk, the NIK-dependent non-canonical RelB subunit of NF-kB (Gringhuis et al., 2009). Another Syk downstream signal recently described, points to the activation of phospholipase C gamma-2, which in turn activates several calcium-dependent and MAPKs-dependent pathways (Xu et al., 2009). One of these calcium-mediated responses involves the calcineurin activation of the nuclear factor of activated T-cells (NFAT), leading to the expression of the cytokines IL-2 and IL-10 and of inflammatory mediators, like cyclooxygenase-2 (COX-2) (Suram et al., 2006; Goodridge et al., 2007). Recently, another calcium-dependent pathway downstream of Dectin-1 and Syk was described. In this pathway, activated calmodulin-dependent kinase II and Pyk2 promote the activation of the ERK–MAPK pathway and CREB, resulting in the generation of an oxidative burst and in the production of IL-2 and IL-10 (Slack et al., 2007; Kelly et al., 2010). The generated reactive oxygen species act through NLRP3 inflammasome and are essential to IL-1β production in response to fungal infections (Gross et al., 2009; Kumar et al., 2009; Said-Sadier et al., 2010).

The Syk-independent pathway downstream Dectin-1 is not fully characterized; however, recent findings suggest that Dectin 1 activation leads to the phosphorylation and activation of RAF1 by Ras proteins, which promotes the phosphorylation of p65, at Ser276 residue, facilitating its acetylation by the histone acetyltransferases CREB-binding protein. Similarly to that described for DC-SIGN, acetylated p65 prolongs and increases the transcription of *IL-10* gene.

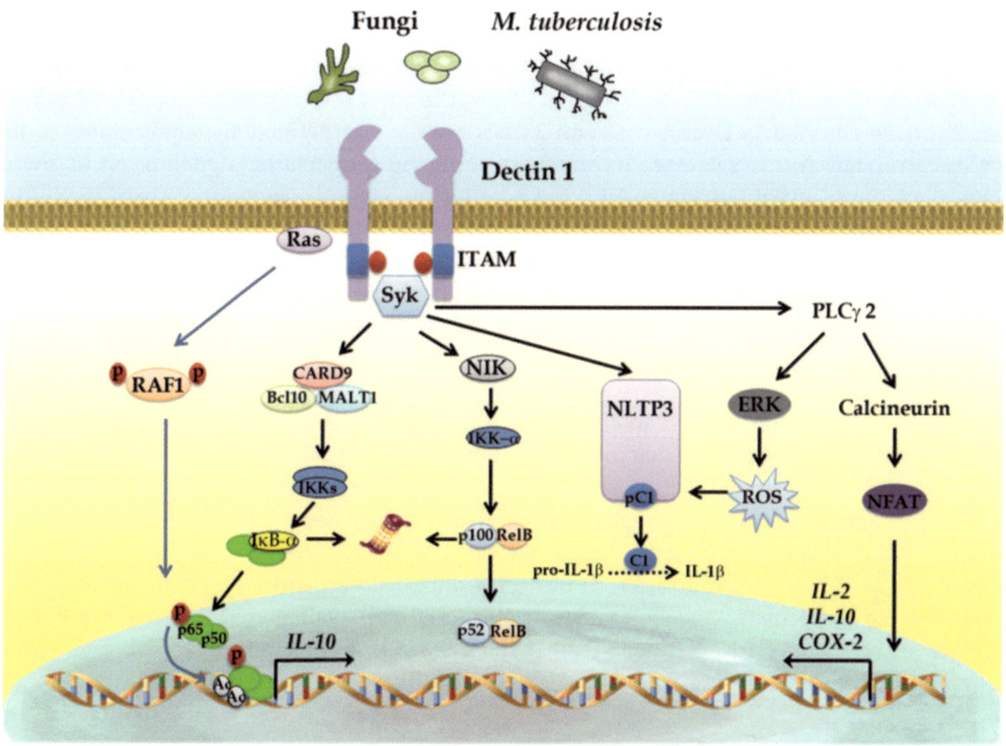

Fig. 4. Signaling through Dectin 1. Recognition of microorganisms by Dectin 1 leads to signal through both, Syk-dependent (black arrows) and Syk-independent pathways (blue arrows). In the Syk-dependent pathway, binding of glucans to Dectin-1 causes the phosphorylation of ITAM-like motifs in its tyrosine residues. Syk is then recruited to the two phosphorylated receptors, leading to the formation of a complex involving CARD9, BCL-10 and MALT1. This activated complex controls NF-κB activation and subsequent expression of cytokines/chemokines, like TNF-α, IL-1β, IL-10, and IL-6. Activation of Syk also leads to the activation of the non-canonical NF-κB pathway, a process mediated by NIK and IKK, and in which RelB-p52 dimers were translocated to nucleus. Another Syk downstream signal leads to activation of PLCγ2, which in turn activates MAPKs-dependent and calcineurin-dependent pathways. Activation of calcineurin promotes the activation of NFAT, leading to the expression of the cytokines IL-2 and IL-10 and COX-2. In turn, activation of ERK, results in the generation of an oxidative burst that acting through the NLRP3 inflammasome, is essential to IL-1β production. In the Syk-independent pathway, Dectin 1 activation leads to the phosphorylation and activation of RAF1 by Ras proteins, leading in turn to the phosphorylation and acetylation of p65. Binding of acetylated p65 to the *Il-10* enhancer, increases the transcription of the gene. C1: caspase 1; pC1: pro-caspase.

2.1.2.1.4 Dectin 2

Dectin 2 was originally found in DCs (Ariizumi et al., 2000), although it is also expressed in tissue macrophages, inflammatory monocytes, B cells, and neutrophils (Fernandes et al., 1999). The receptor has been shown to be involved in recognition of mannan-like or mannan-containing glycoproteins, glycolipids or oligomannosides present in fungi hyphae, being critical for the establishment of Th17 antifungal responses (Sato et al., 2006; Robinson et al., 2009). Furthermore, murine Dectin-2 was also associated with helminth infections by recognition of soluble components derived from the eggs of *Schistosoma mansoni* (Ritter et al., 2010). In contrast to Dectin-1, Dectin-2 does not contain defined signaling motifs in its cytoplasmic tail and is therefore incapable of inducing intracellular signaling on its own. However, the receptor associates with the adaptor molecule Fc receptor γ chain (FcRγ) to transduce intracellular signals, through a Dectin 2-FcRγ-Syk-dependent pathway. FcRγ chain contains an ITAM motif that is dually phosphorylated by Src kinases, promoting the recruitment and activation of Syk. Syk activates, in turn, the NF-κB and MAPKs pathways in a CARD9-dependent or independent fashion, respectively (Saijo et al., 2010) (Figure 5a).

2.1.2.1.5 Mincle

Mincle is a type II transmembrane protein with a highly conserved C-type lectin domain, predominantly expressed in macrophages. It has been implicated in the recognition of *Saccharomyces cerevisiae, C. albicans* and mycobacteria, and was shown to be responsible for specific recognition of α-mannose residues in *Malassezia* species (Bugarcic et al., 2008; Wells et al., 2008; Ishikawa et al., 2009; Yamasaki et al., 2009). Similarly to Dectin-2, it lacks a signaling motif but couples to FcRγ to transduce intracellular signals. Ligation to Mincle of trehalose-6,6-dimycolate, an abundant mycobacterial cell wall glycolipid, was shown to trigger a FcRγ-Syk-CARD9 dependent pathway, leading to protective Th1 and Th17 immune responses (Werninghaus et al., 2009) (Figure 5b).

2.1.2.1.6 BDCA2

BDCA2 is a type II C-type lectin receptor primarily expressed in human plasmacytoid dendritic cells (Dzionek et al., 2001). As endogenous or microbial ligands for BDCA2 have not yet been identified, it is difficult to understand the pathophysiological implications of this CLR. However, it has been shown, by treatment with anti-BDCA-2 monoclonal antibodies, that the receptor crosstalk with other PPARs, namely TLR-9, decreasing the induced IFN-I expression (Jahn et al., 2010). As for Dectin-2 and Mincle, BDCA2 signals through the ITAM motifs of the FcRγ chain. Activation of BDCA2 results in phosphorylation of ITAM motifs of FcRγ, followed by the recruitment and activation of Syk. Activated Syk leads to the formation of a complex, consisting of B cell linker (BLNK), Bruton's tyrosine kinase (BTK) and phospholipase C2 (PLC2), which induces calcium mobilization. This calcium increase appears to be involved in the inhibition of MYD88 adapter recruitment to TLR9 and, thereby, in the reduction of the induced expression of IFN-I, TNF-α and IL-6 (Figure 5c).

2.1.2.1.7 CLEC5A

CLEC5A, also known as Myeloid DNAX activation protein 12 (DAP12)-associating lectin-1 (MDL-1), is a type II C-type lectin receptor expressed in cells of myeloid origin, like monocytes and macrophages, and in human CD66-positive neutrophils (Aoki et al., 2009). Contrarily to other CLRs predominantly involved in fungal and micobacterial recognition, CLEC5A was the first CLRs directly linked to viral recognition. It has been shown that this receptor plays a crucial role in the pathophysiology of dengue virus infection, being directly involved in the production of proinflammatory cytokines by infected macrophages (Chen et

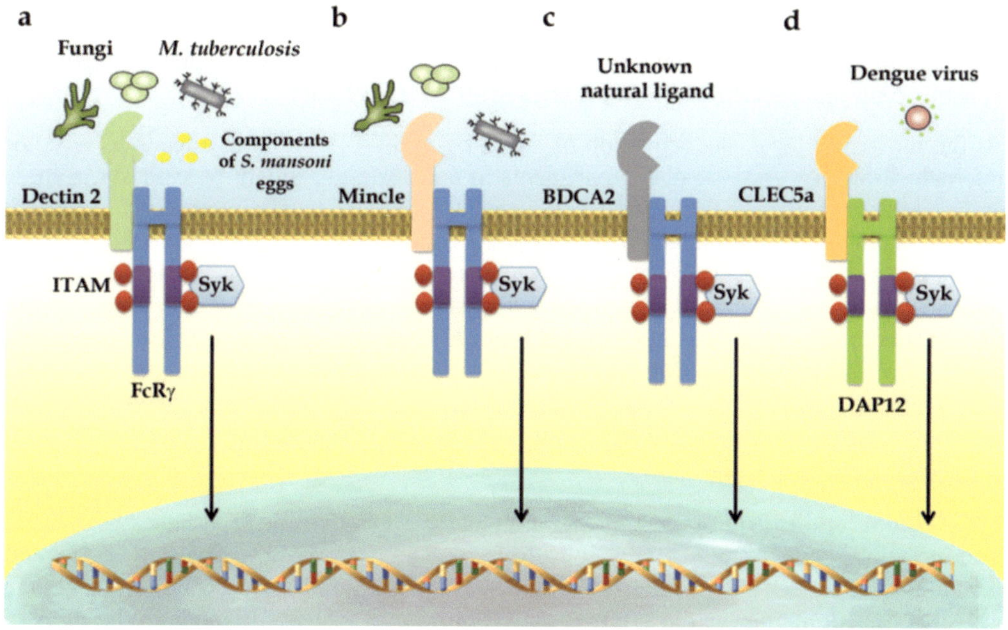

Fig. 5. Signaling through ITAM-coupled C-type lectin receptors. Dectin-2, Mincle and BDCA2 do not contain defined signaling motifs in their cytoplasmic tail being incapable of inducing intracellular signaling on their own. Following ligand binding, these receptors associate with FcRγ leading to recruitment of Syk and subsequent activation of downstream signaling cascades (black arrows). CLEC5a also lacks a citoplasmatic catalytic domain. Recognition of Dengue virions by CLEC5a, results in the association and phosphorylation of DAP12, leading to recruitment of Syk and activation of Syk-dependent downstream signaling.

al., 2008; Watson et al., 2011). CLEC5A has a very short cytoplasmic region lacking a defined signaling motif, yet it transduces intracellular signals trough non-covalent association with the ITAM-bearing adapters DAP10 and DAP12 (Bakker et al., 1999; Inui et al., 2009). DAP10 ITAM motif contains a cytoplasmic sequence that facilitates PI3K recruitment and activation, being possible that it cooperates with DAP12-associated receptors to mediate co-stimulatory signals (Kerrigan & Brown, 2010). Moreover, it was showed that the interaction of dengue virus with CLEC5A causes the phosphorylation of the coupled DAP12 ITAM motif (Chen et al., 2008). Although not formally demonstrated, this molecular event may result in Syk recruitment and activation, followed by downstream signaling that leads to the observed induction of proinflammatory cytokines (Figure 5d)

2.1.2.1.8 DCIR

DCIR was found to be expressed at high levels in blood monocytes, myeloid and plasmacytoid DCs, macrophages and in a less extent in B cells (Bates et al., 1999). Although no endogenous or exogenous specific ligands were yet identified, the receptor was recently shown to play an important role in HIV-1 infection by acting as an attachment factor for the virus (Lambert et al., 2008). DCIR and DC-associated C-type lectin-2 (DCAL-2) are, among the presently identified human CLRs, the only ones containing intracellular immune receptor tyrosine-based inhibition motifs (ITIMs). These ITIMs motifs are responsible, in a phosphatase dependent fashion, for the negative signals that result in repressed activation of neutrophils and dendritic cells (Kanazawa et al., 2002; Richard et al., 2006).

At the molecular level, the activation of DCIR by anti-DCIR antibodies leads to receptor internalization into endosomal compartments in a clathrin-dependent process. As in these endosomal structures are also located TLR8 and TLR9, it is likely that internalized DCIR will interact with them, modulating their signaling. Supporting this hypothesis, recent data shows that the phosphorylation of ITIM promotes the recruitment of the phosphatases SH2-domain-containing protein tyrosine phosphatase 1 (SHP1) or SHP2, which, by an unidentified mechanism, leads to the downregulation of TLR8-induced IL-12 and TNF production in myeloid DCs (Meyer-Wentrup et al., 2009), and to the down-modulation of TLR9-induced IFN and TNF production in plasmacytoid DCs (Meyer-Wentrup et al., 2008).

2.1.3 RIG-I-Like receptors

RIG-I-like receptors (RLRs) constitute a family of three cytoplasmic RNA helicases: retinoic acid-inducible gene I (RIG-I), melanoma differentiation-associated gene 5 (MDA5) and laboratory of genetics and physiology 2 (LGP2). These receptors share a common functional RNA helicase domain near the C terminus (HELICc) that specifically binds to the RNA of viral origin and are, therefore, crucial for antiviral host responses (Yoneyama et al., 2004; Wilkins & Gale, 2010). These responses result from the action of induced inflammatory cytokines and type I interferons over the cells of the innate and adaptive immune system. Inflammatory cytokines primarily promote the recruitment of macrophages and dendritic cells, while type I interferons inhibit viral replication, promote the apoptosis of infected cells and increase the lytic capacity of natural killer cells (Takahasi et al., 2008).

RIG-I is involved in the recognition of a wide variety of RNA viruses belonging to the paramyxovirus and rhabdovirus families, as well as Japanese encephalitis virus, while MDA5 specifically detect, Picornaviruses, such as encephalomyocarditis virus, mengovirus and poliovirus. Somme virus such as dengue virus and West Nile Virus, require, however, the activation of both RIG-I and MDA5 to generate a robust innate immune responses.

Despite structural similarity, RIG-I and MDA5 have been shown to bind distinct types of viral RNAs (Kato et al., 2006). MDA5 preferentially binds long dsRNAs, whilst RIG-I has high affinity for 5'-triphosphate ssRNAs and short dsRNAs without a 5'-triphosphate end (Pichlmair et al., 2006; Kato et al., 2008; Lu et al., 2010). The RIG-I distinction of self from viral ssRNAs is ensured by the predominantly nuclear localization of cellular 5'-triphosphate ssRNAs that even if present in the cytoplasm are normally capped or processed. Recently, the notion that 5'-triphosphate ssRNAs were sufficient to bind to and activate RIG-I was challenged by data obtained with synthetic single-stranded 5′-triphosphate oligoribonucleotides (Schlee et al., 2009). In these experiments, the synthetic 5'-triphosphate ssRNAs were unable to activate RIG-I and only the addition of the synthetic complementary strand resulted in optimal binding and activation of the receptor. The authors hypothesized that this newly data explains how RIG-I detects negative-strand RNA viruses lacking long dsRNA but containing blunt short double strand 5′-triphosphate RNA in the panhandle region of their single-stranded genome.

2.1.3.1 Signaling through RIG-I-Like receptors

RIG-I and MDA5 contain a DExD/H-box helicase domain that recognizes the viral RNA, inducing conformational changes and exposing the caspase-recruitment domains (CARDs) responsible for downstream signaling of these cytoplasmic sensors. CARDs interact with a CARD-containing adaptor, IFN-β promoter stimulator-1 (IPS-1), located in the outer mitochondrial membrane and on peroxisomes (Kawai et al., 2005; Dixit et al., 2010). While peroxisomal IPS-1 induces early expression of interferon-stimulating genes (ISGs) via transcription factor IRF1, mitochondrial IPS-1 induces delayed responses via IRF3/IRF7-controled expression of ISGs and type I interferons. Therefore, signaling through mitochondrial and peroxisomal IPS-1 is essential to an effective antiviral response. From the interaction of IPS-1 with RIG-I and MDA5 CARDs also results the activation of NF-κB, a process that involves the recruitment of TRADD, FADD, caspase-8, and caspase-10 and leads to the induction of proinflammatory cytokines (Takahashi et al., 2006) (Figure 6a). The third member of this cytoplasmic PRRs family, LGP2, similarly to RIG-I and MDA5, possesses a DExD/H-box helicase domain but is devoid of a CARD domain (Yoneyama et al., 2005) and was therefore considered as a negative regulator of RIG-I- and MDA5-mediated signaling (Rothenfusser et al., 2005; Komuro & Horvath, 2006; Saito et al., 2007). Recent *in vivo* experiment showed, however, precisely the opposite, suggesting that LPG2 could contribute to a robust antiviral response, acting as a facilitator of the interaction between viral RNA, RIG-I and MDA5 (Satoh et al., 2010).

2.1.4 NOD-like receptors

Nucleotide-oligomerization domain (NOD)-like receptors (NLRs) are cytosolic sensors of microbial components highly conserved trough evolution. A great number of homologs of these receptors have been described in animals and plants, attesting their importance as ancestral host defense mechanisms. In humans, 23 members of the NLR family were identified, being primarily expressed in immune cells, such lymphocytes, macrophages and dendritic cells, although also found in epithelial and mesothelial cells (Franchi et al., 2009). NLRs contain tree characteristic domains: a) a C-terminal leucine-rich repeat (LRR) domain, responsible for ligand sensing and autoregulation, b) a central nucleotide-binding oligomerization (NOD) domain, required for nucleotide binding and self-oligomerization upon activation and c) a N-terminal effector domain responsible for downstream signal propagation. To date, four different N-terminal domains have been identified: acidic transactivation domain, caspase-recruitment domain (CARD), pyrin domain (PYD), and baculoviral inhibitory repeat (BIR)-like domain (Chen et al., 2009). NOD1 and NOD2, the most studied NLRs, both sense bacterial molecules produced during peptidoglycan synthesis and remodeling. Peptidoglycan is a major component of the bacterial cell wall, formed by alternated residues of *N*-acetylglucosamine (GlcNAc) and *N*-acetylmuramic acid (MurNAc), which are crosslinked by short peptide chains. The bridging aminoacids inside these peptide chains are differentially found in gram-negative and gram-positive bacteria, being responsible for the differential recognition abilities of NOD1 and NOD2 (McDonald et al., 2005). Therefore, NOD2 senses muramyl dipeptide (MDP), which is found in the peptidoglycan of nearly all gram-positive and gram-negative bacteria, while NOD1 sense γ-D-glutamyl-meso-diaminopimelic acid (iE-DAP), an amino acid that is predominantly found in gram-negative bacteria and in some gram-positive bacteria, such as *Listeria monocytogenes* and *Bacillus* spp (Chamaillard et al., 2003).

2.1.4.1 Signaling through NOD-like receptors

The intracellular NLR proteins organize signaling platforms, such as NOD signalosomes and inflammasomes that trigger NF-κB and MAPKs pathways and control the activation of inflammatory caspases (Chen et al., 2009). Upon recognition of their respective ligands, both NOD1 and NOD2 self-oligomerize to recruit and activate the serine-threonine kinase RICK that becomes polyubiquitinated. RICK directly interacts with the regulatory subunit of IKK, the inhibitor of NF-κB kinase γ (IKKγ), promoting the activation of the catalytic subunits IKKα and IKKβ (Inohara et al., 2000). These activated subunits phosphorylate the inhibitor IκB-α, leading to its ubiquitination and subsequent degradation via the proteasome. The released NF-κB translocates to the nucleus, where it promotes the expression of proinflammatory cytokines and chemokines (Masumoto et al., 2006; Werts et al., 2007; Buchholz & Stephens, 2008). Additionally, RICK also promotes the K63-linked polyubiquitination of IKKγ, which facilitates the recruitment of transforming growth factor β-activated kinase (TAK1) (Hasegawa et al., 2008). TAK1 forms a complex with the ubiquitin binding proteins TAK1-binding protein 1 (Tab1), Tab2, and/or Tab3, promoting the phosphorylation of the IKKβ subunit of IKK, that in turn leads to the phosphorylation and degradation of IκB-α. Signaling through NOD1 and NOD2 also results in MAPK activation by a process not fully characterized, but dependent of TAK1 and RICK (Shim et al., 2005) (Figure 6b).

Another process by which NLRs participate in host response to microbial infections is through their involvement in inflammasome formation. Inflammasomes are large protein complexes that includes NLRs proteins, the adapter ASC (apoptosis-associated speck-like protein containing a C-terminal CARD) and pro-caspase-1. This molecular platform is crucial for caspase-1 activation and subsequent processing of pro-IL-1β and pro-IL-18, resulting in the secretion of their mature biologically active forms (Lamkanfi et al., 2007). NLR family members, such as NLRP1, NLRP3 and NLRP4, have shown to be critical factors in the activation of proinflammatory caspase-1 and IL-1β secretion in response to several microbial stimuli (Pedra et al., 2009) (Figure 6b).

NLR signaling: NOD1 senses iE-DAP, an amino acid predominantly found in gram-negative bacteria while NOD2 senses MDP, which is found in the peptidoglycan of nearly all gram-positive and gram negative bacteria. Following recognition of their respective ligands, both NOD1 and NOD2 self oligomerize to recruit and activate RICK, which in turn activates NF-κB via the IKK complex. Signaling through NOD1 and NOD2 also results in MAPK activation by a process not fully characterized but dependent of TAK1 and RICK. Another member of the NLR family constitutes the inflammasome, a multi-protein complex that includes NLRs proteins, the adapter ASC and pro-caspase-1(pC1). In this complex pro-caspase 1 is activated, promoting in turn the maturation of pro-IL-1β cytokine to its bioactive form.

3. Molecular mechanisms by which microorganisms subvert the innate immune system

Common features of pathogenic microorganisms are the exploitation of cytoskeleton and membranous structures to invade/or to gain motility inside the host cell, and also the manipulation of key signaling pathways. In this section, we will specially focus on the mechanisms by which pathogens manipulate signaling pathways in immune cells.

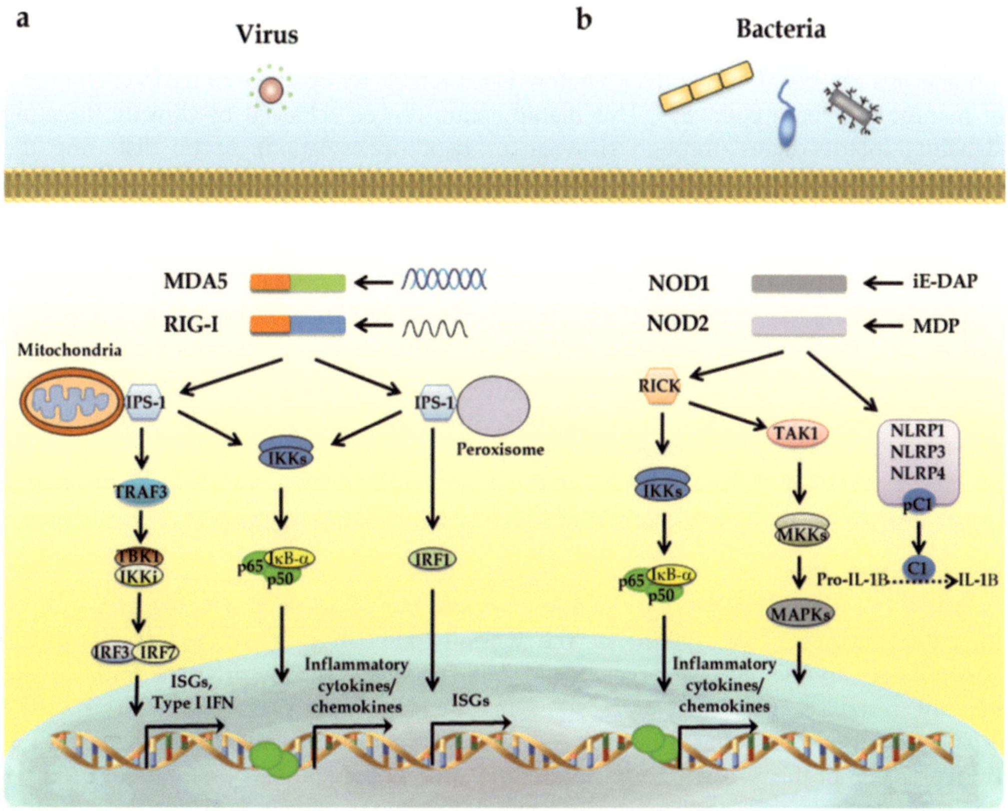

Fig. 6. Signaling through RLRs and NLRs. RLR signaling: RIG-I and MDA5 function as cytosolic sensors of viral RNA, recognizing preferentially 5'-triphosphate ssRNAs and long dsRNAs, respectively. Binding of viral RNAs to these receptors activates signaling through the adaptor protein IPS-1, located in the outer mitochondrial membrane or on in peroxisomes. Mitochondrial IPS-1 leads to activation of NF-κB and IRF3/IRF7 through the IKK complex and TBK1/IKKi, respectively, which results in the production of inflammatory cytokines, type I interferons and interferon-stimulating genes (ISGs). In turn, peroxisomal IPS-1 induces early expression of ISGs via transcription factor IRF1.

As stated in above sections, pattern-recognition receptors confer to mammals an extremely efficient "detection system" of invading microorganism, triggering an intricate signaling network that ultimately orchestrates the establishment of an adequate immune response. However, as part of their pathogenic strategies, several microorganisms evade immune system by circumventing, or distorting, these signaling pathways and creating, therefore, conditions that facilitate their replication and spreading in the host. In the last few years great efforts have been made to understand the molecular mechanisms behind this subversion, and various signaling cascades were identified as main targets of pathogens and virulence factors. When globally analyzed, cascade signals downstream PPRs activation mainly converge to two key signaling pathways: the transcription factor nuclear factor-κB (NF-κB) and the mitogen activated protein kinases (MAPKs). NF-κB is a cornerstone of innate immunity and inflammatory responses, controlling the expression of effector molecules, such as proinflammatory cytokines/chemokines, anti-apoptotic factors and defensins; MAPKs are also signaling cascades intimately connected to the regulation of innumerous aspects of immunity. Therefore is expectable that pathogens try to circumvent, or manipulate, these pathways. This manipulation can be achieved by directly targeting signaling intermediates (through cleavage or dephosphorylation), or by distorting the balance between immunogenic and tolerogenic signals. The later mechanism mostly results from the exploitation of signaling crosstalk between several receptors of innate immune system (Hajishengallis & Lambris, 2011). A classical example is the crosstalk between TLRs and DC-SIGN: while antigen recognition by TLRs triggers a deleterious response, recognition of another antigen of the same pathogen by DC-SIGN negatively modulates the TLR signal, promoting an unresponsive state. In the present chapter we do not intent to cover the general immune evasion strategies of pathogens, but rather focus on the mechanisms by which the microorganisms directly, or by "crosstalk manipulation", interfere with key immune signaling pathways.

3.1 Exploiting CLRs signaling and their crosstalk with other receptors

Mycobacterium tuberculosis, the causative agent of tuberculosis, has been a major world-wide threat for centuries. In 2009 the disease was responsible for 1.8 million deaths and a recent estimative suggests that a third of the world's population is infected (WHO 2010). Macrophages are the primary targets for *M. tuberculosis* and mainly drive the initial innate immune response against the pathogen, while dendritic cells play a central role in the establishment of a subsequent cellular response (Fenton & Vermeulen, 1996; Demangel & Britton, 2000). In this process, DCs capture and process the pathogen, migrate to draining lymph nodes and present the antigenic peptides to naïve T cell, initiating an adaptive immune response. *M. tuberculosis* was shown to modulate the functions of both macrophages and DCs (Balboa et al., 2010; Geijtenbeek et al., 2003), promoting immune conditions that allow a latent infection. Macrophages phagocyte bacteria into phagosomes, which then mature by acquiring low pH, degradative enzymes and reactive oxygen/nitrogen species. Phagosomes fuse with lysosomes to form phagolysosomes, exposing the engulfed microorganism to the lethal action of hydrolases, proteases, super oxide dismutase and lysozymes. However, *M. tuberculosis* escapes death by blocking the maturation of phagosomes and preventing their fusion with lysosomes (Fratti et al., 2003; Hmama et al., 2004). This process was shown to be partially mediated by the binding of mannosylated lipoarabinomannan (ManLAM) to the mannose receptor in macrophages (Kang et al., 2005). ManLAM is a major mannose-containing lipoglycan present in *M. tuberculosis* cell wall that downregulates calmodulin-dependent signal transduction and inhibits sphingosine kinase, preventing the conversion of macrophage sphingosine to sphingosine-1 phosphate (S-1P) (Malik et al., 2003). This arrests the S-1P-dependent increase in Ca^{2+} concentration, disrupting the PI-3K signaling and the subsequent recruitment of Rab5 effector early endosomal antigen 1 (EEA1) to phagosomes (Fratti et al., 2001). EEA1 is crucial for the delivery of lysosomal components from the trans-Golgi network to the phagosome and regulates fusion of phagosomes with lysosomal vesicles. Therefore the MR-mediated phagocytosis of ManLAM-containing *M. tuberculosis* prevents phagosomes to mature and to fuse with lisosomes, allowing bacteria to survive.

One of the most ingenious mechanisms used by microorganisms to escape host immune response is to subvert or disrupt the molecular signaling crosstalk between receptors of the innate immune system (Hajishengallis & Lambris, 2011).

This signaling hijacking frequently leads to an augmented production of immunosuppressive molecules, such as IL-10 and/or to a decreased expression of proinflammatory molecules, such IL-12 and IFN.

Both *M. tuberculosis* and *M. bovis* BCG are able to induce DC maturation through TLR2- and TLR4-mediated signaling (Henderson et al., 1997; Tsuji et al., 2000). However, the concomitant engagement of ManLAM to the C-type lectin receptor DC-SIGN modulates the TLR-induced NF-κB activation, blocking the expression of co-stimulatory molecules CD80, CD83 and CD86 and inducing the production of the immunosuppressive cytokine IL-10 (Geijtenbeek et al., 2003; Gringhuis et al., 2007). Immature mycobacteria-infected DCs and IL-10 dependent blockage of IL-12 production impair the generation of a protective Th1 response, contributing therefore to the establishment of a latent infection. Besides *M. tuberculosis,* other important human pathogens, such *M. leprae, Candida albicans,* measles virus and HIV-1, were shown to explore TLR-DC-SIGN crosstalk to induce the expression of the immunosuppressive cytokine IL-10 (Bergman et al., 2004; Gringhuis et al., 2007; Gringhuis et al., 2009). Similarly to *M. tuberculosis,* HIV-1 activates the Raf-1 pathway through DC-SIGN, modulating TLR signaling, and leading to IL-10 increased production, impairment of TLR-induced dendritic cell maturation and reduced T-cell proliferation. In a process independent of TLR activation, DC-SIGN interacts with HIV-1 envelope glycoprotein gp120, and regulates the gene expression profile of DCs (Hodges et al., 2007). Among the modulated genes, activating transcription factor 3 (ATF3) is of particular importance since it acts as a negative regulator of TLR4-induced expression of proinflammatory cytokines IL-6 and IL-12 (Gilchrist et al., 2006). This is therefore suggestive that DC-SIGN, besides modulating, also represses TLR4 signaling (den Dunnen et al., 2009). Additionally, HIV-1 also exploits the crosstalk between DCIR and TLR8/TLR9 to promote DC infection and to evade host immune response. Binding of the virus to DCIR was shown to down-modulate the production of TLR8-induced IL-12 and TLR9-induced IFN-α, in myeloid and in plasmacytoid DCs, respectively.

In contrast to the above referred mannose-containing pathogens (mycobacteria, *C. albicans* and HIV-1), *Helicobacter pylori* induces IL-10 production and Th1 inhibition, through a Raf-1 independent mechanism. In fact, binding of the fucose-containing LPS Lewis antigens from *Helicobacter pylori* to DC-SIGN actively dissociated the KSR1–CNK–Raf-1 complex from the DC-SIGN signalosome, modifying downstream signal transduction (Gringhuis et al., 2009). Recently, a new form of crosstalk between DC-SIGN and TLRs was described in dendritic cells (Hovius et al., 2008). In this process, *Borrelia burgdorferi* lipoproteins trigger TLR2 activation, while Salp15, a salivary protein from *Ixodes scapularis,* the human vector of *B.burgdorferi,* binds to DC-SIGN and leads to RAF1-mediated MEK activation. MEK-dependent signaling attenuates, in turn, the TLR2-induced proinflammatory cytokine production, by enhancing the decay of *Il6* and *Tnf* mRNA and decreasing IL-12p70 cytokine production. Additionally, this crosstalk synergistically enhances IL-10 production. This immunosuppression reveals to be advantageous for both, the vector and the bacteria, given that it impairs the establishment of an effective adaptive immune response against tick and/or *B. burgdorferi* antigens (Hovius et al., 2008).

In neutrophils, mycobacteria bind, to a yet unidentified C-type lectin receptor (potentially CLEC5A) and induce, via Syk, a crosstalk with the TLR2 adapter molecule MYD88. This results in a rapid and synergistic phosphorylation of Akt and p38 MAK, leading to increased IL-10 production, that in turn contributs to the persistence of high mycobacterial burden (Zhang et al., 2009). Finally, another example of CLRs-TLRs crosstalk that could contribute to the successes of invading pathogens was recently characterized (Goodridge et al., 2007). The C-type lectin receptor Dectin-1 is, in DCs and macrophages, crucial for the detection of the pathogenic fungi *Candida albicans, Aspergillus fumigates* and *Pneumocystis carinii*. Traditionally, regarded as inflammatory stimuli, ligands of Dectin-1 induce significant amounts of the anti-inflammatory cytokine IL-10, conditioning inflammatory cytokine production and Th subsets polarization. The receptor collaborates with TLR2 in NF-κB activation, inducing proinflammatory cytokines, such as IL-6 and TNF-α (Gantner et al., 2003). However, engagement of Dectin 1 was also shown to activate the nuclear factor of activated T-cells (NFAT) that, directly and/or by interference with TLR2, regulates the expression of the immunosuppressive cytokine IL-10 (Goodridge et al., 2007). Moreover, Dectin-1, due to its cytoplasmic adapter ITAM, signals in an autonomous manner, leading to IL-10 production through a calcium-dependent calmodulin (CaM) dependent kinase (CaMK)–Pyk2–ERK signaling pathway (Kelly et al., 2010). Accordingly, the genetic deletion of Dectin-1 only partially blocks inflammatory cytokine production, while severely impairs IL-10 expression (Taylor et al., 2007).

3.2 Exploiting TLRs signaling and their crosstalk with other receptors

Among pattern recognition receptors, TLRs are, by excellence, the orchestrators of innate immunity. However, pathogens might have evolved to interact with, and exploit, TLRs signaling cascades, inducing conflicting signals by distinct pathogen-expressed TLR ligands. TLR2-induced responses represent a paradigm of this TLR-TLR interplay. Signaling through this receptor leads to an overall proinflammatory response, however it also induces the production of substantial levels of the immunosuppressive cytokine IL-10. It was hypothesized that this probably results from the crosstalk between TLR2 and particular co-receptors such CLRs (Zhang et al., 2009).

Several microorganisms exploit TLR2 crosstalk with other TLRs to evade immune system. For example, in macrophages, *C. albicans* was shown to trigger both TLR4 and TLR2 signals. While TLR4 signaling confers protection against infection, TLR2 signaling promotes host susceptibility to invasive candidiasis, through the induction of high levels of IL-10 (Netea et al., 2004). Lipoproteins from *M. tuberculosis* cell wall bind TLR2 and down-regulate the bacterial CpG DnA-TLR9 induced production of IFNα and IFNβ (Simmons et al., 2010). Similarly, in human monocytes, Hepatitis C virus induces TLR2-mediated expression of IL-10, which in turn suppresses TLR9-induced IFNα production by plasmacytoid DCs (Dolganiuc et al., 2006). The pathogens *M. tuberculosis* and *Toxoplasma gondii* promote their survival in macrophages, through TLR2-MYD88-dependent induction of IL-6, IL-10 and granulocyte colony-stimulating factor (GCSF) (El Kasmi et al., 2008). These cytokines, through signal transducer and activator of transcription 3 (STAT3), increase the expression of arginase 1 (ARG1), which by competing with inducible Nitric Oxide Synthase (iNOS) for the common substrate arginine, inhibits the TLR4-mediated production of nitric oxide (NO) (Qualls et al., 2010).

Additionally to interfering with PPR signaling crosstalk, microorganisms also exploit the interplay between other immune receptors, such as TLRs and complement receptors. Normally, complement receptors and TLRs are rapidly activated in response to infection, and their signals synergistically converge to activate ERK and JNK, promoting an effective early innate immune response. However, in macrophages this crosstalk between TLRs and complement receptors is frequently subversive, particularly by reducing the cytokines of IL-12 family (IL-12, IL-23, and IL-27). This decreased cytokine expression translates into a limited polarization of protective Th1 responses (Hawlisch et al., 2005). The molecular mechanisms of this crosstalk are not fully known, but anaphylatoxin receptor C5aR was shown to interfere with TLR-induced cytokine expression, by ERK and PI3K-dependent pathways. The C5aR-ERk-IRF1 pathway preferentially inhibits IL-12p70 production, while the C5aR–PI3k–IRF8 pathway mainly decreases the production of IL-23 (Hawlisch et al., 2005). Several other complement receptors, such as gC1qR, CD46 and CR3, limit the TLR4 and TLR2-induced IL-12 production (Karp et al., 1996; Marth & Kelsall, 1997). HCV core protein has been shown to associate with the putative gC1q receptor expressed in host immune cells, specifically inhibiting TLR-induced production of IL-12. Therefore, engagement of gC1qR on DCs by HCV depresses Th1 immunity and contributes to viral persistence (Waggoner et al., 2007). *L. monocytogenes* and *S. aureus* were also shown to interact with gC1qR, leading probably to a similar evasion mechanism (Braun et al., 2000; Nguyen et al., 2000). Other human pathogens, such as *P. gingivalis, Histoplasma capsulatum* and *B. pertussis* inhibit IL-12 release through CR3-TLR-dependent crosstalk. The fimbriae of *Porphyromonas gingivalis* interacts with complement receptor 3, activating ERK1 and ERK2, and thereby limiting TLR2-induced IL-12 production (Hajishengallis et al., 2007).

These are only some examples of molecular mechanisms by which microorganisms disrupt, or subvert, signaling crosstalk between innate immune receptors, being particularly emphasized in this review the PPRs interplay. This is an exciting and dynamic Immunology field that in last decade brought considerable advances to the understanding of the pathophysiology of several human infectious diseases.

3.3 Direct targeting of signaling intermediates

Another common evasive maneuver used by pathogens is to directly impair signal transduction, through cleavage or dephosphorylation of intermediate molecules in signaling cascades. Cascade signals downstream PPRs activation mainly converge to NF-κB and MAPKs pathways to establish effective immune responses, making the intermediates of these pathways main targets of microorganism hijacking strategies.

Phosphorylation is the most frequent intracellular modification for signal transduction and many pathogens modulate host cell phosphorylation machinery, in order to block or circumvent deleterious signals. *Yersinia* species, causative agents of human diseases, such as bubonic and pneumonic plagues and gastrointestinal disorders, use a wide spectrum of strategies to circumvent immune response. Through a type III secretion system, bacteria can inject into the cytosol of the host cell six different *Yersinia* outer proteins (Yop). These effector proteins interfere with signaling pathways involved in the regulation of the actin cytoskeleton, phagocytosis, apoptosis and the inflammatory response, thus favoring survival of the bacteria (Viboud & Bliska, 2005). The protein YopP/J was shown to be the main antiinflammatory effector protein of *Yersinia,* by inactivating MAPKs and NF-κB pathways. NF-κB pathway inhibition was initially clearly associated to the de-ubiquitinating activity of YopP/J. IκB-α de-ubiquitination impairs its targeting for proteosomal degradation, effectively sequestering NF-κB into the cytoplasm (Zhou et al., 2005). However, this ubiquitin-like protease activity was unable to explain the effects of YopP/J over MAPKs, as ubiquitination is not known to play a direct role in MAPK signaling. Recent data demonstrate that YopP/J has acetyltranferase activity, transferring acetyl moieties to Ser/Thr residues in the activation loop of MKKs and IKKs (Mittal et al., 2006). It was suggested that acetylation competes effectively with phosphorylation at these sites, thereby blocking signal transduction. Vibrio outer protein A (VopA), an YopJ-like protein from *Vibrio parahaemolyticus*, was also shown to selectively inhibit MAPKs signaling by acetylating a conserved lysine in the ATP-binding pocket of MKKs. This not only prevents MKKs activation but also decreases the activity of activated MKKs (Trosky et al., 2007).

Salmonella, another important human pathogen, delivers effector proteins into host cell, suppressing cellular immune response through blockade of NF-κB and MAPKs cascades (McGhie et al., 2009). The effector protein SptP, by its GTPase-activating protein and tyrosine phosphatase activities, reverses MAPKs activation (Murli et al., 2001; Lin et al., 2003) and AvrA, through its acetyltransferase activity toward specific mitogen-activated protein kinase kinases (MAPKKs), potently inhibits JNK (Jones et al., 2008). Other *Salmonella* effector proteins, such SpvC, a phosphothreonine lyase, directly dephosphorylates ERK, JNK and p38 MAPKs (Mazurkiewicz et al., 2008) and Avra and SseL proteins suppress NF-κB activation by impairing IκB-α ubiquitination and degradation (Ye et al., 2007; Le Negrate et al., 2008).

Similarly, as a strategy for repressing innate immunity, *Shigella flexneri* has evolved the capacity to precisely modulate host cell epigenetic "information", interfering with MAPKs and NF-κB pathways at several points. This is mainly drived by the effector protein OspF. OspF is remarkable not only for its biochemistry but also for the fact that is one of the few bacterial effectors that is known to translocate to the host-cell nucleus. At the cytosol level, the protein binds to the ubiquitylated form of the E2 ubiquitin-conjugating enzyme UBCH5B, and independently of IκB phosphorylation, prevents the transfer of ubiquitin to IκB by an E3 ubiquitin–protein ligase (Kim et al., 2005). Additionally, OspF dephosphorylates ERK and p38 MAPKs by either phosphatase (Arbibe et al., 2007) or phosphothreonine lyase (Li et al., 2007) activities. Recent data showed that this protein also manipulates the physical and spatial context of DNA encoding NF-κB-responsive genes (Arbibe et al., 2007). At the host-cell nucleus, OspF dephosphorylates the MAPK ERK2, impairing the activation of mitogen- and stress-activated kinase 1 (MSK1) and MSK2. This prevents subsequent histone phosphorylation, which is necessary for NF-κB-dependent transcription. Therefore, several innate immune-related genes under control of NF-κB remain silent, allowing *S. flexneri* to avoid a deleterious response.

The mechanisms used by microorganisms to modulate NF-κB signaling are diverse and, as exemplified above, a common strategy is to target the steps that lead to IκB degradation. However, several pathogens, such *Toxoplasma gondii* and *Leishmania spp* have evolved distinct processes to block this central signaling pathway. Infection by *T. gondii* provides potent signals for IL-12 production and for induction of strong Th1 immunity, being NF-κB an important player in this process (Caamano & Hunter, 2002). However, at early times of infection (up to 24h) the parasite impairs in macrophages, the NF-κB signaling, limiting the production of IL-12, TNF-α and NO. This blockage was shown to occur independently of infection-induced IKK-dependent degradation of IκB-α, resulting in specific impairment of NF-κB nuclear translocation. The termination of NF-κB signaling was therefore associated with reduced phosphorylation of p65/RelA subunit, an event involved in the ability of NF-κB to translocate to the nucleus and to bind DNA (Shapira et al., 2005).

Regarding *Leishmania*, the infection by this protozoan parasite has long been regarded as the paradigm of a Th2 immune response. Extensive studies have been conducted to disclose the molecular mechanism by which *Leishmania* modulate intracellular signaling events in infected macrophages and dendritic cells. Obtained data indicate that the parasite use an extensive "arsenal" of strategies and virulence factors to alter the host cell signaling, favoring its survival. Infection of macrophages with *L. donovani* promastigotes was shown to increase intracellular ceramide content causing a downregulation of classical PKC activity, up-regulation of calcium independent atypical PKC-zeta and dephosphorylation of ERK. Downregulation of ERK signaling was subsequently found to be associated with the inhibition of activated protein 1 (AP-1) and NF-κB transactivation (Ghosh et al., 2002). Other studies whit the same infection model showed that *Leishmania* alters signal transduction upstream of c-Fos and c-Jun, by inhibiting ERK, JNK and p38 MAP Kinases, resulting in a reduction of AP-1 nuclear translocation (Prive & Descoteaux, 2000). Until recently, little was known about the intervenients and molecular mechanisms behind these immunosuppressive abilities of *Leishmania*. In macrophages infected with *Leishmania mexicana* amastigotes, Cameron and co-workers showed that cysteine peptidase B (CPB) is the virulence factor responsible for proteolytic degradation of NF-κB, ERK and JNK (Cameron et al., 2004). Additionally, CPB is also involved in the activation of host protein tyrosine phosphatase 1B (PTP-1B), inhibition of AP-1 and cleavage of STAT-1α (Abu-Dayyeh et al., 2010). Another *Leishmania* virulence factor, the surface metalloprotease GP63, was shown to cleave host protein tyrosine phosphatases PTP-1B, TCPTP, and SHP-1, resulting in the stimulation of their phosphatase activity and consequent dephosphorylation of key kinases, such as JAK/STAT, IRAK-1 and MAPKs (Gomez et al., 2009). Moreover, GP63 is also responsible for the observed cleavage of NF-κB $p65^{RelA}$ subunit in *L. mexicana* and *L.infantum* -infected macrophages and dendritic cells (Gregory et al., 2008; Neves et al., 2010). From this cleavage results a fragment of approximately 35 kDa that is rapidly translocated into the nucleus where it has some transcriptional activity. It was postulated that the resulting $p35^{RelA}$ fragment may represent an important mediator by which *Leishmania* promastigotes induce several chemokines without inducing other NF-κB-regulated genes, such as iNOS and IL-12 that are detrimental for parasite survival.

Recently, the metalloprotease GP63 was shown to be involved in the decreased general translation observed in macrophages infected with *L.major* (Jaramillo et al., 2011). The parasite protease cleaves the serine/threonine kinase mammalian/ mechanistic target of rapamycin (mTOR), impairing the formation of mTOR complex 1 (mTORC1) and the downstream phosphorylation of translational repressor 4E-binding protein 1/2 (4E-BP1/2). The activity of the translational repressors 4E-BPs is controlled through their phosphorylation state and, in normal conditions, mTORC1 formation leads to hyperphosphorylation of 4E-binding proteins (4E-BPs), causing their dissociation from eukaryotic initiation factor 4F, facilitating this way the translation of mRNA (Gingras et al., 1999). mTORC1, through its downstream targets p70 ribosomal S6 protein kinases 1 and 2 (S6K1/2) and 4E-BPs controls the translation of key innate immune effector molecules, such as type I IFN (Cao et al., 2008; Costa-Mattioli &

Sonenberg, 2008). This cleavage of mTOR by Leishamania GP63 represents, therefore, a survival mechanism where the parasite directly targets the host translational machinery. This strategy is also a common feature of several human viruses. Lytic viruses, such members of the picornavirus group (enterovirus, rhinovirus and aphtovirus) inhibit overall host cellular translation, redirecting the translational apparatus to viral protein synthesis. This effect was shown to be due to the poliovirus 2A protease-mediated cleavage of the translation initiation factor eIF4G (Borman et al., 1997).

Bacillus anthracis, *Chlamydia* and *Escherichia coli* are examples of other human pathogens that directly cleave intermediate molecules from NF-κB and MAPKs signaling cascades. *Bacillus anthracis*, a spore-forming encapsulated gram-positive bacterium kwon to cause anthrax disease, produces innumerous virulence factors critical for the establishment of infection and pathogenesis (Turnbull, 2002). Among these factors, the plasmid-encoded enzymes lethal factor (LF) and oedema factor (OF) are of major importance for the evasion abilities of *B. anthracis*. LF is a particularly selective metalloproteinase that cleaves MKKs at specific sites outside of their catalytic domains, impairing the downstream MAPK activation (Duesbery et al., 1998). In addition, it blocks the p38 MAPK-dependent activation of IRF3 (Dang et al., 2004) and, although not directly affecting NF-κB activity, it causes the downregulation of NF-κB target genes that simultaneously require p38 activity for induction (Park et al., 2002). Consequently, macrophage production of proinflammatory cytokines, such as TNF-α, IL-1β and IL-6, is severely impaired. In turn, OF is an active Ca^{2+} and calmodulin-dependent adenylate cyclase that increases cAMP in the cytosol of host cells (Drum et al., 2002). Raised intracellular levels of cAMP activate PKA, causing downstream inhibition of ERK and JNK pathways, as well as a decreased NADPH oxidase activity, resulting in impaired TNF-α and microbicidal superoxide production (Hoover et al., 1994).

The obligate intracellular bacterial parasite *Chlamydia* is the leading cause of preventable blindness worldwide and urogenital tract infection remains the most prevalent cause of sexually transmitted diseases in developed countries. The parasite avoids host inflammatory response, partially by disrupting the NF-κB signal resultant from the PPR recognition of bacterial component such LPS. This blockage was shown to result from the selective cleavage of the $p65^{RelA}$ subunit of NF-κB by the chlamydial protease-like activity factor (CPAF) (Christian et al., 2010). Similarly, *E.coli* decreases production of proinflammatory cytokines and reduces macrophage bactericidal activity, by targeting NF-κB signal transduction at multiple points. Infection by *E coli* induces a host caspase 3-mediated cleavage of $p65^{RelA}$, by a mechanism not completely defined, but thought to be mediated through the mitochondrial pathway of apoptosis (Albee & Perlman, 2006). In addition, several studies have recently demonstrated that *E. coli* also downregulates NF-κB-mediated gene expression by injecting into host-cells several non-LEE encoded (Nle) effector proteins, such as NleB, NleC and NleE. NleB and NleE prevent IKKβ activation and consequently the degradation of IκB-α, thus limiting p65 translocation to the nucleus (Nadler et al., 2010; Newton et al., 2010) while the zinc-dependent metalloprotease NleC was shown to enzymatically degrade $p65^{RelA}$ and JNK (Yen et al., ; Baruch et al., 2010).

Although more frequent in bacteria, the shutdown of PPR signaling by direct cleavage of cascade intermediates is also a strategy used by some relevant human viral pathogens. As an example, hepatitis C virus-host interactions have revealed several evasion mechanisms used by the virus to control PPRs signaling, providing a molecular basis for viral persistence. In viral infections, recognition of pathogen associated molecular patterns by TLRs and RLRs leads, through independent signaling cascades, to the activation of transcription factors, such as IRF1, IRF3, IRF5, IRF7 and NF-κB. The activity of these transcription factors is crucial for an effective antiviral innate immune response, given that they control the expression of interferon-stimulated genes (ISGs) and type I interferons. Hepatitis C virus (HCV) has evolved to disrupt RLRs signaling by impairing RIG-I pathway, through NS3/4A-mediated cleavage of IPS-1 (Malmgaard, 2004). NS3/4A is formed by a complex of the NS3 and NS4A HCV proteins and has been shown to be an essential viral protein with serine protease activity (Brass et al., 2008). In HCV infection, cleavage of IPS-1 by NS3/4A impairs downstream activation of IRF-3 and NF-κB, blocking the production of IFN-β, as well as the expression of ISGs (Li et al., 2005). This results in a strongly compromised innate immune response, potentiating the propagation of chronic HCV infection.

4. Conclusions

Millenary host-microbe co-evolution has resulted in the development of ingenious strategies by pathogens in order to successfully evade host immune response. Besides the manipulation of host-cell cytoskeleton to gain entry and/or to gain motility in the cell, immune-cell signaling pathways are frequent targets of invading pathogens. Within the past decades, remarkable progress has been made in our understanding on how immune cells sense microorganisms and how microbial effectors counteract innate immune responses. Recognition of conserved microorganism patterns by PRRs activates, in immune cells, an intricate signaling network that culminates in the expression of effector molecules, such as cytokines, chemokines and reactive oxygen species, crucial elements to mount an adequate immune response. A common strategy of pathogens is to disrupt these signaling cascades, by promoting contradictory signals through engagement of distinct PRRs and/or by directly target intermediate components of these signaling pathways. Therefore, understanding the molecular mechanisms used by pathogens to exploit the host signaling networks is of crucial importance for the development of rational interventions in which host response will be redirect to achieve protective immunity.

5. References

Abu-Dayyeh, I., K. Hassani, E. R. Westra, J. C. Mottram & M. Olivier (2010). "Comparative study of the ability of Leishmania mexicana promastigotes and amastigotes to alter macrophage signaling and functions." *Infect Immun* Vol.78, No 6, (Jun), pp. 2438-45, ISSN 1098-5522

Ahren, I. L., E. Eriksson, A. Egesten & K. Riesbeck (2003). "Nontypeable Haemophilus influenzae activates human eosinophils through beta-glucan receptors." *Am J Respir Cell Mol Biol* Vol.29, No 5, (Nov), pp. 598-605, ISSN 1044-1549

Akira, S. (2009). "Pathogen recognition by innate immunity and its signaling." *Proc Jpn Acad Ser B Phys Biol Sci* Vol.85, No 4, 143-56, ISSN 1349-2896

Akira, S., K. Takeda & T. Kaisho (2001). "Toll-like receptors: critical proteins linking innate and acquired immunity." *Nat Immunol* Vol.2, No 8, (Aug), pp. 675-80, ISSN 1529-2908

Akira, S., S. Uematsu & O. Takeuchi (2006). "Pathogen recognition and innate immunity." *Cell* Vol.124, No 4, (Feb 24), pp. 783-801, ISSN 0092-8674

Albee, L. & H. Perlman (2006). "E. coli infection induces caspase dependent degradation of NF-kappaB and reduces the inflammatory response in macrophages." *Inflamm Res* Vol.55, No 1, (Jan), pp. 2-9, ISSN 1023-3830

Aoki, N., Y. Kimura, S. Kimura, T. Nagato, M. Azumi, H. Kobayashi, K. Sato & M. Tateno (2009). "Expression and functional role of MDL-1 (CLEC5A) in mouse myeloid lineage cells." *J Leukoc Biol* Vol.85, No 3, (Mar), pp. 508-17, ISSN 1938-3673

Arbibe, L., D. W. Kim, E. Batsche, T. Pedron, B. Mateescu, C. Muchardt, C. Parsot & P. J. Sansonetti (2007). "An injected bacterial effector targets chromatin access for transcription factor NF-kappaB to alter transcription of host genes involved in immune responses." *Nat Immunol* Vol.8, No 1, (Jan), pp. 47-56, ISSN 1529-2908

Ariizumi, K., G. L. Shen, S. Shikano, R. Ritter, 3rd, P. Zukas, D. Edelbaum, A. Morita & A. Takashima (2000). "Cloning of a second dendritic cell-associated C-type lectin (dectin-2) and its alternatively spliced isoforms." *J Biol Chem* Vol.275, No 16, (Apr 21), pp. 11957-63, ISSN 0021-9258

Bakker, A. B., E. Baker, G. R. Sutherland, J. H. Phillips & L. L. Lanier (1999). "Myeloid DAP12-associating lectin (MDL)-1 is a cell surface receptor involved in the activation of myeloid cells." *Proc Natl Acad Sci U S A* Vol.96, No 17, (Aug 17), pp. 9792-6, ISSN 0027-8424

Balboa, L., M. M. Romero, N. Yokobori, P. Schierloh, L. Geffner, J. I. Basile, R. M. Musella, E. Abbate, S. de la Barrera, M. C. Sasiain & M. Aleman (2010). "Mycobacterium tuberculosis impairs dendritic cell response by altering CD1b, DC-SIGN and MR profile." *Immunol Cell Biol* Vol.88, No 7, (Oct), pp. 716-26, ISSN 1440-1711

Banchereau, J., F. Briere, C. Caux, J. Davoust, S. Lebecque, Y. J. Liu, B. Pulendran & K. Palucka (2000). "Immunobiology of dendritic cells." *Annu Rev Immunol* Vol.18, No, 767-811, ISSN 0732-0582

Banchereau, J. & R. M. Steinman (1998). "Dendritic cells and the control of immunity." *Nature* Vol.392, No 6673, (Mar 19), pp. 245-52, ISSN 0028-0836

Barton, G. M. & J. C. Kagan (2009). "A cell biological view of Toll-like receptor function: regulation through compartmentalization." *Nat Rev Immunol* Vol.9, No 8, (Aug), pp. 535-42, ISSN 1474-1741

Baruch, K., L. Gur-Arie, C. Nadler, S. Koby, G. Yerushalmi, Y. Ben-Neriah, O. Yogev, E. Shaulian, C. Guttman, R. Zarivach & I. Rosenshine (2010). "Metalloprotease type III effectors that specifically cleave JNK and NF-kappaB." *EMBO J* Vol.30, No 1, (Jan 5), pp. 221-31, ISSN 1460-2075

Bates, E. E., N. Fournier, E. Garcia, J. Valladeau, I. Durand, J. J. Pin, S. M. Zurawski, S. Patel, J. S. Abrams, S. Lebecque, P. Garrone & S. Saeland (1999). "APCs express DCIR, a novel C-type lectin surface receptor containing an immunoreceptor tyrosine-based inhibitory motif." *J Immunol* Vol.163, No 4, (Aug 15), pp. 1973-83, ISSN 0022-1767

Bergman, M. P., A. Engering, H. H. Smits, S. J. van Vliet, A. A. van Bodegraven, H. P. Wirth, M. L. Kapsenberg, C. M. Vandenbroucke-Grauls, Y. van Kooyk & B. J. Appelmelk (2004). "Helicobacter pylori modulates the T helper cell 1/T helper cell 2 balance through phase-variable interaction between lipopolysaccharide and DC-SIGN." *J Exp Med* Vol.200, No 8, (Oct 18), pp. 979-90, ISSN 0022-1007

Bogoevska, V., A. Horst, B. Klampe, L. Lucka, C. Wagener & P. Nollau (2006). "CEACAM1, an adhesion molecule of human granulocytes, is fucosylated by fucosyltransferase IX and interacts with DC-SIGN of dendritic cells via Lewis x residues." *Glycobiology* Vol.16, No 3, (Mar), pp. 197-209, ISSN 0959-6658

Borman, A. M., R. Kirchweger, E. Ziegler, R. E. Rhoads, T. Skern & K. M. Kean (1997). "eIF4G and its proteolytic cleavage products: effect on initiation of protein synthesis from capped, uncapped, and IRES-containing mRNAs." *RNA* Vol.3, No 2, (Feb), pp. 186-96, ISSN 1355-8382

Brass, V., J. M. Berke, R. Montserret, H. E. Blum, F. Penin & D. Moradpour (2008). "Structural determinants for membrane association and dynamic organization of the hepatitis C virus NS3-4A complex." *Proc Natl Acad Sci U S A* Vol.105, No 38, (Sep 23), pp. 14545-50, ISSN 1091-6490

Braun, L., B. Ghebrehiwet & P. Cossart (2000). "gC1q-R/p32, a C1q-binding protein, is a receptor for the InlB invasion protein of Listeria monocytogenes." *EMBO J* Vol.19, No 7, (Apr 3), pp. 1458-66, ISSN 0261-4189

Brown, G. D. (2006). "Dectin-1: a signalling non-TLR pattern-recognition receptor." *Nat Rev Immunol* Vol.6, No 1, (Jan), pp. 33-43, ISSN 1474-1733

Brown, G. D. & S. Gordon (2001). "Immune recognition. A new receptor for beta-glucans." *Nature* Vol.413, No 6851, (Sep 6), pp. 36-7, ISSN 0028-0836

Buchholz, K. R. & R. S. Stephens (2008). "The cytosolic pattern recognition receptor NOD1 induces inflammatory interleukin-8 during Chlamydia trachomatis infection." *Infect Immun* Vol.76, No 7, (Jul), pp. 3150-5, ISSN 1098-5522

Bugarcic, A., K. Hitchens, A. G. Beckhouse, C. A. Wells, R. B. Ashman & H. Blanchard (2008). "Human and mouse macrophage-inducible C-type lectin (Mincle) bind Candida albicans." *Glycobiology* Vol.18, No 9, (Sep), pp. 679-85, ISSN 1460-2423

Caamano, J. & C. A. Hunter (2002). "NF-kappaB family of transcription factors: central regulators of innate and adaptive immune functions." *Clin Microbiol Rev* Vol.15, No 3, (Jul), pp. 414-29, ISSN 0893-8512

Cameron, P., A. McGachy, M. Anderson, A. Paul, G. H. Coombs, J. C. Mottram, J. Alexander & R. Plevin (2004). "Inhibition of lipopolysaccharide-induced macrophage IL-12 production by Leishmania mexicana amastigotes: the role of cysteine peptidases and the NF-kappaB signaling pathway." *J Immunol* Vol.173, No 5, (Sep 1), pp. 3297-304, ISSN 0022-1767

Cao, W., S. Manicassamy, H. Tang, S. P. Kasturi, A. Pirani, N. Murthy & B. Pulendran (2008). "Toll-like receptor-mediated induction of type I interferon in plasmacytoid dendritic cells requires the rapamycin-sensitive PI(3)K-mTOR-p70S6K pathway." *Nat Immunol* Vol.9, No 10, (Oct), pp. 1157-64, ISSN 1529-2916

Chamaillard, M., M. Hashimoto, Y. Horie, J. Masumoto, S. Qiu, L. Saab, Y. Ogura, A. Kawasaki, K. Fukase, S. Kusumoto, M. A. Valvano, S. J. Foster, T. W. Mak, G. Nunez & N. Inohara (2003). "An essential role for NOD1 in host recognition of bacterial peptidoglycan containing diaminopimelic acid." *Nat Immunol* Vol.4, No 7, (Jul), pp. 702-7, ISSN 1529-2908

Chen, G., M. H. Shaw, Y. G. Kim & G. Nunez (2009). "NOD-like receptors: role in innate immunity and inflammatory disease." *Annu Rev Pathol* Vol.4, No, 365-98, ISSN 1553-4014

Chen, S. T., Y. L. Lin, M. T. Huang, M. F. Wu, S. C. Cheng, H. Y. Lei, C. K. Lee, T. W. Chiou, C. H. Wong & S. L. Hsieh (2008). "CLEC5A is critical for dengue-virus-induced lethal disease." *Nature* Vol.453, No 7195, (May 29), pp. 672-6, ISSN 1476-4687

Christian, J., J. Vier, S. A. Paschen & G. Hacker (2010). "Cleavage of the NF-kappaB family protein p65/RelA by the chlamydial protease-like activity factor (CPAF) impairs proinflammatory signaling in cells infected with Chlamydiae." *J Biol Chem* Vol.285, No 53, (Dec 31), pp. 41320-7, ISSN 1083-351X

Costa-Mattioli, M. & N. Sonenberg (2008). "RAPping production of type I interferon in pDCs through mTOR." *Nat Immunol* Vol.9, No 10, (Oct), pp. 1097-9, ISSN 1529-2916

Dang, O., L. Navarro, K. Anderson & M. David (2004). "Cutting edge: anthrax lethal toxin inhibits activation of IFN-regulatory factor 3 by lipopolysaccharide." *J Immunol* Vol.172, No 2, (Jan 15), pp. 747-51, ISSN 0022-1767

Demangel, C. & W. J. Britton (2000). "Interaction of dendritic cells with mycobacteria: where the action starts." *Immunol Cell Biol* Vol.78, No 4, (Aug), pp. 318-24, ISSN 0818-9641

den Dunnen, J., S. I. Gringhuis & T. B. Geijtenbeek (2009). "Innate signaling by the C-type lectin DC-SIGN dictates immune responses." *Cancer Immunol Immunother* Vol.58, No 7, (Jul), pp. 1149-57, ISSN 1432-0851

Deng, L., C. Wang, E. Spencer, L. Yang, A. Braun, J. You, C. Slaughter, C. Pickart & Z. J. Chen (2000). "Activation of the IkappaB kinase complex by TRAF6 requires a dimeric ubiquitin-conjugating enzyme complex and a unique polyubiquitin chain." *Cell* Vol.103, No 2, (Oct 13), pp. 351-61, ISSN 0092-8674

Diebold, S. S. (2009). "Activation of dendritic cells by toll-like receptors and C-type lectins." *Handb Exp Pharmacol*, No 188, 3-30, ISSN 0171-2004

Dixit, E., S. Boulant, Y. Zhang, A. S. Lee, C. Odendall, B. Shum, N. Hacohen, Z. J. Chen, S. P. Whelan, M. Fransen, M. L. Nibert, G. Superti-Furga & J. C. Kagan (2010). "Peroxisomes are signaling platforms for antiviral innate immunity." *Cell* Vol.141, No 4, (May 14), pp. 668-81, ISSN 1097-4172

Dolganiuc, A., S. Chang, K. Kodys, P. Mandrekar, G. Bakis, M. Cormier & G. Szabo (2006). "Hepatitis C virus (HCV) core protein-induced, monocyte-mediated mechanisms of reduced IFN-alpha and plasmacytoid dendritic cell loss in chronic HCV infection." *J Immunol* Vol.177, No 10, (Nov 15), pp. 6758-68, ISSN 0022-1767

Drum, C. L., S. Z. Yan, J. Bard, Y. Q. Shen, D. Lu, S. Soelaiman, Z. Grabarek, A. Bohm & W. J. Tang (2002). "Structural basis for the activation of anthrax adenylyl cyclase exotoxin by calmodulin." *Nature* Vol.415, No 6870, (Jan 24), pp. 396-402, ISSN 0028-0836

Duesbery, N. S., C. P. Webb, S. H. Leppla, V. M. Gordon, K. R. Klimpel, T. D. Copeland, N. G. Ahn, M. K. Oskarsson, K. Fukasawa, K. D. Paull & G. F. Vande Woude (1998). "Proteolytic inactivation of MAP-kinase-kinase by anthrax lethal factor." *Science* Vol.280, No 5364, (May 1), pp. 734-7, ISSN 0036-8075

Dzionek, A., Y. Sohma, J. Nagafune, M. Cella, M. Colonna, F. Facchetti, G. Gunther, I. Johnston, A. Lanzavecchia, T. Nagasaka, T. Okada, W. Vermi, G. Winkels, T. Yamamoto, M. Zysk, Y. Yamaguchi & J. Schmitz (2001). "BDCA-2, a novel plasmacytoid dendritic cell-specific type II C-type lectin, mediates antigen capture and is a potent inhibitor of interferon alpha/beta induction." *J Exp Med* Vol.194, No 12, (Dec 17), pp. 1823-34, ISSN 0022-1007

East, L. & C. M. Isacke (2002). "The mannose receptor family." *Biochim Biophys Acta* Vol.1572, No 2-3, (Sep 19), pp. 364-86, ISSN 0006-3002

El Kasmi, K. C., J. E. Qualls, J. T. Pesce, A. M. Smith, R. W. Thompson, M. Henao-Tamayo, R. J. Basaraba, T. Konig, U. Schleicher, M. S. Koo, G. Kaplan, K. A. Fitzgerald, E. I. Tuomanen, I. M. Orme, T. D. Kanneganti, C. Bogdan, T. A. Wynn & P. J. Murray (2008). "Toll-like receptor-induced arginase 1 in macrophages thwarts effective immunity against intracellular pathogens." *Nat Immunol* Vol.9, No 12, (Dec), pp. 1399-406, ISSN 1529-2916

Fenton, M. J. & M. W. Vermeulen (1996). "Immunopathology of tuberculosis: roles of macrophages and monocytes." *Infect Immun* Vol.64, No 3, (Mar), pp. 683-90, ISSN 0019-9567

Fernandes, M. J., A. A. Finnegan, L. D. Siracusa, C. Brenner, N. N. Iscove & B. Calabretta (1999). "Characterization of a novel receptor that maps near the natural killer gene complex: demonstration of carbohydrate binding and expression in hematopoietic cells." *Cancer Res* Vol.59, No 11, (Jun 1), pp. 2709-17, ISSN 0008-5472

Fitzgerald, K. A., E. M. Palsson-McDermott, A. G. Bowie, C. A. Jefferies, A. S. Mansell, G. Brady, E. Brint, A. Dunne, P. Gray, M. T. Harte, D. McMurray, D. E. Smith, J. E. Sims, T. A. Bird & L. A. O'Neill (2001). "Mal (MyD88-adapter-like) is required for Toll-like receptor-4 signal transduction." *Nature* Vol.413, No 6851, (Sep 6), pp. 78-83, ISSN 0028-0836

Franchi, L., N. Warner, K. Viani & G. Nunez (2009). "Function of Nod-like receptors in microbial recognition and host defense." *Immunol Rev* Vol.227, No 1, (Jan), pp. 106-28, ISSN 1600-065X

Fratti, R. A., J. M. Backer, J. Gruenberg, S. Corvera & V. Deretic (2001). "Role of phosphatidylinositol 3-kinase and Rab5 effectors in phagosomal biogenesis and mycobacterial phagosome maturation arrest." *J Cell Biol* Vol.154, No 3, (Aug 6), pp. 631-44, ISSN 0021-9525

Fratti, R. A., J. Chua, I. Vergne & V. Deretic (2003). "Mycobacterium tuberculosis glycosylated phosphatidylinositol causes phagosome maturation arrest." *Proc Natl Acad Sci U S A* Vol.100, No 9, (Apr 29), pp. 5437-42, ISSN 0027-8424

Gantner, B. N., R. M. Simmons, S. J. Canavera, S. Akira & D. M. Underhill (2003). "Collaborative induction of inflammatory responses by dectin-1 and Toll-like receptor 2." *J Exp Med* Vol.197, No 9, (May 5), pp. 1107-17, ISSN 0022-1007

Gazi, U. & L. Martinez-Pomares (2009). "Influence of the mannose receptor in host immune responses." *Immunobiology* Vol.214, No 7, (Jul), pp. 554-61, ISSN 1878-3279

Geijtenbeek, T. B., D. J. Krooshoop, D. A. Bleijs, S. J. van Vliet, G. C. van Duijnhoven, V. Grabovsky, R. Alon, C. G. Figdor & Y. van Kooyk (2000). "DC-SIGN-ICAM-2 interaction mediates dendritic cell trafficking." *Nat Immunol* Vol.1, No 4, (Oct), pp. 353-7, ISSN 1529-2908

Geijtenbeek, T. B., D. S. Kwon, R. Torensma, S. J. van Vliet, G. C. van Duijnhoven, J. Middel, I. L. Cornelissen, H. S. Nottet, V. N. KewalRamani, D. R. Littman, C. G. Figdor & Y. van Kooyk (2000). "DC-SIGN, a dendritic cell-specific HIV-1-binding protein that enhances trans-infection of T cells." *Cell* Vol.100, No 5, (Mar 3), pp. 587-97, ISSN 0092-8674

Geijtenbeek, T. B., R. Torensma, S. J. van Vliet, G. C. van Duijnhoven, G. J. Adema, Y. van Kooyk & C. G. Figdor (2000). "Identification of DC-SIGN, a novel dendritic cell-specific ICAM-3 receptor that supports primary immune responses." *Cell* Vol.100, No 5, (Mar 3), pp. 575-85, ISSN 0092-8674

Geijtenbeek, T. B., S. J. Van Vliet, E. A. Koppel, M. Sanchez-Hernandez, C. M. Vandenbroucke-Grauls, B. Appelmelk & Y. Van Kooyk (2003). "Mycobacteria target DC-SIGN to suppress dendritic cell function." *J Exp Med* Vol.197, No 1, (Jan 6), pp. 7-17, ISSN 0022-1007

Ghosh, S., S. Bhattacharyya, M. Sirkar, G. S. Sa, T. Das, D. Majumdar, S. Roy & S. Majumdar (2002). "Leishmania donovani suppresses activated protein 1 and NF-kappaB activation in host macrophages via ceramide generation: involvement of extracellular signal-regulated kinase." *Infect Immun* Vol.70, No 12, (Dec), pp. 6828-38, ISSN 0019-9567

Gilchrist, M., V. Thorsson, B. Li, A. G. Rust, M. Korb, J. C. Roach, K. Kennedy, T. Hai, H. Bolouri & A. Aderem (2006). "Systems biology approaches identify ATF3 as a negative regulator of Toll-like receptor 4." *Nature* Vol.441, No 7090, (May 11), pp. 173-8, ISSN 1476-4687

Gingras, A. C., B. Raught & N. Sonenberg (1999). "eIF4 initiation factors: effectors of mRNA recruitment to ribosomes and regulators of translation." *Annu Rev Biochem* Vol.68, No, 913-63, ISSN 0066-4154

Gomez, M. A., I. Contreras, M. Halle, M. L. Tremblay, R. W. McMaster & M. Olivier (2009). "Leishmania GP63 alters host signaling through cleavage-activated protein tyrosine phosphatases." *Sci Signal* Vol.2, No 90, ra58, ISSN 1937-9145

Goodridge, H. S., R. M. Simmons & D. M. Underhill (2007). "Dectin-1 stimulation by Candida albicans yeast or zymosan triggers NFAT activation in macrophages and dendritic cells." *J Immunol* Vol.178, No 5, (Mar 1), pp. 3107-15, ISSN 0022-1767

Gregory, D. J., M. Godbout, I. Contreras, G. Forget & M. Olivier (2008). "A novel form of NF-kappaB is induced by Leishmania infection: involvement in macrophage gene expression." *Eur J Immunol* Vol.38, No 4, (Apr), pp. 1071-81, ISSN 0014-2980

Gringhuis, S. I., J. den Dunnen, M. Litjens, M. van der Vlist & T. B. Geijtenbeek (2009). "Carbohydrate-specific signaling through the DC-SIGN signalosome tailors immunity to Mycobacterium tuberculosis, HIV-1 and Helicobacter pylori." *Nat Immunol* Vol.10, No 10, (Oct), pp. 1081-8, ISSN 1529-2916

Gringhuis, S. I., J. den Dunnen, M. Litjens, M. van der Vlist, B. Wevers, S. C. Bruijns & T. B. Geijtenbeek (2009). "Dectin-1 directs T helper cell differentiation by controlling noncanonical NF-kappaB activation through Raf-1 and Syk." *Nat Immunol* Vol.10, No 2, (Feb), pp. 203-13, ISSN 1529-2916

Gringhuis, S. I., J. den Dunnen, M. Litjens, B. van Het Hof, Y. van Kooyk & T. B. Geijtenbeek (2007). "C-type lectin DC-SIGN modulates Toll-like receptor signaling via Raf-1 kinase-dependent acetylation of transcription factor NF-kappaB." *Immunity* Vol.26, No 5, (May), pp. 605-16, ISSN 1074-7613

Gross, O., A. Gewies, K. Finger, M. Schafer, T. Sparwasser, C. Peschel, I. Forster & J. Ruland (2006). "Card9 controls a non-TLR signalling pathway for innate anti-fungal immunity." *Nature* Vol.442, No 7103, (Aug 10), pp. 651-6, ISSN 1476-4687

Gross, O., H. Poeck, M. Bscheider, C. Dostert, N. Hannesschlager, S. Endres, G. Hartmann, A. Tardivel, E. Schweighoffer, V. Tybulewicz, A. Mocsai, J. Tschopp & J. Ruland (2009). "Syk kinase signalling couples to the Nlrp3 inflammasome for anti-fungal host defence." *Nature* Vol.459, No 7245, (May 21), pp. 433-6, ISSN 1476-4687

Hajishengallis, G. & J. D. Lambris (2011). "Microbial manipulation of receptor crosstalk in innate immunity." *Nat Rev Immunol* Vol.11, No 3, (Mar), pp. 187-200, ISSN 1474-1741

Hajishengallis, G., M. A. Shakhatreh, M. Wang & S. Liang (2007). "Complement receptor 3 blockade promotes IL-12-mediated clearance of Porphyromonas gingivalis and negates its virulence in vivo." *J Immunol* Vol.179, No 4, (Aug 15), pp. 2359-67, ISSN 0022-1767

Hasegawa, M., Y. Fujimoto, P. C. Lucas, H. Nakano, K. Fukase, G. Nunez & N. Inohara (2008). "A critical role of RICK/RIP2 polyubiquitination in Nod-induced NF-kappaB activation." *EMBO J* Vol.27, No 2, (Jan 23), pp. 373-83, ISSN 1460-2075

Hashimoto, C., K. L. Hudson & K. V. Anderson (1988). "The Toll gene of Drosophila, required for dorsal-ventral embryonic polarity, appears to encode a transmembrane protein." *Cell* Vol.52, No 2, (Jan 29), pp. 269-79, ISSN 0092-8674

Hawlisch, H., Y. Belkaid, R. Baelder, D. Hildeman, C. Gerard & J. Kohl (2005). "C5a negatively regulates toll-like receptor 4-induced immune responses." *Immunity* Vol.22, No 4, (Apr), pp. 415-26, ISSN 1074-7613

Henderson, R. A., S. C. Watkins & J. L. Flynn (1997). "Activation of human dendritic cells following infection with Mycobacterium tuberculosis." *J Immunol* Vol.159, No 2, (Jul 15), pp. 635-43, ISSN 0022-1767

Hmama, Z., K. Sendide, A. Talal, R. Garcia, K. Dobos & N. E. Reiner (2004). "Quantitative analysis of phagolysosome fusion in intact cells: inhibition by mycobacterial lipoarabinomannan and rescue by an 1alpha,25-dihydroxyvitamin D3-phosphoinositide 3-kinase pathway." *J Cell Sci* Vol.117, No Pt 10, (Apr 15), pp. 2131-40, ISSN 0021-9533

Hodges, A., K. Sharrocks, M. Edelmann, D. Baban, A. Moris, O. Schwartz, H. Drakesmith, K. Davies, B. Kessler, A. McMichael & A. Simmons (2007). "Activation of the lectin DC-SIGN induces an immature dendritic cell phenotype triggering Rho-GTPase activity required for HIV-1 replication." *Nat Immunol* Vol.8, No 6, (Jun), pp. 569-77, ISSN 1529-2908

Hoebe, K., E. Janssen & B. Beutler (2004). "The interface between innate and adaptive immunity." *Nat Immunol* Vol.5, No 10, (Oct), pp. 971-4, ISSN 1529-2908

Hoover, D. L., A. M. Friedlander, L. C. Rogers, I. K. Yoon, R. L. Warren & A. S. Cross (1994). "Anthrax edema toxin differentially regulates lipopolysaccharide-induced monocyte production of tumor necrosis factor alpha and interleukin-6 by increasing intracellular cyclic AMP." *Infect Immun* Vol.62, No 10, (Oct), pp. 4432-9, ISSN 0019-9567

Horng, T., G. M. Barton & R. Medzhitov (2001). "TIRAP: an adapter molecule in the Toll signaling pathway." *Nat Immunol* Vol.2, No 9, (Sep), pp. 835-41, ISSN 1529-2908

Hovius, J. W., M. A. de Jong, J. den Dunnen, M. Litjens, E. Fikrig, T. van der Poll, S. I. Gringhuis & T. B. Geijtenbeek (2008). "Salp15 binding to DC-SIGN inhibits cytokine expression by impairing both nucleosome remodeling and mRNA stabilization." *PLoS Pathog* Vol.4, No 2, (Feb 8), pp. e31, ISSN 1553-7374

Inohara, N., T. Koseki, J. Lin, L. del Peso, P. C. Lucas, F. F. Chen, Y. Ogura & G. Nunez (2000). "An induced proximity model for NF-kappa B activation in the Nod1/RICK and RIP signaling pathways." *J Biol Chem* Vol.275, No 36, (Sep 8), pp. 27823-31, ISSN 0021-9258

Inui, M., Y. Kikuchi, N. Aoki, S. Endo, T. Maeda, A. Sugahara-Tobinai, S. Fujimura, A. Nakamura, A. Kumanogoh, M. Colonna & T. Takai (2009). "Signal adaptor DAP10 associates with MDL-1 and triggers osteoclastogenesis in cooperation with DAP12." *Proc Natl Acad Sci U S A* Vol.106, No 12, (Mar 24), pp. 4816-21, ISSN 1091-6490

Ishikawa, E., T. Ishikawa, Y. S. Morita, K. Toyonaga, H. Yamada, O. Takeuchi, T. Kinoshita, S. Akira, Y. Yoshikai & S. Yamasaki (2009). "Direct recognition of the mycobacterial glycolipid, trehalose dimycolate, by C-type lectin Mincle." *J Exp Med* Vol.206, No 13, (Dec 21), pp. 2879-88, ISSN 1540-9538

Iwasaki, A. & R. Medzhitov (2004). "Toll-like receptor control of the adaptive immune responses." *Nat Immunol* Vol.5, No 10, (Oct), pp. 987-95, ISSN 1529-2908

Iwasaki, A. & R. Medzhitov (2010). "Regulation of adaptive immunity by the innate immune system." *Science* Vol.327, No 5963, (Jan 15), pp. 291-5, ISSN 1095-9203

Jahn, P. S., K. S. Zanker, J. Schmitz & A. Dzionek (2010). "BDCA-2 signaling inhibits TLR-9-agonist-induced plasmacytoid dendritic cell activation and antigen presentation." *Cell Immunol* Vol.265, No 1, 15-22, ISSN 1090-2163

Jaramillo, M., M. A. Gomez, O. Larsson, M. T. Shio, I. Topisirovic, I. Contreras, R. Luxenburg, A. Rosenfeld, R. Colina, R. W. McMaster, M. Olivier, M. Costa-Mattioli & N. Sonenberg (2011). "Leishmania repression of host translation through mTOR cleavage is required for parasite survival and infection." *Cell Host Microbe* Vol.9, No 4, (Apr 21), pp. 331-41, ISSN 1934-6069

Jones, R. M., H. Wu, C. Wentworth, L. Luo, L. Collier-Hyams & A. S. Neish (2008). "Salmonella AvrA Coordinates Suppression of Host Immune and Apoptotic Defenses via JNK Pathway Blockade." *Cell Host Microbe* Vol.3, No 4, (Apr 17), pp. 233-44, ISSN 1934-6069

Kanazawa, N., T. Okazaki, H. Nishimura, K. Tashiro, K. Inaba & Y. Miyachi (2002). "DCIR acts as an inhibitory receptor depending on its immunoreceptor tyrosine-based inhibitory motif." *J Invest Dermatol* Vol.118, No 2, (Feb), pp. 261-6, ISSN 0022-202X

Kang, P. B., A. K. Azad, J. B. Torrelles, T. M. Kaufman, A. Beharka, E. Tibesar, L. E. DesJardin & L. S. Schlesinger (2005). "The human macrophage mannose receptor directs Mycobacterium tuberculosis lipoarabinomannan-mediated phagosome biogenesis." *J Exp Med* Vol.202, No 7, (Oct 3), pp. 987-99, ISSN 0022-1007

Karp, C. L., M. Wysocka, L. M. Wahl, J. M. Ahearn, P. J. Cuomo, B. Sherry, G. Trinchieri & D. E. Griffin (1996). "Mechanism of suppression of cell-mediated immunity by measles virus." *Science* Vol.273, No 5272, (Jul 12), pp. 228-31, ISSN 0036-8075

Kato, H., O. Takeuchi, E. Mikamo-Satoh, R. Hirai, T. Kawai, K. Matsushita, A. Hiiragi, T. S. Dermody, T. Fujita & S. Akira (2008). "Length-dependent recognition of double-stranded ribonucleic acids by retinoic acid-inducible gene-I and melanoma differentiation-associated gene 5." *J Exp Med* Vol.205, No 7, (Jul 7), pp. 1601-10, ISSN 1540-9538

Kato, H., O. Takeuchi, S. Sato, M. Yoneyama, M. Yamamoto, K. Matsui, S. Uematsu, A. Jung, T. Kawai, K. J. Ishii, O. Yamaguchi, K. Otsu, T. Tsujimura, C. S. Koh, C. Reis e Sousa, Y. Matsuura, T. Fujita & S. Akira (2006). "Differential roles of MDA5 and RIG-I helicases in the recognition of RNA viruses." *Nature* Vol.441, No 7089, (May 4), pp. 101-5, ISSN 1476-4687

Kawai, T. & S. Akira (2010)."The role of pattern-recognition receptors in innate immunity: update on Toll-like receptors." *Nat Immunol* Vol.11, No 5, (May), pp. 373-84, ISSN 1529-2916

Kawai, T., K. Takahashi, S. Sato, C. Coban, H. Kumar, H. Kato, K. J. Ishii, O. Takeuchi & S. Akira (2005). "IPS-1, an adaptor triggering RIG-I- and Mda5-mediated type I interferon induction." *Nat Immunol* Vol.6, No 10, (Oct), pp. 981-8, ISSN 1529-2908

Kelly, E. K., L. Wang & L. B. Ivashkiv (2010). "Calcium-activated pathways and oxidative burst mediate zymosan-induced signaling and IL-10 production in human macrophages." *J Immunol* Vol.184, No 10, (May 15), pp. 5545-52, ISSN 1550-6606

Kerrigan, A. M. & G. D. Brown (2010). "Syk-coupled C-type lectin receptors that mediate cellular activation via single tyrosine based activation motifs." *Immunol Rev* Vol.234, No 1, (Mar), pp. 335-52, ISSN 1600-065X

Kim, D. W., G. Lenzen, A. L. Page, P. Legrain, P. J. Sansonetti & C. Parsot (2005). "The Shigella flexneri effector OspG interferes with innate immune responses by targeting ubiquitin-conjugating enzymes." *Proc Natl Acad Sci U S A* Vol.102, No 39, (Sep 27), pp. 14046-51, ISSN 0027-8424

Komuro, A. & C. M. Horvath (2006). "RNA- and virus-independent inhibition of antiviral signaling by RNA helicase LGP2." *J Virol* Vol.80, No 24, (Dec), pp. 12332-42, ISSN 0022-538X

Kumar, H., T. Kawai & S. Akira (2011). "Pathogen recognition by the innate immune system." *Int Rev Immunol* Vol.30, No 1, (Feb), pp. 16-34, ISSN 1563-5244

Kumar, H., Y. Kumagai, T. Tsuchida, P. A. Koenig, T. Satoh, Z. Guo, M. H. Jang, T. Saitoh, S. Akira & T. Kawai (2009). "Involvement of the NLRP3 inflammasome in innate and humoral adaptive immune responses to fungal beta-glucan." *J Immunol* Vol.183, No 12, (Dec 15), pp. 8061-7, ISSN 1550-6606

Lambert, A. A., C. Gilbert, M. Richard, A. D. Beaulieu & M. J. Tremblay (2008). "The C-type lectin surface receptor DCIR acts as a new attachment factor for HIV-1 in dendritic cells and contributes to trans- and cis-infection pathways." *Blood* Vol.112, No 4, (Aug 15), pp. 1299-307, ISSN 1528-0020

Lamkanfi, M., T. D. Kanneganti, L. Franchi & G. Nunez (2007). "Caspase-1 inflammasomes in infection and inflammation." *J Leukoc Biol* Vol.82, No 2, (Aug), pp. 220-5, ISSN 0741-5400

Le Negrate, G., B. Faustin, K. Welsh, M. Loeffler, M. Krajewska, P. Hasegawa, S. Mukherjee, K. Orth, S. Krajewski, A. Godzik, D. G. Guiney & J. C. Reed (2008). "Salmonella secreted factor L deubiquitinase of Salmonella typhimurium inhibits NF-kappaB, suppresses IkappaBalpha ubiquitination and modulates innate immune responses." *J Immunol* Vol.180, No 7, (Apr 1), pp. 5045-56, ISSN 0022-1767

LeibundGut-Landmann, S., O. Gross, M. J. Robinson, F. Osorio, E. C. Slack, S. V. Tsoni, E. Schweighoffer, V. Tybulewicz, G. D. Brown, J. Ruland & C. Reis e Sousa (2007). "Syk- and CARD9-dependent coupling of innate immunity to the induction of T helper cells that produce interleukin 17." *Nat Immunol* Vol.8, No 6, (Jun), pp. 630-8, ISSN 1529-2908

Lemaitre, B., E. Nicolas, L. Michaut, J. M. Reichhart & J. A. Hoffmann (1996). "The dorsoventral regulatory gene cassette spatzle/Toll/cactus controls the potent antifungal response in Drosophila adults." *Cell* Vol.86, No 6, (Sep 20), pp. 973-83, ISSN 0092-8674

Li, H., H. Xu, Y. Zhou, J. Zhang, C. Long, S. Li, S. Chen, J. M. Zhou & F. Shao (2007). "The phosphothreonine lyase activity of a bacterial type III effector family." *Science* Vol.315, No 5814, (Feb 16), pp. 1000-3, ISSN 1095-9203

Li, X. D., L. Sun, R. B. Seth, G. Pineda & Z. J. Chen (2005). "Hepatitis C virus protease NS3/4A cleaves mitochondrial antiviral signaling protein off the mitochondria to evade innate immunity." *Proc Natl Acad Sci U S A* Vol.102, No 49, (Dec 6), pp. 17717-22, ISSN 0027-8424

Lin, S. L., T. X. Le & D. S. Cowen (2003). "SptP, a Salmonella typhimurium type III-secreted protein, inhibits the mitogen-activated protein kinase pathway by inhibiting Raf activation." *Cell Microbiol* Vol.5, No 4, (Apr), pp. 267-75, ISSN 1462-5814

Lu, C., H. Xu, C. T. Ranjith-Kumar, M. T. Brooks, T. Y. Hou, F. Hu, A. B. Herr, R. K. Strong, C. C. Kao & P. Li (2010). "The structural basis of 5' triphosphate double-stranded RNA recognition by RIG-I C-terminal domain." *Structure* Vol.18, No 8, (Aug 11), pp. 1032-43, ISSN 1878-4186

Malik, Z. A., C. R. Thompson, S. Hashimi, B. Porter, S. S. Iyer & D. J. Kusner (2003). "Cutting edge: Mycobacterium tuberculosis blocks Ca2+ signaling and phagosome maturation in human macrophages via specific inhibition of sphingosine kinase." *J Immunol* Vol.170, No 6, (Mar 15), pp. 2811-5, ISSN 0022-1767

Malmgaard, L. (2004). "Induction and regulation of IFNs during viral infections." *J Interferon Cytokine Res* Vol.24, No 8, (Aug), pp. 439-54, ISSN 1079-9907

Marth, T. & B. L. Kelsall (1997). "Regulation of interleukin-12 by complement receptor 3 signaling." *J Exp Med* Vol.185, No 11, (Jun 2), pp. 1987-95, ISSN 0022-1007

Masumoto, J., K. Yang, S. Varambally, M. Hasegawa, S. A. Tomlins, S. Qiu, Y. Fujimoto, A. Kawasaki, S. J. Foster, Y. Horie, T. W. Mak, G. Nunez, A. M. Chinnaiyan, K. Fukase & N. Inohara (2006). "Nod1 acts as an intracellular receptor to stimulate chemokine production and neutrophil recruitment in vivo." *J Exp Med* Vol.203, No 1, (Jan 23), pp. 203-13, ISSN 0022-1007

Mazurkiewicz, P., J. Thomas, J. A. Thompson, M. Liu, L. Arbibe, P. Sansonetti & D. W. Holden (2008). "SpvC is a Salmonella effector with phosphothreonine lyase activity on host mitogen-activated protein kinases." *Mol Microbiol* Vol.67, No 6, (Mar), pp. 1371-83, ISSN 1365-2958

McDonald, C., N. Inohara & G. Nunez (2005). "Peptidoglycan signaling in innate immunity and inflammatory disease." *J Biol Chem* Vol.280, No 21, (May 27), pp. 20177-80, ISSN 0021-9258

McGettrick, A. F. & L. A. O'Neill "Localisation and trafficking of Toll-like receptors: an important mode of regulation." *Curr Opin Immunol* Vol.22, No 1, (Feb), pp. 20-7, ISSN 1879-0372

McGhie, E. J., L. C. Brawn, P. J. Hume, D. Humphreys & V. Koronakis (2009). "Salmonella takes control: effector-driven manipulation of the host." *Curr Opin Microbiol* Vol.12, No 1, (Feb), pp. 117-24, ISSN 1879-0364

Medzhitov, R. (2007). "Recognition of microorganisms and activation of the immune response." *Nature* Vol.449, No 7164, (Oct 18), pp. 819-26, ISSN 1476-4687

Meyer-Wentrup, F., D. Benitez-Ribas, P. J. Tacken, C. J. Punt, C. G. Figdor, I. J. de Vries & G. J. Adema (2008). "Targeting DCIR on human plasmacytoid dendritic cells results in antigen presentation and inhibits IFN-alpha production." *Blood* Vol.111, No 8, (Apr 15), pp. 4245-53, ISSN 0006-4971

Meyer-Wentrup, F., A. Cambi, B. Joosten, M. W. Looman, I. J. de Vries, C. G. Figdor & G. J. Adema (2009). "DCIR is endocytosed into human dendritic cells and inhibits TLR8-mediated cytokine production." *J Leukoc Biol* Vol.85, No 3, (Mar), pp. 518-25, ISSN 1938-3673

Mittal, R., S. Y. Peak-Chew & H. T. McMahon (2006). "Acetylation of MEK2 and I kappa B kinase (IKK) activation loop residues by YopJ inhibits signaling." *Proc Natl Acad Sci U S A* Vol.103, No 49, (Dec 5), pp. 18574-9, ISSN 0027-8424

Murli, S., R. O. Watson & J. E. Galan (2001). "Role of tyrosine kinases and the tyrosine phosphatase SptP in the interaction of Salmonella with host cells." *Cell Microbiol* Vol.3, No 12, (Dec), pp. 795-810, ISSN 1462-5814

Nadler, C., K. Baruch, S. Kobi, E. Mills, G. Haviv, M. Farago, I. Alkalay, S. Bartfeld, T. F. Meyer, Y. Ben-Neriah & I. Rosenshine (2010). "The type III secretion effector NleE inhibits NF-kappaB activation." *PLoS Pathog* Vol.6, No 1, (Jan), pp. e1000743, ISSN 1553-7374

Netea, M. G., R. Sutmuller, C. Hermann, C. A. Van der Graaf, J. W. Van der Meer, J. H. van Krieken, T. Hartung, G. Adema & B. J. Kullberg (2004). "Toll-like receptor 2 suppresses immunity against Candida albicans through induction of IL-10 and regulatory T cells." *J Immunol* Vol.172, No 6, (Mar 15), pp. 3712-8, ISSN 0022-1767

Neves, B. M., R. Silvestre, M. Resende, A. Ouaissi, J. Cunha, J. Tavares, I. Loureiro, N. Santarem, A. M. Silva, M. C. Lopes, M. T. Cruz & A. Cordeiro da Silva (2010). "Activation of phosphatidylinositol 3-kinase/Akt and impairment of nuclear factor-kappaB: molecular mechanisms behind the arrested maturation/activation state of Leishmania infantum-infected dendritic cells." *Am J Pathol* Vol.177, No 6, (Dec), pp. 2898-911, ISSN 1525-2191

Newton, H. J., J. S. Pearson, L. Badea, M. Kelly, M. Lucas, G. Holloway, K. M. Wagstaff, M. A. Dunstone, J. Sloan, J. C. Whisstock, J. B. Kaper, R. M. Robins-Browne, D. A. Jans, G. Frankel, A. D. Phillips, B. S. Coulson & E. L. Hartland (2010). "The type III effectors NleE and NleB from enteropathogenic E. coli and OspZ from Shigella block nuclear translocation of NF-kappaB p65." *PLoS Pathog* Vol.6, No 5, (May), pp. e1000898, ISSN 1553-7374

Nguyen, T., B. Ghebrehiwet & E. I. Peerschke (2000). "Staphylococcus aureus protein A recognizes platelet gC1qR/p33: a novel mechanism for staphylococcal interactions with platelets." *Infect Immun* Vol.68, No 4, (Apr), pp. 2061-8, ISSN 0019-9567

Olynych, T. J., D. L. Jakeman & J. S. Marshall (2006). "Fungal zymosan induces leukotriene production by human mast cells through a dectin-1-dependent mechanism." *J Allergy Clin Immunol* Vol.118, No 4, (Oct), pp. 837-43, ISSN 0091-6749

Oshiumi, H., M. Matsumoto, K. Funami, T. Akazawa & T. Seya (2003). "TICAM-1, an adaptor molecule that participates in Toll-like receptor 3-mediated interferon-beta induction." *Nat Immunol* Vol.4, No 2, (Feb), pp. 161-7, ISSN 1529-2908

Oshiumi, H., M. Sasai, K. Shida, T. Fujita, M. Matsumoto & T. Seya (2003). "TIR-containing adapter molecule (TICAM)-2, a bridging adapter recruiting to toll-like receptor 4 TICAM-1 that induces interferon-beta." *J Biol Chem* Vol.278, No 50, (Dec 12), pp. 49751-62, ISSN 0021-9258

Osorio, F., S. LeibundGut-Landmann, M. Lochner, K. Lahl, T. Sparwasser, G. Eberl & C. Reis e Sousa (2008). "DC activated via dectin-1 convert Treg into IL-17 producers." *Eur J Immunol* Vol.38, No 12, (Dec), pp. 3274-81, ISSN 1521-4141

Park, J. M., F. R. Greten, Z. W. Li & M. Karin (2002). "Macrophage apoptosis by anthrax lethal factor through p38 MAP kinase inhibition." *Science* Vol.297, No 5589, (Sep 20), pp. 2048-51, ISSN 1095-9203

Pedra, J. H., S. L. Cassel & F. S. Sutterwala (2009). "Sensing pathogens and danger signals by the inflammasome." *Curr Opin Immunol* Vol.21, No 1, (Feb), pp. 10-6, ISSN 1879-0372

Pichlmair, A., O. Schulz, C. P. Tan, T. I. Naslund, P. Liljestrom, F. Weber & C. Reis e Sousa (2006). "RIG-I-mediated antiviral responses to single-stranded RNA bearing 5'-phosphates." *Science* Vol.314, No 5801, (Nov 10), pp. 997-1001, ISSN 1095-9203

Prive, C. & A. Descoteaux (2000). "Leishmania donovani promastigotes evade the activation of mitogen-activated protein kinases p38, c-Jun N-terminal kinase, and extracellular signal-regulated kinase-1/2 during infection of naive macrophages." *Eur J Immunol* Vol.30, No 8, (Aug), pp. 2235-44, ISSN 0014-2980

Qualls, J. E., G. Neale, A. M. Smith, M. S. Koo, A. A. DeFreitas, H. Zhang, G. Kaplan, S. S. Watowich & P. J. Murray (2010). "Arginine usage in mycobacteria-infected macrophages depends on autocrine-paracrine cytokine signaling." *Sci Signal* Vol.3, No 135, ra62, ISSN 1937-9145

Richard, M., N. Thibault, P. Veilleux, G. Gareau-Page & A. D. Beaulieu (2006). "Granulocyte macrophage-colony stimulating factor reduces the affinity of SHP-2 for the ITIM of CLECSF6 in neutrophils: a new mechanism of action for SHP-2." *Mol Immunol* Vol.43, No 10, (Apr), pp. 1716-21, ISSN 0161-5890

Ritter, M., O. Gross, S. Kays, J. Ruland, F. Nimmerjahn, S. Saijo, J. Tschopp, L. E. Layland & C. Prazeres da Costa (2010). "Schistosoma mansoni triggers Dectin-2, which activates the Nlrp3 inflammasome and alters adaptive immune responses." *Proc Natl Acad Sci U S A* Vol.107, No 47, (Nov 23), pp. 20459-64, ISSN 1091-6490

Robinson, M. J., F. Osorio, M. Rosas, R. P. Freitas, E. Schweighoffer, O. Gross, J. S. Verbeek, J. Ruland, V. Tybulewicz, G. D. Brown, L. F. Moita, P. R. Taylor & C. Reis e Sousa (2009). "Dectin-2 is a Syk-coupled pattern recognition receptor crucial for Th17 responses to fungal infection." *J Exp Med* Vol.206, No 9, (Aug 31), pp. 2037-51, ISSN 1540-9538

Rogers, N. C., E. C. Slack, A. D. Edwards, M. A. Nolte, O. Schulz, E. Schweighoffer, D. L. Williams, S. Gordon, V. L. Tybulewicz, G. D. Brown & C. Reis e Sousa (2005). "Syk-dependent cytokine induction by Dectin-1 reveals a novel pattern recognition pathway for C type lectins." *Immunity* Vol.22, No 4, (Apr), pp. 507-17, ISSN 1074-7613

Rothenfusser, S., N. Goutagny, G. DiPerna, M. Gong, B. G. Monks, A. Schoenemeyer, M. Yamamoto, S. Akira & K. A. Fitzgerald (2005). "The RNA helicase Lgp2 inhibits TLR-independent sensing of viral replication by retinoic acid-inducible gene-I." *J Immunol* Vol.175, No 8, (Oct 15), pp. 5260-8, ISSN 0022-1767

Said-Sadier, N., E. Padilla, G. Langsley & D. M. Ojcius (2010). "Aspergillus fumigatus stimulates the NLRP3 inflammasome through a pathway requiring ROS production and the Syk tyrosine kinase." *PLoS One* Vol.5, No 4, e10008, ISSN 1932-6203

Saijo, S., S. Ikeda, K. Yamabe, S. Kakuta, H. Ishigame, A. Akitsu, N. Fujikado, T. Kusaka, S. Kubo, S. H. Chung, R. Komatsu, N. Miura, Y. Adachi, N. Ohno, K. Shibuya, N. Yamamoto, K. Kawakami, S. Yamasaki, T. Saito, S. Akira & Y. Iwakura (2010).

"Dectin-2 recognition of alpha-mannans and induction of Th17 cell differentiation is essential for host defense against Candida albicans." *Immunity* Vol.32, No 5, (May 28), pp. 681-91, ISSN 1097-4180

Saito, T., R. Hirai, Y. M. Loo, D. Owen, C. L. Johnson, S. C. Sinha, S. Akira, T. Fujita & M. Gale, Jr. (2007). "Regulation of innate antiviral defenses through a shared repressor domain in RIG-I and LGP2." *Proc Natl Acad Sci U S A* Vol.104, No 2, (Jan 9), pp. 582-7, ISSN 0027-8424

Sato, K., X. L. Yang, T. Yudate, J. S. Chung, J. Wu, K. Luby-Phelps, R. P. Kimberly, D. Underhill, P. D. Cruz, Jr. & K. Ariizumi (2006). "Dectin-2 is a pattern recognition receptor for fungi that couples with the Fc receptor gamma chain to induce innate immune responses." *J Biol Chem* Vol.281, No 50, (Dec 15), pp. 38854-66, ISSN 0021-9258

Satoh, T., H. Kato, Y. Kumagai, M. Yoneyama, S. Sato, K. Matsushita, T. Tsujimura, T. Fujita, S. Akira & O. Takeuchi (2010). "LGP2 is a positive regulator of RIG-I- and MDA5-mediated antiviral responses." *Proc Natl Acad Sci U S A* Vol.107, No 4, (Jan 26), pp. 1512-7, ISSN 1091-6490

Schlee, M., A. Roth, V. Hornung, C. A. Hagmann, V. Wimmenauer, W. Barchet, C. Coch, M. Janke, A. Mihailovic, G. Wardle, S. Juranek, H. Kato, T. Kawai, H. Poeck, K. A. Fitzgerald, O. Takeuchi, S. Akira, T. Tuschl, E. Latz, J. Ludwig & G. Hartmann (2009). "Recognition of 5' triphosphate by RIG-I helicase requires short blunt double-stranded RNA as contained in panhandle of negative-strand virus." *Immunity* Vol.31, No 1, (Jul 17), pp. 25-34, ISSN 1097-4180

Shapira, S., O. S. Harb, J. Margarit, M. Matrajt, J. Han, A. Hoffmann, B. Freedman, M. J. May, D. S. Roos & C. A. Hunter (2005). "Initiation and termination of NF-kappaB signaling by the intracellular protozoan parasite Toxoplasma gondii." *J Cell Sci* Vol.118, No Pt 15, (Aug 1), pp. 3501-8, ISSN 0021-9533

Shim, J. H., C. Xiao, A. E. Paschal, S. T. Bailey, P. Rao, M. S. Hayden, K. Y. Lee, C. Bussey, M. Steckel, N. Tanaka, G. Yamada, S. Akira, K. Matsumoto & S. Ghosh (2005). "TAK1, but not TAB1 or TAB2, plays an essential role in multiple signaling pathways in vivo." *Genes Dev* Vol.19, No 22, (Nov 15), pp. 2668-81, ISSN 0890-9369

Simmons, D. P., D. H. Canaday, Y. Liu, Q. Li, A. Huang, W. H. Boom & C. V. Harding (2010). "Mycobacterium tuberculosis and TLR2 agonists inhibit induction of type I IFN and class I MHC antigen cross processing by TLR9." *J Immunol* Vol.185, No 4, (Aug 15), pp. 2405-15, ISSN 1550-6606

Slack, E. C., M. J. Robinson, P. Hernanz-Falcon, G. D. Brown, D. L. Williams, E. Schweighoffer, V. L. Tybulewicz & C. Reis e Sousa (2007). "Syk-dependent ERK activation regulates IL-2 and IL-10 production by DC stimulated with zymosan." *Eur J Immunol* Vol.37, No 6, (Jun), pp. 1600-12, ISSN 0014-2980

Smeekens, S. P., F. L. van de Veerdonk, J. W. van der Meer, B. J. Kullberg, L. A. Joosten & M. G. Netea (2010). "The Candida Th17 response is dependent on mannan- and beta-glucan-induced prostaglandin E2." *Int Immunol* Vol.22, No 11, (Nov), pp. 889-95, ISSN 1460-2377

Sporri, R. & C. Reis e Sousa (2005). "Inflammatory mediators are insufficient for full dendritic cell activation and promote expansion of CD4+ T cell populations lacking helper function." *Nat Immunol* Vol.6, No 2, (Feb), pp. 163-70, ISSN 1529-2908

Steinman, R. M. (2006). "Linking innate to adaptive immunity through dendritic cells." *Novartis Found Symp* Vol.279, No, 101-9; discussion 109-13, 216-9, ISSN 1528-2511

Suram, S., G. D. Brown, M. Ghosh, S. Gordon, R. Loper, P. R. Taylor, S. Akira, S. Uematsu, D. L. Williams & C. C. Leslie (2006). "Regulation of cytosolic phospholipase A2 activation and cyclooxygenase 2 expression in macrophages by the beta-glucan receptor." *J Biol Chem* Vol.281, No 9, (Mar 3), pp. 5506-14, ISSN 0021-9258

Suzuki, N., S. Suzuki, G. S. Duncan, D. G. Millar, T. Wada, C. Mirtsos, H. Takada, A. Wakeham, A. Itie, S. Li, J. M. Penninger, H. Wesche, P. S. Ohashi, T. W. Mak & W. C. Yeh (2002). "Severe impairment of interleukin-1 and Toll-like receptor signalling in mice lacking IRAK-4." *Nature* Vol.416, No 6882, (Apr 18), pp. 750-6, ISSN 0028-0836

Swantek, J. L., M. F. Tsen, M. H. Cobb & J. A. Thomas (2000). "IL-1 receptor-associated kinase modulates host responsiveness to endotoxin." *J Immunol* Vol.164, No 8, (Apr 15), pp. 4301-6, ISSN 0022-1767

Takahashi, K., T. Kawai, H. Kumar, S. Sato, S. Yonehara & S. Akira (2006). "Roles of caspase-8 and caspase-10 in innate immune responses to double-stranded RNA." *J Immunol* Vol.176, No 8, (Apr 15), pp. 4520-4, ISSN 0022-1767

Takahasi, K., M. Yoneyama, T. Nishihori, R. Hirai, H. Kumeta, R. Narita, M. Gale, Jr., F. Inagaki & T. Fujita (2008). "Nonself RNA-sensing mechanism of RIG-I helicase and activation of antiviral immune responses." *Mol Cell* Vol.29, No 4, (Feb 29), pp. 428-40, ISSN 1097-4164

Takeda, K. & S. Akira (2004). "TLR signaling pathways." *Semin Immunol* Vol.16, No 1, (Feb), pp. 3-9, ISSN 1044-5323

Takeuchi, O. & S. Akira (2010). "Pattern recognition receptors and inflammation." *Cell* Vol.140, No 6, (Mar 19), pp. 805-20, ISSN 1097-4172

Taniguchi, T., K. Ogasawara, A. Takaoka & N. Tanaka (2001). "IRF family of transcription factors as regulators of host defense." *Annu Rev Immunol* Vol.19, No, 623-55, ISSN 0732-0582

Taylor, P. R., G. D. Brown, D. M. Reid, J. A. Willment, L. Martinez-Pomares, S. Gordon & S. Y. Wong (2002). "The beta-glucan receptor, dectin-1, is predominantly expressed on the surface of cells of the monocyte/macrophage and neutrophil lineages." *J Immunol* Vol.169, No 7, (Oct 1), pp. 3876-82, ISSN 0022-1767

Taylor, P. R., S. V. Tsoni, J. A. Willment, K. M. Dennehy, M. Rosas, H. Findon, K. Haynes, C. Steele, M. Botto, S. Gordon & G. D. Brown (2007). "Dectin-1 is required for beta-glucan recognition and control of fungal infection." *Nat Immunol* Vol.8, No 1, (Jan), pp. 31-8, ISSN 1529-2908

Trosky, J. E., Y. Li, S. Mukherjee, G. Keitany, H. Ball & K. Orth (2007). "VopA inhibits ATP binding by acetylating the catalytic loop of MAPK kinases." *J Biol Chem* Vol.282, No 47, (Nov 23), pp. 34299-305, ISSN 0021-9258

Tsuji, S., M. Matsumoto, O. Takeuchi, S. Akira, I. Azuma, A. Hayashi, K. Toyoshima & T. Seya (2000). "Maturation of human dendritic cells by cell wall skeleton of Mycobacterium bovis bacillus Calmette-Guerin: involvement of toll-like receptors." *Infect Immun* Vol.68, No 12, (Dec), pp. 6883-90, ISSN 0019-9567

Turnbull, P. C. (2002). "Introduction: anthrax history, disease and ecology." *Curr Top Microbiol Immunol* Vol.271, No, 1-19, ISSN 0070-217X van Gisbergen, K. P., M. Sanchez-Hernandez, T. B. Geijtenbeek & Y. van Kooyk (2005). "Neutrophils mediate immune modulation of dendritic cells through glycosylation-dependent interactions between Mac-1 and DC-SIGN." *J Exp Med* Vol.201, No 8, (Apr 18), pp. 1281-92, ISSN 0022-1007

van Kooyk, Y. & T. B. Geijtenbeek (2003). "DC-SIGN: escape mechanism for pathogens." *Nat Rev Immunol* Vol.3, No 9, (Sep), pp. 697-709, ISSN 1474-1733

van Vliet, S. J., J. J. Garcia-Vallejo & Y. van Kooyk (2008). "Dendritic cells and C-type lectin receptors: coupling innate to adaptive immune responses." *Immunol Cell Biol* Vol.86, No 7, (Oct), pp. 580-7, ISSN 0818-9641

Viboud, G. I. & J. B. Bliska (2005). "Yersinia outer proteins: role in modulation of host cell signaling responses and pathogenesis." *Annu Rev Microbiol* Vol.59, No, 69-89, ISSN 0066-4227

Waggoner, S. N., C. H. Hall & Y. S. Hahn (2007). "HCV core protein interaction with gC1q receptor inhibits Th1 differentiation of CD4+ T cells via suppression of dendritic cell IL-12 production." *J Leukoc Biol* Vol.82, No 6, (Dec), pp. 1407-19, ISSN 0741-5400

Wang, C., L. Deng, M. Hong, G. R. Akkaraju, J. Inoue & Z. J. Chen (2001). "TAK1 is a ubiquitin-dependent kinase of MKK and IKK." *Nature* Vol.412, No 6844, (Jul 19), pp. 346-51, ISSN 0028-0836

Watson, A. A., A. A. Lebedev, B. A. Hall, A. E. Fenton-May, A. A. Vagin, W. Dejnirattisai, J. Felce, J. Mongkolsapaya, A. S. Palma, Y. Liu, T. Feizi, G. R. Screaton, G. N. Murshudov & C. A. O'Callaghan (2011). "Structural flexibility of the macrophage dengue virus receptor CLEC5A: implications for ligand binding and signaling." *J Biol Chem* Vol.286, No 27, (Jul 8), pp. 24208-18, ISSN 1083-351X

Weis, W. I., M. E. Taylor & K. Drickamer (1998). "The C-type lectin superfamily in the immune system." *Immunol Rev* Vol.163, No, (Jun), pp. 19-34, ISSN 0105-2896

Wells, C. A., J. A. Salvage-Jones, X. Li, K. Hitchens, S. Butcher, R. Z. Murray, A. G. Beckhouse, Y. L. Lo, S. Manzanero, C. Cobbold, K. Schroder, B. Ma, S. Orr, L. Stewart, D. Lebus, P. Sobieszczuk, D. A. Hume, J. Stow, H. Blanchard & R. B. Ashman (2008). "The macrophage-inducible C-type lectin, mincle, is an essential component of the innate immune response to Candida albicans." *J Immunol* Vol.180, No 11, (Jun 1), pp. 7404-13, ISSN 0022-1767

Werninghaus, K., A. Babiak, O. Gross, C. Holscher, H. Dietrich, E. M. Agger, J. Mages, A. Mocsai, H. Schoenen, K. Finger, F. Nimmerjahn, G. D. Brown, C. Kirschning, A. Heit, P. Andersen, H. Wagner, J. Ruland & R. Lang (2009). "Adjuvanticity of a synthetic cord factor analogue for subunit Mycobacterium tuberculosis vaccination requires FcRgamma-Syk-Card9-dependent innate immune activation." *J Exp Med* Vol.206, No 1, (Jan 16), pp. 89-97, ISSN 1540-9538

Werts, C., L. le Bourhis, J. Liu, J. G. Magalhaes, L. A. Carneiro, J. H. Fritz, S. Stockinger, V. Balloy, M. Chignard, T. Decker, D. J. Philpott, X. Ma & S. E. Girardin (2007). "Nod1 and Nod2 induce CCL5/RANTES through the NF-kappaB pathway." *Eur J Immunol* Vol.37, No 9, (Sep), pp. 2499-508, ISSN 0014-2980

Wilkins, C. & M. Gale, Jr. (2010). "Recognition of viruses by cytoplasmic sensors." *Curr Opin Immunol* Vol.22, No 1, (Feb), pp. 41-7, ISSN 1879-0372

Xu, S., J. Huo, K. G. Lee, T. Kurosaki & K. P. Lam (2009). "Phospholipase Cgamma2 is critical for Dectin-1-mediated Ca2+ flux and cytokine production in dendritic cells." *J Biol Chem* Vol.284, No 11, (Mar 13), pp. 7038-46, ISSN 0021-9258

Yamamoto, M., S. Sato, H. Hemmi, K. Hoshino, T. Kaisho, H. Sanjo, O. Takeuchi, M. Sugiyama, M. Okabe, K. Takeda & S. Akira (2003). "Role of adaptor TRIF in the MyD88-independent toll-like receptor signaling pathway." *Science* Vol.301, No 5633, (Aug 1), pp. 640-3, ISSN 1095-9203

Yamamoto, M., S. Sato, H. Hemmi, H. Sanjo, S. Uematsu, T. Kaisho, K. Hoshino, O. Takeuchi, M. Kobayashi, T. Fujita, K. Takeda & S. Akira (2002). "Essential role for TIRAP in activation of the signalling cascade shared by TLR2 and TLR4." *Nature* Vol.420, No 6913, (Nov 21), pp. 324-9, ISSN 0028-0836

Yamamoto, M., S. Sato, K. Mori, K. Hoshino, O. Takeuchi, K. Takeda & S. Akira (2002). "Cutting edge: a novel Toll/IL-1 receptor domain-containing adapter that preferentially activates the IFN-beta promoter in the Toll-like receptor signaling." *J Immunol* Vol.169, No 12, (Dec 15), pp. 6668-72, ISSN 0022-1767

Yamamoto, M., K. Takeda & S. Akira (2004). "TIR domain-containing adaptors define the specificity of TLR signaling." *Mol Immunol* Vol.40, No 12, (Feb), pp. 861-8, ISSN 0161-5890

Yamasaki, S., M. Matsumoto, O. Takeuchi, T. Matsuzawa, E. Ishikawa, M. Sakuma, H. Tateno, J. Uno, J. Hirabayashi, Y. Mikami, K. Takeda, S. Akira & T. Saito (2009). "C-type lectin Mincle is an activating receptor for pathogenic fungus, Malassezia." *Proc Natl Acad Sci U S A* Vol.106, No 6, (Feb 10), pp. 1897-902, ISSN 1091-6490

Ye, Z., E. O. Petrof, D. Boone, E. C. Claud & J. Sun (2007). "Salmonella effector AvrA regulation of colonic epithelial cell inflammation by deubiquitination." *Am J Pathol* Vol.171, No 3, (Sep), pp. 882-92, ISSN 0002-9440

Yen, H., T. Ooka, A. Iguchi, T. Hayashi, N. Sugimoto & T. Tobe "NleC, a type III secretion protease, compromises NF-kappaB activation by targeting p65/RelA." *PLoS Pathog* Vol.6, No 12, e1001231, ISSN 1553-7374

Yoneyama, M., M. Kikuchi, K. Matsumoto, T. Imaizumi, M. Miyagishi, K. Taira, E. Foy, Y. M. Loo, M. Gale, Jr., S. Akira, S. Yonehara, A. Kato & T. Fujita (2005). "Shared and unique functions of the DExD/H-box helicases RIG-I, MDA5, and LGP2 in antiviral innate immunity." *J Immunol* Vol.175, No 5, (Sep 1), pp. 2851-8, ISSN 0022-1767

Yoneyama, M., M. Kikuchi, T. Natsukawa, N. Shinobu, T. Imaizumi, M. Miyagishi, K. Taira, S. Akira & T. Fujita (2004). "The RNA helicase RIG-I has an essential function in double-stranded RNA-induced innate antiviral responses." *Nat Immunol* Vol.5, No 7, (Jul), pp. 730-7, ISSN 1529-2908

Zhang, D., G. Zhang, M. S. Hayden, M. B. Greenblatt, C. Bussey, R. A. Flavell & S. Ghosh (2004). "A toll-like receptor that prevents infection by uropathogenic bacteria." *Science* Vol.303, No 5663, (Mar 5), pp. 1522-6, ISSN 1095-9203

Zhang, X., L. Majlessi, E. Deriaud, C. Leclerc & R. Lo-Man (2009). "Coactivation of Syk kinase and MyD88 adaptor protein pathways by bacteria promotes regulatory properties of neutrophils." *Immunity* Vol.31, No 5, (Nov 20), pp. 761-71, ISSN 1097-4180

Zhou, H., D. M. Monack, N. Kayagaki, I. Wertz, J. Yin, B. Wolf & V. M. Dixit (2005). "Yersinia virulence factor YopJ acts as a deubiquitinase to inhibit NF-kappa B activation." *J Exp Med* Vol.202, No 10, (Nov 21), pp. 1327-32, ISSN 0022-1007

7

Phosphorylation-Regulated Cell Surface Expression of Membrane Proteins

Yukari Okamoto and Sojin Shikano
University of Illinois at Chicago,
USA

1. Introduction

To maintain its functional integrity, a cell senses and reacts to the acute or chronic changes in the environment under physiological and pathological conditions. This typically involves cell surface membrane proteins such as receptors, ion channels, and structural proteins, whose surface expression level is regulated at multiple different steps of their biosynthesis and trafficking. Protein trafficking is mediated by a series of dynamic interactions between the sorting motifs of cargo proteins and the cellular machineries that recognize these motifs. While the constitutive trafficking of many cargo proteins relies on intrinsic sorting signals, post-translational modification of cargo proteins often serves as a key switch that enables the spatio-temporal regulation of their trafficking. Protein phosphorylation is one of the most intensively studied post-translational modifications that control the membrane trafficking. However, molecular mechanisms by which phosphorylation signal regulates the protein localization are diverse and remain not fully understood. The 14-3-3 proteins had been identified to specifically recognize phosphorylated serine or threonine residues, and thus represents one of the most distinct effector molecules that function downstream of the phosphorylation signal by kinases. This chapter will focus on the emerging role of 14-3-3 proteins in the phosphorylation-dependent control of cell surface membrane protein trafficking.

2. Control of cell surface expression by phosphorylation signal

A typical mechanism by which phosphorylation signal controls protein trafficking is that phosphorylation of cargo proteins creates docking sites for the interacting proteins. A well studied example is the internalization of G protein-coupled receptors (GPCRs), where ligand binding induces the conformational change of the receptor and subsequent recruitment of GPCR kinases (GRKs) to the receptor. Receptor phosphorylation by GRKs recruits arrestin that couples the receptor to the adaptor protein of clathrin coat, thereby initiating the internalization of the cargo vesicles (Drake *et al.*, 2006; Tobin, 2008). This way phosphorylation signal leads to the desensitization of ligand stimulus by reducing the cell surface density of GPCRs. On the other hand, phosphorylation signalling can also regulate localization of membrane proteins by attenuating the sorting signal activity. In the neuron, ligand stimulation of N-methyl-D-aspartate (NMDA) receptor leads to receptor phosphorylation by casein kinase II (CK2) at the serine residue within the C-terminal PDZ [postsynaptic density-95 (PSD-95)/Discs large/zona occludens-1] binding motif (IESDV-COOH) of NMDA receptor subunit 2B (NR2B). CK2 phosphorylation disrupts the interaction of NR2B with the PDZ domains of PSD-95 and SAP102 and thereby decreases cell surface NR2B expression (Chung *et al.*, 2004). This represents the regulatory role of phosphorylation in excitatory synaptic function and plasticity.

In contrast to the downregulation of surface expression, the molecular basis for the role of phosphorylation in promoting cell surface trafficking has been less well understood. However, studies in the past decade have revealed that 14-3-3s are the key class of phospho-sensing proteins which mediate cell surface trafficking of various membrane proteins. Here we will review recent findings on the emerging role of 14-3-3 in cell surface protein trafficking, with particular focus on their mechanisms of action and relevant kinases.

2.1 14-3-3 proteins and 14-3-3 binding sites

The 14-3-3 proteins were first discovered in 1967 as brain-rich, acidic protein (Moore & Perez, 1967). The name 14-3-3 refers to the elution and migration profile of these proteins on DEAE-cellulose chromatography and starch gel electrophoresis. They are highly conserved and expressed in all eukaryotic cells, with seven isoforms in mammals (β, γ, ε, ζ, η, τ, σ) and two in yeast (Bmh1 and Bmh2). 14-3-3 proteins participate in fundamental biological processes such as signal transduction, metabolism, protein degradation, and trafficking (Tzivion & Avruch, 2002; van Hemert *et al.*, 2001). All the 14-3-3 proteins, except for the sigma isoform, are able to form stable homo- and heterodimers (Benzinger *et al.*, 2005; Gardino *et al.*, 2006; Wilker *et al.*, 2005). The dimeric structure of the 14-3-3 protein allows it to simultaneously bind two binding sites through an amphipathic ligand-binding groove present in each monomer. In the majority of cases, 14-3-3 proteins recognize phosphorylated peptides in their binding partners. Screening of phosphoserine-oriented peptide libraries has identified two consensus 14-3-3 binding motifs that are present in many of known 14-3-3 binding proteins (Yaffe *et al.*, 1997). These are R-[S/ϕ]-X-pS/pT-X-P (mode I) and R-X-[S/ϕ]-X-pS/pT-X-P (mode II) binding sites where pS/pT is phosphoserine or phosphothreonine, ϕ is an aromatic residue, and X is any residue (typically leucine, glutamate, alanine, and methionine). However, it should be noted that 14-3-3-binding sites in numerous proteins do not conform to these optimal motifs, presumably because other structural features also contribute to the interactions. For instance, proline located at position +2 of the phosphorylation site occurs in only about half of known 14-3-3 binding motifs in mammalian proteins (Johnson *et al.*, 2010).

The C-terminal 14-3-3 binding motifs have recently become a newly recognized group with a distinct mode of interaction (see Table 1). Based on the similarity between the C-terminal 14-3-3 binding motifs of the oAANAT (RRNpSDR-COOH) and H^+-ATPase (QQXYpTV-COOH) proteins, a new mode III consensus for 14-3-3 binding (pSX_{1-2}-COOH) had been proposed (Ganguly *et al.*, 2005). The focal points of this consensus are that the motif is at the extreme C-terminus in contrast to the canonical mode I and II internal binding sites, and that the binding is phosphorylation-dependent. Mode III sequences interact with the same ligand-binding groove of 14-3-3 as do the mode I and mode II motifs (Coblitz *et al.*, 2005). Amino acid selectivity upstream of the phosphorylated residue is conspicuously absent from the proposed mode III motif, presumably due to the discrepancy between the oAANAT and H^+-ATPase motifs. However, upstream arginine residues are preferred for modes I and II 14-3-3 binding as determined by random synthetic peptide library screening

-R-S/ϕ-X-S/T-X-P-	Mode I 14-3-3 binding	
-R-X-S/ϕ-X-S/T-X-P-	Mode II 14-3-3 binding	
	Reported Mode III 14-3-3 binding proteins	Reference
SIRYSGHSL-COOH	Ibα of Ib-IX-V complex	1
MSKARSWTF-COOH	IL-9Rα receptor	2
RRSSV-COOH	TASK-1 channel	3
RRKSV-COOH	TASK-3 channel	4
RGRSWTY-COOH	RGRSWTY	5
RKRSVSL-COOH	GPR15 receptor	6
SYRSSTL-COOH	HAP1A	7
QQSYTV-COOH	Plant plasma membrane H^+-ATPase	8
RRNSDR-COOH	AANAT acetyltransferase	9
RRRQT-COOH	p27Kip1 cyclin kinase inhibitor	10

Table 1. 14-3-3 binding sequences. Consensus mode I and mode II motifs and the reported C-terminal mode III binding sequences are shown. S/T: phosphorylated serine or threonine required for 14-3-3 binding, ϕ: aromatic residue, Reference#1: (Bodnar *et al.*, 1999) , #2: (Sliva *et al.*, 2000), #3 and #4: (O'Kelly *et al.*, 2002):, #5 and #6: (Shikano *et al.*, 2005), #7:(Rong *et al.*, 2007), #8: (Wurtele *et al.*, 2003), #9: (Ganguly *et al.*, 2005), #10: (Fujita *et al.*, 2003).

(Yaffe *et al.*, 1997) and by random C-terminal peptide selection in a cell-based genetic screen (Shikano *et al.*, 2005). Indeed, the majority of the so far identified C-terminal 14-3-3 binding sequences contain arginine residues upstream of the phosphorylated serine or threonine (Table 1). Recent mutagenesis study of the C-terminal 14-3-3 binding site in GPR15 demonstrated the importance of the upstream arginine residue for phosphorylation-dependent 14-3-3 binding (Okamoto & Shikano, 2011). In a crystal structure with 14-3-3, a mode II peptide displayed an arginine in the -4 position from phosphorylated serine (*R*LYHpSLPA) that was looped back to interact with the phosphate on the peptide (Rittinger *et al.*, 1999). These lines of evidence support significant contribution of upstream arginine residues to the 14-3-3 affinity. Thus, Mode III would be better defined as RXXpS/pTX-COOH. For all three modes of 14-3-3 binding, phosphorylation is a prerequisite and arginine residues located upstream of the phospho-serine/threonine are also important for recognition by a number of kinases (Kobe *et al.*, 2005). Thus, the absence of an arginine residue in the C-terminal 14-3-3 binding sequence in plant H^+-ATPase (QQXYpTV-COOH) suggests the possibility that plant and animal differ significantly in kinase recognition. As more C-terminal 14-3-3 binding proteins become available, it would be valuable to revisit the issue of upstream sequence requirements both in terms of 14-3-3 binding *per se* and in terms of kinase recognition.

2.2 Protein kinases that phosphorylate 14-3-3 target sites

Proteomic screens have identified over 200 phosphoproteins which interact with 14-3-3 (Chang *et al.*, 2009; Ichimura *et al.*, 2002; Kakiuchi *et al.*, 2007; Meek *et al.*, 2004; Pozuelo Rubio *et al.*, 2004). Understanding when and how 14-3-3 proteins impact on these targets offers a great opportunity to gain mechanistic insights into many phosphorylation-regulated biological pathways. Since 14-3-3 target sites in proteins must satisfy the specificity requirements for both 14-3-3s and the protein kinases that create the sites in the first place, identification of kinases that phosphorylate 14-3-3 target sites is crucial for elucidating the physiological roles of 14-3-3 binding. Unfortunately, in the majority of cases the identity of the physiologically relevant kinases that phosphorylate the mode I, II or III 14-3-3 binding motifs is still unknown. This is largely due to the high similarities between different serine/threonine protein kinase recognition sites (Table 2) and their likely redundant activities. One reasonable approach to overcome such obstacles and gain better understanding of the physiological kinases for 14-3-3 target proteins would be the global analysis of the actual 14-3-3-binding phosphoproteins. Based on the proteomics data and other available literature on 14-3-3, Johnson *et al.* have recently attempted to define 14-3-3 specificity and identify relevant protein kinases (Johnson *et al.*, 2010). This study points out several features that are distinctive of 14-3-3-binding sequences as compared with other protein phosphorylation sites. For instance, few reported 14-3-3-binding sites have a +1 (relative to phosphorylated serine/threonine) proline residue, which contrasts with phosphoproteomic studies of cell lysates and subcellular fractions where phosphoserine-proline is the most commonly reported phosphorylation motif overall (Ubersax & Ferrell, 2007). This indicates that proline-directed kinases do not phosphorylate 14-3-3-binding sites. Similarly, no reported 14-3-3-binding sites conform to the canonical consensus site for casein kinase II (pS/pT-X-X-D/S/pS), which is probably the second most common type of motif in the entire mammalian phosphoproteome (Salvi *et al.*, 2009). Another interesting notion is that, while the optimal mode I (R-[S/ϕ]-X-pS/pT-X-P) and mode II (R-X-[S/ϕ]-X-pS/pT-X-P) 14-3-3-binding motifs were defined using phosphopeptides, many 14-3-3-binding sites in mammalian proteins (but not in plant 14-3-3-binding proteins) have basic residues in position -5 (and -4) in addition to -3. This creates a motif RXRXXS/T, which is a good target for the basophilic AGC kinase family (cAMP-dependent protein kinases A, cGMP-dependent protein kinases G, and phospholipid-dependent protein kinases) and the calcium/calmodulin-dependent kinase (CaMK) family (Pearson & Kemp, 1991). Indeed, members of these kinase families, including protein kinase A (PKA), protein kinase C (PKC), CaMKI, checkpoint kinases 1 and 2 (Chk1 and 2), Akt/protein kinase B (PKB) and p90 ribosomal S6 kinase (p90Rsk), are all known to phosphorylate sites that mediate 14-3-3 binding (Dougherty & Morrison, 2004). Among these, Akt is one of the most well documented kinases in phosphorylating 14-3-3 client proteins (Mackintosh, 2004).

R-X-R-X-X-S	Akt/PKB
R-X-R-X-X-S	SGK
R-X-R-X-X-S	S6 kinase
K/R-X-X-S	PKA
K/R-X-X-S-X-K/R	PKC
[MVLIF]-X-R-X-X-S-X-X-X-[MVLIF]	CaMK1
R-X-X-S	CaMK2

Table 2. Consensus recognition sequences of major serine/threonine kinases that are known to phosphorylate 14-3-3 binding site.

2.3 Molecular mechanism for the 14-3-3 effects on membrane protein trafficking

The phospho-binding ability of 14-3-3 proteins is reminiscent of other proteins carrying specific modules that recognize phosphorylated sites. Such modules include FHA (Durocher *et al.*, 2000), WD40 (Yaffe & Elia, 2001), Polo-box (Lowery *et al.*, 2005), and BRCT (BRCA-1 C-terminal) repeat domains (Manke *et al.*, 2003), which target serine and threonine phosphorylation, as well as SH2 (Src-homology 2) domains which target phosphorylated tyrosine residues in specific sequence contexts (Bradshaw & Waksman, 2002). These domains are found in a large number of proteins involved in a wide range of signaling processes. 14-3-3s are distinct from those proteins in that 14-3-3s are not modular components of other proteins. They are discrete binding proteins with no intrinsic enzyme activities, except for the nucleoside diphosphate (NDP) kinase-like activity (Yano *et al.*, 1997) and chaperonic activity toward selected substrates (Yano *et al.*, 2006). So, how do 14-3-3s exert their effects? Several excellent reviews discuss different models of 14-3-3 action as masking, scaffolding, or clamping of proteins (Dougherty & Morrison, 2004; Mackintosh, 2004; Mrowiec & Schwappach, 2006). Recruitment of proteins may be regulated by masking of functional signals by 14-3-3 binding. Alternatively, the scaffolding model suggests that 14-3-3 proteins tether different molecules together and form a platform for complex assembly. Clamping describes the idea that 14-3-3 binding alters the functional property of the client protein by stabilizing a certain conformation (Figure 1). It should be noted that combinations of these 'masking', 'scaffolding', and 'clamping' types of 14-3-3 action may occur together. As far as membrane proteins are concerned, very few examples exist where the interactions of 14-3-3 with client proteins are sufficiently well understood to ascribe a particular mode of action. Nevertheless, 14-3-3s may be considered as general switch proteins, of which effect of binding depends on the client protein. In most cases, phosphorylation at a serine or threonine residue activates the switch, and the subsequent binding of 14-3-3 proteins is thought to prevent rapid dephosphorylation. Recent studies have firmly demonstrated that 14-3-3 proteins are involved in controlling cell surface expression level of various cargo membrane proteins. We will discuss the pertinent evidence and hypothetical molecular mechanisms explaining these observations, with attention to the relevant kinases that regulate 14-3-3 binding.

2.3.1 Masking

14-3-3s are implicated in regulating the subcellular localization of many phosphorylated target proteins. The majority of the cases seem to involve the mechanism where 14-3-3 binding blocks the access of other proteins to the sorting signal of target proteins. Such a 'masking' role of 14-3-3 was first implicated in the mitochondria-cytoplasm translocation of the pro-apoptotic protein BAD. BAD interferes with the anti-apoptotic function of Bcl-2 and Bcl-x_L in the mitochondria by binding to those proteins via its BH3 domain (Zha *et al.*, 1996). BH3 domain is located immediately adjacent to the serine^{136}, of which phosphoryaltion by Akt (Datta *et al.*, 1997) is required for the binding of BAD to 14-3-3 in the cytoplasm (Zha *et al.*, 1997). These results suggested that 14-3-3 binding obscures the BH3 domain and prevents the targeting of BAD to mitochondria. A similar mechanism was found for the nuclear-cytoplasmic shuttling of various proteins including tyrosine phosphatase Cdc25C (Kumagai & Dunphy, 1999), transcription factor FKHRL1 (Brunet *et al.*, 1999), glucocorticoidreceptor (Kino *et al.*, 2003), CDK (cyclin-dependent kinase) inhibitor p27

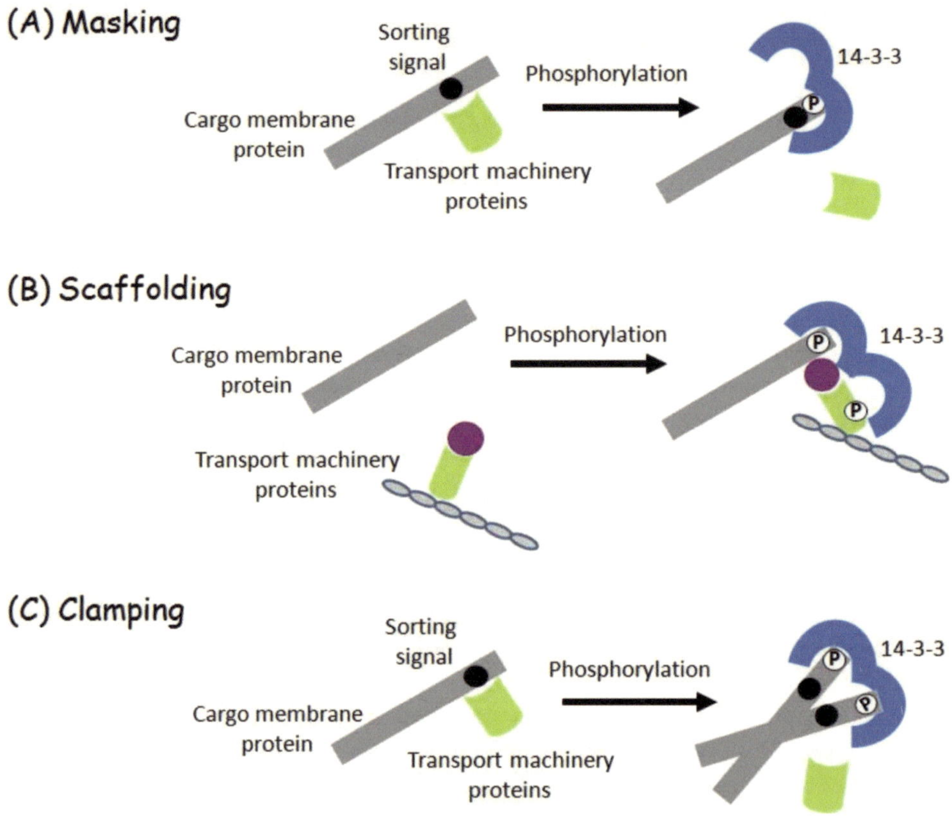

Fig. 1. Hypothetical models for the effects of 14-3-3 binding on the trafficking of membrane proteins. **(A)** *Masking*. Sorting signals (e.g., RXR motif) are physically masked by 14-3-3 binding to the nearby phosphorylated target site (shown as Ⓟ). This blocks the access of transport machineries (e.g., COPI proteins) to the sorting signal. Only the action of one 14-3-3 monomer is drawn here. **(B)** *Scaffolding*. Dimeric 14-3-3 facilitates the interaction of cargo proteins with transport machineries (e.g., motor proteins and microtubules). Binding of 14-3-3 dimer to two different targets might involve phosphorylation-independent and/or outer surface-mediated interaction (see main text). **(C)** *Clamping*. Binding of a 14-3-3 dimer induces a conformation that is unfavorable for sorting signals (e.g., RXR motif), which can be achieved by clustering of targets or relocation from active zones (e.g., proximity to the transmembrane region (see main text)). This could result in reduced accessibility by transport machineries (e.g., COPI proteins).

(Fujita *et al.*, 2003), and catalytic subunit of telomerase TERT (Seimiya *et al.*, 2000). In most cases, 14-3-3 binding promotes the cytoplasmic localization of target proteins. For instance, 14-3-3 binding to serine^{287} of Cdc25C leads to the cytoplasmic retention of Cdc25C. This seems to be due to the occlusion of the closely located bipartite nuclear localization sequence (NLS) at amino acids 298-316 of Cdc25C from importin-α, a receptor for bipartite NLS (Kumagai & Dunphy, 1999). These observations are consistent with the predominant cytoplasmic localization of 14-3-3 proteins at the steady-state, which has led to the hypothesis that they might serve as a universal cytoplasmic anchor that blocks import into the nucleus or other organelles. However, this model of 14-3-3 action is contradicted by the observation that 14-3-3 can also promote the nuclear localization of other binding partners. Seimiya *et al.* found that 14-3-3 binding to human TERT, which requires threonine^{1030}, serine^{1037} and serine^{1041}, lead to the nuclear localization of hTERT. This was attributed to the blocking of the nearby leucine-rich nuclear export sequence (NES) present at residues 970–981 of hTERT, which is otherwise recognized by the nuclear export receptor CRM1 (Seimiya *et al.*, 2000). So, how does 14-3-3 regulate nuclear-cytoplasmic shuttling of different proteins in two opposing directions? One simple hypothesis may be that 14-3-3 itself bears no specific information about protein sorting and the effect of 14-3-3 binding on the subcellular localization of a client protein depends entirely on sorting signals encoded within the client protein and the proximity of those signals to a 14-3-3 binding motif. If so, 14-3-3 binding could possibly affect any other protein sorting pathways.

Consistent with this idea, recent studies report a critical role of 14-3-3 in controlling the cell surface expression level of membrane proteins of different functions, including ion channels, receptors, and adhesion molecules. In many cases, the underlying mechanism seems to involve 'masking' of a short sequence motif, namely a di-arginine or di-basic (RXR) ER localization signal. The RXR-type ER-localisation signals were first identified in ATP-sensitive potassium channels (K_{ATP} channels) (Zerangue *et al.*, 1999). These channels assemble as octameric complexes consisting of four Kir6 channel subunits and four sulphonylurea receptor (SUR) subunits, and each of these subunits carries an RXR signal in the cytoplasmic tail. It is believed that the ER-localisation activity of these motifs is mediated by the retrieval of cargo proteins from the post-ER compartments such as ER-Golgi intermediate compartment (ERGIC) and cis-Glogi through their interaction with the retrograde transport coatomer protein, COPI (Michelsen *et al.*, 2007; Zerangue *et al.*, 1999). Although the RXR motif resembles the C-terminal di-lysine (KKXX) motif which also mediates ER retrieval of membrane protein cargos through direct interaction with COPI (Michelsen *et al.*, 2007), RXR motif is distinct in that it is present almost exclusively in the multimeric cell surface membrane proteins (Michelsen *et al.*, 2005), while KKXX motif is found in the membrane proteins that are resident to the ER such as nucleotide sugar transporters (Jackson *et al.*, 1990, 1993).

Efficient cell surface transport of proteins harbouring RXR motifs will be allowed only when the motif becomes inaccessible to COPI probably by multiple different mechanisms including folding, subunit assembly, post-translational modification, and protein recruitment. 14-3-3 proteins have been implicated in the masking of RXR motif in several instances. The first evidence that 14-3-3 proteins control cell surface transport of membrane protein was shown for two-pore-domain potassium (K_{2P}) channels TASK1 and TASK3 (O'Kelly *et al.*, 2002). These channels bound to 14-3-3 via a mode III C-terminal binding motif (RRSSV-COOH and RRKSV-COOH for TASK1 and TASK3, respectively), where phosphorylation of the penultimate serine and the upstream arginine residues were critically required. In the absence of phosphorylation on the penultimate serine, COPI proteins bind to the adjacent RXR-like sequence (KRR) which shares two arginine residues with the 14-3-3 binding motif. It is thought that phosphorylation switch allows 14-3-3 binding that occludes the partially overlapping RXR-like COPI binding motif. A very recent study by Mant *et al.* reports that PKA, which recognizes a consensus sequence of RXXS/T, phosphorylates the penultimate serine residue in TASK1 and TASK3 channels and promote their expression on cell surface, although it was not shown in the literature whether this phosphorylation actually promotes the binding of 14-3-3 proteins to these sites (Mant *et al.*, 2011).

A similar masking mechanism was suggested for the ER-export of Iip35 isoform of major histocompatibility complex (MHC) class-II-associated invariant chain (O'Kelly *et al.*, 2002). During their assembly in the ER, MHC class-II αβ dimers associate with preformed trimers of the invariant chains (Iip33, Iip35, Iip41, Iip43), to form nonameric $(\alpha\beta Ii)_3$ oligomers. It had been known that phosphorylation of an N-terminal serine[8], present exclusively in the Iip35 cytoplasmic tail (NH_3-MHRRRSRS...), is a prerequisite for efficient ER exit and sorting of class-II/Iip35 complexes to the cell surface. Phosphorylation of serine[8] leads to the 14-3-3 binding to Iip35 and the alanine mutation on this residue inhibits the ER exit of class-II/Iip35 complex (Kuwana *et al.*, 1998). Together with the fact that 14-3-3 and COPI bound to the N-terminal sequence of Iip35 in a mutually exclusive manner, it was concluded that the cell surface transport of MHC class-II complex is promoted by 'masking' effect of 14-3-3 (O'Kelly *et al.*, 2002). However, as Khalil *et al.* pointed out (Khalil *et al.*, 2005), this model seems to require further investigation, since it is not consistent with the fact that, even when associated with a 14-3-3 protein, Iip35 will not leave the ER in the absence of class-II molecules and that mutation of serine[8] to asparagine prevents 14-3-3 protein binding but still allows ER export if class-II molecules are present (Kuwana *et al.*, 1998).

The effect of 14-3-3 binding on promoting cell surface transport was also found by a completely different approach. Shikano *et al.* screened a random peptide library to identify C-terminal peptide signals that would functionally override the ER localization activity of the RXR motif (Shikano *et al.*, 2005). The screening was based on the yeast growth complementation assay using a mutant *Saccharomyces cerevisiae* SGY1528 which cannot survive in low potassium media due to the lack of endogenous potassium uptake transporters. However, SGY1528 growth in low potassium media can be rescued by heterologous expression of mammalian inward rectifying potassium channel Kir2.1, but not when Kir2.1 was artificially fused with the RXR motif (RKR) due to the efficient retention of channel in the ER (Shikano *et al.*, 2005). In the screen, SGY1528 cells were transformed with Kir2.1 constructs where random peptide library of 8-mer sequences was placed at the extreme C-terminus, downstream of the implanted RXR motif. By selecting the transformed cells that survived in low potassium media, the authors searched for the sequence that were able to override the ER localization activity of the RXR motif and restored the surface expression of the chimeric Kir2.1 channel. The screen of about $2x10^6$ clones yielded several sequences that showed robust surface expression of Kir2.1 as tested in mammalian cell. Those sequences shared a minimum consensus of RXXS/TX-COOH and showed strong binding to 14-3-3, which required penultimate serine or threonine and the upstream arginine residues (Shikano *et al.*, 2005). By using one of the identified C-terminal sequences, namely RGRSWTY-COOH, Chung *et al.* investigated the relevant kinases. Using *in vitro* phosphorylation assay using recombinant proteins and *in vivo* studies using reporter Kir2.1 channel bearing both RXR motif and downstream C-terminal RGRSWTY sequence, the authors found that Akt, but not PKA or CamKII, is responsible for direct phosphorylation of the RGRSWTY sequence which recruits 14-3-3 proteins to this site (Chung *et al.*, 2009). Importantly, the extracellular stimulation that activates Akt pathway, such as insulin and platelet-derived growth factor (PDGF), enhanced 14-3-3 binding and promoted the cell surface transport of the reporter Kir2.1 channel (Chung *et al.*, 2009). These results demonstrate that, despite high similarity in the recognition sequences of basophilic serine/threonine protein kinases, specific kinase signaling can modulate membrane protein trafficking through 14-3-3 binding. Further search of the human protein database for this C-terminal 14-3-3-binding motif has identified several candidate membrane proteins that would interact with 14-3-3. These include GPR15, an orphan GPCR that serves for a co-receptor for human immunodeficiency virus (HIV) entry (Farzan *et al.*, 1997). GPR15, which has a C-terminal sequence of RRRKRSVSL-COOH, was indeed confirmed to bind 14-3-3 and this binding absolutely required phosphorylation of penultimate serine^{359}. A recent study by Okamoto and Shikano reported that alanine mutation of the serine^{359} resulted in substantial ER localization of GPR15 and this was mediated by the upstream arginine residues at amino acids 352 and 354, which constitute a COPI-binding RXR motif (Okamoto & Shikano, 2011). These results suggested that mode III binding of 14-3-3 to the receptor C-terminus physically occludes the adjacent RXR motif from the access by COPI, similar to the cases for TASK channels. Thus, a non-biased screening has led to the identification of 14-3-3 function as a key switch that converts phosphorylation signal to the sorting of membrane proteins by modulating the activity of an ER localization signal.

The masking effect of 14-3-3 on the cell surface transport is not restricted to the mode III C-terminal binding. The cytoplasmic tail of ADAM22, a member of ADAM (a disintegrin and metalloprotease domain) protein family, contains two internal 14-3-3 binding sites and three RXR-type ER-localization signals that overlap with both of the 14-3-3 protein binding sites. Mutations in both 14-3-3 binding sites inhibited surface expression of ADAM22, while deletion of both RXR motif and 14-3-3 binding sites restored the surface expression (Godde *et al.*, 2006). Although this study did not investigate COPI interaction with ADAM22, the results suggest the possibility that 14-3-3 binding inhibited the COPI-dependent ER localization activity of RXR signal.

The cell surface transport of a gap junction protein connexin 43 is also promoted by 14-3-3 binding to its internal mode I binding site (RASSRP) (Park *et al.*, 2007). The authors have found that Akt phosphorylates the serine^{373} in this site in the epidermal growth factor (EGF)-stimulated cell. Although not characterized yet, the existence of an overlapping RXR motif (RPR) suggests a possible masking effect of 14-3-3 on the cell surface transport of connexin 43.

NMDA receptors are tetramers composed of homologous subunits (NR1; NR2A-D; NR3A-B (Cull-Candy & Leszkiewicz, 2004). There are multiple NR2 subunits, each with unique spatio-temporal expression patterns, ensuring functional diversity of NMDA receptors. Recently, cerebellar NR2C subunit was found to be directly phosphorylated by Akt at the sequence that conforms to mode I 14-3-3 binding motif (RPRHASLP) (Chen & Roche, 2009). The Akt phosphorylation induced by insulin growth factor (IGF-1) results in the recruitment of 14-3-3 and the increase of cell surface expression of NMDA receptors. Although the causative role of 14-3-3 in surface transport remains unclear in this study, the presence of ER localization signals (Horak & Wenthold, 2009) in the NR1 subunit and the obligatory assembly of NR2 with NR1 for functional NMDA receptor suggest a possible mechanism where Akt-induced 14-3-3 binding to NR2C attenuates the activity of ER localization signals in NR1 by physically occluding them.

Interestingly, 14-3-3s have also been reported to bind directly to the RXR motif itself in a phosphorylation-independent manner. By using an artificial multimer of the distal C-terminus of Kir6.2 channel in pull-down assays from cytosolic cellular extracts, Yuan *et al.* showed that 14-3-3 proteins preferentially bound to the RXR motif (RSRR) in the oligomerized form (Yuan *et al.*, 2003). This interaction was sufficient to allow the exit of a multimeric reporter protein carrying this motif from the ER and promote the subsequent transport to the cell surface. Although the precise role of the 14-3-3 binding in the cell surface transport of octameric K_{ATP} channel consisting of four Kir6.2 and four SUR1 seems to require further investigation (Heusser *et al.*, 2006), this study demonstrated the possibility that 14-3-3 serves for a constitutive check point of ER protein quality control which ensures the cell surface delivery of functional multimeric membrane proteins by preferentially inactivating the ER localisation signals on the properly oligomerized subunits.

2.3.2 Scaffolding

The stable dimeric structure immediately suggests that 14-3-3s might serve as a simple 'scaffold', where two different target proteins bind simultaneously to each monomer of the same 14-3-3 dimer. Indeed, 14-3-3 proteins are often referred to as 'scaffolding proteins' in the literature. However, while 14-3-3s are components of multiprotein complexes (Munday *et al.*, 2000; Pnueli *et al.*, 2001; Widen *et al.*, 2000), the evidence showing the 14-3-3 dimers acting as an intermolecular bridge between two different substrates had been limited to several earlier studies including those reporting the pairings involving Raf-1, namely Raf-1 and Bcr (B-cell receptor) (Braselmann & McCormick, 1995), Raf-1 and A20 (Vincenz & Dixit, 1996) and Raf-1 and PKCζ (Van Der Hoeven *et al.*, 2000). Nevertheless, in the context of membrane protein trafficking, some recent studies implicate the scaffolding role of 14-3-3 in promoting the cell surface expression of membrane proteins.

14-3-3 promotes the ER export of N-cadherin through coupling the N-cadherin/β-catenin/PX-RICS complex to the microtubule-based motor proteins dynein/dynactin (Nakamura *et al.*, 2010). 14-3-3ζ or 14-3-3θ directly interacts in a phosphorylation-dependent manner with a mode I site (RSKSDP) of PX-RICS, a β-catenin-interacting GTPase-activating protein for Cdc42, and this seems to facilitate ER to Golgi trafficking by association of N-cadherin/β-catenin cargo with minus-end motor proteins dynein/dynactin. This results in the increased localization of N-cadherin/β-catenin at cell-cell contact sites. The authors have also shown that CaMKII is responsible for direct phosphorylation of PX-RICS and the subsequent 14-3-3 binding by using *in vitro* phosphorylation assay and siRNA knockdown of CaMKII (Nakamura *et al.*, 2010). It is of note that a similar scaffolding function has been reported for a PDZ protein that couples a cargo receptor and a motor protein which mediates microtubule-based trafficking. PDZ domain of mLin-10 directly interacts with the C-terminal PDZ-binding motif of a neuron-specific plus-end molecular motor KIF-17. mLin-10 also forms a complex with its family members mLin-2 and mLin-7, which in turn interact with PDZ-binding motif of NMDA receptor subunit 2B (Setou *et al.*, 2000).

The interactions between the α3 subunit of the nicotinic acetylcholine receptor (nAChR) and a multi-subunit cytoskeletal-anchoring complex provide another evidence suggesting possible scaffolding role of 14-3-3 (Rosenberg *et al.*, 2008). APC (adenomatous polyposis coli) organizes a multi-protein postsynaptic complex that targets α3nAChRs to synapses. APC interaction with the microtubule plus-end binding protein EB1 is essential for α3nAChR surface membrane insertion and stabilization. 14-3-3 directly interacts with α3 subunit in a phosphorylation-dependent manner and also forms complex with APC. Thus, 14-3-3 proteins may provide for a mechanism by which nAChRs containing only specific subunits are recruited to postsynaptic clusters and may stabilise them there. In both of the above studies, it is still not clear whether and how a 14-3-3 dimer binds to two separate targets with each monomer. Multiple proteomic screenings have revealed that 14-3-3s form complex with a large number of proteins closely involved in vesicular trafficking such as motor proteins, coat proteins, and GTPase regulators (Mrowiec & Schwappach, 2006). Therefore it is conceivable that 14-3-3 proteins do modulate membrane trafficking by serving as a scaffold that connects cargo proteins with cellular transport machineries.

Then, does a 14-3-3 dimer really bind to two different proteins at the same time? It seems somewhat unlikely that there would be frequent occasions where a 14-3-3 dimer binds two separate targets via each canonical ligand-binding groove, unless those target proteins happen to be already close enough to each other and in such position that both of their phosphorylated binding sites can be accommodated by the binding grooves of the same 14-3-3 dimer, whose core is a rigid and unyielding structure (Obsil *et al.*, 2001), but not by the neighboring 14-3-3 dimers. Crystal structure of 14-3-3ζ : AANAT revealed that in addition to the phosphorylation-dependent interaction through its canonical ligand-binding groove, 14-3-3 also makes extensive contacts with AANAT via other regions of the 14-3-3 channel, although these contacts must be insufficient to form a stable complex (Obsil *et al.*, 2001). Moreover, a recent finding by Barry *et al.* demonstrated that 14-3-3 can directly interact with other proteins outside of the canonical binding groove, providing a possible molecular basis for the scaffolding function of 14-3-3 (Barry *et al.*, 2009). 14-3-3ζ undergoes phosphorylation at tyrosine^{179} upon cytokine stimulation and this leads to the binding of Shc protein through its SH2 domain that recognizes phosphorylated tyrosine. This 14-3-3/Shc complex is required for the recruitment of a phosphatidylinositol 3-kinase (PI3K) signaling complex and the regulation of cell survival in response to cytokine. Although this study did not describe whether Shc-bound 14-3-3 proteins bind to any serine/threonine-phosphorylated targets, the result suggests that 14-3-3/Shc scaffolds can act as multivalent signaling nodes for the integration of both phosphoserine/threonine and phosphotyrosine pathways to regulate specific cellular responses. Thus, interaction of 14-3-3 with proteins via non-canonical binding sites of 14-3-3 should contribute to the diversity of its roles in a wide variety of biological pathways including membrane trafficking.

In addition, the propensity of the different 14-3-3 isoforms to form homo- or hetero-dimers may confer additional specificity to the scaffolding roles. Those regions of the 14-3-3 protein which vary between the isoforms are primarily located on the surface of the protein. Therefore, the specificity of interaction of 14-3-3 isoforms with diverse target proteins may involve the outer surface of the protein. For instance, C-termini of 14-3-3 proteins are most divergent and hence most likely to contain isoform-specific structural determinants (Williams *et al.*, 2011). Identification of more protein complexes whose assembly requires 14-3-3 is necessary to gain more mechanistic insights into the scaffolding function of 14-3-3.

2.3.3 Clamping

Another mechanism by which 14-3-3 proteins are thought to exert their effect on their targets are conformational 'clamping'. Clamping can occur when a 14-3-3 dimer binds two sites on the same target protein. A synthetic phosphopeptide with two tandem 14-3-3 consensus motifs binds over 30-fold more tightly than the same peptide containing only a single motif (Yaffe *et al.*, 1997). A number of 14-3-3-binding proteins, including Raf-1 (Muslin *et al.*, 1996), AANAT (Ganguly *et al.*, 2005), ADAM22 (Godde *et al.*, 2006), tyrosine hydroxylase (Toska *et al.*, 2002), and Ndel1 (Johnson *et al.*, 2010), contain two phosphorylated sites that are implicated in 14-3-3 binding, and are separated by polypeptides of various lengths. It has been postulated that one site called the 'gatekeeper' is indispensable for a stable 14-3-3 interaction, whereas a second site 'enhances' the interaction, but has too weak an affinity to bind 14-3-3 alone (Yaffe, 2002). In the case of AANAT, the gatekeeper residue is phosphorylated threonine^{31}. Binding of the gatekeeper leads to binding of a second low-affinity site, in this case phosphorylated serine^{205}, which reflects both the intrinsic affinity of that site and the 'high local concentration induced by its proximity' (Ganguly *et al.*, 2005; Yaffe, 2002). This dual-site binding of AANAT to 14-3-3 provides optimal conformation of the enzyme for high-affinity binding of the substrate arylalkylamine.

14-3-3 clamping can also occur when a 14-3-3 dimer binds two neighboring target proteins. 14-3-3 proteins activate the plant plasma membrane H(+)-ATPase (PMA2) by binding to its C-terminal autoinhibitory domain. This interaction requires phosphorylation of a C-terminal mode III recognition motif as well as an adjacent span of 52 amino acids. X-ray diffraction studies using crystals of 14-3-3 in complex with the entire binding motif of the PMA2 have shown that each 14-3-3 dimer simultaneously binds to two H^+-ATPase molecules. The 3D reconstruction of the purified H(+)-ATPase/14-3-3 complex demonstrated a hexagonal structure consisting of six PMA2 subunits and six 14-3-3 proteins (Ottmann *et al.*, 2007). Thus, a rigid 14-3-3 'clamps' stabilize the dodecameric compex in the active conformation where C-terminal auto-inhibitory domain of PMA would be displaced.

With regards to membrane protein trafficking, clamping activity of 14-3-3 proteins has been much less well understood. This is largely due to the, as yet, small number of studies in this research area and the lack of high-resolution structure of the 14-3-3/target membrane protein complex, which are also true for 'masking' and 'scaffolding' mechanisms. However, the ability of 14-3-3 to change conformation of target proteins suggests the possibility that such conformational change will lead to a new interaction of the target with proteins that are involved in protein trafficking. This way 14-3-3 might indirectly exert the scaffolding function that eventually modulates the protein sorting of the client protein. Alternatively, instead of direct 'masking' of RXR motif by 14-3-3, 14-3-3 binding might force the target protein into the conformation where the RXR motif will be occluded. The RXR motif was previously found to have its functional 'zoning' in relation to the transmembrane region in the context of a reporter CD4 protein (Shikano & Li, 2003). This notion was based on the observation that the ER localization activity of RXR motif was lost when it was positioned proximal to the transmembrane region of CD4. This zoning model was supported by the later study on the gamma-aminobutyric acid type B (GABAB) receptor (Gassmann *et al.*, 2005). An RXR motif (RSRR) is responsible for the ER retrieval of the GABAB1 subunits that were not properly assembled with GABAB2 subunits. It had been thought that coiled-coil interaction of the GABAB1 with GABAB2 will shield RSRR signal on GABAB1. However, closer positioning of RSRR signal to the membrane region drastically reduced its effectiveness and also functional ectopic RSRR signals in GABAB1 were efficiently inactivated by the GABAB2 subunit in the absence of coiled-coil dimerization (Gassmann *et al.*, 2005). These results were consistent with a model in which removal of RSRR from its functionally active zone, rather than its direct shielding by coiled-coil dimerization, triggers cell surface trafficking of GABAB receptors. Thus, it is interesting to speculate that clamping of two different 14-3-3 binding sites, either within the same target protein or in two neighboring proteins, might lead to the placement of RXR motifs in a non-functional zone such as membrane proximity and suppress ER localization of target membrane proteins.

2.3.4 Other mechanisms by which 14-3-3s modulate membrane protein trafficking

Modulation of membrane protein localization by 14-3-3s is not necessarily mediated by their binding to the cargo protein itself. It is not surprising that 14-3-3 binding to any cellular machinery proteins involved in biosynthetic pathways would affect their functions and thereby affect the sorting of the cargo protein.

The best characterized mechanism of this type involves the interaction of 14-3-3 with AS160, a Rab GTPase-activating protein (Rab-GAP). AS160 is an Akt substrate whose phosphorylation contributes to the recruitment of GLUT4 transporters to adipocyte plasma membrane in response to insulin (Watson & Pessin, 2006). It maintains Rab proteins with which it associates in their inactive, GDP-bound states. Several evidences implicate Akt-mediated AS160 phosphorylation and the subsequent 14-3-3 binding in the recruitment of the GLUT4 glucose transporter to the cell surface of adipocytes (Ramm *et al.*, 2006; Watson & Pessin, 2006). Insulin stimulation leads to the recruitment of Akt to the plasma membrane where it gets activated and then phosphorylate AS160 at serine^{341} and threonine^{642}. It is thought that the binding of 14-3-3 to AS160 inhibits its Rab-GAP activity toward substrate Rabs (Ishikura *et al.*, 2007), which stabilizes them at GTP-bound active form and thereby facilitate the trafficking of cargo vesicles. Interestingly, insulin-dependent cell surface transport of GLUT4 is known to be primarily mediated by Akt isoform 2 (Akt2) but not Akt1 in adipocytes (Bae *et al.*, 2003; Cho *et al.*, 2001). A recent study by Gonzalez and McGraw demonstrated that upon insulin stimulation, Akt2 is able to remain associated with plasma membrane longer than Akt1 does and this leads to the Akt2-specific phosphorylation of AS160, which is necessary for GLUT4 trafficking (Gonzalez & McGraw, 2009). It will be interesting to investigate whether and how this Akt isoform-specific phosphorylation of AS160 regulates 14-3-3 binding.

The surface expression of the epithelial sodium channel (ENaC) seems to be regulated in a similar manner that involves AS160-14-3-3 interaction (Liang *et al.*, 2010). Aldosterone stimulation of renal epithelial cell induces the expression of serum- and glucocorticoid-induced kinase (SGK1) that increases the cell surface expression of ENaC and Na absorption (Bhalla *et al.*, 2006). Aldosterone also induces the expression of two 14-3-3 protein isoforms, β and ε (Liang *et al.*, 2006). Liang *et al.* reported that SGK1, which is the downstream kinase of PI3K and shares with Akt the recognition motif of RXRXXS/T (Tessier & Woodgett, 2006), phosphorylates AS160 upon aldosterone stimulation and this recruits the induced 14-3-3 isoforms (Liang *et al.*, 2010), similar to the insulin-induced Akt phosphorylation of AS160. Inhibition of ENaC surface transport by expression of AS160 carrying mutations in SGK1 target sites suggests that 14-3-3 downregulates AS160 function to eventually promote surface transport of ENaC and augment Na absorption in response to aldosterone.

Furthermore, 14-3-3 also controls the surface expression level of ENaC by regulating its degradation machinery. Ubiquitination of ENaC by ubiquitin-E3 protein ligase Nedd4-2 leads to an increased rate of protein degradation. The activation of SGK1 by aldosterone results in the phosphorylation of Nedd4-2 and recruitment of 14-3-3 proteins to Nedd4-2. This 14-3-3 binding inhibits the interaction of Nedd4-2 with ENaC and thereby suppresses the ubiquitin-dependent degradation of ENaC (Ichimura *et al.*, 2005; Liang *et al.*, 2006). This results in the longer stability of ENaC on the cell surface.

14-3-3s also bind to another transport machinery protein, phosphofurin acidic cluster sorting protein (PACS)-2 (Aslan *et al.*, 2009). PACS-2 recognizes an acidic cluster in the cytoplasmic tail of TRPP2 cation channel and localize the channel in the ER through interaction with the COPI complex (Kottgen *et al.*, 2005). The 14-3-3 binding to PACS-2 is dependent on the phosphorylation of PACS-2 by AKt, and this 14-3-3 binding was found to be required for the ER targeting of the PACS-2 substrate cargo protein TRPP2 (Aslan *et al.*, 2009). How binding of PACS-2 to 14-3-3 and COPI cooperate to mediate cargo traffic remains to be determined.

Cell surface expression of membrane proteins could be also regulated by 14-3-3 through modulation of endocytic processes. Recent study demonstrates the interaction of 14-3-3 with transferrin receptor trafficking protein (TTP) (Chiba *et al.*, 2009). TTP specifically promotes the internalization of transferrin receptor (TfR), but not other receptors such as epidermal growth factor receptor (EGFR) and low-density lipoprotein receptor (LDLR), through the clathrin-dependent pathway (Tosoni *et al.*, 2005). 14-3-3 proteins directly bind to TTP in the Akt-dependent manner and this interaction was enhanced by oxidative stress (Chiba *et al.*, 2009). Although the *in vivo* role of this 14-3-3 binding in TTP transport was not shown in this study, it suggests another possible mechanism where kinase signaling utilizes 14-3-3 as an effector molecule to control the cell surface density of proteins.

3. Conclusion

The requirement of phosphorylation for 14-3-3 binding confers 14-3-3 proteins a primary role in regulating protein-protein interactions that are under the control of specific kinases and phosphatases. Accordingly, 14-3-3 constitutes a key player which stands at a point of cross-talk between a plethora of vital biological processes including signaling, metabolism, cell cycle, and protein trafficking. Although available data from biochemical, structural, and bioinformatics studies have provided substantial amounts of information that characterize 14-3-3/client interaction, several fundamental questions regarding 14-3-3 protein biology remain to be addressed.

Although proteomics have revealed over 200 proteins forming complex with 14-3-3, the information regarding kinases and phosphatases that regulate these interactions are limited. The relevant kinase is not known for the majority of 14-3-3-binding proteins and the substantial overlap between the recognition sites of numerous basophilic serine/threonine kinases and 14-3-3 binding sites make their identification very difficult. Equally important but even less well studied are the phosphatases relevant for 14-3-3 binding. When and how the 14-3-3 proteins dissociate from the client proteins is very poorly understood. Does it require dephosphorylation of the 14-3-3 binding site by phosphatases? If so, how can the phosphatases do that when the phosphorylated serine/threonine of a target protein is buried in the amphipathic ligand-binding groove of 14-3-3? Alternatively, does some other signaling event facilitate the release of 14-3-3s from client proteins? In the context of protein trafficking, 14-3-3 is known to modulate sorting of the client proteins by various different mechanisms. In many of the cases where surface membrane transport is promoted by 14-3-3 binding, 14-3-3 seems to impinge on the early step of ER-to-Golgi trafficking. However, except for a very small number of studies (Godde *et al.*, 2006; Okamoto & Shikano, 2011), it is not understood if 14-3-3 proteins remain bound to the client all along the trip to the cell surface or dissociate in any particular step of the vesicular trafficking by phosphatase activity. These problems make us realize that we still do not know enough about the biology of kinases and phosphatases, especially the spatio-temporal regulation of their activities in different cellular compartments.

Another important question that has been long discussed but not fully addressed is whether and how 14-3-3 isoforms play specific roles. The complete sequence conservation in the observed ligand-binding regions of 14-3-3 would support the hypothesis that there may be little isoform specificity in the interaction between 14-3-3 and client proteins; therefore isoform-specific function of 14-3-3 may result either from subcellular localization (Paul *et al.*, 2005; van Hemert *et al.*, 2004) or transcriptional regulation (Liang *et al.*, 2006) of particular isotypes rather than from inherent differences in their ability to bind to particular ligands. However, several findings (Dubois *et al.*, 1997; Gu & Du, 1998; Ichimura *et al.*, 1995) suggest that additional interactions may occur on the outer surface of 14-3-3. This may confer isoform specificity, since residues that are variable between 14-3-3 isoforms are located on the surface of the protein. It is also conceivable that due to the common occurrence of two 14-3-3 binding sites within the target protein, the synergy between the two may also lead to isoform preference of interaction. Studies have shown that 14-3-3 isoforms form hetero-dimers *in vivo* (Alvarez *et al.*, 2003; Liang *et al.*, 2008). The isoform-specific interaction with client proteins may become most relevant in a 'scaffolding' model, where a hetero-dimer consisting of different isoforms would bind to two separate targets via each isoform. Thus, it is likely that its propensity to form homo- and various heterodimeric combinations is crucial for the specificity of 14-3-3 isoform functions. Many apparent conclusions of 14-3-3 function within particular cell types are based on observations of a single isoform, and comparative data among isoforms are still limited. More analysis of the exact combinations of homo- and hetero-dimers of 14-3-3 isoforms that are present within cell compartments and that are involved in interactions with particular proteins will be important.

4. References

Alvarez, D., Callejo, M., Shoucri, R., Boyer, L., Price, G.B. & Zannis-Hadjopoulos, M. (2003) Analysis of the cruciform binding activity of recombinant 14-3-3zeta-MBP fusion protein, its heterodimerization profile with endogenous 14-3-3 isoforms, and effect on mammalian DNA replication in vitro. *Biochemistry*, 42, 7205-7215.

Aslan, J.E., You, H., Williamson, D.M., Endig, J., Youker, R.T., Thomas, L., Shu, H., Du, Y., Milewski, R.L., Brush, M.H., Possemato, A., Sprott, K., Fu, H., Greis, K.D., Runckel, D.N., Vogel, A. & Thomas, G. (2009) Akt and 14-3-3 control a PACS-2 homeostatic switch that integrates membrane traffic with TRAIL-induced apoptosis. *Mol Cell*, 34, 497-509.

Bae, S.S., Cho, H., Mu, J. & Birnbaum, M.J. (2003) Isoform-specific regulation of insulin-dependent glucose uptake by Akt/protein kinase B. *J Biol Chem*, 278, 49530-49536.

Barry, E.F., Felquer, F.A., Powell, J.A., Biggs, L., Stomski, F.C., Urbani, A., Ramshaw, H., Hoffmann, P., Wilce, M.C., Grimbaldeston, M.A., Lopez, A.F. & Guthridge, M.A. (2009) 14-3-3:Shc scaffolds integrate phosphoserine and phosphotyrosine signaling to regulate phosphatidylinositol 3-kinase activation and cell survival. *J Biol Chem*, 284, 12080-12090.

Benzinger, A., Popowicz, G.M., Joy, J.K., Majumdar, S., Holak, T.A. & Hermeking, H. (2005) The crystal structure of the non-liganded 14-3-3sigma protein: insights into determinants of isoform specific ligand binding and dimerization. *Cell Res*, 15, 219-227.

Bhalla, V., Soundararajan, R., Pao, A.C., Li, H. & Pearce, D. (2006) Disinhibitory pathways for control of sodium transport: regulation of ENaC by SGK1 and GILZ. *Am J Physiol Renal Physiol*, 291, F714-721.

Bodnar, R.J., Gu, M., Li, Z., Englund, G.D. & Du, X. (1999) The cytoplasmic domain of the platelet glycoprotein Ibalpha is phosphorylated at serine 609. *J Biol Chem*, 274, 33474-33479.

Bradshaw, J.M. & Waksman, G. (2002) Molecular recognition by SH2 domains. *Adv Protein Chem*, 61, 161-210.

Braselmann, S. & McCormick, F. (1995) Bcr and Raf form a complex in vivo via 14-3-3 proteins. *Embo J*, 14, 4839-4848.

Brunet, A., Bonni, A., Zigmond, M.J., Lin, M.Z., Juo, P., Hu, L.S., Anderson, M.J., Arden, K.C., Blenis, J. & Greenberg, M.E. (1999) Akt promotes cell survival by phosphorylating and inhibiting a Forkhead transcription factor. *Cell*, 96, 857-868.

Chang, I.F., Curran, A., Woolsey, R., Quilici, D., Cushman, J.C., Mittler, R., Harmon, A. & Harper, J.F. (2009) Proteomic profiling of tandem affinity purified 14-3-3 protein complexes in Arabidopsis thaliana. *Proteomics*, 9, 2967-2985.

Chen, B.S. & Roche, K.W. (2009) Growth factor-dependent trafficking of cerebellar NMDA receptors via protein kinase B/Akt phosphorylation of NR2C. *Neuron*, 62, 471-478.

Chiba, S., Tokuhara, M., Morita, E.H. & Abe, S. (2009) TTP at Ser245 phosphorylation by AKT is required for binding to 14-3-3. *J Biochem*, 145, 403-409.

Cho, H., Mu, J., Kim, J.K., Thorvaldsen, J.L., Chu, Q., Crenshaw, E.B., 3rd, Kaestner, K.H., Bartolomei, M.S., Shulman, G.I. & Birnbaum, M.J. (2001) Insulin resistance and a diabetes mellitus-like syndrome in mice lacking the protein kinase Akt2 (PKB beta). *Science*, 292, 1728-1731.

Chung, H.J., Huang, Y.H., Lau, L.F. & Huganir, R.L. (2004) Regulation of the NMDA receptor complex and trafficking by activity-dependent phosphorylation of the NR2B subunit PDZ ligand. *J Neurosci*, 24, 10248-10259.

Chung, J.J., Okamoto, Y., Coblitz, B., Li, M., Qiu, Y. & Shikano, S. (2009) PI3K/Akt signalling-mediated protein surface expression sensed by 14-3-3 interacting motif. *Febs J*, 276, 5547-5558.

Coblitz, B., Shikano, S., Wu, M., Gabelli, S.B., Cockrell, L.M., Spieker, M., Hanyu, Y., Fu, H., Amzel, L.M. & Li, M. (2005) C-terminal recognition by 14-3-3 proteins for surface expression of membrane receptors. *J Biol Chem*, 280, 36263-36272.

Cull-Candy, S.G. & Leszkiewicz, D.N. (2004) Role of distinct NMDA receptor subtypes at central synapses. *Sci STKE*, 2004, re16.

Datta, S.R., Dudek, H., Tao, X., Masters, S., Fu, H., Gotoh, Y. & Greenberg, M.E. (1997) Akt phosphorylation of BAD couples survival signals to the cell-intrinsic death machinery. *Cell*, 91, 231-241.

Dougherty, M.K. & Morrison, D.K. (2004) Unlocking the code of 14-3-3. *J Cell Sci*, 117, 1875-1884.

Drake, M.T., Shenoy, S.K. & Lefkowitz, R.J. (2006) Trafficking of G protein-coupled receptors. *Circ Res*, 99, 570-582.

Dubois, T., Rommel, C., Howell, S., Steinhussen, U., Soneji, Y., Morrice, N., Moelling, K. & Aitken, A. (1997) 14-3-3 is phosphorylated by casein kinase I on residue 233. Phosphorylation at this site in vivo regulates Raf/14-3-3 interaction. *J Biol Chem*, 272, 28882-28888.

Durocher, D., Taylor, I.A., Sarbassova, D., Haire, L.F., Westcott, S.L., Jackson, S.P., Smerdon, S.J. & Yaffe, M.B. (2000) The molecular basis of FHA domain:phosphopeptide binding specificity and implications for phospho-dependent signaling mechanisms. *Mol Cell*, 6, 1169-1182.

Farzan, M., Choe, H., Martin, K., Marcon, L., Hofmann, W., Karlsson, G., Sun, Y., Barrett, P., Marchand, N., Sullivan, N., Gerard, N., Gerard, C. & Sodroski, J. (1997) Two orphan seven-transmembrane segment receptors which are expressed in CD4-positive cells support simian immunodeficiency virus infection. *J Exp Med*, 186, 405-411.

Fujita, N., Sato, S. & Tsuruo, T. (2003) Phosphorylation of p27Kip1 at threonine 198 by p90 ribosomal protein S6 kinases promotes its binding to 14-3-3 and cytoplasmic localization. *J Biol Chem*, 278, 49254-49260.

Ganguly, S., Weller, J.L., Ho, A., Chemineau, P., Malpaux, B. & Klein, D.C. (2005) Melatonin synthesis: 14-3-3-dependent activation and inhibition of arylalkylamine N-acetyltransferase mediated by phosphoserine-205. *Proc Natl Acad Sci U S A*, 102, 1222-1227.

Gardino, A.K., Smerdon, S.J. & Yaffe, M.B. (2006) Structural determinants of 14-3-3 binding specificities and regulation of subcellular localization of 14-3-3-ligand complexes: a comparison of the X-ray crystal structures of all human 14-3-3 isoforms. *Semin Cancer Biol*, 16, 173-182.

Gassmann, M., Haller, C., Stoll, Y., Aziz, S.A., Biermann, B., Mosbacher, J., Kaupmann, K. & Bettler, B. (2005) The RXR-type endoplasmic reticulum-retention/retrieval signal of GABAB1 requires distant spacing from the membrane to function. *Mol Pharmacol*, 68, 137-144.

Godde, N.J., D'Abaco, G.M., Paradiso, L. & Novak, U. (2006) Efficient ADAM22 surface expression is mediated by phosphorylation-dependent interaction with 14-3-3 protein family members. *J Cell Sci*, 119, 3296-3305.

Gonzalez, E. & McGraw, T.E. (2009) Insulin-modulated Akt subcellular localization determines Akt isoform-specific signaling. *Proc Natl Acad Sci U S A*, 106, 7004-7009.

Gu, M. & Du, X. (1998) A novel ligand-binding site in the zeta-form 14-3-3 protein recognizing the platelet glycoprotein Ibalpha and distinct from the c-Raf-binding site. *J Biol Chem*, 273, 33465-33471.

Heusser, K., Yuan, H., Neagoe, I., Tarasov, A.I., Ashcroft, F.M. & Schwappach, B. (2006) Scavenging of 14-3-3 proteins reveals their involvement in the cell-surface transport of ATP-sensitive K+ channels. *J Cell Sci*, 119, 4353-4363.

Horak, M. & Wenthold, R.J. (2009) Different roles of C-terminal cassettes in the trafficking of full-length NR1 subunits to the cell surface. *J Biol Chem*, 284, 9683-9691.

Ichimura, T., Uchiyama, J., Kunihiro, O., Ito, M., Horigome, T., Omata, S., Shinkai, F., Kaji, H. & Isobe, T. (1995) Identification of the site of interaction of the 14-3-3 protein with phosphorylated tryptophan hydroxylase. *J Biol Chem*, 270, 28515-28518.

Ichimura, T., Wakamiya-Tsuruta, A., Itagaki, C., Taoka, M., Hayano, T., Natsume, T. & Isobe, T. (2002) Phosphorylation-dependent interaction of kinesin light chain 2 and the 14-3-3 protein. *Biochemistry*, 41, 5566-5572.

Ichimura, T., Yamamura, H., Sasamoto, K., Tominaga, Y., Taoka, M., Kakiuchi, K., Shinkawa, T., Takahashi, N., Shimada, S. & Isobe, T. (2005) 14-3-3 proteins modulate the expression of epithelial Na+ channels by phosphorylation-dependent interaction with Nedd4-2 ubiquitin ligase. *J Biol Chem*, 280, 13187-13194.

Ishikura, S., Bilan, P.J. & Klip, A. (2007) Rabs 8A and 14 are targets of the insulin-regulated Rab-GAP AS160 regulating GLUT4 traffic in muscle cells. *Biochem Biophys Res Commun*, 353, 1074-1079.

Jackson, M.R., Nilsson, T. & Peterson, P.A. (1990) Identification of a consensus motif for retention of transmembrane proteins in the endoplasmic reticulum. *Embo J*, 9, 3153-3162.

Jackson, M.R., Nilsson, T. & Peterson, P.A. (1993) Retrieval of transmembrane proteins to the endoplasmic reticulum. *J Cell Biol*, 121, 317-333.

Johnson, C., Crowther, S., Stafford, M., Campbell, D.G., Toth, R. & Mackintosh, C. (2010) Bioinformatic and experimental survey of 14-3-3 binding sites. *Biochem J*, 427, 69-78.

Kakiuchi, K., Yamauchi, Y., Taoka, M., Iwago, M., Fujita, T., Ito, T., Song, S.Y., Sakai, A., Isobe, T. & Ichimura, T. (2007) Proteomic Analysis of in Vivo 14-3-3 Interactions in the Yeast Saccharomyces cerevisiae. *Biochemistry*, 46, 7781-7792.

Khalil, H., Brunet, A. & Thibodeau, J. (2005) A three-amino-acid-long HLA-DRbeta cytoplasmic tail is sufficient to overcome ER retention of invariant-chain p35. *J Cell Sci*, 118, 4679-4687.

Kino, T., Souvatzoglou, E., De Martino, M.U., Tsopanomihalu, M., Wan, Y. & Chrousos, G.P. (2003) Protein 14-3-3sigma interacts with and favors cytoplasmic subcellular localization of the glucocorticoid receptor, acting as a negative regulator of the glucocorticoid signaling pathway. *J Biol Chem*, 278, 25651-25656.

Kobe, B., Kampmann, T., Forwood, J.K., Listwan, P. & Brinkworth, R.I. (2005) Substrate specificity of protein kinases and computational prediction of substrates. *Biochim Biophys Acta*, 1754, 200-209.

Kottgen, M., Benzing, T., Simmen, T., Tauber, R., Buchholz, B., Feliciangeli, S., Huber, T.B., Schermer, B., Kramer-Zucker, A., Hopker, K., Simmen, K.C., Tschucke, C.C., Sandford, R., Kim, E., Thomas, G. & Walz, G. (2005) Trafficking of TRPP2 by PACS proteins represents a novel mechanism of ion channel regulation. *Embo J*, 24, 705-716.

Kumagai, A. & Dunphy, W.G. (1999) Binding of 14-3-3 proteins and nuclear export control the intracellular localization of the mitotic inducer Cdc25. *Genes Dev*, 13, 1067-1072.

Kuwana, T., Peterson, P.A. & Karlsson, L. (1998) Exit of major histocompatibility complex class II-invariant chain p35 complexes from the endoplasmic reticulum is modulated by phosphorylation. *Proc Natl Acad Sci U S A*, 95, 1056-1061.

Liang, X., Butterworth, M.B., Peters, K.W. & Frizzell, R.A. (2010) AS160 modulates aldosterone-stimulated epithelial sodium channel forward trafficking. *Mol Biol Cell*, 21, 2024-2033.

Liang, X., Butterworth, M.B., Peters, K.W., Walker, W.H. & Frizzell, R.A. (2008) An obligatory heterodimer of 14-3-3beta and 14-3-3epsilon is required for aldosterone regulation of the epithelial sodium channel. *J Biol Chem*, 283, 27418-27425.

Liang, X., Peters, K.W., Butterworth, M.B. & Frizzell, R.A. (2006) 14-3-3 isoforms are induced by aldosterone and participate in its regulation of epithelial sodium channels. *J Biol Chem*, 281, 16323-16332.

Lowery, D.M., Lim, D. & Yaffe, M.B. (2005) Structure and function of Polo-like kinases. *Oncogene*, 24, 248-259.

Mackintosh, C. (2004) Dynamic interactions between 14-3-3 proteins and phosphoproteins regulate diverse cellular processes. *Biochem J*, 381, 329-342.

Manke, I.A., Lowery, D.M., Nguyen, A. & Yaffe, M.B. (2003) BRCT repeats as phosphopeptide-binding modules involved in protein targeting. *Science*, 302, 636-639.

Mant, A., Elliott, D., Eyers, P.A. & O'Kelly, I.M. (2011) Protein kinase A is central for forward transport of two-pore domain potassium channels K(2P)3.1 and K(2P)9.1. *J Biol Chem*, 286, 14110-14119.

Meek, S.E., Lane, W.S. & Piwnica-Worms, H. (2004) Comprehensive proteomic analysis of interphase and mitotic 14-3-3-binding proteins. *J Biol Chem*, 279, 32046-32054.

Michelsen, K., Schmid, V., Metz, J., Heusser, K., Liebel, U., Schwede, T., Spang, A. & Schwappach, B. (2007) Novel cargo-binding site in the beta and delta subunits of coatomer. *J Cell Biol*, 179, 209-217.

Michelsen, K., Yuan, H. & Schwappach, B. (2005) Hide and run. Arginine-based endoplasmic-reticulum-sorting motifs in the assembly of heteromultimeric membrane proteins. *EMBO Rep*, 6, 717-722.

Moore, B.W. & Perez, V.J. (1967) Specific acidic proteins of the nervous system,. *In: Physiological and Biochemical Aspects of Nervous Integration,*, 343–359.

Mrowiec, T. & Schwappach, B. (2006) 14-3-3 proteins in membrane protein transport. *Biol Chem*, 387, 1227-1236.

Munday, A.D., Berndt, M.C. & Mitchell, C.A. (2000) Phosphoinositide 3-kinase forms a complex with platelet membrane glycoprotein Ib-IX-V complex and 14-3-3zeta. *Blood*, 96, 577-584.

Muslin, A.J., Tanner, J.W., Allen, P.M. & Shaw, A.S. (1996) Interaction of 14-3-3 with signaling proteins is mediated by the recognition of phosphoserine. *Cell*, 84, 889-897.

Nakamura, T., Hayashi, T., Mimori-Kiyosue, Y., Sakaue, F., Matsuura, K., Iemura, S.I., Natsume, T. & Akiyama, T. (2010) The PX-RICS/14-3-3{zeta}/{theta} complex couples N-cadherin/{beta}-catenin with dynein/dynactin to mediate its export from the endoplasmic reticulum. *J Biol Chem*, 285, 16145-16154.

O'Kelly, I., Butler, M.H., Zilberberg, N. & Goldstein, S.A. (2002) Forward transport. 14-3-3 binding overcomes retention in endoplasmic reticulum by dibasic signals. *Cell*, 111, 577-588.

Obsil, T., Ghirlando, R., Klein, D.C., Ganguly, S. & Dyda, F. (2001) Crystal structure of the 14-3-3zeta:serotonin N-acetyltransferase complex. a role for scaffolding in enzyme regulation. *Cell*, 105, 257-267.

Okamoto, Y. & Shikano, S. (2011) Phosphorylation-dependent C-terminal binding of 14-3-3 proteins promotes cell surface expression of HIV co-receptor GPR15. *J Biol Chem*, 286.

Ottmann, C., Marco, S., Jaspert, N., Marcon, C., Schauer, N., Weyand, M., Vandermeeren, C., Duby, G., Boutry, M., Wittinghofer, A., Rigaud, J.L. & Oecking, C. (2007) Structure of a 14-3-3 coordinated hexamer of the plant plasma membrane H+ -ATPase by combining X-ray crystallography and electron cryomicroscopy. *Mol Cell*, 25, 427-440.

Park, D.J., Wallick, C.J., Martyn, K.D., Lau, A.F., Jin, C. & Warn-Cramer, B.J. (2007) Akt phosphorylates Connexin43 on Ser373, a "mode-1" binding site for 14-3-3. *Cell Commun Adhes*, 14, 211-226.

Paul, A.L., Sehnke, P.C. & Ferl, R.J. (2005) Isoform-specific subcellular localization among 14-3-3 proteins in Arabidopsis seems to be driven by client interactions. *Mol Biol Cell*, 16, 1735-1743.

Pearson, R.B. & Kemp, B.E. (1991) Protein kinase phosphorylation site sequences and consensus specificity motifs: tabulations. *Methods Enzymol*, 200, 62-81.

Pnueli, L., Gutfinger, T., Hareven, D., Ben-Naim, O., Ron, N., Adir, N. & Lifschitz, E. (2001) Tomato SP-interacting proteins define a conserved signaling system that regulates shoot architecture and flowering. *Plant Cell*, 13, 2687-2702.

Pozuelo Rubio, M., Geraghty, K.M., Wong, B.H., Wood, N.T., Campbell, D.G., Morrice, N. & Mackintosh, C. (2004) 14-3-3-affinity purification of over 200 human phosphoproteins reveals new links to regulation of cellular metabolism, proliferation and trafficking. *Biochem J*, 379, 395-408.

Ramm, G., Larance, M., Guilhaus, M. & James, D.E. (2006) A role for 14-3-3 in insulin-stimulated GLUT4 translocation through its interaction with the RabGAP AS160. *J Biol Chem*, 281, 29174-29180.

Rittinger, K., Budman, J., Xu, J., Volinia, S., Cantley, L.C., Smerdon, S.J., Gamblin, S.J. & Yaffe, M.B. (1999) Structural analysis of 14-3-3 phosphopeptide complexes identifies a dual role for the nuclear export signal of 14-3-3 in ligand binding. *Mol Cell*, 4, 153-166.

Rong, J., Li, S., Sheng, G., Wu, M., Coblitz, B., Li, M., Fu, H. & Li, X.J. (2007) 14-3-3 protein interacts with Huntingtin-associated protein 1 and regulates its trafficking. *J Biol Chem*, 282, 4748-4756.

Rosenberg, M.M., Yang, F., Giovanni, M., Mohn, J.L., Temburni, M.K. & Jacob, M.H. (2008) Adenomatous polyposis coli plays a key role, in vivo, in coordinating assembly of the neuronal nicotinic postsynaptic complex. *Mol Cell Neurosci*, 38, 138-152.

Salvi, M., Sarno, S., Cesaro, L., Nakamura, H. & Pinna, L.A. (2009) Extraordinary pleiotropy of protein kinase CK2 revealed by weblogo phosphoproteome analysis. *Biochim Biophys Acta*, 1793, 847-859.

Seimiya, H., Sawada, H., Muramatsu, Y., Shimizu, M., Ohko, K., Yamane, K. & Tsuruo, T. (2000) Involvement of 14-3-3 proteins in nuclear localization of telomerase. *Embo J*, 19, 2652-2661.

Setou, M., Nakagawa, T., Seog, D.H. & Hirokawa, N. (2000) Kinesin superfamily motor protein KIF17 and mLin-10 in NMDA receptor-containing vesicle transport. *Science*, 288, 1796-1802.

Shikano, S., Coblitz, B., Sun, H. & Li, M. (2005) Genetic isolation of transport signals directing cell surface expression. *Nat Cell Biol*, 7, 985-992.

Shikano, S. & Li, M. (2003) Membrane receptor trafficking: evidence of proximal and distal zones conferred by two independent endoplasmic reticulum localization signals. *Proc Natl Acad Sci U S A*, 100, 5783-5788.

Sliva, D., Gu, M., Zhu, Y.X., Chen, J., Tsai, S., Du, X. & Yang, Y.C. (2000) 14-3-3zeta interacts with the alpha-chain of human interleukin 9 receptor. *Biochem J*, 345 Pt 3, 741-747.

Tessier, M. & Woodgett, J.R. (2006) Serum and glucocorticoid-regulated protein kinases: variations on a theme. *J Cell Biochem*, 98, 1391-1407.

Tobin, A.B. (2008) G-protein-coupled receptor phosphorylation: where, when and by whom. *Br J Pharmacol*, 153 Suppl 1, S167-176.

Toska, K., Kleppe, R., Armstrong, C.G., Morrice, N.A., Cohen, P. & Haavik, J. (2002) Regulation of tyrosine hydroxylase by stress-activated protein kinases. *J Neurochem*, 83, 775-783.

Tosoni, D., Puri, C., Confalonieri, S., Salcini, A.E., De Camilli, P., Tacchetti, C. & Di Fiore, P.P. (2005) TTP specifically regulates the internalization of the transferrin receptor. *Cell*, 123, 875-888.

Tzivion, G. & Avruch, J. (2002) 14-3-3 proteins: active cofactors in cellular regulation by serine/threonine phosphorylation. *J Biol Chem*, 277, 3061-3064.

Ubersax, J.A. & Ferrell, J.E., Jr. (2007) Mechanisms of specificity in protein phosphorylation. *Nat Rev Mol Cell Biol*, 8, 530-541.

Van Der Hoeven, P.C., Van Der Wal, J.C., Ruurs, P., Van Dijk, M.C. & Van Blitterswijk, J. (2000) 14-3-3 isotypes facilitate coupling of protein kinase C-zeta to Raf-1: negative regulation by 14-3-3 phosphorylation. *Biochem J*, 345 Pt 2, 297-306.

van Hemert, M.J., Niemantsverdriet, M., Schmidt, T., Backendorf, C. & Spaink, H.P. (2004) Isoform-specific differences in rapid nucleocytoplasmic shuttling cause distinct subcellular distributions of 14-3-3 sigma and 14-3-3 zeta. *J Cell Sci*, 117, 1411-1420.

van Hemert, M.J., Steensma, H.Y. & van Heusden, G.P. (2001) 14-3-3 proteins: key regulators of cell division, signalling and apoptosis. *Bioessays*, 23, 936-946.

Vincenz, C. & Dixit, V.M. (1996) 14-3-3 proteins associate with A20 in an isoform-specific manner and function both as chaperone and adapter molecules. *J Biol Chem*, 271, 20029-20034.

Watson, R.T. & Pessin, J.E. (2006) Bridging the GAP between insulin signaling and GLUT4 translocation. *Trends Biochem Sci*, 31, 215-222.

Widen, C., Zilliacus, J., Gustafsson, J.A. & Wikstrom, A.C. (2000) Glucocorticoid receptor interaction with 14-3-3 and Raf-1, a proposed mechanism for cross-talk of two signal transduction pathways. *J Biol Chem*, 275, 39296-39301.

Wilker, E.W., Grant, R.A., Artim, S.C. & Yaffe, M.B. (2005) A structural basis for 14-3-3sigma functional specificity. *J Biol Chem*, 280, 18891-18898.

Williams, D.M., Ecroyd, H., Goodwin, K.L., Dai, H., Fu, H., Woodcock, J.M., Zhang, L. & Carver, J.A. (2011) NMR spectroscopy of 14-3-3zeta reveals a flexible C-terminal extension: differentiation of the chaperone and phosphoserine-binding activities of 14-3-3zeta. *Biochem J*, 437, 493-503.

Wurtele, M., Jelich-Ottmann, C., Wittinghofer, A. & Oecking, C. (2003) Structural view of a fungal toxin acting on a 14-3-3 regulatory complex. *Embo J*, 22, 987-994.

Yaffe, M.B. (2002) How do 14-3-3 proteins work?-- Gatekeeper phosphorylation and the molecular anvil hypothesis. *FEBS Lett*, 513, 53-57.

Yaffe, M.B. & Elia, A.E. (2001) Phosphoserine/threonine-binding domains. *Curr Opin Cell Biol*, 13, 131-138.

Yaffe, M.B., Rittinger, K., Volinia, S., Caron, P.R., Aitken, A., Leffers, H., Gamblin, S.J., Smerdon, S.J. & Cantley, L.C. (1997) The structural basis for 14-3-3:phosphopeptide binding specificity. *Cell*, 91, 961-971.

Yano, M., Mori, S., Niwa, Y., Inoue, M. & Kido, H. (1997) Intrinsic nucleoside diphosphate kinase-like activity as a novel function of 14-3-3 proteins. *FEBS Lett*, 419, 244-248.

Yano, M., Nakamuta, S., Wu, X., Okumura, Y. & Kido, H. (2006) A novel function of 14-3-3 protein: 14-3-3zeta is a heat-shock-related molecular chaperone that dissolves thermal-aggregated proteins. *Mol Biol Cell*, 17, 4769-4779.

Yuan, H., Michelsen, K. & Schwappach, B. (2003) 14-3-3 dimers probe the assembly status of multimeric membrane proteins. *Curr Biol*, 13, 638-646.

Zerangue, N., Schwappach, B., Jan, Y.N. & Jan, L.Y. (1999) A new ER trafficking signal regulates the subunit stoichiometry of plasma membrane K(ATP) channels. *Neuron*, 22, 537-548.

Zha, J., Harada, H., Osipov, K., Jockel, J., Waksman, G. & Korsmeyer, S.J. (1997) BH3 domain of BAD is required for heterodimerization with BCL-XL and pro-apoptotic activity. *J Biol Chem*, 272, 24101-24104.

Zha, J., Harada, H., Yang, E., Jockel, J. & Korsmeyer, S.J. (1996) Serine phosphorylation of death agonist BAD in response to survival factor results in binding to 14-3-3 not BCL-X(L). *Cell*, 87, 619-628.

8

Regulation of Retrotransposition of Long Interspersed Element-1 by Mitogen-Activated Protein Kinases

Yukihito Ishizaka[1,*], Noriyuki Okudaira[1] and Tadashi Okamura[2]
[1]*Department of Intractable Diseases, National Center for Global Health and Medicine,*
[2]*Division of Animal Model, Department of Infections Diseases, National Center for Global Health and Medicine*
Japan

1. Introduction

Our genome contains a higher amount of endogenous retroelements (~42 %) than mouse (~37 %) or fruit fly (~3.6 %) (1-3). Long interspersed element-1 (L1) is the most abundant of transposable elements, comprising ~17% of the genome (1-4). L1 is an autonomous endogenous retroelement that has evolved in a single, unbroken lineage for the past 40 million years in primates (5). A single human cell has more than 5×10^5 copies of L1 (2,4), and most of them are functionally defective (6). However, 80 to 100 copies of L1 are competent for retrotransposition (L1-RTP) (7), and approximately 10 % of these are highly active for "copy and paste" (7). L1 is actively expressed in embryonal stem cells (8) and L1-RTP is induced in oocytes or early embryonic development (9-11). L1-RTP occurring in germ lines would function an intrinsic factor responsible for allelic variants among individuals (12,13). However, aberrant L1-RTP alternates critical gene structures, leading to the development of inborn errors (14). At the moment, at least 17 genetic diseases have been reported as sporadic cases of inheritable disorders caused by aberrant insertion of L1 (14). On the other hand, recent observations suggest that L1-RTP occurs in somatic cells. Strikingly, it was shown that copy numbers of L1 is increased in human brain tissues (15,16). Aberrant L1 insertions have been detected in c-*myc* gene and the *APC* gene in breast carcinoma and colon carcinoma, respectively (17,18). Moreover recent analysis demonstrated that L1 is frequently mobilized in human lung cancers and pancreatic carcinomas (19,20). These observations indicate that it is important to understand the mode of L1-RTP, but little is known about the cellular factors for the induction of L1-RTP in somatic cells. We herein summarize our current understanding of L1-RTP induction, with an emphasis on mitogen-activated protein kinases (MAPKs), which are activated by environmental compounds, and we discuss their roles in genome shuffling.

* Corresponding Author

2. Biology of L1-RTP

L1, a non-long terminal repeat (non-LTR)-type endogenous retroelement, encodes two proteins: open reading frames 1 and 2 (ORF1 and 2) (3). ORF1 is a cytoplasmic 40 kDa protein that is present within ribonucleoprotein complexes (21-23). ORF1 associates in *cis* with L1-mRNA (24) and functions as a chaperone of L1-mRNA (25). ORF2 is a protein of about 150 kDa with dual activities as reverse transcriptase (RT) (26) and an endonuclease (27). ORF2 recognizes the 5′-TTAAAA hexanucleotide in the genome and induces a nick between 3′-AA and TTTT in the complementary strand (28,29). It has been proposed that the first-strand DNA is synthesized by target site-primed reverse transcription (3,29). ORF1 and 2 complete the entire process of L1-RTP and are competent for the induction of retrotransposition of *Alu*, a non-autonomous retroelements (30, 31).

3. Reported triggers of L1-RTP

As to the environmental factors that induce L1-RTP in somatic cells, Farkash *et al.* reported that gamma irradiation at 4.5 Gy induced L1-RTP (32). Independently, Deiniger's group reported that heavy metals of such as mercury, cadmium and nickel also induced L1-RTP (33,34). They also reported that nickel-induced L1-RTP is induced by a post-transcriptional mechanism (34). As to an environmental carcinogen, Stribinskis and Ramos found that benzo[*a*]pyrene (B[*a*]P) induced L1-RTP (35). An extensive analysis revealed that aryl hydrocarbon receptor (AhR), which serves as a receptor for such environmental pollutants as 2,3,7,8-tetrachlorodibenzo-*p*-dioxin (TCDD) (36), was required for the B[*a*]P-induced L1-RTP (35). Because TCDD, a non-genotoxic hydrocarbon carcinogen, did not induce L1-RTP, it was proposed that as one of the its mechanisms an AhR-dependent cellular response converts B[*a*]P into an active genotoxic compound, which in turn induces L1-RTP (35). Although the exact modes of L1-RTP are unclear, these studies inspired us to investigate the possibility that various environmental compounds can induce L1-RTP.

4. Induction of L1-RTP by an environmental compound and identification of p38 as a pivotal cellular factor

First, we found that 6-formylindolo[3,2-*b*]carbazole (FICZ), a tryptophan photoproduct, induced L1-RTP (37). FICZ is highly active, and even picomolar comncentratiom of the compound induced L1-RTP. In mammalian cells, six groups of MAPKs, namely extracellular signal-regulated protein kinase (ERK)1/2, ERK5, JNK, p38, ERK3/4 and ERK7/8, are identified, and are activated by intracellular and extracellular stimuli (38). Among these, cellular signal cascades of ERK1/2, p38 and JNK have been well characterized, because of the availability of inhibitors, including PD98,059, SB202190 and SP600125, respectively. Using these MAPK inhibitors, we found that FICZ-induced L1-RTP was dependent on p38 (37). Interestingly, the compound induced phosphorylation of cyclic-AMP responsive element binding protein (CREB), and the down-regulation of endogenous CREB by short interference RNA (siRNA) attenuated the induction of L1-RTP by FICZ. Moreover, a transfection-back experiment of cDNA that encoded a siRNA-resistant CREB restored the induction of L1-RTP. These data indicate that the induction of L1-RTP by FICZ depended on p38-CREB-dependent signaling. Intriguingly, L1-RTP by FICZ was not dependent on AhR, although FICZ is a candidate physiological ligand of AhR (39). In contrast, L1-RTP by FICZ was dependent on AhR nuclear translocator 1 (ARNT1), a binding partner of AhR (40).

AhR and ARNT1 are members of the basic helix-loop-helix/per-arnt-sim (bHLH/PAS) family, which are transcription factors involved in a variety of biological functions (41). Recently, it was shown that the bHLH/PAS family is functionally linked with environmental adaptation of living organisms (42). When AhR binds environmental compounds, it forms a heterodimer with ARNT1, which is recruited from the cytoplasm to chromatin and recognizes a xenobiotic responsive element (XRE) (36). It has been shown that the chromatin recruitment of ligand-bound AhR depends on the nuclear localization signal of ARNT1 (43), but there are no reports showing that ARNT1 functions as a receptor for environmental compounds. A cellular factor that cooperates with ARNT1 in FICZ-induced L1-RTP has yet to be identified.

5. MAPKs required for L1-RTP by FICZ

To explore the involvement of MAPKs in L1-RTP, we extended our experiments to explore whether environmental carcinogens induce L1-RTP. In two-stage chemical carcinogenesis, it has been shown that skin tumors develop by treatment with 7,12-dimethylbenz[*a*]anthracene (DMBA) plus 12-*O*-tetradecanoylphorbol-13-acetate (TPA) (44). DMBA functions as an initiator and activates *H-ras* gene, whereas TPA functions as a tumor promoter thorough non-genotoxic effects (45). However, how TPA induces tumor progression remains to be clarified. We first analyzed whether L1-RTP is involved under skin carcinogenesis. When transgenic mice harboring human L1 as a transgene (hL1-EGFP mouse) were subjected to DMBA/TPA-induced skin carcinogenesis, L1-RTP was frequently observed in the DMBA/TPA-induced skin tumors (46). Interestingly, *in vitro* experiments revealed that both DMBA and TPA were active for the induction of L1-RTP. On the other hand, *in vivo* experiments, in which hL1-EGFP mice were transiently treated with DMBA or TPA suggested that L1-RTP in the skin tumors was attributable to the effects of the repeated treatment with TPA. Notably, we observed that the mode of L1-RTP by DMBA and TPA was different. DMBA-induced L1-RTP was dependent on both AhR and ARNT1, whereas TPA-induced L1-RTP required neither protein. Instead, it depended on ERK1/2 and epidermal growth factor receptor (EGFR). Since Balmain *et al* (44) originally reported on DMBA/TPA-induced two-stage carcinogenesis, a major issue of cancer research is to clarify the mechanism of the TPA-induced tumor promotion. Using genetically-engineered mice, it has been proven that TPA-induced tumor promotion depends on ERK1/2 and EGFR (47,48). Interestingly, TPA-induced L1-RTP was shown to be dependent on these molecules, suggesting that the genome shuffling by L1-RTP is linked with the mode of TPA-dependent tumor promotion.

6. MAPKs are involved in the induction of L1-RTP by carcinogens

Given that environmental compounds seemed to induce L1-RTP by involving different cellular proteins, we investigated other carcinogens such as B[*a*]P and 3-methylcholanthrene (3-MC). Consistent with a previous report (35), B[*a*]P induced L1-RTP in an AhR-dependent manner (46). Additionally, 3-MC also induced L1-RTP in an AhR-dependent manner (46). However, we found that the L1-RTP was induced even when siRNA against *ARNT1* was transfected into the cell (Fig. 1b, lanes 9 and 18). The siRNA clearly suppressed the mRNA expression of *CYP1A1* (Fig. 1c, lanes 11 and 12), indicating that the siRNA effectively abrogated the function of endogenous ARNT1 protein. These data support the idea that ARNT1 is dispensable for the induction of L1-RTP by these compounds. Because it has been

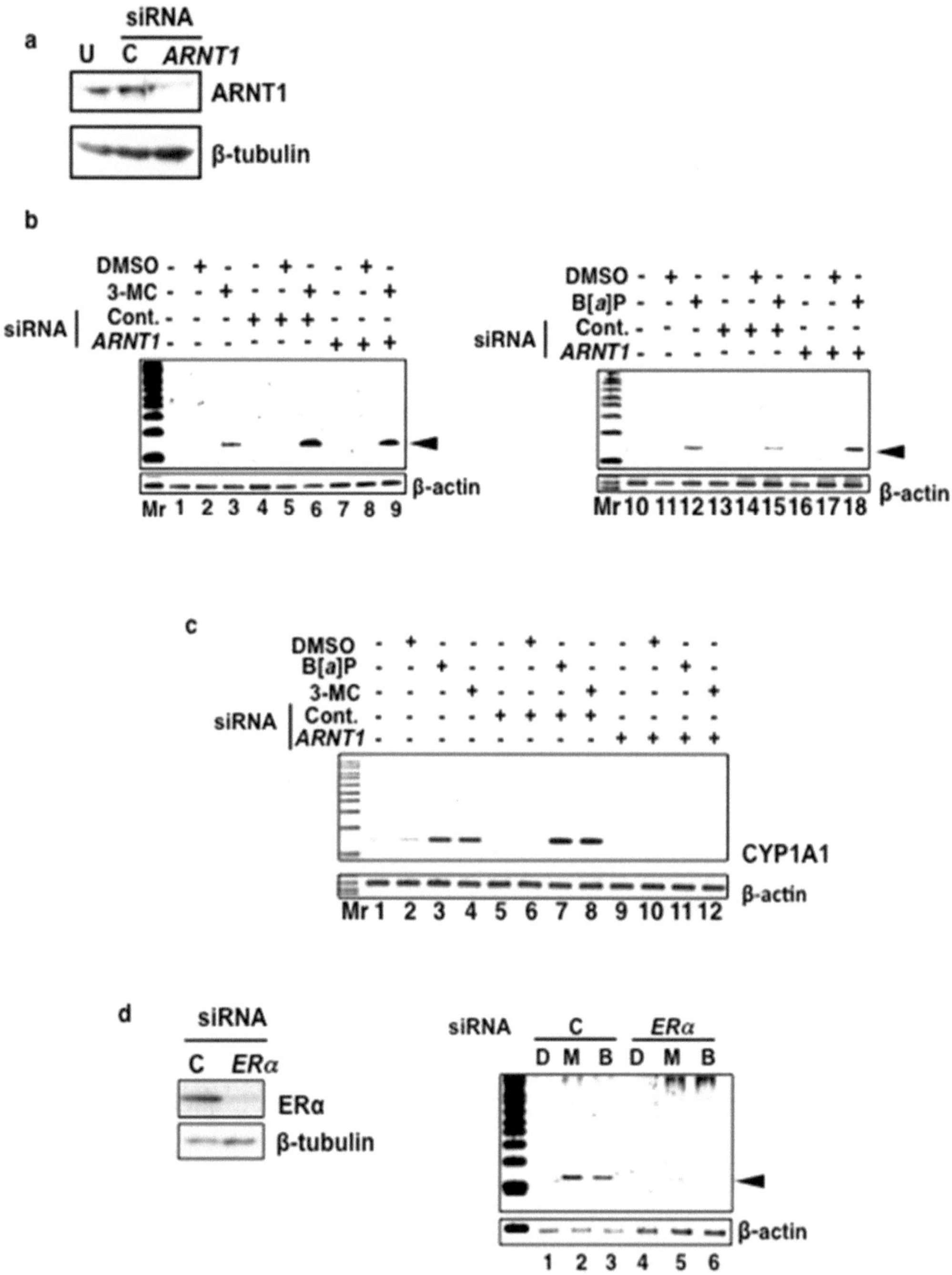

Fig. 1. B[*a*]P and 3-MC induced L1-RTP depending on AhR and ERα, but not on ARNT1. An L1-RTP assay was performed according to the procedures described (37,46). Briefly, HuH-7 cells from a human hepatoma cell line were transfected with pEF06R on day 0, then treated with 0.5 μg/mL puromycin for two days (days 1-3). The cells were then trypsinized and

replated for treatment with the compounds. Two days after the addition of 3 μM B[*a*]P or 1 μM 3-MC, the cells were harvested and their DNA extracted. No cytotoxicity was caused by 3 μM B[*a*]P or 1 μM 3-MC (data not shown). For the PCR-based assay, a spliced form of *EGFP* cDNA (140 bp in length) was amplified by PCR with primers specific for the separated exons of *EGFP* cDNA. The amplified DNA was then loaded onto an agarose gel and detected after staining with SYBR Green. As an internal control, the same samples were used as templates for the amplification of β-actin. **a**. Effects of *ARNT1* siRNA on the down-regulation of endogenous ARNT1. Western blot analysis was performed on day 2 after the transfection of *ARNT1* siRNA. U, untreated; C, control siRNA; *ARNT1*, *ARNT1* siRNA. **b**. L1-RTP caused by B[*a*]P and 3-MC was independent of ARNT1. The PCR-based assay of the effects of *ARNT1* siRNA is shown. HuH-7 cells were transfected with pEF06R on day 0 and then selected from days 1-3. On day 3, the cells were trypsinized, replated, and further transfected with control (Cont.) or *ARNT1* siRNA. On day 4, the cells were again divided into three groups and treated with DMSO, 3-MC (left panel), or B[*a*]P (right panel). After two days, DNA was extracted and subjected to a PCR-based assay. The arrowhead indicates the PCR-amplified band corresponding to the induction of L1-RTP. **c**. *ARNT1* siRNA effectively blocked the mRNA expression of *CYP1A1*, which was induced by the compounds. HuH-7 cells were first transfected with control or *ARNT1* siRNAs. On day 2 after transfection, the cells were trypsinized, replated, and treated with B[*a*]P or 3-MC. RT-PCR analysis was performed on day 2 after the addition of the compounds. **d**. ERα is required for the induction of L1-RTP by B[*a*]P or 3-MC. In this experiment, MCF-7 cells from a human breast carcinoma cell line were used. Using a similar experimental protocol, the effect of *ERα* siRNA on the induction of L1-RTP was examined. As an internal control, β-actin was amplified.

Cellular factors		Inducers				
		FICZ	B[*a*]P	3-MC	DMBA	TPA
AhR		-	○	○	○	-
ARNT1		○	-	-	○	-
ERα		N.T.	○	○	-	-
MAPK	SB202190	○	○	○	-	-
	SP600125	-	○	○	-	-
	PD98,059	N.T.	N.T.	N.T.	-	○

○, dependent; -, independent; N.T., not tested.
The induction of L1-RTP was examined by a PCR-based assay (see legend for Fig. 1).

Table 1. Summary of cellular factors required for L1-RTP by environmental compounds

shown that AhR forms a complex with estrogen receptor α (ERα) (49), we further tested the involvement of ERα in the induction of L1-RTP. Interestingly, the transfection of *ERα* siRNA attenuated L1-RTP induced by these compounds (Fig. 1d, lanes 5 and 6). In addition, we found that CREB was definitely phorphorylated (Fig. 2a, lane 4), and checked the effects of MAPK inhibitors on the induction of L1-RTP by 3-MC. As shown in Fig. 2b, SB202190 attenuated the induction of L1-RTP (lane 8), whereas SP600125 did not (lane 10). To further identify a candidate substrate of p38, we examined the effects of *CREB* siRNA. The transfection of *CREB* siRNA abrogated the induction of L1-RTP by 3-MC (Fig. 2c). These data suggest that L1-RTP by 3-MC is induced by the cooperative function of AhR and ERα depending on a signal cascade involving the p38-CREB pathway. Our data also indicate that the induction of L1-RTP by B[*a*]P is dependent on p38 and JNK (Fig. 2d, lanes 8 and 10).

Although further study is required, our current understanding is that various environmental compounds induce L1-RTP by combinations of the bHLH/PAS family and MAPKs (Table 1). L1-RTP was differentially induced by FICZ, DMBA, B[*a*]P, 3-MC and TPA. Most of the compounds examined, with the exception of DMBA, depended on MAPKs. Moreover, the L1-RTP by carcinogens depended on AhR, whereas FICZ did not. It is important to collect more information about chemical compounds active in the induction of L1-RTP and to elucidate the involvement of MAPKs.

It has been proposed that L1-RTP is controlled at the transcriptional and post-transcriptional levels. *In vitro* experiments revealed that that the expression of L1 is tightly regulated by methytlation of CpG in the region of 5′-LTR. In normal somatic cells, the 5′-LTR of L1 is methylated at CpG (50,51), but it is hypomethylated in transformed cells (52). It has been consistently reported that treatment with B[*a*]P induced the hypomethylation of CpG in HeLa cells (53). Moreover, it was reported that L1-5′UTR has a ubiquitously active antisense promoter that encodes small interfering RNAs, which effectively suppressed the retrotransposition of L1 (54). These observations indicate that epigenetic alternation of the 5′-UTR was proposed as the activation mode of L1-UTR by the compound. However, the following *in vitro* experiments suggested the presence of another regulatory system of L1-RTP. A reporter construct was transfected into cultured cells, and treatment with the compounds increased the frequency of L1-RTP. Because the reporter construct (*e.g.*, pEF06R, which carries *EGFP* cDNA as a reporter gene) contained a potent CMV promoter (32,34), L1-mRNA was strongly expressed when it was transfected into cultured cells. Even under such conditions, remarkable effects on the induction of L1-RTP were detectable by additing inducers such as FICZ, B[*a*]P, and 3-MC (37,46). Data indicate the presence of an additional regulatory system in which cellular proteins regulate the induction of L1-RTP. One possible mode of regulation is the chromatin recruitment of ORF1.

7. The chromatin recruitment of ORF1 is MAPK-dependent

Because it has been postulated that ORF1 is present in the cytoplasm (21-23) and carcinogen-induced L1-RTP was dependent on AhR, it is plausible that ORF1 is functionally associated with the bHLH/PAS family. To prove this, we evaluated the association of ORF1 and AhR by an immunoprecipitation followed by Western blot analysis with a polyclonal antibody to human ORF1. Intriguingly, ORF1 and AhR were associated even under normal conditions (Okudaira N, submitted). More importantly, we detected that recruitment of ORF1 into the chromatin-rich fraction was coupled with L1-RTP. As reported, chromatin recruitment of ORF1 was induced by FICZ in a MAPK-dependent manner. It is interesting that the chromatin recruitment of ORF1 was induced by FICZ, although FICZ-induced L1-RTP was not dependent on AhR (37). Interestingly, ARNT1 was associated with ORF1 when FICZ was added to the culture medium (37). Although the precise role of the MAPK is unclear, these data suggest that the chromatin recruitment of ORF1 is the important regulatory step in L1-RTP, where at least p38 is involved as a crucial cellular factor.

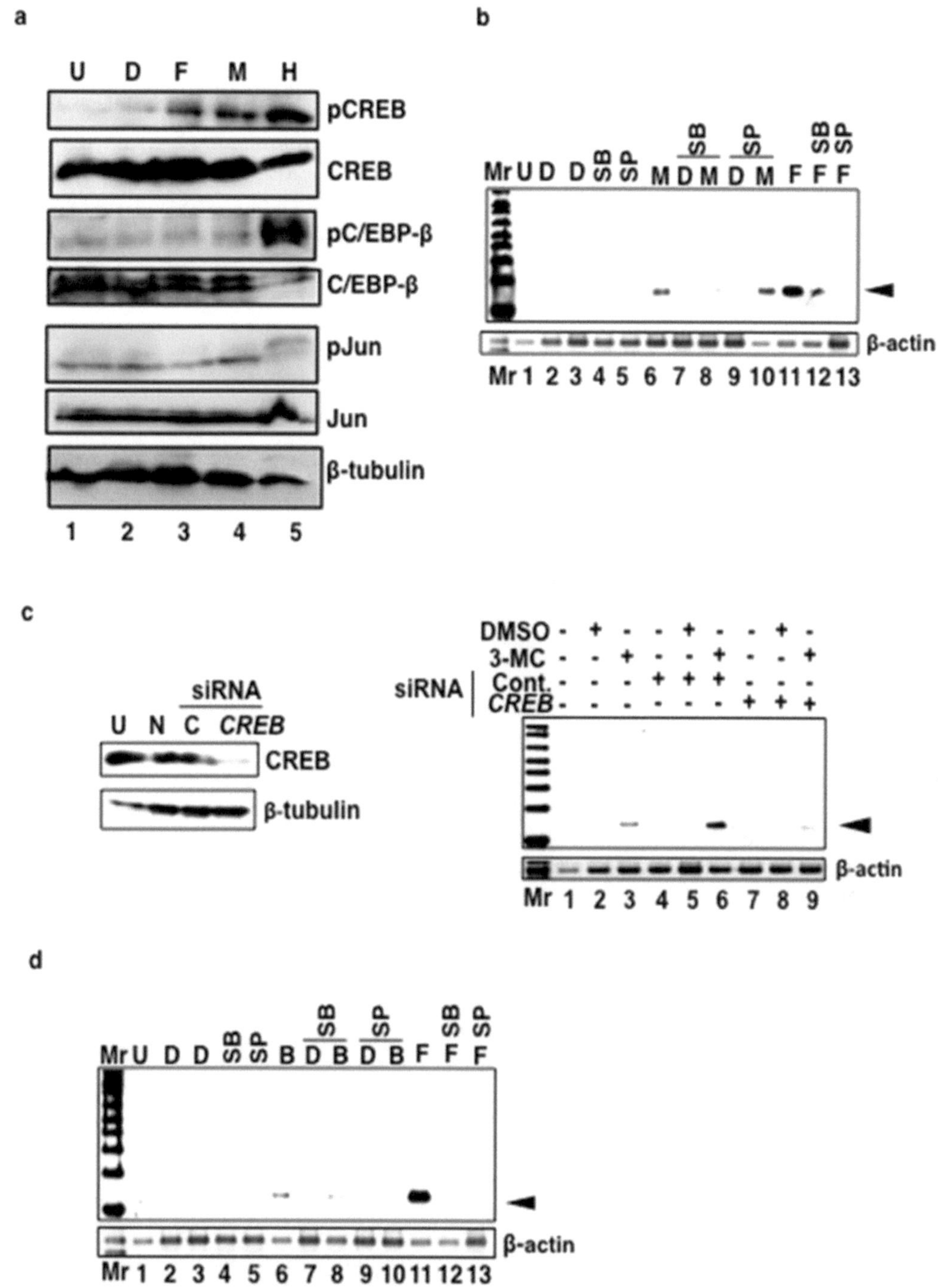

Fig. 2. MAPK is required for the induction of L1-RTP by B[*a*]P or 3-MC. **a**. Phosphorylation of MAPK substrates induced by 3-MC. HuH-7 cells were analyzed on day 2 after the addition of the compound. U, untreated; D, dimethylsulfoxide (DMSO); F, FICZ; M, 3-MC; H, H_2O_2. **b**. Effects of MAPK inhibitors on L1-RTP induced by 3-MC. SB and SP are

SB202190 and SP600125, respectively. U, untreated; D, DMSO; F, FICZ; M, 3-MC; H, H_2O_2. Of note, L1-RTP caused by 3-MC was attenuated by SB (lane 8), but not by SP (lane 10). The arrowhead indicates the induction of L1-RTP. **c**. CREB is required for L1-RTP induced by 3-MC. Left panel: Western blot analysis detected efficient down-regulation of the endogenous protein by *CREB* siRNA. U, untreated; N, non-transfected; C, control siRNA; *CREB*, *CREB* siRNA. Right panel: PCR-based assay after the transfection of *CREB* siRNA. *CREB* siRNA attenuated L1-RTP induced by 3-MC (lane 9). **d**. Effects of MAPK inhibitors on L1-RTP induced by B[*a*]P. Reagents similar to those described in Fig. 2b were used. L1-RTP by B[*a*]P was attenuated by both SB (lane 8) and SP (lane 10).

8. Roles of MAPK on L1-RTP

It has been supposed that the increase of transposable elements coupled with evolution (7). Even in *Candida albicans*, an L1-like structure is present as a functional gene (55). On the other hand, the bHLH/PAS family, which has a variety of biological functions including the metabolism of xenobiotics, maintenance of the circadian rhythm, cellular responses to hypoxia, and neuronal differentiation (41,42), is also well conserved from lower species to mammals (56). Interestingly, *AhR* homologs are also present in the genomes of *Drosophila melanogaster* and *Caenorhabditis elegans* (56). Although no direct evidence on the functional relationship between these two biological phenomena has been claimed, our observation is the first to demonstrate the functional link of these biological events. Moreover, data suggest that MAPKs are involved in the bHLH/PAS-dependent L1-RTP. MAPKs are involved in cellular response to intracellular and extracellular stress (38,57), and it is plausible that MAPKs mediate various stresses in the induction of L1-RTP, resulting in genome shuffling. Random mutagenesis by L1-RTP may give emerging novel organisms to survive in altered environments.

It is important to clarify the roles of MAPKs in the induction of L1-RTP. At present, at least two functions of MAPKs can be postulated. As explained, environmental compounds activate MAPKs, by which the chromatin recruitment of ORF1 is induced as a necessary step in L1-RTP. ORF1 functions in *cis* with L1- mRNA and functions as a chaperon of L1-mRNA (24,25). Using MAPK inhibitors, we observed that L1-RTP was abrogated concomitantly with the reduced chromatin recruitment of ORF1. These observations suggest that MAPK activation drives the mobilization of ORF1 to chromatin, by which retroelements are translocated to chromtin.

Another possible role for MAPKs is related to the activity of the APOBEC family. It has been proposed that APOBEC family functions as innate restriction factors that suppress the activity of endogenous retroelements (58). Originally, it was postulated that the APOBEC family inhibits HIV-1 infection by editing C to T via deaminase activity (58). Vif, a gene product of HIV-1, degrades APOBEC proteins, causing infected cells to become permissive for HIV-1 infection (59). We previously showed that all members of the APOBEC family exhibit inhibitory activity toward L1-RTP (60). However, it was recently postulated that the APOBEC has dual activity (61) and inhibits the activity of RT (62). In *in vitro* experiments in which APOBEC3G were added to the reaction of RT in the synthesis of viral DNA, APOBEC interfered with elongation of the viral DNA (62). Interestingly, it has been shown that C/EBP-β bound APOBEC3G and attenuate the inhibitory activity of APOBEC3G (63). Moreover, it was demonstrated that the mutation of serine at 228 (S228), the phosphorylation of which is correlated with the cytoplasmic localization of the molecule (63), abolished both binding and inhibitory activity on APOBEC3G (64). Given that C/EBP-β is a substrate of p38

(38), a plausible model is that p38 augments the blocking activity of C/EBP-β on APOBEC3G via phosphorylation.

9. Further implications

Ataxia telangiectasis mutated (ATM), a phosphoinositide 3-kinase, has a functional link with L1-RTP (16). In an intriguing recent observation, the copy number of L1 increased in the brain tissues of patients with ATM (16). L1-RTP is consistently increased in the brain tissue of *ATM*-knock out mouse. Although these observations suggest that ATM functions as a negative regulator of L1-RTP, Gasior *et al.* originally reported that ATM was required for the induction of L1-RTP (65). Because of controversial observations regarding the role of ATM in L1-RTP, we focused on MAPKs in the current study.

Recent observations revealed that genome shuffling by L1-RTP in human somatic cells is a source of interindividual genomic heterogeneity (12,13). In addition, independent research groups reported that L1-RTP is frequently induced in tumors (19,20), suggesting the involvement of L1-RTP in the development of carcinogenesis. Importantly, L1 proteins are active on the retrotransposition of *Alu* (30,31), a non-autonomous retoroelement. On the other hand, it has been shown that *Alu* induces genomic instability via non-allelic homologous recombination (66). Thus, it is important to understand the activation mechanisms of L1. Our current observations support the idea that the chromatin recruitment of ORF1, which is controlled by cooperative regulation by members of the bHLH/PAS family and MAPKs, is a critical step in the regulation of L1-RTP. If this is the case, L1-RTP induction in the genome is selectively determined by cellular factors. Because AhR is a transcription factor that recognizes specific nucleotide element (36), carcinogens possibly induce L1-RTP in the genomes in the vicinity of the *cis*-element.

As observed in the analysis of L1-RTP by B[*a*]P and 3-MC, L1-RTP was not induced via the classical pathway controlled by both AhR and ARNT1. Our data suggest that L1-RTP is not necessarily induced by genotoxic activities of these compounds, further implying that L1-RTP is a novel type of genomic instability by which cellular cascades activated by environmental compounds lead to genome shuffling and generate stable phenotypes of the affected cells. The suppression of L1-RTP in somatic cells by targeting MAPK activity may be a novel strategy to protect the development of intractable diseases that include carcinogenesis.

10. Acknowledgments

We are grateful to Dr. Elena. T. Luning Prak (University of Pennsylvania Medical Center) for pEF06R. This work was supported in parts by a research grant for the Log-range Research Initiative (LRI) from Japan Chemical Industry Association (JCIA), and The Grant for National Center for Global Health and Medicine 21A-104) and a Grant-in-Aid for Research from the Ministry of Health, Labour, and Welfare of Japan (109156296). Mr. Noriyuki Okudaira is an applicant supported by Grant-in-Aid from the Tokyo Biochemical Research Foundation.

All authors declare that they have no conflict of interest for the current work.

11. References

[1] Haig H Kazazian Jr, "Mobile elements: drivers of genome evolution", *Science* 303 (2004): 1626-32.

[2] Eric S Lander *et al.*, "Initial sequencing and analysis of the human genome", *Nature* 409 (2001): 860-921.
[3] Norbert Bannert and Reinhard Kurth, "Retroelements and the human genome: new perspectives on an old relation", *Proc Acad Natl Sci USA* 101 (2004): 14572-79.
[4] John L Goodier and Haig H Kazazian Jr, "Retrotransposons revisited: the restraint and rehabilitation of parasites", *Cell* 135 (2008): 23-35.
[5] Hameed Khan, Arian Smit and Stephane Boissinot, "Molecular evolution and tempo of amplification of human LINE-1 retorotransposons since the origin of primates", *Genome Res:* 16 (2006) 78-87.
[6] Donna M Sassaman *et al.*, "Many human L1 elements are capable of retrotransposition", *Nat Genet* 16 (1997): 37-43.
[7] Brook Brouha *et al*, "Hot L1s account for the bulk of retrotransposition in the human population", *Proc Natl Acad Sci USA* 100 (2003): 5280-85.
[8] Ioannis Georgiou *et al.*, "Retrotransposon RNA expression and evidence for retrotransposition events in human oocytes", *Hum Mol Genet* 18 (2009): 1221-28.
[9] Angela Macia *et al.*, "Epigenetic control of retrotransposon expression in human embryonic stem cells", *Mol Cell Biol* 31 (2011): 300-16.
[10] Jose AJM van den Hurk *et al.*, "L1 retrotransposition can occur early in human embryonic development", *Hum Mol Genet* 16 (2007): 1587-92.
[11] Hiroki Kano *et al.*, "L1 retrotransposition occurs mainly in embryogenesis and creates somatic mosaicism", *Genes Dev* 23 (2009): 1303-12.
[12] Christine R Beck *et al.*, "LINE-1 retrotransposition activity in human genomes", *Cell* 141 (2010): 1159-70.
[13] Lisa CR Huang *et al.*, "Mobile interspersed repeats are major structural variants in the human genome", *Cell* 141 (2010): 1171-82.
[14] Daria V Babushok and Haig H Kazazian Jr, "Progress in understanding the biology of the human mutagen LINE-1", *Hum Mutat* 28 (2007): 527-39.
[15] Kenneth J Baillie *et al.*, "Somatic retrotransposition alters the genetic landscape of the human brain", *Nature* 479 (2011): 534-7.
[16] Nicole G Coufal *et al.*, "Ataxia telangiectasia mutated (ATM) modulates long interspersed element-1 (L1) retrotransposition in human neural stem cells", *Proc Acad Natl Sci USA* 108 (2011): 20382-7.
[17] Buzzy Morse *et al.*, "Insertional mutagenesis of the myc locus by a LINE-1 sequence in a human breast carcinoma", *Nature* 333 (1988): 87-90.
[18] Yoshio Miki *et al.*, "Disruption of the APC gene by a retrotransposable insertion of L1 sequence in a colon cancer", *Cancer Res* 52 (1992): 643-5.
[19] Rebecca C Iskow *et al.*, "Natural mutagenesis of human genomes by endogenous retrotransposons", *Cell* 141 (2010):1253-61.
[20] David T Ting *et al.*, "Aberrant overexpression of satellile repeats in pancreatic and other epithelial cancers", *Science* 331 (2011) 593-6.
[21] Bratthauer GL and Fanning TG, "Active LINE-1 retrotransposons in human testicular cancer", *Oncogene* 3 (1992): 507-10.
[22] Hirohiko Hohjoh and Mxine F Singer, "Cytoplasmic ribonucleoprotein complexes containing human LINE-1 protein and RNA", *EMBO J* 15 (1996): 630-9.
[23] John L Goodier *et al.*, "LINE-1 ORF1 protein localizes in stress granules with other RNA-binding proteins, including components of RNA interference RNA-induced silencing complex", *Mol Cell Biol* 27 (2007): 6469-83.
[24] Wei Wei *et al.*, "Human L1 retrotransposition: *cis* preference versus trans complememntation", *Mol Cell Biol* 21 (2001): 1429-39.

[25] Sandra L Martin, "Nucleic acid chaperone properties of ORF1p from the non-LTR retrotransposon, LINE-1", *RNA Biol* 7 (2010): 706-11.

[26] Qinghua Feng *et al.*, "Human L1 retrotransposison encodes a conserved endonuclease required for retrotransposition", *Cell* 87 (1996) 905-16.

[27] Stephen ML Mathias e*t al.*, "Reverse transcriptase encoded by a human transposable element", *Science* 254 (1991): 1808-10.

[28] Jerzy Jurka, "Sequence patterns indicate an enzymatic involvement in integration of mammalian retroposons", *Proc Natl Acad Sci USA* 94 (1997): 1872-7.

[29] Nicolas Gilbert *et al.*, "Multiple fates of L1 retrotransposition intermediates in cultured human cells", *Mol Cell Biol* 25 (2005): 7780-95.

[30] Marie Dewannieux, Cecile Esnault and Thierry Heidmann, "LINE-mediated retrotransposition of marked Alu sequences", *Nat Genet* 35 (2003): 41-8.

[31] Nichola Wallace *et al.*, "LINE-1 ORF1 protein enhances Alu SINE retrotrasposition", *Gene* 419 (2008): 1-6.

[32] Evan A Farkash *et al.*, "Gamma radiation increases endonuclease-dependent L1 retrotransposition in a cultured cell assay", *Nucleic Acids Res* 34 (2006): 1196-1204.

[33] Shubha P Kale *et al.*, "Heavy metals stimulate human LINE-1 retrotransposition", *Int J Environ Res Public Health* 2 (2005): 14-23.

[34] Mohammed El-Sawy *et al.*, "Nickel stimulates L1 retrotransposition by a post-transcriptional mechanism", *J Mol Biol* 354 (2005): 246-57.

[35] Vilius Stribinskis and Kenneth S Ramos, "Activation of human long interspersed nuclear element 1 retrotransposition by benzo(*a*)pyrene, an ubiquitous environmental carcinogen", *Cancer Res* 66 (2006): 2616-20.

[36] Timothy V Beischlag *et al.*, "The aryl hydrocarbon receptor complex and the control of gene expression", *Crit Rev Eukaryot Gene Expr* 18 (2008): 207-50.

[37] Noriyuki Okudaira *et al.*, "Induction of long interspersed nucleotide element-1 (L1) retrotransposition by 6-formylindolo[3,2-*b*]carbazole, a tryptophan photoproduct", *Proc Natl Acad Sci USA* 107 (2010): 18487-92.

[38] Marie Cargnello and Philippe P Roux, "Activation and function of the MAPKs and their substrates, the MAPK-activated protein kinases", *Micobiol Mol Biol Rev* 75 (2011): 50-83.

[39] Emma Wincent *et al.*, "The suggested physiologic aryl hydrocarbon receptor activator and cytochrome P4501 substrate 6-formylindolo[3,2-b]carbazole is present in humans", *J Biol Chem* 284 (2009): 2690-6.

[40] Emilly C Hoffman *et al.*, "Cloning of a factor required for activity of the Ah (dioxin) receptor", *Science* 252 (1991): 954-958.

[41] Robyn J Kewley *et al.*, "The mammalian basic helix-loop-helix/PAS family of transcriptional regulators", *Int J Biochem Cell Biol* 36 (2004): 189-204.

[42] Brian E McIntosh *et al.*, "Mammalian Per-Arnt-Sim proteins in environmental adaptation", *Annu Rev Physiol* 72 (2010): 625-45.

[43] Hideaki Eguchi *et al.*, "A nuclear localization signal of human aryl hydrocarbon receptor nuclear translocator/hypoxia-inducible factor 1beta is a novel bipartite type recognized by the two components of nuclear pore-targeting complex", *J Biol Chem* 272 (1997): 17640-17647.

[44] Alian Balmain *et al.*, "Activation of the mouse cellular Harvey-ras gene in chemically induced benign skin papillomas", *Nature* 307 (1984): 658-60.

[45] Mark A Nelson *et al.*, "Detection of mutant Ha-*ras* in chemically initiated mouse skin epidermis before the development of benign tumors", *Proc Natl Acad Sci USA* 89 (1992): 6398-6402.

[46] Noriyuki Okudaira *et al.*, "Involvement of retrotransposposition of long interspersed nucleotide element-1 in skin tumorigenesis induced by 7,12-dimethylbenz[*a*]anthracene and 12-O-tetradecanoylphorbol-13-acetate, *Cancer Sci* 102 (2011): 2000-6.
[47] Christine Bourcier *et al.*, "p44 mitogen-activated protein klinase (extracellular signal-regulated kinase 1)-dependent signaling contributes to epithelial skin carcinogenesis", *Cancer Res* 66 (2006): 2700-7.
[48] Llanos M Casanova *et al.*, "A critical role for *ras*-mediated, epidermal growth factor receptor-dependent angiogenesis in mouse skin carcinogenesis", *Cancer Res* 62 (2006): 3402-7.
[49] Fumiaki Ohtake *et al.*, "Modulation of oestrogen receptor signaling by association with the activated dioxin receptor", *Nature* 423 (2993): 545-50.
[50] Kikumi Hata and Yoshiyuki Sakaki, "Identification of critical CpG sites for repression of L1 transcription by DNA methylation", *Gene* 189 (1997): 227-34.
[51] David M Woodcock *et al.*, "Asymmetric methylation in the hypermethylated CpG promoter region of the human L1 retrotransposon", *J Biol Chem* 272 (1997): 7810-6.
[52] Andrea R Florl *et al.*, "DNA methylation and expression of LINE-1 and HERV-K provirus sequences in urothelial and renal cell carcinomas", *Br J Cancer* 80 (1999): 1312-21.
[53] Ivo Teneng *et al.*, "Reactivation of L1 retrotransposon by benzo(*a*)pyrene involved complex genetic and epigenetic regulation, *Epigenetics* 6 (2011): 355-67.
[54] Nuo Yang and Haig H Kazazian Jr, "L1 retrotransposition is suppressed by endogenously encoded small interfering RNAs in human cultured cells", *Nat Struct Mol Biol* 13 (2006): 763-71.
[55] Chung Dong, Russell T Poulter and Jeffrey S Han, "LINE-like retrotransposition in *Saccharomyces cerevisiae*", *Genetics* 181 (2009): 301-11.
[56] Mark E Hahn, "Aryl hydrocarbon receptors: diversity and evolution", *Chem-Biol Interact* 141 (2002): 131-60.
[57] Helmuth Gehart *et al.*, "MAPK signaling in cellular metabolism: stress or wellness?", *EMBO Reports* 11 (2010): 834-40.
[58] Ya-Lin Chiu and Warner C Greene, "The APOBECG3 cytidine deaminase: an innate defensive network opposing exogenous retroviruses and endogenous retroelements", *Annu Rev Immunol* 26 (2008): 317-53.
[59] Ann M Sheehy *et al.*, "Isolation of a human gene that inhibits HIV-1 infection and is suppressed by the viral Vif protein", *Nature* 418 (2002): 646-50.
[60] Masanobu Kinomoto *et al.*, "All APOBEC3 family proteins differentially inhibit LINE-1 retrotransposition", *Nucl Acids Res* 35 (2007): 2955-64.
[61]Cecile Esnault *et al.*, "Dual inhibitory effects of APOBEC family proteins on retrotransposition of mammalian endogenous retroviruses", *Nucl Acids Res* 34 (2006): 1522-31.
[62] Kate N Bishop *et al.*, "APOBEC3G inhibits elongation of HIV-1 reverse transcripts", *PLoS Pathogens* 4 (2008): e1000231.
[63] Shigemi M Kinoshita and Shizuka Taguchi, "NF-IL6 (C/EBPβ) induces HIV-1 replication by inhibiting cytidine deaminase APOBEC3G", *Proc Natl Acad Sci USA* 105 (2008): 15022-27.
[64] Martina Buck *et al.*, "Nuclear export of phosphorylated C/EBPβ mediates the inhibition of albumin expression by TNF-α", *EMBO J* 20 (2001): 6712-23.
[65] Stephen L. Gasior, "The human LINE-1 retrotransposon creates DNA double-strand breaks", *J Mol Biol* 357 (2006): 1383-93.
[66] Victoria P Belancio, Astrid M Roy-Engel and Prescott L Deininger, "All y'all need to know 'bout' retroelements in cancer", *Sem Cancer Biol* 20 (2010): 200-10.

Connexins as Substrates for Protein Kinases and Phosphoprotein Phosphatases

José L. Vega[1,2], Elizabeth Baeza[1], Viviana M. Berthoud[3] and Juan C. Sáez[1,4]
[1]*Departamento de Fisiología, Pontificia Universidad Católica de Chile, Santiago,*
[2]*Laboratorio de Fisiología Experimental (EPhyL), Universidad de Antofagasta, Antofagasta, Chile,*
[3]*Department of Pediatrics, Section of Hematology, Oncology, University of Chicago, IL,*
[4]*Instituto Milenio, Centro Interdisciplinario de Neurociencias de Valparaíso, Valparaíso,*
[3]*USA,*
[1,2,4]*Chile*

1. Introduction

1.1 Mammalian connexins

Connexins are protein subunits expressed by cordates that form gap junction channels (GJCs) and hemichannels (HCs) (Goodenough, 1974; Makowski et al., 1977). A GJC is formed by the head-to-head docking of two HCs, each contributed by one of the two contacting cells (Meşe et al., 2007). Each HC is an oligomeric assembly of six identical (homomeric) or six different (heteromeric) Cx subunits (Sáez et al., 2005). GJCs and HCs subserve different functions; while GJCs communicate the cytoplasm of contacting cells, HCs provide a pathway for communication between the intracellular and extracellular compartments (Bruzzone and Dermietzel, 2006). Although both types of channels are permeable to ions and small molecules, GJCs and HCs composed of the same Cx subtype are likely to present differences in permeability and regulatory properties (Sáez et al., 2003; Meşe et al., 2007; Sáez et al., 2010).

The family of connexin genes has 20 members in the mouse genome and 21 members in the human genome (Eiberger et al., 2001; Willecke et al., 2002; Söhl and Willecke, 2003; 2004). Most Cx genes have a similar structure and contain the protein coding region as a single exon (Willecke et al., 2002; Söhl and Willecke, 2003; 2004; Pfenniger et al., 2011). Cxs were initially denoted according to the tissue of origin or the apparent size of a polypeptide as determined by SDS-PAGE. Shortly thereafter, it became clear that such designations were inappropriate, because many of these proteins are expressed in more than one tissue (Beyer et al., 1987) and their apparent molecular mass may vary with electrophoresis conditions (Green et al., 1988). Therefore, a standard nomenclature was developed to distinguish members of this family. The current nomenclature uses the abbreviated symbol "Cx" (for connexin) followed by a suffix that indicates the molecular mass of the Cx amino acid sequence (in kDa) predicted from its cDNA. In some cases, a prefix is added to indicate the species of origin. Hydropathicity plots of the Cx amino acid sequences have been used to predict their membrane topology. These analyses predicted the presence of four hydrophobic domains, three hydrophilic cytoplasmic domains (the amino and carboxyl termini and an intracellular loop) and two extracellular loops (Heynkes et al., 1986; Paul, 1986; Beyer et al., 1987). This topology was supported by experiments that studied the binding of site-specific antibodies and protease sensitive sites (Zimmer et al., 1987; Hertzberg et al., 1988; Milks et al., 1988; Yancey et al., 1989; Zhang and Nicholson, 1994; Quist et al., 2000). The cytoplasmic loop and the carboxyl terminus vary extensively in length and amino acid composition and probably contain most of the regulatory sites of GJCs and HCs.

1.2 Mammalian protein kinases and phosphoprotein phosphatases

Most Cxs contain putative phosphorylation sites (Lampe and Lau, 2004). As with all phosphoproteins, their phosphorylation state will depend on the activities of protein kinases and phosphoprotein phosphatases. Mammalian cells express several different types of protein kinases and phosphoprotein phosphatases with more than 500 putative kinase genes in the human and mouse genome (Manning et al., 2002; Caenepeel et al., 2004). Protein kinases and phosphoprotein phosphatases have been subdivided according to their substrate specificities, activators, cofactors and/or amino acid sequence homology. It would be beyond the scope of this chapter to attempt to review them here and thus, we will briefly summarize the characteristics of the kinases and phosphatases that have most frequently been studied as possible effectors of the phosphorylation state of connexins.

1.3 Serine/threonine protein kinases

cAMP- and cGMP-dependent protein kinases (PKA and PKG, respectively) can be activated by increasing the concentration of the corresponding cyclic nucleotide (e.g., treatment with membrane permeable analogs of cAMP or cGMP such as 8-Bromo-cAMP and 8-Bromo-cGMP or forskolin, which activates adenylyl cyclase). The CAMKII isoenzymes are activated by binding of Ca^{2+}/calmodulin but other protein binding partners can also regulate their activity (Griffith, 2004). Casein kinase I (CK1) is a family of monomeric serine/threonine kinases that are constitutively active. This family shows a strong preference for pre-phosphorylated substrates. Several inhibitors for members of this family have been described including CKI-7 and IC261 (Perez et al., 2011). Protein kinase C (PKC) has several isoforms that have been subdivided in three subtypes: conventional, novel and atypical. They differ in their activation by Ca^{2+}, binding of diacylglycerol (DAG) and in their response to phorbol esters. Conventional PKCs bind Ca^{2+} and DAG. Novel PKCs lack amino acids involved in Ca^{2+} binding, but bind DAG. The catalytic activity of atypical PKCs is independent of Ca^{2+} and DAG; these PKC isoforms do not bind phorbol esters (Newton, 1995). The phorbol ester tumor promoter, 12-O-tetradecanoylphorbol 13-acetate (TPA) and 1-oleoyl-2-acetyl-sn-glycerol (OAG), an analog of diacylglycerol have been commonly used as activators of PKC. MAPKs are subdivided in three subfamilies: the extracellular signal-regulated kinases (ERKs), the c-Jun amino-terminal kinases (JNKs) and the p38 MAPKs. They are activated by protein kinase cascades [MKKK-MKK(or MEK for ERKs)-MAPK], although MKK-independent activation of p38α has been reported (Johnson and Lapadat, 2002). Finally, cyclin-dependent kinases (Cdks) constitute a family of serine/threonine kinases that regulate proliferation, differentiation, senescence and apoptosis. In post-mitotic neurons, all Cdks, with the exception of Cdk5, are silenced.

1.4 Tyrosine kinases

The tyrosine kinases can be divided in two groups: receptor tyrosine kinases (RPTKs; e.g, growth factor receptors, ephrin receptors) and non-receptor (cytoplasmic) tyrosine kinases (NRPTKs; e.g., Src, FAK, JAK). RPTKs can be further subdivided into 20 subfamilies and NRPTKs into 10 subfamilies. In the case of RPTKs, ligand-induced oligomerization and conformational changes result in tyrosine autophosphorylation of the receptor subunits which activates the catalytic activity and mediate the specific binding of cytoplasmic signaling proteins containing Src homology-2 (SH2) and protein tyrosine-binding domains. The NRPTK, c-Src, contains an SH2 domain through which it can bind to specific tyrosine autophosphorylation sites in ligand-stimulated RPTKs and mediate mitogenic signaling. c-Src can also be activated by binding to proline-rich sequences in target proteins through its SH3 domain or by dephophorylation of Tyr527 (Blume-Jensen and Hunter, 2001). The viral form of Src, v-Src, is constitutively active and oncogenic. It contains a shorter sequence at the carboxyl terminus that lacks Tyr527, which is required for inactivation. v-Src has been extensively studied in relation to connexins for its effects on gap junction function.

1.5 Serine/threonine phosphoprotein phosphatases

The phosphoserine/phosphothreonine protein phosphatases have been classified in three subfamilies (PPM, FCP and PPP). Members of the PPP (PP1, PP2A and PP2B) and PPM (PP2C) subfamilies which use a metal ion-catalyzed reaction account for most of the serine/phosphothreonine phosphatase activity *in vivo* (Barford et al., 1998). Several phosphatase inhibitors with different specificities are available including calyculin A (which inhibits PP1 and PP2A), cyclosporine A (an inhibitor of PP2B), FK506 (an inhibitor of PP2B) and okadaic acid (which inhibits PP1).

1.6 Phosphotyrosine phosphatases

The phosphotyrosine phosphatases (PTPs) have been classified in class I-IV based on the amino acid sequence of their catalytic domains (class I-III are cysteine-based PTPs and class IV are aspartic-based PTPs). The cysteine-based family can be subdivided in classical PTPs, dual-specificity PTPs, cdc25 PTPs, and low-molecular weight PTPs. Classical PTPs can be further subdivided into transmembrane receptor-like enzymes and intracellular non-receptor PTPs. Eighty one of the 107 PTP genes in the human genome are active protein phosphatases (Alonso et al., 2004).

2. Methods used to demonstrate that connexins are phosphoproteins

The most frequently used experimental approaches to demonstrate that a particular Cx is a phosphoprotein include metabolic labeling of cultured cells with ^{32}P followed by immunoprecipitation and alkaline phosphatase treatment, phosphoamino acid analysis (Sáez et al. 1986; Takeda et al., 1989; Musil et al., 1990; Crow et al., 1990; Sáez et al., 1990; Lau et al.,

1992; Goldberg and Lau, 1993; Kurata and Lau, 1994; Doble et al., 1996; Warn-Cramer et al., 1996; Mikalsen et al., 1997; Cheng and Louis, 1999) or two-dimensional phosphopeptide mapping (Sáez et al., 1990; Kurata and Lau, 1994; Díez et al., 1995; Loo et al., 1995; Warn-Cramer et al., 1996; Berthoud et al., 1997; Díez et al., 1998; Kanemitsu et al., 1998) *in vitro* phosphorylation assays using fusion proteins or synthetic peptides containing the putative phosphorylation site(s) and purified protein kinases (Sáez et al., 1990; Loo et al., 1995; Warn-Cramer et al., 1996; Berthoud et al., 1997; Kanemitsu et al., 1998; Shah et al., 2002; O'Brien et al., 2004; Ouyang et al., 2005; Yogo et al., 2006; Alev et al., 2008; Morel et al., 2010); treatment of cultured cells with specific protein kinase or phosphoprotein phosphatase activators or inhibitors to alter ^{32}P incorporation or the immunoblot pattern of connexins (Lau et al., 1992; Husøy et al., 1993; Guan et al., 1996; Berthoud et al., 1997; Cruciani et al., 1999; Duthe et al., 2000; Li and Nagy, 2000; Sirnes et al., 2009; Morley et al., 2010); overexpression or knockdown of a specific protein kinase or phosphoprotein phosphatase (Kanemitsu et al., 1998; Lampe et al., 1998; Doble et al., 2000; Lin et al., 2001; Chu et al., 2002; Petrich et al., 2002; Doble et al., 2004; Peterson-Roth et al., 2009; Ai et al., 2011); mass spectrometry (MS) analyses of immunoprecipitated connexins or *in vitro* phosphorylated fusion proteins containing a Cx intracellular domain (Cooper et al., 2000; Yin et al., 2000; TenBroek et al., 2001; Cooper and Lampe, 2002; Cameron et al., 2003; Axelsen et al., 2006; Locke et al., 2006; Solan et al., 2007; Shearer et al., 2008; Locke et al., 2009; Wang and Schey, 2009; Huang et al., 2011) and more recently, luminescence resonance energy transfer (Bao et al., 2007). Mutagenesis of the identified phosphorylation sites has been used to determine the functional consequences of their phosphorylation/dephosphorylation in cultured cells as well as *in vivo* after transfection or knock-in of a phosphosite-directed mutant Cx (Lampe et al., 1998; Remo et al., 2011).

3. Metabolic labeling with ^{32}P

The first reports that demonstrated a particular Cx to be a phosphoprotein using metabolic labeling with ^{32}P showed phosphorylation of Cx32 in hepatocytes (treated with phorbol esters, OAG, forskolin or cAMP analogs)((Sáez et a., 1986; Takeda et al., 1989; Sáez et al., 1990) and phosphorylation of Cx43 in uninfected and Rous sarcoma virus (RSV)-transformed fibroblasts (Crow et al., 1990). Phosphoamino acid analysis indicated that hepatocyte Cx32 and Cx43 in uninfected fibroblasts were phosphorylated on seryl residues (Takeda et al., 1989; Crow et al., 1990; Sáez et al., 1990), but Cx43 was also phosphorylated in tyrosyl residues in RSV-transformed fibroblasts (Crow et al., 1990). Using metabolic labeling with ^{32}P, other studies described that EGF-induced phosphorylation of Cx43 on serine residues in T51B cells through activation of mitogen-activated protein kinase (MAPK) (Lau et al., 1992; Warn-Cramer et al., 1996), FGF-2 induced phosphorylation of Cx43 in cardiomyocytes (Doble et al., 1996), tyrosine phosphorylation of Cx43 in early passage hamster embryo fibroblast (Mikalsen et al., 1997), phosphorylation of Cx56 by PKC and Cx49 by casein kinase 1 (CK1) in lens fiber cells (Berthoud et al., 1997; Cheng and Louis, 1999). In some cases, the specific phosphorylation site has been identified in reconstituted connexons expressed in *Xenopus laevis* oocytes. Using this approach, it has been demonstrated that v-Src induces tyrosine phosphorylation of Cx43 but not Cx32 (Swenson et al., 1990), and that serine368 of Cx43 (but not serine372) is directly phosphorylated by PKC (Bao et al., 2004a; 2004b).

3.1 *In vitro* phosphorylation

Another widely used approach to identify putative phosphorylation sites is *in vitro* phosphorylation assays. In this case, a polypeptide, fusion protein or synthetic peptide (corresponding to a fragment of the connexin that includes the putative phosphorylation site(s)) is incubated with a purified protein kinase in the presence of [γ-^{32}P]ATP and its ability to be a substrate for that protein kinase is evaluated by the incorporation of ^{32}P. Sáez and collaborators (1990) also performed *in vitro* kinase assays using the catalytic subunits of PKA, PKC or CaMK II and purified gap junctions or synthetic peptides as substrates, and compared their two-dimensional pattern of phosphopeptides with those obtained from metabolically labeled cells. Using glutathione S-transferase (GST) fusion proteins of Cx56 containing the carboxyl terminus or the intracellular loop, *in vitro* phosphorylation of Cx56 by PKC and PKA have been demonstrated in serine118 (in the intracellular loop) and serine493 (in the carboxyl terminus)(Berthoud et al., 1997).

Phosphorylation of Cx43 is among the best characterized. Polypeptides, fusion proteins and several synthetic peptides containing putative phosphorylation sites within the carboxyl terminus of Cx43 have been used to carry out *in vitro* phosphorylation and identify phosphorylation sites. These experiments have demonstrated that Cx43 is a substrate of p34^{cdc2} kinase (cell division cycle 2 kinase also known as cyclin dependent kinase 1) which mediates phosphorylation of Cx43 on Ser255 and possibly Ser262 (Kanemitsu et al., 1998). Cx43 is also a substrate for PKC and PKA. Kinetic analyses of wild type and mutant (S364P and S365N) Cx43 peptides (containing amino acid residues 359-376) *in vitro* phosphorylated by PKA and PKC have suggested that phosphorylation of Ser364 may be required for subsequent phosphorylation by PKC (Shah et al., 2002). *In vitro* phosphorylation of Ser365, Ser368, Ser369, and Ser373 by PKA has been described using a His-tagged Cx43-CT (containing amino acid residues E227-I382)(Yogo et al., 2006).

Other studies have shown *in vitro* phosphorylation of perch Cx35 by PKA and mouse Cx36 by CaMKII using fusion proteins containing the carboxyl terminus or the intracellular loop (O'Brien et al., 2004; Ouyang et al., 2005; Alev et al., 2008). A polypeptide containing the polymorphic variants S319 and P319 of the carboxyl terminus of human Cx37 (amino acid residues 233-333) was *in vitro* phosphorylated by glycogen synthase kinase-3β (Morel et al., 2010). *In vitro* kinase assays have also been used to demonstrate that phosphorylation of Cx32 by PKC prevents its proteolysis by calpains (Elvira et al., 1993).

Analyses of two dimensional maps of mixes of tryptic phosphopeptides from a connexin immunoprecipitated after metabolic labeling and from a (poly)peptide after *in vitro* phosphorylation together with phosphopeptide sequencing have been used often to identify the phosphorylated sites of the immunoprecipitated connexin and changes in their phosphorylation state under different experimental conditions.

4. Pharmacological modulation of phosphoprotein phosphatases

Changes in the phosphorylation state of Cxs can be induced by activating or inhibiting a specific intracellular phosphoprotein phosphatase. This type of approach allows identification of the protein phosphatases involved in the effects observed.

Using this approach, it has been demonstrated that treatment of V79 fibroblasts with several phosphoprotein phosphatase inhibitors (i.e., calyculin A, cyclosporin A or FK506) does not change the immunoblot pattern of Cx43 (Husøy et al., 1993; Cruciani et al., 1999). However, the dephosphorylation of immunoprecipitated Cx43 from TPA-exposed V79 cells is more efficiently reduced by PP2A than by PP1, PP2B or PP2C inhibitors (Cruciani et al., 1999). In WB-F344 cells, a rat liver epithelial cell line, calyculin A prevents the dephosphorylation of Cx43 induced by 18β-glycyrrhetinic acid (Guan et al., 1996). However, in primary cultures of astrocytes, calyculin A had little effect on hypoxia-induced Cx43 dephosphorylation; in this cell type, inhibition of PP2B with cyclosporin A or FK506 reduced Cx43 dephosphorylation after hypoxia (Li and Nagy, 2000). Calyculin A significantly retarded the loss of channel activity seen in ventricular myocytes in ATP-deprived conditions; conversely, stimulation of endogenous PP1 activity by treatment with p-nitrophenyl phosphate or 2,3-butanedione monoxime (a dephosphorylating chemical agent) induced a reversible interruption of cell-to-cell communication (Duthe et al., 2000; 2001).

The effect of okadaic acid on Cx43 also varies depending on cell type. It inhibits dephosphorylation of Cx43 in untreated and EGF-treated T51B rat liver epithelial cells and prevents the dephosphorylation of Cx43 induced by 18β-glycyrrhetinic acid in WB-F344 rat liver epithelial cells (Lau et al., 1992; Guan et al., 1996). Okadaic acid also significantly retards the loss of gap junction channel activity seen in ventricular myocytes in ATP-deprived conditions (Duthe et al., 2000; 2001). In other cell types, it has little or no effect on the immunoblot pattern of Cx43 (Berthoud et al., 1992; Husøy et al., 1993; Cruciani et al., 1999), and has little effect on hypoxia-induced Cx43 dephosphorylation in primary cultures of astrocytes (Li and Nagy, 2000). Altogether these results suggest the involvement of different protein phosphatases in the phosphorylation state of Cx43 in different cell types under various experimental conditions.

5. Genetic activation or inhibition of a protein kinase or phosphatase

In some studies, changes in the phosphorylation state of Cxs have been induced by genetic manipulation through chemical-induced mutagenesis of genomic DNA or transfection with mammalian expression vectors and/or infection with virus containing cDNAs coding for a protein of interest. These methods can be used to modify the kinase activity using cDNAs encoding active or dominant negative mutant forms of a specific kinase. Lampe et al. used the FT210 cell line which contains a temperature-sensitive mutant of $p34^{cdc2}$/cyclin B kinase to demonstrate that the formation of the phosphoform of Cx43 present in mitotic cells was dependent on the activity of this kinase. However, the two-dimensional tryptic phosphopeptide map of immunoprecipitated Cx43 from mitotic cells had many major and minor tryptic phosphopeptides that could not be attributed to direct $p34^{cdc2}$/cyclin B kinase phosphorylation of the Cx43CT (Lampe et al., 1998). Doble et al. (2000) used transient tranfection and adenoviral infection of truncated or dominant-negative forms of PKCε to demonstrate that this kinase is required for Cx43 phosphorylation in cardiomyocytes (Doble et al., 2000).

The mechanism by which v-Src affects Cx43 phosphorylation and function has been extensively explored. Several studies have shown that expression of v-Src in mammalian fibroblasts leads to phosphorylation of Cx43 in tyrosyl residues (Crow et al., 1990). Mutants of Cx43 and v-Src SH2 and SH3 domains have been used to demonstrate that the SH2 and

SH3 domains of v-Src interact with Cx43; the SH3 domain binds to a proline-rich motif and the SH2 domain binds to a phosphorylated tyrosyl residue in the carboxyl terminus of Cx43 (Kanemitsu et al., 1997). Two specific phosphorylation sites for v-Src have been identified in Cx43, Tyr247 and Tyr265, by stably re-expressing wild type or mutant Cx43 with v-Src in Cx43 knockout cells (Lin et al., 2001). Moreover, using a triple serine-to-alanine mutant at the MAPK sites (S255/279/282A) it has been shown that phosphorylation of Cx43 by MAPK is not required for v-Src-induced disruption of gap junctional intercellular communication (Lin et al., 2006).

Several studies have been carried out on cardiac cells. Phosphorylation of Cx43 in Ser262 regulates DNA synthesis in cardiomyocytes forming cell-cell contact (Doble et al., 2004). Expression of an activated mutant of mitogen-activated protein kinase kinase 7 (a JNK-specific upstream activator) in cultured cardiomyocytes and in the heart *in vivo* demonstrated that Cx43 expression is regulated by JNK, although this effect may not be mediated by direct phosphorylation of Cx43 (Petrich et al., 2002). Transgenic mice with cardiac-specific overexpression of a constitutively active form of calcineurin (a calcium-dependent serine/threonine phosphatase) showed differences in the distribution of Cx43 in the ventricles, and Cx43 was mainly present in the nonphosphorylated form (Chu et al., 2002). Overexpression of p21-activated kinase 1 (PAK1, an activator of PP2A) increased PP2A activity and induced dephosphorylation of Cx43 in rabbit myocytes and Cx43-overexpressing HEK293 cells (Ai et al., 2011).

6. Genetic modification of a phosphosite-specific mutant connexin

A more recent approach is the generation of connexin knock-in mice in which the coding region of the wild type protein is replaced by DNA encoding a phosphosite-specific mutant. The only available report to date using this approach showed that mice in which Cx43 was replaced by a Cx43 mutant at the CK1 sites in which serines 325/328/330 were replaced with phosphomimetic glutamic acids (S3E) were resistant to gap junction remodeling and less susceptible to the induction of arrhythmias. In contrast, mice in which a Cx43 mutant with serines 325/328/330 mutated to non-phosphorylatable alanines (S3A) was knocked-in in place of Cx43 had severe alterations in gap junction formation and function, and had a proarrhythmic phenotype (Remo et al., 2011). This report shows a mechanistic link between the phosphorylation state of Cx43 and arrhythmic susceptibility (Remo et al., 2011).

7. Phosphospecific antibodies

Antibodies that recognize a specific phosphorylated (or dephosphorylated) site in a connexin have been developed. These have been extensively used to identify the state of phosphorylation of the phosphosite they recognize and to determine associated changes in connexin distribution in cells under different physiological and pathological conditions. Using this approach, it has been described that ischemic preconditioning prevents the changes in the phosphorylation state of Cx43 observed in a model of ischemia/reperfusion in pig hearts (Schulz et al., 2003). It has also been reported that PKC phosphorylates Cx43 in Ser368 (Solan et al., 2003), and that scratch wounding of primary human keratinocytes causes a PKC-dependent increase in phosphorylation at this site in cells adjacent to the scratch (Richards et al., 2004). Leykauf et al. used a specific antibody against PSer279-

PSer282 of Cx43 to demonstrate that different phosphorylated forms of Cx43 coexist at the plasma membrane (Leykauf et al., 2003). Two antibodies recognizing the same phosphosites were used to show that EGF and activation of its receptor with quinones induce phosphorylation of Cx43 in these serine residues (Abdelmohsen et al., 2003; Leykauf et al., 2003). Using an antibody that specifically recognizes Cx43 phosphorylated at serines 325, 328 and/or 330 (PS325/328/330), Lampe and colleagues showed that while Cx43 relocalizes to the lateral edges in ischemic hearts, Cx43 phosphorylated at these residues remained mostly at the intercalated disk (Lampe et al., 2006). An antibody that recognizes dephosphorylated Ser364/Ser365 and binds preferentially to Golgi-localized Cx43 in cultured cells has been used to demonstrate conformational changes in Cx43 (Sosinsky et al., 2007). Other studies have described that phosphorylation of connexin 43 at Ser262 is associated with a cardiac injury-resistant state (Srisakuldee et al., 2009).

Phosphospecific antibodies have been used in combination with PKC or MEK inhibitors to determine the protein kinase pathway involved in the effects observed. Sirnes et al. reported that TPA induces phosphorylation of Ser255 and Ser262 of Cx43 in a MAPK-dependent manner (Sirnes et al., 2009). A MAPK-dependent phosphorylation of serines 255, 262 and 279/282 of Cx43 has also been demonstrated using phosphospecific antibodies and a MEK inhibitor in follicles exposed to luteinizing hormone (Norris et al., 2008). In MC3T3-E1 osteoblasts, treatment with fibroblast growth factor 2 induces a PKCδ-dependent increase in phosphorylation at Ser368 of Cx43 (Niger et al., 2010). Solan and Lampe used several anti-Cx43 phosphospecific antibodies that recognize Src, MAPK or PKC sites and LA-25 cells (which express a temperature-sensitive v-Src) grown at the permissive and non-permissive temperatures to show that distinct tyrosine and serine residues are phosphorylated in response to v-Src activity (Solan and Lampe, 2008). Li et al. used antibodies that specifically recognize PSer110 and PSer276 in Cx35 to demonstrate that the level of phosphorylation of these serines depends on PKA activity and regulates photoreceptor coupling in zebrafish retina (Li et al., 2009).

8. Mass spectrometry analyses

Another technique that has been used to identify putative phosphorylation sites is mass spectrometry (MS) analysis of connexins isolated from tissue or cultured cells or *in vitro* phosphorylated (poly)peptides. For this purpose, the immunoprecipitated/isolated connexin or *in vitro* phosphorylated polypeptide is digested with a protease or a mix of proteases, the sample is enriched in phosphopeptides and subjected to MS. This technique is highly sensitive and it does not require the use of radioactivity.

The first studies using this technique to identify phosphorylation sites in Cxs were reported several years ago (Cooper et al., 2000; Yin et al., 2000). Cooper et al. showed that *in vitro* phosphorylation of the carboxyl terminus of Cx43 with $p34^{cdc2}$/cyclin B kinase resulted in phosphorylation of Ser255 using liquid chromatography coupled with tandem mass spectrometry (LC-MS/MS) (Cooper et al., 2000). Yin et al. demonstrated that lens Cx45.6 is phosphorylated in the chicken lens *in vivo* at Ser363 using nanoelectrospray and tandem mass spectrometry (Yin et al., 2000).

Several studies using mass spectrometry analysis have been performed on Cx43. Ser364 was identified as a phosphorylation site in Cx43 using matrix-assisted laser desorption/ionization-time of flight (MALDI-TOF MS) and LC-MS/MS (TenBroek et al., 2001). MALDI-TOF MS in combination with metabolic labeling of normal rat kidney (NRK) epithelial cells (in the presence and absence of a casein kinase 1 inhibitor) and *in vitro* phosphorylation of Cx43CT fusion proteins with casein kinase 1δ (CK1δ) have been used to determine that serines 325, 328 or 330 are potential sites of CK1 phosphorylation in these cells (Cooper and Lampe, 2002). Cameron et al. (2003) used MALDI-TOF MS to identify Ser255 of Cx43 as the preferred site for big MAPK 1 (BMK1)/ERK5 phosphorylation. This finding was further supported by the lack of phosphorylation of GST fusion proteins containing mutant carboxyl termini of Cx43 in which Ser255 had been mutated to alanine (S255A and S255A/S279A/ S282A). Axelsen et al. (2006) reported the time course of changes in phosphorylation of Cx43 immunopurified from perfused rat hearts under non-ischemic and ischemic conditions. These authors identified thirteen phosphorylation sites using MALDI MS and LC-MS/MS in non-ischemic conditions and detected site-specific changes

	Connexin	Cell Type	Kinase or Phosphatase	Identification site	References
Metabolic labeling	Cx32	Rat hepatocytes and purified liver gap junction	cAMP-PK, PKC, Ca2+/CaM-PK-II	Ser233	Sáez et al., 1990. Eur J Biochem 192:263-73.
	Cx43	Rat liver epithelial cells (T51B)	-	Serine residues	Lau et al., 1992. Mol Biol Cell 3:865-74.
	Cx43	Rat neonatal cardiomyocytes	-	Serine residues	Doble et al., 1996. Circ Res 79:647-658
	Cx43	in vitro	MAPK	Ser255, Ser279, Ser282	Warn-Cramer et al., 1996. J Biol Chem 271:3779-86
	Cx43	Hamster fibroblast	-	Tyrosine residues	Mikalsen et al., 1997. FEBS Lett 401:271-5.
	Cx49	Sheep lens fiber cell	casein kinase 1	Ser/Thr	Cheng and Louis, 1999. Eur J Biochem 263:276-86.
	Cx43	Expression in Xenopus	PKC	Ser368	Bao et al., 2004. J Biol Chem 279:20058-66.
	Cx43	Expression in Xenopus	PKC	Ser368	Bao et al., 2004 Am J Physiol Cell Physiol 286: C647-C654
	Cx43	Expression in Xenopus	pp60v-src	Tyr265	Swenson et al., 1990. Cell Regul 1:989-1002
Kinase assays	Cx56	chicken lens primary cultures	PKC	Ser118	Berthoud et al., 1997. Eur J Biochem 244:89-97.
	Cx43	Rat-1 fibroblast	p34cdc2/cyclin B	amino acids 241-264	Kanemitsu et al., 1998. Cell Growth Differ 9:13-21.
	Cx43	Mouse fibroblast (L929)	PKA-PKC	Ser364	Shah et al., 2002. Mol Cell Biochem 238:57-68
	Cx43	Rat granulosa cells	PKA	Ser365, Ser368, Ser369, and Ser373	Yogo et al., 2006. J Reprod Dev 52:321-8
	Cx35	HeLa Cells	PKA	cytoplasmic domain	O'Brien et al., 2004. J Neurosci 24:5632-5642
	Cx35	HeLa Cells	PKA	intracel loop & cytoplasmic domain	Ouyang et al., 2005. Brain Res Mol Brain Res 135:1-11.
	Cx36	GST-Cx36 fusion protein	CaMKII	cytoplasmic domains	Alev et al., 2008. Proc Natl Acad Sci U S A 105:20964-9
	Cx37	HeLa and SK-HEP-1 cells	GSK-3β	Ser319 and Pro319	Morel et al., 2010.Carcinogenesis 31:1922-1931.
Pharmacological Modulation	Cx43	Hamster embryo cells and lung fibroblasts	PKC	-	Husoy et al., 1993 Carcinogenesis. 1993 14(11):2257-65.
	Cx43	Rat liver epithelial cells	PP1 and PP2A	-	Guan et al., 1996 Mol Carcinog 16, 157-164
	Cx43	Hamster fibroblast	PP1, PP2A, PP2B and PP2C	-	Cruciani et al., 1999 Exp Cell Res 252: 449-463
	Cx43	Astrocytes	PP1, PP2A and PP2B	-	Li and Nagy, 2000 Eur J Neurosci 12, 2644-2650.
	Cx43	neonatal rat cardiomyocytes	serine/threonine protein kinases	-	Duthe et al., 2000 Gen Physiol Biophys 19: 441-449.
	Cx43	neonatal rat cardiomyocytes	PP1 and PP2A	-	Duthe et al., 2001 Am J Physiol Cell Physiol. 2001 281:C1648-56.
	Cx43	Madin Darby canine kidney (MDCK) cells	PKC, cAMP- or cGMP-dependent PK	-	Berthoud et al., 1992 Eur J Cell Biol 57: 40-50
Genetic Modulation	Cx43	Rat1 fibroblasts	p34cdc2 kinase	Ser255	Lampe et al., 1998. J Cell Sci 111:833-41.
	Cx43	neonatal rat cardiomyocytes	PKC epsilon	-	Doble et al., 2000 Circ Res 86:293-301
	Cx43	Cx43 knockout mouse cell line.	v-Src	Y247, Y265	Lin et al., 2001 J Cell Biol 154:815-27.
	Cx43	neonatal rat cardiomyocytes	c-jun	-	Petrich et al., 2002 Circ Res 91:640-7.
	Cx43	Transgenic Mouse hearts	PP3	-	Chu et al., 2002 Cardiovasc Res 54: 105-116.
	Cx43	embryonic fibroblasts	v-Src	Tyrosine residues	Peterson-Roth et al. 2009 Cancer Res 69:3619-3624
	Cx43	left ventricular myocytes	p21-activated kinase 1 and PP2A	-	Al et al., 2011 Cardiovasc Res. 92(1):106-14.
	Cx43	Cx43 germline knock-in mice	-	Ser325, Ser 328, Ser330	Remo et al., 2011 Circ Res. 108(12):1459-66.
	Cx43	in vitro	v-Src	Y265	Kanemitsu et al., 1997 J Biol Chem 272:22824-31.
	Cx43	neonatal rat cardiomyocytes	PKC	Ser262	Doble et al., 2004 Journal of Cell Science 117:507-514
	Cx43	Cx43 knockout mouse fibroblasts	V-Src	-	Lin et al., 2006 Cell Commun Adhes 13:199-216.
Phosphospecific antibodies	Cx43	rabbit lens ephitelial cells	PKC-gamma	-	Lin et al., 2003
	Cx43	rat kidney epithelial cells	PKC	Ser368	Solan et al., 2003
	Cx43	liver epithelial cells	ERK1, ERK2	Ser279, Ser282	Abdelmohsen et al., 2003
	Cx43	Pig hearts	PKCα, p38MAPKα, and p38MAPKβ	-	Schulz et al., 2003 FASEB J 17:1355-7
	Cx43	Rat liver epithelial cells	MAPK	-	Leykauf et al., 2003 Cell Tissue Res 311:23-30
	Cx43	Humna keratinocyte	PKC	Ser368	Richards et al., 2004 J Cell Biol. 167:555-62.
	Cx43	Cx43 knockout mouse fibroblasts and heart	casein kinase 1	Ser325, Ser328, Ser330	Lampe et al., 2006 J Cell Sci 119:3435-3442
	Cx43	Rat kidney epithelial cells	vSrc	Y247, Y265, Ser262, Ser279/282, Ser368	Solan and Lampe, 2008 Cell Commun Adhes 15:75-84.
	Cx43	Mouse ovarian follicles	MAPK	Ser255, Ser262, Ser279/282	Norris et al., 2008 Development 135:3229-3238
	Cx35	Zebrafish retina	PKA	Ser110, Ser276	Li et al., 2009 J Neurosci 29:15178-15186.
	Cx43	Rat liver epithelial cell	PKC and MAPK	Ser368, Ser255, Ser262	Simes et al., 2009 Biochem Biophys Res Commun 382:41-45.
	Cx43	Rat hearts	PKC	Ser262	Srisakuldee et al., 2009 Cardiovasc res 83:672-81.
	Cx43	MC3T3 osteoblasts	PKC delta	Ser368	Niger et al., 2010 BMC Biochemistry 11:14
Mass spectrometry analyses	Cx45.6	Avian lens primary cell cultures	casein kinase II	Ser363	Yin et al., 2000 J Biol Chem 275:6850-6.
	Cx43	mouse fibroblasts	PKA	Ser364	TenBroek et al., 2001 J Cell Biol 155:1307-18
	Cx43	rat kidney cells	casein kinase 1 (gamma)	Ser325, Ser328, Ser330	Cooper and Lampe, 2002 J Biol Chem 277:44962-8.
	Cx43	HEK-293	BMK1/ERK5	Ser255	Cameron et al., 2003 J Biol Chem 278:18682-8.
	Cx43	isolated perfused rat hearts	PKA, PKCα, PKCε, PKG, AMP-dependent,	30 residues	Axelsen et al. 2006 J Mol Cell Cardiol 40:790-8.
	Cx26, Cx32	HeLa Cells	-	-	Locke et al. 2006 FASEB J 20:1221-3.
	Cx43	NRK-E51, MDCK	PKC	Ser365	Solan et al. 2007 J Cell Biol 179:1301-9.
	Cx44, Cx49	Bovine Lens Fiber	-	Ser and Thr residues in the C-tail of Cx44 and Ser residues in Cx49	Shearer et al. 2008 Invest Ophthalmol Vis Sci 49:1553-1562.
	Cx46, Cx50	Bovine lens	-	9 residues in Cx46 and 18 residues in Cx50	Wang and Schey 2009 Exp Eye Res 89:898-904
	Cx26	in vitro	-	Thr123, Thr177, Ser183, Thr186, Tyr233, Tyr235, Tyr240	Locke et al 2009 Biochem. J 424:385-398
	Cx43	in vitro	CaMKII	Ser296, Ser365, Ser369, Ser373, Ser244, Ser306	Huang et al. 2011 J Proteome Res 10:1098-109
LRET	Cx43	purified WT Cx43	PKC	-	Bao et al., 2007. Proc Natl Acad Sci U S A 104:4919-24.

Table 1. Techniques used for identification of connexins as substrates for protein kinases and phosphoprotein phosphatases.

in Cx43 phosphorylation during the course of ischemia. Phosphorylation of Ser365 has also been demonstrated in Cx43 immunoprecipitated from NRK cells using liquid chromatography coupled to electrospray ionization tandem mass spectrometry (LC/ESI MS/MS) (Solan et al., 2007). Fifteen putative phosphorylation sites on Cx43 have also been identified after *in vitro* phosphorylation of a GST fusion protein containing the Cx43CT with CaMK II by high-resolution mass spectrometry (Huang et al., 2011).

Post-translational modification by phosphate has also been identified by mass spectrometry in Cx26 and Cx32; Cx26 is phosphorylated in the intracellular loop and the second extracellular loop, and Cx32 is phosphorylated in the amino and carboxyl termini (Locke et al., 2006). Two studies have used mass spectrometry to identify phosphorylation sites in the bovine lens fiber connexins, Cx44 and Cx49. While phosphorylation sites were identified only on the carboxyl terminus of Cx44, phosphosites were identified in both the intracellular loop and carboxyl terminus of Cx49 (Shearer et al., 2008; Wang and Schey, 2009).

9. Luminescence resonance energy transfer

Another recent approach used to evaluate the functional effect(s) of phosphorylation of Cxs is the generation of hemichannels of known composition, stoichiometry that can be assessed by luminescence resonance energy transfer (LRET)(Bao et al., 2007). This method uses terbium ions (Tb^{3+}), which have a long lifetime emission as donor and fluorescein as acceptor. The technique is based on the detection of LRET between Cx43 subunits labeled with Tb^{3+} and those labeled with fluorescein. The composition of the HCs can be determined based on the number of acceptor-labeled monomers per HC. Using HC of known composition, Bao and colleagues have determined that in a Cx43 HC all six subunits have to be phosphorylated by PKC at Ser368 to abolish sucrose permeability, although the HC pore still has a sizable diameter and allows permeation of smaller molecules (Bao et al., 2007).

10. Conclusions and future directions

In summary, connexins are substrates for various protein kinases and phosphoprotein phosphatases. Several of the phosphorylation sites have been identified, and the effect of phosphorylation at many of these sites on connexin channel activity has been studied. In some cases, pathophysiological conditions that alter their phosphorylation state have been reported. Although significant progress has been made in the area of connexin phosphorylation, there are many associated aspects that require further investigation.

A question that remains unanswered is whether all connexins are phosphoproteins. Does phosphorylation affect connexin channel function in all members? Does phosphorylation at a specific site induce consistent functional changes in gap junction channels and hemichannels? Or, can phosphorylation at a specific site induce changes in one channel type, and not in the other?. Because phosphorylation has been implicated in several steps of the connexin's life cycle, it is also important to determine which phosphorylation events are associated with proper trafficking to the plasma membrane, formation of gap junctional plaques or internalization and degradation. Are connexins sorted/targeted to different compartments depending on their cohort of phosphorylated sites? Where do these phosphorylation events take place? Since some hierarchy in the phosphorylation events has been shown for Cx43, it is interesting to know whether changes in phosphorylation are also associated with other post-translational modifications. Do these have a hierarchical sequence? Because connexins and changes in the activity of protein kinases/phosphoprotein phosphatases have been associated with disease, it would be important to know how the phosphorylation state of connexins is affected in disease. What are the intracellular signals and mechanisms of regulation of phosphorylation/dephosphorylation of connexins? What are the endogenous activators of the protein kinases/phosphoprotein phosphatases involved? Although the answers to some of these questions are known for some of the phosphorylation sites identified, especially in the case of Cx43, these questions have not been addressed for most connexins.

11. References

Abdelmohsen K, Gerber PA, Montfort von C, Sies H, Klotz L-O (2003) Epidermal growth factor receptor is a common mediator of quinone-induced signaling leading to phosphorylation of connexin-43: role of glutathione and tyrosine phosphatases. J Biol Chem 278:38360–38367.

Ai X, Jiang A, Ke Y, Solaro RJ, Pogwizd SM (2011) Enhanced activation of p21-activated kinase 1 in heart failure contributes to dephosphorylation of connexin 43. Cardiovasc Res 92:106–114.

Alev C, Urschel S, Sonntag S, Zoidl G, Fort AG, Höher T, Matsubara M, Willecke K, Spray DC, Dermietzel R (2008) The neuronal connexin36 interacts with and is phosphorylated by CaMKII in a way similar to CaMKII interaction with glutamate receptors. Proc Natl Acad Sci USA 105:20964–20969.

Alonso A, Sasin J, Bottini N, Friedberg I, Friedberg I, Osterman A, Godzik A, Hunter T, Dixon J, Mustelin T (2004) Protein tyrosine phosphatases in the human genome. Cell 117:699–711.

Axelsen LN, Stahlhut M, Mohammed S, Larsen BD, Nielsen MS, Holstein-Rathlou N-H, Andersen S, Jensen ON, Hennan JK, Kjølbye AL (2006) Identification of ischemia-regulated phosphorylation sites in connexin43: A possible target for the antiarrhythmic peptide analogue rotigaptide (ZP123). J Mol Cell Cardiol 40:790-798.

Bao X, Altenberg GA, Reuss L (2004a) Mechanism of regulation of the gap junction protein connexin 43 by protein kinase C-mediated phosphorylation. Am J Physiol, Cell Physiol 286:C647–C654.

Bao X, Lee SC, Reuss L, Altenberg GA (2007) Change in permeant size selectivity by phosphorylation of connexin 43 gap-junctional hemichannels by PKC. Proc Natl Acad Sci USA 104:4919–4924.

Bao X, Reuss L, Altenberg GA (2004b) Regulation of purified and reconstituted connexin 43 hemichannels by protein kinase C-mediated phosphorylation of Serine 368. J Biol Chem 279:20058–20066.

Barford D, Das AK, Egloff MP (1998) The structure and mechanism of protein phosphatases: insights into catalysis and regulation. Annu Rev Biophys Biomol Struct 27:133–164.

Berthoud VM, Beyer EC, Kurata WE, Lau AF, Lampe PD (1997) The gap-junction protein connexin 56 is phosphorylated in the intracellular loop and the carboxy-terminal region. Eur J Biochem 244:89–97.

Berthoud VM, Ledbetter ML, Hertzberg EL, Sáez JC (1992) Connexin43 in MDCK cells: regulation by a tumor-promoting phorbol ester and Ca^{2+}. Eur J Cell Biol 57:40–50.

Beyer EC, Paul DL, Goodenough DA (1987) Connexin43: a protein from rat heart homologous to a gap junction protein from liver. J Cell Biol 105:2621–2629.

Blume-Jensen P, Hunter T (2001) Oncogenic kinase signalling. Nature 411:355–365.

Bruzzone R, Dermietzel R (2006) Structure and function of gap junctions in the developing brain. Cell Tissue Res 326:239–248.

Caenepeel S, Charydczak G, Sudarsanam S, Hunter T, Manning G (2004) The mouse kinome: discovery and comparative genomics of all mouse protein kinases. Proc Natl Acad Sci USA 101:11707–11712.

Cameron S, Malik S, Akaike M (2003) Regulation of epidermal growth factor-induced connexin 43 gap junction communication by big mitogen-activated protein kinase1/ERK5 but not ERK1/2 kinase activation. J Biol Chem 278:18682-18688.

Cheng HL, Louis CF (1999) Endogenous casein kinase I catalyzes the phosphorylation of the lens fiber cell connexin49. Eur J Biochem 263:276–286.

Chu G, Carr AN, Young KB, Lester JW, Yatani A, Sanbe A, Colbert MC, Schwartz SM, Frank KF, Lampe PD, Robbins J, Molkentin JD, Kranias EG (2002) Enhanced myocyte contractility and Ca^{2+} handling in a calcineurin transgenic model of heart failure. Cardiovasc Res 54:105–116.

Cooper CD, Lampe PD (2002) Casein kinase 1 regulates connexin-43 gap junction assembly. J Biol Chem 277:44962–44968.

Cooper CD, Solan JL, Dolejsi MK, Lampe PD (2000) Analysis of connexin phosphorylation sites. Methods 20:196–204.

Crow DS, Beyer EC, Paul DL, Kobe SS, Lau AF (1990) Phosphorylation of connexin43 gap junction protein in uninfected and Rous sarcoma virus-transformed mammalian fibroblasts. Mol Cell Biol 10:1754–1763.

Cruciani V, Kaalhus O, Mikalsen SO (1999) Phosphatases involved in modulation of gap junctional intercellular communication and dephosphorylation of connexin43 in hamster fibroblasts: 2B or not 2B? Exp Cell Res 252:449–463.

Díez JA, Elvira M, Villalobo A (1995) Phosphorylation of connexin-32 by the epidermal growth factor receptor tyrosine kinase. Ann N Y Acad Sci 766:477–480.

Díez JA, Elvira M, Villalobo A (1998) The epidermal growth factor receptor tyrosine kinase phosphorylates connexin32. Mol Cell Biochem 187:201–210.

Doble BW, Chen Y, Bosc DG, Litchfield DW, Kardami E (1996) Fibroblast growth factor-2 decreases metabolic coupling and stimulates phosphorylation as well as masking of connexin43 epitopes in cardiac myocytes. Circ Res 79:647–658.

Doble BW, Dang X, Ping P, Fandrich RR, Nickel BE, Jin Y, Cattini PA, Kardami E (2004) Phosphorylation of serine 262 in the gap junction protein connexin-43 regulates DNA synthesis in cell-cell contact forming cardiomyocytes. J Cell Sci 117:507–514.

Doble BW, Ping P, Kardami E (2000) The epsilon subtype of protein kinase C is required for cardiomyocyte connexin-43 phosphorylation. Circ Res 86:293–301.

Duthe F, Dupont E, Verrecchia F, Plaisance I, Severs NJ, Sarrouilhe D, Hervé JC (2000) Dephosphorylation agents depress gap junctional communication between rat cardiac cells without modifying the connexin43 phosphorylation degree. Gen Physiol Biophys 19:441–449.

Duthe F, Plaisance I, Sarrouilhe D, Hervé JC (2001) Endogenous protein phosphatase 1 runs down gap junctional communication of rat ventricular myocytes. Am J Physiol, Cell Physiol 281:C1648–C1656.

Eiberger J, Degen J, Romualdi A, Deutsch U, Willecke K, Söhl G (2001) Connexin genes in the mouse and human genome. Cell Commun Adhes 8:163–165.

Elvira M, Díez JA, Wang KK, Villalobo A (1993) Phosphorylation of connexin-32 by protein kinase C prevents its proteolysis by μ-calpain and m-calpain. J Biol Chem 268:14294–14300.

Goldberg GS, Lau AF (1993) Dynamics of connexin43 phosphorylation in pp60$^{v\text{-}src}$-transformed cells. Biochem J 295: 735–742.

Goodenough DA (1974) Bulk isolation of mouse hepatocyte gap junctions. Characterization of the principal protein, connexin. J Cell Biol 61:557–563.

Green CR, Harfst E, Gourdie RG, Severs NJ (1988) Analysis of the rat liver gap junction protein: clarification of anomalies in its molecular size. Proc R Soc Lond B, Biol Sci 233:165–174.

Griffith LC (2004) Regulation of calcium/calmodulin-dependent protein kinase II activation by intramolecular and intermolecular interactions. J Neurosci 24:8394–8398.

Guan X, Wilson S, Schlender KK, Ruch RJ (1996) Gap-junction disassembly and connexin 43 dephosphorylation induced by 18β-glycyrrhetinic acid. Mol Carcinog 16:157–164.

Hertzberg EL, Disher RM, Tiller AA, Zhou Y, Cook RG (1988) Topology of the Mr 27,000 liver gap junction protein. Cytoplasmic localization of amino- and carboxyl termini and a hydrophilic domain which is protease-hypersensitive. J Biol Chem 263:19105–19111.

Heynkes R, Kozjek G, Traub O, Willecke K (1986) Identification of a rat liver cDNA and mRNA coding for the 28 kDa gap junction protein. FEBS Lett 205:56–60.

Huang RY-C, Laing JG, Kanter EM, Berthoud VM, Bao M, Rohrs HW, Townsend RR, Yamada KA (2011) Identification of CaMKII phosphorylation sites in connexin43 by high-resolution mass spectrometry. J Proteome Res 10:1098–1109.

Husøy T, Mikalsen SO, Sanner T (1993) Phosphatase inhibitors, gap junctional intercellular communication and [^{125}I]-EGF binding in hamster fibroblasts. Carcinogenesis 14:2257–2265.

Johnson GL, Lapadat R (2002) Mitogen-activated protein kinase pathways mediated by ERK, JNK, and p38 protein kinases. Science 298:1911–1912.

Kanemitsu MY, Jiang W, Eckhart W (1998) Cdc2-mediated phosphorylation of the gap junction protein, connexin43, during mitosis. Cell Growth Differ 9:13–21.

Kanemitsu MY, Loo LW, Simon S, Lau AF, Eckhart W (1997) Tyrosine phosphorylation of connexin 43 by v-Src is mediated by SH2 and SH3 domain interactions. J Biol Chem 272:22824–22831.

Kurata WE, Lau AF (1994) p130$^{gag\text{-}fps}$ disrupts gap junctional communication and induces phosphorylation of connexin43 in a manner similar to that of pp60$^{v\text{-}src}$. Oncogene 9:329–335.

Lampe PD, Cooper CD, King TJ, Burt JM (2006) Analysis of Connexin43 phosphorylated at S325, S328 and S330 in normoxic and ischemic heart. J Cell Sci 119:3435–3442.

Lampe PD, Kurata WE, Warn-Cramer BJ, Lau AF (1998) Formation of a distinct connexin43 phosphoisoform in mitotic cells is dependent upon p34^{cdc2} kinase. J Cell Sci: 833–841.

Lampe PD, Lau AF (2004) The effects of connexin phosphorylation on gap junctional communication. Int J Biochem Cell Biol 36:1171–1186.

Lau AF, Kanemitsu MY, Kurata WE, Danesh S, Boynton AL (1992) Epidermal growth factor disrupts gap-junctional communication and induces phosphorylation of connexin43 on serine. Mol Biol Cell 3:865–874.

Leykauf K, Dürst M, Alonso A (2003) Phosphorylation and subcellular distribution of connexin43 in normal and stressed cells. Cell Tissue Res 311:23–30.

Li H, Chuang AZ, O'Brien J (2009) Photoreceptor coupling is controlled by connexin 35 phosphorylation in zebrafish retina. J Neurosci 29:15178–15186.

Li WE, Nagy JI (2000) Connexin43 phosphorylation state and intercellular communication in cultured astrocytes following hypoxia and protein phosphatase inhibition. Eur J Neurosci 12:2644–2650.

Lin R, Martyn KD, Guyette CV, Lau AF, Warn-Cramer BJ (2006) v-Src tyrosine phosphorylation of connexin43: regulation of gap junction communication and effects on cell transformation. Cell Commun Adhes 13:199–216.

Lin R, Warn-Cramer BJ, Kurata WE, Lau AF (2001) v-Src phosphorylation of connexin 43 on Tyr247 and Tyr265 disrupts gap junctional communication. J Cell Biol 154:815–827.

Locke D, Bian S, Li H, Harris AL (2009) Post-translational modifications of connexin26 revealed by mass spectrometry. Biochem J 424:385–398.

Locke D, Koreen IV, Harris AL (2006) Isoelectric points and post-translational modifications of connexin26 and connexin32. FASEB J 20:1221–1223.

Loo LW, Berestecky JM, Kanemitsu MY, Lau AF (1995) pp60^{src}-mediated phosphorylation of connexin 43, a gap junction protein. J Biol Chem 270:12751–12761.

Makowski L, Caspar DL, Phillips WC, Goodenough DA (1977) Gap junction structures. II. Analysis of the x-ray diffraction data. J Cell Biol 74:629–645.

Manning G, Whyte DB, Martinez R, Hunter T, Sudarsanam S (2002) The protein kinase complement of the human genome. Science 298:1912–1934.

Meşe G, Richard G, White TW (2007) Gap junctions: basic structure and function. J Invest Dermatol 127:2516–2524.

Mikalsen SO, Husøy T, Vikhamar G, Sanner T (1997) Induction of phosphotyrosine in the gap junction protein, connexin43. FEBS Lett 401:271–275.

Milks LC, Kumar NM, Houghten R, Unwin N, Gilula NB (1988) Topology of the 32-kd liver gap junction protein determined by site-directed antibody localizations. EMBO J 7:2967–2975.

Morel S, Burnier L, Roatti A, Chassot A, Roth I, Sutter E, Galan K, Pfenniger A, Chanson M, Kwak BR (2010) Unexpected role for the human Cx37 C1019T polymorphism in tumour cell proliferation. Carcinogenesis 31:1922–1931.

Morley M, Jones C, Sidhu M, Gupta V, Bernier SM, Rushlow WJ, Belliveau DJ (2010) PKC inhibition increases gap junction intercellular communication and cell adhesion in human neuroblastoma. Cell Tissue Res 340:229–242.

Musil LS, Beyer EC, Goodenough DA (1990) Expression of the gap junction protein connexin43 in embryonic chick lens: molecular cloning, ultrastructural localization, and post-translational phosphorylation. J Membr Biol 116:163–175.

Newton AC (1995) Protein kinase C: structure, function, and regulation. J Biol Chem 270:28495–28498.

Niger C, Hebert C, Stains JP (2010) Interaction of connexin43 and protein kinase C-delta during FGF2 signaling. BMC Biochem 11:14.

Norris RP, Freudzon M, Mehlmann LM, Cowan AE, Simon AM, Paul DL, Lampe PD, Jaffe LA (2008) Luteinizing hormone causes MAP kinase-dependent phosphorylation and closure of connexin 43 gap junctions in mouse ovarian follicles: one of two paths to meiotic resumption. Development 135:3229–3238.

O'Brien J, Nguyen HB, Mills SL (2004) Cone photoreceptors in bass retina use two connexins to mediate electrical coupling. J Neurosci 24:5632–5642.

Ouyang X, Winbow VM, Patel LS, Burr GS, Mitchell CK, O'Brien J (2005) Protein kinase A mediates regulation of gap junctions containing connexin35 through a complex pathway. Brain Res Mol Brain Res 135:1-11.

Paul DL (1986) Molecular cloning of cDNA for rat liver gap junction protein. J Cell Biol 103:123-134.

Pérez DI, Gil C, Martínez A (2011) Protein kinases CK1 and CK2 as new targets for neurodegenerative diseases. Med Res Rev 31:924-954.

Peterson-Roth E, Brdlik CM, Glazer PM (2009) Src-Induced cisplatin resistance mediated by cell-to-cell communication. Cancer Res 69:3619-3624.

Petrich BG, Gong X, Lerner DL, Wang X, Brown JH, Saffitz JE, Wang Y (2002) c-Jun N-terminal kinase activation mediates downregulation of connexin43 in cardiomyocytes. Circ Res 91:640-647.

Pfenniger A, Wohlwend A, Kwak BR (2011) Mutations in connexin genes and disease. Eur J Clin Invest 41:103-116.

Quist AP, Rhee SK, Lin H, Lal R (2000) Physiological role of gap-junctional hemichannels. Extracellular calcium-dependent isosmotic volume regulation. J Cell Biol 148:1063-1074.

Remo BF, Qu J, Volpicelli FM, Giovannone S, Shin D, Lader J, Liu F-Y, Zhang J, Lent DS, Morley GE, Fishman GI (2011) Phosphatase-resistant gap junctions inhibit pathological remodeling and prevent arrhythmias. Circ Res 108:1459-1466.

Richards TS, Dunn CA, Carter WG, Usui ML, Olerud JE, Lampe PD (2004) Protein kinase C spatially and temporally regulates gap junctional communication during human wound repair via phosphorylation of connexin43 on serine368. J Cell Biol 167:555-562.

Sáez JC, Berthoud VM, Brañes MC, Martínez AD, Beyer EC (2003) Plasma membrane channels formed by connexins: their regulation and functions. Physiol Rev 83:1359-1400.

Sáez JC, Nairn AC, Czernik AJ, Spray DC, Hertzberg EL, Greengard P, Bennett MV (1990) Phosphorylation of connexin 32, a hepatocyte gap-junction protein, by cAMP-dependent protein kinase, protein kinase C and Ca^{2+}/calmodulin-dependent protein kinase II. Eur J Biochem 192:263-273.

Sáez JC, Retamal MA, Basilio D, Bukauskas FF, Bennett MVL (2005) Connexin-based gap junction hemichannels: gating mechanisms. Biochim Biophys Acta 1711:215-224.

Sáez JC, Spray DC, Nairn AC, Hertzberg EL, Greengard P, Bennett MVL (1986) cAMP increases junctional conductance and stimulates phosphorylation of the 27kDa principal gap junction polypeptide. Proc Natl Acad Sci USA 83:2473-2477.

Sáez JC, Schalper KA, Retamal MA, Orellana JA, Shoji KF, Bennett MVL (2010) Cell membrane permeabilization via connexin hemichannels in living and dying cells. Exp Cell Res 316:2377-2389.

Schulz R, Gres P, Skyschally A, Duschin A, Belosjorow S, Konietzka I, Heusch G (2003) Ischemic preconditioning preserves connexin 43 phosphorylation during sustained ischemia in pig hearts in vivo. FASEB J 17:1355-1357.

Shah MM, Martinez A-M, Fletcher WH (2002) The connexin43 gap junction protein is phosphorylated by protein kinase A and protein kinase C: *In vivo* and *in vitro* studies. Mol Cell Biochem 238:57-68.

Shearer D, Ens W, Standing K, Valdimarsson G (2008) Posttranslational modifications in lens fiber connexins identified by off-line-HPLC MALDI-quadrupole time-of-flight mass spectrometry. Invest Ophthalmol Vis Sci 49:1553-1562.

Sirnes S, Kjenseth A, Leithe E, Rivedal E (2009) Interplay between PKC and the MAP kinase pathway in Connexin43 phosphorylation and inhibition of gap junction intercellular communication. Biochem Biophys Res Commun 382:41-45.

Solan JL, Fry MD, TenBroek EM, Lampe PD (2003) Connexin43 phosphorylation at S368 is acute during S and G2/M and in response to protein kinase C activation. J Cell Sci 116:2203–2211.

Solan JL, Lampe PD (2008) Connexin 43 in LA-25 cells with active v-src is phosphorylated on Y247, Y265, S262, S279/282, and S368 via multiple signaling pathways. Cell Commun Adhes 15:75–84.

Solan JL, Marquez-Rosado L, Sorgen PL, Thornton PJ, Gafken PR, Lampe PD (2007) Phosphorylation at S365 is a gatekeeper event that changes the structure of Cx43 and prevents down-regulation by PKC. J Cell Biol 179:1301–1309.

Sosinsky GE, Solan JL, Gaietta GM, Ngan L, Lee GJ, Mackey MR, Lampe PD (2007) The C-terminus of connexin43 adopts different conformations in the Golgi and gap junction as detected with structure-specific antibodies. Biochem J 408:375–385.

Söhl G, Willecke K (2003) An update on connexin genes and their nomenclature in mouse and man. Cell Commun Adhes 10:173–180.

Söhl G, Willecke K (2004) Gap junctions and the connexin protein family. Cardiovasc Res 62:228–232.

Srisakuldee W, Jeyaraman MM, Nickel BE, Tanguy S, Jiang Z-S, Kardami E (2009) Phosphorylation of connexin-43 at serine 262 promotes a cardiac injury-resistant state. Cardiovasc Res 83:672–681.

Swenson KI, Piwnica-Worms H, McNamee H, Paul DL (1990) Tyrosine phosphorylation of the gap junction protein connexin43 is required for the pp60$^{v\text{-}src}$-induced inhibition of communication. Cell Regul 1:989–1002.

Takeda A, Saheki S, Shimazu T, Takeuchi N (1989) Phosphorylation of the 27-kDa gap junction protein by protein kinase C in vitro and in rat hepatocytes. J Biochem 106:723–727.

TenBroek EM, Lampe PD, Solan JL, Reynhout JK, Johnson RG (2001) Ser364 of connexin43 and the upregulation of gap junction assembly by cAMP. J Cell Biol 155:1307–1318.

Wang Z, Schey KL (2009) Phosphorylation and truncation sites of bovine lens connexin 46 and connexin 50. Exp Eye Res 89:898–904.

Warn-Cramer BJ, Lampe PD, Kurata WE, Kanemitsu MY, Loo LW, Eckhart W, Lau AF (1996) Characterization of the mitogen-activated protein kinase phosphorylation sites on the connexin-43 gap junction protein. J Biol Chem 271:3779–3786.

Willecke K, Eiberger J, Degen J, Eckardt D, Romualdi A, Güldenagel M, Deutsch U, Söhl G (2002) Structural and functional diversity of connexin genes in the mouse and human genome. Biol Chem 383:725–737.

Yancey SB, John SA, Lal R, Austin BJ, Revel JP (1989) The 43-kD polypeptide of heart gap junctions: immunolocalization, topology, and functional domains. J Cell Biol 108:2241–2254.

Yin X, Jedrzejewski PT, Jiang JX (2000) Casein kinase II phosphorylates lens connexin 45.6 and is involved in its degradation. J Biol Chem 275:6850–6856.

Yogo K, Ogawa T, Akiyama M, Ishida-Kitagawa N, Sasada H, Sato E, Takeya T (2006) PKA implicated in the phosphorylation of Cx43 induced by stimulation with FSH in rat granulosa cells. J Reprod Dev 52:321–328.

Zhang JT, Nicholson BJ (1994) The topological structure of connexin 26 and its distribution compared to connexin 32 in hepatic gap junctions. J Membr Biol 139:15–29.

Zimmer DB, Green CR, Evans WH, Gilula NB (1987) Topological analysis of the major protein in isolated intact rat liver gap junctions and gap junction-derived single membrane structures. J Biol Chem 262:7751–7763.

10

Protein Phosphatases Drive Mitotic Exit

Megan Chircop
Children's Medical Research Institute, The University of Sydney, Australia

1. Introduction

Mitosis is the final stage of the cell cycle that results in the formation of two independent daughter cells with an equal and identical complement of chromosomes (Figure 1). This requires a complex series of events such as nuclear envelope breakdown, spindle formation, equal chromosome segregation, packaging of chromosomes into daughter nuclei and constriction of the plasma membrane at the cell equator, which is subsequently abscised to generate two independent daughter cells. For mitosis to be successful, these events need to occur in a strict order and be spatiotemporally controlled, which is primarily mediated by protein phosphorylation (Dephoure et al., 2008). In human cells more than one thousand proteins show increased phosphorylation during mitosis (Dephoure et al., 2008). These phosphorylation events are mediated by mitotic protein kinases such as cyclin-dependent kinases (Cdks), Auroras, Polo-like kinases (Plks), Mps1, Neks and NimA (Ma and Poon, 2011). In mammalian cells, the majority of phosphorylation events and thus mitotic progression is driven by the activity of Cdk1, which is the main subtype of Cdks (Dephoure et al., 2008). Its activity during mitosis is due to binding cyclin B1 and phosphorylation of a residue in the T-loop.

Mitotic exit involves two stages: (1) membrane ingression, which begins during anaphase following chromosome segregation and involves the breakdown of mitotic structures including the mitotic spindle. It also involves the physical constriction of the cell membrane between segregating chromosomes at the cell equator to generate a thin intracellular bridge between nascent daughter cells. This is followed by (2) membrane abscission at a specific location along the intracellular bridge to generate two independent daughter cells (Figure 1). During mitotic exit, cells also decondense their chromosomes and re-assemble interphase structures such as the nuclear envelope and endoplasmic reticulum. Again, these events need to occur in a strict ordered sequence and requires the reversal of Cdk1-mediated phosphorylation events. Cdk1 is inactivated upon anaphase and is largely dependent on proteasomal-mediated degradation of cyclin B1 by the anaphase promoting complex/cyclosome (APC^{Cdc20}) (Peters, 2006). However, downregulation of Cdk is not sufficient for mitotic exit in human cells. Thus, mitotic phosphatases are also thought to contribute to both the inactivation of Cdk1 at the onset of anaphase and to the mitotic exit process in higher eukaryotes. Consistent with this idea, in the early stages of mitotic exit, Cdk1 is transiently inhibited by phosphorylation prior to the degradation of cyclin B1 (D'Angiolella et al., 2007). It is possible that the transient phosphorylation of Cdk1 is also due to inhibition of the Cdc25C phosphatase by the PP2A phosphatase, which is the same phosphatase that keeps Cdc25C inactive during interphase (Forester et al., 2007). However, recent evidence indicates that the Cdc14B phosphatase dephosphorylates Cdc25C resulting in its inhibition and consequent phosphorylation of Cdk1 (Tumurbaatar et al., 2011). Moreover, Cdk substrates are dephosphorylated in an ordered sequence from anaphase to cytokinesis (Bouchoux and Uhlmann, 2011). Thus, mitotic exit further depends on the activation of protein phosphatase(s). Indeed, mitotic exit is blocked in cells lacking Cdk1 activity when protein phosphatase activity is suppressed (Skoufias et al., 2007).

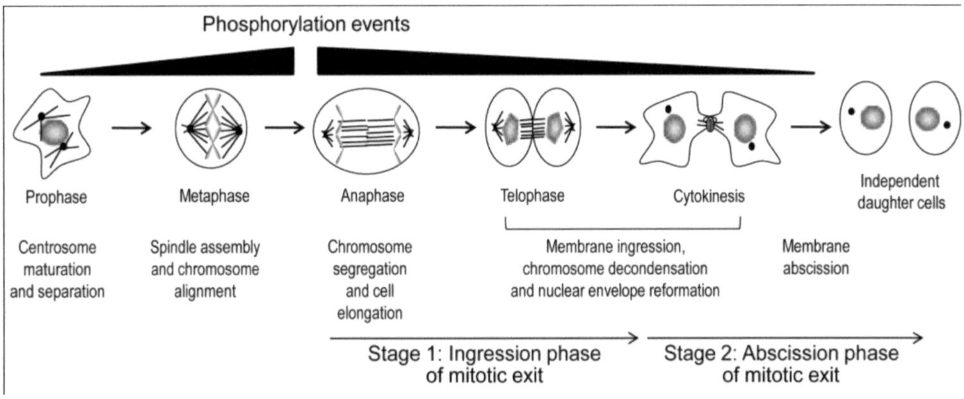

Fig. 1. Schematic illustration of the stages of mitosis. The relative abundance of phosphorylation events is shown above each mitotic stage and the major cellular events occurring at each stage are shown below. These events are known to be regulated by phosphorylation/dephosphorylation. Mitotic exit begins during anaphase and involves two sequential stages: (1) membrane ingression that generates a cleavage furrow followed by (2) membrane abscission of the intracellular bridge that connects the two nascent daughter cells. Chromosomes/nuclei shown in blue. Midbody shown in red.

Although there is a large body of knowledge about the phosphoproteins and protein kinases involved in mitosis and how they are regulated, the specific dephosphorylation events and the involvement of specific phosphatases in mitosis has only recently become appreciated. Studies are now revealing how the timely execution of mitotic events depends on the delicate interplay between protein kinases and phosphatases. To date, most reviews have focused on the role of protein dephosphorylation at the mitotic spindle and specifically how it regulates chromosome alignment (metaphase) and segregation (anaphase) (Bollen et al., 2009, De et al., 2009). This chapter will focus on providing an updated overview of the protein dephosphorylation events that occur during the later stages of mitosis (anaphase - cytokinesis) that contribute to driving mitotic exit and the generation of two independent daughter cells. More specifically, this chapter will provide insights into the protein phosphatases responsible for these dephosphorylation events and how they are regulated in mammalian cells.

2. Mitotic phosphatases in mammalian cells

In *Saccharomyces cerevisiae* (Shou et al., 1999, Visintin et al., 1999) and *Schizosaccharomyces pombe* (Cueille et al., 2001, Trautmann et al., 2001), mitotic exit and co-ordination of the final stage of mitosis, cytokinesis, are driven by the dual serine-threonine and tyrosine-protein phosphatase, Cdc14. Thus the action of Cdc14 is, in part, to counteract Cdk activity by dephosphorylating Cdk substrates (Visintin et al., 1998). Cdc14 is tightly regulated both spatially and temporally (Stegmeier and Amon, 2004, Queralt and Uhlmann, 2008) as well as being a part of several feedback loops that contribute to a rapid metaphase-anaphase transition (Holt et al., 2008). We have gained a detailed molecular picture of the way that the Cdc14 phosphatase orchestrates mitotic exit in yeast (reviewed in (Stegmeier and Amon, 2004, Queralt and Uhlmann, 2008)). However, much less is known about the protein dephosphorylation events and the responsible phosphatases that reverse Cdk phosphorylation and thus drive mitotic exit in eukaryotes. Homologues of Cdc14 exist in most if not all eukaryotes, but they do not seem to have the same central function in late mitosis as in budding yeast (Trautmann and McCollum, 2002). In *Caenorhabditis elegans*, depletion of CeCDC-14 by RNAi causes defects in cytokinesis; however, this is most likely due to failure to form an intact central spindle (Gruneberg et al., 2002). The human genome encodes two Cdcl4 homologues, Cdc14A and Cdc14B and both can rescue Cdc14 yeast phenotypes (Queralt and Uhlmann, 2008), suggesting functional conservation. However, neither Cdc14A nor Cdc14B are required for mitotic exit in higher eukaryotes (Berdougo et al., 2008) although they do seem to be required to generally dephosphorylate Cdk targets (Mocciaro and Schiebel, 2010). This indicates that they have overlapping functions or that additional mitotic exit phosphatases are required. Instead, recent reports suggest that Cdc14s might act by reversing the activating phosphorylations on Cdc25 phosphatases, thereby indirectly contributing to the regulation of Cdk activity in human cells (Krasinska et al., 2007, Vazquez-Novelle et al., 2010, Tumurbaatar et al., 2011). A survey of phosphatase contribution to cell cycle progression in Drosophila failed to identify a specific candidate for a mitotic exit phosphatase (Chen et al., 2007), suggesting that more than one phosphatase may act redundantly, or that its involvement in mitotic exit is not the only function of the phosphatase. Recent efforts into identifying phosphatases other than Cdc14 that drive mitotic exit have revealed the serine-threonine calcium- and calmodulin-activated phosphatase, calcineurin (CaN or PP2B) (Chircop et al., 2010a), the protein tyrosine phosphatase containing domain 1 (Ptpcd-1) (Zineldeen et al., 2009), PP1 (Wu et al., 2009), PP2A (Mochida et al., 2009, Schmitz et al., 2010) and oculocerebrorenal syndrome of Lowe 1 (OCRL1) (Ben El et al., 2011) as being required for mitotic exit in mammalian cells (Table 1).

2.1 Cdc14A and Cdc14B

Although the roles of human Cdc14A and Cdc14B are poorly understood, Cdc14A has been linked to centrosome separation and cytokinesis (Kaiser et al., 2002, Yuan et al., 2007), while Cdc14B participates in centrosome duplication and microtubule stabilization (Cho et al., 2005).

2.1.1 Cdc14A

The role of Cdc14A in cytokinesis has been linked to the membrane abscission stage. Ectopically expressed *Xenopus* Cdc14A localizes to the midbody of cytokinetic cells. *Xenopus* oocytes overexpressing wild-type or phosphatase-dead Cdc14A arrests cells in late stage cytokinesis, whereby the nascent daughter cells are connected by a thin intracellular bridge. Neither central spindle formation, nor the re-localization of passenger proteins and centralspindlin complexes to the midbody are affected. Instead targeting of the essential midbody abscission components, exocyst and SNARE complexes to the midbody, are disrupted in these cells (Krasinska et al., 2007), indicating that Cdc14 midbody localization and more specifically its phosphatase activity is required for abscission.

Phosphatase	Substrates	Function(s)	References
Cdc14A			
	Unidentified Cdk substrates	Centrosome separation, chromosome segregation and cytokinesis	(Kaiser et al., 2002, Mailand et al., 2002, Yuan et al., 2007)
	Cdc25	Inhibition of Cdk1	(Krasinska et al., 2007)
Cdc14B			
	N.D.	Stabilisation and bundling of MTs	(Cho et al., 2005)
	SIRT2	Downregulation of SIRT2 deacetylase activity by promoting its degradation	(Dryden et al., 2003)
Ptpcd-1			
	Unidentified Cdk substrates	Cytokinesis	(Zineldeen et al., 2009)
PP1			
	I2	Chromosome segregation	(Wu et al., 2009)
	Moesin	Cell shape changes for anaphase elongation	(Kunda et al., 2011)
	AIB1	Relocate AIB1 to chromatin for transcription	(Ferrero et al., 2011)
	PNUTS	Chromosome decondensation	(Landsverk et al., 2005)
	B-type lamins	Targeting ER to chromatin and nuclear envelop reformation	(Steen et al., 2000, Ito et al., 2007)
	Histone H3	Chromsomal reorganisation and nuclear envelope reformation	(Vagnarelli et al., 2011)
PP2A			
	Unidentified Cdk substrates	Mitotic exit	(Schmitz et al., 2010, Burgess et al., 2010)
CaN (PP2B)			
	Dynamin II	Membrane abscission	(Chircop et al., 2010a, Chircop et al., 2010b)
OCRL			
	PI(4,5)P2	Cleavage furrow formation and membrane ingression	(Ben El et al., 2011)

Table 1. The substrates and function of mitotic phosphatases required for mitotic exit in mammalian cells. N.D. not determined.

Biochemical studies in human HeLa cells suggests that Plk1 regulates the phosphatase activity of Cdc14A during mitosis (Yuan et al., 2007). Plk1 interacts with and phosphorylates Cdc14A resulting in release of Cdc14 auto-inhibited phosphatase activity *in vitro*. This is likely to occur during anaphase. Indeed, overexpression of a phospho-mimetic mutant of Cdc14A in HeLa cells results in aberrant chromosome alignment with delay in prometaphase (Yuan et al., 2007). This suggests that Cdc14A activity is associated with metaphase-anaphase progression and chromosome segregation.

2.1.2 Cdc14B

Although Cdc14B is not required for mitotic exit in mammalian cells, it does appear to play a role in mitosis during the latter stages. The SIRT2 protein is a NAD-dependent deacetylase (NDAC) that is a member of the SIR2 gene family with roles in chromatin structure, transcriptional silencing, DNA repair, and control of cellular life span. SIRT2 abundance and phosphorylation status increase upon mitotic entry. During late stages of mitosis, Cdc14B, but not Cdc14A, mediates SIRT2 dephosphorylation, which in turn targets it for degradation by the 26S proteasome (Dryden et al., 2003). Cells stably overexpressing wildtype SIRT2 but not missense mutants lacking NDAC activity have a prolonged mitotic phase (Dryden et al., 2003). Thus, Cdc14B may contribute to chromatin changes during mitotic exit such as chromosome decondensation by targeting SIRT2 for destruction.

2.2 Ptpcd-1

Of all the phosphatases implicated in mammalian cell mitotic exit to date, the dual-specificity phosphatase, Ptpcd-1, is structurally the most related to Cdc14 (Zineldeen et al., 2009). It is suggested to be a functional isozyme of mammalian Cdc14A (Zineldeen et al., 2009). Like Cdc14A, Ptpcd-1 associates with and co-localises with Plk1 at the midbody of cells in cytokinesis. Both overexpression of Ptpcd-1 and Plk1 cause cytokinesis failure and multinucleatc cell formation. No Ptpcd-1 substrates have yet been identified, however like Cdc14B its function is most likely regulated by Plk1. Ptpcd-1 possesses four Plk1 consensus phosphorylation sites and its overexpression could not rescue cytokinesis failure induced by Plk1 depletion (Zineldeen et al., 2009), suggesting that it lies downstream of Plk1. In support of this idea, the yeast homolog of Plk1, Cdc5, regulates Cdc14 phosphorylation and its subcellular localization for mitotic exit (Visintin et al., 2003, Visintin et al., 2008). Based on their midbody co-localization, it is possible that this Plk1/Ptpcd-1 signalling pathway contributes to membrane abscission.

2.3 PP1 and PP2A

The phosphatase inhibitor, okadaic acid, can induce mitotic entry in interphase cells (Yamashita et al., 1990) and this mitotic state can be maintained if Cdk1 activity is inhibited (Skoufias et al., 2007). Okadaic acid inhibits the activity of the protein phosphatase (PP)1 and PP2A. Consequently, both phosphatases have been implicated in reversing mitotic phosphorylation events in *Xenopus* egg extracts (Wu et al., 2009, Mochida et al., 2009). Not surprisingly, both phosphatases are inactivated during mitosis and their reactivation is important for mitotic exit (Wu et al., 2009, Mochida et al., 2009). However, both phosphatases have distinct substrates and are regulated via different mechanisms, which is in line with these enzymes being structurally diverse (Virshup and Shenolikar, 2009).

2.3.1 PP1

PP1 is activated at the metaphase-anaphase transition by a mechanism involving both inactivation of Cdk1 and proteasome-dependent degradation of an unknown protein (Mochida and Hunt, 2007, Skoufias et al., 2007). During early stages of mitosis, PP1 activity is suppressed and this supression is maintained through dual inhibition by Cdk1 phosphorylation and the binding of inhibitor-1 (I1) (Wu et al., 2009). Protein kinase A phosphorylates I1, mediating its binding to PP1. Partial PP1 activation is achieved during anaphase following a drop in Cdk1 levels due to cyclin B degradation. This shifts the Cdk1/PP1 ratio in favour of PP1 allowing auto-dephosphorylation of PP1 at its Cdk1-mediated phosphorylation site. PP1 subsequently mediates the dephosphorylation of I2 at the I2 activating site, resulting in dissociation of the PP1-I2 inhibitor complex. This results in full activation of PP1 and initiation of mitotic exit. During anaphase, when the outer kinetochore is dis-assembled, I2 levels drop and this may also contribute to the up-regulation of PP1 (Li et al., 2007, Wang et al., 2008). PP1 itself participates in outer kinetochore dis-assembly and chromosome segregation and this may be due to its ability to dephosphorylate Aurora B substrates (Emanuele et al., 2008). Thus, several feedback loops exist and involve protein kinases to initiate and maintain PP1 activity for mitotic exit.

PP1 plays roles in several mitotic events that need to occur in a sequential order and include cell elongation, chromosome segregation, chromosome decondensation and nuclear envelope re-formation. At the onset of mitosis, the cell rounds up and forms a stiff, rounded metaphase cortex. Moesin, the sole *Drosophila* Ezrin-Radixin-Moesin (ERM)-family protein which functions to regulate actin dynamics and cytoskeleton organization (Fehon et al., 2010), plays a critical role in this process and is dependent on the phospho-form of moesin (Kunda and Baum, 2009, Roch et al., 2010, Roubinet et al., 2011). Consequently, dephosphorylation of moesin at the cell poles is required to dismantle this rigid cortex to allow for anaphase elongation and cytokinesis. An RNAi screen for phosphatases involved in the temporal and spatial control of moesin identified PP1 as the responsible phosphatase (Kunda et al., 2011). Overexpression of phosphomimetic-moesin and PP1 depletion blocks proper anaphase elongation of the cell (Kunda et al., 2011).

PP1 is involved in the first step of nuclear envelope re-formation by stimulating the targeting of endoplasmic reticulum to chromatin (Ito et al., 2007). The assembly of the nuclear lamina depends on the dephosphorylation of B-type lamins, which is catalyzed by a PP1/AKAP149 complex that is associated with the nuclear envelope (Steen et al., 2000).

The role of PP1 in chromosome decondensation involves its association with Repo-Man and the PP1 nuclear targeting subunit (PNUTS). During anaphase, a Repo-Man/PP1 complex forms following Repo-Man dephosphorylation. Repo-Man targets the complex to chromosomes to allow PP1 to mediate the dephosphorylation of histone H3 (Vagnarelli et al., 2011). This contributes to the loss of chromosome architecture (Vagnarelli et al., 2006, Trinkle-Mulcahy and Lamond, 2006, Trinkle-Mulcahy et al., 2006). During telophase, PP1 targets PNUTS to the reforming nuclei following the assembly of nuclear membranes concomitant with chromatin decondensation. Here, PNUTS enhances *in vitro* chromosome decondensation in a PP1-dependent manner (Landsverk et al., 2005). Thus, targeting of PNUTS to the reforming nuclei in telophase may be part of a signalling event promoting chromatin decondensation as cells re-enter interphase.

Finally, PP1 appears to play a role in the initiation of transcription upon entry into the next cell cycle, as it is responsible for reversing the inhibitory Cdk1-mediated phosphorylation events of the transcription factor, AIB1 (Ferrero et al., 2011). AIB1 phosphorylation does not appear to affect its transcriptional activity but instead excludes it from condensed chromatin during mitosis to prevent its access to the promoters of AIB1-dependent genes. Its dephosphorylation by PP1 would presumably allow AIB1 to relocate to decondensed chromatin upon entry into the next cell cycle to re-initiate gene transcription.

2.3.2 PP2A

PP2A forms a complex with B-type regulatory subunits and these subunits contribute to PP2A localisation and substrate specificity. As such, PP2A-B55α and PP2A-B55δ were considered strong mitotic exit phosphatase candidates since these B55 regulatory subunits are substrate specifiers for Cdk substrates (Janssens et al., 2008). Indeed, both have been shown to be regulators of mitotic exit in human cells (Mochida et al., 2009, Schmitz et al., 2010). In *Xenopus* egg extracts, PP2A-B55δ is negatively regulated by the kinase greatwall (MASTL in humans) during early mitotic stages to allow the accumulation of mitotically phosphorylated proteins. This is achieved by greatwall-mediated phosphorylation of the small protein ARPP-19, which converts it into a potent PP2A inhibitor (Burgess et al., 2010). PP2A activation induced by MASTL knockdown leads to premature mitotic exit in human cells (Burgess et al., 2010). How PP2A is reactivated once Cdk1 activity decreases to drive mitotic exit remains unclear. Presumably greatwall needs to be inactivated and this is likely to involve an as yet unidentified phosphatase. Alternatively or in addition to this, PP2A may be activated via auto-phosphorylation in a similar manner to PP1 (Wu et al., 2009). Moreover, the identification of PP2A substrates and the role of PP2A for mitotic exit remain key questions for future investigation.

2.4 CaN (PP2B)

Endocytosis is thought to shut down during mitosis then resume during the final stage, cytokinesis (Schweitzer et al., 2005). Endocytosis is required for cytokinesis (Feng et al., 2002) and thought to contribute to the pool of recycling endosomes that are eventually delivered to the site of abscission. Here, they are proposed to (i) provide extra total cell surface area, an increase of at least 25% is required to complete division (Boucrot and Kirchhausen, 2007), (ii) deliver critical cytokinetic proteins to the abscission site (Low et al., 2003), and/or (iii) be directly involved in compound fusion, whereby numerous vesicles fuse with the plasma membrane during abscission to separate the daughter cells (Low et al., 2003, Gromley et al., 2005, Goss and Toomre, 2008, Prekeris and Gould, 2008).

The calcium- and calmodulin-dependent phosphatase, calcineurin (CaN) is an excellent candidate phosphatase for restarting endocytosis during cytokinesis, since it initiates endocytosis in neurons (Liu et al., 1994). In support of this idea, the fission yeast CaN gene is required for cytokinesis and the CaN inhibitors cyclosporin A (CsA) and FK506 block yeast cytokinesis (Yoshida et al., 1994). CaN is required for calcium-induced mitotic exit in cytostatic factor-arrested *Xenopus* oocytes (Mochida and Hunt, 2007, Nishiyama et al., 2007). CaN is upregulated in *Xenopus* oocytes from metaphase of meiosis II. An increase in cytoplasmic calcium upon fertilisation triggers meiosis II exit in these oocytes, which involves calmodulin-activating kinase-dependent activation of the APC. APC-mediated inactivation of Cdk is not sufficient to drive Cdk substrate dephosphorylation and meiotic exit in these oocytes and thus activation of CaN is likely to occur in parrallel to drive this process. A recent report has indicated that CaN is also required for completion of abscission in human cells (Chircop et al., 2010a). During cytokinesis CaN locates to two 1.1 μm diameter flanking midbody rings (FMRs) that reside on either side of the 1.6 μm diameter γ-tubulin midbody ring (MR) within the centre of the intracellular bridge. The endocytic protein, dynamin II (dynII), is mitotically phosphorylated by Cdk1/cyclin B1 upon mitotic entry and co-localises with CaN at the FMRs during cytokinesis (Chircop et al., 2010a, Chircop et al., 2010b). CaN inhibition by CsA, dynII depletion, phospho-mimetic dynII phosphopeptides and small molecule dynamin inhibitors lead to aborted cytokinesis and multinucleation (Joshi et al., 2010, Chircop et al., 2010a, Chircop et al., 2010b, Chircop et al., 2011). At the FMRs, a calcium influx activates CaN resulting in dephosphorylation of dynII. This is one of the last molecular events known to occur prior to abscission. Thus, it is possible that CaN-mediated dynII dephosphorylation may be the trigger for cellular abscission to complete cytokinesis.

In the brain, for clathrin-mediated endocytosis (CME) and activity-dependent bulk endocytosis (ADBE), CaN not only targets dynI but also α-adaptin, epsin and eps15 (Cousin and Robinson, 2001). Like dynII, epsin and α-adaptin are mitotically phosphorylated (Chen et al., 1999, Kariya et al., 2000, Dephoure et al., 2008). Thus, they represent additional potential CaN substrates during cytokinesis. Once dephosphorylated they may contribute to the recruitment of the dephosphorylated form of dynII to the abscission site or directly in CME within the intracellular bridge.

2.5 Oculocerebrorenal syndrome of Lowe 1 (OCRL), an inositol 5-phosphatase

Generation of the cleavage furrow during the membrane ingression stage of cytokinesis involves an actin-myosin II contractile ring. At the cleavage furrow, the phosphoinositide phosphatidylinositol 4,5-bisphosphate (PI(4,5)P2) plays an important role in this process by recruiting and regulating essential proteins of the cytokinesis machinery (Janetopoulos and Devreotes, 2006). PI(4,5)P2 mis-regulation blocks cleavage furrow formation leading to generation of a multinucleated cell (Emoto et al., 2005, Field et al., 2005, Wong et al., 2005). In Drosophila, the localization of PI(4,5)P2 is restricted at the cleavage furrow by the Drosophila ortholog of human oculocerebrorenal syndrome of Lowe 1 (OCRL1) (Ben El et al., 2011), an inositol 5-phosphatase mutated in the X-linked disorder oculocerebrorenal Lowe syndrome. Depletion of this phosphatase results in cytokinesis failure due to mis-localization of several essential cleavage furrow components to giant cytoplasmic vacuoles that are rich in PI(4,5)P2 and endocytic markers (Ben El et al., 2011). dOCRL is associated with endosomes and mediates PI(4,5)P2 dephosphorylation on internal membranes to restrict this phosphoinositide at the plasma membrane and thereby regulate cleavage furrow formation and ingression.

3. Conclusion

Here, the role of Cdc14A and Cdc14B, PP1, PP2A-B55, CaN, Ptpcd-1 and OCRL in regulating and driving mitotic exit in mammalian cells was reviewed. It is clear that we have only scraped the surface in our investigations into understanding the role and regulation of protein phophatases in mitosis. The identification of all protein phosphatases involved in driving mitotic exit in mammalian cells, their relevant substrates and function as well as how their action is spatio- and temporal regulated *in vivo* remain key questions for future investigation.

Understanding how protein dephosphorylation regulates mitotic exit and how the responsible protein phosphatases are regulated will provide an improved understanding to how two independent daughter cells are generated. Mitotic exit failure results in aneuploidy, which leads to genomic instability and thus contributes to the initiation and progression of tumourigenesis. Thus, an understanding of the molecular pathways that drive mitotic exit may highlight molecular targets for the development of new anti-cancer chemotherapeutic agents. In line with this idea, a recent publication has identified the CaN substrate, dynII, as a molecular target for the treatment of cancer (Chircop et al., 2010a, Chircop et al., 2010b). Inhibitors of dynII possess anti-cancer properties due to their ability to cause cytokinesis failure and subsequent cell growth arrest or apoptotic cell death (Joshi et al., 2010). It will be interesting to pursue the development of other targeted inhibitors to determine if they also possess anti-cancer properties as well as being useful molecular tools to unravel the signalling pathways required for mitotic exit in mammalian cells.

4. References

Ben El KK, Roubinet C, Solinet S, Emery G, Carreno S. The inositol 5-phosphatase dOCRL controls PI(4,5)P2 homeostasis and is necessary for cytokinesis. Curr Biol 2011;21:1074-1079.

Berdougo E, Nachury MV, Jackson PK, Jallepalli PV. The nucleolar phosphatase Cdc14B is dispensable for chromosome segregation and mitotic exit in human cells. Cell Cycle 2008;7:1184-1190.

Bollen M, Gerlich DW, Lesage B. Mitotic phosphatases: from entry guards to exit guides. Trends Cell Biol 2009;19:531-541.

Bouchoux C, Uhlmann F. A quantitative model for ordered Cdk substrate dephosphorylation during mitotic exit. Cell 2011;147:803-814.

Boucrot E, Kirchhausen T. Endosomal recycling controls plasma membrane area during mitosis. Proc Natl Acad Sci U S A 2007;104:7939-7944.

Burgess A, Vigneron S, Brioudes E, Labbe JC, Lorca T, Castro A. Loss of human Greatwall results in G2 arrest and multiple mitotic defects due to deregulation of the cyclin B-Cdc2/PP2A balance. Proc Natl Acad Sci U S A 2010;107:12564-12569.

Chen F, Archambault V, Kar A, Lio' P, D'Avino PP, Sinka R, Lilley K, Laue ED, Deak P, Capalbo L and others. Multiple protein phosphatases are required for mitosis in Drosophila. Curr Biol 2007;17:293-303.

Chen H, Slepnev VI, Di Fiore PP, De Camilli P. The interaction of epsin and Eps15 with the clathrin adaptor AP-2 is inhibited by mitotic phosphorylation and enhanced by stimulation-dependent dephosphorylation in nerve terminals. J Biol Chem 1999;274:3257-3260.

Chircop M, Malladi CS, Lian AT, Page SL, Zavortink M, Gordon CP, McCluskey A, Robinson PJ. Calcineurin activity is required for the completion of cytokinesis. Cell Mol Life Sci 2010a;67:3725-3737.

Chircop M, Perera S, Mariana A, Lau H, Ma MP, Gilbert J, Jones NC, Gordon CP, Young KA, Morokoff A and others. Inhibition of dynamin by dynole 34-2 induces cell death following cytokinesis failure in cancer cells. Mol Cancer Ther 2011;10:1553-1562.

Chircop M, Sarcevic B, Larsen MR, Malladi CS, Chau N, Zavortink M, Smith CM, Quan A, Anggono V, Hainsa PG and others. Phosphorylation of dynamin II at serine-764 is associated with cytokinesis. Biochim Biophys Acta 2010b;1813:1689-1699.

Cho HP, Liu Y, Gomez M, Dunlap J, Tyers M, Wang Y. The dual-specificity phosphatase CDC14B bundles and stabilizes microtubules. Mol Cell Biol 2005;25:4541-4551.

Cousin MA, Robinson PJ. The dephosphins: dephosphorylation by calcineurin triggers synaptic vesicle endocytosis. Trends Neurosci 2001;24:659-665.

Cueille N, Salimova E, Esteban V, Blanco M, Moreno S, Bueno A, Simanis V. Flp1, a fission yeast orthologue of the s. cerevisiae CDC14 gene, is not required for cyclin degradation or rum1p stabilisation at the end of mitosis. J Cell Sci 2001;114:2649-2664.

D'Angiolella V, Palazzo L, Santarpia C, Costanzo V, Grieco D. Role for non-proteolytic control of M-phase-promoting factor activity at M-phase exit. PLoS One 2007;2:e247.

De WP, Montani F, Visintin R. Protein phosphatases take the mitotic stage. Curr Opin Cell Biol 2009;21:806-815.

Dephoure N, Zhou C, Villen J, Beausoleil SA, Bakalarski CE, Elledge SJ, Gygi SP. A quantitative atlas of mitotic phosphorylation. Proc Natl Acad Sci U S A 2008;105:10762-10767.

Dryden SC, Nahhas FA, Nowak JE, Goustin AS, Tainsky MA. Role for human SIRT2 NAD-dependent deacetylase activity in control of mitotic exit in the cell cycle. Mol Cell Biol 2003;23:3173-3185.

Emanuele MJ, Lan W, Jwa M, Miller SA, Chan CS, Stukenberg PT. Aurora B kinase and protein phosphatase 1 have opposing roles in modulating kinetochore assembly. J Cell Biol 2008;181:241-254.

Emoto K, Inadome H, Kanaho Y, Narumiya S, Umeda M. Local change in phospholipid composition at the cleavage furrow is essential for completion of cytokinesis. J Biol Chem 2005;280:37901-37907.

Fehon RG, McClatchey AI, Bretscher A. Organizing the cell cortex: the role of ERM proteins. Nat Rev Mol Cell Biol 2010;11:276-287.

Feng B, Schwarz H, Jesuthasan S. Furrow-specific endocytosis during cytokinesis of zebrafish blastomeres. Exp Cell Res 2002;279:14-20.

Ferrero M, Ferragud J, Orlando L, Valero L, Sanchez Del PM, Farras R, Font de MJ. Phosphorylation of AIB1 at Mitosis Is Regulated by CDK1/CYCLIN B. PLoS One 2011;6:e28602.

Field SJ, Madson N, Kerr ML, Galbraith KA, Kennedy CE, Tahiliani M, Wilkins A, Cantley LC. PtdIns(4,5)P2 functions at the cleavage furrow during cytokinesis. Curr Biol 2005;15:1407-1412.

Forester CM, Maddox J, Louis JV, Goris J, Virshup DM. Control of mitotic exit by PP2A regulation of Cdc25C and Cdk1. Proc Natl Acad Sci U S A 2007;104:19867-19872.

Goss JW, Toomre DK. Both daughter cells traffic and exocytose membrane at the cleavage furrow during mammalian cytokinesis. J Cell Biol 2008;181:1047-1054.

Gromley A, Yeaman C, Rosa J, Redick S, Chen CT, Mirabelle S, Guha M, Sillibourne J, Doxsey SJ. Centriolin anchoring of exocyst and SNARE complexes at the midbody is required for secretory-vesicle-mediated abscission. Cell 2005;123:75-87.

Gruneberg U, Glotzer M, Gartner A, Nigg EA. The CeCDC-14 phosphatase is required for cytokinesis in the Caenorhabditis elegans embryo. J Cell Biol 2002;158:901-914.

Holt LJ, Krutchinsky AN, Morgan DO. Positive feedback sharpens the anaphase switch. Nature 2008;454:353-357.

Ito H, Koyama Y, Takano M, Ishii K, Maeno M, Furukawa K, Horigome T. Nuclear envelope precursor vesicle targeting to chromatin is stimulated by protein phosphatase 1 in Xenopus egg extracts. Exp Cell Res 2007;313:1897-1910.

Janetopoulos C, Devreotes P. Phosphoinositide signaling plays a key role in cytokinesis. J Cell Biol 2006;174:485-490.

Janssens V, Longin S, Goris J. PP2A holoenzyme assembly: in cauda venenum (the sting is in the tail). Trends Biochem Sci 2008;33:113-121.

Joshi S, Perera S, Gilbert J, Smith CM, Gordon CP, McCluskey A, Sakoff JA, Braithwaite A, Robinson PJ, Chircop (nee Fabbro) M. The dynamin inhibitors MiTMAB and OcTMAB induce cytokinesis failure and inhibit cell proliferation in human cancer cells. Mol Cancer Ther 2010;9:1995-2006.

Kaiser BK, Zimmerman ZA, Charbonneau H, Jackson PK. Disruption of centrosome structure, chromosome segregation, and cytokinesis by misexpression of human Cdc14A phosphatase. Mol Biol Cell 2002;13:2289-300.

Kariya K, Koyama S, Nakashima S, Oshiro T, Morinaka K, Kikuchi A. Regulation of complex formation of POB1/epsin/adaptor protein complex 2 by mitotic phosphorylation. J Biol Chem 2000;275:18399-18406.

Krasinska L, de Bettignies G, Fisher D, Abrieu A, Fesquet D, Morin N. Regulation of multiple cell cycle events by Cdc14 homologues in vertebrates. Exp Cell Res 2007;313:1225-1239.

Kunda P, Baum B. The actin cytoskeleton in spindle assembly and positioning. Trends Cell Biol 2009;19:174-179.

Kunda P, Rodrigues NT, Moeendarbary E, Liu T, Ivetic A, Charras G, Baum B. PP1-Mediated Moesin Dephosphorylation Couples Polar Relaxation to Mitotic Exit. Curr Biol 2011.

Landsverk HB, Kirkhus M, Bollen M, Kuntziger T, Collas P. PNUTS enhances in vitro chromosome decondensation in a PP1-dependent manner. Biochem J 2005;390:709-717.

Li M, Satinover DL, Brautigan DL. Phosphorylation and functions of inhibitor-2 family of proteins. Biochemistry 2007;46:2380-2389.

Liu JP, Sim ATR, Robinson PJ. Calcineurin Inhibition of Dynamin-I GTPase Activity Coupled to Nerve-Terminal Depolarization. Science 1994;265:970-973.

Low SH, Li X, Miura M, Kudo N, Quinones B, Weimbs T. Syntaxin 2 and endobrevin are required for the terminal step of cytokinesis in mammalian cells. Dev Cell 2003;4:753-759.

Ma HT, Poon RY. How protein kinases co-ordinate mitosis in animal cells. Biochem J 2011;435:17-31.

Mailand N, Lukas C, Kaiser BK, Jackson PK, Bartek J, Lukas J. Deregulated human Cdc14A phosphatase disrupts centrosome separation and chromosome segregation. Nat Cell Biol 2002;4:317-22.

Mocciaro A, Schiebel E. Cdc14: a highly conserved family of phosphatases with non-conserved functions? J Cell Sci 2010;123:2867-2876.

Mochida S, Hunt T. Calcineurin is required to release Xenopus egg extracts from meiotic M phase. Nature 2007;449:336-340.

Mochida S, Ikeo S, Gannon J, Hunt T. Regulated activity of PP2A-B55 delta is crucial for controlling entry into and exit from mitosis in Xenopus egg extracts. Embo J 2009;28:2777-2785.

Nishiyama T, Yoshizaki N, Kishimoto T, Ohsumi K. Transient activation of calcineurin is essential to initiate embryonic development in Xenopus laevis. Nature 2007;449:341-345.

Peters JM. The anaphase promoting complex/cyclosome: a machine designed to destroy 20456. Nat Rev Mol Cell Biol 2006;7:644-656.

Prekeris R, Gould GW. Breaking up is hard to do - membrane traffic in cytokinesis. J Cell Sci 2008;121:1569-1576.

Queralt E, Uhlmann F. Cdk-counteracting phosphatases unlock mitotic exit. Curr Opin Cell Biol 2008;20:661-668.

Roch F, Polesello C, Roubinet C, Martin M, Roy C, Valenti P, Carreno S, Mangeat P, Payre F. Differential roles of PtdIns(4,5)P2 and phosphorylation in moesin activation during Drosophila development. J Cell Sci 2010;123:2058-2067.

Roubinet C, Decelle B, Chicanne G, Dorn JF, Payrastre B, Payre F, Carreno S. Molecular networks linked by Moesin drive remodeling of the cell cortex during mitosis. J Cell Biol 2011;195:99-112.

Schmitz MH, Held M, Janssens V, Hutchins JR, Hudecz O, Ivanova E, Goris J, Trinkle-Mulcahy L, Lamond AI, Poser I and others. Live-cell imaging RNAi screen identifies PP2A-B55alpha and importin-beta1 as key mitotic exit regulators in human cells. Nat Cell Biol 2010;12:886-893.

Schweitzer JK, Burke EE, Goodson HV, D'Souza-Schorey C. Endocytosis resumes during late mitosis and is required for cytokinesis. J Biol Chem 2005;280:41628-41635.

Shou W, Seol JH, Shevchenko A, Baskerville C, Moazed D, Chen ZW, Jang J, Charbonneau H, Deshaies RJ. Exit from mitosis is triggered by Tem1-dependent release of the protein phosphatase Cdc14 from nucleolar RENT complex. Cell 1999;97:233-44.

Skoufias DA, Indorato RL, Lacroix F, Panopoulos A, Margolis RL. Mitosis persists in the absence of Cdk1 activity when proteolysis or protein phosphatase activity is suppressed. J Cell Biol 2007;179:671-685.

Steen RL, Martins SB, Tasken K, Collas P. Recruitment of protein phosphatase 1 to the nuclear envelope by A-kinase anchoring protein AKAP149 is a prerequisite for nuclear lamina assembly. J Cell Biol 2000;150:1251-1262.

Stegmeier F, Amon A. Closing mitosis: the functions of the Cdc14 phosphatase and its regulation. Annu Rev Genet 2004;38:203-232.

Trautmann S, McCollum D. Cell cycle: new functions for Cdc14 family phosphatases. Curr Biol 2002;12:R733-R735.

Trautmann S, Wolfe BA, Jorgensen P, Tyers M, Gould KL, McCollum D. Fission yeast Clp1p phosphatase regulates G2/M transition and coordination of cytokinesis with cell cycle progression. Curr Biol 2001;11:931-940.

Trinkle-Mulcahy L, Andersen J, Lam YW, Moorhead G, Mann M, Lamond AI. Repo-Man recruits PP1 gamma to chromatin and is essential for cell viability. J Cell Biol 2006;172:679-692.

Trinkle-Mulcahy L, Lamond AI. Mitotic phosphatases: no longer silent partners. Curr Opin Cell Biol 2006;18:623-631.

Tumurbaatar I, Cizmecioglu O, Hoffmann I, Grummt I, Voit R. Human Cdc14B promotes progression through mitosis by dephosphorylating Cdc25 and regulating Cdk1/cyclin B activity. PLoS One 2011;6:e14711.

Vagnarelli P, Hudson DF, Ribeiro SA, Trinkle-Mulcahy L, Spence JM, Lai F, Farr CJ, Lamond AI, Earnshaw WC. Condensin and Repo-Man-PP1 co-operate in the regulation of chromosome architecture during mitosis. Nat Cell Biol 2006;8:1133-1142.

Vagnarelli P, Ribeiro S, Sennels L, Sanchez-Pulido L, de Lima AF, Verheyen T, Kelly DA, Ponting CP, Rappsilber J, Earnshaw WC. Repo-Man coordinates chromosomal reorganization with nuclear envelope reassembly during mitotic exit. Dev Cell 2011;21:328-342.

Vazquez-Novelle MD, Mailand N, Ovejero S, Bueno A, Sacristan MP. Human Cdc14A phosphatase modulates the G2/M transition through Cdc25A and Cdc25B. J Biol Chem 2010;285:40544-40553.

Virshup DM, Shenolikar S. From promiscuity to precision: protein phosphatases get a makeover. Mol Cell 2009;33:537-545.

Visintin C, Tomson BN, Rahal R, Paulson J, Cohen M, Taunton J, Amon A, Visintin R. APC/C-Cdh1-mediated degradation of the Polo kinase Cdc5 promotes the return of Cdc14 into the nucleolus. Genes Dev 2008;22:79-90.

Visintin R, Craig K, Hwang ES, Prinz S, Tyers M, Amon A. The phosphatase Cdc14 triggers mitotic exit by reversal of Cdk-dependent phosphorylation. Mol Cell 1998;2:709-18.

Visintin R, Hwang ES, Amon A. Cfi1 prevents premature exit from mitosis by anchoring Cdc14 phosphatase in the nucleolus. Nature 1999;398:818-23.

Visintin R, Stegmeier F, Amon A. The role of the polo kinase Cdc5 in controlling Cdc14 localization. Mol Biol Cell 2003;14:4486-4498.

Wang W, Stukenberg PT, Brautigan DL. Phosphatase inhibitor-2 balances protein phosphatase 1 and aurora B kinase for chromosome segregation and cytokinesis in human retinal epithelial cells. Mol Biol Cell 2008;19:4852-4862.

Wong R, Hadjiyanni I, Wei HC, Polevoy G, McBride R, Sem KP, Brill JA. PIP2 hydrolysis and calcium release are required for cytokinesis in Drosophila spermatocytes. Curr Biol 2005;15:1401-1406.

Wu JQ, Guo JY, Tang W, Yang CS, Freel CD, Chen C, Nairn AC, Kornbluth S. PP1-mediated dephosphorylation of phosphoproteins at mitotic exit is controlled by inhibitor-1 and PP1 phosphorylation. Nat Cell Biol 2009;11:644-651.

Yamashita K, Yasuda H, Pines J, Yasumoto K, Nishitani H, Ohtsubo M, Hunter T, Sugimura T, Nishimoto T. Okadaic acid, a potent inhibitor of type 1 and type 2A protein phosphatases, activates cdc2/H1 kinase and transiently induces a premature mitosis-like state in BHK21 cells. Embo J 1990;9:4331-4338.

Yoshida T, Toda T, Yanagida M. A Calcineurin-Like Gene Ppb1(+) in Fission Yeast - Mutant Defects in Cytokinesis, Cell Polarity, Mating and Spindle Pole Body Positioning. J Cell Sci 1994;107:1725-1735.

Yuan K, Hu H, Guo Z, Fu G, Shaw AP, Hu R, Yao X. Phospho-regulation of HsCdc14A By Polo-like kinase 1 is essential for mitotic progression. J Biol Chem 2007;282:27414-27423.

Zineldeen DH, Shimada M, Niida H, Katsuno Y, Nakanishi M. Ptpcd-1 is a novel cell cycle related phosphatase that regulates centriole duplication and cytokinesis. Biochem Biophys Res Commun 2009;380:460-466.

11

Regulations and Functions of ICK/MAK/MOK – A Novel MAPK-Related Kinase Family Linked to Human Diseases

Zheng "John" Fu
University of Virginia School of Medicine, Department of Medicine, USA

1. Introduction

The RCK (the tyrosine kinase gene *v-ros* cross-hybridizing kinase) family within the CMGC (CDK/MAPK/GSK3/CLK) group of the human kinome consists of ICK/MRK (Intestinal cell kinase/MAK-related kinase) (Abe, Yagi et al. 1995; Togawa, Yan et al. 2000), MAK (male germ cell-associated kinase) (Matsushime, Jinno et al. 1990), and MOK (MAPK/MAK/MRK-overlapping kinase) (Miyata, Akashi et al. 1999) (Fig. 1). In the N-terminal catalytic domain, they all share significant sequence homology with MAPK (mitogen-activated protein kinase) and contain a MAPK-like TXY motif in the activation T-loop. However, they display significant divergence in the composition of their C-terminal non-catalytic domains which may determine their functional specificity and confer distinct regulatory mechanisms.

The biological functions of the ICK/MAK/MOK family have been elusive until recently. MAK is highly expressed in testis, however the MAK null mouse is viable and fertile, suggesting the existence of functional redundancy or compensation for the lack of MAK in testis (Shinkai, Satoh et al. 2002). In the MAK-null retina, photoreceptors exhibit elongated cilia and progressive degeneration, suggesting that MAK is required for regulation of ciliary length and retinal photoreceptor survival (Omori, Chaya et al. 2010). Exome sequencing has identified multiple point mutations in the kinase domain (Fig. 1) or an Alu-insertion in exon 9 of MAK as potential causes of retinitis pigmentosa (RP), a genetically heritable and autosomal recessive disease (Ozgul, Siemiatkowska et al. 2011; Tucker, Scheetz et al. 2011). MAK is a co-activator for androgen receptor (AR) in prostate cancer cells and is required for AR-mediated signaling and cell proliferation (Xia, Robinson et al. 2002; Ma, Xia et al. 2006). A loss-of-function point mutation R272Q of ICK (Fig. 1) has been recently identified as the causative mutation in a neonatal lethal multiplex human syndrome ECO (endocrine-cerebro-osteodysplasia), implicating a key role for ICK in development of multiple organ systems (Lahiry, Wang et al. 2009). Using shRNA knockdown, we have shown that suppression of ICK expression in intestinal epithelial cells markedly impaired cell proliferation and G1 cell cycle progression (Fu, Kim et al. 2009). Furthermore, ICK deficiency led to a significant decrease in the mTORC1 (mammalian target of rapamycin complex 1) activity, concomitant with reduced expression of specific mTORC1 downstream targets cyclinD1 and c-Myc (Fu, Kim et al. 2009). These results suggest that ICK may target

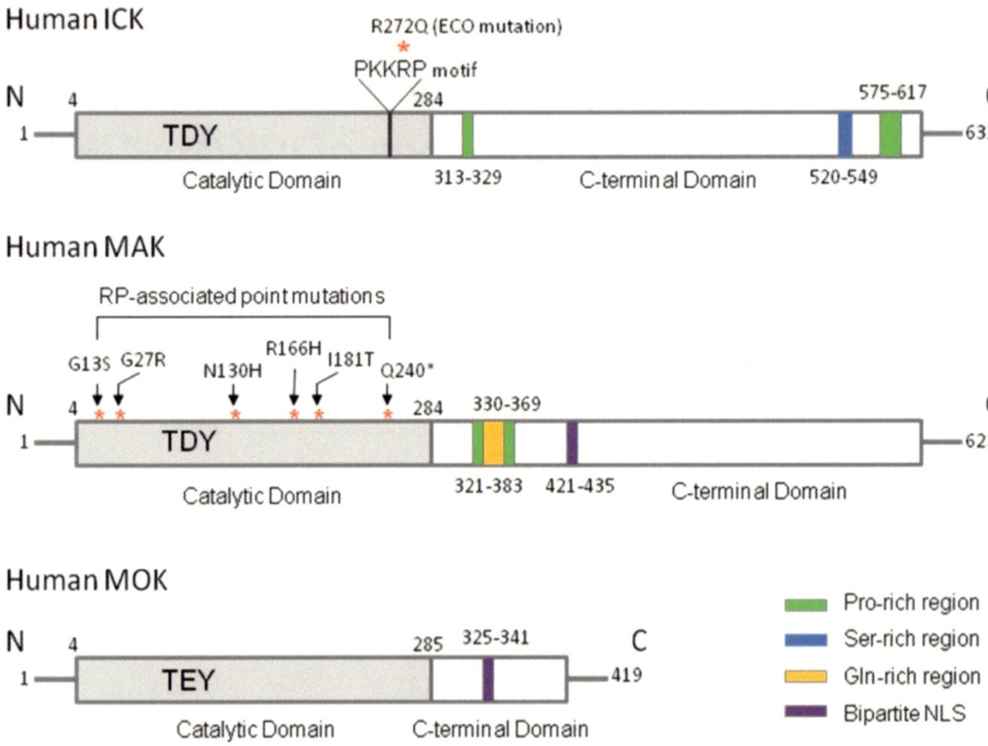


```
ICK     MNRYTTIRQLGDGTYGSVLLGRSIESGELIAIKKMKRKFYSWEECMNLREVKSLKKLN-H 59
MAK     MNRYTTMRQLGDGTYGSVLMGKSNESGELVAIKRMKRKFYSWDECMNLREVKSLKKLN-H 59
MOK     MKNYKAIGKIGEGTFSEVMKMQSLRDGNYYACKQMKQRFESIEQVNNLREIQALRRLNPH 60
        *:.*.:: ::*:**:..*:  :* ..*:  * *:**::* * ::  ****:::*::** *

ICK     ANVVKLKEVIRE--NDHLYFIFEYMKENLYQLIKERNKLFPESAIRNIMYQILQGLAFIH 117
MAK     ANVIKLKEVIRE--NDHLYFIFEYMKENLYQLMKDRNKLFPESVIRNIMYQILQGLAFIH 117
MOK     PNILMLHEVVFDRKSGSLALICELMDMNIYELIRGRRYPLSEKKIMHYMYQLCKSLDHIH 120
        .*:: *:**: :  .. * :* * *. *:*:*:: *.  :.*. * : ***: :.* .**

ICK     KHGFFHRDLKPENLLCMGPELVKIADFGLAREIRSKPPYTDYVSTRWYRAPEVLLRSTNY 177
MAK     KHGFFHRDMKPENLLCMGPELVKIADFGLARELRSQPPYTDYVSTRWYRAPEVLLRSSVY 177
MOK     RNGIFHRDVKPENILIK-QDVLKLGDFGSCRSVYSKQPYTEYISTRWYRAPECLLTDGFY 179
        ::*:****:****:*    :::*:.*** .*.: *: ***:*:********* **  .  *

ICK     SSPIDVWAVGCIMAEVYTLRPLFPGASEIDTIFKICQVLGTPKKTDWPEGYQLSSAMNFR 237
MAK     SSPIDVWAVGSIMAELYMLRPLFPGTSEVDEIFKICQVLGTPKKSDWPEGYQLASSMNFR 237
MOK     TYKMDLWSAGCVFYEIASLQPLFPGVNELDQISKIHDVIGTPAQKILTK-FKQSRAMNFD 238
        :  :*:*:.*.:: *:  *:*****..*:* * ** :*:*** :.  .: :: : :***

ICK     WPQCVPNNLKTLIPNASSEAVQLLRDMLQWDPKKRPTASQALRYPYF 284
MAK     FPQCVPINLKTLIPNASNEAIQLMTEMLNWDPKKRPTASQALKHPYF 284
MOK     FPFKKGSGIPLLTTNLSPQCLSLLHAMVAYDPDERIAAHQALQHPYF 285
        :*     .:  * .* * :.:.*:  *: :**.:* :* ***::***
```

Fig. 1. Schematic illustration of structural organization and features of human ICK, MAK and MOK; Sequence alignment of their catalytic domains. ECO, endocrine-cerebro-osteodysplasia syndrome; RP, retinitis pigmentosa the mTORC1 signaling pathway to regulate cell proliferation and cell cycle progression. The biological functions of MOK are the least understood in the ICK/MAK/MOK family, except that a previous study indicated that MOK may be involved in growth arrest and differentiation in the intestinal epithelium (Uesaka and Kageyama 2004).

In this chapter, the current knowledge about the regulations and functions of this novel group of serine/threonine protein kinases will be reviewed. Furthermore, and more importantly, the many "unknowns" about the biology of ICK/MAK/MOK will be identified and discussed, the answers to which should provide new insights into their unique regulatory mechanisms, diverse biological substrates and physiological functions.

2. ICK signaling cascade

ICK was separately cloned from a rat heart cDNA library (Abe, Yagi et al. 1995) and a mouse small intestinal crypt cDNA library (Togawa, Yan et al. 2000) by using degenerate oligonucleotide primers recognizing sequences from highly conserved subdomains of serine-threonine kinases. ICK, named after its cloning origin the intestine, is actually a ubiquitously expressed Ser/Thr protein kinase. Northern analysis with specific ICK probes detected ICK mRNAs in most mouse, rat and human tissues examined (Abe, Yagi et al. 1995; Togawa, Yan et al. 2000). ICK and MAK contain nearly identical N-terminal catalytic domains (87% identity) but more divergent C-terminal noncatalytic domains. The ICK/MAK catalytic domain is related to both mitogen-activated protein kinases (MAPKs) and cyclin-dependent protein kinases (CDKs), with a MAPK-like TDY motif in the activation loop and the CDK-like regulatory sites $T^{14}Y^{15}$, but lacking the PSTAIRE cyclin-binding motif found in most CDKs. ICK and MAK are also conserved from yeast to humans. *Saccharomyces cerevisiae* has one closely related kinase, Ime2p (inducer of meiosis) that is a meiosis-specific homolog of human CDK2 and required for timing meiotic S phase (Foiani, Nadjar-Boger et al. 1996; Clifford, Stark et al. 2005). *Caenorhabditis elegans* has one homolog DYF5, a dye-filling defective mutant identified from a forward genetic screen that plays an important role in the control of cilia length and the docking and undocking of kinesin-2 motors from IFT (intraflagellar transport) particles (Burghoorn, Dekkers et al. 2007). *Danio rerio* (zebrafish) also has one homologous gene whose in situ expression at the basal level was detected in the retinal photoreceptor cell layer (http://zfin.org).

2.1 Regulation of activity by the TDY motif phosphorylation through a pair of yin-yang regulators

So far, we have established ICK as the prototype for a new group of kinases with MAPK-like regulation at TDY motifs (Fu, Schroeder et al. 2005). By mass spectrometry, we have shown that ICK can be specifically phosphorylated in the TDY motif *in vivo*. ICK requires an intact and doubly phosphorylated TDY motif for maximum activity. Autophosphorylation on Tyr-159 in the TDY motif only confers basal kinase activity. Full activation of ICK requires additional phosphorylation of Thr-157 in the TDY motif. Furthermore, we have identified PP5 (protein phosphatase 5) and CCRK (cell cycle related kinase) as a pair of *yin-yang* regulators for Thr-157 phosphorylation (Fu, Larson et al. 2006).

CCRK (Cell Cycle-Related Kinase): Since the catalytic domain of ICK is similar to those of both ERK2 (extracellular signal-regulated kinase) and CDK2 (cyclin-dependent protein kinase 2), we tested whether the ERK2 activator MEK1/2 (MAPK/ERK kinase 1/2) and CDK2

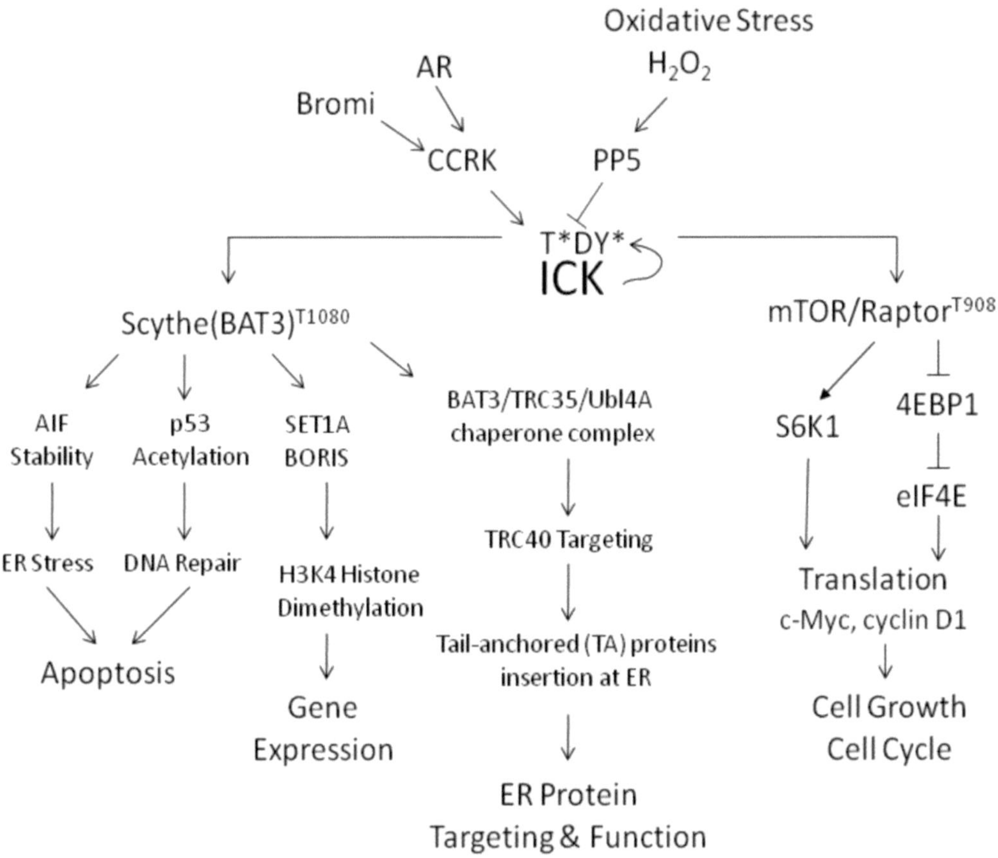

Fig. 2. Working model of the ICK signaling cascade. CCRK, cell cycle-related kinase; PP5, protein phosphatase 5; Bromi, broad-minded; AR, androgen receptor; mTOR, mammalian target of rapamycin; Raptor, regulatory associated protein of mTOR; Scythe/BAT3, HLA-b-associated transcript 3.

activator CAK (Cdk activating kinase) activates ICK. Surprisingly, neither MEK1/2 nor the Cdk7 complex (CDK7/cyclin H/MAT1) phosphorylates ICK in the T-loop, instead our data implicated ICK as a physiologic downstream target of CCRK (Fu, Schroeder et al. 2005; Fu, Larson et al. 2006). CCRK is most closely related to yeast CAK based on sequence homology, however it is a point of controversy as to whether CCRK has the intrinsic CAK activity (Liu, Wu et al. 2004; Wohlbold, Larochelle et al. 2006). CCRK may support a role for ICK in the regulation of proliferation and/or apoptosis. CCRK was identified in a large scale siRNA screen for suppressors of apoptosis (MacKeigan, Murphy et al. 2005). CCRK was also shown to be important for cell growth in HeLa, HCT116 and U2OS cells (Liu, Wu et al. 2004; Wohlbold, Larochelle et al. 2006). CCRK is a novel candidate oncogene in human glioblastoma (Ng, Cheung et al. 2007), colon cancer (An, Ng et al. 2010) and hepatocellular carcinoma (Feng, Cheng et al. 2011). The heart expresses a splice variant of CCRK, which promotes cardiac cell growth and survival; differs from the generic isoform in terms of protein-protein interactions, substrate specificity and regulation of the cell cycle; and is down-regulated significantly in heart failure (Qiu, Dai et al. 2008). Recently, CCRK was shown to interact with Broad-minded (Bromi) to control cilia assembly and mammalian Sonic hedgehog (Shh) signaling transduction (Ko, Norman et al. 2010). The endogenous CCRK protein level was significantly reduced in Bromi mutant embryos and fibroblasts, while the CCRK mRNA level was unaffected, suggesting that Bromi promotes CCRK stability. CCRK was also implicated as a downstream mediator of AR (androgen receptor) signaling that drives hepatocarcinogenesis through a β-catenin and TCF (T cell receptor)-dependent pathway (Feng, Cheng et al. 2011). Ligand-bound AR is able to up-regulate CCRK transcription and protein expression through direct binding to the AR-responsive element of its promoter. What remains unknown is how the activity of CCRK is regulated independent of its expression level. Given CCRK is localized in both nuclear and cytoplasmic compartments, it will be important to know whether CCRK is differentially regulated and performs distinct biological functions in two different locations. It is also worth pointing out that a previous gel filtration study demonstrated the presence of at least two different CAK activities in human cells, with the second CAK activity detected at 30-40 KDa resembling the biochemical properties of Cak1p (Kaldis and Solomon 2000). This observation raises the possibility that a "small" CAK other than CCRK may be an upstream activator of ICK and MAK.

PP5 (Protein Phosphatase 5): PP5 plays important roles in cell cycle checkpoints, DNA damage response and proliferation (Golden, Swingle et al. 2008; Hinds and Sanchez 2008). Inhibition of PP5 expression results in a marked antiproliferative effect through the activation of the p53-dependent G1 checkpoint (Zuo, Dean et al. 1998). PP5 is required for both ATM (Ataxia telangiectasia mutated) and ATR (Ataxia telangiectasia and Rad3-related protein) checkpoint signaling, operant in S and G2/M (Ali, Zhang et al. 2004; Zhang, Bao et al. 2005). Hydrogen peroxide treatment induces activation of PP5 leading to the negative regulation of ASK1 (apoptosis signal-regulating kinase 1) and inhibition of apoptosis (Morita, Saitoh et al. 2001). PP5 dephosphorylates two functional sites in DNA-PKc (DNA-dependent protein kinase, catalytic subunit) that are required for functions in the DNA repair of double strand breaks (Wechsler, Chen et al. 2004). PP5 also dephosphorylates Raf1 (proto-oncogene c-Raf) at Ser-338 to inhibit Raf1 activity and its downstream signaling to MEK and ERK (von Kriegsheim, Pitt et al. 2006). Similarly, PP5 can inactivate ICK by dephosphorylating the essential phospho-threonine residue within the T-loop (Fu, Larson et al. 2006). We also showed that hydrogen peroxide treatment induces activation of the endogenous PP5 to negatively regulate ICK phosphorylation in the T-loop (Fu, Larson et al. 2006). Identifying ICK as a new downstream target of PP5 leads to our hypothesis that PP5 may modulate some branch of checkpoint signaling in response to stress and DNA-damage through the inactivation of ICK.

2.2 Regulation of activity by nuclear targeting

Prior studies from us and others have shown that GFP-tagged ICK is predominantly nuclear (Yang, Jiang et al. 2002; Fu, Schroeder et al. 2005). Our studies further established that the catalytic domain, but not the C-terminal domain, of ICK is required for nuclear localization. Neither the kinase activity nor the TDY phosphorylation appears to be necessary. Instead, an intact subdomain XI is required as well as the conserved arginine, R272, in the PKKRP motif and its interacting networks including W184 and E169. Loss of nuclear localization was associated with a significant reduction in its catalytic activity, suggesting that nuclear targeting is important for the maximal activation of ICK, consistent with the predominant nuclear localization of its upstream activator CCRK.

It still remains elusive how the endogenous ICK is distributed in cells. Our unpublished data implicate the presence of endogenous ICK signals in the cytoplasm. Given that ICK has multiple splicing variants, it is possible that different isoforms of ICK may exhibit differential subcellular localization. Although CCRK is mainly localized to the nucleus, it is also present in the cytoplasm (Liu, Wu et al. 2004; An, Ng et al. 2010; Ko, Norman et al. 2010), thus capable of phosphorylating and activating cytoplasmic ICK as well. It remains to be determined whether cytoplasmic ICK and nuclear ICK have distinct biological activities.

2.3 The ICK substrate phosphorylation consensus

In order to identify putative substrates for ICK, a positional scanning peptide array method was used to determine the sequence specificity surrounding the ICK phosphorylation site (Fu, Larson et al. 2006). The phosphorylation consensus for ICK is [R]-[P]-[X]-[S/T]-[P/A/T/S], with the strongest selection for arginine at P-3 and proline at P-2. A preference for proline at P+1 was observed and was expected, given the similarity of ICK to ERK2 in the catalytic domain. However, the selection for proline at P+1 position is not absolutely stringent because alanine, threonine, and serine were also selected albeit less well. Despite some similarities, the ICK phosphorylation consensus is distinct from that of ERK2 due to the lack of absolute stringency for proline at P+1 or that of CDK2 due to the lack of a strong preference for basic residues (K/R) at P+3.

Due to the difficulty in obtaining a large quantity of highly purified and active full-length ICK protein, we were only able to use the catalytic domain of ICK as the kinase source in the peptide library scan (Fu, Larson et al. 2006). Therefore, despite our recent successes in using this consensus motif to identify several candidate physiological substrates for ICK (see Fig. 2), there still is the possibility that the full-length ICK including the long C-terminal domain may add additional features or modifications to the current substrate consensus sequence for ICK.

Given that ICK and MAK are essentially identical in the catalytic domain, we anticipate that this consensus sequence for ICK may also be useful for selecting putative substrates and/or phosphorylation sites for MAK. For example, we have identified Scythe/BAT3 as a candidate physiological substrate for ICK (Fu, Larson et al. 2006). Scythe/BAT3 is especially enriched in testis and abundant in male germ cells (Wang and Liew 1994), an expression pattern very similar to that of MAK (Matsushime, Jinno et al. 1990). In addition, both MAK and Scythe/BAT3 mRNAs increase dramatically in the mouse testis at around 14 to 20 days after birth. These correlations suggest an interesting hypothesis that Scythe may be a direct substrate downstream of MAK to function in some aspects of spermatogenesis.

2.4 Candidate physiological substrates

Scythe/BAT3 (HLA-b-associated transcript 3): We have established, *in vitro*, that ICK can phosphorylate Scythe (Fu, Larson et al. 2006), an important mediator of apoptosis and proliferation during mammalian development (Desmots, Russell et al. 2005). Scythe was originally identified as a novel reaper-binding apoptotic regulator in *Drosophila melanogaster* (Thress, Henzel et al. 1998; Thress, Evans et al. 1999). Recently, Scythe was identified as the key interacting partner of the human small glutamine-rich TPR-containing protein (hSGT) that is required for progression through cell division (Winnefeld, Grewenig et al. 2006). Findings from studies of Scythe-deficient mice indicated that Scythe is a novel and essential regulator of p53-mediated responses to genotoxic stress by controlling DNA-damage induced acetylation of p53 (Sasaki, Gan et al. 2007). As a possible link of ICK to colon cancer, Scythe was a newly identified candidate tumor suppressor gene in colon cancer cells (Ivanov, Lo et al. 2007). Scythe is also essential for selective elimination of defective proteasomal substrate as a ubiquitin-like protein (Minami, Hayakawa et al. 2010) and acts as a tansmembrane domain (TMD)-selective chaperon that effectively channels tail-anchored (TA) proteins into the ER membrane (Mariappan, Li et al. 2010). Using the ICK phosphorylation consensus sequence R-P-X-S/T, we reported that an *in vivo* phosphorylation site, Thr-1080, in Scythe is a major ICK phosphorylation site *in vitro* (Fu, Larson et al. 2006). Functions are yet to be defined for this ICK phosphorylation site in Scythe-regulated biological events such as proliferation, apoptosis and DNA damage control in response to genotoxic stress.

Raptor (Regulatory associated protein of mTOR): The serine/threonine protein kinase mammalian target of rapamycin (mTOR) is the core catalytic component of two structurally and functionally distinct protein complexes, mTOR complex 1 (mTORC1) and mTOR complex 2 (mTORC2), which collectively integrates nutrient, hormonal, and energy signal inputs to control cell growth, proliferation and survival (Bhaskar and Hay 2007; Hall 2008; Laplante and Sabatini 2009). mTORC1, when activated by growth factors and nutrients, stimulate cell growth and proliferation by phosphorylating two key regulators of mRNA translation and ribosome biogenesis, S6K1 (ribosomal protein S6 kinase) and 4EBP1 (eukaryotic initiation factor 4E-binding protein 1) (Hara, Yonezawa et al. 1997; Fingar, Richardson et al. 2004; Proud 2004; Ma and Blenis 2009). In addition to the catalytic subunit mTOR, mTORC1 also contains four associated components, Raptor, mLST8/GβL, PRAS40, and Deptor (Kim, Sarbassov et al. 2002; Loewith, Jacinto et al. 2002; Kim, Sarbassov et al. 2003; Kim and Sabatini 2004; Vander Haar, Lee et al. 2007; Wang, Harris et al. 2008; Peterson, Laplante et al. 2009). Raptor plays an important role as a scaffolding protein to recruit substrates S6K1 and 4EBP1 to mTOR (Nojima, Tokunaga et al. 2003). Upon growth factor stimulation, Raptor binding to substrates can be enhanced by the dissociation of the competitive inhibitor PRAS40 (the proline-rich Akt substrate of 40KDa) from mTORC1 (Fonseca, Smith et al. 2007; Sancak, Thoreen et al. 2007; Wang, Harris et al. 2008; Nascimento and Ouwens 2009). Raptor can also positively regulate mTOR activity in response to nutrient sufficiency by directly interacting with Rag family GTPases to induce mTORC1 re-localization to an intracellular vesicular compartment containing RheB, a Ras-like GTP-binding protein that activates mTOR via an unknown mechanism (Hanrahan and Blenis 2006; Kim, Goraksha-Hicks et al. 2008; Sancak, Peterson et al. 2008).

Recently, multiple phosphorylation sites of Raptor have been identified, several of which are critical for the regulation of mTORC1 activity in response to insulin, nutrients or energy stress. Phosphorylation of Raptor Ser-722 and Ser-792 by AMPK is required for the inhibition of mTORC1 and cell cycle arrest induced by energy stress (Gwinn, Shackelford et al. 2008). RSK mediated phosphorylation of Ser-719/721/722 enhances mTORC1 activity stimulated by Ras/MAPK pathway (Carriere, Cargnello et al. 2008). Phosphorylation of Ser-863 by either mTOR or ERK1/2 promotes mTORC1 activation in response to various stimuli including growth factors, nutrients and cellular energy (Wang, Lawrence et al. 2009; Foster, Acosta-Jaquez et al. 2010; Carriere, Romeo et al. 2011). Taken together, a plethora of emerging evidence indicates that the complex phosphorylation status of Raptor is tightly associated with the activity of mTORC1. Our data indicated that ICK associates with Raptor and phosphorylates Raptor at Thr-908 (Fu, Kim et al. 2009). More importantly, knockdown the ICK expression significantly reduced the phosphorylation of S6K1 at Thr-389 targeted by the mTORC1, suggesting that ICK is an upstream regulator of the mTORC1 activity in the regulation of cell growth and proliferation (Fu, Kim et al. 2009).

2.5 Role of ICK in ECO and multiple organ development during embryogenesis

Protein kinases comprise one of the largest and most abundant gene families in humans. Both inherited germ-line and somatic mutations in kinase genes have been associated with many human diseases including developmental and metabolic disorders and neoplastic malignancies (Lahiry, Torkamani et al. 2010). The human endocrine-cerebro-osteodysplasia (ECO) syndrome is a newly identified congenital neonatal-lethal disorder whose clinical manifestations include osteodysplasia, cerebral anomalies, and endocrine gland hypoplasia. A homozygous missense mutation R272Q in ICK was identified as the causative mutation for ECO. ICK, named after its cloning origin the intestine, is a misnomer because it is highly conserved and ubiquitously expressed in human tissues, which may explain why the R272Q mutation in ICK causes developmental defects in multiple organs. Previously we have established that the R272A mutation in ICK impairs the nuclear targeting and the catalytic activity of ICK; this result was confirmed with the R272Q mutation as well. Many of the malformations observed in ECO involve a defect in apoptosis, especially the cleft lip and palate, syndactyly, prolonged persistence of fusion of the eyelids, and unfused urogenital folds. ECO-affected infants also develop phenotypes observed with Scythe deficiency, including hydrocephalus, dilated and hypoplastic kidneys. Given the role of scythe in apoptosis, these observations support the hypothesis that ICK targets scythe to regulate apoptosis during mammalian organ development.

2.6 Role of ICK in the intestinal epithelium

A delicate balance of cell renewal, differentiation and cell death is crucial to maintain the gastrointestinal tissue architecture that forms the basis for the normal function of the gut. An early study using in situ hybridization showed that the expression of ICK mRNA was localized specifically to the crypt compartment of the small intestine (Togawa, Yan et al. 2000). The crypt is the compartment of the intestinal epithelium where stem cells, progenitor cells and rapidly replicating transit-amplifying cells reside raising the hypothesis that ICK may play a role in epithelial replication, lineage specification and cell fate determination in crypt epithelium. Using a conditional ICK knockout mouse model, we are currently addressing whether intestine-specific ablation of the ICK gene affects cell proliferation, differentiation, migration and lineage allocation in the intestinal epithelium during normal development and homeostasis, and whether the ICK expression is important for the expansion and proliferation of the intestinal stem cell population and their progenitors in the restoration of the normal epithelial architecture after mucosal injury.

Recently, we reported that suppression of ICK expression in cultured colorectal carcinoma and intestinal epithelial cell lines by short hairpin RNA (shRNA) interference significantly impaired cellular proliferation and induced features of gene expression characteristic of colonic or enterocytic differentiation (Fu, Kim et al. 2009). Downregulation of ICK altered expression of cell cycle regulators (cyclin D1, c-Myc, and $p21^{Cip1/WAF1}$) of G1-S transition, consistent with the G1 cell cycle delay induced by ICK shRNA. ICK deficiency also led to a significant decrease in the expression and/or activity of S6K1, indicating that disrupting ICK function downregulates the mTORC1 signaling pathway. Our prior studies also provided biochemical evidence that ICK interacts with the mTOR/Raptor complex in cells and Raptor is an *in vitro* substrate for ICK (Fu, Kim et al. 2009). Recently, we investigated whether and how ICK targets Raptor to regulate the activity of mTORC1. Our results indicate that ICK is able to promote mTORC1 activation through phosphorylation of Raptor Thr-908 (Wu D et al., J Biol Chem, in press).

2.7 Role of ICK in cardiac development and hypertrophy

Abe, S and colleagues (Abe, Yagi et al. 1995) observed the ICK/MRK protein signals in the cytosol of cardiomyocytes at day 11 rat embryos, and the ICK/MRK immunohistological staining appeared to be weaker and in a patchy, speckled pattern in adult rat hearts, suggesting downregulation of the ICK/MRK signals during cardiac development. Furthermore, the intensity of the ICK/MRK staining and the number of ICK positive cardiomyocytes were both increased in hypertrophic hearts with experimentally induced stenosis of the abdominal aorta, implicating that the ICK/MRK expression is inducible by external stress such as pressure overload.

3. MAK signaling cascade

In 1990, MAK was first isolated from a human genomic DNA library by using weak cross-hybridization with a tyrosine kinase gene (v-*ros*) in Professor Masabumi Shibuya's laboratory. This gene was designated as MAK (male germ cell-associated kinase) because it is highly expressed in testicular germ cells. In contrast to the ubiquitous expression pattern of ICK, MAK expression is more restricted. MAK mRNAs are enriched in testis and expressed in male germ cells during and after meiosis (Matsushime, Jinno et al. 1990). MAK expression was also detected in prostate and retina. MAK was identified as an androgen-inducible gene in LNCaP prostate epithelial cells and as a co-activator of androgen receptor signaling in prostate cancer (Ma, Xia et al. 2006). Recently, MAK expression was detected in cilia of the retina where it was suggested to be involved in photoreceptor cell survival (Omori, Chaya et al. 2010).

3.1 cell cycle-dependent localization and regulation in the TDY motif

The subcellular localization of MAK is dynamic during cell cycle (Wang and Kung 2011). MAK displays uniform localization in the nucleus during interphase, and associates with mitotic spindles and centrosomes at metaphase and anaphase. This dynamic nuclear localization of MAK is associated with its cell cycle-related role (see 3.4). Similar to ICK, MAK also requires an intact and dually phosphorylated TDY motif for full activation. CCRK, but not MEK, is the upstream activating kinase for MAK in the TDY motif (Wang and Kung 2011). More interestingly, although the expression level of MAK remained constant, the TDY-dual phosphorylation level oscillated during cell cycle (Wang and Kung

2011). It increased at S phase, peaked at G2 to early M phase, and decreased at late M phase. It is not clear, however, whether this oscillation of the TDY-dual phosphorylation of MAK during cell cycle is associated with the expression and/or activity levels of CCRK. The high level TDY-phosphorylation of MAK at G2/M does provide the molecular basis for an important role of MAK during the metaphase-anaphase transition (see 3.4).

3.2 MAK in testis and spermatogenesis

Northern blot analysis revealed two discrete transcripts (2.6 and 3.8 kb) of the *mak* gene that are mainly expressed in germ cells at and/or after the pachytene stage (Matsushime, Jinno et al. 1990). Since these two *mak* transcripts display differential temporal expression patterns during spermatogenesis, it was speculated that they may have distinct physiological functions in germ cells differentiation. Subsequent studies from Professor Shibuya's lab identified two MAK protein products that are mainly localized in the cytoplasm and a phosphorylated 210-KDa protein as a candidate physiological substrate for MAK. The true identify of this MAK-associated 210-KDa protein still remains a mystery.

In 2002, phenotypic analysis of the MAK knockout mouse was reported by Yoichi Shinkai and colleagues. Overall, MAK-deficient mice developed normally with no gross abnormalities (Shinkai, Satoh et al. 2002). Surprisingly, most of the MAK null mice were fertile, suggesting no major defects in spermatogenesis in the absence of MAK gene in mice. The only mild phenotype in MAK-deficient male mice is reduced sperm motility. These data suggest that MAK is not essential for spermatogenesis and male fertility, raising the possibility that ICK may compensate for the role of MAK in spermatogenesis. Our studies indicate that both ICK and MAK, proteins are abundantly expressed in mouse testis, and in primary spermatocytes and sertoli cell lines (Fu Z et al., unpublished data), providing further molecular basis for the speculation that ICK and MAK may have redundant biological functions in testis. Knockout of both MAK and ICK genes in mice will be required to test this hypothesis.

3.3 MAK in retina, ciliogenesis and retinitis pigmentosa

A potential role of MAK and ICK in regulating cilia structure and functions has long been speculated based on the observations that loss of functions of their homologs in *Chlamydomonas reinhardtii* (*LF4p*) (Berman, Wilson et al. 2003), in *Caenorhabditis elegans* (*Dyf-5*) (Burghoorn, Dekkers et al. 2007), and in *Leishmania Mexicana* (*LmxMPK9*) (Bengs, Scholz et al. 2005) causes elongated cilia or flagella to various degrees. During a microarray screening to identify photoreceptor cell-specific genes involved in the conversion of photoreceptors to amacrine-like cells, Takahisa Furukawa and colleagues found a retina-specific isoform of MAK cDNA containing a 75-bp in-frame insertion to the originally reported form of MAK in testis. Using the same MAK-null mice that display no major phenotype in spermatogenesis, Furukawa's lab investigated whether MAK has any important functions in retina where MAK is predominantly expressed in photoreceptor cells and localized in the photoreceptor connecting cilia and outer segment axonemes. The MAK-null retina appears to be normal until the completion of retinogenesis at postnatal day 14, suggesting that MAK gene is not essential for cell fate determination in retina. However, photoreceptor cells in the MAK-null retina do exhibit progressive degeneration associated with two major hallmarks: elongated cilia and aberrant outer-segment disk formation. The role of MAK in regulating cilia length was also confirmed in serum starved NIH 3T3 cells where MAK is mainly localized in the nuclei and in the cilia as well. Both the kinase activity and the C-terminal region of MAK are essential for the regulation of the cilia length. However, only the C-terminal noncatalytic domain, but not the kinase activity, of MAK is required for its ciliary localization.

Retinitis pigmentosa 1 (RP1) was implicated as a candidate physiological substrate for MAK in regulating ciliary structure and organization in that RP1 induces ciliary elongation and reduces the effect of MAK overexpression, and furthermore MAK physically interacts with RP1 and directly phosphorylates RP1 *in vitro* (Omori, Chaya et al. 2010). Scythe may be another candidate substrate for MAK related to its role in RP for the following reasons. Scythe was identified from a yeast two-hybrid screen using ICK kinase domain as the bait and was confirmed to be an *in vitro* and *in vivo* substrate for ICK (Fu, Larson et al. 2006). ICK and MAK are essentially identical in the kinase domain, therefore Scythe maybe a common substrate for both. RP is characterized by apoptotic death of photoreceptor cells and scythe is known to be important for regulating apoptosis and is abundantly expressed in retina.

MAK was found in two cell types involved in sensory transduction, photoreceptors and olfactory receptors as well as epithelial of the respiratory tract and choroid plexus (Bladt and Birchmeier 1993). Interestingly, Furukawa and his colleagues also noted the reduced MAK expression in respiratory epithelia of the nasal cavity and in epididymal sperm cells of the testis, yet the ciliary length of neither cell types differ in wild-type and MAK-KO mice (Omori, Chaya et al. 2010). One possible explaination for this retina-specific phenotype in MAK-KO mice is the tissue-specific functional redundancy/compensation. Our studies indicate that both ICK and MAK are expressed in respiratory systems and testis, but very little of ICK or MOK protein is detected in retina where MAK is highly expressed (Fu Z et al., unpublished data), consistent with the notion that functional redundancy may exist between MAK, ICK and MOK in testis and lung but not in retina.

The essential role of MAK in supporting the biological functions of retina was further substantiated recently by similar findings from two independent studies (Ozgul, Siemiatkowska et al. 2011; Tucker, Scheetz et al. 2011). Exome sequencing and cis-regulatory mapping identified six missense point mutations of highly conserved residues within the catalytic domain of MAK as a cause of retinitis pigmentosa (RP) (Fig. 1). Also by exome sequencing, a 353-bp Alu repeat insertion was found to disrupt the correct splicing of exon 9 of *Mak* gene, thereby preventing mature retinal cells from expressing the correct MAK isoform in retina. In either case, a lack of active form of MAK gene product in retina was implicated as a cause of RP. Interestingly, as pointed out in (Ozgul, Siemiatkowska et al. 2011), all of the MAK mutations identified in the retinal isoform should also be present in testis isoform, yet there were no reports of infertility in the identified male objects, consistent with the phenotype of the *Mak* knockout mice. This finding again raises the possibility that ICK may be able to compensate for the lack of MAK in testis where both genes are abundantly expressed.

3.4 MAK in prostate cancer, AR signaling and mitosis

In contrast to ICK, expression of its closely related kinase MAK is more restricted. However, in addition to testis and retina, MAK expression was also detected in prostate (Xia,

Robinson et al. 2002). MAK was identified as an androgen-inducible co-activator of androgen receptor (AR) in prostate cancer cells (Ma, Xia et al. 2006). Similar to the role of ICK in intestinal epithelial cells, MAK is also required for prostate epithelial cell replication (Ma, Xia et al. 2006). A recent study from Kung's lab indicated that over-expression of MAK in prostate cancer cells caused mitotic defects that are independent of AR signaling but are associated with deregulation of the APC/C(CDH1) (Wang and Kung 2011).

Unpublished data from Kung's lab and our own data have suggested that AR is not the substrate for MAK *in vitro* or *in vivo*. Given the tight physical association of MAK with AR complex in cells and the significant biological effect of MAK knockdown on AR signaling events, it is quite possible that an AR-associated protein within the AR complex serves as the direct target of MAK in regulating the AR signaling.

Recently, Kung's lab demonstrated that MAK is over-expressed in prostate cancer cells and causes mitotic defects such as centrosome amplification and lagging chromosomes via deregulation of APC/C^{CDH1}, thus providing an AR-independent mechanism to promote prostate cancer development (Wang and Kung 2011). This report also indicated that CDH1 is an *in vitro* substrate for MAK and MAK can negatively regulate APC/C^{CDH1} through phosphorylation of CDH1 at the same CDK-dependent sites. It requires further studies to determine whether CDH1 is a true physiologic substrate for MAK and how MAK and CDK coordinate to target the same sites to regulate APC/C^{CDH1} activities.

4. MOK signaling cascade

MOK was identified through *in silico* computer screening of the GENBANK EST database using MAP kinase consensus sequences as probes (Miyata, Akashi et al. 1999). MOK encodes a protein of 419 (human) and 420 (mouse) amino acids, containing the conserved kinase subdomains I-XI and the TEY motif in the activation loop. Structurally, MOK belongs to the MAP kinase superfamily, and is closely related to MAK and ICK/MRK, thus termed as MOK (MAPK/MAK/MRK overlapping kinase). Although MOK shares highest homologies to MAK and ICK, especially in the catalytic domains (41-43% identity), it also displays certain structural features that are distinct from MAK and ICK. MAK and ICK have the TDY motif in the activation loop, as compared with the TEY motif for MOK. Similar to Cdk2, MAK and ICK possess $T^{14}Y^{15}$ motif in the N-terminal end of their catalytic domains, while MOK exhibits $T^{14}F^{15}$ motif instead in the same position. More significantly, the C-terminal noncatalytic domain of MOK is much shorter than that of either MAK or ICK (Fig. 1) and show very little sequence homology to MAK and ICK or any other known protein kinases.

Interestingly, MOK was also isolated from a blast search to analyze homologies to the RAGE (renal cell carcinoma antigen)-gene family that encodes antigens of human renal carcinoma cells recognized by autologous cytolytic T lymphocytes (Eichmuller, Usener et al. 2002). Sequence alignment indicates MOK is identical to RAGE-1, -2, -3 at the 3'-region, but is completely different at the 5'-region, suggesting that MOK and RAGE genes may be the splicing products from the same gene or MOK may be aberrantly inserted into RAGE genes by translocation. What is the molecular genetic basis for this observation and whether MOK is involved in RAGE-gene family associated tumorigenesis are interesting questions that will motivate further studies.

4.1 Regulation of MOK activity in the TEY motif and by TPA

MOK possesses protein kinase activity towards exogenous substrates for MAPK such as c-Jun, MBP, cyclin B1 and c-Myc and undergoes autophosphorylation (Miyata, Akashi et al. 1999). Similar to ICK and MAK, an intact TEY motif in the activation loop of MOK is essential for its kinase activity (Miyata, Akashi et al. 1999). The autokinase activity of MOK was almost completely abolished when the TEY motif was mutated to AEF, although it is not clear whether the Thr and/or the Tyr in the TEY motif are the phosphor-acceptor sites by autophosphorylation.

Similar to ICK and MAK, MOK could not be significantly activated by many extracellular stimuli (serum, anisomycin, and hyperosmotic shock) that stimulate MAP kinases (Miyata, Akashi et al. 1999). TPA (phorbol 12-myristate 13-acetate), however, at a concentration of 100 ng/ml, was able to stimulate the MOK activity up to about threefold, albeit at a much longer time point after treatment (15-20 min) than that required to maximally activate ERK1/2 and p38 MAPK (5 min) (Miyata, Akashi et al. 1999). These results suggest that the activation mechanism of MOK by TPA is different from that of MAP kinases. The identity of the upstream activating kinase for MOK remains unknown.

4.2 Regulation of MOK activity and/or functional specificity by subcellular localization

Recombinant MOK is predominantly cytoplasmic (Miyata, Akashi et al. 1999). In our subcellular fractionation studies, we observed both cytoplasmic and nuclear localization of endogenous MOK in human cell lines, suggesting that MOK may shuttle between these two compartments (Fu, Z et al., unpublished data). The molecular size (48 KDa) of MOK allows its nuclear entry by diffusion. In addition, there is a consensus bipartite NLS located in its C-terminal domain (Fig. 1), possibly permitting selective nuclear targeting as well. Further studies are required to elucidate the molecular mechanisms underlying its subcellular distribution and to address whether MOK targets different subsets of substrates in different locations and thus performs distinct biological functions.

4.3 Regulation of MOK stability by chaperone proteins

Work from Professor Eisuke Nishida's lab suggested specific association of MOK with a set of molecular chaperones including HSP90, HSP70 and Cdc37 (Miyata, Ikawa et al. 2001). Inhibition of HSP90 chaperone activity caused rapid degradation of MOK through proteasome-dependent pathways, suggesting that chaperone association is required to stabilize MOK. Interestingly, in the same study both MAK and ICK/MRK were also reported to associate with HSP90, although the biological effects of HSP90 association with MAK and ICK/MRK were not examined. In our unpublished studies, we also observed robust association of chaperones (HSP70 and Cdc37) with ICK and its upstream activating kinase CCRK. These results taken together suggest that ICK/MAK/MOK may require the presence of chaperone proteins in the same protein complexes for assistance in folding and stabilization, a biochemical property that is strikingly different from classic MAP kinases that do not specifically associate with chaperones (Miyata, Ikawa et al. 2001).

4.4 MOK in testis and germ cell development

MOK (T/STK30, testis-derived serine/threonine kinase 30) was isolated from adult testis using a PCR-based strategy to identify novel protein kinases expressed in germ cells

(Gopalan, Centanni et al. 1999). T/STK30 transcripts are most abundantly expressed in testis and ovary, and were not detected in a sterile mutant testis that lacks germ cells, further demonstrating that T/STK30 expression in testis is restricted to germ cells. By in situ hybridization, Donovan P.J. and colleagues further demonstrated that T/STK30 transcripts were detected in pachytene spermatocytes and round spermatids, but not in spermatogonia or testicular somatic cells. Similarly, T/STK30 is highly expressed in female germ cells, but not in the surrounding somatic cells that are mostly proliferating. These results, taken together, suggest T/STK30 is involved in some aspects of germ cell differentiation and maturation. Functional studies (targeted gene disruption and/or over-expression) will be required to address the role of MOK (T/STK30) in mammalian gametogenesis.

4.5 MOK in intestinal cell differentiation

Although MOK is closely related to ICK in the N-terminal catalytic domain, they differ significantly in the structural organization of the C-terminal non-catalytic domain. While MOK mRNA appears to be restricted to the crypt compartment of the small intestine, MOK protein was detected in the upper crypt and lower villus epithelial cells (Uesaka and Kageyama 2004). In HT-29 cells, MOK activity was reported to be elevated by sodium butyrate, which is known to inhibit growth and induce differentiation of HT-29 cells (Uesaka and Kageyama 2004). This observation suggests a possible role for MOK in the regulation of intestinal epithelial differentiation. Does MOK play an important role in the induction of growth arrest and differentiation in the intestinal epithelium, which could be directly opposite to the role of ICK? What are the upstream modulators and the downstream physiological substrates of MOK? Does MOK crosstalk with ICK during the regulation of gastrointestinal proliferation and differentiation? If so, what is the molecular basis for this interaction? These questions about MOK and its relationship to ICK remain to be clarified.

5. Conclusion and significant questions

The ICK/MAK/MOK family shares significant sequence and structural homology to MAPKs and CDKs. They all contain a TXY motif in the activation loop that is required for full activity. ICK and MAK also possess a CDK-like TY motif, but so far it is unknown whether these sites are phosphorylated *in vivo* and be able to regulate the catalytic activity. The regulatory mechanisms of ICK/MAK/MOK appear to be very different from that of MAPKs. Unlike classic MAPKs, they are not acutely activated by growth factors or stress. The upstream activating kinase for ICK and MAK is CCRK, not MEK or CDK7 complex. Both ICK and MAK have a long C-terminal non-catalytic domain with postulated functions in protein-protein and protein-DNA interactions, enabling them to operate through signaling pathways distinct from that of classic MAPKs.

Emerging evidence strongly suggest that this group of kinases have important biological functions in mammalian development and human diseases. They are involved in regulating many fundamental biological processes including cell proliferation, differentiation, apoptosis and cell cycle. Yet the molecular basis underlying these regulations is still largely elusive.

Even though tremendous progress has been made to elucidate the regulations and functions of ICK/MAK/MOK since their discoveries some 20 years ago, many significant questions, such as a few named below, are yet to be answered in order to fully understand the biology of this kinase family in mammalian development and human diseases.

Q1: What is the molecular basis to determine the substrate and/or signaling specificity for ICK/MAK/MOK? This information may be stored in their C-terminal non-catalytic domains since they share extensive homology in their N-terminal catalytic domains. The recent data showing that the C-terminal domain of MAK is essential for its ciliary localization and function in regulating ciliary length provided further support to this notion.

Q2: What are the upstream stimuli or environmental cues that activate ICK/MAK/MOK? Are the expression and activity levels of ICK/MAK/MOK regulated during development or stress?

Q3: Do ICK/MAK/MOK have different subsets of physiologic substrates in different subcellular compartments and thus regulate distinct biological processes?

Q4: Is there a functional redundancy between ICK and MAK in testis? More specifically, does ICK compensate for the lack of MAK in spermatogenesis in MAK KO mice? Do MAK and ICK signal through scythe or different substrates to regulate germ cell development during spermatogenesis?

Q5: Does MOK have an opposite function to that of ICK or MAK in the intestine and testis given the existing evidence seem to indicate that ICK and MAK are pro-proliferation and MOK is pro-differentiation?

Q6: How the ICK, MAK and MOK genes are regulated at the transcription level in a specific biological context during development and diseases? Currently very little is known on this topic. ICK and FBX9 are divergently transcribed from a bi-directional promoter that contains functional sites for β-catenin/TCF7L2 and FOXA (Sturgill, Stoddard et al. 2010). MAK transcripts can be down-regulated by retinol during spermatogonial proliferation phase of spermatogenesis (Wang and Kim 1993). Cdx2, a caudal-related homeobox transcription factor, interacts with the MOK promoter and induces expression of MOK transcripts (Uesaka and Kageyama 2004).

6. Acknowledgement

I am indebted to Professor Thomas W. Sturgill (University of Virginia) for introducing me to the world of ICK/MAK/MOK. Fu, Z is currently supported by the NIH grant DK082614.

7. References

Abe, S., T. Yagi, et al. (1995). "Molecular cloning of a novel serine/threonine kinase, MRK, possibly involved in cardiac development." *Oncogene* 11(11): 2187-2195.

Ali, A., J. Zhang, et al. (2004). "Requirement of protein phosphatase 5 in DNA-damage-induced ATM activation." *Genes Dev* 18(3): 249-254.

An, X., S. S. Ng, et al. (2010). "Functional characterisation of cell cycle-related kinase (CCRK) in colorectal cancer carcinogenesis." *Eur J Cancer* 46(9): 1752-1761.

Bengs, F., A. Scholz, et al. (2005). "LmxMPK9, a mitogen-activated protein kinase homologue affects flagellar length in Leishmania mexicana." *Mol Microbiol* 55(5): 1606-1615.

Berman, S. A., N. F. Wilson, et al. (2003). "A novel MAP kinase regulates flagellar length in Chlamydomonas." *Curr Biol* 13(13): 1145-1149.

Bhaskar, P. T. and N. Hay (2007). "The two TORCs and Akt." *Dev Cell* 12(4): 487-502.

Bladt, F. and C. Birchmeier (1993). "Characterization and expression analysis of the murine rck gene: a protein kinase with a potential function in sensory cells." *Differentiation* 53(2): 115-122.

Burghoorn, J., M. P. Dekkers, et al. (2007). "Mutation of the MAP kinase DYF-5 affects docking and undocking of kinesin-2 motors and reduces their speed in the cilia of Caenorhabditis elegans." *Proc Natl Acad Sci U S A* 104(17): 7157-7162.

Carriere, A., M. Cargnello, et al. (2008). "Oncogenic MAPK signaling stimulates mTORC1 activity by promoting RSK-mediated raptor phosphorylation." *Curr Biol* 18(17): 1269-1277.

Carriere, A., Y. Romeo, et al. (2011). "ERK1/2 phosphorylate Raptor to promote Ras-dependent activation of mTOR complex 1 (mTORC1)." *J Biol Chem* 286(1): 567-577.

Clifford, D. M., K. E. Stark, et al. (2005). "Mechanistic insight into the Cdc28-related protein kinase Ime2 through analysis of replication protein A phosphorylation." *Cell Cycle* 4(12): 1826-1833.

Desmots, F., H. R. Russell, et al. (2005). "The reaper-binding protein scythe modulates apoptosis and proliferation during mammalian development." *Mol Cell Biol* 25(23): 10329-10337.

Eichmuller, S., D. Usener, et al. (2002). "mRNA expression of tumor-associated antigens in melanoma tissues and cell lines." *Exp Dermatol* 11(4): 292-301.

Feng, H., A. S. Cheng, et al. (2011). "Cell cycle-related kinase is a direct androgen receptor-regulated gene that drives beta-catenin/T cell factor-dependent hepatocarcinogenesis." *J Clin Invest* 121(8): 3159-3175.

Fingar, D. C., C. J. Richardson, et al. (2004). "mTOR controls cell cycle progression through its cell growth effectors S6K1 and 4E-BP1/eukaryotic translation initiation factor 4E." *Mol Cell Biol* 24(1): 200-216.

Foiani, M., E. Nadjar-Boger, et al. (1996). "A meiosis-specific protein kinase, Ime2, is required for the correct timing of DNA replication and for spore formation in yeast meiosis." *Mol Gen Genet* 253(3): 278-288.

Fonseca, B. D., E. M. Smith, et al. (2007). "PRAS40 is a target for mammalian target of rapamycin complex 1 and is required for signaling downstream of this complex." *J Biol Chem* 282(34): 24514-24524.

Foster, K. G., H. A. Acosta-Jaquez, et al. (2010). "Regulation of mTOR complex 1 (mTORC1) by raptor Ser863 and multisite phosphorylation." *J Biol Chem* 285(1): 80-94.

Fu, Z., J. Kim, et al. (2009). "Intestinal cell kinase, a MAP kinase-related kinase, regulates proliferation and G1 cell cycle progression of intestinal epithelial cells." *Am J Physiol Gastrointest Liver Physiol* 297(4): G632-640.

Fu, Z., K. A. Larson, et al. (2006). "Identification of yin-yang regulators and a phosphorylation consensus for male germ cell-associated kinase (MAK)-related kinase." *Mol Cell Biol* 26(22): 8639-8654.

Fu, Z., M. J. Schroeder, et al. (2005). "Activation of a nuclear Cdc2-related kinase within a mitogen-activated protein kinase-like TDY motif by autophosphorylation and cyclin-dependent protein kinase-activating kinase." *Mol Cell Biol* 25(14): 6047-6064.

Golden, T., M. Swingle, et al. (2008). "The role of serine/threonine protein phosphatase type 5 (PP5) in the regulation of stress-induced signaling networks and cancer." *Cancer Metastasis Rev* 27(2): 169-178.

Gopalan, G., J. M. Centanni, et al. (1999). "A novel mammalian kinase, T/STK 30, is highly expressed in the germ line." *Molecular Reproduction and Development* 52(1): 9-17.

Gwinn, D. M., D. B. Shackelford, et al. (2008). "AMPK phosphorylation of raptor mediates a metabolic checkpoint." *Mol Cell* 30(2): 214-226.

Hall, M. N. (2008). "mTOR-what does it do?" *Transplant Proc* 40(10 Suppl): S5-8.

Hanrahan, J. and J. Blenis (2006). "Rheb activation of mTOR and S6K1 signaling." *Methods Enzymol* 407: 542-555.

Hara, K., K. Yonezawa, et al. (1997). "Regulation of eIF-4E BP1 phosphorylation by mTOR." *J Biol Chem* 272(42): 26457-26463.

Hinds, T. D., Jr. and E. R. Sanchez (2008). "Protein phosphatase 5." *Int J Biochem Cell Biol* 40(11): 2358-2362.

Ivanov, I., K. C. Lo, et al. (2007). "Identifying candidate colon cancer tumor suppressor genes using inhibition of nonsense-mediated mRNA decay in colon cancer cells." *Oncogene* 26(20): 2873-2884.

Kaldis, P. and M. J. Solomon (2000). "Analysis of CAK activities from human cells." *Eur J Biochem* 267(13): 4213-4221.

Kim, D. H. and D. M. Sabatini (2004). "Raptor and mTOR: subunits of a nutrient-sensitive complex." *Curr Top Microbiol Immunol* 279: 259-270.

Kim, D. H., D. D. Sarbassov, et al. (2002). "mTOR interacts with raptor to form a nutrient-sensitive complex that signals to the cell growth machinery." *Cell* 110(2): 163-175.

Kim, D. H., D. D. Sarbassov, et al. (2003). "GbetaL, a positive regulator of the rapamycin-sensitive pathway required for the nutrient-sensitive interaction between raptor and mTOR." *Mol Cell* 11(4): 895-904.

Kim, E., P. Goraksha-Hicks, et al. (2008). "Regulation of TORC1 by Rag GTPases in nutrient response." *Nat Cell Biol* 10(8): 935-945.

Ko, H. W., R. X. Norman, et al. (2010). "Broad-minded links cell cycle-related kinase to cilia assembly and hedgehog signal transduction." *Dev Cell* 18(2): 237-247.

Lahiry, P., A. Torkamani, et al. (2010). "Kinase mutations in human disease: interpreting genotype-phenotype relationships." *Nat Rev Genet* 11(1): 60-74.

Lahiry, P., J. Wang, et al. (2009). "A multiplex human syndrome implicates a key role for intestinal cell kinase in development of central nervous, skeletal, and endocrine systems." *Am J Hum Genet* 84(2): 134-147.

Laplante, M. and D. M. Sabatini (2009). "mTOR signaling at a glance." *J Cell Sci* 122(Pt 20): 3589-3594.

Liu, Y., C. Wu, et al. (2004). "p42, a novel cyclin-dependent kinase-activating kinase in mammalian cells." *J Biol Chem* 279(6): 4507-4514.

Loewith, R., E. Jacinto, et al. (2002). "Two TOR complexes, only one of which is rapamycin sensitive, have distinct roles in cell growth control." *Mol Cell* 10(3): 457-468.

Ma, A. H., L. Xia, et al. (2006). "Male germ cell-associated kinase, a male-specific kinase regulated by androgen, is a coactivator of androgen receptor in prostate cancer cells." *Cancer Res* 66(17): 8439-8447.

Ma, X. M. and J. Blenis (2009). "Molecular mechanisms of mTOR-mediated translational control." *Nat Rev Mol Cell Biol* 10(5): 307-318.

MacKeigan, J. P., L. O. Murphy, et al. (2005). "Sensitized RNAi screen of human kinases and phosphatases identifies new regulators of apoptosis and chemoresistance." *Nat Cell Biol* 7(6): 591-600.

Mariappan, M., X. Li, et al. (2010). "A ribosome-associating factor chaperones tail-anchored membrane proteins." *Nature* 466(7310): 1120-1124.

Matsushime, H., A. Jinno, et al. (1990). "A novel mammalian protein kinase gene (mak) is highly expressed in testicular germ cells at and after meiosis." *Mol Cell Biol* 10(5): 2261-2268.

Minami, R., A. Hayakawa, et al. (2010). "BAG-6 is essential for selective elimination of defective proteasomal substrates." *J Cell Biol* 190(4): 637-650.

Miyata, Y., M. Akashi, et al. (1999). "Molecular cloning and characterization of a novel member of the MAP kinase superfamily." *Genes Cells* 4(5): 299-309.

Miyata, Y., Y. Ikawa, et al. (2001). "Specific association of a set of molecular chaperones including HSP90 and Cdc37 with MOK, a member of the mitogen-activated protein kinase superfamily." *J Biol Chem* 276(24): 21841-21848.

Morita, K., M. Saitoh, et al. (2001). "Negative feedback regulation of ASK1 by protein phosphatase 5 (PP5) in response to oxidative stress." *EMBO J* 20(21): 6028-6036.

Nascimento, E. B. and D. M. Ouwens (2009). "PRAS40: target or modulator of mTORC1 signalling and insulin action?" *Arch Physiol Biochem* 115(4): 163-175.

Ng, S. S., Y. T. Cheung, et al. (2007). "Cell cycle-related kinase: a novel candidate oncogene in human glioblastoma." *J Natl Cancer Inst* 99(12): 936-948.

Nojima, H., C. Tokunaga, et al. (2003). "The mammalian target of rapamycin (mTOR) partner, raptor, binds the mTOR substrates p70 S6 kinase and 4E-BP1 through their TOR signaling (TOS) motif." *J Biol Chem* 278(18): 15461-15464.

Omori, Y., T. Chaya, et al. (2010). "Negative regulation of ciliary length by ciliary male germ cell-associated kinase (Mak) is required for retinal photoreceptor survival." *Proc Natl Acad Sci U S A* 107(52): 22671-22676.

Ozgul, R. K., A. M. Siemiatkowska, et al. (2011). "Exome sequencing and cis-regulatory mapping identify mutations in MAK, a gene encoding a regulator of ciliary length, as a cause of retinitis pigmentosa." *Am J Hum Genet* 89(2): 253-264.

Peterson, T. R., M. Laplante, et al. (2009). "DEPTOR is an mTOR inhibitor frequently overexpressed in multiple myeloma cells and required for their survival." *Cell* 137(5): 873-886.

Proud, C. G. (2004). "mTOR-mediated regulation of translation factors by amino acids." *Biochem Biophys Res Commun* 313(2): 429-436.

Qiu, H., H. Dai, et al. (2008). "Characterization of a novel cardiac isoform of the cell cycle-related kinase that is regulated during heart failure." *J Biol Chem* 283(32): 22157-22165.

Sancak, Y., T. R. Peterson, et al. (2008). "The Rag GTPases bind raptor and mediate amino acid signaling to mTORC1." *Science* 320(5882): 1496-1501.

Sancak, Y., C. C. Thoreen, et al. (2007). "PRAS40 is an insulin-regulated inhibitor of the mTORC1 protein kinase." *Mol Cell* 25(6): 903-915.

Sasaki, T., E. C. Gan, et al. (2007). "HLA-B-associated transcript 3 (Bat3)/Scythe is essential for p300-mediated acetylation of p53." *Genes Dev* 21(7): 848-861.

Shinkai, Y., H. Satoh, et al. (2002). "A testicular germ cell-associated serine-threonine kinase, MAK, is dispensable for sperm formation." *Mol Cell Biol* 22(10): 3276-3280.

Sturgill, T. W., P. B. Stoddard, et al. (2010). "The promoter for intestinal cell kinase is head-to-head with F-Box 9 and contains functional sites for TCF7L2 and FOXA factors." *Mol Cancer* 9: 104.

Thress, K., E. K. Evans, et al. (1999). "Reaper-induced dissociation of a Scythe-sequestered cytochrome c-releasing activity." *EMBO J* 18(20): 5486-5493.

Thress, K., W. Henzel, et al. (1998). "Scythe: a novel reaper-binding apoptotic regulator." *EMBO J* 17(21): 6135-6143.

Togawa, K., Y. X. Yan, et al. (2000). "Intestinal cell kinase (ICK) localizes to the crypt region and requires a dual phosphorylation site found in map kinases." *J Cell Physiol* 183(1): 129-139.

Tucker, B. A., T. E. Scheetz, et al. (2011). "Exome sequencing and analysis of induced pluripotent stem cells identify the cilia-related gene male germ cell-associated kinase (MAK) as a cause of retinitis pigmentosa." *Proc Natl Acad Sci U S A* 108(34): E569-576.

Uesaka, T. and N. Kageyama (2004). "Cdx2 homeodomain protein regulates the expression of MOK, a member of the mitogen-activated protein kinase superfamily, in the intestinal epithelial cells." *FEBS Lett* 573(1-3): 147-154.

Vander Haar, E., S. I. Lee, et al. (2007). "Insulin signalling to mTOR mediated by the Akt/PKB substrate PRAS40." *Nat Cell Biol* 9(3): 316-323.

von Kriegsheim, A., A. Pitt, et al. (2006). "Regulation of the Raf-MEK-ERK pathway by protein phosphatase 5." *Nat Cell Biol* 8(9): 1011-1016.

Wang, L., T. E. Harris, et al. (2008). "Regulation of proline-rich Akt substrate of 40 kDa (PRAS40) function by mammalian target of rapamycin complex 1 (mTORC1)-mediated phosphorylation." *J Biol Chem* 283(23): 15619-15627.

Wang, L., J. C. Lawrence, Jr., et al. (2009). "Mammalian target of rapamycin complex 1 (mTORC1) activity is associated with phosphorylation of raptor by mTOR." *J Biol Chem* 284(22): 14693-14697.

Wang, L. Y. and H. J. Kung (2011). "Male germ cell-associated kinase is overexpressed in prostate cancer cells and causes mitotic defects via deregulation of APC/C(CDH1)." *Oncogene.*

Wang, R. and C. C. Liew (1994). "The human BAT3 ortholog in rodents is predominantly and developmentally expressed in testis." *Mol Cell Biochem* 136(1): 49-57.

Wang, Z. Q. and K. H. Kim (1993). "Retinol differentially regulates male germ cell-associated kinase (mak) messenger ribonucleic acid expression during spermatogenesis." *Biol Reprod* 49(5): 951-964.

Wechsler, T., B. P. Chen, et al. (2004). "DNA-PKcs function regulated specifically by protein phosphatase 5." *Proc Natl Acad Sci U S A* 101(5): 1247-1252.

Winnefeld, M., A. Grewenig, et al. (2006). "Human SGT interacts with Bag-6/Bat-3/Scythe and cells with reduced levels of either protein display persistence of few misaligned chromosomes and mitotic arrest." *Exp Cell Res* 312(13): 2500-2514.

Wohlbold, L., S. Larochelle, et al. (2006). "The cyclin-dependent kinase (CDK) family member PNQALRE/CCRK supports cell proliferation but has no intrinsic CDK-activating kinase (CAK) activity." *Cell Cycle* 5(5): 546-554.

Xia, L., D. Robinson, et al. (2002). "Identification of human male germ cell-associated kinase, a kinase transcriptionally activated by androgen in prostate cancer cells." *J Biol Chem* 277(38): 35422-35433.

Yang, T., Y. Jiang, et al. (2002). "The identification and subcellular localization of human MRK." *Biomol Eng* 19(1): 1-4.

Zhang, J., S. Bao, et al. (2005). "Protein phosphatase 5 is required for ATR-mediated checkpoint activation." *Mol Cell Biol* 25(22): 9910-9919.

Zuo, Z., N. M. Dean, et al. (1998). "Serine/threonine protein phosphatase type 5 acts upstream of p53 to regulate the induction of p21(WAF1/Cip1) and mediate growth arrest." *J Biol Chem* 273(20): 12250-12258.

12

Protein Kinases and Protein Phosphatases as Participants in Signal Transduction of Erythrocytes

Ana Maneva[1] and Lilia Maneva-Radicheva[2]
[1]*Medical University, Pharmaceutical Faculty, Department of Chemistry and Biochemistry, Plovdiv,*
[2]*Medical University, Medical Faculty, Department of Chemistry and Biochemistry, Sofia, Bulgaria*

1. Introduction

Signal transduction is defined as the transfer of a signal starting from a primary messenger with a ligand that binds to specific receptors on the cell membrane. The signal then reaches the effector molecule(s) through a cascade involving various protein kinases and/or protein phosphatases as well as other molecules such as adaptor proteins, anchoring proteins, amplifier proteins, etc. Although erythrocytes are nucleus-free cells, they use elements of cell signal pathways that help them to maintain membrane integrity, ion transport, and metabolism. Aging, some "stress" conditions, and various diseases also may generate cell signals into erythrocytes (Minetti и Low, 1997; Antonelou et al. 2010; Berzosa et al. 2011; Pantaleo et al. 2010). Molecules from both erythrocyte membranes and from cytosol may take part in signal transduction. Membrane components involved in signal transduction include receptors, heterotrimeric G-proteins, adaptor and anchoring proteins, and some proteins of the Ras-superfamily. The erythrocyte intracellular signaling cascade is divided into protein kinases and phosphatases, second messengers and small GTPases, which could further be defined as cell surface and intracellular activities (Pasini et al. 2006). In general, the mechanisms of signal transduction in erythrocytes can be summarized in the following groups: 1) specific mechanisms triggering changes in the activity of erythrocyte enzymes, 2) specific mechanisms inducing changes in the activity of transport systems, 3) signal transduction related to membrane association/dissociation of cytoskeletal and integral proteins, 4) "stressful" conditions may induce activation of specific cellular signals into erythrocytes. Phosphorylation of the erythrocyte proteins by protein kinases and phosphatases is the key mechanism for controlling erythrocyte functions.

2. Binding of extracellular ligands to receptors on erythrocytes membrane initiates various processes of phosphorylation of erythrocyte proteins

2.1 Thyroid stimulating hormone

Thyroid stimulating hormone (TSH) binds to a specific TSH receptor (TSHR) which activates adenylate cyclase and increases cAMP levels in thyroid cells. Recent studies have reported that TSHR and Na/K-ATPase are localized on the membranes of both erythrocytes and erythrocyte ghosts. TSHR responds to TSH treatment by increasing intracellular cAMP levels from two to tenfold. The authors suggest a novel cell signalling pathway, potentially active in local circulatory control (Balzan et al. 2009).

2.2 Parathyroid hormone

Parathyroid hormone (PTH) operates with G-protein dependent receptors and has been shown to decrease erythrocyte deformability in a Ca^{2+}-dependent manner (Bogin et al. 1986). In pseudohypoparathyroidism, PTH resistance results from impairment caused by a deficiency of Gsα-signaling cascade due to a metylation defect of the GNAS gene and decrease in erythrocyte Gsα activity (Zazo et al. 2011).

2.3 Insulin

Insulin stimulates erythrocyte glycolysis, Na+/H+-antiport and Na+/K+-ATPase (Rizvi et al.1994), as well as membrane-associated NO-synthase (Bhattacharaya al. 2001). It has been shown that insulin uses signaling pathways involving both ζ-PKC and phospatidylinositol 3 kinase (PI3K), which leads to the activation of Na^+/H^+-antiports (Sauvage et al. 2000). Insulin is also used in the MAPK-signaling cascade of phosphorylation and activation of protein NHE1, responsible for Na^+/H^+ - antiport (Sartori et al. 1999).

Participation of C-peptide in insulin signaling pathways has also been discussed, including involvement of G-protein and Ca-dependent phosphatase. It has been reported that insulin restores the activity of Na+/K +-ATPase (De La Tour et al. 1998). The decrease in Na^+/K^+ adenosine triphosphatase (ATPase) in erythrocytes of type 1 diabetes is thought to play a role in the development of long-term complication. Infusion of insulin may restore this enzyme activity in red cells (Djemli-Shipkolye et al. 2000).

A reduced erythrocyte insulin receptor binding and tyrosine kinase activity was measured in hypertensive subjects with hyperinsulinemia (Corry et al. 2002). Erythrocytes from normal individuals showed increased pH and increased sodium influx (NHE1) after insulin stimulation. In contrast, insulin had no effect on NHE1 activity of eryhthrocyte from obese individuals (Kaloyianni et al. 2001). Insulin activation of insulin receptor kinase in erythrocytes is not altered in non-insulin-dependent diabetes and not influenced by hyperglycemia (Klein et al. 2000).

2.4 Insulin-like growth factor I

Erythrocytes possess receptors for insulin-like growth factor I (IGF-I). Binding to these receptors is dependent on cell age (Polychronakos et al. 1983). Acromegalic patients with higher plasma IGF-I and insulin levels presented lower IGF-I specific binding and affinity than normal adults. Growth hormone (GH)-deficient children showed higher IGF-I binding without significant affinity alterations than normal prepubertal children (el-Andere et al. 1995).

2.5 Leptin

The specific binding of leptin on erythrocytes is established by Scatchard analysis (Tsuda, 2006). NHE1 (Na^+/H^+-exchanger) activity increases in the presence of leptin but significantly less in the obese than in the control group. Since NHE1 activity is associated with insulin resistance and hypertension, the activation of this antiport by leptin may represent a link between adipose tissue hypertrophy and cardiovascular complication of obesity (Konstantinou-Tegou et al. 2001). It is possible to assume that leptin, similarly to insulin, might be able to activate NHE1 through MAPK activation (Sartori et al. 1999; Bianchini et al. 1991).

2.6 Adrenaline, Noradrenaline and DOPA

The β-2-adrenergic receptor coupled to the G-protein binds cateholamines and activates adenylate cyclase in human erythrocytes (Horga et al. 2000). The functional beta-receptor response depends to a large extent on Ca^{2+} concentrations (Horga et al. 2000). According to Muravyov et al. (2010) a crosstalk between adenylyl cyclase signaling pathway and Ca^{2+} regulatory mechanism exists. The potent beta-adrenergic agonist, isoproterenol (2 microM), epinephrine (10 microM) and norepinephrine (10 μM) stimulated the cAMP-dependent protein kinase in erythrocyte membranes, 38 +/- 7%, 31 +/- 6%, and 30 +/- 6%, respectively (Tsukamoto and Sonenberg, 1979). Micromolar concentration of noradrenaline (1 μM) increases the ^{32}P intake in band 2 with 70%, and with 40% in band 3 (Nelson et al. 1979).

Adrenaline and noradrenaline are both found to stimulate the erythrocyte Na^{+}/H^{+}-antiport (Perry et al. 1991; Paajaste and Nikinmar, 1991). Adrenaline stimulates Na^{+}/H^{+}-antiport through activation of NHE1 transport system with participation of PKC, since the effect has been reported to potentiate in the presence of phorbol ester (PKC activator), and being inhibited from calphostin (PKC inhibitor), respectively (Bourikas et al. 2003).

A specific erythrocyte DOPA transport protein was found who is also capable to transports choline. Its functions are regulated by insulin (Azoui et al. 1996).

Several catecholamines (phenylephrine, dobutamine and dopamine) inhibit the Cl-removal-activated Ca^{2+} entry into erythrocytes, thus preventing increase of cytosolic Ca^{2+} activity, subsequent cell shrinkage and activation of erythrocyte scramblase. The catecholamines thus counteract erythrocyte phosphatidylserine exposure and subsequent clearance of erythrocytes from circulating blood (Lang et al. 2005).

Exposure of RBCs to adrenaline resulted in a concentration-dependent increase in RBC filterability and authors supposed that adrenergic agonists may improve passage of erythrocytes through microvasculature (Muravyov et al. 2010). Rasmussen et al (1975) reported that the same low doses of adrenaline and isoproterenol induce a decrease of erythrocyte deformability.

2.7 Adenosine

Adenosine binds to the erythrocyte adenosine type1 receptors (A1AR) and adenosine type 2 receptors (A2AR), (Lu et al. 2004; Zhang et al. 2011). A1ARs are functionally coupled with pertussis toxin-sensitive G proteins and inhibit the activity of adenylate cyclase. A1ARs bind to erythrocyte membrane cytoskeletal protein 4.1G, which can inhibit A1-receptor action (Lu et al. 2004). A1ARs activation can also trigger the release intercellular Ca^{2+} (Lu et al. 2004). Increased adenosine levels promoted sickling, hemolysis and damage to multiple tissues in SCD transgenic mice and promoted sickling of human erythrocytes. (Zhang et al. 2011).

2.8 Prostaglandins (PG)

Binding of PGE2 to receptors coupled to the G-protein activates phospholipase C which in turn catalizes phospholipide turnover (Minetti et al. 1997) and/or stimulates a Ca^{2+}-dependent K^+ chanell in human erythrocytes and alters cell volume and filterability (Li et al. 1996). The PGE1 receptor coupled to the G-protein activates adenylate cyclase in human erythrocytes and increases erythrocyte deformality (Dutta-Roy et al. 1991). Prostacyclin (PGI) binding to human erythrocyte receptors stimulate cAMP synthesis and ATP release (Sprague et al. 2008).

2.9 Thyroid and steroid hormones

Lipophilic hormones take part in signal transduction also with "non-genomic" effects using signals starting from the plasma membrane (Falkenstein et al. 2000). Their role in erythrocyte signal transduction pathways and their activities have not been fully explored. Similar results have been reported for thyroid hormones (Angel et al.1989; Botta and Farias, 1985), and for estrogens (Gonçalves et al. 2001). Effect of thyroid hormones is probably related to calmodulin-dependent activation of erythrocyte membrane ATPase (Lawrence et al. 1993), that is inhibited in the presence of retinoic acid (Smith et al. 1989).

In vitro beta-estradiol 10^{-5} M decreased erythrocyte aggregation in blood samples of postmenopausal women undergoing hormone therapy, which could prevent high blood viscosity and, consequently, cardiovascular events (Gonçalves et al. 2001).

Results from X-ray diffraction studies revealed that cortisol and estradiol bind into the erythrocyte membrane bilayer, and exert opposite effect over Na^+/K^+-ATPase activity: cortisol diminishes its activity by 24%, but estradiol increases it by 18% (Golden et al. 1999). Other published *in vitro* studies showed that aldosterone stimulates Na^+/K^+-ATPase activity in human erythrocyte membranes (Hamlyn and Duffy, 1978). One possible explanation is that the incorporation leads to conformational changes and reorganization in the active center of the enzyme molecule of Na^+/K^+-ATPase. In addition to the delayed genomic steroid actions, increasing evidence for rapid, nongenomic steroid effects has been demonstrated for virtually all groups of steroids, and transmission by so far hypothetical specific membrane receptors is very likely. Nongenomic effects on cellular function involve conventional second messenger cascades (Falkenstein et al., 2000). Plasma selenium as well as plasma and erythrocyte gluthatione peroxidase activity increase with estrogen during the menstrual cycle. The mechanism is still unknown (Ha and Smith, 2003).

2.10 Cytokines

Biochemical evidence is provided for the presence of endothelin (ET) receptor subtype B in syckle and normal red cells. It was found that ET-1, PAF (Platelet Activating Factor), RANTES and IL-10 induce a significant increase in red cell density. These data suggest that activation of the Gardos channel is functionally coupled to receptors as C-X-C(PAF), C-C (RANTES) and ET receptors type B and the cell volume regulation or erythrocyte hydratation state might be altered by activation of the Gardos channel by cytokine in vivo (Rivera et al. 2002).

Human red cells bind specifically IL-8 (a neutrophil activating chemokine) with IL-8RA and IL-8RB receptors. Red cell absorbtion of IL-8 may function to limit stimulation of leucocytes by IL-8 released into blood (Darbonne et al. 1991). IL-8 released after acute myocardial infarction is mainly bound to erythrocytes (de Winter et al. 1997).

ICAM-4 belongs to the intercellular adhesion molecules and is an erythrocyte membrane component. ICAM-4 interacts specifically with platelet alphaIIbeta 3 integrin. RBCs are considerated passively entrapped in fibrin polymers during trombosis through ICAM-4 (Hermand et al. 2003).

2.11 Thrombin

It is assumed that thrombin induces a signal that stimulates the formation of cAMP and PGE1 via activation of Gs-dependent adenylate cyclase and Ca^{2+}-independent PKC in eryhtrocyte progenitors. Signaling pathway is inhibited by amiloride and by PKC inhibitors such as GF-109203X, Go 6976 and staurosporine (Haslauer et al. 1998). However, whether or not these signals are also valid for mature erythrocytes has not been studied yet.

2.12 Lactoferrin

Lactoferrin (Lf) is a metal-binding glycoprotein with antioxidative (Cohen et al., 1992), anti-inflammatory, immunomodulatory (Legrand et al., 2004), anticancerogenic (Thotathil and Jameson, 2007) anti-bacterial (Weinberg, 2007), antiviral (Mistry et al., 2007), antiatherogenic (Kajikawa et al. 1994), and antithrombotic properties (Levy-Toledano et al., 1995).

Our previous studies showed that Lf binding with erythrocyte membrane receptors (Taleva et al., 1999) results in stimulation of glycolysis, antioxidative protection (Maneva et al., 2003) and activation of Na^+/K^+-ATPase activity (Maneva et al., 2007). Lf–receptor interaction might intervene in short–term effects of regulation, involving processes of changes in association, phosphorylation and oxidation of the membrane proteins.

Lf (10-50 nM) decreased the ATP content from 9 to 57% depending on the concentration used. There is a negative correlation found between the concentration of added Lf and formed ATP: $y = 2.550 - 0.0076.x$, $r = - 0.993$, $p <0.001$, $n = 5$ (Figure 1). The reason for the decrease of ATP in the presence of Lf could be the processes of phosphorylation and/or ion transport activated by Lf. Lf is an activator of protein phosphorylation (Maekawa et al. 2002; Curran et al. 2006) and an activator of ion transport (Sun et al. 1991). Such a decrease in ATP content was found by other authors (Assouline-Cohen and Beitner, 1999; Boadu & Sager, 2000) due to stimulation of vital processes of erythrocytes.

Lf could stimulate glycolysis by interfering with phosphorylation processes: 1) There is evidence that insulin activates the Na^+/H^+ - antiport in erythrocytes by PI3K-dependent signaling pathway (Sauvage et al. 2000) leading to activation of glycolysis (Madshus, 1988). Like insulin, Lf also activates Na^+/H^+ - antiport (Sun et al. 1991) and stimulates the lactate formation (Maneva et al. 2003). Therefore, it could be suggested that Lf uses the same PI3K - dependent mechanism for glycolysis activation (Boivin, 1988); 2) Phosphorylation of tyrosin residues in band 3 (Boivin, 1988) leads to dissociation of the complex with glycolytic enzymes, leading to their activation (Low et al. 1993). Lf could also stimulate the glycolytic enzymes by activating Src-kinase-dependent phosphorylation of band 3. There is data showing that Lf is an activator of tyrosine phosphorylation by Src-kinases (Takayama & Mizumachi, 2001).

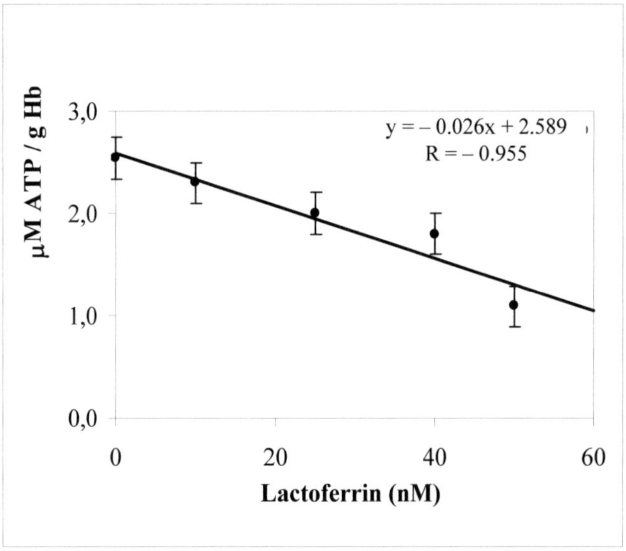

Fig. 1. Correlation and regression analysis of the relation between the concentration of Lf and content in erythrocyte ATP (unpublished data).

Sigma-Aldrich test kit has been used in this experiment. The method is based on using a reaction from oxidative phosphorylation (OP) in glycolysis where ATP is produced. OP is two–stage process involving the enzymes glyceraldehyde-3-phosphate dehydrogenase (GAPDH), and 3-phosphoglycerate kinase. The output products of OP are 3-phosphoglycerate, $NADH^+$ and ATP. The test uses GAPDH in reverse direction: a conversion of 1,3 bisphosphoglycerate into glyceraldehyde-3-phosphate, where the oxidation of $NADH^+$ to NAD^+ is directly proportional to the amounts of the ATP produced. The results are presented in µmols ATP/g.Hb.

Several possible mechanisms could be discussed in regard to how Lf binding to the receptors on the erythrocyte membrane might lead to Na^+/K^+-ATPase activation (Maneva al. 2007) via changes in levels of phosphorylation of membrane and cytosolic proteins: 1) Changes in the levels of phosphorylation of tyrosine residues in erythrocyte membrane modulates not only the activities of band 3 (Brunati et al. 2000), but also the activity of α–subunit of Na^+/K^+-ATPase (Done et al. 2002). α-subunit of Na^+/K^+-ATPase is phosphorylated by tyrosine kinases, as the target of phosphorylation is known to be the 537 tyrosne residue in the molecule. This phosphorylation is proved to be a prerequisite for the participation of Na^+/K-ATPase in signal transduction in experiments with renal cells (Done et al. 2002). Lf stimulates tyrosine phosphorylation of proteins (Takayama & Mizumachi, 2001). 2) It is known that Lf activates casein kinase 2 (CK2), and is phosphorylated by the latter (Maekawa et al. 2002). CK2 phosphorylates and activates a wide spectrum of erythrocyte membrane proteins associated in multiprotein complexes with Na^+/K^+-ATPase (Wei and Tao, 1993); 3) Non-receptor tyrosine kinase Src takes part in signal pathway that regulates Na-pump activity in other cell types (Haas et al. 2002). Lf is also able to activate Src (Takayama and Mizumashi, 2001). 4) Stimulating effect of Lf could be indirect as Lf activates Na^+/H^+ antiport (Sun et al. 1991). By this mechanism erythrocytes are loaded with Na^+ thus activating Na^+/K^+-ATPase which exports Na^+ out of the erythrocyte cell.

2.13 Polyamines

Membrane receptors for polyamines were proved to exist on erythrocyte membranes (Moulinoux et al. 1984). After binding with their receptors, polyamines enhance the protein-protein interactions among the cytoskeletal proteins and membrane lipid bilayer (Bratton, 1994; Farmer et al. 1985). Polyamine also inhibits transbilayer movement of plasma membrane phospholipids in the erythrocyte ghost (Bratton, 1994).

Polyamines are CK2 protein kinase activators (Leroy et al.1997). It was found that positively charged betaine and polyamines inhibit the activity of membrane Na^+/K^+-ATPase in erythrocytes. It has been demonstrated that the effect of polyamines is based on the competitive inhibition mechanism (Kanbak et al. 2001). A similar mechanism is most likely responsible for the poor ion transport in cancer patients whose tumor cells secrete polyamines (Villano et al. 2001). Raised erythrocyte polyamine levels are estimated in patients with diabetes type 1 and in non-insulin dependent diabetes mellitus with great vessel disease and albuminuria (Seghieri et al. 1997).

2.14 Caffeine

Caffeine binds to erythrocytes membrane proteins (Sato et al. 1990) and could induce changes in the erythrocyte protein complex's formation, thus modulating the activity of enzymes involved in signal transduction. Caffeine could interfere with the phosphorylation processes in erythrocytes as an inhibitor of erythrocyte CK2 (Lecomte et al. 1980), and PI3K (Buckley, 1977) as it was reported in other cell types. Methylxanthines are capable for inhibition of cyclin-dependent protein kinases found in both cytosol and erythrocyte membranes (Biovin, 1988).

Processes of phosphorylation-dephosphorylation of erythrocyte membrane and plasma proteins provide different levels of interaction and participate in maintaining the integrity of the erythrocyte membrane. They also exert control over important metabolic processes in erythrocytes. The literature data shows that signals in erythrocytes lead to changes in phosphorylation and association of integral membrane proteins and other intracellular proteins. Cytosolic protein kinases and protein phosphatases were found in erythrocyte membrane.

3. Protein kinases

3.1 Protein serine/threonine kinases

3.1.1 cAMP dependent protein kinases

Type 1 isoform of cAMP dependent protein kinase is localized in the membrane, while type 2 is located in the cytosol (Dreytuss et al. 1978). cAMP-protein kinases-dependent phosphorylation of isoforms of membrane Ca^{2+}-ATPase was found. Isoform 1 is shown to be better substrate in comparison to isoforms 2 and 4. Isoform 1 is susceptible to degradation by calpain (thiol-dependent protease) (Guerini et al. 2003).

In all cells, increasing in cAMP are regulated by the activity of phosphodiesterases (PDEs) (Sheppard and Tsien, 1975). In erythrocytes, activation of either beta adrenergic receptors (beta (2) AR) or the prostacyclin receptor (IPR) results in increases in cAMP and ATP release

(Sprague et al. 2001). Receptor-mediated increases in cAMP are tightly regulated by distinct PDEs associated with each signaling pathway, as shown by the finding that selective inhibitors of the PDEs localized to each pathway potentiate both increases in cAMP and ATP release (Adderley et al. 2010).

Adenylyl cyclase and cAMP are components of a signal-transduction pathway relating red blood cells (RBC) deformation to ATP release from human and rabbit RBCs (Sprague et al. 2008). Exposure of RBC to catecholamines (epinephrine, phenylephrine, an agonist of α1-adrenergic receptors, clonidine, an agonist of α2-adrenergic receptors and isoproterenol, an agonist of β-adrenergic receptors) led to change in the RBC microrheological properties. Forskolin (10 μM), an adenylate cyclase stimulator, increases the RBC deformability (RBCD). A somewhat more significant deformability rise appeares after RBC incubation with dibutyril-AMP. Red blood cell aggregation (RBCA) is significantly decreased under these conditions. All drugs having PDE inhibitory activity increase red cell deformability (Muravyov et al. 2009).

Ca^{2+} entry increase is accompanied by red cell aggregation rise, while adenylyl cyclase-cAMP system stimulation led to red cell deformability increase and its aggregation lowered. The crosstalk between two intracellular signaling systems is probably connected with phosphodiesterase activity. It was found that all four PDE inhibitors: IBMX, vinpocetine, rolipram, pentoxifylline decreased red cell aggregation significantly and, quite the contrary, they increased red cell deformability (Muravyov et al. 2010)

Erythrocytes are oxygen sensors and modulators of vascular tone (Ellsworth et al. 2009). It has become evident that erythrocytes participate in the regulation of vascular caliber in the microcirculation via release of the potent vasodilator, adenosine triphosphate (ATP). The regulated release of ATP from erythrocytes occurs via a defined signaling pathway and requires increases in cyclic 3',5'- adenosine monophosphate (cAMP) (Adderley et al. 2010) . Heterotrimeric G protein Gi is involved in a signal transduction pathway for ATP release from erythrocytes (Olearczyk et al. 2010). Insulin inhibits human erythrocyte cAMP accumulation and ATP release. The targets of insulin action are phosphodiesterase 3 and phosphoinositide 3-kinase (Hanson et al. 2010). TSH signalling pathway is cAMP-dependent and probably it is potentially active in local circulatory control (Balzan et al. 2009).

3.1.2 cGMP-pependent protein kinases

Petrov et al. (1994) supposed that human erythrocytes possess membrane and soluble guanylate-cyclase activity stimulated by atrial natriuretic peptide III (ANP-III) and that activation of Na^+/H^+ exchange by this peptide is mediated by cGMP (Petrov and Lijnen, 1996). There are no reported data considering participation of protein kinases and phosphatases in the signal pathway.

3.1.3 Casein kinase 1 and 2

Membrane proteins of human erythrocytes can be phosphorylated not only by membrane casein kinase but also by cytosolic casein kinases, resembling casein kinase 1 and 2 (CK1 and CK2), respectively. CK1 and CK2 phosphorylate serine and threonine residues in target proteins (Boivin, 1988). CK2 inactivation mediated by phosphatidylinositol-4, 5-bisphosphate, a substrate for phospholipase C which catalyzes formation of lipid mediators IP3 and DAG

(Chauhan et al. 1993) has been found. It was established that CK2 is inactivated by insulin (Sommerson et al. 1987).

An increased phosphorylation of the membrane proteins, promoted by the okadaic acid (strong inhibitor of P-Ser/Thr-protein phosphatase(s)), is accompanied by a release of CK from the membrane into the cytosol. Such an intracellular translocation might provide a feedback mechanism for the regulation of the CK catalyzed phosphorylation of membrane proteins in human erythrocytes (Bordin et al. 1994).

The membrane mechanical stability of erythrocytes is exclusively regulated by phosphorylation of β-spectrin by membrane bound CK1. Increased phosphorylation of β-spectrin decreased membrane mechanical stability while decreased phosphorylation increased membrane mechanical stability (Manno et al. 1995).

CK2 isolated from erythrocyte membrane and cytosolic fractions exhibited the same subunit composition (αα1) and the ability to utilize ATP and GTP as phosphate donors. Both kinases were found to catalize the phosphorylation of several erythrocyte membrane cytoskeletal proteins (spectrin, ankyrin, adductin, protein 4.1 and protein 4.9). Unlike CK1, CK2 did not phosphorylate band 3. Spermine, spermidine, and putrescine stimulated to varying degrees the activities of erythrocyte CK2, whereas heparin inhibited the kinase activities (Wei and Tao, 1993).

The phosphorylation sites of calmodulin are important for its ability to activate the human erythrocyte Ca^{2+}-ATPase. Phosphorylation of mammalian calmodulin on serine/threonine residues by casein kinase 2 decreased its affinity for Ca^{2+}-ATPase by two fold. In contrast, tyrosine phosphorylation of calmodulin by the insulin-receptor kinase did not significantly alter calmodulin-stimulated Ca^{2+}-ATPase activity (Sacks et al. 1996).

The COP9 signalosome (CSN) is a multimeric complex that is conserved from yeast to man (Bech-Otschir et al., 2002). Immunoprecipitation and far-western blots reveal that CK2 and PKD are associated with CSN. The COP9 signalosome (CSN) purified from human erythrocytes possesses kinase activity that phosphorylates proteins such as c-Jun and p53 with consequence for their ubiquitin (Ub)-dependent degradation. (Uhle et al., 2003)

3.1.4 MAPK (Mitogen Activated Protein Kinases)

It is well known that signaling pathways involving MAPKs are associated with control of cell growth and proliferation, but erythrocytes are mature highly differentiated cells. There are only a few known participants in the MARK-signaling pathways so far. Immunoblot with antiMAPK antibody revealed the two erythrocyte forms of MAPK–p44 (ERK1) and p42 (ERK2). Insulin and okadaic acid (inhibitor of serine/threonine protein phosphatases) stimulate MAPK activity. Insulin enhances the erythrocyte Na^+/H^+-exchanger through MAPK activation (Sartori et al. 1999; Bianchini et al. 1991). Membranes of human erythrocytes contain several proteins of the Ras superfamily (Ikeda et al. 1988; Damonte et al. 1990). One of them, RhoA, was detected in both cytosol and membrane fraction of the erythrocytes. Cytosolic Rho bound specifcally to the cytoplasmic surface of the erythrocyte membrane. The translocation of Rho to the membrane was absolutely GTP -dependent at low Mg^{2+} concentration (Boukharov et al. 1998).

3.2 Protein tyrosine kinases

Erythrocyte membrane associated tyrosine kinase activity has been established (Zylinska et al. 2002), together with surprisingly high levels of phosphorylated tyrosine in erythrocytes (Phan-Dinh-Tuy et al. 1983).

Two tyrosine kinases that are able to phosphorylate band 3 have been found. Involvement of the Src-kinases family in regulation of erythrocyte membrane transport has been suggested. That family represents an important class of non-receptor protein kinases participating in the regulation of cell communications, proliferation, differentiation, migration and survival (Minetti et al. 2004). Subsequent two-stage phosphorylation of band-3 from Syk- and Lyn-tyrosin kinase was found. (Brunati et al. 2000).

3.3 Ca^{2+} dependent phosphorylation

Since mature RBCs lack intracellular calcium stores, elevation in intracellular calcium must stem from calcium influx. At low intracellular Ca^{2+}, efflux of potassium and water predominates, leading to changes in erythrocyte rheology (Andrews et al. 2002). At higher Ca^{2+} content, activation of kinases and phosphatases was observed (Minetti et al.1996; Cohen and Gascard, 1992). Ca^{2+} ions are involved in regulation of phospholipase C, the enzyme generating inositol-1,4,5-triphosphate (InsP3), phosphatidylinositol-3-kinase (PI3K), the enzyme metabolizing InsP3 to InsP4 and in the regulation of protein kinase C (Carafoli, 1994; Carafoli, 2002).

3.3.1 Protein kinase C (PKC)

There are four isoforms of PKC established in the erythrocyte: α, ζ, ι and μ (PKD) (Govekar and Zingle, 2001). PKC translocates from cytosol to membrane - a process shown to be initiated by phorbol esters and diacylglycerols (Cohen and Foley, 1986). PKD and CK2 are associated with COP9 signalosome (CSN) (Uhle et al., 2003).

PKC phosphorylates serine residues in band 3, band 4.1 and band 4.8 (Govekar and Zingle, 2001). It has been reported that after stimulation with phorbol ester (PMA), PKC-α translocates towards the erythrocyte membrane (Govekar and Zingde, 2001), where it phosphorylates serine residues in band 2, and 4.1 and band 4.9 which in turn leads to thorough rearrangement of the cytoskeleton membrane network (Giraud et al. 1988; Ceolotto et al. 1998). PKC phosphorylates membrane Na^+/K^+-ATPase (Wright et al. 1993) and the carboxyl terminus of the plasma membrane Ca^{2+}-ATPase in human erythrocytes (Wang et al. 1991; Wright et al.1993; Smallwood et al.1988) and the effect of phosphorylation on the activity of the both enzymes depends on the isoenzyme form of protein kinase C.

The interactions between membrane, peripheral and cytoskeleton proteins are responsible for the maintenance of erythrocyte deformability and some of these interactions are modulated by PKC activity. Correlation was established between cytoskeleton proteins, PKC activity, band 3 phosphorylation degrees and erythrocyte deformability (de Oliveira et al. 2008). PKC participates in the processes of phosphorylation of band 3 that regulates its activity as a transport protein (Ceolotto et al. 1998). Protein kinase C mediates erythrocyte "programmed cell death" following glucose depletion (Klarl et al. 2006).

In erythrocytes, lead acetate stimulates the phosphorylation of membrane cytoskeletal proteins by a mechanism dependent on protein kinase C. Since levels of calcium or diacylglycerols did not increase, it appears that lead may activate the enzyme by a direct interaction (Belloni-Olivi et al. 1996).

Calpain is a Ca^{2+}-dependent thiol protease that translocates towards the erythrocytes membrane and activates with increase of intracellular calcium concentration (Glasser et al. 1994). Calmodulin binds proteins as calpain substrates (Wang et al. 1989). It is known that calpain causes transition from Ca^{2+}-dependent PKC form to Ca^{2+}-independent one, followed by decrease in its activity (Saido et al. 1994).

3.3.2 Calmodulin (CaM)

The increase in cellular Ca^{2+} results in reversible formation of Ca^{2+}/CaM complex that binds to a number of enzymes modulating their activity (Carafoli, 2002).

Calmodulin is a ubiquitous protein whose activity is regulated through phosphorylation and specific Ca^{2+}-binding (Benaim and Villalobo, 2002). Ca^{2+}-ATPase interacts with the carboxy-terminal half of calmodulin, which is the region that contains the majority of the phosphorylation sites in calmodulin (Bzdega and Kosk-Kosicka, 1992). CaM is phosphorylated by serine/threonine kinases such as CK2 and PKA (Benaim and Villalobo, 2002). Phosphorylation of a tyrosine residue in CaM (Tyr99) increases the affinity of its binding to the target proteins thus increasing their activity. A similar stimulating effect of CaM has been reported for Ca^{2+}-ATPase (Kosk-Kosicka et al. 1990), myosin light chain kinase (MLCK), Ca-CaM dependent kinase (CaM kinase II) and Ca-CaM dependent protein phosphatase 2B (calcineurin) (Corti et al. 1999). Elevated intracellular Ca^{2+}, in association with the Ca^{2+}-binding protein, calmodulin, stimulates erythrocytes phosphofructokinase (PFK) activity. This activation involves the detachment of the enzyme from erythrocyte membranes, which has been described as an important mechanism of glycolysis regulation on these cells (Zancan and Sola-Penna, 2005).

CaM is an inhibitor of PI3P-kinase (Villalonga et al. 2002), Na^+/H^+ antiport (Yingst et al. 1992) and of erythrocyte Na^+/K^+-ATPase (Yingst at al. 1992). The latter effect could be exerted via inducing of signal pathway involving CaM-dependent kinase that activates membrane phospholipase A2(PLA2), which in turn inhibits Na-pump (Okafor et al. 1997).

4. Protein phosphatases

Erythrocyte phosphatases that dephosphorylate phosphoserine, phosphothreonine and phosphotyrosine residues in different proteins were found. Cytoplasmic phosphatases, low-molecular-weight acid phosphatases, and neutral membrane-associated tyrosine phosphatases have been reported (Clari et al. 1987; Graham et al. 1974).

Protein phosphatases originated from two ancestor genes, one serving as the prototype for the phosphotyrosine phosphatases family (the PTP family), and the other for phosphoserine/phosphothreonine phosphatases (the PPP and PPM family). The PPP family groups the PP1, PP2, PP2B, PP4, PP5, PP6 and PP7 enzymes, whereas PP2C and bacterial enzymes like SpoII or PrpC belong to the PPM family (Batford, 1996; Kiener et al. 1987).

Band 3 is a target for tyrosine phosphatases SHP-1 and SHP-2 (Bordin et al. 2002) and PTP1B activities (Zipser and Kosower, 1996). Ca^{2+} promotes erythrocyte band 3 phosphorylation via dissociation of phosphotyrosine phosphatase from band 3 (Ziper et al. 2002)

SHP-2 is an essential soluble protein tyrosine phosphatase (PTP) containing two spectrin homologous domains - SH2. SHR-2 domain participates in multiple signaling pathways of growth factors (GF) and cytokines, and plays an important role in the release of signals from the cell surface to the nucleus. SH2 is the anchor (docking) place where the protein tyrosine kinases, e.g. Src family are attracted to and attached to, thus facilitating phosphorylation-dephosphorylation cycle. SHR-2 is located in the erythrocytes cytosol and it is translocated to the erythrocyte membrane when there is increased tyrosine phosphorylation of the transmembrane protein band 3, induced by PTP inhibitors such as pervanadate, and N-ethylmaleimide. Band 3 is both anchoring protein and substrate for the SHP-2 (Bordin et al. 2002). SHP-1 and SHP-2 ensure dephosphorylation of band 3 in different conditions: SHP-2 through interaction of its SH2 domain(s) to p-tyr protein is regulated by the band 3 tyr-phosphorylation level; SHP-1 dephosphorylates tyr-8, tyr-21 and tyr-904 and may be involved by simple membrane rearrangement (Bragadin et al. 2007).

Protein phosphatase-1 α (PP1alpha) is a selective substrate of peroxynitrite activated src family kinase frg and tyrosine phosphorylation of PP1alpha correspond to inhibition of its enzyme activity. The final effect of peroxynitrite is the ampliphication of tyrosine dependent signaling, a finding of general interest in nitrite oxide related pathophysiology (Mallozzi et al. 2005)

PTP1B is localized at the erythrocyte membrane associated with band 3. It is activated by Mg^{2+} and inhibited by Mn^{2+} and vanadate ions (VO_3^-), (Zipser and Kosower, 1996). PTP1B, unlike the other enzymes examined, was quantitatively conserved during erythrocyte aging (Minetti et al. 2004) and erythrocytes may undergo in vivo activation of the Ca^{2+}-dependent calpain system that proteolytically regulates PTP1B activity (Ciana et al. 2004).

The activities of Na^+/H^+ exchanger and Na-K-2Cl cotransporter in rat erythrocytes are regulated by protein phosphatases PP1 and PP2 and stimulated when protein dephosphorylation is inhibited (Ivanova et al. 2006).

5. Control over glycolytic enzymes

Glycolytic erythrocyte enzymes that form a complex with band 3 are under regulation by the processes of phosphorylation and oxidation (Campanella et al. 2005), and pyruvate kinase (PK) - by phosphorylation/dephosphorylation (Kiner and Westhead, 1980). Binding of phosphofructokinase (PPK), glyceraldehyde-3-phosphate dehydrogenase (GAPDH) and aldolase to band 3 lead to their reversible inhibition (Low et al. 1993), which is neutralized by tyrosine kinase phosphorylation of band 3 resulting in enzymes activation. (Harrison et al. 1991). $p72^{syk}$ and $p56/53^{lyn}$ tyrosine kinases are also involved in band 3 phosphorylation (Hubert et al. 2000), and phosphorylation/dephosphorylation cycle is maintained by protein phosphatases PTP1B and SHP-2 (Bordin et al. 2002; Zipser and Kosower, 1996; Minetti et al. 2004).

PK turns to an inactive form after phosphorylation by cAMP-dependent protein kinase, though it is unclear why since erythrocytes do not undergo gluconeogenesis. PK is activated by erythrocyte protein phosphatases (Kiener and Weathead, 1980).

CaM induces dimerisation of phosphofructokinase (PFK) and physiological levels of concentration of intracellular Ca^{2+} stimulates its catalytic activity (Marinho-Carvalho et al., 2006). However, results obtained from other authors showed that Ca^{2+}/calmodulin protein kinase (CaM-kinase) phosphorylation inhibits phosphofructokinase (PFK) in sheep heart (Mahrenholz et al. 1991)

Ferricyanide activates glycolysis in erythrocytes in two ways. Ferricyanide is the only known non-physiological extracellular agent who induces signal transduction leading to activation of cytoplasmic protein tyrosin kinase that phosphorylates enzyme-binding site in band 3, which in turn leads to release and activation of glycolytic enzymes GAPDH (glyceraldehyde-3-phosphate-dehydrogenase), PPK (phosphofructokinase) and aldolase (Low at al. 1990). It is proved that ferricyanide is a stimulator of glycolysis as end acceptor of electrons produced in glyceraldehyde-3-phosphate dehydrogenase reaction. Thus, ferricyanide is capable of recovering the level of oxidated NAD^+ which is important for maintaining the rate of glycolysis (Orringer and Roer, 1979). There are experiment-based evidence for the existence of erythrocyte *e-* transport transmembrane chain related to transferrin receptor (Orringer and Roer, 1979; Goldenberg et al. 1990), that is known to reduce $K_3Fe(CN)_6$ and removes protons from inside the cell through the cell membrane (Low et al. 1987). Our own experiments showed that $K_3Fe(CN)_6$ inhibits ^{59}FeLF binding to erythrocytes (Maneva et al. 2003) as most probably Lf and $K_3Fe(CN)_6$ compete for one and the same electron formed in oxidative phosphorylation in glycolysis; Lf is the physisological activator of the signal pathway used by ferricyanide that leads to tyrosine phosphorylation of band 3 and consequent glycolysis activation.

6. Modulation of components of erythrocyte membrane skeleton

In terms of proteins, the RBC membrane is a complex network of transporters, cytoskeletal molecules, and membrane-bound enzymes. The ability of transmembrane receptor proteins to change their cytoskeleton associations in response to ligand binding looks like a key mechanism for cell signaling through erythrocyte membrane. The erythrocyte membrane skeleton has 3 main components: spectrin molecules forming tetramers $\alpha_2\beta_2$, short actin oligomers containing 12-15 monomers, and band 4.1. A variety of membrane-associated enzymes, including several kinases (protein kinase A, protein kinase C, cdc kinase, casein kinase 1) are thought to regulate interactions within the network through induction of phosphorylation, methylation, myristoylation, palmytoylation, or farnesylation (Pasini et al. 2010).

Spectrin binding to actin is initiated from aductin and is significantly inhibited in the presence of Ca^{2+} and CaM. CaM is able to form weak bonds with spectrine and in the presence of Ca^{2+} could influence binding of protein 4.1 to actin, and probably spectrin phosphorylation as well (Manno et al. 1995).

Band 4.1 is mainly involved in membrane skeleton organization due to its ability to initiates spectrin/actin association. Band 4.1 binds to the membrane in at least two sites: one high affinity site localized on glycophorins, and another low-affinity site associated with band 3. Phosphorylation of Band 4.1 by PKC was found to modulate its binding to band 3. Most probably, this increases the flexibility of the membrane in response to mechanical stress as was reported for some other models where the elevated levels of membrane phosphorylation made cells more deformable and more resistant to mechanical stress (Danilov and Cohen, 1989). Association of band 4.1 with glycophorin is important for the maintenance of cell shape and is regulatated by polyphosphoinositide co-factor. Indeed, when a change in erythrocyte shape was found, a change in the phosphoinositide content was also found (Danilov et al. 1990). Band 4.1 phosphorylation regulates the assembling of spectrin, actin, and band 4.1 as phosphorylation diminish affinity of band 4.1 to spectrin. Band 4.1 and band 4.9 could be phosphorylated in different sites by PKC and by cAMP-dependent protein kinases (Horne et al.1990).

Cytoskeletal protein band 4.1G binds to the third intercellular loop of the A1 adenosine receptor and inhibits receptor action. (Lu et al., 2004)

7. Modulation of band 3

The two most important integral proteins in human erythrocyte memranes are glycophorin A and band 3. Band 3 is multifunctional protein whose N- and C-ends are localized on the cytoplasmic surface of the membrane and crosses the cell membrane 14 times. The C-terminal end is 52 kD domain (residues from 360 to 919) of band 3 and is responsible for anion transport through the membrane, while 40 kD from N-terminal end of the cytosolic domain plays a critical role in the binding of bilayer to spectrin-based skeletal net. Cytoplasmic domain of band 3 serves as a center of membrane organization, interacting with proteins, such as ankyrin, protein 4.1 and 4.4, hemoglobin, some glycolytic enzymes, tyrosine phosphatases, tyrosine kinase p72 (^{syk}), and Na^+/K^+ - ATPase. There are about 1 million copies of band 3 per cell and they are presented as dimeric and tetrameric forms. Approximately 40-60% of band 3 is associated with spectrin-stabilizing cytoskeleton (van den Akker et al. 2010).

Band 3 (AE1) is a member of the family of anion-transportng proteins that maintain Cl^-/HCO3 $^-$ exchange. It refers mainly to erythtocytes and plays a role in CO_2 transport among the tissues and lungs (Saldanha et al. 2007). Erythrocyte PKC phosphorylates serine residues in band 3 (Govekar and Zingle, 2001).

Band 3 modifications normally occur during physiological red blood cell (RBC) senescence and some pathological condition in humans (Santos-Silva et al. 1998). Band 3-tyrosine phosphorylation might be induce in "stress" conditions: "oxidative stress", inhibition of the phosphoprotein phosphatase activity by vanadate or by *thiol group blockers* (diamide, N-ethylmaleimide, etc.), protein tyrosine kinase (PTK) activation from hypertonic NaCl solution, or by intracellular increase of Ca^{2+} levels (Hecht & Zick, 1992). Erythrocyte thiol status is an intrinsic regulator of phospho tyrosine residues levels in band 3 by oxidation/reduction of band 3–associated phosphotyrosine protein phosphatases (PTP) (Zipser et al. 1997). It is considered that the increase in phosphorylation induced by Ca^{2+} ions, and including significant inhibition of PTP activity by dissociation of PTPase from band 3 may also play an important part in signal transduction pathways in some pathophysiological conditions accompanied with increased levels of intracellular Ca^{2+} (Minetti and Low, 1997).

Abnormal band 3 tyrosine phosphorylation has been observed in a number of red cell disorders (Terra et al. 1998). Hyper-phosphorylated band 3 showed a manifest tendency to cluster, indicating a change in its interactions with the cytoskeletal network. Irreversible band 3 tyrosine phosphorylation leads to membrane vesiculation in G6PD deficient red cells. Syk kinase inhibition largely prevents red cell membrane lysis and vesiculation, strongly suggesting a functional role of band 3 tyrosine phosphorylation in the red cell membrane destabilization (Pantaleo et al. 2011; Bordin et al. 2005).

Based on similarity in oligosaccharide components of band 3 and Lf (Ando et al. 1996), a competition for binding sites could be assumed (Beppu et al. 2000; Eda et al. 1996) together with restructuring of already established complexes. Our experiments revealed that Lf activates erythrocyte glycolysis (Maneva et al. 2003). This fact might be explained with some conformational changes leading to the release of glycolytic enzymes from their complex with band 3. Studies of the same investigators show that Lf, and erythrocyte band 3 occupy the same places on monocyte leukemic cell line THP-1 (Eda et al. 1996). Lf could also activate tyrosine phosphorylation of band 3 since Lf is already proved to be able to increase tyrosine phosphorylation of membrane proteins (Tanata et al. 1998).

8. Changes in phosphorylation in oxidative stress

The alteration of red cell thiol status affects the cell phosphotyrosine status and oxidative stress involves inhibition of PTP. Erythrocyte thiol alkylation by N-ethylmaleimide results in irreversible PTP inhibition and irreversible phosphorylation (Zipser et al. 1997).

Oxidation of erythrocyte membrane by diamide leads to formation of disulfide bonds and following conformational changes in band 3. Most probably those changes lead to the opening of cryptic sites that become accessible for the binding of anti-band 3 antibodies (Turrini et al. 1994). Macrophages recognize oxidatively damaged autologous erythrocytes, and cell surface fibronectin of macrophages enhances the recognition (Beppu et al. 1991). Ca^{2+} signaling including Ca^{2+} influx, calmodulin activation, and myosin light chain phosphorylation are involved in the fibronectin stimulation of the recognition of macrophages for oxidized erythrocytes (Beppu et al. 2000).

Pervanadate, N-ethylmaleimide and diamide strongly increase the two-stage phosphorylation of tyrosine residues of band 3 by Syk-kinase and Src-family (probably Lyn-kinase). An intriguing fact is that there are different mechanisms by which osmotic and oxidative stress activate PTK Syk: oxidative stress leads to autophosphorylation, but osmotic one-to SH-2 domain phosphorylation. It was demonstrated that the same agents strongly enhance interaction between SHP-2 and band 3, which take part simultaneously with the translocation of this phosphatase from cytosol to erythrocyte membrane. These events are most likely mediated by Src-phosphorylation, since both translocation of SHP-2 and phospatase interaction with the erythrocyte membrane are prevented by PP2, a specific Src inhibitor. SHP-2 binds to band 3 via its SH2 region(s). Authors suggest that the attraction of cytosolic SHP-2 to band 3 preceeds the next dephosphorylation of the transmembrane protein (Bordin et al. 2002).

Peroxynitrite ($ONOO^-$) is a product of the reaction of nitric oxide and superoxide anion. It is able to nitrate protein tyrosine. If this modification occurs on phosphotyrosine kinase, substrates can down-regulate cell signaling. $ONOO^-$ at low concentrations is a stimulator of both band 3 tyrosine phosphorylation and erythrocyte glycolysis, but higher concentratons $ONOO^-$ induces cross-linking of membrane proteins, inhibition of band 3 phosphorylation, nitration of tyrosines in cytosolic domain of band 3, and irreversible inhibition of lactate production (Mallozi et al. 1997).

Deoxygenation and increase in intracellular Mg^{2+} content induce phosphorylation of tyrosine residues in human erythrocytes (Barbul et al. 1999).

Increasing in erythrocyte volume is combined with stimulation in the activities of the two PTK- p72 (syk) and p56 (lyn) that phosphorylate band 3 (Musch et al. 1999).

Oxidation of methionine residues in CaM prevents its activating effect on membrane Ca^{2+}-ATPase (Gaop et al. 2001).

The erythrocyte has a pool of flavonic compounds, which are considered a buffer, maintaining the antioxidative activity of the erythrocyte (Fiorani et al., 2003) Quercetin and resveratrol (piceatannol) take part in the regulation of the phosphorylation of band 3 in the presence of the physiological oxidant peroxinitrite. Quercetin decreases the Syk activity and particularly prevents mediated by the free radical peroxynitrite ($ONOO_{\cdot}$) phosphotyrosine phosphatase inhibition. Restraverol (whose anlogue is piceatannol) has another mechanism of action - it enables the mediated by peroxynitrite stimulation of tyrosine phosphorylation by another phosphotyrosine protein kinase - Lyn (Maccaglia et al., 2003).

Pleiotropic effects of resveratrol include antioxidant activity and inhibition of cyclooxygenase with decrease of PGE(2) formation. In erythrocytes, oxidation and PGE(2) activate Ca^{2+}-permeable cation channels. The Ca^{2+}-entry leads to activation of Ca^{2+}-sensitive K^{+}channels with subsequent cell shrinkage and cell membrane scrambling with phosphatidylserine (PS) exposure at the erythrocyte surface. Cell shrinkage and phosphatidylserine exposure are hallmarks of suicidal erythrocyte death or eryptosis. Eryptotic cells adhere to the vascular wall, thus compromising microcirculation, and are cleared from circulating blood, thus leading to anemia. Resveratrol is a potent inhibitor of suicidal erythrocyte death during energy depletion, oxidative stress, and isoosmotic cell shrinkage. The nutrient could thus counteract anemia and impairment of microcirculation under conditions with excessive eryptosis (Qadri al. 2009). However, in some cases, flavonoids have been suggested to work as prooxidants (Kitagawa et al. 2004)

9. Eryptosis (Apoptosis)

Two signaling pathways converge to trigger apoptosis: (1) formation of PGE2 leads to activation of Ca^{2+}-permeable cation channels; (2) the PLA2 (Phospholipase A 2) mediated release of PAF (platelet–activating factor) activates a sphingomielinase, leading to formation of ceramide. Increased cytosolic Ca^{2+} activity and enhanced cermide levels lead to membrane scrambling with subsequent PS exposure. Ca^{2+}-activated Ca^{2+}-sensitive K^{+} channels leading to cellular KCl loss and cell shrinkage. Ca^{2+} stimulates the protease calpain, resulting in degradation of cytoskeleton (Foller et al. 2008; Lang et al. 2010).

Most triggers of eryptosis, except oxidative stress, are effective without activation of caspases. The involvement of Fas/caspase 8/caspase 3-dependent signaling in erythrocytes leads to PS externalization, a central feature of erythrophagocytosis and erythrocyte biology. The oxidatively stressed red cell recapitulated apoptotic events, including translocation of Fas into rafts, formation of a Fas-associated complex, and activation of caspases 8 and 3. The ROS (Radical Oxygen Species) scavenger N-acetylcysteine inhibits eryptosis (Mandal et al. 2005).

Protein kinase C mediates erythrocyte "programmed cell death" following glucose depletion (Klarl et al. 2006).

Eryptosis is stimulated in a wide variety of diseases including sepsis, haemolytic uremic syndrome, malaria, sickle-cell anemia, beta-thalassemia, glucose-6-phosphate dehydrogenase (G6PD)-deficiency, phosphate depletion, iron deficiency, and Wilson's disease. Excessive eryptosis is observed in erythrocytes lacking the cGMP-dependent protein kinase-1 (cGK1) or cAMP activated protein kinase (AMPK). Moreover, eryptosis is elicited by osmotic shock, oxidative stress, and energy depletion, as well as a wide variety of endogenous mediators and xenobiotics. Inhibitors of eryptosis include erythropoietin, nitric oxide NO, catecholamines, and high concentrations of urea (Lang et al. 2006).

10. Effect of modulators of phosphorylation on activity of erythrocyte glycolysis and sodium pump

Modulation of cellular signals from other agents is a leading pharmaceutical approach for developing a therapeutic strategy for many diseases. Of significant practical importance is to address the right questions and to find the appropriate answers in regard to whether certain agents would exert synergic or opposite effects over key cellular functions.

Erythrocyte membranes mediate the relation between ion transport and glycolysis. Na^+/K^+-ATPase is not only transport system but also participates in ouabain - initiated signal transduction (Haas et al. 2002; Xie and Cai, 2003; Mohammadi et al.2001), and inititated cell signals in cardiac myocytes that are independent from changes of intracellular concentrations of Na^+ and Ca^{2+} (Liu et al. 2000). It was demonstrated that glyceraldehyde-3-phosphate-dehydrogenase interacts with cytoplasmic surface of $Na^+/K+$-ATPase, and that interaction is shown to be inhibited by ouabain (Fossel and Solomon, 1978). The α-subunit of Na^+/K^+-ATPase and band 3 form a heterodimeric structure (Martin and Sachs,. 1992). Similar structural integration allows shared control of both glycolysis and ion transport, and possible interference of modulators of ion transport in the control of glycolysis might be suggested (Fossel and Solomon, 1978). To verify our hypothesis we studied the effect of the same modulators of phosphorylation on both glycolysis (by measuring the amounts of its final product lactate) and $Na^+/K+$-ATPase activity.

10.1 Methods

10.1.1 Isolation of the erythrocytes

Heparinized fresh drawn blood from healthy donors was centrifuged at 2,000 ×g for 5 min at 4°C and the pellet was resuspended in 4 volumes of phosphate buffered saline (PBS) pH 7.4. After three washes at 1,800, 1,500, 1,300 xg, the erythrocytes were isolated by density separation (Cohen et al., 1976). The erythrocyte fraction was resuspended in PBS pH 7.4 to obtain the same concentration as in the fresh blood. Cell concentration was counted in Burker's camera by a Standard KF2 microscope (Carl Zeiss, Jena, Germany). The suspension did not contain other cell species.

10.1.2 Preparation of the erythrocyte membranes

Five milliliters of packed RBCs were mixed with 15 ml cold PBS (0.144 g/l $KH_2PO_4.7H_2O$, 9.0 g/l NaCl, and 0.795 g/l $Na_2HPO_4.7H_2O$), pH 7.4, and centrifuged at 5,900 ×*g* for 10 min at 4°C. The supernatant was discarded, and cells were washed with 15 ml cold PBS, centrifuged as above, and resuspended in 5 ml of 5 mM Na_2HPO_4, pH 8.0, for hypotonic cell lysis. Lysed cells were centrifuged at 25,000 ×*g* for 15 min at 4°C, after which the supernatant was gently aspirated and discarded. The RBC membrane pellet was repeatedly washed (4–5 times) with 5 mM Na_2HPO_4 until the pellet appeared white (indicating removal of Hb), and membranes were used for further experiments. The protein content of hemoglobin-free pellets was determined according to Bradford (1976), with human serum albumin as a calibrator. Samples were diluted to protein contents of 1.5 g/L.

10.1.3 Hemoglobin measurement

The hemoglobin (Hb) concentration of the erythrocyte suspension was determined according to Beutler (1975).

10.1.4 Cell treatment

The 50 μl erythrocyte suspension (2×10^7 cells/ml) was incubated 30 min at 25°C. Samples were performed in quadruplicates. In order to estimate the effect of cell signal modulators, samples were incubated either in the presence or in the absence of the agents.

10.1.5 Chemicals

All chemicals were purchased from Sigma-Aldrich Co, St Louis, MO USA. The mentioned below concentrations were chosen according the producer's prescription and data existing in the literature about their optimal effect: Go6976 (50 nM) and Go6983 (50 nM) – inhibitors of protein serine kinase PKC. Go6976 is an inhibitor of the classic isoforms of PKC and PKD, while Go6983 inhibits only the classic isoforms. Both inhibitors were used simultaneously to exclude the effect of PKD; Caffeine (20 mM) was used as a phosphodiesterase inhibitor; Okadaic acid (20 nM) and Calyculin A (50 nM) as inhibitors of serine/treonine protein phosphatases; N-(6-Aminohexyl)-5-chloro-1-naphtalene-sulfonamide (W-7) (30 μM) as a calmodulin antagonist.

10.1.6 Determination of lactate content in erythrocytes

After centrifugation for 10 min at 2,000 ×*g* erythrocytes were resuspended in 0.4 ml 10% trichloroacetic acid (TCA). For obtaining of total precipitation the samples were cooled on ice 10 min and then centrifuged again at the same conditions. 0.1 ml from the supernatant was used further according the prescription of the test kit reagent obtained from "SIGMA Diagnostics" (St. Louis, USA). The method was based on the reaction of pyruvate oxidation in the presence of NADH and lactate dehydrogenase. To estimate lactate, the reaction was carried out with excess NAD^+ and the OD of NADH was measured at 340 nm.

50 μl of erythrocyte suspension (2×10^7 cells/ml) was incubated for 30 min at 25°C in the presence (samples) or absence (control) of the chosen agents. After 10 min of centrifugation at 2,000 ×g, erythorcytes were resuspended in 0.4 ml of 10% TCA. For total sedimentation, erythrocytes were cooled down for 10 min and the centrifugation step was repeated at the same conditions. 0.1 ml from the supernatant was added to a preliminary prepared solution of NAD^+ in 2.0 ml glycine buffer, 4.0 ml ddH_2O and 0.1 ml lactate dehydrogenase. Instead of

supernatant, 0.1 ml 10% TCA was added as a blank. After 30 min of incubation at room temperature, the extinction was read at OD=340 nm. Results were calculated using a standart curve and equation according to kit instruction. Results were presented in µg/ml.

10.1.7 Erythrocyte membrane ATPase activity

Here, a method based on enzyme kinetics was used for evaluation of ATPase activity based on the amounts of generatated NAD^+ in glycolytic reactions. The following reactions were explored: 1) conversion of phosphoenolpyruvate (PEP) to pyruvate and ATP by pyruvate kinase; 2) interconversion of pyruvate and lactate with concomitant interconversion of NADH and NAD+ catalyzed by lactate dehydrogenase (Vásárhelyi et al., 1997).

Samples (50 µl) with protein concentration 1.5 g/l of erythrocytes membranes were added to 450 µl of the following solution with final concentration per liter: 100 mmol of NaCl, 20 mmol of KCl, 2.5 mmol of $MgCl_2$, 0.5 mmol of EGTA, 50 mmol of Tris-HCl, pH 7.4, 1.0 mmol of ATP, 1.0 mmol of phosphoenolpyruvate, 0.16 mmol of NADH, 5 kU of pyruvate kinase, 12 kU of lactate dehydrogenase (all purchased from Sigma). Finally, the tested modulators were added in the samples in above mentioned concentrations. After 5 min, 5 µl of 10 mmol/l ouabain was added to inhibit the ouabain-sensitive ATPase activity. The change in absorbance was measured at OD=340 nm by a twin test (i.e., combination of two assays in one cuvette); Rate A (i.e., slope of total ATPase activity), 80–280 s; Rate B (i.e., slope of ouabain-resistant ATPase activity), 400–600 s. The difference between the two slopes was proportional to the Na^+/K^+-ATPase activity (Vásárhelyi *et al.* 1997). One unit of ATPase activity (1U) equals to one µmol oxidized NADH for 1 min. Calculations are based on the fact that NADH solutition with concentration 1 mg/ml has 0.80 extinction at OD=340 nm (E_{340} = 0.80) (Vásárhelyi et al. 1997).

10.2 Results

PKC inhibitors (Go6876 and Go6983), protein phosphatase inhibitors (OA and calyculin A), caffeine, and W-7 increased reliable lactate formation. Calyculin A inhibits reliable Na^+/K^+-ATPase (Table 1 and 2). From all studied modulators only Go6976 and Go6983, which are PKCs inhibitors, stimulated reliably Na^+/K^+-ATPase, with 194% and 84%, respectively (Table 1 and 2; Figure 2 and 3).

Agents (n = 6)	Lactate µmol/g.Hb x ± SD	p
No agents (control)	5.15 ± 0.19	-
Go6976	8.23 ± 0.63	<0.001
Go6983	8.21 ± 3.24	<0.05
Caffeine	8.07 ± 0.65	<0.001
OA	7.84 ± 1.45	<0.001
Calyculin A	7.23 ± 0.37	<0.001
W-7	11.12 ± 0.30	<0.001

Table 1. Effect of modulators of phosphorylation on lactate formation

10.3 Discussion

10.3.1 Effect of PKC inhibitors

Four isoforms of PKC in erythrocytes have been found: α, ζ, ι and μ (PKD). Erythrocyte PKC phosphorylates serine residues in band 3, band 4.1 and band 4.8 (Govekar and Zingle, 2001). Go6976 and Go6983 are used to exclude the involvement of PKD in cell signals. Go6976 inhibits classical isoforms of PKC and PKD, but Go6983 inhibits classical isoforms only. The absence of differences in the effects of Go6976 and Go6983 on the lactate formation (Table 1) excludes the involvement of PKD in the control of glycolysis and may indicate involvement of classical PKCα as a negative regulator of glycolysis. It was demonstrated that αPKC translocated from the interior of erythrocytes to the membrane as a result of various stimuli (Govekar and Zingle, 2001).

Go6976 (inhibitor of classic isoforms and PKD), and also PKD "switch-off" inhibitor Go6983, reliably activate ATPase activity. This presents PKC as a negative regulator of Na^+/K^+-ATPase. The effect of stimulation with Go6976 is almost twice higher than stimulation with Go6983 (Table 2 and Fig.3). This probably suggests a specific function of PKD as an inhibitor of Na^+/K^+-ATPase. PKD has a catalytic domain, which shows more similarity to CaM-dependent kinases than to PKC. PKD efficiently phosphorylated synthetic substrates of Ca^{2+}/CaM -dependent kinase II, but does not catalyze phosphorylation of substrates

Agents (n = 6)	Na^+/K^+ ATPase [U/g. Protein]	p
No agents (control)	4.44 ± 0.98	-
Go6976	13.04 + 2.11	<0.001
Go6983	8.18 + 3.24	<0.05
Caffeine	5.68 + 1.57	>0.1
OA	5.18 ± 0.61	>0.1
Calyculin A	2.71 ± 0.98	<0.02
W-7	4.81 + 0.86	>0.1

Table 2. Effect of modulators of phosphorylation on Na^+/K^+-ATPase

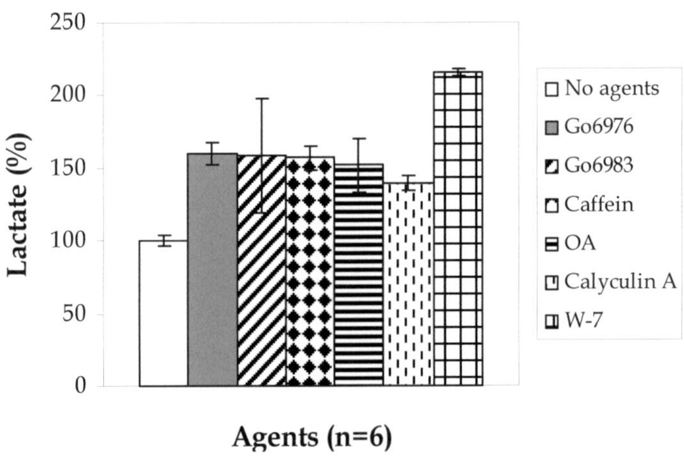

Fig. 2. Effect of modulators of phosphorylation on lactate formation (control=100%).

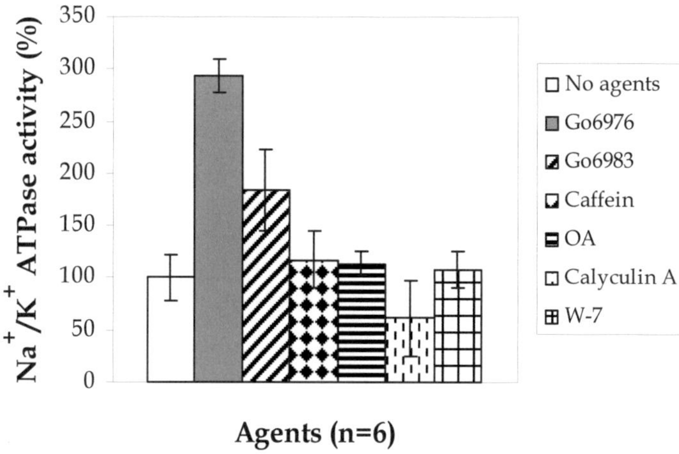

Fig. 3. Effect of modulators of phosphorylation on Na^+/K^+-ATPase (control=100 %)

typical of PKC (Ron et al., 1999). CaM inhibits the sodium pump in erythrocytes (Okafor et al. 1997; Yingst et al. 1992), and because of similarities with the effects of CaM, PKD participation in erythrocyte phosphorylation could be assumed, leading to a decrease in activity of Na^+/K^+-ATPase. It is also proven an inhibitory effect of PKD on Na^+/H^+ -antiport in other cell types (Haworth and Sinnet-Smithh, 1999). It could be speculated that due to the antiport inhibition by PKD, the activity of Na^+/K^+-ATPase is being reduced.

10.3.2 Effect of caffeine

The mechanisms by which caffeine stimulates erythrocyte glycolysis (Table 1 and Fig.2) may include modulation of erythrocyte protein phosphorylation (Boivin, 1988) or changes in the conformation of the erythrocyte membrane proteins that facilitate the access of enzymes phosphorylating band 3 (Sato et al. 1990). It is known that methylxanthines are able to inhibit cyclic nucleotide-dependent protein kinases present in the cytosol and the erythrocyte membrane (Boivin, 1988). This inhibition may be due to competition for ATP (Boivin, 1988) and may be involved in the regulation of PK, whose active form is the dephosphorylated one (Kiener and Weathead, 1980; Garrillo et al., 2000; Nakashima et al., 1982). Caffeine can exert an indirect effect on lactate formation in erythrocytes through stimulation of Na^+/K^+-ATPase (Gusev et al. 2000), but our results showed that caffeine has no reliable effect on the activity of erythrocyte Na^+/K^+-ATPase (Table 2 and Fig 3). The identified differences may be due to the fact that in that study, erythrocytes from Rana temporaria were examined (Gusev et al, 2000).

10.3.3 Effect of protein phosphatases inhibitors

Okadaic acid (OA) is an inhibitor of serine/treonine phosphatases (Bordin et al. 1993). The incentive effect of OA on glycolysis (Table 1 and Fig.2) may be indirect, as a consequence of activation of MAPK-dependent way that increases Na^+/H^+-antiport (Sartori et al. 1999). It is known that one of the mechanisms for glycolysis stimulation is the activation of proton export through this antiport (Madshus, 1988). An interesting fact is that OA and various growth factors (GF), e.g. TGF and EGF, exert a stimulatory effect on Na^+/H^+-antiport (NHE-1) through phosphorylation of identical serine residues in its molecule (Sardet et al. 1991). Calyculin A and OA are likely to cause stimulation of glycolysis by using pathways involving protein phosphatase type II (PP2), Table 1 and Fig 2. Protein phosphatase type I (PP1) is highly sensitive to calyculin A (CalA), but not to OA. Protein phosphatase type II (PP2) is highly sensitive to both inhibitors, CalA and OA (Bize et al. 1999). OA has no effect on erythrocyte sodium pump (Table 2 and Fig.3), which probably means that phosphatase type II (PP2) is not involved in cellular signals engaged with the control of Na^+/K^+-ATPase, but Calyculin A inhibits reliably (with about 40%) the pump activity, which may be due to the involvement of PP1 in cell signals leading to activation of Na^+/K^+-ATPase (Table2 and Fig.3).

10.3.4 Effect of W-7

As calmodulin antagonist W-7 restores reduced physiological ability of erythrocytes to change their shape when loaded with Ca^{2+} (Murakami et al. 1986) and also exerts a vasodilator effect (Beresewiz, 1989). The beneficial effect of W-7 on erythrocyte rheology may be due to its stimulatory effect on glycolysis and improvement of erythrocyte bioenergetics (Table 1 and Fig.2). The stimulatory effect of W-7 on glycolysis may be due to blocking the inhibitory effect of calmodulin on Na^+/H^+-antiport (Yingst et al. 1992), or to decrease in the processes of phosphorylation/dephosphorylation, regulated by Ca^{2+}calmodulin (Benaim and Villalobo, 2002, Corti et al. 1999). W-7 increases by 8% the activity of Na^+/K^+-ATPase but the effect lacks statistical significance (Table 2 and Fig.3) in the absence of Ca^{2+}. There is evidence in the literature for an inhibitory effect of calmodulin on erythrocyte Na^+/K^+-ATPase, which occurs at 2 μM Ca^{2+} in the incubation medium (Yingst et al. 1992). The difference with our results could be explained probably with these specific experimental conditions. In our study, no extra ammounts of calcium were added into incubation medium. Erythrocytes have about 80 nM intracellular Ca^{2+} (Astarie et al. 1992). Perhaps the inhibitory effect of calmodulin (resp. stimulation with W-7) on the sodium pump acts as a regulatory mechanism when an increase in intracellular calcium content is presented, and primarily affects Ca^{2+}-binding capacity of calmodulin (Astarie et al. 1992).

10.4 Conclusions

Processes of phosphorylation-dephosphorylation of erythrocyte membrane and plasma proteins provide different levels of interaction, and also participate in maintaining the integrity of erythrocyte membrane and exerting control over important metabolic processes in erythrocytes. Summary of the existing data in literature shows they are more comprehensive in terms of the effect of primary messengers that bind to membrane receptors on erythrocytes and the consequent biological effects of the ligand-receptor interaction. Down-stream signaling pathways, though, where a major role in phosphorylation processes is played by protein kinases and protein phosphatases, remain less clear. Expansion of our knowledge and better understanding of signal transduction in erythrocytes will enable possible control to be exerted over their impact in maintenance of vascular tone, procoagulant activity and antioxidant status, erythrocyte aging, senescence, eryptosis (apoptosis), and erythrophagocytosis. Our results demonstrate that erythrocytes use different cell signals for regulation of glycolysis and activity of the sodium pump. They also reveal that different classes of PKC most likely taking part in regulation of glycolysis and Na^+/K^+-ATPase. Protein phosphatase inhibitors have opposite effects in the control of glycolysis and sodium pump, which indicates involvement of different protein phosphatases in cellular signal transduction that controls ion transport and cell energetics. Caffeine and W-7 have a reliable effect on the stimulation of glycolysis, but not on sodium pump.

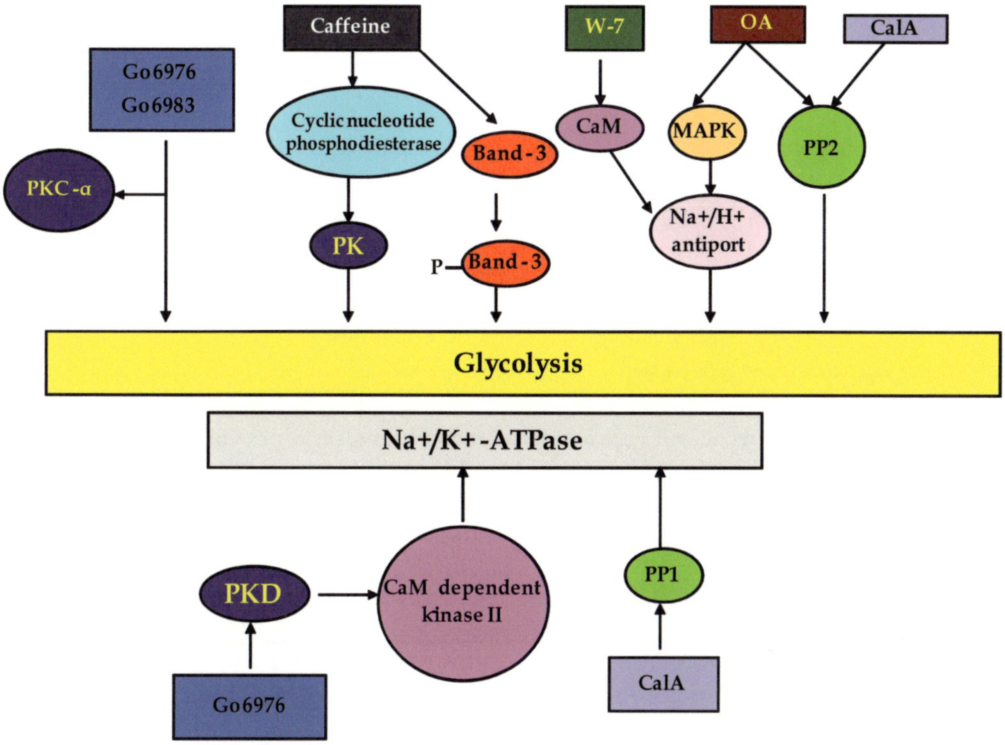

Fig. 4. Erythrocytes use different processes of phosphorylation in regulation of glycolysis and sodium pump.
(A). Glycolysis. Go6976 and Go6983 are inhibitors of PKC and stimulators of glycolysis. The target for this effect is PKC-α, which is probably a negative regulator of glycolysis. Caffeine increases the formation of lactate as: a) an activator of phosphodiesterase and inhibitor of c.AMP-dependent phosphorylation processes that keeps pyruvate kinase in dephosphorylated active form; b) induces conformational changes or activates phosphorylation of band 3. W-7 as a CaM antagonist blocks its inhibitory effect on Na^+/H^+-antiport, which activates glycolysis. Okadaic acid (OA) stimulates glycolysis as an inhibitor of protein phosphatase PP2, which is probably a negative regulator of glycolysis, and as a stimulator of MAPK-dependent phosphorylation of Na^+/H^+-antiport. Calyculin A (CalA) activates glycolysis as PP2 inhibitor that turns the proteins involved in glycolysis to their active dephosphorylated form. (B) Na^+/K^+-ATPase. PKD is likely a negative regulator of the sodium pump using CaM-dependent pathway since its specific inhibitor Go6976 significantly increases the activity of Na^+/K^+-ATPase. Cal A inhibits the activity of the pump reliably, which may be due to the involvement of PP1 in cell signals leading to activation of Na^+/K^+-ATPase.

11. Acknowledgments

The authors would like to thank Dr. Ilian Radichev (Staff Scientist at Sanford Research/USD), for his valuable comments, and to acknowledge the helpful assistance of Jacob Ellefson (Research Associate at Sanford Research/USD) on improving the language quality of this work.

12. References

Adderley, S.P.; Sprague, R.S.; Stephenson, A.H.; Hanson, M.S. Regulation of cAMP by phosphodiesterases in erythrocytes.(2010). *Pharmacological Reports*, vol. 62, № 3, pp 475-482

Ando, K.; Kikugawa, K.; Beppu, M. (1996). Binding of anti-band 3 autoantibody to sialylated poly-N-acetyllactosaminyl sugar chains of band 3 glycoprotein on polyvinylidene difluoride membrane and sepharose gel: further evidence for anti-band 3 autoantibody binding to the sugar chains of oxidized and senescent erythrocytes. *Journal of Biochemistry*, vol. 119, № 4, pp. 639-647

Andrews, D.A; Yang, L.; Low, P.S. (2002). Phorbol ester stimulates a protein kinase C-mediated agatoxin-TK-sensitive calcium permeability pathway in human red blood cells. *Blood*, vol.100, № 9, pp. 3392-3399

Angel, R.C.; Botta, J.A.; Farias, R.N. (1989). High affinity L-triiodothyronine binding to right-side-out and inside-out vesicles from rat and human erythrocyte membrane. *Journal of Biological Chemistry*, vol. 264, №32, pp.19143-19146

Antonelou, M.H; Kriebardis, A.G.; Papassideri, I.S. (2010). Aging and death signalling in mature red cells: from basic science to transfusion practice. *Blood Transfusion*, № 8, Suppl 3, 39-47

Assouline-Cohen, M.; Beitner, R. (1999). Effects of Ca2+ on erythrocyte membrane skeleton-bound phosphofructokinase, ATP levels, and hemolysis. *Molecular Genetics and Metabolism*, Vol. 66, № 1, pp. 56-61

Astarie, C.; Pernollet, M.G.; Del Pino, M.; Levenson, J.; Simon, A.; Devynck, M.A. (1992). Control of the erythrocyte free Ca2+ concentration in essential hypertension. *Hypertension*, vol.19, pp. 167-174

Azoui, R.; Cuche, J.L.; Renaud, J.F.; Safar, M.; Daher, G. (1996). A dopamine transporter in human erythrocytes: modulation by insulin. *Experimental Physiology* vol. 81, № 3, pp. 421- 434

Balzan, S.; Del Carratore, R., Nicolini, G.; Forini, F.; Lubrano, V.; Simili, M.; Benedetti, P.A.; Levrasi, G. (2009). TSH induces co-localization of TSH receptor and Na/K-ATPase in human erythrocytes. *Cell Biochemistry and Function*, vol. 27, № 5, pp. 259-263

Barbul, A.; Zipser, Y.; Nachles, A.; Korenstein, R. (1999). Deoxygenation and elevation of intracellular magnesium induce tyrosine phosphorylation of band 3 in human erythrocytes. *FEBS Letters*, vol. 455, № (1-2), pp. 87-91

Barford, D. (1996). Molecular mechanisms of the protein serine/threonine phosphatases. *Trends in Biochemical Sciences*, vol. 21, № 11, pp. 407–412

Bech-Otschir, D.; Seeger, M.; Dubiel, W. (2002). The COP9 signalosome: at the interface between signal transduction and ubiquitin-dependent proteolysis. *Journal of Cell Science*, vol. 115, (Pt. 3), pp. 467-473

Belloni-Olivi, L.; Annadata, M.; Goldstein, G.W.; Bressler, J. P. (1996). Phosphorylation of membrane proteins in erythrocytes treated with lead. *Biochemical Journal*, vol. 315, (Pt.2), pp. 401-406

Benaim, G.; Villalobo, A. (2002). Phosphorylation of calmodulin. Functional implications. *European Journal of Biochemistry*, vol. 269, № 15, pp. 3619-3631

Beppu, M.; Ando, K.; Saeki, M.; Yokoyama, N.; Kikugawa, K. (2000). Binding of oxidized Jurkat cells to THP-1 macrophages and antiband 3 IgG through sialylated poly-N-acetyllactosaminyl sugar chains. *Archivs of Biochemistry and Biophysics*, vol. 384, № 2, pp. 368-374

Beppu, M.; Masa H.; Hora, M.; Kikugawa, K (1991). Augmentation of macrophage recognition of oxidatively damaged erythrocytes by substratum-bound fibronectin and macrophage surface fibronectin. *FEBS Letters*, vol. 295, №(1-3), pp. 135-140

Beresewiz, A.. (1989), Anti-ischemic and membrane stabilizing activity of calmodulin inhibitors. *Basic Research in Cardiology*, vol. 84, № 6, pp. 631-645

Berzosa, C.; Gómez-Trullén, E.M.; Piedrafita, E.; Cebrián, I.; Martínez-Ballarín, E.; Miana-Mena, F.J.; Fuentes-Broto, L.; García, J.J. (2011). Erythrocyte membrane fluidity and indices of plasmatic oxidative damage after acute physical exercise in humans. *European Journal of Applied Physiology*, vol. 111, № 6, pp. 1127-1133

Beutler, E. (1975) Hemoglobin estimation. In: *Red Cell Metabolism. A Manual of Biochemical Methods. Basic techniques and equipment.* 2-nd Ed. (Grune and Stratton, ed.), New York, pp.112-114

Bhattacharaya, S.; Chakraborty, P.S.; Basu, R.S.; Kahn, N.N.; Sinha, A.K. (2001). Purification and properties of insulin-activated nitric oxide synhase from human erythrocyte membrane. *Archives of Physiology and Biochemistry*, vol. 109, № 5, pp.441-449

Bianchini, I.; Woodside, M.; Sardet, C.; Pouyssegur, J.; Takai, A.; Grinstein, S. (1991). Okadaic acid, a phosphatase inhibitor, induces activation and phosphorylation of the Na+/H+ antiport. *Journal of Biological Chemistry*, vol. 266, № 23, pp. 15406-15413

Bize, I.; Güvenç, B.; Robb, A.; Buchbinder, G.; Brugnara, C. (1999). Serine/ threonine protein phosphatases and regulation of K-Cl cotransport in human erythrocytes. *American Journal of Physiology*, vol. 277, (5 pt 1), pp. C926-C936

Boadu, E.; Sager, G. (2000). ATPase activity and transport by a cGMP transporter in human erythrocyte ghosts and proteoliposome-reconstituted membrane extracts. *Biochimica et Biophysica Acta*, vol. 1509, № (1-2), pp.467-474

Bogin, E.; Earon, Y.; Blum, M. (1986). Effect of parathyroid hormone and uremia on erythrocyte deformability. *Clinica Chimica Acta*, vol. 161, № 3, pp. 293-299

Boivin, P. (1988). Role of phosphorylation of red blood cell membrane proteins. *Biochemical Journal*, vol. 256, № 3, pp. 689-695

Bordin, L.; Brunati, A.M.; Donella-Deana A.; Baggio, B.; Toninello A.; Clari G. (2002). Band 3 is an anchor protein and a target for SHP-2 tyrosine phosphatase in human erythrocytes. *Blood*, vol. 100, № 1, pp. 276-282

Bordin, L.; Clari G.; Baggio, B.; Gambaro, G.; Moret, V. (1994). Relationship between membrane protein phosphorylation and intracellular translocation of casein kinase in human erythrocytes. *Biochemical and Biophysical Research Communication*, vol. 203, № 1 pp. 681-685

Bordin, L.; Clari, G.; Bellato, M.; Tessarin, C.; Moret, V. (1993). Effect of okadaic acid on membrane protein phosphorylation in human erythrocytes. *Biochemical and Biophysical Research Communication*, vol. 195, № 2, pp. 723-729

Bordin, L.; Zen, F.; Ion-Popa, F.; Barbetta, M.; Baggio, B.; Clari, G. (2005). Band 3 tyr-phosphorylation in normal and glucose-6-phosphate dehydrogenase-deficient human erythrocytes. *Molecular Membrane Biology*, vol. 22, № 5, pp. 411- 420

Botta, J.A.; Farías, R. N. (1985). Solubilization of L-triiodothyronine binding site from human erythrocyte membrane. *Biochemical and Biophysical Reseach Communication*, vol. 133, № 2, pp. 442-448

Boukharov, A.A.; Cohen, C.M. (1998). Guanine nucleotide-dependent translocation of RhoA from cytosol to high affinity membrane binding sites in human erythrocytes. *Biochemical Journal*, vol. 330, (Pt 3), pp. 1391-1398

Bourikas, D.; Kaloyianni, M.; Bougolia, M.; Zolota, Z.; Koliakos, G. (2003). Modulation of the Na+-H+ antiport activity by adrenaline on erythrocytes from normal and obese patients. *Molecular Endocrinology*, vol. 205, № 1-2, pp. 141-150

Bradford, M.M. (1976). A rapid and sensitive method for quantitation of microgram quantities of protein-binding. *Analalitical Biochemistry*, vol. 72, pp. 248-254

Bragadin, M.; Ion-Popa, F.; Clari, G.; Bordin, L. (2007). SHP-1 tyrosine phosphatase in human erythrocytes. *Annual of the National Academy of Sciences of the United States of America*, vol. 1095, pp. 193-203

Bratton D.L. (1994). Polyamine inhibition of trans bilayer movement of plasma membrane phospholipids in the erythrocyte ghost. *Journal of Biological Chemistry*, vol. 269, № 36, pp. 22517-22523

Brunati, A.M.; Bordin, L.; Clari, G.; James, P.; Quadroni, M.; Baritono, E.; Pinna, L.A.; Donella-Deana, A. (2000). Sequential phophorylation of protein bind 3 by Syk and Lyn tyrosine kinases in intact human erythrocytes: identification of primary and secondary phosphorylation sites. *Blood*, vol. 96, № 4, pp. 1560-1567

Buckley J.T. (1977). Properties of human erythrocyte phosphatidylinositol kinase and inhibition by adenosine, ADP and related compounds. *Biochim Biophys Acta*. Vol. 498, № 1, pp. 1-9

Bzdega, T.; Kosk-Kosicka, D. (1992). Regulation of the erythrocyte Ca^{2+}-ATPase by mutant calmodulins with Glu-Ala substitutions in the Ca^{2+}-binding domains, *Journal of Biological Chemistry*, vol. 267, № 7, pp. 4394-4397

Campanella, M.E.; Chu, H.; Low, P.S. (2005). Assembly and regulation of a glycolytic enzyme complex on the human erythrocyte membrane. *Proceedings of the National Academy of Sciences of the United States of America*, vol. 102 , № 7, pp. 2402-2407

Carafoli, E. (1994). Plasma membrane calcium ATPase: 15 years of work on the purified enzyme, *Federation of American Societies for Experimental Biology (FASEB) Journal*, vol. 8, № 13, pp. 993-1002.

Carafoli, E. (2002) Calcium signaling: a tale for all seasons. *Proceedings of the National Academy of Sciences of the United States of America*, vol. 99, № 3, pp. 1115-1122

Ceolotto, G.; Corelin, P.; Clari, G.; Semplicini, A.; Canessa, M. (1998). Protein kinase C and insulin regulation of red blood cell Na+/H+ exchange and cytosolic free calcium in human. *American Journal of Hypertension*, vol. 11, (3 Pt 1), pp. 81-87

Chauhan, V.P.; Singh, S.S.; Chauhan, A.; Brockerhoff. H. (1993). Magnesium protects phosphatidylinositol-4,5- bisphosphate-mediated inactivation of casein kinase I in erythrocyte membrane. *Biochimica et Biophysica Acta,* vol.1177, № 3, pp.318-321

Ciana, A,; Minetti, G,; Balduini, C. (2004). Phosphotyrosine phosphatases acting on band 3 in human erythrocytes of different age: PTP1B processing during cell ageing. *Bioelectrochemistry.* Vol. 62, № 2, pp.169-173

Clari, G.; Brunati, A. M.; Moret, V. (1987). Membrane-bound phosphotyrosil-protein phosphatase activity in human erythrocytes. Dephosphorylation of band 3. *Biochemical and Biophysical Research . Communication*, vol. 142, № 2, pp. 587-594

Cohen NS, Ekholm JE, Luthra MG, Hanahan DJ. (1976). Biochemical characterisation of density-separated human erythrocytes. *Biochimica et Biophysica Acta,* vol. 419, pp. 229-237

Cohen, C.M.; Foley, S.F. (1986). Organization of the spectrin-actin-band 4.1 ternary complex and its regulation by band 4.1 phosphorylation. In: *Membrane and cytoskeletal-membrane association*, pp. 212-222, Alan R.Liss, Inc.N.Y

Cohen, C.M.; Gascard. P. (1992). Regulation and posttranslational modification of erythrocyte membrane- skeletal proteins. *Seminars in Hematology,* vol. 29, № 4, pp. 244-292

Cohen, M.S.; Mao, J.; Rasmussen, J.; Serodi, J.; C.; Brittigan, B.E. (1992). Interaction of lactoferrin and lipopolysacharide (LPS). Effect on the antioxidant property of lactoferrin and the ability of LPS to prime human neutrophils for enchanced superoxide formation. *Journal of Infection Disease,* vol. 166, № 6, pp. 1375- 1378

Gonçalves, I.; Saldanha, C.; Martins e Silva, J. (2001) Beta-estradiol effect on erythrocyte aggregation-a controlled in vitro study. *Clinical Hemorheology and Microcirculation,* vol. 25, № (3-4), pp. 127-134

Corry, D.H.; Joolhar, F.S.; Hori, M.T.; Tuck, M.L. (2002). Decreased erythrocyte insulin binding in hypertensive subjects with hyperinsulinemia. *American Journal of Hypertension*, vol. 15, (4Pt1), pp. 296-301

Corti, C.; Leclerc L'Hostis, E.; Quadroni, M.; Schmid, H.; Durussel, I.; Cox, J.; Dainese Hatt, P.; James, P.; Carafoli, E. (1999). Tyrosine phosphorylation modulates the interaction of calmodulin with its target proteins. *European Journal of Biochemistry,* vol. 262, № 3, pp.790-802

Curran C.S., Demick K.P. and Mansfield J.M. (2006). Lactoferrin activates macrophages via TLR4-dependent and -independent signaling pathways. *Cell Immunol.* vol.242, №1, pp. 23-30

Damonte, G.; Sdraffa, A.; Zocchi E.; Guida, L.; Polvani, C.; Tonetti, M.; Benatti, U.; Boquet, P.; De Flora, A. (1990). Multiple small molecular weight guanine nucleotide-binding proteins in human erythrocyte membranes. *Biochemical and Biophysical Research Communication*, Vol. 166, № 3 pp. 1398-13405

Danilov, Y.N.; Cohen, C.M. (1989). Wheat germ agglutinin but not concanavalin A modulates protein kinase C-mediated phosphorylation of red cell skeletal proteins. *FEBS Letters*, vol. 257, № 2, pp. 431-434

Danilov, Y.N.; Fennell, R.; Ling, E.; Cohen, C.M. (1990). Selective modulation of band 4.1 binding to erythrocyte membranes by protein kinase C. *Journal of Biological Chemistry,* vol. 265, № 5, pp.2556-2562

Darbonne, W.C.; Rice, G.C.; Mohler, M.A.; Apple, T.; Hebert, C.A.; Valente, A.J.; Baker, J.B. (1991). Red blood cells are a sink for interleukin 8, a leukocyte chemotaxin. *Journal of Clinical .Investigation,* vol. 88, № 4, pp. 1362-1369

De La Tour, D.D.; Raccah, D.; Jannot M..F.; Coste T.; Rougerie C.; Vague P. (1998). Erythrocyte Na+/K+ ATPase activity and disease: relationship with C-peptide level. *Diabetologia,* vol. 41, № 9, pp. 1080-1084

de Oliveira, S.; Silva-Herdade, A.S.; Saldanha, C. (2008). Modulation of erythrocyte deformability by PKC activity. *Clinical Hemorheology and Microcirculation,* vol. 39, № (1-4), pp.363-373

de Winter, R.J.; Manten, A.; de Jong, Y.P.; Adams, R.; van Deventer, S.J.; Lie, K.L. (1997). Interleukin-8 released after acute myocardial infarction is mainly bound to erythrocytes. *Heart,* vol. 78, № 6, pp.598-602

Djemli-Shipkolye, A.; Gallice, P.; Coste, T.; Jannot, M.F.; Tsimaratos, M.; Raccah, D.; Vague, P. (2000). The effects ex vivo and in vitro of insulin and C-peptide on Na/K adenosine triphosphatase activity in red blood cell membranes of type 1 diabetic patients. *Metabolism,* vol. 49, № 7, pp.868-872

Done´, S.C.; Leibiger, I.B.; Efendiev, R.; Katz, A.I.; Leibiger, B.; Berggren, P.O.; Pedemonte, C.H.; Bertorello A.M. (2002). Tyrosine 537 within the Na, K-ATPase-Subunit Is Essential for AP-2 Binding and Clathrin-dependent Endocytosis. *Journal of Biological Chemistry,* vol. 277, № 19, 17108–17111

Dreytuss, G.; Schwartz, K.J.; Blotet E.R. (1978). Compartmentalization of cyclic AMP-dependent protein kinases in human erythrocytes. *Proceedings of the National Academy of Sciences of the United States of America,* vol. 75, № 12, pp. 5926-5930

Dutta-Roy, A.K.; Kahn, N.N.; Sinha, A.K. (1991). Interaction of receptors for prostaglandin E1/prostacyclin and insulin in human erythrocytes and platelets. *Life Sciences,* vol. 49, № 16, pp.1129-1139.

Eda, S.; Kikugawa, K.; Beppu, M. (1996). Binding characteristics of human lactoferrin to the human monocytic leukemia cell line THP-1 differentiated into macrophages. *Biological & Pharmaceutical Bulletin,* vol. 19, № 2 pp.167-175

el-Andere, W.; Lerário, A.C.; Netto, D.G.; Wajchenberg, B.L. (1995). Erythrocyte insulin-like growth factor-I receptor evaluation in normal subjects, acromegalics, and growth hormone-deficient and insulin-dependent diabetic children. *Metabolism,* vol. 44, № 7, pp. 923-928

Ellsworth, M.L.; Ellis, C.G.; Goldman, D.; Stephenson, A.H.; Dietrich, H.H.; Sprague RS. (2009). Erythrocytes: oxygen sensors and modulators of vascular tone. *Physiology (Bethesda).* Vol. 24, 107-116

Falkenstein,E.; Tillmann, H-C.; Christ, M.; Feuring, M.; Wehling, M. (2000). Multiple actions of steroid hormones - a focus on rapid, nongenomic effects. *Pharmacological Reviews,* vol. 52, № 4, pp. 513–555

Farmer, B.T. 2nd; Harmon, T.M.; Butterfield, D.A.(1985). ESR studies of the erythrocyte membrane skeletal protein network: influence of the state of aggregation of spectrin on the physical state of membrane protein, bilayer lipids, and cell surface carbohydrates. *Biochimica et Biophysica Acta,* vol. 821, № 3, pp. 420-430

Fiorani, M.; Accorsi, A.; Cantoni, O. (2003). Human red blood cells as a natural flavonoid reservoir. *Free Radical Research,* vol. 37, №12, pp.1331-1338

Föller, M.; Huber, S.M., Lang, F. (2008). Erythrocyte programmed cell death. *IUBMB Life.* vol. 60, № 10, pp. 661- 668

Fossel, E.T.; Solomon, A.K. (1978). Ouabain-sensitive interaction between human red cell membrane and glycolytic enzyme complex in cytosol. *Biochimica et Biophysica Acta,* vol. 510, № 1, pp. 99-111

Gaop J. ; Yao, Y.; Squier, T. C. (2001). Oxidatively modified calmodulin binds to the plasma membrane Ca-ATPase in a nonproductive and conformationally disordered complex. *Biophysical Journal,* vol. 80, № 4, pp. 1791-1801

Garrillo, J.J.; Ibares, B.; Esteban-Gamboa, A.; Felin, J.E. (2000). Involvement of both phosphatidylinositol 3-kinase and p44/p42 mitogen-activated protein kinase pathways in the short-term regulation of pyruvate kinase L by insulin. *Endocrynology,* vol. 142, № 3, pp. 1057-1064

Giraud, F.; Gascard P.; Sulpice J. C. (1988). Stimulation of polyphosphoinositide turnover upon activation of protein kinases in human erythrocytes. *Biochimica et Biophysyca Acta,* vol. 968, № 3, pp. 367-378.

Glasser, T.; Schwartz-Benmier, N.; Barnoy, S.; Barak, S.; Eshhar, Z.; Kosower, N.S.. (1994). Calpain (Ca2+-dependent thiol protease) in erythrocytes from young and old individuals. *Proceedings of the National Academy of Sciences of the United States of America,* vol. 91, № 17, pp. 7879-7893

Golden, G.A.; Mason, R.P.; Tulenko, T.N.; Zubenko, G.S.; Rubin, R.T. (1999). Rapid and opposite effects of cortisol and estradiol on human erythrocyte Na+/K+ -ATPase activity: intercalation into the cell membrane. *Life Sciences,* vol. 65, № 12, pp. 1247-1255

Goldenberg, H.; Dodel, B.; Seidl, D. (1990). Plasma membrane Fe2-transferrin reductase and iron uptake in K562 cells are not directly related. *European Journal of Biochemistry,* vol. 192, № 2, pp. 475-480

Gonçalves,, I.; Saldanha, C.; Martins e Silva, J. (2001). Beta-estradiol effect on erythrocyte aggregation - a controlled in vitro study. *Clinical Hemorheology and Microcirculation,* vol. 25, № (3-4), pp. 127-134

Govekar, R. B.; Zingde, S.M. (2001). Protein kinase C isoforms in human erythrocytes. *Annals of Hematology,* vol. 80, № 9, pp. 531-534

Graham, C.; Auruch, J.; Fairbanks, G. (1974). Phosphoprotein phosphatase of the human erythrocyte. *Biochemical and Biophysical Research Communication,* vol. 72, № 2, pp. 701-708

Guerini, D.; Pan, B.; Carafoli, E. (2003). Expression, purification, and characterization of isoform 1 of the plasma membrane Ca2+ pump: focus on calpain sensitivity. *Journal of Biological Chemistry,* vol. 278, № 40, pp. 38141-38148.

Gusev, G.P.; Agalakova, N.I. (2000). Na, K-Pump activation by isoproterenol, methylxanthines, and iodoacetate in erythrocytes of the frog Rana temporaria. *Zhurnal evoliutsionnoi biokhimii i fiziologii,* vol. 36, № 2 , pp.106-111

Ha, E.J.; Smith, A.M. (2003). Plasma selenium and plasma and erythrocyte gluthatione peroxidase activity increase with estrogen dunig the menstrual cycle. *Journal of the American College of Nutrition,* vol. 22, № 1, pp.43-51

Haas, M.; Wang, H.; Tian, J.; Xie, Z. (2002). Src-mediated inter-receptor cross-talk between the Na+/K+-ATPase and the epidermal growth factor receptor relays the signal from ouabain to mitogen-activated protein kinases. *Journal of Biological Chemistry*, vol. 277, № 21, pp. 18694-18702

Hamlyn, J.M.; Duffy, T. (1978). Direct stimulation of human erythrocyte membrane (Na+ +K+) ATPase activity in vitro by physiological concentration of d-aldosterone. *Biochemical and Biophysical .Research Communication.*, vol.84, №2, pp.458-464

Hanson, M.S.; Stephenson, A. H.; Bowles, E.A.; Sprague, R.S. (2010). Insulin inhibits human erythrocyte cAMP accumulation and ATP release: role of phosphodiesterase 3 and phosphoinositide 3-kinase. *Experimentsl Biology and Medicine (Maywood)*, vol. 235, № 2, pp. 256-262

Harrison, M.L.; Rathinavelu, P.; Arese, P.; Geahlen, R.L.; Low, P.S. (1991). Role of band 3 tyrosine phosphorylation in the regulation of erythrocyte glycolysis. *Journal of Biological Chemistry*, vol. 266, № 7, pp. 4106-4111

Haslauer, M.; Baltensperger, K.; Porzig, H. (1998). Thrombin and phorbol esters potentiate Gs-mediated c.AMP formation in intact human erythroid progenitors via two synergistic signaling pathways converging on adenylate cyclase type VII. *Molecular Pharmacology*, vol. 53, № 5, pp. 837-845

Haworth, RS.; Sinnet-Smithh, J. (1999). Protein kinase D inhibits plasma membrane Na+/H+ exchanger activity. *American Journal of Physiology*, vol. 277, № (6 pt 1), pp. C1202-C1209

Hecht, D.; Zick, Y. (1992). Selective inhibition of protein tyrosine phosphatase activities by H2O2 and vanadate in vitro. *Biochemical and Biophysical Research Communication*, vol. 188, № 2, pp.773-779

Hermand, P.; Gane, P.; Huet, M.; Jallu, V.; Kaplan, C.; Sonneborn, H.H.; Cartron, J.P.; Bailly, P. (2003) Red cell ICAM-4 is a novel ligand for platelet-activated alpha IIbbeta 3 integrin. *Journal of Biological Chemistry*, vol. 278, № 7, pp. 4892-4898

Horga, J.F,; Gisbert, J.; De Agustín, J.C.; Hernández, M.; Zapater, P. (2000). A beta-2-adrenergic receptor activates adenylate cyclase in human erythrocyte membranes at physiological calcium plasma concentrations. *Blood Cells, Molecules and Diseases*, vol. 26 № 3, pp. 223-228

Horne, W.C.; Prinz, W.C.; Tang, E.K. (1990). Identification of two cAMP-dependent phosphorylation sites on erythrocyte protein 4.1.*Biochimica et Biophysica Acta*, vol. 1055, № 1, pp. 87-92

Hubert, E.M.; Musch, M.W.; Goldstein, L. (2000). Inhibition of volume stimulated taurine efflux and tyrosine kinase activity in the skate red blood cells. *Pflugers Archiv*, vol. 440, № 1, pp.132-139

Ikeda, K.; Kikuchi, A.; Takai, Y. (1988). Small molecular weight GTP-binding proteins in human erythrocyte ghosts. *Biochem Biophys Res Commun.*, Vol. 156, № 2, pp. 889-897.

Ivanova, T.I.; Agalakova, N.I.; Gusev, G.P (2006) Activation of sodium transport in rat erythrocytes by inhibition of protein phosphatases 1 and 2A. *Comparative Biochemistry and Physiology: Part B - Biochemistry and Molecular Biology*, vol. 145, № 1 pp.60-67

Kajikawa, M.; Ohta, T.; Takase, M.; Kawase, K.; Shimamura, S.; Matsuda. I (1994). Lactoferrin inhibits cholesterol accumulation in macrophages mediated by acetylated or oxidized low-density lipoproteins. *Biochimica et Biophysica Acta*, vol. 1213, № 1, pp. 82-90.

Kaloyianni, M.; Bourikas, D.; Koliakos, G. (2001). The effect of insulin on Na+-H+ antiport activity of obese and normal subjects erythrocytes. *Cellular Physiology and Biochemistry*, vol. 11, № 5, pp.253-258

Kanbak, G.; Akynz, M.; Jnab, M. (2001). Preventive effect of betaine in ethanol-induced membrane lipid composition and membrane ATPases. *Archives of Toxicology*, vol. 75, № 1, pp. 59-61

Kiener, P.A.; Carroll, D.; Roth, B.J.; Westhead, E.W. (1987). Purification and characterization of a high molecular weight type I phosphoprotein posphatase from human erythrocytes. *Journal of Biological Chemistry*, vol. 262, № 5, pp. 2016-2024

Kiener P.A.; Westhead, E.W. (1980). Dephosphorylation and reactivation of phosphorylated pyruvate kinase by a cytosolic phosphoprotein phosphatases from human erythrocytes. *Biochemical and Biophysical Research Communication*, vol. 96, № 2, pp. 551-557

Kitagawa, S.; Sakamoto, H.; Tano, H. (2004). Inhibitory effects of flavonoids on free radical-induced hemolysis and their oxidative effects on hemoglobin. *Chemical & Pharmaceutical Bulletin (Tokyo)*, vol. 52, № 8, pp.999-1001.

Klarl, B.A.; Lang, P.A.; Kempe, D.S.; Niemoeller, O.M.; Akel, A.; Sobiesiak, M.; Eisele, K.; Podolski, M.; Huber, S.M.; Wieder, T.; Lang, F. (2006). Protein kinase C mediates erythrocyte "programmed cell death" following glucose depletion. *Americal Journal of Physiology- Cell Physiology*, vol. 290, № 1, pp. C244-C253.

Klein, H.H.; Muller, R.; Drenckhan, M.; Schutt, M.; Batge, B.; Fehm, H.L. (2000). Insulin activation of insulin receptor kinase in erythrocytes is not altered in non-insulin-dependent diabetes and not influenced by hyperglycemia. *Journal of Endocrinology*, vol. 166, № 2, pp. 275-281

Konstantinou-Tegou, A.; Kaloyianni, M.; Bourikas, D.; Koliakos, G. (2001). The effect of leptin on Na (+)-H(+) antiport (NHE1)activity of obese and normal subjects erythrocytes. *Molecular and Cellular Endocrinology*, vol. 25, № 183, pp. 11-18

Kosk-Kosicka, D.; Bzdega, T.; Johnson, J. D. (1990). Fluorescence studies on calmodulin binding to erythrocyte Ca2+'-ATPase in different oligomerization states, *Biochemistry*, vol. 29, № 7, pp. 1875 - 1879

Lang, F.; Gulbins, E.; Lang, P. A.; Zappulla, D.; Föller, M. (2010). Ceramide in suicidal death of erythrocytes. *Cellular Physiology and Biochemistry*, vol.26, № 1, pp. 21-28.

Lang, F.; Lang, K. S.; Lang, P. A.; Huber, S. M.; Wieder, T. (2006). Mechanisms and significance of eryptosis. *Antioxidants &Redox Signal*ing, vol.8, № (7-8), pp.1183-1192

Lang, P.A.; Kempe, D.S.; Akel, A.; Klarl, B.A.; Eisele, K.; Podolski, M.; Hermle, T.; Niemoeller, O.M.; Attanasio, P.; Huber, S.M.; Wieder, T.; Lang, F.; Duranton, C. (2005). Inhibition of erythrocyte "apoptosis" by catecholamines. *Naunyn Schmiedeberg's Archives of Pharmacology*, vol. 372, № 3, pp. 228-235

Lawrence, W. D.; Deziel M.R.; Davis, P. J.; Schoenl, M.; Davis, F.B.; Blas, S. D. (1993). Thyroid hormone stimulates release of calmodulin-enhancing activity from human erythrocyte membranes in vitro. *Clinical Science*, (London). Vol. 84, № 2, pp. 217-223

Lecomte, M.C.; Galand, C.; Biovin, P. (1980). Inhibition of human erythrocyte casein kinase by methylxanthines. Study of inhibition mechanism. *FEBS Letters*, vol. 116, № 1, pp.45-47

Legrand, D.; Elass, E.; Pierce, A.; Mazurier, J. (2004). Lactoferrin and host defence: an overview of its immuno-modulating and anti-inflammatory properties. *Biometals,* Vol. 17, № 3 pp. 225-229

Leroy, D.; Filhol, O.; Delcros, J. G.; Pares, S.; Chambaz, E. M.; Cochet, C. (1997). Chemical features of the protein kinase CK2 polyamine binding site. *Biochemistry,* vol. 36, № 6, pp. 1242-1250

Lévy-Toledano, S.; Grelac, F.; Caen, J.P.; Maclouf, J. (1995). KRDS, a peptide derived from human lactotransferrin, inhibits thrombin-induced thromboxane synthesis by a cyclooxygenase-independent mechanism. *Journal of Thrombosis and Haemostasis,* Vol. 73, № 5, pp. 857-861

Li, Q.; Jungmann, V.; Kiyatkin, A.; Low, P.S (1996). Prostaglandin E2 stimulates a Ca2+-dependent K+ channel in human erythrocytes and alters cell volume and filterability. *Journal of Biological Chemistry,* vol. 271, № 31, pp. 18651-18656.

Liu, J.; Tian, J.; Haas, M.; Shapiro, J.I.; Askari, A.; Xie, Z. (2000). Ouabain interaction with cardiac Na+/K+-ATPase initiates signal cascades independent of changes in intracellular Na+ and Ca2+ concentrations. *J Biol Chem.* Vol. 275, № 36, pp. 27838-27844.

Low, P.S.; Ceahlen, R.L.; Mehler, E.; Harrison, M.L. (1990). Extracellular control of erythrocyte metabolism mediated by a cytoplasmic tyrosine kinase. *Biomedica Biochimica. Acta,* vol. 140, № (2-3), pp. 135-140

Low, H.; Grebing, C.; Lindgren, A.; Tally, M.; Sun, I. L.; Crane, F.L. (1987). Involvement of transferrin in the reduction of iron by the transplasma membrane electron transport system. *Journal of Bioenergetics and Biomembranes,* vol. 19, № 5, pp. 535-549

Low, P.S.; Rathinavelu, P.; Harrison, M. L. (1993). Regulation of glycolysis via reversible enzyme binding to the membrane protein, band 3. *Journal of Biological Chemistry,* vol. 268, № 20, pp.14627-14631

Lu, D.; Yan, H.; Othman, T.; Turner, C. P.; Woolf, T.; Rivkees, S.A. (2004). Cytoskeletal protein band 4.1G binds to the third intercellular loop of the a1 adenosine receptor and inhibits receptor action. *Biochemical Journal,* vol. 377, (Pt 1), pp. 51-59

Maccaglia, A.; Mallozzi, C.; Minetti, M. (2003). Differential effects of quercetin and resveratrol on Band 3 tyrosine phosphorylation signalling of red blood cells. *Biochemical and Biophyscal Research Communication,* vol. 305, № 3, pp. 541-547

Madshus, I.H. (1988). Regulation of intracellular pH in eukaryotic cells. *Biochemical Journal ,* vol. 250, № 1, pp.1-8

Maekawa, T.; Fujihara, M.; Ohtsuki. K. (2002). Characterization of human lactoferricin as a potent protein kinase CK2 activator regulated by A-kinase in vitro. *Biological & Pharmaceutical Bulletin,* vol. 25, 118-121

Mahrenholz, A.M.; Lan, L.; Mansour, T. E. (1991). Phosphorylation of heart phosphofructokinase by Ca2+/calmodulin protein kinase. *Biochemical and Biophysical Research Commununication,* vol. 174, № 3 , pp. 1255-1259.

Mallozzi, C.; De Franceschi, L.; Brugnara, C.; Di Stasi, A. M. (2005). Protein phosphatase 1alpha is tyrosine-phosphorylated and inactivated by peroxynitrite in erythrocytes through the src family kinase fgr. *Free Radical Biology & Medicine,* vol. 38, № 12, pp. 1625-1636

Mallozi, C.; Di Stasi, A.M.; Minetti, M. (1997). Peroxynitrite modulates tyrosine-dependent signal transduction pathway of human erythrocyte band 3. *FASEB Journal,* vol. 11, № 14, 1281-1290

Mandal, D.; Mazumder, A.; Das, P.; Kundu, M.; Basu, J. (2005) Fas-, caspase 8-, and caspase 3-dependent signaling regulates the activity of the aminophospholipid translocase and phosphatidylserine externalization in human erythrocytes. *Journal of Biological Chemistry,* vol. 280, № 47, pp. 39460-39467

Maneva, A.; Angelova-Gateva, P.; Taleva, B.; Maneva-Radicheva, L. (2007). Lactoferrin stimulates erythrocyte Na^+/K^+ Adenosine Triphosphatase: Effect of some modulators of membrane phosphorylation. *Zeitschruft fur Naturforschung C,* vol. 62, № (11-12), pp. 897- 905

Maneva, A.; Taleva, B.; Maneva, L. (2003). Lactoferrin-Protector against oxidative stress and regulation of glycolysis in human erythrocytes. *Zeitschruft fur Naturforschung C,* vol. 58, № (3-4), pp. 256-262

Manno, S.; Takakuwa, Y.; Nagao, K.; Mohandas, N. (1995). Modulation of erythrocyte membrane mechanical function by beta-spectrin phosphorylation and dephosphorylation. *Journal of Biological Chemistry,* vol. 270, № 10, pp. 5659-5965

Marinho-Carvalho, M.M.; Zancan, P.; Sola-Penna, M. (2006). Modulation of 6-phosphofructo-1-kinase oligomeric equilibrium by calmodulin: formation of active dimers. *Molecular Genetics and Metabolism,* vol. 87, № 3, pp. 253-261

Martin, D.W.; Sachs, J.R. (1992). Cross-linking of the erythrocyte (Na^+, K^+) –ATPase. Chemical cross-linkers induce α-subunit-band 3 heterodimers and do not induce α-subunit homodimers. *Journal of Biological Chemistry,* vol. 267 № 33, pp. 23922-23929

Minetti, G.; Ciana, A.; Baldini, C. (2004). Differential sorting of tyrosine kinases and phosphotyrosine phosphatases acting on band 3 during vesiculation of human erythrocytes. *Biochemical Journal,* vol. 377, (Pt 2), pp. 487-497

Minetti, G.; Low, P.S. (1997). Erythrocyte signal transduction pathways and their possible functions. *Current Opinion in Hematology,* vol. 4, № 2, pp. 116-121

Minetti, G.; Piccinini, G.; Balduini C, Seppi C, Brovelli A. (1996). Tyrosine phosphorylation of band 3 in Ca2+/ A23187-treated human erythrocytes. *Biochemical Journal,* vol. 320, (Pt. 2), pp. 445-450

Mistry, N.; Drobni, P.; Näslund, J.; Sunkari, V.G.; Jenssen, H.; Evander, M. (2007). The anti-papillomavirus activity of human and bovine lactoferricin. *Antiviral Research,* vol. 75, № 3, pp. 258-265.

Mohammadi, K.; Kometiani P.; Xie, Z.; Askari, A. (2001). Role of protein kimase C in signal pathways that link Na^+/K^+-ATPase to ERK1/2. *Journal of Biological Chemistry,* vol. 276, № 45, pp. 42050-42056

Moulinoux, J.P.; Calve, M.; Quemener, V.; Quash, G. (1984). In vitro studies on the entry of polyamines into normal red blood cells. *Biochimie,* vol. 66, № 5, pp.385-393

Murakami, J.; Maeda, N.; Kon, K.; Shiga, T. (1986), A contribution of calmodulin to cellular deformability of calcium-loaded human erythrocytes. *Biochimica et Biophysica Acta,* vol. 863, No 1, pp. 23-32

Muravyov, A. V.; Tikhomirova, I. A.; Maimistova, A. A.; Bulaeva, S. V.; Zamishlayev, A. V.; Batalova, E. A. (2010). Crosstalk between adenylyl cyclase signaling pathway and Ca2+ regulatory mechanism under red blood cell microrheological changes. *Clinical Hemorheology and Microcirculation,* vol. 45, No (2-4), 337- 345

Muravyov, A.V.; Tikhomirova, I. A.; Maimistova, A. A.; Bulaeva, S.V. (2009). Extra- and intracellular signaling pathways under red blood cell aggregation and deformability changes. *Clinical Hemorheology and Microcirculation*, vol. 43, No 3, pp.223-232.

Musch, M.W.; Hubert, E. M.; Goldstein, L (1999). Volume expansion stimulates p72(syk) and p56(lyn) in skate erythrocytes. *Journal of Biological Chemistry*, vol. 274, No 12, pp. 7923-7928

Nakashima, K.; Fujii, S.; Kaku, K.; Kaneko, T. (1982). Calcium-calmodulin dependent phosphorylation of erythrocyte pyruvate kinase. *Biochimical and Biophysical Research Commununication*, Vol. 104, No 1, pp. 551-557

Nelson, M.J.; Ferrell J. E.; Huestis, W.H. (1979). Adrenergic stimulation of membrane protein phosphorylation in human erythrocytes. *Biochimica et Biophysica Acta*, vol. 558, No 1, pp. 136-140

Okafor, M.C.;.Schiebinger. R.J. Yingst, D.R. (1997). Evidence for a calmodulin-dependent phospholipase A2 that inhibits Na-K-ATPase. *American .Journal of Physiology*, vol. 272, (4 Pt 1), pp. C1365-C1372

Olearczyk, J.J.; Stephenson, A.H.; Lonigro, A.J.; Sprague, R.S. (2004). Heterotrimeric G protein Gi is involved in a signal transduction pathway for ATP release from erythrocytes. *American Journal of Physiology - Heart and Circulatory Physiology*, vol. 286, No 3, H940-H945

Orringer, E.P.; Roer, M.E. (1979). An ascorbate-mediated transmembrane-reducing system of the human erythrocyte. *Journal of Clinical Investigation*, vol. 63, No 1, pp. 53-58

Paajaste, M.; Nikinmar, M. (1991). Effect of noradrenaline on the methemoglobin concentration of rainbow trout red cells. *Journal of Experimental Zoology*, vol. 260, No 1, pp. 28-32

Pantaleo, A.; De Franceschi, L.; Ferru, E.; Vono, R.; Turrini, F. (2010). Current knowledge about the functional roles of phosphorylative changes of membrane proteins in normal and diseased red cells. *Journal of Proteomics*. Vol. 73, No 3, pp. 445-455

Pantaleo, A.; Ferru, E.; Carta F.; Mannu, F.; Simula, L. F.; Khadjavi, A.; Pippia, P.; Turrini, F. (2011). Irreversible AE1 Tyrosine Phosphorylation Leads to Mem brane Vesiculation in G6PD Deficient Red Cells . *PLoS ONE*, vol. 6, No 1, e15847, 1-8

Pasini, E.M.; Kirkegaard, M.; Mortensen, P.; Lutz, H.U.; Thomas, A.W.; Mann, M. (2006). In-depth analysis of the membrane and cytosolic proteome of red blood cells. *Blood*, vol. 108, No 3, pp. 791-801

Pasini, E.M.; Mann, M.; Thomas, A.W. (2010). Red blood cell proteomics. *Transfusion Clinique et Biologique*, vol. 17, No 3, pp. 151–164

Perry, S.F.; Wood, C.M.; Thomas, S.; Walsch, P.J. (1991). Adrenergic stimulation of carbon dioxide excretion by trout red blood cells in vitro is mediated by activation of Na+/H+ exchange. *Journal of Experimental Biology*, vol. 157, pp. 367-380

Petrov, V.; Amery, A.; Lijnen, P. (1994). Role of cyclic GMP in atrial-natriuretic-peptide stimulation of erythrocyte Na+/H+ exchange. *European Journal of Biochemistry*, vol. 221 No, pp. 195-1999

Petrov, V,; Lijnen P. (1996) Regulation of human erythrocyte Na+/H+ exchange by soluble and particulate guanylate cyclase. *American Journal of Physiology*; vol. 271, (5 Pt 1), pp. C1556-C1564

Phan-Dinh-Tuy, F.; Henry, J.; Rosenfeld, C.; Kahn, A.(1983), High tyrosine kinase activity in normal non proliferating cells. *Nature*, vol. 305, No 5933, pp. 435-440

Polychronakos, C.; Guyda, H. J.; Posner B. I. (1983). Receptors for the insulin-like growth factors on human erythrocytes. *Journal of Clinical Endocrinology and Metabolism*, vol. 57, No 2, pp. 436- 438.

Qadri, S.M; Föller, M.; Lang, F. (2009). Inhibition of suicidal erythrocyte death by resveratrol. *Life Sciences*, vol. 85, No (1-2), pp.33-38

Rasmussen, H.; Lake, W.; Allen, J.E. (1975). The effect of catecholamines and prostaglandins upon human and rat erythrocytes. *Biochimica et Biophys Acta*, vol. 411, No 1, pp. 63-73.

Rivera, A..; Jarolim, P.; Brugnara , C. (2002). Modulation of Gardos channel activity by cytokines in sickle erythrocytes. *Blood*, vol. 99, No 1, pp.357-603

Rizvi, S.I.; Incerpi, S.; Luly, P. (1994). Insulin modulation of Na/H antiport in rat red blood cells. *Indian Journal of Biochemistry and Biophysics*, vol. 31, No 2, pp. 127-130

Ron, D.; Kazanietz, M.G. (1999). New insight into the regulation of protein kinase C and novel phorbol ester receptors. *FASEB Journal*, vol. 13, No 13, pp. 1658-1676

Sacks, D.B.; Lopez, M.M.; Li, Z.; Kosk-Kosicka, D. (1996). Analysis of phosphorylation and mutation of tyrosine residues of calmodulin on its activation of the erythrocyte Ca (2+)-transporting ATPase. *European Journal of Biochemistry*, vol. 239, No 1, pp. 98-104.

Saido, T.C.; Sorimachi, H.; Suzuki, K. (1994). Calpain: new perspectives in molecular diversity and physiological/pathological involvement. *FASEB Journal*, vol. 8, No 11, pp. 814- 822

Saldanha, C.; Silva, A.S.; Gonçalves, S.; Martins-Silva, J. (2007). Modulation of erythrocyte hemorheological properties by band 3 phosphorylation and dephosphorylation. *Clinical Hemorheology and Microcirculation*, vol. 36, No 3, pp.183-194.

Santos-Silva, A.; Castro, E.M.; Teixeira, N.A.; Guerra, F.C; Quintanilha, A. (1998). Erythrocyte membrane band 3 profile imposed by cellular aging by activated neutrophils and by neutrophilic elastase. *Clinica Chimica Acta*, vol. 275, № 2, pp.185-196

Sardet, C.; Farounoux, P.; Poreysseger, J. (1991). Alpha-thrombin, epidermal growth factor, and okadaic acid activate the Na^+/H^+ exchanger, NHE-1, phosphorylating a set of common sites. *Journal of Biological Chemistry*, vol. 266, No 29, pp. 19166-19171

Sartori, M.; Ceolotto, G.; Semplieini, A. (1999). MAPKinase and regulation of the sodium-proton exchanger in human red cell. *Biochemica et Biophysica Acta*, vol. 1421, No 1, pp. 140-148

Sato, Y.; Miura, T. and Suzuki, Y. (1990). Interaction of pentoxifylline with human erythrocytes. I. Interaction of xanthine derivatives with human erythrocyte ghosts. *Chemical Pharmaceutical Bulletin (Tokyo)*, vol. 38, No 2, pp. 552-554

Sauvage, M.; Maziere, P.; Fathalah, H.; Girard, E.. (2000). Insulin stimulates NHE1 activity by sequential activation of phosphatidylinositol3-kinase and PKC zeta in human erythrocytes. *European Journal of Biochemistry*, vol. 267, No 4, pp. 955-962

Seghieri, G.; Anichini, R.; Ciuti, M.; Gironi, A.; Bennardini, F.; Franconi, F. (1997). Raised erythrocyte polyamine levels in non-insulin -dependent diabetes mellitus with great vessel disease and albuminuria. *Diabetes Research and Clinical Practice*, vol. 37, No 1, pp. 15-20

Sheppard, H.; Tsien, W.H. (1975). Alteration in the hydrolytic activity, inhibitor sensitivity and molecular size of the rat erythrocyte cyclic AMP-phosphodiesterase by calcium and hypotonic sodium chloride. *Journal of Cyclic Nucleotide Research,* vol. 1, No 4, pp. 237-242

Smallwood, J.I.; Gugi, B.; Rasmussen, H. (1988). Regulation of erythrocyte Ca2+ pump activity by protein kinase C. *Journal of Biological Chemistry,* vol. 263, No 5, pp. 2195-2202

Smith, T.J.; Davis, F.B.; Davis, P.J. (1989). Retinoic Acid Is a Modulator of Thyroid Hormone Activation of Ca2+- ATPase in the Human Erythrocyte Membrane. *Journal of Biological Chemistry,* vol. 264, No 2, pp. 687-689

Sommerson, J.; Mulligan, J.A.; Lozeman, F.J.; Krebs, E.G. (1987). Activation of casein kinase in response to insulin and to epidermal growth factor. *Proceedings of National Academy of Sciences of the United States of America,* vol. 84, No 24, pp. 8834-8838

Sprague, R.S.; Bowles, E.A.; Hanson, M.S.; DuFaux, E.A.; Sridharan, M.; Adderley, S.; Ellsworth, M.L.; Stephenson, A.H. (2008). Prostacyclin analogs stimulate receptor-mediated cAMP synthesis and ATP release from rabbit and human erythrocytes. *Microcirculation.* vol. 15, No 5, pp. 461-471

Sprague, R. S.; Ellsworth, M. L.; Stephenson, A H.; Lonigro, A. J. (2001). Participation of cAMP in a signal-transduction pathway relating erythrocyte deformation to ATP release *American Journal of Physiology- Cell Physiology,* vol. 281,No 4, C1158–C1164

Sun, I.L.; Crane, F.L.; Morré, D.J.; Löw, H.; Faulk, W.P. (1991). Lactoferrin activates plasma membrane oxidase and Na+/H+ antiport activity. *Biochemical and Biophysical Reseach Communication,* vol. 176, № 1, pp. 498-504

Takayama, Y.; Mizumachi, K. (2001). Effects of lactoferrin on collagen gel contractile activity and myosin light chain phosphorylation in human fibroblasts. *FEBS Letters,* vol. 508, No 1, pp.111-116.

Taleva, B.; Maneva, A.; Sirakov. L. (1999). Essential metal ions alter the lactoferrin binding to the erythrocyte plasma membrane receptors. *Biological Trace Element Research,* vol. 68, No 1, pp. 2-23

Tanaka, T.; Omata, Y.; Isamida, T.; Saito, A.., Shimazaki, K.; Yamauchi, K.; Suzuki, N. (1998). Growth inhibitory effect of bovine lactoferrin to Toxoplasma gonadii achyzoites in murine mactophages: tyrosine phosphorylation in murine macrophages induced by bovine lactoferrin. *Journal of Veterinary .Medicine,* vol. 60, № 3, pp. 369-371

Terra, H.T., Saad, M.J.; Carvalho, C.R.; Vicentin, D.L.; Costa, F.F.; Saad, S.T. (1998). Increased tyrosine phosphorylation of band 3 in hemoglobinopathies. *American Journal of Hematology,* vol. 58, No 3, pp. 224–230.

Thotathil, Z.; Jameson, M. B. (2007). Early experience with novel immunomodulators for cancer treatment. *Expert Opinion on Investigational Drugs,* Vol. 16, No 9, pp. 1391-1403.

Tsuda, K. (2006). Leptin receptor and membrane microviscosity of erythrocytes in essential hypertension. *American Joournal of Hypertension,* vol. 19, No 8, 874-875.

Tsukamoto, T.; Sonenberg, M. (1979). Catecholamine regulation of human erythrocyte membrane protein kinase. *Journal of Clinical Investation,* vol. 64, № 2, pp. 534-540.

Turrini, F.; Mannu, F.; Cappadero, M.; Ulliers, D.; Giribaldi, G.; Arese, P. (1994). Binding of naturally occurring antibodies to oxidatively modified erythrocyte band 3. *Bichimica et Biophysica Acta,* vol. 1190, No 2, pp. 287-303

Uhle, S.; Medalia, O.; Waldron, R.; Dumdey, R.; Henklein, P.; Bech-Otschir, D.; Huang, X.; Berse, M.; Sperling, J.; Schade R.; Dubiel, W. (2003). Protein kinase CK2 and protein kinase D are associated with the COP9 signalosome *The EMBO Journal*, vol. 22, No. 6, pp. 1302-1312

van den Akker, E.; Satchwell, T. J.; Williamson, R.C.; Toye, A. M. (2010). Band 3 multiprotein complexes in the red cell membrane; of mice and men. *Blood Cells Mol Dis*. Vol. 45, No 1, pp.1-8.

Vásárhelyi, B.; Szabó, T.; Vér, Á.; Tulassay, T. (1997). Measurement of Na^+/K^+-ATPase activity with an automated analyzer. *Clinical Chemistry,* vol. 43, No. 10, pp. 1986-1987

Villalonga, P.; Lopez-Alcala, C.; Chiloeches, A.; Gil, J.; Marais, R.., Bachs, O.; Agell N. (2002). Calmodulin prevents activation of Ras by PKC in 3T3 fibroblasts. *Journal of Biological Chemistry,* vol. 277, No 40, pp. 37929-37935

Villano, P.J.; Gallice, P.M.; Nicoara, A.E.; Honore, S.G.; Owczarczak, K.; Favre, R.G.; Briand, C.M. (2001). Polyamine secreted by cancer cells possibly account for the impairment of the human erythrocyte sodium pump. *Cell and Molecular Biology,* vol. 47, No 2, pp. 305-312

Wang, K.K.W.; Villalobo, A.; Roufogalis, B.D. (1989). Calmodulin binding proteins as calpain substrates. *Biochemical Journal,* vol. 262, No 3, pp. 693-706

Wang, K.K.; Wright, L.C.; Machan, C.L.; Allen, B.G.; Conigrave, A.D.; Roufogalis, B.D. (1991). Protein kinase C phosphorylates the carboxyl terminus of the plasma membrane Ca(2+)-ATPase from human erythrocytes. *Journal of Biological Chemistry,* vol. 266, No 14, pp. 9078-9085

Wei, T.; Tao, M. (1993). Human erythrocyte casein kinase II: characterization and phosphorylation of membrane cytoskeletal proteins. *Archives of Biochemistry and Biophysics,* Vol. 307, No 1, pp. 206-216

Weinberg, E.D. (2007). Antibiotic properties and applications of lactoferrin. *Current Pharmieutical Design,* vol. 13, No 8, pp. 801-811

Wright, L.C.; Chen, S.; Roufogalis. B.D. (1993). Regulation of the activity and phosphorylation of the plasma membrane Ca(2+)-ATPase by protein kinase C in intact human erythrocytes. *Archives of Biochemistry and Biophysics,* vol. 306, No 1, pp. 277-284.

Yingst, D.R.; Ye-Hu, J.; Chen, H.; Barrett, V. (1992). Calmodulin increases Ca^{2+}-dependent inhibition of Na, K-ATPase in human red blood cells. *Archives of Biochemistry and Biophysics,* vol. 295, pp. 49-54

Zancan, P.; Sola-Penna, M. (2005). Calcium influx: a possible role for insulin modulation of intracellular distribution and activity of 6-phosphofructo-1-kinase in human erythrocytes. *Molecular Genetics and Metabolism,* Vol. 86, No 3, pp. 392-400

Zazo, C.;Thiele, S.; Martín, C.; Fernandez-Rebollo, E.; Martinez-Indart, L.; Werner, R.; Garin, I.; Group, S.P.; Hiort, O.; Perez de Nanclares, G. (2011). Gsα activity is reduced in erythrocyte membranes of patients with psedohypoparathyroidism due to epigenetic alterations at the GNAS locus. *Journal of Bone and Mineral Research,* Vol. 26, No 8, pp.1864-1870.

Zhang, Y.; Dai, Y.; Wen, J.; Zhang, W.;Grenz, A.; Sun, H.; Tao, L.; Lu, G.; Alexander, D.C.; Milburn, M.V.; Carter-Dawson, L.; Lewis, D.E.; Zhang, W.; Eltzschig, H.K.;

Kellems, R.E.; Blackburn, M.R.;Juneja, H.S.; Xia, Y. (2011). Detrimental effects of adenosine signaling in sickle cell disease. *Nat Med.* Vol. 17, No 1, pp 79-86.

Zipser, Y.; Kosower. N.S. (1996). Phosphotyrosine phosphatase associated with band 3 protein in the human erythrocyte membrane. *Biochemical Journal*, vol. 314, (Pt.3), pp. 881-887

Zipser, Y.; Piade, A.; Barbul, A.; Korenstein, R.; Kosower, N.S. (2002). Ca2+ promotes erythrocyte band 3 tyrosine phosphorylation via dissociation of phosphotyrosine phosphatase from band 3. *Biochemical Journal*, vol. 368, (Pt 1), pp. 137- 144

Zipser, Y.; Plade, A.; Kosower, N.S. (1997). Erythrocyte thiol status regulates band phosphotyrosine level via oxidation/reduction of band 3 associated phosphotyrosine phosphatase. *FEBS Letters*, vol. 406, No (1-2), pp. 126-130

Zylinska, L.; Sobolewska, B.; Gulczynska, E.; Ochedalski, T.; Soszynski, M. (2002). Protein kinases activities in erythrocyte membranes of asphyxiated newborns. *Clinical Biochemistry*, vol. 35, No 2, pp. 93-98

Xie, Z.; Cai, T. (2003) Na^{+}-K^{+}- ATPase-mediated signal transduction: from protein interaction to cellular function. *Molecular Interventions*, vol. 3, No 3, pp.157-168

13

Role of Protein Kinase Network in Excitation-Contraction Coupling in Smooth Muscle Cell

Etienne Roux[1,2], Prisca Mbikou[3] and Ales Fajmut[4]
[1] *Univ. de Bordeaux, Adaptation Cardiovasculaire à l'ischémie, Pessac,*
[2]*INSERM, Adaptation Cardiovasculaire à l'ischémie, Pessac,*
[3]*Institute of Biomedical Technologies, Auckland University of Technology, Auckland,*
[4]*University of Maribor, Medical Faculty, Faculty of Natural Sciences and Mathematics and Faculty of Health Sciences,*
[1,2]*France*
[3]*New Zealand*
[4]*Slovenia*

1. Introduction

The aim of this chapter is to present a review of the main protein kinases involved in the signalling pathways between the stimulation of smooth muscle cell and the resulting dynamic contraction.

As in striated muscle cells, contraction in smooth muscle cells (SMC) is primarily triggered by intracytosolic Ca^{2+} ($[Ca^{2+}]_i$) increase. However, by contrast with striated muscle, in SMCs $[Ca^{2+}]_i$ increase generates contraction by activation of the myosin light chain kinase (MLCK) via the formation of the Ca^{2+}-calmodulin-MLCK complex. Activated MLCK phosphorylates the 20kDa regulatory light chain (MLC_{20}) of the thick filament. Phosphorylated MLC_{20} allows myosin to bind to actin, and this phosphorylation is critical for SMC contraction, since its inhibition generally abolishes agonist-dependent contraction, whereas relaxation is induced by MLC_{20} dephosphorylation by myosin light chain phosphatase (MLCP). Hence, excitation-contraction coupling in SMCs critically depends on Ca^{2+}-dependent MLCK activation and the balance of MLCK/MLCP activity. Moreover, it has been shown that these two major enzymes can be modulated by several protein kinases such as protein kinase A (PKA), protein kinase C (PKC) and Rho kinase (RhoK). As a consequence, these protein kinases indirectly modulate the activity of the thick filament of myosin. Additionally, actin-myosin interaction can be modulated by proteins associated to the thin filament of actin such as caldesmon and calponin, which modulation depends on their phosphorylation by several protein kinases. It appears then that in SMCs, the excitation-contraction coupling is determined by interacting signalling pathways involving various protein kinases, so that the canonical MLCK/MLCP enzymatic balance is embedded in a complex network of protein kinases acting both on the thick and thin filaments of the contractile apparatus. The chapter will present the functional structure of the contractile apparatus of SMC, and detail its activation by the Ca^{2+}-calmodulin-MLCK complex and down regulation by MLCP, and the action of the main PK that have been shown to modulate the sensitivity of the contractile apparatus to Ca^{2+}. The resultant behaviour of the SMC stimulated by contractile agonists not only depends on the structure of the regulatory network that modulates the contractile apparatus but also on the dynamics of the reactions. Mathematical modelling of this signalling network is of great help to decipher how different protein kinases involved in this network participate to the time-dependent behaviour of the contractile system. Several theoretical models have been developed in this sense. The chapter will present the general principles of these models, their predictions and how they help in understanding the role of PK in the time course of the contractile response of SCMs.

In this chapter, examples are taken from airway smooth muscle cells, though this chapter is not limited to this tissue, and describes general mechanisms present in other SMC types.

2. Ca^{2+} signalling and the contractile apparatus

2.1 Ca^{2+} signalling in airway smooth muscle cells

In smooth muscle cells, as in heart and skeletal muscle, $[Ca^{2+}]_i$ is the primary intracellular messenger that generates contraction. Stimulation of smooth muscle cells by contractile agonists results in an increase in $[Ca^{2+}]_i$, which in turn activates the contractile apparatus. However, the mechanisms by which Ca^{2+} stimulates the formation of actin-myosin bridges critically differ in smooth muscle cells from striated muscle cells (see below). Though Ca^{2+} signalling is not primarily triggered via direct protein kinase activation by contractile agonists, several agents that contribute to the Ca^{2+} signal can be modulated by protein kinases and, on the other hand, activation of several protein kinases involved in the modulation of the contractile apparatus is Ca^{2+} sensitive. Hence, this section will present a general overview of the mechanisms of Ca^{2+} signalling in smooth muscle cells, including in airways.

$[Ca^{2+}]_i$ value in basal conditions is maintained around 100 nM, a low value compared to extracellular medium, around 1-2 mM, and intracellular organelles like the sarcoplasmic reticulum, by active mechanisms of Ca^{2+} efflux through the plasma membrane and Ca^{2+} pumping into intracellular Ca^{2+} stores. When cells are stimulated, $[Ca^{2+}]_i$ is increased via different mechanisms which relative importance depends on smooth muscle cell types and/or contractile agonists. Basically, the general mechanisms of $[Ca^{2+}]_i$ increase involve either intracellular Ca^{2+} influx or intracellular Ca^{2+} release from internal Ca^{2+} stores, or both (Sanders, 2001; Somlyo *et al.*, 1994). Voltage-dependent Ca^{2+} occurs via L-type voltage-operated Ca^{2+} channel (VOC). Normal resting membrane potential in airway myocytes is around -60 mV (Roux *et al.*, 2006), and is highly dependent on basal K^+ conductance. Some K^+ channels are active at rest, and contribute to the resting K^+ conductance and hence resting membrane potential. Closure of K^+ channels tends to depolarize the plasma membrane which in turn induces extracellular Ca^{2+} influx via VOC. By contrast, additional activation of K^+ channels tends to hyperpolarize the plasma membrane, to inhibit VOC-dependent Ca^{2+} influx and hence to induce relaxation. Voltage-dependent Ca^{2+} influx and subsequent contraction is called the electromechanical coupling. In parallel to voltage-operated Ca^{2+} entry, Ca^{2+} influx can be generated by voltage-independent membrane channels. Receptor-operated channels can be opened by direct binding of the agonist on the membrane receptor, like P2X purinergic receptors (Mounkaila *et al.*, 2005). Another voltage-independent source of Ca^{2+} entry is Ca^{2+} influx through store-operated Ca^{2+} channels (SOC). These channels are activated by emptying of intracellular Ca^{2+} stores (Marthan, 2004).

The other origin of $[Ca^{2+}]_i$ increase is Ca^{2+} release from intracellular organelles (Sanders, 2001). The major one from which Ca^{2+} is released upon contractile stimulation is the sarcoplasmic reticulum (SR). Ca^{2+} release from the sarcoplasmic reticulum, with internal Ca^{2+} concentration in the millimolar range, occurs via two main types of sarcoplasmic receptors, the inositol-trisphosphate receptors ($InsP_3Rs$) and the ryanodine receptors (RyRs). $InsP_3Rs$ are activated by $InsP_3$, which is produced from phosphatidylinositol phosphate by phospholipase C (PLC) upon stimulation by contractile agonists (Roux *et al.*, 1998). Activation of contraction via primary $InsP_3$ production and $InsP_3R$-operated Ca^{2+} release, a consequence of enzymatic activation of PLC, is called the pharmacomechanical coupling, in opposition with the electromechanical coupling described above. RyRs are physiologically activated by an increase in Ca^{2+} concentration on the cytosolic face of the RyR, or by direct mechanical coupling between L-type VOC, and it has been shown that cyclic ADP ribose is a co-agonist of RyR (Prakash *et al.*, 1998). RyR activation by Ca^{2+} self-amplifies $[Ca^{2+}]_i$ increase, whatever its initial mechanism. This Ca^{2+}-induced Ca^{2+} release (CICR), however, is not observed in all smooth muscle cells. In human bronchial smooth muscle, for example, RyR, though present and functional, do not participate in the acetylcholine-induced Ca^{2+} response (Hyvelin *et al.*, 2000a).

Basal maintenance of low $[Ca^{2+}]_i$ and Ca^{2+} removal form the cytosol upon and after stimulation is due to active mechanisms that either extrude Ca^{2+} in the extracellular medium or uptake Ca^{2+} into intracellular Ca^{2+} stores. Ca^{2+} extrusion is mainly due to the activity of the plasma membrane Ca^{2+} ATPase (PMCA), and the Na^+-Ca^{2+} exchanger (NCX) (Sanders, 2001).The main mechanisms of Ca^{2+} uptake from the cytosol are Ca^{2+} pumping back into the SR by sarcoendoplasmic Ca^{2+} ATPase (SERCA), Ca^{2+} uptake into the mitochondria. Also, several Ca^{2+}-binding proteins can buffer cytosolic Ca^{2+} and hence decrease $[Ca^{2+}]_i$ (Gunter *et al.*, 2000; Roux *et al.*, 2004).

The Ca^{2+} signal is not only the overall increase in the cellular Ca^{2+} pool and subsequent $[Ca^{2+}]_i$ but also, and most importantly, the pattern of the dynamic change in $[Ca^{2+}]_i$. Indeed, the time-dependent variation in the shape of $[Ca^{2+}]_i$ is the actual "Ca^{2+} signal", since Ca^{2+} binding to signalling protein, e. g., calmodulin, depends on cytosolic Ca^{2+} concentration. Overall $[Ca^{2+}]_i$ measurements generally show a transient increase in $[Ca^{2+}]_i$ (Ca^{2+} peak) followed by a progressive decay to a steady-state value (Ca^{2+} plateau) and/or by Ca^{2+} oscillations (Bergner *et al.*, 2002; Hyvelin *et al.*, 2000b; Kajita *et al.*, 1993; Liu *et al.*, 1996). Changes in $[Ca^{2+}]_i$ are not uniform within the cytosol, and studies have evidenced the role of local change in Ca^{2+} signalling (Prakash *et al.*, 2000). Ca^{2+} signalling should hence be defined as spatiotemporal changes in $[Ca^{2+}]_i$, from which depends the activity of the contractile apparatus, and other cell functions. The shape of this Ca^{2+} signal critically depends on the dynamics of Ca^{2+} fluxes between intra- and extracellular media and also between intracellular Ca^{2+} compartments. The dynamic relationship between the calcium signal and MLCK activity and its consequence on force development will be discussed in more detail in the section "Theoretical modelling of PK and ASMC contraction".

2.2 Smooth muscle contractile apparatus

2.2.1 Components and general organization

The contractile apparatus of smooth muscle is basically constituted of thick filaments of myosin II and thin filament of actin and associated proteins, the thin/thick filament ratio being about 20/1 to 30/1 (Kuo *et al.*, 2003; Somlyo *et al.*, 1983). These filaments are not organized in sarcomeres and do not form well individualized myofibrils. Thick filaments are anchored on dense bodies in the cell and dense area on the plasma membrane and actin filaments are positioned between thick filaments. Dense bodies and filaments are connected by non-contractile intermediate filaments that constitute an intracellular network.

Thick filaments are 1.5 µm long and 12-14 nm in diameter, and are composed of polymerized monomers of myosin. Each monomer of myosin is formed by the association of 2 identical heavy chains (MHC) complexed to 2 pairs of light chains (MLC). Each of the 2 heavy chains has a C-terminal extended α-helix twisted to form a single tail and an N-terminal head. 4 isoforms of MHC have been described in smooth muscle cells, corresponding to 4 alternatively-spliced variants derived from the single smooth muscle myosin heavy chain gene MYH11, differing in their amino-terminal and carboxy-terminal portions (Hong *et al.*, 2011). Polymerized and monomeric myosins are in equilibrium and the stimulation of the myocyte increases myosin polymerization. The distal portion of the myosin head has the ATPase enzymatic activity required for its motor function, and a binding site for actin. A pair of light chains, a 17 kDa one and a 20 kDa one, is complexed with each of the head of the myosin heavy chain. The role of MLC_{17} is unclear, and it is thought to contribute to the stability of the molecule of myosin. Phosphorylation of the so-called regulatory MLC_{20} is required for actin-myosin binding, and hence phosphorylation/dephosphorylation of MLC_{20} regulates actin-myosin cross bridge and contraction. 2 residues, located in the amino-terminal portion of the protein, threonine 18 and serine 19, can be phosphorylated by MLCK. The major site of phosphorylation by MLCK in intact myocyte is ser19, though thr18 can also be phosphorylated by MLCK. It has been shown that Rho kinase can also directly phosphorylate MLC_{20}, but, at least in airway smooth muscle, Rho kinase regulates phosphorylated/dephosphorylated MLC_{20} ratio by acting on MLCP (Mbikou *et al.*, 2011). It has been shown that PKC may also phosphorylate MLC_{20} but on different residues, serine 1, serine 2 and threonine 9 (Barany *et al.*, 1996). However, it is generally admitted that PKC contributes to smooth muscle contraction mainly via MLCP inhibition.

Smooth muscle thin filaments are formed by a double helix of F-actin complexed with tropomyosin, caldesmon and calponin. Globular smooth muscle α-actin molecules assemble into filamentous polymers of F-actin, and 2 parallel strands of F-actin rotate on each other to form a double helix structure. F-actin filaments are about 1 µm long and 5-7 nmn in diameter. As in striated muscle, tropomyosin binds to F-actin in the furrow formed by the double helix of actin, and contributes to actin filament stabilization (Shah *et al.*, 2001). In contrast with striated muscle, here is no troponin on the thin filament of actin in smooth muscle, and other proteins, caldesmon and calponin, are associated with actin. Caldesmon and calponin are regulatory proteins that can be phosphorylated by several kinases and their phosphorylation modulates the formation of the actin-myosin bridge. Regulation of contraction by caldesmon and calponin is detailed below.

2.2.2 Actin-mysosin bridge cycling

Basically, muscle contraction is based on cycling attachment and detachment between thick filaments of myosin and thin filaments of actin. Interaction between actin and myosin, which corresponds to the formation of the actin-myosin crossbridge, is triggered by the head of myosin. As seen above, in smooth muscle, phosphorylation of MLC_{20} is the key event that allows the formation of the actin-myosin bridge, a step required for contraction to occur (Harnett *et al.*, 2003; Wingard *et al.*, 2001). MLC_{20} phosphorylation induces a conformational change of the neck of myosin so that the head of myosin can bind to actin. This allows the rotation of the head of myosin responsible for the "sliding" of actin and myosin filaments. MLC_{20} phosphorylation also activates the ATPase activity of the myosin head, followed by actin and myosin detachment. When MLC_{20} remains phosphorylated all along the crossbridge cycle, crossbridge cycling is rapid, and associated with high ATP consumption. However, sustained contraction can occur even if $[Ca^{2+}]_i$ and subsequent MLC_{20} phosphorylation decrease (Mbikou *et al.*, 2006), and this is generally attributed to the fact that if dephosphorylation of MLC_{20} occurs after the attachment of myosin on actin, crossbridge cycle goes on but at a slower rate, in particular in the stage where dephosphorylated myosin detaches actin. These maintained dephosphorylated crossbridges that cycle at a slow rate are termed latch-bridges.

Hence, 2 types of crossbridges cycling can occur in smooth muscles, a fast, phosphorylated one (during which MLC_{20} remains phosphorylated) and a slow, partially dephosphorylated one (during which MLC_{20} is phosphorylated after actin-myosin attachment), and, accordingly, a 4-state model of the contractile apparatus has been proposed, initially by Hai and Murphy (Hai *et al.*, 1988a). This model is presented in Figure 1. The first state corresponds to unbound actin and unphosphorylated MLC_{20}, the second one to unbound actin and phosphorylated MLC_{20}, the third one to actin bond to phosphorylated MLC_{20}, and the fourth one to actin bond to dephosphorylated MLC_{20}. Transition between state 1 and 2, and 3 and 4, is reversible and depends on MLC_{20} phosphorylation/dephosphorylation. Transition between state 2 and 3 is reversible and corresponds to the phosphorylated crossbridge cycling. Rate constant from state 1 to 4 is very slow, and it can be considered that transition from state 4 to 1 is almost irreversible. State 1-2-3-4-1 cycling corresponds to the latch-bridge cycling.

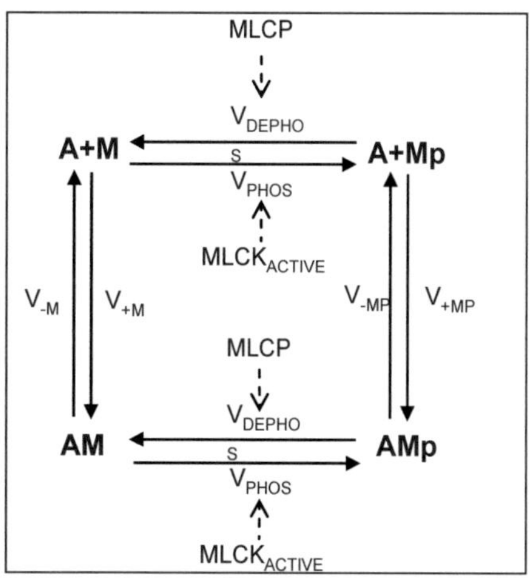

Fig. 1. **4 state model of smooth muscle contractile apparatus.** A, actin; M, unphosphorylated myosin; Mp, phosphorylated myosin; AM, unphosphorylated actin-myosin bridge; AMp, phosphorylated actin-myosin bridge; MLCK, myosin light chain kinase; MLCK, myosin light chain phosphatase; CaM, calmodulin. V, rate constants of the reactions.

The main enzyme by which MLC_{20} is phosphorylated is MLCK. Dephosphorylation of phosphorylated MLC_{20} is catalysed by MLCP. Though other enzymes can contribute to smooth muscle contraction, muscle contraction critically depends on the balance between MLCK and MLCP activity.

2.3 Modulation of MLC_{20} phosphorylation: The MLCK/MLCP balance

2.3.1 MLCK

MLCK has a ubiquitous distribution and is present in non-muscular cells as well as cardiac, skeletal and smooth muscle cells. Several genes encode for distinct isoforms of MLCK. In humans, MYLK2 and MYLK3 genes encode for the MLCK isoforms expressed in skeletal and cardiac myocytes, respectively, and the smooth muscle isoform is encoded by the MYLK1 gene. This gene has 2 initiation sites generating a short isoform corresponding to the smooth muscle MLCK and a long isoform corresponding to nonmuscle MLCK. The domain structure of smooth muscle mammalian MLCK is shown in Figure 2. MLCK has 2 CaM-binding domains, one at its aminoterminal portion, composed of 3 DFRxxL motifs, and the other at its carboxyterminal portion, before the IgT domain located at the N-terminus. The DFrxxL domain is also a binding site to actin, and Ca^{2+}-CaM binding to this domain results in weakened actin binding. Some results suggest that Igl/Ig2 domains bind to actin. IgT domain is a binding site to myosin. The kinase domain is the catalytic portion of MLCK and includes a binding site for ATP and for MLC_{20}. The role of the other domains is unknown (Hong *et al.*, 2011). Hence, MLCK interacts with myosin on two sites. The catalytic core of MLCK interacts with the aminoterminal portion of MLC_{20}, which allows phosphorylation MLC_{20}. The IgT domain of MLCK may also interact with myosin at the head-neck junction. It is thought that IgT binding to myosin near the MLC_{20} increases the catalytic activity of MLCK. The concentration of MLCK is likely to vary from muscle to muscle, but typical values range from 1 to 8 μM, which is low compared with the typical concentration of myosin, classically ranging from 50 to 100 μM.

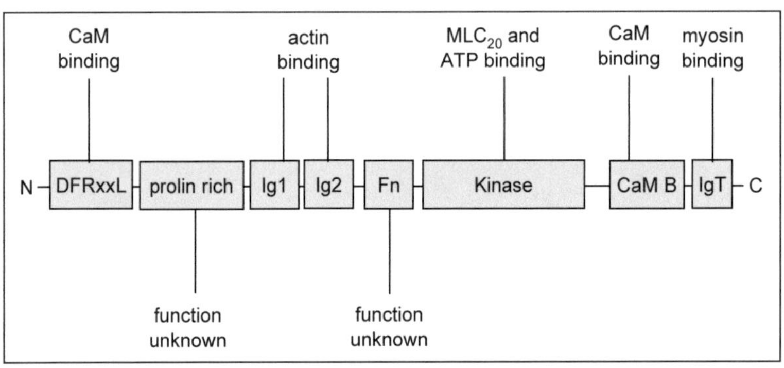

Fig. 2. **Structural scheme of myosin light chain kinase (MLCK) and its different domains.** In the absence of calmodulin (AcM) binding, the catalytic site (Kinase) is masked by the N-terminal CaM binding and IgT domains and hence inactive. Binding of Ca2+-CaM complex induces a conformational change and unmasks the kinase domain hence active.

The major substrate of MLCK in smooth muscle is the smooth muscle myosin, but recent results showed that non muscle myosin, which is also expressed in smooth muscle myosin, can be a substrate for MLCK (Yuen *et al.*, 2009). MLCK phosphorylates MLC_{20} on serine 19 and, with lower enzymatic activity, on threonine 18 (Deng *et al.*, 2001). MLCK activity is essential for airway smooth muscle contraction, as it has been evidenced by pharmacological inhibition of MLCK in rat trachea (Mbikou *et al.*, 2006) and knockout mice (Zhang *et al.*, 2010).

The most important regulator of MLCK activity is the Ca^{2+}-CaM complex. MLCK unbound to Ca^{2+}-CaM is in an auto-inhibitory state. The IgT domain and the adjacent CaM binding domain constitute an auto-inhibitory sequence that masks the catalytic core. Ca^{2+}-CaM binding to the auto-inhibitory site induces a conformational change. This displacement of the auto-inhibitory sequence unmasks the catalytic core and thereby reveals the enzymatic activity of MLCK. On the other hand, phosphorylation of MLCK on serine 512 by several kinases such as protein kinase A in the C-terminal CaM binding domain decreases MLCK affinity for the Ca^{2+}CaM complex, thereby inactivating MLCK even in the presence of Ca^{2+}, a mechanism responsible for MLCK desensitization (Stull *et al.*, 1993). CaM is a rather small protein (16700 Da) consisting of a single polypeptidic chain of 148 to 149 aminoacids. 3 distinct domains can be identified: an N-terminal domain, a central helix domain, and a C-terminal domain. Both amino- and carboxy-terminal domain possesses 2 binding sites for Ca^{2+}. The 4 sites have a high affinity for Ca^{2+}, though the affinity of the C-terminal sites is higher. These sites have an EF-hand structure. Ca^{2+} binding induces a conformational change of the F-helix of the EF-hand motif. In this conformation CaM binds to several proteins and modulates their activity. Amongst almost 40 different proteins that have been shown to be regulated by CaM, the most important in smooth muscle are MLCK and CaM-kinase II.

2.3.2 MLCP

MLCP is a holoenzyme composed of three subunits: the 38 kDa phosphatase catalytic subunit PP1cδ, the 110 kDa regulatory subunit MYPT1, and a 20 kDa noncatalytic subunit, M-20, which role remains unclear (Hartshorne *et al.*, 1999; Hartshorne *et al.*, 1998). PP1cδ is a serine/threonine phosphatase that dephosphorylates MLC_{20} on serine 19 and threonine 18. Disruption of the quaternary structure of MLCP, either by dissociation of the regulatory subunit MYPT1 to PP1c or by disruption of the interactions between the three subunits, decreases MLCP activity toward MLC_{20} dephosphorylation.

PP1c alone is capable of dephosphorylating the MLC_{20}, but its activity is potentiated when complexed to MYPT1. Indeed, MYPT1s targets PP1c to its substrate, MLC_{20}, and confers substrate specificity to the phosphatase. A recent study by crystallography has revealed a basic structure which potentiates the catalytic activity of the holoenzyme (Terrak *et al.*, 2004). Structurally, MYPT1 subunit possesses a C-terminal leucine-zipper domain and a binding site of protein kinase G. Four N-terminal amino acids, KVKF, give a strong affinity for PP1c. MYPT1 also has six repeated domains of ankyrin, which serve as anchoring sites for protein-protein interactions. Several kinases are able to phosphorylate MYPT1 at various sites: Rho kinase, MYPT kinase (also known as Zip-like kinase), integrin-like kinase, myotonic dystrophy kinase (DMPK). PKC also has been shown to phosphorylate ankyrin repeated domains and inhibit the interaction between MYPT1 and PP1c (Toph *et al.*, 2000). In addition to this negative regulation of MLCP, MYPT1 phosphorylation is also associated with an upregulation. Recent studies have reported that protein kinases dependent on cyclic nucleotides such as PKA and PKG are able to phosphorylate MYPT1 on serine 692, 695 and 852. The S695 and S852 residues are adjacent to two residues, T696 and T853, that are phosphorylation sites of MYPT1 by RhoK. Phosphorylation at S695 by PKG or PKA prevents RhoK phosphorylation at S696 and vice versa (Wooldridge *et al.*, 2004), and it is supposed that phosphorylation at S852 has a similar effect on T853 phosphorylation. Thus the phosphorylation of MYPT1 by PKA and PKG protein antagonizes the inhibition of MLCP induced by RhoK (Grassie *et al.*, 2011).

CPI-17 is another regulation pathway of MLCP activity (Ito *et al.*, 2004). CPI-17 is a 17 kDa protein composed of 147 aminoacids that can inhibit the activity of the MLCP holoenzyme as well as that of isolated PP1c. Phosphorylation of CPI-17 at threonine 38 enhances its inhibitory potency about 1000 fold. The proposed mechanism is that CPI-17 binds at the PP1c active site, resulting in formation of a complex of MYPT1/PP1C/CPI-17 (Eto *et al.*, 2004). The major kinase responsible for in *vivo* CPI-17 phosphorylation is PKC, especially PKCα and PKCδ. *In vitro* studies have shown that other kinases can phosphorylate CPI-17 at T38, including Rho kinase, protein kinase N (PKN), MYPT-kinase, integrin-like kinase and PKA (Ito *et al.*, 2004). The scheme of the regulation of MLCP by kinase activity is given in Figure 3.

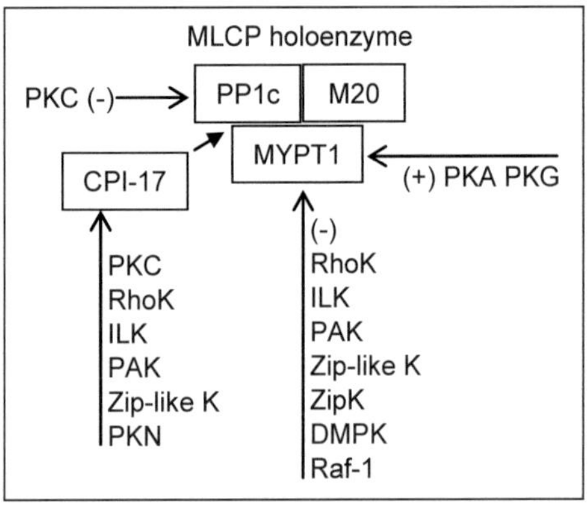

Fig. 3. **Regulation of MLCP activity by protein kinases.** DAP kinase, death-associated protein kinase;; DMPK, myotonic dystrophy protein kinase; ILK, integrin-linked kinase; MLCP, myosin light chain phosphatase; MYPT1, myosin phosphatase target subunit 1 of MLCP; M20, 20 kDa non-catalytic subunit of MLCP; PAK, p21-activated protein kinase; PKA, protein kianse A; PKC, protein kinase C; PKG, protein kinase G; PKN, protein kinase N; PLA2, phospholipase A2; PP1c, catalytic subunit of type1 protein phosphatase; RhoK, Rhokinase; ZipK; zipper-interacting protein kinase (adapted from (Hirano *et al.*, 2003)).

2.4 Thin filament-associated regulation of AM bridge cycling

2.4.1 Caldesmon

Caldesmon is a small protein of 87 kDa associated with the thin filaments of actin and involved in the regulation of smooth muscle contraction. Structurally, caldesmon possesses 4 distinct domains and can bind to actin, tropomyosin, myosin and calmodulin. The fourth domain located at the C-terminus is the most important since it is the binding domain to actin, to myosin and to Ca^{2+}-binding proteins. Caldesmon interacts stoichiometrically with tropomyosin to consolidate the actin filaments (Gusev, 2001), and contributes to the sensitivity of the myofilament to Ca^{2+}. In the absence of Ca^{2+}, the fourth domain interacts closely with actin and inhibits the ATPase activity. $[Ca^{2+}]_i$ increase induces a conformational change of the actin-tropomyosin-caldesmon complex, allowing the ATPase activity of myosin. In addition, it allows an interaction between the thin and thick filaments as it is able to bind to both actin and myosin and this property seems to be involved in the sustained phase of the contractile response (Marston *et al.*, 1991). In vitro studies have shown that caldesmon can be a substrate for several kinases (Kordowska *et al.*, 2006).

2.4.2 Calponin

Calponin is a molecule of approximately 34 kDa related to the skeletal muscle troponin T. Calponin is present in smooth muscle at a concentration almost equivalent to that of tropomyosin. Calponin binding to F-actin disrupts actin-myosin interaction and hence negatively regulates the contraction of smooth muscle. Calponin inhibits the sliding of actin filament on myosin and the ATPase activity of myosin. The inhibitory effect of calponin is suppressed by high $[Ca^{2+}]_i$ values, but only in the presence of calmodulin. Phosphorylation and dephosphorylation of calponin plays a key role in the regulation of smooth muscle contraction. Indeed, the inhibitory effect of calponin is abolished when calponin is phosphorylated by CaM kinase II or PKC, and is restored when calponin is dephosphorylated by phosphatase 2A (Winder *et al.*, 1998; Winder *et al.*, 1990). Based on these properties, 3 physiological roles have been identified for calponin: (1) maintenance of relaxation at rest when $[Ca^{2+}]_i$ is low; (2) Ca^{2+}-dependent contribution, involving calmodulin, to contraction when $[Ca^{2+}]_i$ is increased and, (3) Ca^{2+}-independent contribution to contraction via PKC-dependent phosphorylation.

3. Role of protein kinases in the modulation of excitation-contraction coupling

3.1 General overview

3.1.1 Modulation of Ca^{2+} signalling

Though protein kinases modulate smooth muscle contraction by acting mainly downstream of the Ca^{2+} signal, some of them have been shown to act upstream of the Ca^{2+} signal and regulate the activity of structures involved in Ca^{2+} signal encoding. It has been shown by *in vitro* studies that PKA can phosphorylate $InsP_3$ receptors on serine 1755 and serine 1589 (Ferris *et al.*, 1991a). Studies using purified receptor protein reconstituted in liposomes have also evidenced that PKC and CaM kinase II can phosphorylate $InsP_3$ receptors (Ferris *et al.*, 1991b). Ryanodine receptors also can be phosphorylated by PKA, PKC and CaM kinase II (Coronado *et al.*, 1994). These phosphorylations may play an important role in smooth muscle physiology since Ca^{2+} release from SR plays a critical role in Ca^{2+} homeodynamics in airway myocyte (Ay *et al.*, 2004; Roux *et al.*, 2004). Indeed, contractile agonists such as acetylcholine act via $InsP_3$ production but also activate PKC (Mbikou *et al.*, 2006). On the other hand, PKA stimulation is a important pathway involved in bronchorelaxation since β_2-adrenergic stimulation, a major pharmacological treatment of asthma, acts mainly via Gs protein stimulation, adenylate cyclase activation, cAMP production and PKA activation (Anderson, 2006; Johnson, 2006). Extracellular ATP has also been shown to induce airway relaxation via PKA activation (Mounkaila *et al.*, 2005). Protein kinase may also modulate Ca^{2+} homeostasis by acting on voltage-dependent Ca^{2+} entry. As explained above, K^+ channels critically contribute to the maintenance of myocyte membrane voltage, and it has been shown in vascular smooth muscle that PKA and PKC can phosphorylate several voltage-dependent K^+ channels, Ca^{2+}-activated K^+ channels and ATP-dependent K^+ channels, thus regulating the electromechanical coupling (Ko *et al.*, 2008). Phosphorylation by PKA activates these K^+ channels and hence tends to hyperpolarize the plasma membrane and limit extracellular Ca^{2+} entry, whereas phosphorylation by PKC has an opposite effect. The effective contribution of these mechanisms in airway in unclear, since it has been shown that in airways activation of Ca^{2+}-activated K^+ channels does not seem to have a significant effect on β-adrenergic relaxation (Corompt *et al.*, 1998).

3.1.2 Modulation of the Ca2+ sensitivity of the contractile apparatus

The signalling pathways capable of modulating the contraction of a given Ca^{2+} signal constitute the modulation of the sensitivity of the contractile apparatus to Ca^{2+}. As detailed previously, the importance of contraction depends on the balance between MLCK and MLCP activity, which is under control of several signalling pathways. In parallel with the canonical MLCK-dependent MLC_{20} phosphorylation, in several smooth muscles, MLC can be directly phosphorylated by Ca^{2+}-independent kinases like ZIP kinases and Integrin-linked kinase (ILK), which can also inhibit MLCP (Deng *et al.*, 2002; Huang *et al.*, 2006; Murthy, 2006; Niiro *et al.*, 2001). Rho kinase, activated by small GTPase (Rho), which seems to be activated both by Ca^{2+}-dependent and Ca^{2+}-independent pathways, can phosphorylate directly MLC and, by inhibition of MLCP, indirectly increase MLC phosphorylation and hence regulate smooth muscle contraction (Bai *et al.*, 2006; Murthy, 2006; Schaafsma *et al.*, 2006). Also, contractile agonists may, in parallel to MLCK activation, inhibit MLCP activity. PKC can inhibit MLCP, and hence modulate contraction (Bai *et al.*, 2006). By contrast, relaxant agonists, such as β_2-agonists in airways, may act via MLCP activation and/or MLCK inhibition (Janssen *et al.*, 2004; Johnson, 1998). So, in addition with the Ca^{2+}-calmodulin-MLCK pathway, other Ca^{2+}-dependent and independent enzymatic pathways regulate the contractile apparatus in smooth muscles, which status depends on the balance between MLCP/MLCK phosphorylation and dephosphorylation. Additionally, smooth muscle contraction can be modulated independently from MLC_{20} phosphorylation via caldesmon and calponin. The major kinases involved in smooth muscle contraction are listed below.

3.2 RhoK

Rho-associated kinase (Rho-kinase), originally identified as an effector of the small GTPase Rho, has been shown to play a major role in many processes including cell migration, neuronal polarisation, cytokinesis and cell contraction. It is a serine/threonine kinase structurally related to myotonic dystrophy kinase as there is 72% homology in the kinase domains. Different studies support the idea that Rho-kinase exists as a dimer resulting from parallel association through the coil-coil domain (Shimizu *et al.*, 2003). There are two Rho-kinase members, Rho-kinase I/ROKβ/p160ROCK and Rho-kinase II /ROKα/Rho kinase, which share 65% sequence identity and 95% sequence similarity at the amino acid level (Riento *et al.*, 2003b). The kinase domain is highly conserved between these two proteins (83% identical), suggesting that they may have similar substrate specificity. The consensus phosphorylation sequence for Rho-kinase is R/KXS/T or R/KXXS/T (X is any amino acid). The Rho-kinase protein is composed of three domains: a N-terminal kinase domain, a central coiled-coil domain containing a Rho binding site (RhoBD), and a C-terminal pleckstrin homology-like domain (PH-like domain) containing a Cys-rich region similar to the C1 domain of protein kinase C. The C-terminal region including the RhoBD plus the PH domain has been shown to directly interact with the kinase domain to inhibit its activity (Amano *et al.*, 1999). The interaction between Rho•GTP and the RhoBD releases this autoinhibition and thus activates the kinase (Ishizaki *et al.*, 1996). In addition to Rho other small as GTPases such Rnd3/RhoE, Gem and Rad can bind Rho-kinase outside of the Rho-binding region and inhibit its function (Komander *et al.*, 2008; Riento *et al.*, 2003a). Arachidonic acid has been also shown to activate Rho-kinase via its PH domain (Araki *et al.*, 2001). Although both Rho-kinase I and Rho-kinase II proteins are ubiquitously expressed in most tissues, higher levels of Rho-kinase II are found in brain and muscles whereas higher levels of Rho-kinase I are found in non-neuronal tissues such as liver, lung and testis (Leung *et al.*, 1996; Nakagawa *et al.*, 1996). Functional differences have been reported between Rho-kinase I and Rho-kinase II. For instance Rho-kinase I seems to be essential for the formation of stress fibres, whereas Rho-kinase II seems important for phagocytosis and cell contraction, both being dependent on MLC phosphorylation (Wang *et al.*, 2009; Yoneda *et al.*, 2005). Binding tests revealed that RhoE preferentially binds Rho-kinase I, but not Rho-kinase II, whereas MYPT1 binds only Rho-kinase II (Komander *et al.*, 2008; Wang *et al.*, 2009).

There has been a great deal of interest in the involvement of the Rho/ROCK signalling pathway in excitation–contraction coupling (Iizuka *et al.*, 1999; Mbikou *et al.*, 2011; Yoshii *et al.*, 1999). Activation of the Rho/Rho-kinase pathways is likely due to the stimulation of the G-coupled protein receptor. Stimulation of M_3 muscarinic receptors agonist for instance activates the Gq, G_{12} and G_{13} α-subunits which cause the cascade activation of GEFs, RhoA and Rho-kinase leading to Ca^{2+} sensitization (Hirano *et al.*, 2004; Somlyo *et al.*, 2003). It is well-known that Rho/Rho-Kinase signalling modulates Ca^{2+} sensitivity of the smooth muscle likely either by inhibition of the MLCP activity or by direct phosphorylation of the MLC_{20}. Precisely, Rho-kinase has been shown to phosphorylate the MYPT1 subunit of the MLCP at two inhibitory sites, Thr696 and Thr853, by that causing the dissociation between the MLCP and MLC_{20} (Feng *et al.*, 1999; Kawano *et al.*, 1999; Kimura *et al.*, 1996; Velasco *et al.*, 2002). Phosphorylation of MYPT1 by Rho-kinase alters the MYPT1-PP1c interaction, which decreases MLCP activity toward MLC_{20}. Rho-kinase is also able to directly phosphorylate Ser19 and Thr18 in non-muscle cells (Ueda *et al.*, 2002), but the direct contribution of Rho-kinase to phospho-MLC_{20} levels *in vivo* is not yet proven. Moreover, Rho kinase exerts an inhibitory activity toward PP1c subunit of MLCP upon the phosphorylation of the latter (Eto *et al.*, 1995; Eto *et al.*, 1997). Taken together, the mechanisms triggered by Rho-kinase activity enhance the level of phosphorylated MLC_{20} and, consequently, myosin ATPase activity and therefore contraction. *In vitro* tests have shown that Rho-kinase is also able to phosphorylate CPI17 (MLCP inhibitor protein), thus inducing the inhibition of MLCP and thereby raising the level of phosphor MLC20 (Amano *et al.*, 2010).

3.3 PKC

The protein kinase C (PKC) family is the largest serine/threonine-specific kinase family known. It embraces a large family of enzymes that differ in structure, cofactor requirements and function (Nishizuka, 1995). Depending on their cofactor requirements, the homologous group of PKC can be divided into three groups as follows: the group of conventional (c)PKC isoforms (α, β_I, β_{II} and γ), that require Ca^{2+} and diacylglycerol (DAG) to become activated; the group of novel (n)PKC isoforms (δ, ε, ζ, θ and μ) that require only DAG; and group of the atypical (a)PKC isoforms, namely ζ, ι and λ (the mouse homologue of human PKCι), that require neither Ca^{2+} nor DAG. The general structure of a PKC molecule consists of a catalytic domain in N-terminal and a regulatory pseudosubstrate in C-terminal, both framing 3 distinct sites able to bind specifically ATP, Ca^{2+} or phosphatidyl serine. The pseudosubstrate region is a small sequence of amino acids that binds the substrate-binding cavity in the catalytic domain, thus keeping the enzyme inactive. The activity of PKC is controlled by its compartmentalization within the cell. All PKC family members possess a phosphatidylserine binding domain for membrane interaction. The expression and distribution of PKC isoforms is tissue- and species–specific. Some isoforms (e.g. PKCα) are ubiquitously expressed in tissues whereas others seem to be restricted to certain tissues (Webb *et al.*, 2000). In ASM, the protein and mRNA expression of PKC isoforms differs depending on the specie. In human trachealis for instance, there is expression of the conventional α, β_I, and β_{II} PKC isoforms as well as novel (δ, ε, ζ, θ) and atypical (ζ) (Webb *et al.*, 1997) while canine trachealis does not express the α isoform and bovine ASM expresses the θ-PKC variants (Webb *et al.*, 2000).

It is well known that the stimulation of a Gq protein-coupled receptor, for instance by cholinergic agonist, induces the elevation of the cytosolic Ca^{2+} concentration and the production of DAG by phospholipase C. Both Ca^{2+} and DAG bind to the C2 and C1 domain, respectively, and recruit PKC to the membrane (Bell *et al.*, 1991; Huang, 1989). This interaction with the membrane results in release of the pseudosubstrate from the catalytic site and activation of the enzyme(Lester *et al.*, 1990).

PKC may modulate the sensitivity of the contractile apparatus to Ca^{2+}, since it can inhibit MLCP activity. However, the effectiveness of PKC contribution to airway contraction remains controversial. PKC has been shown to be involved in force maintenance in human airways (Rossetti *et al.*, 1995). However, other studies do not implicate PKC in agonist-induced Ca^{2+} sensitization and point to other effectors such as the small GTP-binding protein p21rho (Akopov *et al.*, 1998; Itoh *et al.*, 1994; Otto *et al.*, 1996; Yoshii *et al.*, 1999). Our recent studies have shown that PKC contributes to the sustained phase, but not to the initial phase, of cholinergic-induced contraction in rat airways (Mbikou *et al.*, 2006).

3.4 PKA

Cyclic-AMP-dependent protein kinase (PKA) is an ubiquitous mammalian enzyme which catalyzes Ser/Thr phosphorylation in protein substrates that in turn control a wide range of cellular functions including gene regulation, cell cycle, metabolism and cell death (Shabb, 2001). This tetrameric holoenzyme comprises two catalytic (C) subunits that possess kinase activity and two inhibitory regulatory (R) subunits, each including two tandem cAMP binding domains, *i.e.* CBD-A and CBD-B (Johnson *et al.*, 2001). cAMP is the essential second messenger that activates PKA (Berman *et al.*, 2005; McNicholl *et al.*). In the absence of cAMP, the R-subunit and the C-subunit create a complex that blocks substrate access and thus prevents the kinase activity. cAMP binding to the R-subunits releases these inhibitory interactions and unleashes the C-subunit, allowing substrate phosphorylation. There are three isoforms of C (Cα, Cβ, and Cγ) and two major isoforms of R (R_I and R_{II}) that are further distinguished into subforms (α and β) (Zhao *et al.*, 1998). The physiological importance of these isozyme variations is not fully understood, but anchoring proteins (AKAPs) for R_{II} give it a unique cellular distribution (Scott *et al.*, 1994) R_I and R_{II} show sequence homology in their cAMP-binding and pseudosubstrate domains but differ extensively in their dimerization domains as well as in the sequence connecting the dimerization and pseudosubstrate domains. All known R-subunit isoforms share a common organization that consists of a dimerization domain at the NH_2 terminus followed by an autoinhibitor site and two-tandem cAMP-binding domains noted CBD-A and B (Taylor *et al.*, 2005). While the portion of the R subunit COOH-terminal to the inhibition site is responsible for high affinity binding of the C-subunit and cAMP, the remaining NH_2 terminus serves as an adaptor for binding to kinase anchoring proteins (Scott *et al.*, 1994) and is responsible for *in vivo* subcellular localization and targeting of PKA.

The ordered sequential mechanism of PKA activation is described as follows: cAMP binds first to CBD-B, making site CBD-A accessible to a second molecule of cAMP, which in turn causes the release of the active C-subunit (Kim *et al.*, 2007; Su *et al.*, 1995). In other words CBD-B functions as a gatekeeper for CBD-A, whereas the latter acts as the central controlling unit of the PKA system and provides the primary interfaces with the C-subunit (McNicholl *et al.*). The structures of the R-subunit in its active and inhibited states have also demonstrated that although CBD-A and CBD-B play clearly distinct roles in the activation of PKA, they both share similar allosteric features.

PKA activation is closely dependent on the cytosolic cAMP level which is itself regulated by G proteins via an enzyme named adenylate cyclase. Stimulation of GPCRs, by muscarinic receptor agonist for instance, can rise up or drop off AMPC production depending on the type of GPCRs. Indeed, GPCRs that activate the Gαi subunits inhibit cAMP production whereas GPCRs that activate the Gαs subunits activate cAMP production through this specific sequence: 1) activated Gαs subunit interacts with the adenylate cyclase 2) adenylate cyclase quickly converts ATP into cAMP, 3) AMPc molecule activate the PKA. Activation of PKA is involved in airway smooth muscle relaxation (Zhou *et al.*, 1992), which may involve phosphorylation of a number of effector proteins that cause either reduction of $[Ca^{2+}]_i$ and/or reduction of MLCK sensitivity to Ca^{2+}-calmodulin (de Lanerolle *et al.*, 1991).

Studies have demonstrated that cAMP-dependent signaling pathway activation prevents or reverses the ASM contraction indirectly, via the inhibition of $InsP_3$ receptor (InsP3R) of the sarcoplasmic reticulum, reducing the $[Ca^{2+}]_i$. The InsP3R are responsible for mobilizing Ca^{2+} from sarcoplasmic reticulum in response to agonist binding. PKA has been shown to mediate the phosphorylation of the InsP3R which consequently reduces the ability of Inositol(1,4,5)triphosphate to release Ca^{2+} from membrane vesicles (Schramm *et al.*, 1995). The mechanism by which PKA-induced phosphorylation decreases $insP_3$-induced Ca^{2+} release has not been determined, but known consequences of receptor/channel phosphorylation include altered agonist affinity as well as altered function.

3.5 CaMkinase II

Calcium-calmodulin-dependent protein kinase II (CaMKII) is an oligomeric serine/threonine-specific protein kinase which belongs to a family of enzymes regulated by the calcium-calmodulin complex similarly to the MLCK. Increases in the cytosolic Ca^{2+} concentration following the stimulation modulate the function of many intracellular proteins (Zhou *et al.*, 1994). One of the most important intracellular acceptors of the Ca^{2+} signal is calmodulin (CaM), which exerts a modulating influence on the function of Ca^{2+}/CaM-dependent protein kinases. Among them, the CaMKII shows a broad substrate specificity and has been thought to be a multifunctional protein kinase (Cohen *et al.*, 1992; Colbran *et al.*, 1989a; Colbran *et al.*, 1989b). Four genes encode related but distinct isoforms of CaM kinase II (α, β, γ, and δ). It was originally isolated from brain tissues (Fukunaga *et al.*, 1982) (Goldenring *et al.*, 1983; Kuret *et al.*, 1984) and preparations of CaM kinase II purified from rat forebrain consist of the α (50 kDa) and β (60 kDa) subunits whose cDNAs have been cloned and sequenced (Kolb *et al.*, 1998). Non neuronal tissues express mostly the isoform γ, and δ but in such a low level that it makes difficult the purification of the enzyme; so most biochemical and physical data on the enzyme have been established with CaMkinase II-α and -β/β′ from mammalian brain. In smooth muscle, CaM kinase II was first isolated as caldesmon kinase (Ikebe *et al.*, 1990b) with a molecular mass of the major subunit of 56 kDa. Isolated smooth-muscle CaM kinase II has enzymological properties similar to that of brain; however, smooth-muscle CaMKII is a tetramer according to its native molecular mass rather than a decamer or octamer as are the brain enzymes (Zhou *et al.*, 1994).

The CaMKII subunits are thought to assemble to holoenzyme through their C-terminal association domains (Fahrmann *et al.*, 1998). The linear structure of the CaMKII core consists of a catalytic/autoregulatory domain (A) containing a variable region V1, a conserved linker (B), and an association domain that contains two highly conserved sequences (C and D) as well as multiple variable regions (V2-V4). Function of each variable region has been identified as follows: V1 contains insert implicated in SR-membrane targeting; V2 possesses a functional nuclear localization signal, as well as a site of autophosphorylation; Insert X within V3 is rich in proline residues and conforms to a SH_3-binding sequence. The variable regions are diversely expressed in the different subunits. For instance the α and δ_G isoforms are the smallest catalytically competent CaMKII products because they contain the A–D core sequences, but no inserts (Hudmon *et al.*, 2002). The most prominent proteins phosphorylated by the Type II CaM kinase are its own subunits (Bennett *et al.*, 1983; Kennedy *et al.*, 1983; Miller *et al.*, 1985; Miller *et al.*, 1986). Some studies suggest that kinase activity decrease after autophosphorylation (Kuret *et al.*, 1984; LeVine *et al.*, 1985; Yamauchi *et al.*, 1985) while others suggest that it increases (Shields *et al.*, 1984) or becomes autonomous (Saitoh *et al.*, 1985). Despite these discrepancies, the most accepted theory is that the autophosphorylation allows for the activation of the catalytic domain (Hanley *et al.*, 1987). The Ca2+-CaM complex interacts with and promotes autophosphorylation of each subunit of the CaMKII.

The implication of the CAMKII in the artery smooth muscle reactivity has been extensively investigated whereas little is known regarding its role in the airways smooth muscle. In contracted cultured ASM cells from bovine, studies demonstrated that CaMKII is responsible for the phosphorylation of the MLCK (Stull *et al.*, 1993). Biochemistry data showed that MLCK is phosphorylated by CaMKII (Hashimoto *et al.*, 1990; Ikebe *et al.*, 1990a) at a specific serine near the calmodulin-binding domain; and this phosphorylation brings about the reduction of the affinity of MLCK for the Ca^{2+}•CaM complex (Stull *et al.*, 1990). In another hand the phosphorylation of MLCK by CaMKII have been shown to decrease the

Ca^{2+} sensitivity of MLC_{20} phosphorylation (Tansey *et al.*, 1994; Tansey *et al.*, 1992). Taken together, these findings would suggest that the net effect of the CaMKII is the relaxation of the ASM. However, enzymatic tests revealed two key elements which exclude this hypothesis: 1) the rate of phosphorylation of MLCK likely by the CaMKII is first of all slower than the rates of increase in cytosolic Ca^{2+} concentrations, 2) and also slower than the rate of phosphorylation of the MLC in intact tracheal smooth muscle cells in culture (Tansey *et al.*, 1994). Therefore, the CaMKII activity is thought not to affect significantly the reactivity of the bovine ASM. Moreover, other studies showed that the CAMKII do not play a role in the profile of the contractile response upon stimulation in intact ASM from rat or cow (Liu *et al.*, 2005; Mbikou *et al.*, 2011; Sakurada *et al.*, 2003).

3.6 PI3K

The phosphatidy inositol 3-kinases (PI3K) superfamily draws together all the enzymes capable of phosphorylating specifically the hydroxyl group of a membrane phospholipid called phosphatidyl inositol (PtdIns). Cloning approaches revealed the existence of eight distinct PI3K genes expressing eight isoforms in human and mouse genomes. Based on their domain structure, lipid substrate specificity and associated regulatory subunits, these isoforms have been divided into three main classes as follows: class I including p110α, p110β, p110δ and p110γ; class II including PI3K-C2α, PI3K-C2β and PI3K-C2γ, and the class III consisting of the sole enzyme Vps34. The PI3K phosphorylates the PtdIns into four possible products, PtdIns(3)phosphate, PtdIns(3,4)biphosphate, PtdIns(3,5) biphosphate and PtdIns(3,4,5)triphosphate, which are involved in a wide range of cellular functions, including cell growth, proliferation, motility, differentiation, survival and intracellular trafficking (Fry, 2001; Fry, 1994; Katso *et al.*, 2001; Rameh *et al.*, 1999).

Purified PI3K is a heterodimer of 85 and 110 kDa subunits (Carpenter *et al.*, 1990; Fry *et al.*, 1992; Morgan *et al.*, 1990; Shibasaki *et al.*, 1991). Analysis of the primary sequence of p85 reveals a multidomain protein which contains a number of non-catalytic domains, a Src homology region 3(SH3), and a region with significant sequence similarity to the product of the breakpoint cluster region gene BCR (Otsu *et al.*, 1991). In vascular smooth muscle, PI3K appears to play a role in the regulation of contraction as experiments showed that the specific isoform PI3K-C2α is necessary for Rho/RhoKinase-dependent MLCP inhibition and consequently for the MLC20 phosphorylation and the contraction (Yoshioka *et al.*, 2007). However, in the ASM, the PI3K does not regulate the agonist-induced contractile response (Mbikou *et al.*, 2006).

4. Theoretical modelling of PK and ASMC contraction

4.1 Interest and general priniples

Since the contractile pattern of SMC in response to contractile agonists not only depends on the structure of the regulatory network that modulates the contractile apparatus but also on the dynamics of the reactions, understanding of the mechanisms underlying this contractile profile is quite impossible by a non-mathematical intuitive approach. In the following sections, we will present the concepts of our recent mathematical modelling of isometric contraction and force development in airway smooth muscle cells based on the 4-state latch bridge model (Hai *et al.*, 1988a) and upgraded by the unique description of MLCK and

MLCP regulatory pathways (Fajmut *et al.*, 2008; Fajmut *et al.*, 2005a; Fajmut *et al.*, 2005b; Fajmut *et al.*, 2005c; Mbikou *et al.*, 2011; Mbikou *et al.*, 2006).

The latch state was first described by the mathematical model by Hai and Murphy in 1988 (Hai *et al.*, 1988a). They introduced the model of isometric contraction based on the 4-state kinetic scheme of actomyosin crossbridges, in which to actin bound and unbound to myosin (phosphorylated and dephosphorylated) represent four different states. Even nowadays, that model represents the reference in modelling of smooth muscle contraction.

Huxley (Huxley, 1957) pioneered the modelling of smooth muscle contraction. His model from 1957 was based on the sliding filament theory. Until the occurrence of Hai and Murphy's model in 1988 (Hai *et al.*, 1988b), relatively small number of models of smooth muscle contraction was developed compared to striated muscles. In 1986 Gestrelius and Börgström (Gestrelius *et al.*, 1986) introduced new concepts in modelling. They first considered viscoelastic properties of the filaments and the cytoskeleton. The later model and Huxley's model enabled the study of nonisometric contraction and took into account mostly the dynamics and mechanics of the filaments and myosin crossbridges. Although the original Hai and Murphy's model (Hai *et al.*, 1988a) enabled only the prediction of isometric force its advantage was in considering the regulatory mechanisms that drive smooth muscle contraction. Several authors have later upgraded it to enable the prediction of nonisometric contraction. In 1997 Yu et al. (Yu *et al.*, 1997) expanded it to simulate the nonisometric contractions of smooth muscles, added length dynamics and assumed length-dependent bonding and unbonding rates to be distributed according to the Gaussian distribution. In 1999 Fredberg et al. (Fredberg *et al.*, 1999) and Mijailovich et al. (Mijailovich *et al.*, 2000) integrated the latch regulation scheme of Hai and Murphy with Huxley's sliding filament model of muscle contraction for the studies of the effects of length fluctuations on the dynamically evolving cross-bridge distributions, simulating those that occur in airway smooth muscle during breathing. The later model has been recently upgraded by simple description of Ca^{2+}-dependent regulation of MLCK activity (Bates *et al.*, 2009; Wang *et al.*, 2008). Our approach in the modelling of smooth muscle contraction is to upgrade the 4-state kinetic description of Hai and Murphy (Hai *et al.*, 1988a) with the signalling pathways that regulate the MLC phosphorylation and dephosphorylation. This was for a long time the missing part in the modelling of smooth muscle contraction.

Development of force in smooth muscles is achieved by interactions between myosin cross-bridges and actin filaments. To describe theses interactions Hai and Murphy proposed a kinetic scheme shown in Figure 1 (Hai *et al.*, 1988a). The scheme was later upgraded by Rembold and Murphy (Rembold *et al.*, 1990) with the consideration of the attachment of dephosphorylated myosin to actin with a very slow rate. Many authors ignore this interaction, since the rate of attachment of dephosphorylated myosin is 150-fold lower than that of phosphorylated myosin (0.002 s^{-1} and 0.3 s^{-1}, respectively) (Trybus, 1996). In our studies we consider it in the modelling (Fajmut *et al.*, 2008; Fajmut *et al.*, 2005a; Mbikou *et al.*, 2011; Mbikou *et al.*, 2006).

The four different states of the myosin cross-bridges are presented in Figure 1. : A+M - detached, dephosphorylated, A+MP - detached, phosphorylated, AMP - attached, phosphorylated, and AM - attached, dephosphorylated, the last one termed also a latch bridge. The corresponding reaction velocities of phosphorylation/dephosphorylation and attachment/detachment of myosin cross bridges, the relevant variables of mathematical modelling, are indicated together with sites of action of enzymes MLCK and MLCP responsible for phosphorylation and dephosphorylation of MLCs, respectively. In the model, the magnitude of stress in smooth muscles is proportional to the concentration of myosin cross-bridges associated with actin filaments (AMP and AM), whereby myosin cross-bridges in the state AMP generate stress and cross-bridges in the state AM maintain stress in smooth muscles. v are either the velocities of crossbridge attachment/detachment or the velocities of phosphorylation/ dephosphorylation, which are regulated by MLCK and MLCP.

Hai and Murphy's model considered a very simple semi-theoretical description of the Ca^{2+}-dependent MLCK activation, and thus first coupled Ca^{2+} signalling pathway with the contraction. In this chapter we will present an upgrade from that model in the sense of purely theoretical and more detailed description of Ca^{2+}-dependent MLCK activation and functioning as well as MLCP regulation of force development. Our new concepts will be compared with the old ones. As a basic model scheme, we take the four-state model, in which we incorporate explicitly the description of both enzymatic reactions. The activity of MLCK is under the influence of transmitting Ca^{2+} signal and, hence, Ca^{2+} directly affects the force development. On the other hand, the action of MLCP is considered either independent of other signalling pathways or being under influence of Rho-kinase (RhoK), which phosphorylates MLCP and regulates its activity and catalytic properties.

4.2 Modelling of MLCK activity

According to the generally accepted view, MLCK is activated by the Ca^{2+}-CaM complex (Kamm *et al.*, 2001). In general, an increase in $[Ca^{2+}]_i$ is considered to initiate the binding of four Ca^{2+} ions to CaM, which is finally followed by association of complex Ca_4CaM with MLCK (Dabrowska *et al.*, 1982; Smith *et al.*, 2000). In this chapter we will present three different approaches to deal with interactions between Ca^{2+}, CaM and MLCK: a three-state model proposed by Kato et al. in 1984 (Kato *et al.*, 1984), our eight-state model proposed by Fajmut et al. in 2005 (Fajmut *et al.*, 2005b) and a semi-theoretical approach proposed by Rembold and Murphy in 1990 (Rembold *et al.*, 1990). The kinetic scheme of a three-state model is presented in Figure 4.

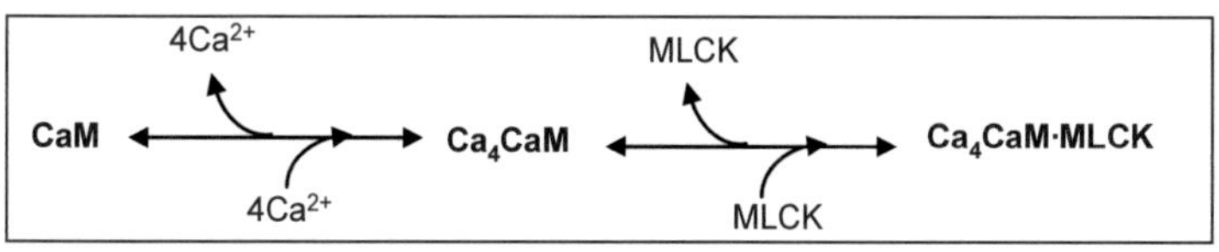

Fig. 4. **The three-state kinetic scheme of MLCK activation proposed by Kato *et al.*** (Kato *et al.*, 1984).

It assumes four independent and equivalent Ca^{2+} binding sites on CaM and considers a minimum number of possible states for CaM in the activation of MLCK. According to the kinetic scheme, MLCK is activated in two steps. The first step assumes simultaneous binding of four Ca^{2+} ions to CaM. In the second step, the intermediate complex of CaM with four Ca^{2+} ions bound (Ca_4CaM) is associated with MLCK. The final product Ca_4CaM-MLCK represents the active form of MLCK in the sense that it is able to phosphorylate MLCs. It should be noted, that Kato's kinetic scheme (Kato *et al.*, 1984) (Figure 4) reflects a somewhat older and simplified view on the interactions between Ca^{2+}, CaM and MLCK. Newer experimental studies show that not only Ca_4CaM complex but also Ca_2CaM complexes and Ca^{2+}-free CaM interact with MLCK (Bayley *et al.*, 1996; Johnson *et al.*, 1996). It has been also shown that CaM and MLCK can interact without the presence of Ca^{2+} (Wilson *et al.*, 2002) and that the velocity of Ca^{2+} binding to CaM is different with respect to N and C terminals on CaM (Brown *et al.*, 1997; Johnson *et al.*, 1996; Minowa *et al.*, 1984; Persechini *et al.*, 2000).

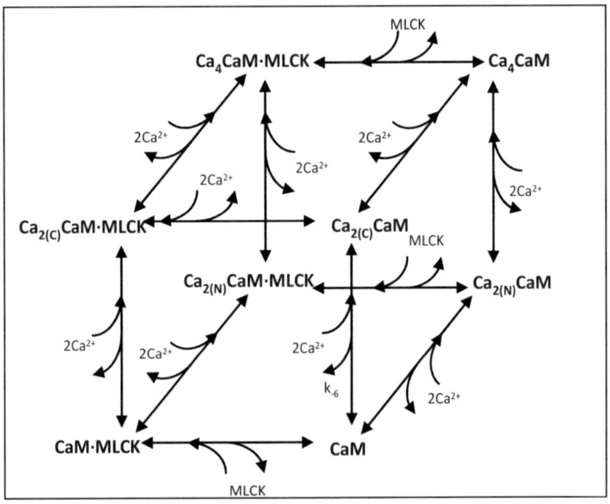

Fig. 5. **The eight-state kinetic scheme of MLCK activation proposed by Fajmut et al.** (Fajmut *et al.*, 2005a).

On the basis of these experimental findings we proposed more detailed and profound kinetic model of MLCK activation that considers either six (Fajmut *et al.*, 2005c) or eight (Fajmut *et al.*, 2005a) different states of CaM with respect to binding sites for Ca^{2+} on C and N terminals and a binding site for MLCK. The corresponding eight-state kinetic scheme is presented in Figure 5.

Binding reactions and reaction components involved can easily be recognised from the scheme. The complex $Ca_4CaM \cdot MLCK$ represents the active form of MLCK and can be applied in the 4-state kinetic scheme of force development as indicated in Figure 1.

In contrast to theoretical models of MLCK activation described here, a semi-theoretical approach was proposed by Hai and Murphy (Hai *et al.*, 1988a). This approach was based on the expression fitted to the experimentally determined dependence of MLC phosphorylation on $[Ca^{2+}]_i$. The corresponding expression was built into the mathematical model of force development proposed by them (Hai *et al.*, 1988a). The predictions of all abovementioned models for the steady state relative amount of the active MLCK (A), i.e., the ratio between $Ca_4CaM \cdot MLCK$ and total MLCK versus $[Ca^{2+}]_i$, are presented in Figure 6.

From the comparison of the model results (lines on the left panel) with the measurements (open circles) (Geguchadze *et al.*, 2004) in Figure 6 one can conclude that predictions of the eight-state model (Fajmut *et al.*, 2005a) (full line), most properly describes the process of MLCK activation. The other two models give either too sensitive (dotted line) (Rembold *et al.*, 1990) or too insensitive (dashed line) (Kato *et al.*, 1984) responses to $[Ca^{2+}]$ and they also

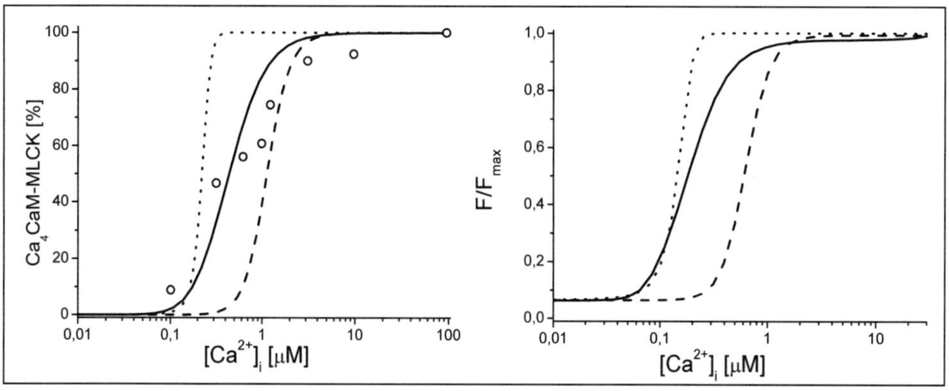

Fig. 6. **Model prediction of MLCK activation.** *Left panel:* Relative amount of the active MLCK (*A*) in dependence on $[Ca^{2+}]$ according to: semi-theoretical approach by Rembold and Murphy (Rembold *et al.*, 1990) (dotted line); eight-state model by Fajmut et al. (Fajmut *et al.*, 2005a) (full line); three-state model by Kato et al. (Kato *et al.*, 1984) (dashed line). Experimental results of Geguchadze et al. (Geguchadze *et al.*, 2004) (open circles). *Right panel:* The corresponding steady state force dependencies on $[Ca^{2+}]$ according to: Rembold and Murphy (Rembold *et al.*, 1990) (curve 1); Fajmut et al. (Fajmut *et al.*, 2005a) (curve 2); Kato et al. (Kato *et al.*, 1984) (curve 3).

do not have strong support by other experimental evidence (Gallagher *et al.*, 1993; Geguchadze *et al.*, 2004). The corresponding steady state force dependency on $[Ca^{2+}]$ (right panel) shows similar behaviour. Rembold and Murphy's model gives sensitive and Kato's model gives insensitive response to $[Ca^{2+}]$ and again don't have a strong support by experiments (Sieck *et al.*, 1998). These findings speak in favour of the necessity for more complex description of interactions between Ca^{2+}, CaM and MLCK in Ca^{2+}-signal transduction pathway. Another important property, which has not been implicated in this steady state analysis, is time dependency. In the past the interactions between Ca^{2+}, CaM and MLCK were considered as very fast (Kasturi *et al.*, 1993; Torok *et al.*, 1994) and thus modelled as being in the steady state. The reason for that were the results obtained in *in vitro* experiments, which exhibited very fast kinetics between Ca^{2+}, CaM and MLCK. However, experiments from 2001 performed by Wilson et al. (Wilson *et al.*, 2002) revealed that this might not be the case *in vivo*. Their experiments showed that at low $[Ca^{2+}]_i$ MLCK is present in complexes with Ca^{2+}-free CaM as well as with Ca_2CaM, in which Ca^{2+} is bound to the C-terminal of CaM (Wilson *et al.*, 2002). It has been suggested that this is a consequence of the increased affinity of CaM for Ca^{2+} by the presence of MLCK in the complex (Wilson *et al.*, 2002). In accordance with these findings the Ca^{2+}-CaM activation of MLCK appears at sufficiently high Ca^{2+} levels, when in addition to the C-terminal binding sites, the N-terminal binding sites for Ca^{2+} on CaM are saturated. However, the transition from CaM-MLCK and Ca_2CaM·MLCK complexes is much slower compared to the transition from Ca^{2+}- and MLCK- free CaM state. Our model simulations show for typical physiological total amounts of CaM and MLCK (10 μM and 2 μM, respectively) in smooth muscle cells that for $[Ca^{2+}]_i = 0.3$ μM approximately one third of MLCK is in complexes with Ca_2CaM, one third is Ca^{2+}-CaM-free and one third is in active form. Moreover, the half saturation time for achieving the final active MLCK form for $[Ca^{2+}]_i = 0.1$ μM is 1.5 s and for $[Ca^{2+}]_i = 0.5$ μM it is 0.5 s. These results of the eight-state model show that the processes of activation/deactivation of MLCK are not as fast as proposed by earlier models (Kato *et al.*, 1984) and some *in vitro* experiments on isolated CaM and MLCK (Kasturi *et al.*, 1993; Torok *et al.*, 1994). Moreover, the half-saturation time of MLCK activation/deactivation is of the same order of magnitude as the typical periods of oscillatory Ca^{2+} signals in smooth muscle cells (Mbikou *et al.*, 2006; Perez *et al.*, 2005), thus the processes of MLCK activation significantly contribute to decoding of oscillatory Ca^{2+} signal into a rather steady developed force already at the cellular level (Fajmut *et al.*, 2008; Fajmut *et al.*, 2005b; Mbikou *et al.*, 2011; Mbikou *et al.*, 2006) and add a small delay in force development after $[Ca^{2+}]_i$ increase.

4.3 Modelling of the MLCK/MLP balance and Ca^{2+}-contraction coupling

In our models (Fajmut *et al.*, 2008; Fajmut *et al.*, 2005b; Mbikou *et al.*, 2011; Mbikou *et al.*, 2006) we showed that the transduction of the Ca^{2+} signal from its appearance in the cytosol as a time-dependent variation of concentration to the development of force in smooth muscle cells is decoded mainly by the interactions between Ca^{2+}, CaM and MLCK, and is further translated to force by the balance between the phosphorylation and dephosphorylation of MLC, whereby both processes are regulated by MLCK and MLCP, respectively. The abundance of actomyosin crossbridges either phosphorylated or in the latch state is reflected in the magnitude of developed force. Our model results point out that a complete description of MLCK activation by Ca^{2+}-CaM is necessary for the relevant prediction of Ca^{2+}-contraction coupling and that 4-state latch bridge model upgraded with Ca^{2+}-CaM-dependent MLCK activation well describes the fast phase (first few minutes) of force development in the isometric contraction of rat tracheal rings (Fajmut *et al.*, 2008; Mbikou *et al.*, 2011; Mbikou *et al.*, 2006).

Essentially, the last and the most elaborated version of the model describing Ca^{2+}-contraction coupling consists of three parts (Mbikou *et al.*, 2011). The first one describes the activation of MLCK by Ca^{2+}-CaM complexes, considers binding of MLCK to Ca^{2+}-free CaM as well as to various Ca^{2+}-CaM complexes, predicts the concentration of the active form of MLCK, i.e. the complex $Ca_4CaM{\cdot}MLCK$, in dependence of $[Ca^{2+}]_i$. The kinetic scheme of these interactions is presented in Figure 5.

The second part describes the regulation of MLCP activity and represents an essential upgrade from the original description of Hai and Murphy (Hai *et al.*, 1988a), in which dephosphorylation velocity was taken as a linear function of AMP and MP. In our first models (Fajmut *et al.*, 2008; Fajmut *et al.*, 2005b; Mbikou *et al.*, 2006) we treated dephosphorylation process with two parallel enzymatic reactions of Michaelis-Menten type with AMP and MP as the substrates for MLCP and with MLCP·MP and MLCP·AMP as the intermediate complexes. All these models considered constant total amount of the enzyme and one Ca^{2+} independent catalytic activity of MLCP. In accordance with recent experimental results indicating that the activity of MLCP is under the influence of Rho-Kinase (RhoK) (Somlyo *et al.*, 2000), we modelled RhoK-dependent MLCP regulation (Mbikou *et al.*, 2011). RhoK phosphorylates MLCP and thus modifies its enzymatic properties. It decreases the rate constant of enzyme-substrate breakdown (k_{cat}) and increases the Michaelis constant (K_M) (Feng *et al.*, 1999; Ichikawa *et al.*, 1996; Lukas, 2004). The mechanisms by which RhoK is itself activated have not yet been fully determined. However, according to our present experimental results, RhoK is likely to be activated by the Ca^{2+} signal (Mbikou *et al.*, 2011). Thus, we consider the transition of RhoK from inactive state into active state to be Ca^{2+}-dependent. On these statements, RhoK activity has been modelled according to the kinetic scheme, represented in Figure 7 (Mbikou *et al.*, 2011).

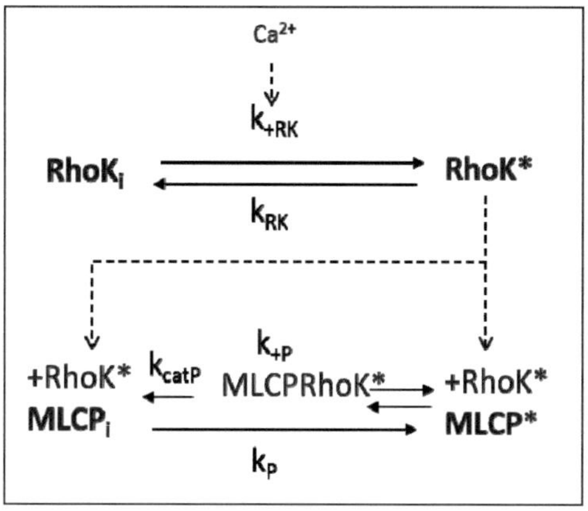

Fig. 7. **Kinetic scheme of RhoK dependent MLCP phosphorylation and partial inhibition proposed by Mbikou et al. (Mbikou *et al.*, 2011)**. $RhoK_i$, inactive RhoK; RhoK*, active Rhok. $MLCP_i$, inactive MLCP; MLCP*, active MLCP.

RhoK can be either in an inactive ($RhoK_i$) or active ($RhoK^*$) state. k_{+RK} and k_{-RK} are the corresponding on- and off- rate constants. k_{+RK} depends on the Ca^{2+} response, that is the $[Ca^{2+}]_i$ above baseline. Modelling of MLCP phosphorylation by $RhoK^*$ is based on the steady-state Michaelis-Menten enzyme kinetics, whereby $MLCP^*RhoK^*$ is the intermediate complex and k_{catP} is the rate constant for the breakdown of this intermediate complex into product, i.e. $MLCP_i$. k_{+P} and k_{-P} are the corresponding overall rate constants for MLCP phosphorylation and dephosphorylation, respectively. The values of the rate constant of MLCP-substrate breakdown (k_{cat}) and the Michaelis constant of MLCP (K_M) depend on the net effective form of MLCP, which is dependent on the fraction of unphosphorylated and inhibited MLCP ($[MLCP]^*/[MLCP]_i$).

The third part of the modelling represents the well-known 4-state actomyosin latch bridge model (Hai *et al.*, 1988a). Links between all three parts of the model are the active form of MLCK and the net effective form of MLCP, which both modulate the rate of MLC phosphorylation and dephosphorylation.

In our experimental and theoretical study (Mbikou *et al.*, 2006), the version of the model without RhoK-dependent regulation of MLCP was first developed and analysed. The model was applied to the studies of the effect of different calcium signals to the amplitude and the velocity of force developed in airway smooth muscles. It was shown that the velocity and magnitude of the force that develops in several seconds after cholinergic stimulation are determined by the following signal parameters: the amplitude and the frequency of the oscillating Ca^{2+} signal as well as the plateau - but not the peak - in the biphasic Ca^{2+} signal, which comprises a peak followed by a decline to a plateau phase (Mbikou *et al.*, 2006). On the other hand, the increased frequency of oscillating Ca^{2+} signal is translated into the increase of force magnitude (Fajmut *et al.*, 2008; Mbikou *et al.*, 2006). One main physiological implication of that model (Mbikou *et al.*, 2006) was the prediction of the temporal delay of force generation with respect to Ca^{2+} transient. Figure 8 (left panel) presents the time dependent relative MLC phosphorylation (*p*) (full line) and relative MLCK activity (*A*) (dotted line) after biphasic $[Ca^{2+}]_i$ with a peak and a plateau with the characteristic values (peak: 0.6 μM, plateau: 0.2 μM, baseline: 0.15 μM) as well as the time dependent relative force development (*F*) as predicted by the model (Mbikou *et al.*, 2006) (right panel).

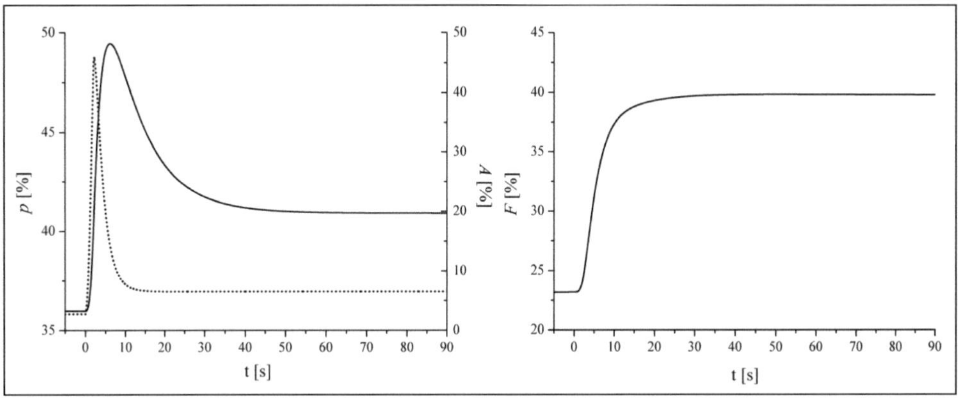

Fig. 8. **Predicted MLCK activity and MLC20 phosphorylation.** *Left panel:* Time dependent relative MLC phosphorylation (*p*) (full line) and relative MLCK activity (*A*) (dotted line) after biphasic $[Ca^{2+}]_i$ with a peak and plateau as predicted by the model (Mbikou *et al.*, 2006). *Right panel:* Time dependent relative force development (*F*) as predicted by the model (Mbikou *et al.*, 2006). *F* is defined relatively to the force obtained at supramaximal steady $[Ca^{2+}]_i$.

This time-delay originates also from the process of MLCK activation/inactivation. Namely, the process of Ca^{2+}-CaM-dependent MLCK activation contributes significantly to the time delay, in contrast with other studies (Kasturi *et al.*, 1993; Torok *et al.*, 1994), hypothesizing that the process of MLCK activation is extremely fast and is not likely to contribute more than a few milliseconds to the overall delay in force development (Sieck *et al.*, 1998). Slow force generation in our model is also a consequence of the slow-rate of MLCK activation/inactivation kinetics. Additionally, the kinetics of formation of actomyosin cross bridges explains the delay between MLC phosphorylation and force development, and the Hill-shaped time course of isometric contraction.

In another purely theoretical study (Fajmut *et al.*, 2008) we confirmed that upon biphasic Ca^{2+}-signal transduction through the system, MLCK controls amplitude more than duration, whereas MLCP tends to control both. These general characteristic regulatory properties of kinases and phosphatases were previously described by Heinrich et al. and Hornberg et al. (Heinrich *et al.*, 2002; Hornberg *et al.*, 2005) in other signalling processes.

In our most recent work (Mbikou *et al.*, 2011), the model was applied to the studies of RhoK contribution to the early phase of the Ca^{2+}-contraction coupling in airway smooth muscle. For this purpose, the simulation of RhoK inhibitor Y27632 was simulated by the model. Theoretical results of early stress development agreed with experimental results, which showed an evident drop in the stress development after RhoK inhibition in the early phase of contraction, whereby the shape and the characteristic time of stress development did not change significantly. The model further predicted that maximal RhoK activation and subsequent MLCP inactivation occur in less than 10 s, i.e., before the short time maximal contraction is achieved (Mbikou *et al.*, 2011).

An essential property of our models coupling Ca^{2+} and contraction is the explicit consideration of MLC phosphorylation and dephosphorylation steps described by Michaelis-Menten kinetics. This permits simulations of the effects that variations in enzyme contents and their catalytic properties exert on their velocities of phosphorylation/dephosphorylation, signal transduction and development of force. In this sense our models provide an upgrade with respect to the original Hai and Murphy's model as well as other models describing smooth muscle contraction.

4.4 Modelling of thin filament-associated regulation of contraction

An additional upgrade to the original Hai and Murphy's model has been proposed by Hai and Kim in 2005 (Fajmut *et al.*, 2005c) to address some experimental data that could not be explained by the four-state model. Because in phorbol ester-induced force development constant myosin phosphorylation could not be explained by the original 4-state model, the authors proposed and postulated a thin-filament-regulated latch-bridge model that includes two latch-bridge 4-state cycles, one of which is identical to the original Hai and Murphy's model, and the other one is the ultraslow 4-state cycle with lower cross-bridge cycling rates (Hai *et al.*, 2005). The model is able to fit phorbol ester-induced contractions at constant myosin phosphorylation. This was achieved by shifting cross bridges from the regular to the ultraslow cross-bridge cycle. It was also proposed that PKC activation leads to the thin-filament-based inhibition of actomyosin ATPase activity in ultraslow cycle, however, authors did not specify the target, the signalling pathway and the mechanism of this regulation. They hypothesized about calponin and caldesmon - the thin-filament-based regulatory proteins - being the candidates for the inhibition of actomyosin ATPase activity caused by PKC, because both proteins can exist in the unphosphorylated and phosphorylated form (Gerthoffer *et al.*, 1994).

In the early 1990s it was believed that the actin bound proteins, calponin and caldesmon, have large modulatory role in the latch state, however, in the late 1990s, the discovery of the significant regulatory role of myosin light chain phosphatase in smooth muscle contraction draw attention away from the thin-filament-based regulatory proteins (Paul, 2009). There is evidence that regulation of the response to a given $[Ca^{2+}]_i$, that is, regulation of the 'Ca^{2+} sensitivity' via modulating phosphatase activity, is as important as regulation of $[Ca^{2+}]_i$ in the control of contractility (Paul, 2009). PKC-mediated phosphorylation of CPI-17 has been postulated as a mechanism of PKC-mediated inhibition of MLCP (Hirano *et al.*, 2003). However, according to Hai and Kim (Hai *et al.*, 2005) this mechanism can potentially explain only the initial increase in the MLC phosphorylation but not the continued force development after myosin phosphorylation has already reached steady state.

In our experimental studies (Mbikou *et al.*, 2011; Mbikou *et al.*, 2006) we also observed the biphasic force development similar to that observed by phorbol ester-induced contraction (Hai *et al.*, 2005). The analysis of the time course of isometric contraction and MLC phosphorylation showed that the contractile response of rat tracheal rings to cholinergic stimulation developed in two distinct phases (Mbikou *et al.*, 2006). The first, short-time contractile response, which represents 70 % of the total contraction, was associated with a fast $[Ca^{2+}]_i$ peak followed by a plateau with, in some cases, with superimposed $[Ca^{2+}]_i$ oscillations, in correlation with fast and transient MLC phosphorylation, and Hill-shaped force development (Mbikou *et al.*, 2006) as shown in our model simulations presented in

Figure 8. The first fast phase was then followed by the second slow phase in which progressive increase of force reached plateau after 30 minutes (Mbikou *et al.*, 2006). Our model (Mbikou *et al.*, 2006) of force development generated by Ca^{2+}-dependent MLCK activation properly predicted the short-time contractile response. Simulations with the existing model showed that the long-time contractile response might be explained either by a long-term increase in oscillation frequency or $[Ca^{2+}]_i$ plateau. However, recordings of $[Ca^{2+}]_i$ responses to cholinergic stimulation for several minutes did not support such a hypothesis (Perez *et al.*, 2005). A possible explanation was, in parallel with the activation of MLCK, the inactivation of MLCP, which may be due to the action of PKC or RhoK (Mizuno *et al.*, 2008). For RhoK we have recently shown that it is implicated only in the fast phase of force development (Mbikou *et al.*, 2011). However, incorporation of progressive slow inactivation of MLCP in our modelling predicts a time course of isometric contraction similar to the experimental one, and explains the second increase in MLC phosphorylation and a slow phase in force development (Mbikou *et al.*, 2006). But, it does not explain a decrease in MLC20 phosphorylation associated with maximal force observed after 30 minutes (Mbikou *et al.*, 2006).

5. Conclusion

In airway smooth muscle as in other smooth muscles, actin-myosin cross bridge cycling critically depends on the phosphorylation of MLC_{20} and hence on MLCK/MLCP balance. MLCK, which activity is modulated by the Ca^{2+} signal through the formation of the Ca^{2+}-calmodulin-MLCK complex, is the most important kinase in airway myocyte contraction and the contractile properties of airway smooth muscle cells are lost when MLCK is inactivated or deleted. However, though activation of MLCK is indispensable, contraction of airway smooth muscle, both in its amplitude and time-course, is modulated by a network of kinases that can act upstream the Ca^{2+} signal, modulating the Ca^{2+} signal itself, or downstream, modulating the sensitivity of the contractile apparatus to Ca^{2+}. The main targets of the protein kinases acting on the decoding of the Ca^{2+} signal are MLCP and MLCK, though direct MLC_{20} phosphorylation, in parallel to MLCK, may be possible. Indeed, MLCP and MLCK have several sites of phosphorylation and their enzymatic activity depends on whether these sites are phosphorylated or not. These phosphorylations may up- or downregulate MLCP and MLCK activity. Stimulation of airways by contractile agonists such as acetylcholine activates kinases such as Rho kinase and PKC that inhibit MLCP activity and hence increase the sensitivity of the contractile apparatus to Ca^{2+}. By contrast, β_2-adrenergic stimulation, a major relaxant pathway, activates PKA which inhibits MLCK and favours MLCP activity. Additionally, contraction may be modulated by phosphorylation of caldesmon and calponin, proteins associated with the thin filament of actin. The MLCK/MLCP balance is hence embedded in a network of protein kinases. The resultant contractile behaviour of the SMC depends on the dynamics of the reaction of this regulatory network, and mathematical modelling is essential to decipher how the different protein kinases determine the time-dependent variation of the contractile status of the airway smooth muscle cell.

6. Acknowledgements

This work was supported by a Proteus Hubert-Curien partnership. The authors gratefully acknowledge a grant from the Cultural Service of the French Embassy in Slovenia and the French Institute Charles Nodier for the stay of A. Fajmut in Bordeaux.

7. References

Akopov SE, Zhang L, Pearce WJ (1998). Regulation of Ca2+ sensitization by PKC and rho proteins in ovine cerebral arteries: effects of artery size and age. *Am J Physiol* 275(3 Pt 2): H930-939.

Amano M, Chihara K, Nakamura N, Kaneko T, Matsuura Y, Kaibuchi K (1999). The COOH terminus of Rho-kinase negatively regulates rho-kinase activity. *J Biol Chem* 274(45): 32418-32424.

Amano M, Nakayama M, Kaibuchi K (2010). Rho-kinase/ROCK: A key regulator of the cytoskeleton and cell polarity. *Cytoskeleton (Hoboken)* 67(9): 545-554.

Anderson GP (2006). Current issues with beta2-adrenoceptor agonists: pharmacology and molecular and cellular mechanisms. *Clin Rev Allergy Immunol* 31(2-3): 119-130.

Araki S, Ito M, Kureishi Y, Feng J, Machida H, Isaka N, *et al.* (2001). Arachidonic acid-induced Ca2+ sensitization of smooth muscle contraction through activation of Rho-kinase. *Pflugers Arch* 441(5): 596-603.

Ay B, Prakash YS, Pabelick CM, Sieck GC (2004). Store-operated Ca2+ entry in porcine airway smooth muscle. *Am J Physiol Lung Cell Mol Physiol* 286(5): L909-917.

Bai Y, Sanderson MJ (2006). Modulation of the Ca2+ sensitivity of airway smooth muscle cells in murine lung slices. *Am J Physiol Lung Cell Mol Physiol.*

Barany K, Barany M (1996). Myosin Light Chains. In: Barany M (ed)^(eds). *Biochemistry of Smooth Muscle Contraction*, edn. San Diego: Academic Press. p^pp 21-35.

Bates JHT, Bullimore SR, Politi AZ, Sneyd J, Anafi RC, Lauzon A-M (2009). Transient oscillatory force-length behavior of activated airway smooth muscle. *American Journal of Physiology - Lung Cellular and Molecular Physiology* 297(2): L362-L372.

Bayley PM, Findlay WA, Martin SR (1996). Target recognition by calmodulin: Dissecting the kinetics and affinity of interaction using short peptide sequences. *Protein Sci.* 5: 1215-1228.

Bell RM, Burns DJ (1991). Lipid activation of protein kinase C. *J Biol Chem* 266(8): 4661-4664.

Bennett MK, Erondu NE, Kennedy MB (1983). Purification and characterization of a calmodulin-dependent protein kinase that is highly concentrated in brain. *J Biol Chem* 258(20): 12735-12744.

Bergner A, Sanderson MJ (2002). Acetylcholine-induced calcium signaling and contraction of airway smooth muscle cells in lung slices. *J Gen Physiol* 119(2): 187-198.

Berman HM, Ten Eyck LF, Goodsell DS, Haste NM, Kornev A, Taylor SS (2005). The cAMP binding domain: an ancient signaling module. *Proc Natl Acad Sci U S A* 102(1): 45-50.

Brown SE, Martin SR, Bayley PM (1997). Kinetic control of the dissociation pathway of calmodulin-peptide complexes. *J. Biol. Chem.* 272(6): 3389-3397.

Carpenter CL, Duckworth BC, Auger KR, Cohen B, Schaffhausen BS, Cantley LC (1990). Purification and characterization of phosphoinositide 3-kinase from rat liver. *J Biol Chem* 265(32): 19704-19711.

Cohen A, Boeijinga JK, van Haard PM, Schoemaker RC, van Vliet-Verbeek A (1992). Gastrointestinal blood loss after non-steroidal anti-inflammatory drugs. Measurement by selective determination of faecal porphyrins. *Br J Clin Pharmacol* 33(1): 33-38.

Colbran RJ, Schworer CM, Hashimoto Y, Fong YL, Rich DP, Smith MK, *et al.* (1989a). Calcium/calmodulin-dependent protein kinase II. *Biochem J* 258(2): 313-325.

Colbran RJ, Smith MK, Schworer CM, Fong YL, Soderling TR (1989b). Regulatory domain of calcium/calmodulin-dependent protein kinase II. Mechanism of inhibition and regulation by phosphorylation. *J Biol Chem* 264(9): 4800-4804.

Corompt E, Bessard G, Lantuejoul S, Naline E, Advenier C, Devillier P (1998). Inhibitory effects of large Ca2+-activated K+ channel blockers on beta-adrenergic- and NO-donor-mediated relaxations of human and guinea-pig airway smooth muscles. *Naunyn Schmiedebergs Arch Pharmacol* 357(1): 77-86.

Coronado R, Morrissette J, Sukhareva M, Vaughan DM (1994). Structure and function of ryanodine receptors. *Am J Physiol* 266(6 Pt 1): C1485-1504.

Dabrowska R, Hinkins S, Walsh MP, Hartshorne DJ (1982). The binding of smooth-muscle myosin light chain kinase to actin. *Biochem. Biophys. Res. Commun.* 107: 1524-1531.

de Lanerolle P, Paul RJ (1991). Myosin phosphorylation/dephosphorylation and regulation of airway smooth muscle contractility. *Am J Physiol* 261(2 Pt 1): L1-14.

Deng JT, Sutherland C, Brautigan DL, Eto M, Walsh MP (2002). Phosphorylation of the myosin phosphatase inhibitors, CPI-17 and PHI-1, by integrin-linked kinase. *Biochem J* 367(Pt 2): 517-524.

Deng JT, Van Lierop JE, Sutherland C, Walsh MP (2001). Ca2+-independent smooth muscle contraction. a novel function for integrin-linked kinase. *J Biol Chem* 276(19): 16365-16373.

Eto M, Kitazawa T, Brautigan DL (2004). Phosphoprotein inhibitor CPI-17 specificity depends on allosteric regulation of protein phosphatase-1 by regulatory subunits. *Proc Natl Acad Sci U S A* 101(24): 8888-8893.

Eto M, Ohmori T, Suzuki M, Furuya K, Morita F (1995). A novel protein phosphatase-1 inhibitory protein potentiated by protein kinase C. Isolation from porcine aorta media and characterization. *J Biochem* 118(6): 1104-1107.

Eto M, Senba S, Morita F, Yazawa M (1997). Molecular cloning of a novel phosphorylation-dependent inhibitory protein of protein phosphatase-1 (CPI17) in smooth muscle: its specific localization in smooth muscle. *FEBS Lett* 410(2-3): 356-360.

Fahrmann M, Mohlig M, Schatz H, Pfeiffer A (1998). Purification and characterization of a Ca2+/calmodulin-dependent protein kinase II from hog gastric mucosa using a protein-protein affinity chromatographic technique. *Eur J Biochem* 255(2): 516-525.

Fajmut A, Brumen M (2008). MLC-kinase/phosphatase control of Ca2+ signal transduction in airway smooth muscles. *J Theor Biol* 252(3): 474-481.

Fajmut A, Brumen M, Schuster S (2005a). Theoretical model of the interactions between Ca2+, calmodulin and myosin light chain kinase. *FEBS Lett.* 579: 4361-4366.

Fajmut A, Dobovisek A, Brumen M (2005b). Mathematical modeling of the relation between myosin phosphorylation and stress development in smooth muscles. *J. Chem. Inf. Model.* 45(6): 1610-1615.

Fajmut A, Jagodic M, Brumen M (2005c). Mathematical modeling of the myosin light chain kinase activation. *J. Chem. Inf. Model.* 45(6): 1605-1609.

Feng J, Ito M, Ichikawa K, Isaka N, Nishikawa M, Hartshorne DJ, *et al.* (1999). Inhibitory phosphorylation site for Rho-associated kinase on smooth muscle myosin phosphatase. *J Biol Chem* 274(52): 37385-37390.

Ferris CD, Cameron AM, Bredt DS, Huganir RL, Snyder SH (1991a). Inositol 1,4,5-trisphosphate receptor is phosphorylated by cyclic AMP-dependent protein kinase at serines 1755 and 1589. *Biochem Biophys Res Commun* 175(1): 192-198.

Ferris CD, Huganir RL, Bredt DS, Cameron AM, Snyder SH (1991b). Inositol trisphosphate receptor: phosphorylation by protein kinase C and calcium calmodulin-dependent protein kinases in reconstituted lipid vesicles. *Proc Natl Acad Sci U S A* 88(6): 2232-2235.

Fredberg JJ, Inoyue DS, Mijalovich SM, Butler JP (1999). Perturbed equilibrium of myosin binding in airway smooth muscle and its implications in bronchospasm. *Am. J. Respir. Crit. Care Med.* 159(3): 959-967.

Fry MJ (2001). Phosphoinositide 3-kinase signalling in breast cancer: how big a role might it play? *Breast Cancer Res* 3(5): 304-312.

Fry MJ (1994). Structure, regulation and function of phosphoinositide 3-kinases. *Biochim Biophys Acta* 1226(3): 237-268.

Fry MJ, Panayotou G, Dhand R, Ruiz-Larrea F, Gout I, Nguyen O, *et al.* (1992). Purification and characterization of a phosphatidylinositol 3-kinase complex from bovine brain by using phosphopeptide affinity columns. *Biochem J* 288 (Pt 2): 383-393.

Fukunaga K, Yamamoto H, Matsui K, Higashi K, Miyamoto E (1982). Purification and characterization of a Ca2+- and calmodulin-dependent protein kinase from rat brain. *J Neurochem* 39(6): 1607-1617.

Gallagher P, Herring B, Trafny A, Sowadski J, Stull J (1993). A molecular mechanism for autoinhibition of myosin light chain kinases. *J. Biol. Chem.* 268(35): 26578-26582.

Geguchadze R, Zhi G, Lau KS, Isotani E, Persechini A, Kamm KE, *et al.* (2004). Quantitative measurements of Ca2+/calmodulin binding and activation of myosin light chain kinase in cells. *FEBS Lett.* 557(1-3): 121-124.

Gerthoffer W, Pohl J (1994). Caldesmon and calponin phosphorylation in regulation of smooth muscle contraction. *Can J Physiol Pharmacol* 72(11): 1410-1414.

Gestrelius S, Borgstrom P (1986). A dynamic model of smooth muscle contraction. *Biophys J* 50(1): 157-169.

Goldenring JR, Gonzalez B, McGuire JS, Jr., DeLorenzo RJ (1983). Purification and characterization of a calmodulin-dependent kinase from rat brain cytosol able to phosphorylate tubulin and microtubule-associated proteins. *J Biol Chem* 258(20): 12632-12640.

Grassie ME, Moffat LD, Walsh MP, MacDonald JA (2011). The myosin phosphatase targeting protein (MYPT) family: a regulated mechanism for achieving substrate specificity of the catalytic subunit of protein phosphatase type 1delta. *Arch Biochem Biophys* 510(2): 147-159.

Gunter TE, Buntinas L, Sparagna G, Eliseev R, Gunter K (2000). Mitochondrial calcium transport: mechanisms and functions. *Cell Calcium* 28(5-6): 285-296.

Gusev NB (2001). Some properties of caldesmon and calponin and the participation of these proteins in regulation of smooth muscle contraction and cytoskeleton formation. *Biochemistry (Mosc)* 66(10): 1112-1121.

Hai C-M, Kim HR (2005). An expanded latch-bridge model of protein kinase C-mediated smooth muscle contraction. *J. Appl. Physiol.* 98(4): 1356-1365.

Hai CM, Murphy RA (1988a). Cross-bridge phosphorylation and regulation of latch state in smooth muscle. *Am J Physiol Cell Physiol* 254(1): C99-106.

Hai CM, Murphy RA (1988b). Regulation of shortening velocity by cross-bridge phosphorylation in smooth muscle. *Am J Physiol* 255(1 Pt 1): C86-94.

Hanley RM, Means AR, Ono T, Kemp BE, Burgin KE, Waxham N, *et al.* (1987). Functional analysis of a complementary DNA for the 50-kilodalton subunit of calmodulin kinase II. *Science* 237(4812): 293-297.

Harnett KM, Biancani P (2003). Calcium-dependent and calcium-independent contractions in smooth muscles. *Am J Med* 115 Suppl 3A: 24S-30S.

Hartshorne DJ, Hirano K (1999). Interactions of protein phosphatase type 1, with a focus on myosin phosphatase. *Mol Cell Biochem* 190(1-2): 79-84.

Hartshorne DJ, Ito M, Erdodi F (1998). Myosin light chain phosphatase: subunit composition, interactions and regulation. *J Muscle Res Cell Motil* 19(4): 325-341.

Hashimoto Y, Soderling TR (1990). Phosphorylation of smooth muscle myosin light chain kinase by Ca2+/calmodulin-dependent protein kinase II: comparative study of the phosphorylation sites. *Arch Biochem Biophys* 278(1): 41-45.

Heinrich R, Neel B, Rapoport T (2002). Mathematical models of protein kinase signal transduction. *Mol Cell* 9(5): 957-970.

Hirano K, Derkach DN, Hirano M, Nishimura J, Kanaide H (2003). Protein kinase network in the regulation of phosphorylation and dephosphorylation of smooth muscle myosin light chain. *Mol Cell Biochem* 248(1-2): 105-114.

Hirano K, Hirano M, Kanaide H (2004). Regulation of myosin phosphorylation and myofilament Ca2+ sensitivity in vascular smooth muscle. *J Smooth Muscle Res* 40(6): 219-236.

Hong F, Haldeman BD, Jackson D, Carter M, Baker JE, Cremo CR (2011). Biochemistry of smooth muscle myosin light chain kinase. *Arch Biochem Biophys* 510(2): 135-146.

Hornberg JJ, Bruggeman FJ, Binder B, Geest CR, de Vaate AJMB, Lankelma J, *et al.* (2005). Principles behind the multifarious control of signal transduction. ERK phosphorylation and kinase/phosphatase control. *FEBS Journal* 272(1): 244-258.

Huang J, Mahavadi S, Sriwai W, Hu W, Murthy KS (2006). Gi-coupled receptors mediate phosphorylation of CPI-17 and MLC20 via preferential activation of the PI3K/ILK pathway. *Biochem J* 396(1): 193-200.

Huang KP (1989). The mechanism of protein kinase C activation. *Trends Neurosci* 12(11): 425-432.

Hudmon A, Schulman H (2002). Neuronal CA2+/calmodulin-dependent protein kinase II: the role of structure and autoregulation in cellular function. *Annu Rev Biochem* 71: 473-510.

Huxley AF (1957). Muscle structure and theories of contraction. *Prog Biophys Biophys Chem* 7: 255-318.

Hyvelin JM, Martin C, Roux E, Marthan R, Savineau JP (2000a). Human isolated bronchial smooth muscle contains functional ryanodine/caffeine-sensitive Ca-release channels. *Am J Respir Crit Care Med* 162(2 Pt 1): 687-694.

Hyvelin JM, Roux E, Prevost MC, Savineau JP, Marthan R (2000b). Cellular mechanisms of acrolein-induced alteration in calcium signaling in airway smooth muscle. *Toxicol Appl Pharmacol* 164(2): 176-183.

Ichikawa K, Ito M, Hartshorne DJ (1996). Phosphorylation of the large subunit of myosin phosphatase and inhibition of phosphatase activity. *J Biol Chem* 271(9): 4733-4740.

Iizuka K, Yoshii A, Samizo K, Tsukagoshi H, Ishizuka T, Dobashi K, *et al.* (1999). A major role for the rho-associated coiled coil forming protein kinase in G-protein-mediated Ca2+ sensitization through inhibition of myosin phosphatase in rabbit trachea. *Br J Pharmacol* 128(4): 925-933.

Ikebe M, Reardon S (1990a). Phosphorylation of smooth myosin light chain kinase by smooth muscle Ca2+/calmodulin-dependent multifunctional protein kinase. *J Biol Chem* 265(16): 8975-8978.

Ikebe M, Reardon S, Scott-Woo GC, Zhou Z, Koda Y (1990b). Purification and characterization of calmodulin-dependent multifunctional protein kinase from smooth muscle: isolation of caldesmon kinase. *Biochemistry* 29(51): 11242-11248.

Ishizaki T, Maekawa M, Fujisawa K, Okawa K, Iwamatsu A, Fujita A, *et al.* (1996). The small GTP-binding protein Rho binds to and activates a 160 kDa Ser/Thr protein kinase homologous to myotonic dystrophy kinase. *EMBO J* 15(8): 1885-1893.

Ito M, Nakano T, Erdodi F, Hartshorne DJ (2004). Myosin phosphatase: structure, regulation and function. *Mol Cell Biochem* 259(1-2): 197-209.

Itoh T, Suzuki A, Watanabe Y (1994). Effect of a peptide inhibitor of protein kinase C on G-protein-mediated increase in myofilament Ca(2+)-sensitivity in rabbit arterial skinned muscle. *Br J Pharmacol* 111(1): 311-317.

Janssen LJ, Tazzeo T, Zuo J (2004). Enhanced myosin phosphatase and Ca(2+)-uptake mediate adrenergic relaxation of airway smooth muscle. *Am J Respir Cell Mol Biol* 30(4): 548-554.

Johnson DA, Akamine P, Radzio-Andzelm E, Madhusudan M, Taylor SS (2001). Dynamics of cAMP-dependent protein kinase. *Chem Rev* 101(8): 2243-2270.

Johnson JD, Snyder C, Walsh M, Flynn M (1996). Effects of myosin light chain kinase and peptides on Ca2+ exchange with the N- and C-terminal Ca2+ binding sites of calmodulin. *J. Biol. Chem.* 271(2): 761-767.

Johnson M (1998). The beta-adrenoceptor. *Am J Respir Crit Care Med* 158(5 Pt 3): S146-153.

Johnson M (2006). Molecular mechanisms of beta(2)-adrenergic receptor function, response, and regulation. *J Allergy Clin Immunol* 117(1): 18-24; quiz 25.

Kajita J, Yamaguchi H (1993). Calcium mobilization by muscarinic cholinergic stimulation in bovine single airway smooth muscle. *Am J Physiol* 264(5 Pt 1): L496-503.

Kamm KE, Stull JT (2001). Dedicated myosin light chain kinases with diverse cellular functions. *J. Biol. Chem.* 276(7): 4527-4530.

Kasturi R, Vasulka C, Johnson J (1993). Ca2+, caldesmon, and myosin light chain kinase exchange with calmodulin. *J. Biol. Chem.* 268(11): 7958-7964.

Kato S, Osa T, Ogasawara T (1984). Kinetic model for isometric contraction in smooth muscle on the basis of myosin phosphorylation hypothesis. *Biophys. J.* 46(1): 35-44.

Katso R, Okkenhaug K, Ahmadi K, White S, Timms J, Waterfield MD (2001). Cellular function of phosphoinositide 3-kinases: implications for development, homeostasis, and cancer. *Annu Rev Cell Dev Biol* 17: 615-675.

Kawano Y, Fukata Y, Oshiro N, Amano M, Nakamura T, Ito M, *et al.* (1999). Phosphorylation of myosin-binding subunit (MBS) of myosin phosphatase by Rho-kinase in vivo. *J Cell Biol* 147(5): 1023-1038.

Kennedy MB, McGuinness T, Greengard P (1983). A calcium/calmodulin-dependent protein kinase from mammalian brain that phosphorylates Synapsin I: partial purification and characterization. *J Neurosci* 3(4): 818-831.

Kim C, Cheng CY, Saldanha SA, Taylor SS (2007). PKA-I holoenzyme structure reveals a mechanism for cAMP-dependent activation. *Cell* 130(6): 1032-1043.

Kimura K, Ito M, Amano M, Chihara K, Fukata Y, Nakafuku M, *et al.* (1996). Regulation of myosin phosphatase by Rho and Rho-associated kinase (Rho-kinase). *Science* 273(5272): 245-248.

Ko EA, Han J, Jung ID, Park WS (2008). Physiological roles of K+ channels in vascular smooth muscle cells. *J Smooth Muscle Res* 44(2): 65-81.

Kolb SJ, Hudmon A, Ginsberg TR, Waxham MN (1998). Identification of domains essential for the assembly of calcium/calmodulin-dependent protein kinase II holoenzymes. *J Biol Chem* 273(47): 31555-31564.

Komander D, Garg R, Wan PT, Ridley AJ, Barford D (2008). Mechanism of multi-site phosphorylation from a ROCK-I:RhoE complex structure. *EMBO J* 27(23): 3175-3185.

Kordowska J, Huang R, Wang CL (2006). Phosphorylation of caldesmon during smooth muscle contraction and cell migration or proliferation. *J Biomed Sci* 13(2): 159-172.

Kuo KH, Herrera AM, Seow CY (2003). Ultrastructure of airway smooth muscle. *Respir Physiol Neurobiol* 137(2-3): 197-208.

Kuret J, Schulman H (1984). Purification and characterization of a Ca2+/calmodulin-dependent protein kinase from rat brain. *Biochemistry* 23(23): 5495-5504.

Lester DS, Doll L, Brumfeld V, Miller IR (1990). Lipid dependence of surface conformations of protein kinase C. *Biochim Biophys Acta* 1039(1): 33-41.

Leung T, Chen XQ, Manser E, Lim L (1996). The p160 RhoA-binding kinase ROK alpha is a member of a kinase family and is involved in the reorganization of the cytoskeleton. *Mol Cell Biol* 16(10): 5313-5327.

LeVine H, 3rd, Sahyoun NE, Cuatrecasas P (1985). Calmodulin binding to the cytoskeletal neuronal calmodulin-dependent protein kinase is regulated by autophosphorylation. *Proc Natl Acad Sci U S A* 82(2): 287-291.

Liu C, Zuo J, Pertens E, Helli PB, Janssen LJ (2005). Regulation of Rho/ROCK signaling in airway smooth muscle by membrane potential and [Ca2+]i. *Am J Physiol Lung Cell Mol Physiol* 289(4): L574-582.

Liu X, Farley JM (1996). Frequency modulation of acetylcholine-induced Ca(++)-dependent Cl- current oscillations are mediated by 1, 4, 5-trisphosphate in tracheal myocytes. *J Pharmacol Exp Ther* 277: 796-804.

Lukas TJ (2004). A signal transduction pathway model prototype I: from agonist to cellular endpoint. *Biophys. J.* 87(3): 1406-1416.

Marston SB, Redwood CS (1991). The molecular anatomy of caldesmon. *Biochem J* 279 (Pt 1): 1-16.

Marthan R (2004). Store-operated calcium entry and intracellular calcium release channels in airway smooth muscle. *Am J Physiol Lung Cell Mol Physiol* 286(5): L907-908.

Mbikou P, Fajmut A, Brumen M, Roux E (2011). Contribution of Rho kinase to the early phase of the calcium-contraction coupling in airway smooth muscle. *Exp Physiol* 96(2): 240-258.

Mbikou P, Fajmut A, Brumen M, Roux E (2006). Theoretical and experimental investigation of calcium-contraction coupling in airway smooth muscle. *Cell Biochem Biophys* 46(3): 233-252.

McNicholl ET, Das R, SilDas S, Taylor SS, Melacini G Communication between tandem cAMP binding domains in the regulatory subunit of protein kinase A-Ialpha as revealed by domain-silencing mutations. *J Biol Chem* 285(20): 15523-15537.

Mijailovich SM, Butler JP, Fredberg JJ (2000). Perturbed equilibria of myosin binding in airway smooth muscle: bond-length distributions, mechanics, and ATP metabolism. *Biophys J* 79(5): 2667-2681.

Miller SG, Kennedy MB (1985). Distinct forebrain and cerebellar isozymes of type II Ca2+/calmodulin-dependent protein kinase associate differently with the postsynaptic density fraction. *J Biol Chem* 260(15): 9039-9046.

Miller SG, Kennedy MB (1986). Regulation of brain type II Ca2+/calmodulin-dependent protein kinase by autophosphorylation: a Ca2+-triggered molecular switch. *Cell* 44(6): 861-870.

Minowa O, Yagi K (1984). Calcium binding to tryptic fragments of calmodulin. *J. Biochem. (Tokyo)* 56: 1175-1182.

Mizuno Y, Isotani E, Huang J, Ding H, Stull JT, Kamm KE (2008). Myosin light chain kinase activation and calcium sensitization in smooth muscle in vivo. *Am J Physiol Cell Physiol* 295(2): C358-364.

Morgan SJ, Smith AD, Parker PJ (1990). Purification and characterization of bovine brain type I phosphatidylinositol kinase. *Eur J Biochem* 191(3): 761-767.

Mounkaila B, Marthan R, Roux E (2005). Biphasic effect of extracellular ATP on human and rat airways is due to multiple P2 purinoceptor activation. *Respir Res* 6(1): 143.

Murthy KS (2006). Signaling for contraction and relaxation in smooth muscle of the gut. *Annu Rev Physiol* 68: 345-374.

Nakagawa O, Fujisawa K, Ishizaki T, Saito Y, Nakao K, Narumiya S (1996). ROCK-I and ROCK-II, two isoforms of Rho-associated coiled-coil forming protein serine/threonine kinase in mice. *FEBS Lett* 392(2): 189-193.

Niiro N, Ikebe M (2001). Zipper-interacting protein kinase induces Ca(2+)-free smooth muscle contraction via myosin light chain phosphorylation. *J Biol Chem* 276(31): 29567-29574.

Nishizuka Y (1995). Protein kinase C and lipid signaling for sustained cellular responses. *FASEB J* 9(7): 484-496.

Otsu M, Hiles I, Gout I, Fry MJ, Ruiz-Larrea F, Panayotou G, *et al.* (1991). Characterization of two 85 kd proteins that associate with receptor tyrosine kinases, middle-T/pp60c-src complexes, and PI3-kinase. *Cell* 65(1): 91-104.

Otto B, Steusloff A, Just I, Aktories K, Pfitzer G (1996). Role of Rho proteins in carbachol-induced contractions in intact and permeabilized guinea-pig intestinal smooth muscle. *J Physiol* 496 (Pt 2): 317-329.

Paul RJ (2009). Regulation of smooth muscle contraction/relaxation: paradigm shifts and quantifying arrows. *The Journal of Physiology* 587(14): 3413-3414.

Perez JF, Sanderson MJ (2005). The frequency of calcium oscillations induced by 5-HT, ACH, and KCl determine the contraction of smooth muscle cells of intrapulmonary bronchioles. *J Gen Physiol* 125(6): 535-553.

Persechini A, Yano K, Stemmer PM (2000). Ca2+ binding and energy coupling in the calmodulin-myosin light chain kinase complex. *J. Biol. Chem.* 275(6): 4199-4204.

Prakash YS, Kannan MS, Walseth TF, Sieck GC (1998). Role of cyclic ADP-ribose in the regulation of [Ca2+]i in porcine tracheal smooth muscle. *Am J Physiol* 274(6 Pt 1): C1653-1660.

Prakash YS, Pabelick CM, Kannan MS, Sieck GC (2000). Spatial and temporal aspects of ACh-induced [Ca2+]i oscillations in porcine tracheal smooth muscle. *Cell Calcium* 27(3): 153-162.

Rameh LE, Cantley LC (1999). The role of phosphoinositide 3-kinase lipid products in cell function. *J Biol Chem* 274(13): 8347-8350.

Rembold CM, Murphy RA (1990). Latch-bridge model in smooth-muscle - [Ca-2+]i can quantitatively predict stress. *Am. J. Physiol.* 259(2): C251-C257.

Riento K, Guasch RM, Garg R, Jin B, Ridley AJ (2003a). RhoE binds to ROCK I and inhibits downstream signaling. *Mol Cell Biol* 23(12): 4219-4229.

Riento K, Ridley AJ (2003b). Rocks: multifunctional kinases in cell behaviour. *Nat Rev Mol Cell Biol* 4(6): 446-456.

Rossetti M, Savineau JP, Crevel H, Marthan R (1995). Role of protein kinase C in nonsensitized and passively sensitized human isolated bronchial smooth muscle. *Am J Physiol* 268(6 Pt 1): L966-971.

Roux E, Hyvelin JM, Savineau JP, Marthan R (1998). Calcium signaling in airway smooth muscle cells is altered by in vitro exposure to the aldehyde acrolein. *Am J Respir Cell Mol Biol* 19(3): 437-444.

Roux E, Marhl M (2004). Role of sarcoplasmic reticulum and mitochondria in ca(2+) removal in airway myocytes. *Biophys J* 86(4): 2583-2595.

Roux E, Noble PJ, Noble D, Marhl M (2006). Modelling of calcium handling in airway myocytes. *Prog Biophys Mol Biol* 90(1-3): 64-87.

Saitoh T, Schwartz JH (1985). Phosphorylation-dependent subcellular translocation of a Ca2+/calmodulin-dependent protein kinase produces an autonomous enzyme in Aplysia neurons. *J Cell Biol* 100(3): 835-842.

Sakurada S, Takuwa N, Sugimoto N, Wang Y, Seto M, Sasaki Y, *et al.* (2003). Ca2+-dependent activation of Rho and Rho kinase in membrane depolarization-induced and receptor stimulation-induced vascular smooth muscle contraction. *Circ Res* 93(6): 548-556.

Sanders KM (2001). Invited review: mechanisms of calcium handling in smooth muscles. *J. Appl. Physiol.* 91(3): 1438-1449.

Schaafsma D, Boterman M, de Jong AM, Hovens I, Penninks JM, Nelemans SA, *et al.* (2006). Differential Rho-kinase dependency of full and partial muscarinic receptor agonists in airway smooth muscle contraction. *Br J Pharmacol* 147(7): 737-743.

Schramm CM, Chuang ST, Grunstein MM (1995). cAMP generation inhibits inositol 1,4,5-trisphosphate binding in rabbit tracheal smooth muscle. *Am J Physiol* 269(5 Pt 1): L715-719.

Scott JD, McCartney S (1994). Localization of A-kinase through anchoring proteins. *Mol Endocrinol* 8(1): 5-11.

Shabb JB (2001). Physiological substrates of cAMP-dependent protein kinase. *Chem Rev* 101(8): 2381-2411.

Shah V, Bharadwaj S, Kaibuchi K, Prasad GL (2001). Cytoskeletal organization in tropomyosin-mediated reversion of ras-transformation: Evidence for Rho kinase pathway. *Oncogene* 20(17): 2112-2121.

Shibasaki F, Homma Y, Takenawa T (1991). Two types of phosphatidylinositol 3-kinase from bovine thymus. Monomer and heterodimer form. *J Biol Chem* 266(13): 8108-8114.

Shields SM, Vernon PJ, Kelly PT (1984). Autophosphorylation of calmodulin-kinase II in synaptic junctions modulates endogenous kinase activity. *J Neurochem* 43(6): 1599-1609.

Shimizu T, Ihara K, Maesaki R, Amano M, Kaibuchi K, Hakoshima T (2003). Parallel coiled-coil association of the RhoA-binding domain in Rho-kinase. *J Biol Chem* 278(46): 46046-46051.

Sieck GC, Han Y-S, Prakash YS, Jones KA (1998). Cross-bridge cycling kinetics, actomyosin ATPase activity and myosin heavy chain isoforms in skeletal and smooth respiratory muscles. *Comp. Biochem. Physiol. Part B: Biochem. Mol. Biol.* 119(3): 435-450.

Smith L, Stull JT (2000). Myosin light chain kinase binding to actin filaments. *FEBS Lett.* 480: 298-300.

Somlyo AP, Somlyo AV (2003). Ca2+ sensitivity of smooth muscle and nonmuscle myosin II: modulated by G proteins, kinases, and myosin phosphatase. *Physiol Rev* 83(4): 1325-1358.

Somlyo AP, Somlyo AV (1994). Signal transduction and regulation in smooth muscle. *Nature* 372(6503): 231-236.

Somlyo AP, Somlyo AV (2000). Signal transduction by G-proteins, rho-kinase and protein phosphatase to smooth muscle and non-muscle myosin II. *J Physiol* 522 Pt 2: 177-185.

Somlyo AP, Somlyo AV, Kitazawa T, Bond M, Shuman H, Kowarski D (1983). Ultrastructure, function and composition of smooth muscle. *Ann Biomed Eng* 11(6): 579-588.

Stull JT, Hsu LC, Tansey MG, Kamm KE (1990). Myosin light chain kinase phosphorylation in tracheal smooth muscle. *J Biol Chem* 265(27): 16683-16690.

Stull JT, Tansey MG, Tang DC, Word RA, Kamm KE (1993). Phosphorylation of myosin light chain kinase: a cellular mechanism for Ca2+ desensitization. *Mol Cell Biochem* 127-128: 229-237.

Su Y, Dostmann WR, Herberg FW, Durick K, Xuong NH, Ten Eyck L, *et al.* (1995). Regulatory subunit of protein kinase A: structure of deletion mutant with cAMP binding domains. *Science* 269(5225): 807-813.

Tansey MG, Luby-Phelps K, Kamm KE, Stull JT (1994). Ca(2+)-dependent phosphorylation of myosin light chain kinase decreases the Ca2+ sensitivity of light chain phosphorylation within smooth muscle cells. *J Biol Chem* 269(13): 9912-9920.

Tansey MG, Word RA, Hidaka H, Singer HA, Schworer CM, Kamm KE, *et al.* (1992). Phosphorylation of myosin light chain kinase by the multifunctional calmodulin-dependent protein kinase II in smooth muscle cells. *J Biol Chem* 267(18): 12511-12516.

Taylor SS, Kim C, Vigil D, Haste NM, Yang J, Wu J, *et al.* (2005). Dynamics of signaling by PKA. *Biochim Biophys Acta* 1754(1-2): 25-37.

Terrak M, Kerff F, Langsetmo K, Tao T, Dominguez R (2004). Structural basis of protein phosphatase 1 regulation. *Nature* 429(6993): 780-784.

Toph A, Kiss E, Gergely P, Walsh MP, Hartshorne DJ, Erdodi F (2000). phosphorylation of MYPT1 by protein kinase C attenuates interaction with PP1 catalytic subunit and 20 kDa light chain of myosin. *FEBS Lett* 484: 113-117.

Torok K, Trentham DR (1994). Mechanism of 2-chloro-(epsilon-amino-Lys75)-[6-[4-(N,N-diethylamino)phenyl]-1,3,5-triazin-4-yl]calmodulin interactions with smooth muscle myosin light chain kinase and derived peptides. *Biochem.* 33: 12807-12820.

Trybus KM (1996). In: Barany M (ed)^(eds). *Biochemistry of smooth muscle contraction*, edn. San Diego, CA: Academic Press. p^pp 37-46.

Ueda K, Murata-Hori M, Tatsuka M, Hosoya H (2002). Rho-kinase contributes to diphosphorylation of myosin II regulatory light chain in nonmuscle cells. *Oncogene* 21(38): 5852-5860.

Velasco G, Armstrong C, Morrice N, Frame S, Cohen P (2002). Phosphorylation of the regulatory subunit of smooth muscle protein phosphatase 1M at Thr850 induces its dissociation from myosin. *FEBS Lett* 527(1-3): 101-104.

Wang I, Politi A, Tania N, Bai Y, Sanderson M, Sneyd J (2008). A mathematical model of airway and pulmonary arteriole smooth muscle. *Biophys J* 94(6): 2053-2064.

Wang Y, Zheng XR, Riddick N, Bryden M, Baur W, Zhang X, *et al.* (2009). ROCK isoform regulation of myosin phosphatase and contractility in vascular smooth muscle cells. *Circ Res* 104(4): 531-540.

Webb BL, Hirst SJ, Giembycz MA (2000). Protein kinase C isoenzymes: a review of their structure, regulation and role in regulating airways smooth muscle tone and mitogenesis. *Br J Pharmacol* 130(7): 1433-1452.

Webb BL, Lindsay MA, Seybold J, Brand NJ, Yacoub MH, Haddad EB, *et al.* (1997). Identification of the protein kinase C isoenzymes in human lung and airways smooth muscle at the protein and mRNA level. *Biochem Pharmacol* 54(1): 199-205.

Wilson DP, Sutherland C, Walsh MP (2002). Ca2+ activation of smooth muscle contraction. Evidence for the involvement of calmodulin that is bound to the triton-insoluble fraction even in the absence of Ca2+. *J. Biol. Chem.* 277(3): 2186-2192.

Winder SJ, Allen BG, Clement-Chomienne O, Walsh MP (1998). Regulation of smooth muscle actin-myosin interaction and force by calponin. *Acta Physiol Scand* 164(4): 415-426.

Winder SJ, Walsh MP (1990). Smooth muscle calponin. Inhibition of actomyosin MgATPase and regulation by phosphorylation. *J Biol Chem* 265(17): 10148-10155.

Wingard CJ, Nowocin JM, Murphy RA (2001). Cross-bridge regulation by Ca(2+)-dependent phosphorylation in amphibian smooth muscle. *Am J Physiol Regul Integr Comp Physiol* 281(6): R1769-1777.

Wooldridge AA, MacDonald JA, Erdodi F, Ma C, Borman MA, Hartshorne DJ, *et al.* (2004). Smooth muscle phosphatase is regulated in vivo by exclusion of phosphorylation of threonine 696 of MYPT1 by phosphorylation of Serine 695 in response to cyclic nucleotides. *J Biol Chem* 279(33): 34496-34504.

Yamauchi T, Fujisawa H (1985). Self-regulation of calmodulin-dependent protein kinase II and glycogen synthase kinase by autophosphorylation. *Biochem Biophys Res Commun* 129(1): 213-219.

Yoneda A, Multhaupt HA, Couchman JR (2005). The Rho kinases I and II regulate different aspects of myosin II activity. *J Cell Biol* 170(3): 443-453.

Yoshii A, Iizuka K, Dobashi K, Horie T, Harada T, Nakazawa T, *et al.* (1999). Relaxation of contracted rabbit tracheal and human bronchial smooth muscle by Y-27632 through inhibition of Ca2+ sensitization. *Am J Respir Cell Mol Biol* 20(6): 1190-1200.

Yoshioka K, Sugimoto N, Takuwa N, Takuwa Y (2007). Essential role for class II phosphoinositide 3-kinase alpha-isoform in Ca2+-induced, Rho- and Rho kinase-dependent regulation of myosin phosphatase and contraction in isolated vascular smooth muscle cells. *Mol Pharmacol* 71(3): 912-920.

Yu SN, Crago PE, Chiel HJ (1997). A nonisometric kinetic model for smooth muscle. *Am J Physiol Cell Physiol* 272(3): C1025-1039.

Yuen SL, Ogut O, Brozovich FV (2009). Nonmuscle myosin is regulated during smooth muscle contraction. *Am J Physiol Heart Circ Physiol* 297(1): H191-199.

Zhang WC, Peng YJ, Zhang GS, He WQ, Qiao YN, Dong YY, *et al.* (2010). Myosin light chain kinase is necessary for tonic airway smooth muscle contraction. *J Biol Chem* 285(8): 5522-5531.

Zhao J, Hoye E, Boylan S, Walsh DA, Trewhella J (1998). Quaternary structures of a catalytic subunit-regulatory subunit dimeric complex and the holoenzyme of the cAMP-dependent protein kinase by neutron contrast variation. *J Biol Chem* 273(46): 30448-30459.

Zhou HL, Newsholme SJ, Torphy TJ (1992). Agonist-related differences in the relationship between cAMP content and protein kinase activity in canine trachealis. *J Pharmacol Exp Ther* 261(3): 1260-1267.

Zhou ZL, Ikebe M (1994). New isoforms of Ca2+/calmodulin-dependent protein kinase II in smooth muscle. *Biochem J* 299 (Pt 2): 489-495.

14

Regulation of Na^+/H^+ Exchanger Isoform 3 by Protein Kinase A in the Renal Proximal Tubule

Adriana Castello Costa Girardi[1] and Luciene Regina Carraro-Lacroix[2]
[1]*University of São Paulo Medical School*
[2]*Hospital for Sick Children*
[1]*Brazil*
[2]*Canada*

1. Introduction

One of the major functions of the kidneys is to maintain the volume and composition of the body fluids constant despite wide variation in the daily intake of water and solutes. To accomplish this task, the activities of a number of transport proteins along the nephron are tightly regulated.

The nephron is the functional unit of the kidneys. Each human kidney contains approximately 1.2 millions of nephrons. At the beginning of each nephron, in the glomerulus, the blood is filtered: cells and most proteins are retained, whereas water and small solutes pass from the glomerular capillaries to the Bowman's capsule. As the glomerular filtrate leaves Bowman's capsule and enters the renal tubule, it flows sequentially through the proximal tubule, the loop of Henle, the distal tubule, and the collecting duct. Along this course, greater part of the glomerular filtrate is transported across and between the tubule cells and reenters the blood (reabsorption), whereas some is secreted from the blood into the luminal fluid (secretion). The formation of urine involves the sum of these three major processes: ultrafiltration of plasma by the glomerulus, reabsorption of water and solutes from the ultrafiltrate, and secretion of solutes into the tubular fluid. Although 180 liters of plasma is filtered by the human glomeruli each day, less than 1% of water, sodium chloride and variable amounts of other solutes are excreted in the urine. By the processes of reabsorption and secretion the renal tubule modulates the volume and composition of the urine. Consequently, the tubules precisely control the volume, composition, and pH of the body fluids.

The renal proximal tubule is responsible for reabsorption of the majority of the filtered sodium, bicarbonate, chloride and water. Na^+/H^+ exchange is the predominant mechanism for absorption of Na^+ and secretion of H^+ across the apical membrane of proximal tubule cells (Alpern, 1990). Apical membrane Na^+/H^+ exchange also has a major role in mediating chloride reabsorption in the proximal tubule through its combined activity with a Cl^-/base exchanger and by creating an increase in luminal chloride concentration that favors the diffusion of the anion from the tubular lumen to the blood (Warnock and Yee, 1981; Aronson and Giebisch, 1997). The sodium/proton exchanger isoform 3 (NHE3) represents the major topic of this chapter; therefore the properties of the NHE family will be briefly discussed bellow.

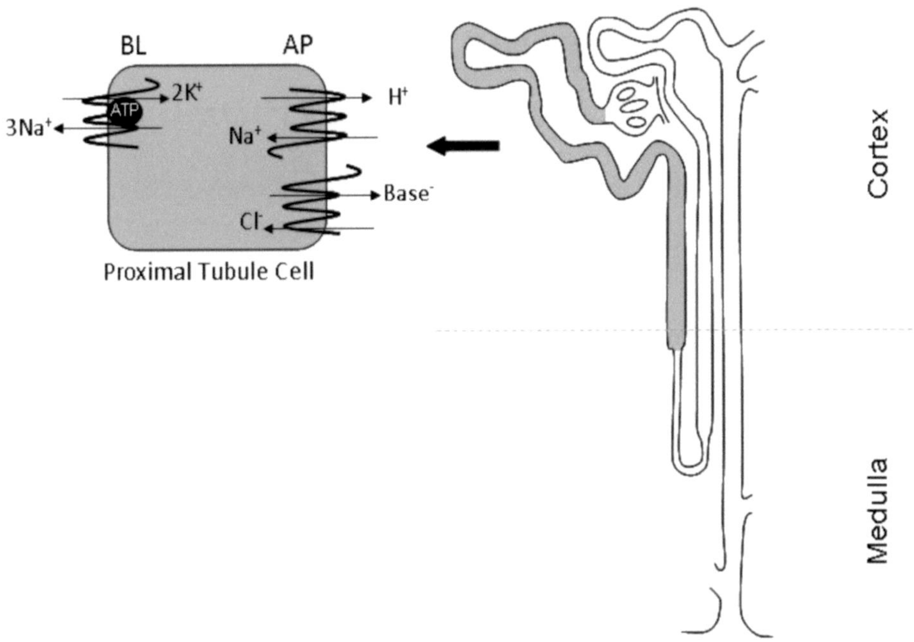

Fig. 1. Schematic illustration of the nephron depicting the most important transport mechanisms involved with NaCl reabsorption in proximal tubule. The inset shows Na^+/K^+-ATPase, Na^+/H^+ exchanger and Cl^-/Base exchanger localization in proximal tubule. BL = basolateral; AP = apical.

The mammalian Na^+/H^+ exchanger (NHE) gene family (SLC9) consists of secondary active transporters that mediate the electroneutral exchange of intracellular protons for extracellular sodium (Aronson, 1985). The transport activity of this protein is crucial to regulation of intracellular pH and cellular volume. In polarized epithelia, Na^+/H^+ exchangers are also involved in transepithelial NaHCO3 and NaCl transport.

All members of the NHE family share a common structural feature. They consist of two major portions, an N-terminal transmembrane domain and a large cytoplasmic C-terminal domain. The N-terminal portion of all known isoforms is predicted to span the plasma membrane twelve times. This domain is responsible for the Na^+/H^+ exchange transport function (Pouyssegur, 1994). The C-terminal portion is mainly hydrophilic and it is the portion through which the activity of the exchanger is regulated.

Thus far, five sodium proton exchangers (NHE1, NHE2, NHE3, NHE4 and NHE8) have been identified in plasma membrane of renal tubular cells (Biemesderfer et al., 1992; Biemesderfer et al., 1993a; Amemiya et al., 1995; Chambrey et al., 1997; Chambrey et al., 1998). Of these, NHE3, the most abundant NHE isoform in renal tissue, is confined to the apical membrane of proximal tubule and thin and thick ascending limb. Several lines of evidence strongly support the conclusion that NHE3 is the principal NHE isoform responsible for apical membrane Na^+/H^+ exchange in the proximal tubule. First, studies

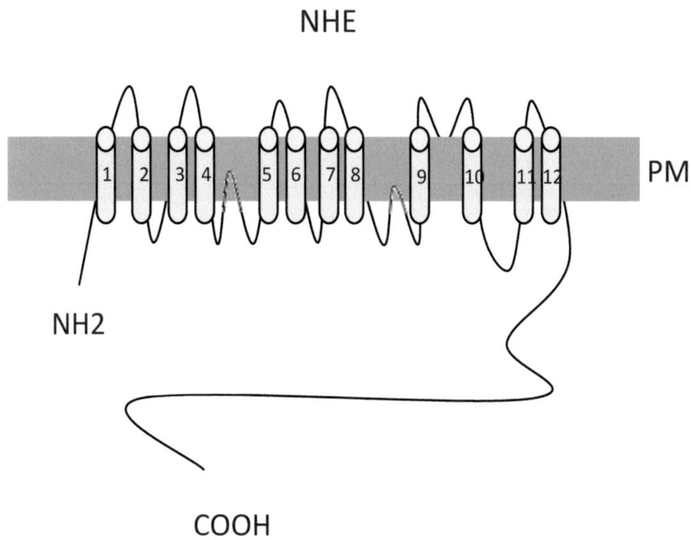

Fig. 2. Secondary structure of sodium/proton exchangers (NHE). All NHE isoforms have a membrane topology of 10-12 transmembrane segments. The carboxy-terminal region at the cytoplasmic site of NHE3 has consensus phosphorylation sites for PKA.

using isoform-specific antibodies have demonstrated that NHE3 is expressed on the brush border membrane of proximal tubule cells (Biemesderfer et al., 1993b; Biemesderfer et al., 1997). Second, the pattern of sensitivity to different inhibitors of Na^+/H^+ exchange in renal brush border vesicles and of bicarbonate absorption in microperfused proximal tubules is most consistent with the properties of NHE3 among the known NHE isoforms (Vallon et al., 2000). Third, micropuncture analysis on NHE3 knockout mice revealed a remarkable reduction of fluid and bicarbonate reabsorption in renal proximal tubule further supporting the concept that NHE3 is the isoform that accounts for most Na^+/H^+ exchange in this nephron segment (Lorenz et al., 1999; Wang et al., 1999). Indeed, mice lacking NHE3 display hypotension and had a mild hyperchloremic metabolic acidosis (Schultheis et al., 1998). Despite the reduced salt reabsortive capacity in the renal proximal tubule, the NHE3 deficient mice grows well when fed a normal sodium diet, mostly due to reduced glomerular filtration rate and increased sodium and bicarbonate reabsorption in the distal nephron. However, if these animals are subjected to dietary salt restriction the adaptative mechanisms are not sufficient to fully compensate for the large defect on proximal reabsorption and they may die from hypovolemic shock (Ledoussal et al., 2001).

Given the important role of NHE3 in mediating $NaHCO_3$ and NaCl reabsorption in the proximal tubule, this transporter is subject to acute and chronic regulation in response to a variety of conditions and humoral factors affecting acid-base or salt balance. In this chapter we will focus on the regulation of NHE3 by protein kinase A and the implications of this regulatory mechanism on renal function under physiological and pathophysiological conditions.

2. Regulation of NHE3 activity by hormones that activate cAMP-dependent protein kinase A (PKA) in renal proximal tubule

The signal transduction cascade mediating the acute effect of NHE3 agonists and antagonists involves multiple pathways. One of the best studied regulatory mechanisms affecting NHE3 activity is the inhibition resulting from protein kinase A (PKA) activation. Hormones activating cAMP-dependent PKA have been shown to reduce sodium and bicarbonate reabsorption in renal proximal tubule by inhibiting NHE3 transport activity. Table 1 presents a summary of hormones and molecular mechanisms associated with inhibition of NHE3 by PKA. These hormones act via G-protein coupled receptors (GPCR) expressed in the apical membrane of the renal proximal tubule (Felder et al., 1984; Muff et al., 1992; Marks et al., 2003; Schlatter et al., 2007; Crajoinas et al., 2011) . GPCR signal transduction occurs through coupling to heterotrimeric G proteins on the intracellular side of the membrane. Heterotrimeric G proteins contain three subunits referred as Gα, Gβ, and Gγ. Upon ligand binding, the GPCR undergoes a conformational change that promotes the exchange of bound guanosine diphosphate (GDP) from the Gα subunit for guanosine triphosphate (GTP). The G protein Gα subunit bound to GTP can then dissociate from the Gβγ dimer and initiate the intracellular signaling cascade that leads to cAMP dependent PKA activation that further elicits NHE3 inhibition (Fig. 3).

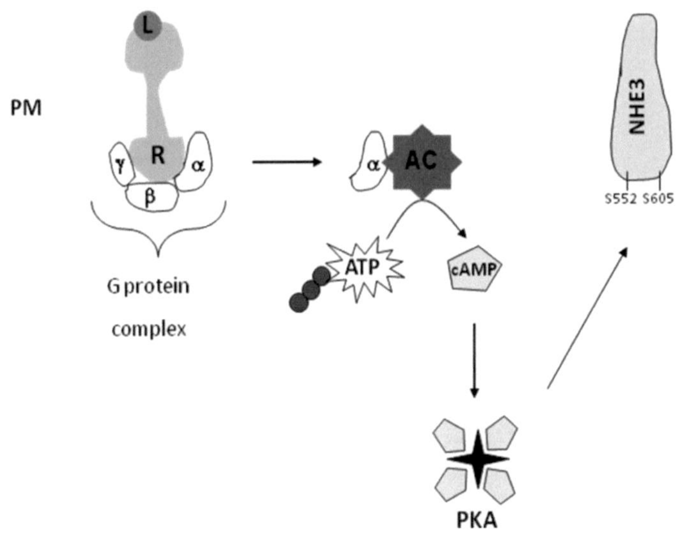

Fig. 3. Downstream effect of a receptor coupled to adenylyl cyclase. The ligand (L) binds to its receptor coupled to protein G (α_S) and adenylyl cyclise (AC) is activated. The activated enzyme converts ATP to cAMP, which activates protein kinase A (PKA) which, in turn, is ready to phosphorylate NHE3 at the specific PKA consensus sites.

2.1 Parathyroid hormone

The kidney is a principal target organ for the action of the parathyroid hormone (PTH). PTH is primarily involved in modulation of serum calcium and phosphate homeostasis but also acts on the proximal tubule, the thick ascending limb, and the distal convoluted tubule to alter urinary electrolyte and fluid excretion. The inhibitory effect of PTH on renal proximal

Hormone/Condition	Associated Mechanism	References
Parathyroid hormone	↓ of NHE3 affinity for protons due to phosphorylation of the exchanger ↓ of NHE3 maximum velocity due to a decrement of the number of NHE3 molecules at the plasma membrane ↓ of NHE3 promoter activity, NHE3 mRNA and protein abundance	(Fan et al., 1999; Collazo et al., 2000; Zhang et al., 1999; Girardi et al., 2000; Bezerra et al., 2008)
Dopamine	↓ NHE3-mediated Na^+/H^+ exchange due to increased endocytosis caused by increased PKA-mediated NHE3 phosphorylation	(Gomes & Soares-da-Silva, 2004; Hu et al., 2001; Bacic et al., 2003)
Glucagon-like peptide 1	↓ NHE3 activity due to increased NHE3 phosphorylation	(Crajoinas et al., 2011; Carraro-Lacroix et al., 2009)
Glucagon	Acutely - ↓ NHE via a PKA-dependent pathway Chronically - ↑ NHE3 mRNA and protein expression at the plasma membrane	(Amemiya et al., 2002)
Guanylins	↓ NHE3-mediated Na^+/H^+ due to increased levels of NHE3 PKA-dependent phosphorylation and reduction of the exchanger at the plasma membrane	(Amorim et al., 2006; Lessa LM, Girardi AC, and Malnic G, unpublished observations)
High Salt Diet	↓ NHE3-mediated Na^+/H^+ due to higher NHE3 phosphorylation on serine 552, redistribution from microvilli to the intermicrovillar region together with its regulatory partner dipeptidyl peptidase IV	(Yang et al., 2008)
Hypertension	↓ NHE3-mediated Na^+/H^+ due to higher NHE3 phosphorylation on serine 552, redistribution of the transporter to the intermicrovillar region	(Magyar et al., 2000; Panico et al., 2009; Crajoinas et al., 2010)
Heart Failure	↑ NHE3-mediated sodium reabsorption due to increased renal cortical NHE3 mRNA and protein levels and lower levels of NHE3 phosphorylation at serine 552	(Lutken et al., 2009; Inoue et al., 2012)

Table 1. Major factors and hormones that inhibit NHE3 activity via cAMP-dependent PKA activation.

tubule NHE3 activity has been consistently reported by several laboratories on both *in vivo* (Bank and Aynediian, 1976; Girardi et al., 2000) and *in vitro* studies (Pollock et al., 1986; Mrkic et al., 1992; Mrkic et al., 1993). Experiments performed *in vitro* and *ex vivo* provide evidence that the acute inhibition of NHE3 by PTH is mediated by molecular mechanisms the include reduction of the transporter's apparent affinity for protons in consequence of the direct phosphorylation of the exchanger followed by reduction of its maximum velocity due to a decrement of the number of NHE3 molecules expressed at the plasma membrane (Fan et al., 1999; Collazo et al., 2000). Consistent with a decrease of NHE3 surface expression in response to PTH, studies carried out by the McDonough laboratory have shown that reduction of NHE3 activity in response to acute treatment with this hormone is a consequence of NHE3 redistribution from the apical microvilli to the base of the intermicrovillar region of the proximal tubule brush border (Zhang et al., 1999).

The chronic effect of PTH on NHE3 regulation has also been evaluated (Girardi et al., 2000; Bezerra et al., 2008). Long term inhibition of NHE3 by PTH is associated with a reduction on NHE3 protein and mRNA levels. PTH also provokes a mild inhibitory effect on NHE3 promoter that seems to be PKA-dependent (Bezerra et al., 2008).

2.2 Dopamine

The intrarenal dopamine natriuretic system is critical for mammalian sodium homeostasis. Numerous studies have demonstrated that dopamine remarkably increases urinary sodium excretion mainly by inhibiting tubular sodium reabsorption. The inhibitory effect of dopamine on NHE3 transport activity is mediated mainly via the dopamine D1 receptor and stimulation of adenylyl cyclase/PKA system and phospholipase C/PKC (Gomes and Soares-da-Silva, 2004). The underlying molecular mechanisms by which dopamine decreases NHE3-mediated Na^+/H^+ exchange in renal proximal tubule involves increased endocytosis and is associated with increased PKA-mediated NHE3 phosphorylation (Hu et al., 2001; Bacic et al., 2003).

2.3 Glucagon-like peptide-1

Glucagon-like peptide-1 (GLP-1) is produced by posttranslational modification of the proglucagon gene product in the intestinal L-cells, predominantly localized in the colon and ileum (Holst, 1997; Drucker, 2005). This incretin hormone plays an important role on the maintenance of systemic glucose homeostasis by stimulating insulin secretion and improving insulin sensitivity (Drucker, 2005). Numerous reports in the literature have demonstrated that GLP-1 also exerts renoprotective actions. In this regard, continuous administration of GLP-1 induces diuresis and natriuresis in both humans (Gutzwiller et al., 2004; Gutzwiller et al., 2006) and experimental animal models (Moreno et al., 2002; Yu et al., 2003).

The molecular mechanisms underlying the renal actions of GLP-1 seems to involve increases of GFR and RPF and decrease of NHE3-mediated Na^+/H^+ exchange in the renal proximal tubule (Carraro-Lacroix et al., 2009). Recent studies by our group have demonstrated that binding of GLP-and/or the GLP-1R agonist exendin-4 to its receptor in the renal proximal tubule activates the cAMP/PKA signaling pathway, leading, in turn, to phosphorylation of the PKA consensus sites located at the C-terminal region of the exchanger (Carraro-Lacroix et al., 2009; Crajoinas et al., 2011). Increased NHE3 phosphorylation levels induced by GLP-1 was not accompanied by a decrease of NHE3 expression at the microvillar microdomain of the brush border, suggesting that the mechanism by which GLP-1 inhibits NHE3 activity does not involve subcellular redistribution of the exchanger between the subcompartiments of the renal proximal tubule brush border.

2.4 Glucagon

Glucagon is a 29-amino-acid pancreatic peptide produced by the α-cells present at the periphery of the islets of Langerhans (Baum et al., 1962) and its major function is the maintenance of plasma glucose homeostasis between meals and during fasting. Glucagon binding to its receptor primarily activates adenylyl cyclase and increases cAMP (Pohl et al., 1971; Rodbell et al., 1971). The tissue distribution of the glucagon receptor is broad, with higher levels of expression in liver and kidney (Svoboda et al., 1993; Dunphy et al., 1998).

In the kidney, glucagon affects renal glomerular filtration, renal blood flow, and decreases renal tubular sodium reabsorption (Pullman et al., 1967). Part of the acute natriuretic action of glucagon are mediated by inhibition of NHE3 in the renal proximal tubule via a cAMP/PKA-dependent pathway. Interestingly, in vitro studies using OKP cells have shown that glucagon acutely inihibits and chronically stimulates NHE3 activity (Amemiya et al., 2002).

2.5 Guanylins

The guanylin and uroguanylin are endogenous ligants of the Escherichia coli heat-stable enterotoxin (STa) receptor, guanylate cyclase C (Currie et al., 1992; Fonteles et al., 1998) and are known to be involved in a control system that regulates salt balance in response to oral salt intake. Both guanylin and uroguanylin are synthesized in the intestine and in the kidney and have already been identified in several animal species, including as mammals, fishes and birds (Forte, 2004).

The renal effects of uroguanylin are much more pronounced than the ones produced by guanylin and include natriuresis, kaliuresis, diuresis and increased excretion of cGMP (Forte et al., 1996; Greenberg et al., 1997; Fonteles et al., 1998). In the renal proximal tubule, uroguanylin significantly inhibits NHE3 transport function (Amorim et al., 2006). Ongoing studies by the Malnic laboratory have demonstrated that the mechanism by which uroguanyn inhibits NHE3 involves increased levels of NHE3 phosphorylation followed by retrieval of the exchanger from the plasma membrane (unpublished observations, Lessa LM, Girardi AC, and Malnic G)). The mechanism by which NHE3 is phosphorylated by PKA in response to coupling of uroguanylin to its receptor in the renal proximal tubule possibly involves a crosstalk mechanism between cGMP and cAMP pathways.

3. The Na^+/H^+ exchanger regulatory factor NHERF

Although a large number of hormones reported to affect NHE3 share the same signal pathways, the molecular mechanisms by which they regulate NHE3 may differ greatly among them. The identification of regulatory proteins that interact with NHE3 has unraveled some aspects of the molecular mechanisms underlying this transporter regulation.

The first NHE3 regulatory factor was isolated and characterized byWeinman and Shenolikar (Weinman et al., 1995; Weinman et al., 2000a; Weinman et al., 2000b). These investigators demonstrated that the presence of this cofactor was essential for PKA-mediated inhibition of NHE3. This protein was cloned and termed NHERF-1 (Na/H exchanger regulatory factor). Subsequently, in an attempt to identify proteins that interact with NHE3, Yun and coworkers used the C-terminus of NHE3 as a bait in a yeast two-hybrid screen and isolated E3KARP (exchanger-3 kinase A regulatory protein, or NHERF-2) . NHERF-1 and NHERF-2 are highly homologous proteins (52% sequence identity for the human orthologs) (Yun et al., 1997). Physical association of NHERF-1 and NHERF-2 with NHE3 has been demonstrated by binding assays using fusion proteins or by co-precipitation experiments using transfected cells overexpressing NHE3.

NHERF-1 and NHERF-2 are both members of a family of proteins that contain two tandem PDZ domains (that are conserved modules that mediate protein-protein interaction) and a C-terminal ezrin-radixin-moesin (ERM) binding domain which anchors the proteins to the actin cytoskeleton through ezrin. Lamprecht and Yun have proposed a model whereby the complex NHERF/ezrin acts as a functional AKAP (A kinase anchoring protein) for NHE3, serving as a structural link between not only NHE3 and the cytoskeleton but also between NHE3 and PKA, since ezrin is capable of binding the RII regulatory subunit of PKA (Dransfield et al., 1997; Lamprecht et al., 1998). The current model suggests that NHERF-1 is required for cAMP-dependent regulation of NHE3 and acts as an adapter to link NHE3 to ezrin, which then serves as a PKA anchoring protein. Upon activation of PKA by hormones and/or factors that increase intracellular cAMP, PKA phosphorylates serine residues in the C-terminal hydrophilic domain of NHE3 (Fig. 4). Biochemical experiments with brush border vesicles isolated from NHERF-1 knockout mice corroborate with this model (Weinman et al., 2003). These studies showed that NHERF-1 is crucial for PKA-mediated phosphorylation and inhibition of NHE3 transport activity. Moreover, NHE3 expression was not affected in these animals, showing that NHERF-1 plays an essential role on NHE3 modulation but it is not required for expression or apical targeting of the transporter in the proximal tubule (Weinman et al., 2003).

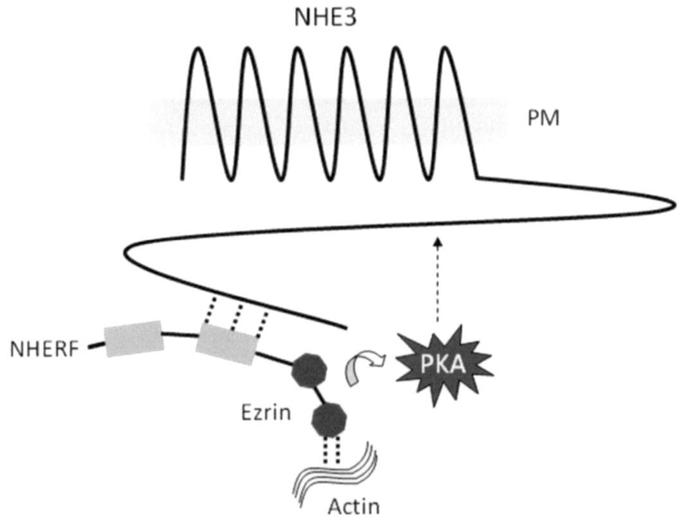

Fig. 4. Model of NHERF requirement for PKA-mediated NHE3 inhibition. NHERF facilitates cAMP-dependent regulation of NHE3 by interacting with the cytoskeleton to target PKA to phosphorylation sites within the cytoplasmic domain of NHE3.

To date, four members of the NHERF family of PDZ domain proteins have been described. These proteins bind to a variety of membrane transporters regulating their cell surface expression, protein interactions as well as the formation of signaling complexes.

4. NHE3 phosphorylation at the PKA consensus sites

Moe and coworkers were the first to demonstrate that the cytoplasmic domain of NHE3 is a substrate for PKA in vitro (Moe et al., 1995). There are multiple putative consensus motifs for PKA on the NHE3 C-terminal region. Three of these sites (S552, S605 and S634) are highly conserved among species. Based on that, transfection studies using truncation of the NHE3 C-terminal domain and site direct mutagenesis of the above mentioned serine residues were carried out to evaluate whether PKA directly phosphorylates one or more of these consensus sites in vivo (Cabado et al., 1996; Kurashima et al., 1997; Zhao et al., 1999). Kurashima and coworkers have shown that both serines 605 and 634 are important for PKA-mediated inhibition of NHE3, although only serine 605 is phosphorylated in vivo (Kurashima et al., 1997). The studies by Zhao and colleagues confirmed that phosphorylation of serine 605 is increased by cAMP/PKA. These investigators also found that PKA directly phosphorylates serine 552 and that both 552 and 605 residues appear to be critical for inhibition of NHE3 by PKA (Zhao et al., 1999).

Years later, the Aronson laboratory generated phosphospecific antibodies directed to the PKA consensus sites S552 and S605 of rat NHE3 (Kocinsky et al., 2005). These reagents are of great value, since they have enable investigators to evaluate the phosphorylation state of these two residues in endogenous NHE3 under basal conditions, under a variety of physiological and pharmacological maneuvers and in disease states. Indeed, increments in the phosphorylation status of NHE3 at serines 552 and 605 have been shown to occur in response to dopamine (Kocinsky et al., 2005), PTH (Kocinsky et al., 2007), and GLP-1 (Crajoinas et al., 2011).

Studies by Kocinsky and colleagues have also demonstrated that serine 552 is phosphorylated to a much greater extent than serine 605 in baseline in vivo. Moreover, these investigators found that when NHE3 is phosphorylated at serine 552, it mainly resides at the intermicrovillar domain of the brush border (Kocinsky et al., 2005). This observation is consistent with a decrease on NHE3-mediated Na^+/H^+ exchange in the renal proximal tubule, since within the intermicrovillar subcompartment of the brush border, this transporter must have very limited assess to the tubular fluid.

The precise mechanism by which NHE3 phosphorylation leads to NHE3 inhibition remains obscure. Although phosphorylation of NHE3 at the 552 and 605 residues are necessary for the PKA-dependent inhibitory effect, phosphorylation of NHE3 at these PKA consensus sites precedes transport inhibition (Kocinsky et al., 2007) indicating that phosphorylation per se is not sufficient to inhibit NHE3 activity. The current body of data suggests that PKA phosphorylation may ultimately result in inhibition of NHE3 by modulating NHE3 subcellular trafficking, interaction with regulatory proteins, or localization within the plasma membrane.

5. Pathophysiological Implications of NHE3 Phosphorylation by Protein Kinase A

As mentioned above, NHE3 is phosphorylated on serine 552 under basal conditions by the adenylyl cyclase/cAMP-activated-protein kinase A (PKA) and the endogenous levels of phosphorylation is often affected as part of acute NHE3 regulation. Interestingly, baseline levels of NHE3 phosphorylation at this residue may also be associated with chronic regulation of NHE3 activity.

5.1 High Salt Diet

During high salt diet the kidneys increase sodium and volume excretion to match intake. IYang and colleagues have demonstrated that three weeks of high salt diet (4%) doubled the levels of NHE3 phosphorylation at the serine 552 (Yang et al., 2008), possibly contributing to the natriuretic effect triggered by high sodium load.

5.2 Hypertension

Mice lacking Na^+/H^+ exchanger NHE3 are hypotensive and hypovolemic, underscoring the importance of the transporter on blood pressure control.

We have recently assessed in vivo NHE3 transport activity and defined the mechanisms underlying NHE3 regulation before and after development of hypertension in spontaneously hypertensive rats (SHR). By means of in vivo stationary microperfusion, we found that NHE3-mediated bicarbonate reabsorption is higher in the proximal tubule of 5-week-old pre-hypertensive spontaneously hypertensive rat (SHR) and lower in 14-week-old SHR compared to age-matched normotensive rats (Wistar Kyoto, WKY). Higher NHE3 activity in young pre-hypertensive SHR is associated with lower phosphorylation levels (serine 552) and increased NHE3 expression at the microvillar brush border. During the hypertensive stage, NHE3 was found to be mainly confined to the intermicrovillar region and the relative abundance of NHE3 phosphorylated on serine 552 is increased compared to normotensive animals (Fig. 5) (Crajoinas et al., 2010).

5.3 Heart failure

Heart failure (HF) is associated with sodium and water retention and extracellular volume expansion. A principal site of renal salt and water reabsorption is the proximal tubule. We therefore hypothesized that NHE3, the major apical transcellular pathway for sodium reabsorption in the proximal tubule, might be upregulated in heart failure. To test this hypothesis, we employed both stationary in vivo microperfusion and pH-dependent sodium uptake to verify whether NHE3 activity would be altered in the proximal tubule of an experimental model of heart failure (Antonio et al., 2009). Our data demonstrated that heart failure rats display enhanced NHE3-mediated sodium reabsorption in the proximal tubule which may contribute to extracellular volume expansion and edema. In addition to increased renal cortical NHE3 expression at both protein and mRNA levels, we have also observed that the levels of NHE3 phosphorylation at serine 552 in renal cortical membranes of heart failure rats are lower than in Sham animals. Thus, the molecular mechanisms mediating enhanced sodium reabsorption in the renal proximal tubule of heart failure rats also involves posttranslational covalent modification of NHE3 (Fig. 5) (Inoue et al., 2012).

6. Conclusion

PKA activation plays an important role in inhibiting the activity NHE3, the major apical transcellular pathway for sodium reabsorption in the proximal tubule. Similarly, PKA inhibition may be involved in NHE3 stimulation, as suggested by the studies carried out in young hypertensive and in heart failure animals. The elucidation of the mechanisms by which phosphorylation of NHE3 at the PKA consensus sites leads to inhibition of NHE3-mediated Na^+/H^+ exchange in the renal proximal tubule may also unravel important molecular underpinnings that lead to the development and/or progression of primary kidney diseases and other conditions that affect the kidneys.

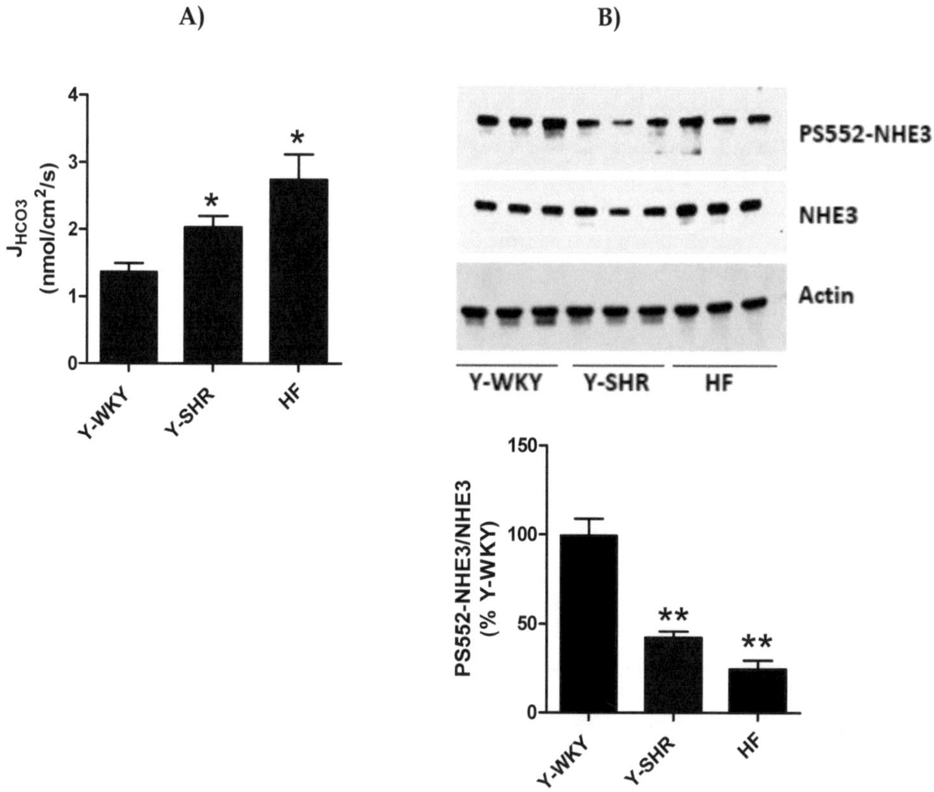

Fig. 5. NHE3 activity negatively correlates with the levels of endogenous phosphorylation of the PKA consensus site, serine 552, located at the C-terminus region of the exchanger Adapted from (Crajoinas et al., 2010; Inoue et al., 2012). **(A)** Stationary microperfusion was employed to measure NHE3-mediated bicarbonate reabsorption (J_{HCO3-}) in the proximal tubules of 5-week-old Wistar Kyoto rats (Y-WKY, n = 5 rats, 12 tubules), 5-week old SHR (Y-SHR, n = 5 rats, 12 tubules, and heart failure rats (HF, n = 5 rats, 15 tubules). Data are expressed as means ± SE. *P < 0.05 *vs.* Y-WKY. **(B)** *Top*: Representative immunoblot of phosphorylated and total NHE3 expression in renal cortical membranes isolated from Y-WKY, Y-SHR or HF rats. Equivalent samples (15 µg of protein for NHE3 and 5 µg for PS552-NHE3 and actin) of renal cortical membranes were prepared for immunoblot analysis. The membranes were incubated with monoclonal antibodies against phosphorylated NHE3 at serine 552 (PS552-NHE3 (1:1,000)), total NHE3 (1:1,000) or anti-actin (1:50,000). *Bottom: Left* - Graphical representation of the phosphorylation ratio of NHE3 at Serine 552 to total NHE3 in renal cortical membranes (PS552-NHE3/NHE3). Values are means ± SE. n = 3/group, **P < 0.001 *vs.* Y-WKY.

7. Acknowledgments

This work was supported by Fundação de Amparo à Pesquisa do Estado de São Paulo, São Paulo, SP, Brazil (Grant 2007/52945-8).

8. References

Alpern RJ (1990) Cell mechanisms of proximal tubule acidification. *Physiol Rev* 70:79-114.

Amemiya M, Kusano E, Muto S, Tabei K, Ando Y, Alpern RJ and Asano Y (2002) Glucagon acutely inhibits but chronically activates Na(+)/H(+) antiporter 3 activity in OKP cells. *Exp Nephrol* 10:26-33.

Amemiya M, Loffing J, Lotscher M, Kaissling B, Alpern RJ and Moe OW (1995) Expression of NHE-3 in the apical membrane of rat renal proximal tubule and thick ascending limb. *Kidney International* 48:1206-1215.

Amorim JB, Musa-Aziz R, Lessa LM, Malnic G and Fonteles MC (2006) Effect of uroguanylin on potassium and bicarbonate transport in rat renal tubules. *Can J Physiol Pharmacol* 84:1003-1010.

Antonio EL, Dos Santos AA, Araujo SR, Bocalini DS, Dos Santos L, Fenelon G, Franco MF and Tucci PJ (2009) Left ventricle radio-frequency ablation in the rat: a new model of heart failure due to myocardial infarction homogeneous in size and low in mortality. *J Card Fail* 15:540-548.

Aronson PS (1985) Kinetic properties of the plasma membrane Na+-H+ exchanger. *Annual Review of Physiology* 47:545-560.

Aronson PS and Giebisch G (1997) Mechanisms of chloride transport in the proximal tubule. *Am J Physiol* 273:F179-192.

Bacic D, Kaissling B, McLeroy P, Zou L, Baum M and Moe OW (2003) Dopamine acutely decreases apical membrane Na/H exchanger NHE3 protein in mouse renal proximal tubule. *Kidney Int* 64:2133-2141.

Bank N and Aynediian HS (1976) A micropuncture study of the effect of parathyroid hormone on renal bicarbonate reabsorption. *J Clin Invest* 58:336-344.

Baum J, Simons BE, Jr., Unger RH and Madison LL (1962) Localization of glucagon in the alpha cells in the pancreatic islet by immunofluorescent technics. *Diabetes* 11:371-374.

Bezerra CN, Girardi AC, Carraro-Lacroix LR and Reboucas NA (2008) Mechanisms underlying the long-term regulation of NHE3 by parathyroid hormone. *Am J Physiol Renal Physiol* 294:F1232-1237.

Biemesderfer D, Pizzonia J, Abu-Alfa A, Exner M, Reilly R, Igarashi P and Aronson PS (1993a) NHE3: a Na+/H+ exchanger isoform of renal brush border. *American Journal of Physiology* 265:F736-742.

Biemesderfer D, Pizzonia J, Abu-Alfa A, Exner M, Reilly R, Igarashi P and Aronson PS (1993b) NHE3: a Na+/H+ exchanger isoform of renal brush border. *Am J Physiol* 265:F736-742.

Biemesderfer D, Reilly RF, Exner M, Igarashi P and Aronson PS (1992) Immunocytochemical characterization of Na(+)-H+ exchanger isoform NHE-1 in rabbit kidney. *American Journal of Physiology* 263:F833-840.

Biemesderfer D, Rutherford PA, Nagy T, Pizzonia JH, Abu-Alfa AK and Aronson PS (1997) Monoclonal antibodies for high-resolution localization of NHE3 in adult and neonatal rat kidney. *Am J Physiol* 273:F289-299.

Cabado AG, Yu FH, Kapus A, Lukacs G, Grinstein S and Orlowski J (1996) Distinct structural domains confer cAMP sensitivity and ATP dependence to the Na+/H+ exchanger NHE3 isoform. *J Biol Chem* 271:3590-3599.

Carraro-Lacroix LR, Malnic G and Girardi AC (2009) Regulation of Na+/H+ exchanger NHE3 by glucagon-like peptide 1 receptor agonist exendin-4 in renal proximal tubule cells. *Am J Physiol Renal Physiol* 297:F1647-1655.

Chambrey R, Achard JM, St John PL, Abrahamson DR and Warnock DG (1997) Evidence for an amiloride-insensitive Na+/H+ exchanger in rat renal cortical tubules. *Am J Physiol* 273:C1064-1074.

Chambrey R, Warnock DG, Podevin RA, Bruneval P, Mandet C, Belair MF, Bariety J and Paillard M (1998) Immunolocalization of the Na+/H+ exchanger isoform NHE2 in rat kidney. *American Journal of Physiology* 275:F379-386.

Collazo R, Fan L, Hu MC, Zhao H, Wiederkehr MR and Moe OW (2000) Acute regulation of Na+/H+ exchanger NHE3 by parathyroid hormone via NHE3 phosphorylation and dynamin-dependent endocytosis. *J Biol Chem* 275:31601-31608.

Crajoinas RO, Lessa LM, Carraro-Lacroix LR, Davel AP, Pacheco BP, Rossoni LV, Malnic G and Girardi AC (2010) Posttranslational mechanisms associated with reduced NHE3 activity in adult vs. young prehypertensive SHR. *Am J Physiol Renal Physiol* 299:F872-881.

Crajoinas RO, Oricchio FT, Pessoa TD, Pacheco BP, Lessa LM, Malnic G and Girardi AC (2011) Mechanisms mediating the diuretic and natriuretic actions of the incretin hormone glucagon-like peptide-1. *Am J Physiol Renal Physiol* 301:F355-363.

Currie MG, Fok KF, Kato J, Moore RJ, Hamra FK, Duffin KL and Smith CE (1992) Guanylin: an endogenous activator of intestinal guanylate cyclase. *Proc Natl Acad Sci U S A* 89:947-951.

Dransfield DT, Yeh JL, Bradford AJ and Goldenring JR (1997) Identification and characterization of a novel A-kinase-anchoring protein (AKAP120) from rabbit gastric parietal cells. *Biochem J* 322 (Pt 3):801-808.

Drucker DJ (2005) Biologic actions and therapeutic potential of the proglucagon-derived peptides. *Nat Clin Pract Endocrinol Metab* 1:22-31.

Dunphy JL, Taylor RG and Fuller PJ (1998) Tissue distribution of rat glucagon receptor and GLP-1 receptor gene expression. *Mol Cell Endocrinol* 141:179-186.

Fan L, Wiederkehr MR, Collazo R, Wang H, Crowder LA and Moe OW (1999) Dual mechanisms of regulation of Na/H exchanger NHE-3 by parathyroid hormone in rat kidney. *J Biol Chem* 274:11289-11295.

Felder RA, Blecher M, Calcagno PL and Jose PA (1984) Dopamine receptors in the proximal tubule of the rabbit. *Am J Physiol* 247:F499-505.

Fonteles MC, Greenberg RN, Monteiro HS, Currie MG and Forte LR (1998) Natriuretic and kaliuretic activities of guanylin and uroguanylin in the isolated perfused rat kidney. *Am J Physiol* 275:F191-197.

Forte LR, Fan X and Hamra FK (1996) Salt and water homeostasis: uroguanylin is a circulating peptide hormone with natriuretic activity. *Am J Kidney Dis* 28:296-304.

Forte LR, Jr. (2004) Uroguanylin and guanylin peptides: pharmacology and experimental therapeutics. *Pharmacol Ther* 104:137-162.

Girardi AC, Titan SM, Malnic G and Reboucas NA (2000) Chronic effect of parathyroid hormone on NHE3 expression in rat renal proximal tubules. *Kidney Int* 58:1623-1631.

Gomes P and Soares-da-Silva P (2004) Dopamine acutely decreases type 3 Na(+)/H(+) exchanger activity in renal OK cells through the activation of protein kinases A and C signalling cascades. *Eur J Pharmacol* 488:51-59.

Greenberg RN, Hill M, Crytzer J, Krause WJ, Eber SL, Hamra FK and Forte LR (1997) Comparison of effects of uroguanylin, guanylin, and Escherichia coli heat-stable enterotoxin STa in mouse intestine and kidney: evidence that uroguanylin is an intestinal natriuretic hormone. *J Investig Med* 45:276-282.

Holst JJ (1997) Enteroglucagon. *Annu Rev Physiol* 59:257-271.

Hu MC, Fan L, Crowder LA, Karim-Jimenez Z, Murer H and Moe OW (2001) Dopamine acutely stimulates Na+/H+ exchanger (NHE3) endocytosis via clathrin-coated vesicles: dependence on protein kinase A-mediated NHE3 phosphorylation. *J Biol Chem* 276:26906-26915.

Inoue BH, Dos Santos L, Pessoa TD, Antonio EL, Pacheco BP, Savignano FA, Carraro-Lacroix LR, Tucci PJ, Malnic G and Girardi AC (2012) Increased NHE3 abundance and transport activity in renal proximal tubule of rats with heart failure. *Am J Physiol Regul Integr Comp Physiol* 302:R166-174.

Kocinsky HS, Dynia DW, Wang T and Aronson PS (2007) NHE3 phosphorylation at serines 552 and 605 does not directly affect NHE3 activity. *Am J Physiol Renal Physiol* 293:F212-218.

Kocinsky HS, Girardi AC, Biemesderfer D, Nguyen T, Mentone S, Orlowski J and Aronson PS (2005) Use of phospho-specific antibodies to determine the phosphorylation of endogenous Na+/H+ exchanger NHE3 at PKA consensus sites. *Am J Physiol Renal Physiol* 289:F249-258.

Kurashima K, Yu FH, Cabado AG, Szabo EZ, Grinstein S and Orlowski J (1997) Identification of sites required for down-regulation of Na+/H+ exchanger NHE3 activity by cAMP-dependent protein kinase. phosphorylation-dependent and -independent mechanisms. *J Biol Chem* 272:28672-28679.

Lamprecht G, Weinman EJ and Yun CH (1998) The role of NHERF and E3KARP in the cAMP-mediated inhibition of NHE3. *J Biol Chem* 273:29972-29978.

Ledoussal C, Lorenz JN, Nieman ML, Soleimani M, Schultheis PJ and Shull GE (2001) Renal salt wasting in mice lacking NHE3 Na+/H+ exchanger but not in mice lacking NHE2. *Am J Physiol Renal Physiol* 281:F718-727.

Lorenz JN, Schultheis PJ, Traynor T, Shull GE and Schnermann J (1999) Micropuncture analysis of single-nephron function in NHE3-deficient mice. *Am J Physiol* 277:F447-453.

Lutken SC, Kim SW, Jonassen T, Marples D, Knepper MA, Kwon TH, Frokiaer J and Nielsen S (2009) Changes of renal AQP2, ENaC, and NHE3 in experimentally induced heart failure: response to angiotensin II AT1 receptor blockade. *Am J Physiol Renal Physiol* 297:F1678-1688.

Magyar CE, Zhang Y, Holstein-Rathlou NH and McDonough AA (2000) Proximal tubule Na transporter responses are the same during acute and chronic hypertension. *Am J Physiol Renal Physiol* 279:F358-369.

Marks J, Debnam ES, Dashwood MR, Srai SK and Unwin RJ (2003) Detection of glucagon receptor mRNA in the rat proximal tubule: potential role for glucagon in the control of renal glucose transport. *Clin Sci (Lond)* 104:253-258.

Moe OW, Amemiya M and Yamaji Y (1995) Activation of protein kinase A acutely inhibits and phosphorylates Na/H exchanger NHE-3. *J Clin Invest* 96:2187-2194.

Moreno C, Mistry M and Roman RJ (2002) Renal effects of glucagon-like peptide in rats. *Eur J Pharmacol* 434:163-167.

Mrkic B, Forgo J, Murer H and Helmle-Kolb C (1992) Apical and basolateral Na/H exchange in cultured murine proximal tubule cells (MCT): effect of parathyroid hormone (PTH). *J Membr Biol* 130:205-217.

Mrkic B, Tse CM, Forgo J, Helmle-Kolb C, Donowitz M and Murer H (1993) Identification of PTH-responsive Na/H-exchanger isoforms in a rabbit proximal tubule cell line (RKPC-2). *Pflugers Arch* 424:377-384.

Muff R, Fischer JA, Biber J and Murer H (1992) Parathyroid hormone receptors in control of proximal tubule function. *Annu Rev Physiol* 54:67-79.

Panico C, Luo Z, Damiano S, Artigiano F, Gill P and Welch WJ (2009) Renal proximal tubular reabsorption is reduced in adult spontaneously hypertensive rats: roles of superoxide and Na+/H+ exchanger 3. *Hypertension* 54:1291-1297.

Pohl SL, Birnbaumer L and Rodbell M (1971) The glucagon-sensitive adenyl cyclase system in plasma membranes of rat liver. I. Properties. *J Biol Chem* 246:1849-1856.

Pollock AS, Warnock DG and Strewler GJ (1986) Parathyroid hormone inhibition of Na+-H+ antiporter activity in a cultured renal cell line. *Am J Physiol* 250:F217-225.

Pouyssegur J (1994) Molecular biology and hormonal regulation of vertebrate Na+/H+ exchanger isoforms. *Renal Physiology & Biochemistry* 17:190-193.

Pullman TN, Lavender AR and Aho I (1967) Direct effects of glucagon on renal hemodynamics and excretion of inorganic ions. *Metabolism* 16:358-373.

Rodbell M, Birnbaumer L, Pohl SL and Krans HM (1971) The glucagon-sensitive adenyl cyclase system in plasma membranes of rat liver. V. An obligatory role of guanylnucleotides in glucagon action. *J Biol Chem* 246:1877-1882.

Schlatter P, Beglinger C, Drewe J and Gutmann H (2007) Glucagon-like peptide 1 receptor expression in primary porcine proximal tubular cells. *Regul Pept* 141:120-128.

Schultheis PJ, Clarke LL, Meneton P, Miller ML, Soleimani M, Gawenis LR, Riddle TM, Duffy JJ, Doetschman T, Wang T, Giebisch G, Aronson PS, Lorenz JN and Shull GE (1998) Renal and intestinal absorptive defects in mice lacking the NHE3 Na+/H+ exchanger. *Nat Genet* 19:282-285.

Svoboda M, Ciccarelli E, Tastenoy M, Robberecht P and Christophe J (1993) A cDNA construct allowing the expression of rat hepatic glucagon receptors. *Biochem Biophys Res Commun* 192:135-142.

Vallon V, Schwark JR, Richter K and Hropot M (2000) Role of Na(+)/H(+) exchanger NHE3 in nephron function: micropuncture studies with S3226, an inhibitor of NHE3. *Am J Physiol Renal Physiol* 278:F375-379.

Wang T, Yang CL, Abbiati T, Schultheis PJ, Shull GE, Giebisch G and Aronson PS (1999) Mechanism of proximal tubule bicarbonate absorption in NHE3 null mice. *Am J Physiol* 277:F298-302.

Warnock DG and Yee VJ (1981) Chloride uptake by brush border membrane vesicles isolated from rabbit renal cortex. Coupling to proton gradients and K+ diffusion potentials. *J Clin Invest* 67:103-115.

Weinman EJ, Minkoff C and Shenolikar S (2000a) Signal complex regulation of renal transport proteins: NHERF and regulation of NHE3 by PKA. *Am J Physiol Renal Physiol* 279:F393-399.

Weinman EJ, Steplock D, Donowitz M and Shenolikar S (2000b) NHERF associations with sodium-hydrogen exchanger isoform 3 (NHE3) and ezrin are essential for cAMP-mediated phosphorylation and inhibition of NHE3. *Biochemistry* 39:6123-6129.

Weinman EJ, Steplock D and Shenolikar S (2003) NHERF-1 uniquely transduces the cAMP signals that inhibit sodium-hydrogen exchange in mouse renal apical membranes. *FEBS Lett* 536:141-144.

Weinman EJ, Steplock D, Wang Y and Shenolikar S (1995) Characterization of a protein cofactor that mediates protein kinase A regulation of the renal brush border membrane Na(+)-H+ exchanger. *J Clin Invest* 95:2143-2149.

Yang LE, Sandberg MB, Can AD, Pihakaski-Maunsbach K and McDonough AA (2008) Effects of dietary salt on renal Na+ transporter subcellular distribution, abundance, and phosphorylation status. *Am J Physiol Renal Physiol* 295:F1003-1016.

Yu M, Moreno C, Hoagland KM, Dahly A, Ditter K, Mistry M and Roman RJ (2003) Antihypertensive effect of glucagon-like peptide 1 in Dahl salt-sensitive rats. *J Hypertens* 21:1125-1135.

Yun CH, Oh S, Zizak M, Steplock D, Tsao S, Tse CM, Weinman EJ and Donowitz M (1997) cAMP-mediated inhibition of the epithelial brush border Na+/H+ exchanger, NHE3, requires an associated regulatory protein. *Proc Natl Acad Sci U S A* 94:3010-3015.

Zhang Y, Norian JM, Magyar CE, Holstein-Rathlou NH, Mircheff AK and McDonough AA (1999) In vivo PTH provokes apical NHE3 and NaPi2 redistribution and Na-K-ATPase inhibition. *Am J Physiol* 276:F711-719.

Zhao H, Wiederkehr MR, Fan L, Collazo RL, Crowder LA and Moe OW (1999) Acute inhibition of Na/H exchanger NHE-3 by cAMP. Role of protein kinase a and NHE-3 phosphoserines 552 and 605. *J Biol Chem* 274:3978-3987.

15

cGMP-Dependent Protein Kinase in the Regulation of Cardiovascular Functions

Yuansheng Gao, Dou Dou, Xue Qin, Hui Qi and Lei Ying
[1]*Department of Physiology and Pathophysiology, Peking University Health Science Center,*
[2]*Key Laboratory of Molecular Cardiovascular Science (Peking University), Ministry of Education,*
China

1. Introduction

Cyclic GMP-dependent protein kinase (PKG) was discovered in 1970 in lobster muscle (Kuo & Greengard, 1970). It is a serine/threonine protein kinase specifically activated by cyclic guanosine monophosphate (cGMP). PKG is a ubiquitous intracellular second messenger mediating the biological effects of cGMP elevating agents including nitric oxide (NO), natriuretic peptides, and guanylin (an intestinal peptide involved in intestinal fluid regulation). It is now well recognized that PKG plays a central role in a broad range of physiological processes, such as contractility and proliferation of smooth muscle and cardiac myocytes, platelet aggregation, synaptic plasticity and learning, behavior, intestinal chloride reabsorption, renin secretion, and endochondral ossification (Francis et al., 2010; Hofmann et al., 2009; Lincoln et al., 2001; Lohmann & Walter, 2005). This chapter will focus on the role of PKG in the regulation of cardiovascular functions under physiological and pathophysiological conditions.

2. PKG structure and tissue distribution

In mammalian cells PKG exists as two types, PKG-I and PKG-II, respectively. They are encoded by two separated genes *prkg1* and *prkg2*. The human *prkg1* gene is located on chromosome 10 at p11.2 - q11.2 and has 15 exons. The NH_2 terminus (the first 100 residues) of PKG I is encoded by two alternative exons that produce the isoforms PKG Iα and PKG Iβ. The human *prkg2* gene is located on chromosome 4 at q13.1 - q21.1 and has 19 exons. PKG-I and PKG-II is composed of two identical subunits of the homodimer about 75-80 kDa and 84-86 kDa, respectively and shares common structural features. Each subunit of the enzyme consists of a regulatory domain and a catalytic domain. The regulatory domain is composed of an N-terminal domain and a cGMP binding domain. The N-terminal domain mediates homodimerization, suppression of the kinase activity in the absence of cGMP, and interactions with other proteins including protein substrates. The cGMP binding domain contains a high and a low cGMP affinity binding sites. The two cGMP-binding sites interact allosterically. Binding of cGMP releases the inhibition of the catalytic center by the N-terminal autoinhibitory/pseudosubstrate domain and allows the phosphorylation of target proteins. The catalytic domain contains a MgATP and a target protein-binding site, which catalyze the phosphate transfer from ATP to the hydroxyl group of a serine/threonine side chain of the target protein. When stimulated with cGMP, the phosphotransferase activity increases by 3- to 10-fold (Francis et al., 2010; Hofmann et al., 2009).

PKG-I is predominantly localized in the cytoplasm (except in the platelets where it is with the membrane). PKG-II is anchored to the plasma membrane by N-terminal myristoylation. In general, PKG-I and PKG-II are expressed in different cell types. PKG-I exists at high concentrations in all types of smooth muscle cells (~0.1μM) including vascular smooth muscle cells and at lower levels in vascular endothelium and cardiomyocytes. The enzyme has also been detected in other cell types such as fibroblasts, certain types of renal cells and leukocytes, and in specific regions of the nervous system. Platelets express predominantly PKG Iβ while both PKG Iα and PKG Iβ isoforms are present in smooth muscle, including uterus, blood vessels, intestines, and trachea. PKG-II is expressed in several brain nuclei, intestinal mucosa, kidney, chondrocytes and the lung but not in cardiac and vascular myocytes (Francis et al., 2010; Hofmann et al., 2009).

Existing research results show that PKG-I is the major type of the enzymes in the cardiovascular system involved in the regulation of vascular tone, regulation of vascular smooth muscle cells and myocardial cells proliferation and phenotypic modulation, and inhibiting platelet aggregation. Both PKG-Iα and PKG-Iβ can be specially activated by cGMP, with the former is about 10 times more sensitive to cGMP than the latter. PKG can also be activated by cAMP, although more than 100 times less potent than cGMP. The main role of PKG-II is phosphorylation in the intestinal mucosa of cystic fibrosis transmembrane conductance regulator, regulation of intestinal chloride ion/fluid secretion, inhibition of renin secretion in the kidney, and the regulation of bone tissue and bone endochondral bone growth (Francis et al., 2010; Hofmann et al., 2009; Lincoln et al., 2001; Lohmann & Walter, 2005).

3. PKG function in the cardiovascular system

3.1 Blood vessels

3.1.1 Vasodilatation

PKG is involved in vasodilatation caused by cGMP elevating agents including endothelium-derived NO, ANP, CNP, and exogenous nitrovasodilators (Gao, 2010; Hofmann et al., 2009). In certain vessel types such as ovine perinatal pulmonary artery and vein (Dhanakoti et al., 2000; Gao et al., 1999) as well as porcine coronary artery and vein (Qin et al., 2007; Qi et al., 2007) relaxation caused by nitrovasodilators is primarily mediated by PKG. Studies show that the expression and activity of PKG can be modulated by physiological variables such as oxygenation (Gao et al., 2003).

Activation of calcium activated potassium (BK) channels has been implicated as a mechanism for PKG-mediated relaxation of vascular smooth muscle in a number of vessel types including cerebral artery (Robertson et al., 1993), coronary artery (White et al., 2000), and pulmonary artery (Barman et al., 2003), which leads to increased membrane polarization and thus decreased Ca^{2+} influx and vasodilatation. PKG may stimulate BK channels by direct phosphorylation of the α-subunit at serine 1072 (Fukao et al., 1999) or through activation of protein phosphatase 2A (Zhou et al., 1996). In ovine basilar arterial smooth muscle cells PKG has been shown to play a larger role in the regulation of BK activity in fetal than in adult myocytes, indicating a developmental changes in the role of PKG (Lin et al., 2005).

PKG may also modulate Ca^{2+} release from the inositol-trisphosphate receptor (IP_3R) of the sarcoplasmic/endoplasmic reticulum (SERCA) through phosphorylation of IP_3R-associated cGMP kinase substrate (IRAG), a 125-kDa protein that resides in the SERCA membrane in a trimeric complex with PKG Iβ and IP_3R. In aortic smooth muscle cells of mice expressing a mutated IRAG protein that does not interact with the IP_3R the inhibition of cGMP on hormone-induced increases in $[Ca^{2+}]_i$ and contractility are blunted (Geiselhöringer et al., 2004). NO-, ANP-, and cGMP-dependent relaxation of aortic vessels is also attenuated in IRAG-knockout mice (Desch et al., 2010).

Increasing evidence has pointed to Ca^{2+} desensitization through interference with RhoA and Rho kinase (ROK) signaling as a key mechanism for PKG-mediated vasodilatation (A.P. Somlyo & A.V. Somlyo, 2003). PKG may phosphorylate RhoA at Ser188, resulting in increased extraction of Rho A from cell membranes and thus reduced activation of this small GTPase protein and attenuated vasocontractility (Loirand et al., 2006). PKG may suppress the inhibitory effect of ROK on myosin light chain phosphatase (MLCP) by phosphorylation of the regulatory subunit of MLCP, myosin phosphatase targeting subunit (MYPT1), at Ser695 and Ser852, which leads to decreased phosphorylation of MYPT1 at Thr696 and Thr853 by ROK, increased activity of MLCP, decreased phosphorylation of myosin light chain (MLC), and diminished vasoreactivity (Wooldridge et al., 2004; Gao et al., 2007 &2008). The effect of PKG on MLCP requires its binding to the leucine zipper domain in the C-terminal of MYPT1. The expression of the leucine zipper domain in MYPT1 is modulated by various physiological and pathophyiological conditions (Chen et al., 2006; Dou et al., 2010; Payne et al., 2006), which may alter the action of PKG on MLCP. Studies also show that PKG/MYPT1 signaling plays a greater role in mediating relaxation of proximal arteries induced by NO than that of distal arteries in coronary vasculature (Ying et al., 2011).

A number of PKG substrates not mentioned above may also be targeted by PKG and involved in PKG-mediated vasodilatation, such as phosphodiesterase 5 (PDE5), phospholamban, and RGS (regulator of G-protein signaling) proteins (Schlossmann & Desch, 2009). It is worth noting that cGMP may affect vasodilatation by PKG-dependent and independent mechanism. Global PKG-knockout causes only a slight hypertension in young mice whereas in the adult the basal blood pressure of the PKG-knockout mice is not different from the control (Pfeifer et al. 1998), indicating other mechanisms may take place to compensate the lose of PKG in maintaining a normal blood pressure.

3.1.2 Phenotype modulation and antiproliferation action

Vascular smooth muscle cells (VSMCs) exist in either a differentiated, contractile or a dedifferentiated, synthetic phenotype. A normal PKG activity appears critical to maintain vascular smooth muscle cells in a contractile and differentiated state. Repetitively passaged VSMCs of the rat aorta do not express PKG and exist in the synthetic phenotype. Transfection of PKG Iα cDNA induces a morphologic change of VSMCs consistent with the contractile phenotype, which is prevented by the inhibition of PKG (Dey et al., 2005).

Myocardin is a smooth muscle and cardiac muscle-specific transcriptional coactivator of serum response factor (SRF) while E26-like protein-1 (Elk-1) is a SRF/myocardin transcription antagonist. PKG-I has been shown to decrease Elk-1 activity by sumo modification of Elk-1, thereby increasing myocardin-SRF activity on SMC-specific gene expression and keeping the cells in a contractile phenotype (Choi et al., 2010). In VSMCs of ovine fetal pulmonary veins hypoxia-induced reduction in PKG protein expression is closely correlated with the repressed expression of VSMC phenotype markers, along with a reduced expression of myocardin and increased expression of Elk-1. It is postulated that the increased expression of Elk-1 resulting from the downregulation of PKG under hypoxia displaces myocardin from SRF and thereby leads to suppression of SMC marker genes and activation of expressions of genes related with the synthetic phenotype (Zhou et al., 2007 & 2009). The PKG-dependent modulation of phenotypes of VSMCs appears to need the cysteine-rich LIM-only protein CRP4 to act as a scaffolding protein that promotes cooperation between SRF and other transcription factors and cofactors since PKG stimulation of the SM-α-actin promoter is suppressed when CRP4 is deficient in PKG binding (Zhang et al., 2007).

PKG has been reported to exert anti- and pro-atherogenic effects in vascular smooth muscle. In coronary and cerebral arterial smooth muscle cells (El-Mowafy et al., 2008; Luo et al., 2011) the proliferation induced by vascular mitogens was inhibited by the cGMP elevating agent or PKG I transfection. However, 8-Br-cGMP stimulated proliferation of aortic SMCs from the wide-type mice but not from PKG I-deficient mice (Wolfsgruber et al., 2003). The contradictory effects may have in part resulted from differences in PKG activation levels (i.e., basal activation vs. hyperactivation). For instance, PKG at low activation levels prevents apoptosis whereas high-level activation causes apoptosis of aortic SMCs of the mice (Wong & Fiscus, 2010).

3.2 Heart

3.2.1 Cardiac contractility

A critical role for PKG in the negative inotropic effect caused by NO and cGMP has been demonstrated in myocardial preparations from PKG I-deficient juvenile mice, from the cardiomyocyte-specific knockout adult mice (Wegener et al., 2002), and in rat ventricular myocytes with the PKG inhibitor (Layland et al., 2002). The Ca^{2+} current of rat ventricular cells is inhibited by cGMP and a catalytically active fragment of PKG (Mery et al., 1991). In murine cardiac myocytes overexpressing PKG I the basal and stimulated activities of L-type Ca^{2+} channels are inhibited by NO and the cGMP analog (Schroder et al. 2003). Hence, PKG may exert its negative inotropic effect action by reducing $[Ca^{2+}]_i$ through inhibiting the activity of Ca^{2+} channels. Recent studies suggest that PKG I-mediated inhibition of L-type Ca^{2+} channels of cardiac myocytes may result from the phosphorylation of Ca_v 1.2 α_{1c} and β_2 subunits (Yang et al., 2007).

PKG may also reduce $[Ca^{2+}]_i$ of cardiomyocytes through phosphorylation of phospholamban, which leads to an increased activity of the sarcoplasmic reticulum Ca^{2+}-ATPase (SERCA) and thereby an increased Ca^{2+} uptake from the cytosol. Indeed, phospholamban of the rabbit cardiac myocytes is phosphorylated by cGMP in a manner sensitive to the inhibition of PKG. Moreover, the inhibitory effect of contractility of the cardiac myocytes caused by cGMP is prevented by the inhibition of PKG or SERCA (Zhang et al., 2005). C-type natriuretic peptide and the cGMP analog have been found to cause positive inotropic and lusitropic responses of murine hearts, which are associated with an increased phosphorylation of phospholamban (Wollert et al., 2003). These observations are in vary with those obtained in studies by Zhang et al. discussed above. The underlying reasons for the different inotropic effects remain to be determined (Wollert et al., 2003).

In intact cardiomyocytes of the rat, the negative inotropic and relaxant effects of DEA/NO, an NO donor, occur without significant changes in the amplitude or kinetics of the intracellular Ca^{2+} transient. The effect is diminished in the presence of the inhibitor of soluble guanylyl cyclase (sGC) or PKG, indicating a PKG-dependent Ca^{2+} desensitization of the myofilaments. Meanwhile, hearts treated with DEA/NO showed a significant increase in troponin I phosphorylation (Layland et al., 2002). The PKG may reduce the Ca^{2+} sensitivity of cardiac myofilaments through phosphorylation of cardiac troponin I (cTnI) at the same sites (Ser23/24) as those phosphorylated by PKA (Layland et al., 2005). Studies suggest that cardiac Troponin T may serve as an anchoring protein for PKG to facilitate preferential and rapid cTnI phosphorylation (Yuasa et al., 1999).

3.2.2 Antihypertrophy

An increased left ventricular mass has been recognized as an independent risk factor that correlates closely with cardiovascular risk and has strong prognostic implications. In the mice administration of sildenafil, which elevates cGMP level by inhibiting PDE5, suppresses the development of cardiac hypertrophy caused by chronic pressure overload and can even reverse pre-established cardiac enlargement. These effects are associated with an increased activity of PKG I (Takimoto et al. 2005). Mice with myocyte-specific PDE5 gene overexpression develop more severe cardiac hypertrophy and PKG activation is inhibited as compared to controls in response to pressure overload. Under such a cardiomyopathic state, the suppression of PDE5 expression/activity in myocytes enhanced PKG activity and reversed all previously amplified maladaptive responses (Zhang et al., 2010). In contrast to many studies which indicate an antihypertrophic role for PKG, Lukowski et al. have found that total PKG I-knockout and myocyte-specific rescue of PKG expression (in the context of global gene silencing) did not affect isoproterenol and stress-induced development of cardiac hypertrophy in mice (Lukowski et al., 2010). It is suspected that the lack of differences between controls and PKG I-deficient mice may be in part due to that PKG I-targeted cascades have not been activated under the experimental conditions (Kass & Takimoto, 2010).

RGSs are GTPase-accelerating proteins that promote GTP hydrolysis by the alpha subunit of heterotrimeric G proteins, thereby accelerating signal termination in response to GPCR stimulation (Schlossmann & Desch, 2009). Among more than 30 RGS proteins, RGS4 is richly expressed in murine coronary myocytes. In cultured cardiac myocytes, atrial natriuretic peptide stimulated PKG-dependent phosphorylation of RGS4 and association of RGS4 with the alpha subunit of Gq protein. Mice lacking guanylyl cyclase-A (GC-A), a natriuretic peptide receptor, have pressure-independent cardiac hypertrophy, reduced expression and phosphorylation of RGS4 in the hearts compared with wild-type mice. The RGS4 overexpression in GC-A-KO mice reduced cardiac hypertrophy and suppressed the augmented cardiac expressions of hypertrophy-related genes. These results suggest that GC-A activation may counteract cardiac hypertrophy via RGS4 in a PKG-dependent mechanism (Tokudome et al., 2008). It appears that ANP-cGMP-PKG-RGS signaling is involved in β-adrenergic but not angiotensin II (Ang II)–induced (Gs vs. Gaq mediated) cardiomyocyte hypertrophy. ANP attenuated Ang II-stimulated Ca^{2+} currents of cardiomycytes but had no effect on isoproterenol stimulation. The effect of ANP on Ang II stimulation was eliminated in cardiomyocytes of mice deficient in GC-A, in PKG I, or in RGS2. Furthermore, cardiac hypertrophy induced by Ang II but not by β- adrenoreceptor was exacerbated in mice with cardiomyocyte-restricted GC-A deletion (Klaiber et al., 2010).

Multiple subclasses of transient receptor potential (TRP) channels are expressed in the heart. These channels, especially the TRPC subclass, have been implicated being involved in the regulation of the cardiac hypertrophic response, most likely coordinating signaling within local domains or through direct interaction with Ca^{2+}-dependent regulatory proteins. Overexpression of TRPC6 in mice lacking GC-A exacerbated cardiac hypertrophy while the blockade of TRPC channels attenuated the cardiac hypertrophy. ANP inhibited agonist-evoked Ca^{2+} influx of murine cardiomyocytes. The inhibitory effects of ANP were abolished by PKG inhibitors or by substituting an alanine for threonine 69 in TRPC6, suggesting that PKG-dependent phosphorylation of TRPC6 at threonine 69 is a critical target of antihypertrophic effects elicited by ANP (Kinoshita et al., 2010)

3.2.3 Cardioprotective action against ischemia-reperfusion injury

In isolated murine heart and cardiomyocytes elevation of cGMP by the activators of soluble or particulate guanylyl cyclase, by the inhibitor of PDE5, or by the cGMP analog elicits potent protection against myocardial ischemia-reperfusion injury and reduces cardiomyocyte necrosis and apoptosis. These effects are accompanied by an increased PKG activity and attenuated by PKG inhibitors or by selective knockdown of PKG in cardiomyocytes (Das, 2008 et al.; Gorbe et al., 2010). It is generally recognized that ischemia/reperfusion injury arises primarily from the opening of the mitochondrial permeability transition pore (mPTP) in the first minutes of reperfusion. cGMP-PKG signaling may prevent opening of mPTP via activation of the mitochondrial K_{ATP} channels, direct phosphorylation of an unknown protein on the mitochondrial outer membrane, and upregulation of the antiapoptotic protein Bcl-2 (Costa et al.,2008; Deschepper, 2010). Glycogen synthase kinase 3β (GSK-3β) plays a central role in transferring cardio protective signals downstream to target(s) that act at or in proximity to the mPTP. Phosphorylation and inhibition of GSK-3β has also been demonstrated being involved in PKG-mediated cardioprotective action (Das et al., 2008; Juhaszova et al., 2009; Xi et al., 2010).

3.3 eNOS activity and endothelial permeability

PKG I has been detected within a range of 0.15 to 0.5 μg/mg cellular protein in adult artery and vein endothelial cells (ECs) and in microvascular ECs (Diwan et al., 1994; Draijer et al., 1995; MacMillan-Crow et al., 1994). There are only limited studies on the role of PKG in the regulation of endothelial function, which is related to eNOS activity and endothelial permeability (Butt et al., 2000; Draijer et al., 1995; Moldobaeva et al., 2006; Rentsendorj et al., 2008). Studies using recombinant human eNOS suggest that the enzyme can be phosphorylated at Ser^{1177}, Ser^{633}, Thr^{495} and activated by PKG II in a manner independent of Ca^{2+} and calmodulin (Butt et al., 2000). Cyclic GMP analog inhibits an increase in $[Ca^{2+}]_i$ and endothelial permeability caused by thrombin in cultured ECs expressing PKG I but not those lacking PKG expressing (Draijer et al., 1995). In human pulmonary artery endothelial cells infected with adenovirus encoding PKG Iβ the cGMP analog prevents the increase in endothelial permeability caused by H_2O_2. The barrier protection effect was not affected by inhibition of the expression of VASP, a PKG substrate (Moldobaeva et al., 2006; Rentsendorj et al., 2008).

3.4 Anti-platelet aggregation action

Substantial evidence supports a critical role for PKG in mediating the anti-platelet aggregation action caused by cGMP elevating agents such as endothelium-derived NO (EDNO) and exogenous nitrovasodilators (Walter & Gambaryan, 2009; Dangel et al., 2010). PKG Iβ is the predominate isoform of the enzyme in platelets. The concentration of PKG Iβ in human platelets is 3.65μM, which is higher than that of any other cell type examined (Antl et al., 2007; Eigenthaler et al., 1992). In PKG-deficient murine platelets the inhibition of the cGMP analog on granule secretion, aggregation and adhesion is severely affected (Massberg et al. 1999; Schinner et al., 2011). The effect of PKG may be in part mediated by IRAG. IRAG is abundantly expressed in platelets and constitutively formed in a macrocomplex with PKGIβ and the $InsP_3R$. PKGIβ phosphorylates IRAG at Ser664 and Ser677 in intact platelets, resulting in attenuated release of Ca^{2+} from the sarcoplasmic reticulum evoked by IP_3. Targeted deletion of the IRAG-$InsP_3RI$ interaction in $IRAG^{\Delta12/\Delta12}$ mutant mice causes a loss of NO/cGMP-dependent inhibition of $[Ca^{2+}]_i$ increase and platelet aggregation. The preventive effect of NO on arterial thrombosis in the injured carotid artery was observed in wide-type platelets but not in $IRAG^{\Delta12/\Delta12}$ mutants (Antl et al. 2007).

Vasodilator-stimulated phosphoprotein (VASP) belongs to the Ena-VASP protein family. It is associated with filamentous actin formation and may play a widespread role in cell adhesion and motility. In VASP-deficient mice, the inhibitory effect of NO on platelet adhesion is impaired. Under physiologic conditions, platelet adhesion to endothelial cells was enhanced in VASP null mutants. Under pathophyiological conditions, the loss of VASP augments platelet adhesion to the postischemic intestinal microvasculature, to the atherosclerotic endothelium of ApoE-deficient mice, and to the subendothelial matrix of blood vessels (Massberg et al. 2004). In VASP-deficient mice, although cGMP-mediated inhibition of platelet aggregation is impaired, cGMP-dependent inhibition of agonist-induced increases in cytosolic calcium concentrations and granule secretion is preserved (Aszódi et al., 1999).

Although it is a currently prevailing concept that PKG signaling inhibits platelet function, some studies show that activation of NO-cGMP-PKG pathway promotes platelet aggregation (Blackmore, 2011; Li et al., 2003; Zhang et al., 2011). In PKG knockout mice platelet responses to von Willebrand factor (vWF) or low doses of thrombin are impaired and bleeding time is prolonged. Human platelet aggregation induced by these agents is also diminished by PKG inhibitors but enhanced by cGMP (Li et al., 2003). A defect in platelet aggregation in response to low doses of collagen or thrombin also occurs in platelet-specific sGC-deficient mice (Zhang et al., 2011). It appears that cGMP at low concentrations promotes while at higher concentrations inhibits platelet aggregation (Blackmore et al., 2011; Li et al., 2003).

4. PKG and cardiovascular diseases

4.1 Hypertension

Global deletion of eNOS (Huang et al., 1995), sGC (Friebe et al., 2007), or PKG I (Pfeifer et al. 1998) results in hypertension in mice. About 80% of the mice that are deficient in PKG I died at age of 8-week. Those lived to adulthood showed no significant difference in blood pressure from the wild type animals, indicating compensatory mechanisms are functioning (Pfeifer et al. 1998). Loss of PKG I abolishes NO- and cGMP-dependent relaxations of smooth muscle (Pfeifer et al. 1998). In mice with a selective mutation in the N-terminal protein interaction domain of PKG Iα also results in reduced vasodilator response to EDNO and cGMP and increased systemic blood pressure, suggesting that the hypertension results from a diminished response of blood vessels to cGMP (Michael et al., 2008). Vascular reconstitution of PKG Iα or PKG Iβ in PKG I-deficient mice restores the diminished vasodilatation to NO and cGMP and normalizes the elevated blood pressure, (Weber et al., 2007). In spontaneously hypertensive rats (SHR) cardiomyocytes PKG-I expression is decreased, making the NO/cGMP-dependent regulation on calcium transient in cardiomyocytes weakened, and promoting cardiac hypertrophy (Mazzetti et al., 2001).

Abnormality in the renin-angiotensin-aldosterone system is an important etiologic event in the development of hypertension. Renal renin mRNA levels under stimulatory (low-salt diet plus ramipril) and inhibitory (high-salt diet) conditions were elevated in PKG II deficient mice. The deletion of PKG II abolishes the attenuation of forskolin-stimulated renin secretion caused by 8-Br-cGMP in cultured renal juxtaglomerular cells. Activation of PKG by 8-Br-cGMP decreased renin secretion from the isolated perfused rat kidney of the wild-type mice but not that of PKG II-/- mice. These findings suggest that PKG II exerts an inhibitory effect on renin secretion (Wagner et al., 1998). Mice deficient in PKG II display no elevated blood pressure, suggesting that PKG II is not critically involved in the regulation of overall systemic blood pressure (Hofmann et al., 2009).

4.2 Atherosclerosis

In an animal model of late-stage atherosclerosis obtained by feeding 8-week-old rabbits with hypercholesterol diet for 50 weeks the protein levels of sGC and PKG I of the aorta were reduced. These changes were most prominent in the neointimal layer. Phosphorylation of VASP at Ser239, a specific indicator of PKG activity, was also reduced. The preferential down-regulation of cGMP/PKG signaling in neointima suggests a direct connection of these changes to neointimal proliferation and vascular dysfunction occurred in atherosclerosis (Melichar et al., 2004). It seems that the decreased PKG expression occurred only at late-stage atherosclerosis, as the protein level of PKG was unaltered in Watanabe heritable hyperlipidemic rabbits of three month old (Warnholtz et al., 2002). Thrombospondin-1 and osteopontin are extracellular matrix (ECM) proteins involved in the development of atherosclerosis. PKG may exert its anti- atherosclerotic effect in part through these two ECM proteins, since their expression could be marked reduced by PKG I (Dey et al., 1998). Interestingly, postnatal ablation of PKG I selectively in the VSMCs of mice reduced atherosclerotic lesion area, which would suggest that smooth muscle PKG I promotes atherogenesis (Wolfsgruber et al., 2003).

4.3 Diabetic vascular disease

High glucose exposure has been found to reduce the protein and mRNA levels of PKG I as well as PKG I activity in cultured rat VSMCs. PKG I protein levels were decreased in femoral arteries from diabetic mice. Glucose-mediated decrease in PKG I levels was inhibited by the superoxide scavenger or NAD(P)H oxidase inhibitors. High glucose exposure increased the protein levels and phosphorylated levels of p47phox (an NADPH oxidase subunit) in VSMCs, associated with increased superoxide production. The suppressed PKG expression and increased superoxide production were prevented by transfection of cells with siRNA-p47phox, suggesting that NADPH oxidase-derived superoxide may mediate the high glucose-induced downregulation of PKG occurred in diabetic blood vessels (Liu et al., 2007). Studies also show that activation of PKG by expression of constitutively active PKG suppressed high glucose-induced VSMC proliferation and inhibited first gap phase (G1) to synthesis phase (S) phase progression of the cell cycle. These changes were accompanied with reduced glucose-induced cyclin E expression and cyclin E-cyclin-dependent kinase 2 activity as well as inhibition of glucose-induced phosphorylation of retinoblastoma protein (Rb) and p27 degradation. It suggests that PKG may inhibit VSMC proliferation through attenuation of cyclin E expression and increase in p27 protein stability, which leads to decreased CDK 2 activity and reduced Rb phosphorylation, thereby resulting in cell cycle arrest and cell growth inhibition (Wang & Li, 2009). Increased activity of transforming growth factor-β (TGF-β) is implicated in the development of diabetic macrovascular fibroproliferative remodeling. High glucose was found to stimulate the expression of thrombospondin1 (TSP1), a major activator of transforming growth factor-β (TGF-β), and to stimulate TGF-β activation in primary murine aortic SMCs. These effects were inhibited by overexpression of constitutively active PKG. Since PKG is downregulated in diabetic vasculature, it is likely that the downregulation of PKG action may relieve its suppression on TSP1 expression and TGF-β activity, thereby leading to augmented vascular remodeling in diabetes (Wang et al., 2010).

4.4 Pulmonary arterial hypertension

PKG expression and/or activity are/is reduced in animal models of pulmonary arterial hypertension (PAH) induced by ligation of the ductus arteriosus of fetal lambs (Resnik et al., 2006) and in caveolin-1 (Cav-1) knockout mice (Zhao et al., 2009). Cav-1, a 21-kDa integral membrane protein, is an intracellular physiological inhibitor of eNOS activity. Mice deficient in Cav-1 led to chronic eNOS activation and PAH. Activation of eNOS in Cav-1$^{-/-}$ lungs resulted in an impaired PKG activity through tyrosine nitration, probably at Tyr345 or Tyr549 of the catalytic domain of human PKG Iα. The PAH phenotype in Cav-1$^{-/-}$ lungs could be rescued by overexpression of PKG 1α. The treatment of these mice with either a superoxide scavenger or an eNOS inhibitor reverses their pulmonary vascular pathology and PAH phenotype, suggesting that an increased peroxynitrite formed from chronic overproduction of NO and superoxide may result in tyrosine nitration and loss of activity of PKG. Clinically, lung tissues from patients with idiopathic PAH have been found to display reduced Cav-1 expression, increased eNOS activation, and PKG nitration (Zhao et al., 2009). In ovine fetal pulmonary veins hypoxic exposure also causes peroxynitrite-mediated PKG nitration, reduced PKG activity, and suppressed dilator response to 8-Br-cGMP (Negash et al., 2007).

An upregulated ROK activity is implicated in a number of cardiovascular diseases including PAH (Satoh et al., 2011). ROK augments vasoconstriction primarily by inhibiting MLCP activity through phosphorylation of the regulatory subunit MYPT1 at Thr696 and Thr853, which leads to increased Ca^{2+}-sensitization of smooth muscle. The effect of ROK can be counteracted by the stimulatory action of PKG through phosphorylation of MYPT1 at Ser695 and Ser852. Pulmonary arteries from fetuses exposed to chronic intrauterine hypoxia (CH) displayed thickening vessel walls and diminished relaxant response to 8-Br-cGMP, two important characteristics of newborn PAH (Bixby et al., 2007; Gao et al., 2007). Rp-8-Br-PET-cGMPS, a specific PKG inhibitor, attenuated relaxation to 8-Br-cGMP in control vessels to a greater extent than in CH vessels while Y-27632, a ROCK inhibitor, potentiated 8-Br-cGMP-induced relaxation of CH vessels and had only a minor effect in control vessels. The specific activity of PKG was decreased while ROK activity was increased in CH vessels as compared with the controls. The phosphorylation of MYPT1 at Thr696 and Thr853 was inhibited by 8-Br-cGMP to a lesser extent in CH vessels than in controls. The difference was eliminated by Y-27632. These data indicate that the attenuated PKG-mediated relaxation in pulmonary arteries exposed to chronic hypoxia in utero is due to inhibition of PKG activity and due to enhanced ROCK activity. Increased ROCK activity may inhibit PKG action through increased phosphorylation of MYPT1 at Thr696 and Thr853 (Gao et al., 2007). In contrast to pulmonary arteries, relaxation of pulmonary veins of fetuses exposed to (CH) displayed no changes in the thickness of vessel walls and relaxant response to 8-Br-cGMP. In these veins phosphorylation of MYPT1 at Thr696 by ROK and at Ser695 by PKG was diminished as compared with control veins, suggesting that CH attenuates both PKG action and ROK action on MYPT1, resulting in an unaltered response to cGMP (Gao et al., 2008).

Bone morphogenetic proteins (BMPs) are members of the TGF-β superfamily. Mutations in the BMP type II receptor (BMPR-II) are responsible for the majority of cases of heritable PAH. Dysfunction in BMP signaling is implicated in idiopathic PAH and in a number of experimental models of PAH (Toshner et al., 2010). Studies found that PKG I may regulate the activation of BMP receptor and receptor-regulated Smad, a key mediator for BMP signaling, at the plasma membrane and regulate the expression of BMP target genes in the nucleus. These mechanisms may enable PKG I to compensate for the aberrant cellular responses to BMP caused by mutations in BMPRII found in PAH patients. Indeed, the overexpression of PKG I restores normal BMP responsiveness in cells expressing signaling deficient PAH mutant receptors such as the mutant BMPRII-Q657ins16 (Schwappacher et al., 2009; Thomson et al., 2000).

4.5 Nitrate tolerance

Nitroglycerine (NTG) is a widely used vasodilator in the treatment of angina pectoris and acute heart failure. It is converted inside the cell to NO or an NO-related intermediate and causes vasodilatation in a cGMP-dependent fashion. The effectiveness of NTG is often diminished when it is continuously used for a period of time, termed nitrate tolerance. The underlying mechanisms include an increased production of reactive oxygen species (ROS), impairment of biotransformation of NTG by aldehyde dehydrogenase, desensitization of sGC, upregulation of phosphodiesterases, and downregulation PKG activity (Münzel et al., 2005). In human arteries and veins, nitrate tolerance is associated with decreased PKG activity (Schulz et al., 2002). In the arteries of rats and rabbits, nitrate tolerance induced by low-dose NTG is associated with decreased PKG activity, while the tolerance induced by high-dose NTG is associated with decreased PKG protein level and activity (Mülsch *et al.*, 2001). In porcine coronary arteries nitrate tolerance induced by NTG at low concentrations is prevented by the scavenger of ROS. However the tolerance induced by NTG at higher concentrations is not affected by the scavenger of ROS and shows cross-tolerance to the NO donor and 8-Br-cGMP. Meanwhile, the protein and mRNA levels of PKG are reduced. It seems that the tolerance induced by NTG at higher concentrations may be due to suppression of PKG expression resulting from sustained activation of the enzyme (Dou et al., 2008). A diminished expression and activity of PKG was also observed in pulmonary veins of newborn lambs after prolonged exposure to the NO donor (Gao et al., 2004).

Activation of MLCP is a key mechanism for vasodilatation induced by nitrovasodilators such as NTG and NO. MLCP is a heterotrimer, composed of a catalytic subunit PP1cδ, a regulatory subunit MYPT1, and a subunit with unknown function. The regulatory subunit MYPT1 exists as isoform either with or without leucine zipper domain in its C-terminal [MYPT1 (LZ+) and MYPT1 (LZ−), respectively]. The presence of leucine zipper is necessary for PKG binding to MYPT1 and for PKG-mediated stimulatory effect on MLCP. Studies consistently demonstrate that the expression of MYPT1 (LZ+) determines the sensitivity to cGMP-mediated vasodilatation (Lee et al., 2007; A.P. Somlyo & A.V. Somlyo, 2003). Nitrate tolerance induced under *in vitro* conditions in porcine coronary arteries and induced under *in vivo* preparations in murine aorta show a decreased protein levels of MYPT1 (LZ+) but not of PP1cδ. The decrease in the MYPT1 (LZ+) protein level of coronary artery can also be induced by the NO donor and 8-Br-cGMP in a manner sensitive to the inhibitors of sGC and PKG, respectively. The tolerance to NTG in porcine coronary artery and mouse aorta is ameliorated by proteasome inhibitors. Therefore a downregulation of MYPT1 (LZ+) caused by increased proteasome-dependent degradation may contribute to development of nitrate tolerance (Dou et al., 2010).

5. Conclusion

Overwhelming evidence, obtained by genetic manipulation and pharmacological tools, under both *in vivo* and *in vitro* conditions, suggests that PKG is the primary enzyme in mediating vasodilatation, antiproliferation of vascular smooth muscle, and anti-platelet aggregation action induced by endogenous and exogenous nitrovasodilators via cGMP elevation (Francis et al., 2010; Gao, 2010; Hofmann et al., 2009; Walter & Gambaryan, 2009). Studies also support a barrier protection effect in the vascular endothelium (Moldobaeva et al., 2006; Rentsendorj et al., 2008). Increasing evidence also suggests that PKG exerts negative inotropic and antihypertrophic actions in the heart (Takimoto et al. 2005; Yang et al., 2007; Zhang et al., 2010) as well as a cardioprotective action against ischemia-reperfusion injury (Das et al., 2008; Juhaszova et al., 2009; Xi et al., 2010). Despite substantial progress has been made in elucidating the role of PKG in the regulation of cardiovascular functions there are many aspects remain to be explored. For instance, the developing and ageing aspects for the role of PKG, the gender difference, and the heterogeneity in the role of PKG in different vasculatures. Also, the roles of many PKG substrates in the regulation of cardiovascular activities remain to be defined (Schlossmann & Desch, 2009).

Dysfunction in NO-cGMP signaling is a common initiator and independent predictor of cardiovascular events (Vanhoutte et al., 2009). An impaired PKG action has been implicated in various cardiovascular disorders such as hypertension, atherosclerosis, diabetic vascular disease, pulmonary arterial hypertension, and nitrate tolerance (Francis et al., 2010; Gao, 2010; Hofmann et al., 2009). Cardiovascular alterations are a long-term process comprising functional and structural changes with remarkable complexities, which undoubtedly make the dissection of the role of PKG rather challenging. However, a better understanding of its role and the underlying mechanism will be of great therapeutic significance.

6. References

Antl, M.;, von Bruhl, M.L.; Eiglsperger, C.; Werner, M.; Konrad, I.; Kocher, T.; Wilm, M.; Hofmann, F.; Massberg, S.; Schlossmann, J. (2007). IRAG mediates NO/cGMP-dependent inhibition of platelet aggregation and thrombus formation. *Blood* Vol.109, No.2, (January 2007), pp. 552-559.

Aszódi, A.; Pfeifer, A.; Ahmad, M.; Glauner, M.; Zhou, X.H.; Ny, L.; Andersson, K.E.; Kehrel, B.; Offermanns, S.; Fässler, R. (1999). The vasodilator- stimulated phosphoprotein (VASP) is involved in cGMP- and cAMP-mediated inhibition of agonist- induced platelet aggregation, but is dispensable for smooth muscle function. *EMBO J.* (January 1999), Vol.18, No.1, pp. 37-48.

Barman, S.A.; Zhu, S.; Han, G.; White, R.E. (2003). cAMP activates BK_{Ca} channels in pulmonary arterial smooth muscle via cGMP-dependent protein kinase. *Am. J. Physiol. Lung Cell Mol. Physiol.,* (June 2003), Vol.284, No.6, pp. L1004-L1011.

Bixby, C.E.; Ibe, B.O.; Abdallah, M.F.; Zhou, W.; Hislop, A.A.; Longo, L.D., Raj, J.U. (2007). Role of platelet-activating factor in pulmonary vascular remodeling associated with chronic high altitude hypoxia in ovine fetal lambs. *Am. J. Physiol. Lung Cell Mol. Physiol.,* (December 2007), Vol.293, No. 6, pp.L1475-82.

Blackmore, P.F. (2011). Biphasic effects of nitric oxide on calcium influx in human platelets. *Thromb. Res.,* (January 2011), Vol.127, No.1 pp.e8-e14.

Butt, E.; Bernhardt, M.; Smolenski, A.; Kotsonis, P; Frohlich, L.G.; Sickmann, A.; Meyer, H.E.; Lohmann, S.M.; Schmidt, H.H. (2000). Endothelial nitric-oxide synthase (type III) is activated and becomes calcium independent upon phosphorylation by cyclic nucleotide-dependent protein kinases. *J. Biol. Chem.,* (February 2000), Vol.275, No.7, pp.5179-5187.

Chen, F.C.; Ogut, O.; Rhee, A.Y.; Hoit, B.D.; Brozovich, F.V. (2006). Captopril prevents myosin light chain phosphatase isoform switching to preserve normal cGMP-mediated vasodilatation. *J. Mol. Cell Cardiol.,* (September 2006), Vol.41, No.3, pp.488-495.

Choi, C.; Sellak, H.; Brown, F.M.; Lincoln, T.M. (2010). cGMP-dependent protein kinase and the regulation of vascular smooth muscle cell gene expression: possible involvement of Elk-1 sumoylation. *Am. J. Physiol. Heart Circ. Physiol.,* (November 2010), Vol.299, No. 5, pp.H1660-H1670.

Costa, A.D.; Pierre, S.V.; Cohen, M.V.; Downey, J.M.; Garlid, K.D. (2008). cGMP signalling in pre- and post-conditioning: the role of mitochondria. *Cardiovasc. Res.,* (January 2008), Vol.77, No.2, pp.344-352.

Dangel, O.; Mergia, E.; Karlisch, K.; Groneberg, D.; Koesling, D.; Friebe, A. (2010). Nitric oxide-sensitive guanylyl cyclase is the only nitric oxide receptor mediating platelet inhibition. *J. Thromb. Haemost.,* (June 2010), Vol.8, No. 6, pp.1343-1352.

Das, A.; Xi, L.; Kukreja, R.C. (2008). Protein Kinase G-dependent Cardioprotective Mechanism of Phosphodiesterase-5 Inhibition Involves Phosphorylation of ERK and GSK3β. *J. Biol. Chem.,* (October 2008), Vol.283, No. 43, pp.29572-29585.

Desch, M.; Sigl, K.; Hieke, B.; Salb, K.; Kees, F.; Bernhard, D.; Jochim, A.; Spiessberger, B.; Höcherl, K.; Feil, R.; Feil, S.; Lukowski, R.; Wegener, J.W.; Hofmann, F.;

Schlossmann, J. (2010). IRAG determines nitric oxide- and atrial natriuretic peptide-mediated smooth muscle relaxation. *Cardiovasc. Res.*, (June 2010), Vol.86, No.3, pp.496-505.

Deschepper, C.F. (2010). Cardioprotective actions of cyclic GMP: lessons from genetic animal models. *Hypertension*, (February 2010), Vol.55, No. 2, pp.453-458.

Dey, N.B.; Boerth, N.J.; Murphy-Ullrich, J.E.; Chang, P.L.; Prince, C.W.; Lincoln, T.M. (1998). Cyclic GMP-dependent protein kinase inhibits osteopontin and thrombospondin production in rat aortic smooth muscle cells. Circ. Res., (Feb 1998), Vol.82, No.2, pp.139-146.

Dey, N.B.; Foley, K.F.; Lincoln, T.M.; Dostmann, W.R. (2005). Inhibition of cGMP-dependent protein kinase reverses phenotypic modulation of vascular smooth muscle cells. *J. Cardiovasc. Pharmacol.*, (May 2005), Vol.45, No.5, pp404-413.

Dhanakoti, S.; Gao, Y.; Nguyen, M.Q.; Raj, J.U. (2000). Involvement of cGMP-dependent protein kinase in the relaxation of ovine pulmonary arteries to cGMP and cAMP. *J. Appl. Physiol.*, (May 2000), Vol.88, No.5, pp.1637-1642.

Diwan, A.H.; Thompson, W.J.; Lee, A.K.; Strada, S.J. (1994). Cyclic GMP-dependent protein kinase activity in rat pulmonary microvascular endothelial cells. *Biochem. Biophys. Res. Commun.*, (July 1994), Vol.202, No.2, pp.728-735.

Dou, D.; Ma, H.; Zheng, X.; Ying, L.; Guo, Y.; Yu, X.; Gao, Y. (2010). Degradation of leucine zipper-positive isoform of MYPT1 may contribute to development of nitrate tolerance. *Cardiovasc. Res.*, (April 2010), Vol.86, No.1, pp.151-159.

Dou, D.; Zheng, X.; Qin, X.; Qi, H.; Liu, L.; Raj, J.U.; Gao, Y. (2008). Role of cGMP-dependent protein kinase in development of tolerance to nitroglycerine in porcine coronary arteries. *Br. J. Pharmacol.*, (February 2008), Vol.153, No.3, pp.497-507.

Draijer, R.; Vaandrager, A.B.; Nolte, C.; de Jonge, H.R.; Walter, U.; van Hinsbergh, V.W. (1995). Expression of cGMP-dependent protein kinase I and phosphorylation of its substrate, vasodilator-stimulated phosphoprotein, in human endothelial cells of different origin. *Circ. Res.*, (November 1995), Vol.77, No.5, pp.897-905.

Eigenthaler, M.; Nolte, C.; Halbrugge, M.; Walter, U. (1992). Concentration and regulation of cyclic nucleotides, cyclic-nucleotide-dependent protein kinases and one of their major substrates in human platelets. Estimating the rate of cAMP-regulated and cGMP-regulated protein phosphorylation in intact cells. *Eur. J. Biochem.*, (April 1992), Vol.205, No. 2, pp.:471-481.

El-Mowafy, A.M.; Alkhalaf, M.; El-Kashef, H.A. (2008). Resveratrol reverses hydrogen peroxide-induced proliferative effects in human coronary smooth muscle cells: a novel signaling mechanism.*Arch. Med. Res.*, (February 2008), Vol.39, No.2, pp.155-161.

Francis, S.H.; Busch, J.L.; Corbin, J.D.; Sibley, D. (2010). cGMP-dependent protein kinases and cGMP phosphodiesterases in nitric oxide and cGMP action. *Pharmacol. Rev.*, (September 2010), Vol.62, No.3, pp.525-563.

Friebe, A.; Mergia, E.; Dangel, O.; Lange, A.; Koesling, D. (2007). Fatal gastrointestinal obstruction and hypertension in mice lacking nitric oxide-sensitive guanylyl cyclase. *Proc. Natl. Acad. Sci. USA*, (May 2007), Vol.104, No.18, pp.7699-7704.

Fukao, M.; Mason, H.S.; Britton, F.C.; Kenyon, J.L.; Horowitz, B.; Keef, K.D. (1999). Cyclic GMP-dependent protein kinase activates cloned BKCa channels expressed in mammalian cells by direct phosphorylation at serine 1072. *J. Biol. Chem.*, (April 1999), Vol.274, No.16, pp.10927-10935.

Gao, Y. (2010). The multiple actions of NO. Pflugers Arch.- Eur. J. Physiol., (May 2010), Vol.459, No.6, pp.829-839.

Gao, Y.; Dhanakoti, S.; Tolsa, J.-F.; Raj, J.U. (1999). Role of protein kinase G in nitric oxide and cGMP-induced relaxation of newborn ovine pulmonary veins. *J. Appl. Physiol.* (September 1999), Vol.87, No.3, pp.993-998.

Gao, Y.; Dhanakoti, S.; Trevino, E.M.; Sander, F.C.; Portuga, A.M.; Raj, J.U. (2003). Effect of oxygen on cyclic GMP-dependent protein kinase-mediated relaxation in ovine fetal pulmonary arteries and veins. *Am. J. Physiol. Lung Cell Mol. Physiol.*, (September 2003), Vol.285, No.3, pp.L611-L618.

Gao, Y.; Dhanakoti, S; Trevino, E.M.; Wang, X.; Sander, F.C.; Portuga, A.D.; Raj, J.U. (2004). Role of cGMP-dependent protein kinase in development of tolerance to nitric oxide in pulmonary veins of newborn lambs. *Am J Physiol Lung Cell Mol Physiol.*, (April 2004), Vol.286, No.4, ppL786–L792.

Gao, Y.; Portugal, A.D.; Negash, S.; Zhou, W.; Longo, L.D.; Usha Raj, J. (2007). Role of Rho kinases in PKG-mediated relaxation of pulmonary arteries of fetal lambs exposed to chronic high altitude hypoxia. *Am J Physiol Lung Cell Mol Physiol.*, (March 2007), Vol.292, No.3, pp.L678-L684.

Gao, Y.; Portugal, A.D.; Liu, J.; Negash, S.; Zhou, W.; Tian, J.; Xiang, R.; Longo, L.D.; Raj, J.U. (2008). Preservation of cGMP-induced relaxation of pulmonary veins of fetal lambs exposed to chronic high altitude hypoxia: role of PKG and Rho kinase. *Am. J. Physiol. Lung Cell Mol. Physiol.*, (November 2008), Vol.295, No.5, pp.L889-L896.

Geiselhöringer, A.; Werner, M.; Sigl, K.; Smital, P.; Wörner, R.; Acheo, L.; Stieber, J.; Weinmeister, P.; Feil, R.; Feil, S.; Wegener, J.; Hofmann, F.; Schlossmann, J. (2004). IRAG is essential for relaxation of receptor-triggered smooth muscle contraction by cGMP kinase. *EMBO J.*, (October 2004), Vol.23, No.21, pp.4222-4231.

Gorbe, A.; Giricz, Z.; Szunyog, A.; Csont, T.; Burley, D.S.; Baxter, G.F.; Ferdinandy, P. (2010). Role of cGMP-PKG signaling in the protection of neonatal rat cardiac myocytes subjected to simulated ischemia/reoxygenation. *Basic. Res. Cardiol.*, (September 2010), Vol.105, No.5, pp.643-650.

Hofmann, F.; Bernhard, D.; Lukowski, R.; Weinmeister, P. (2009). cGMP regulated protein kinases (cGK). *Handb. Exp. Pharmacol.*, Vol.191, 137-162.

Huang, P.L.; Huang, Z.; Mashimo, H.; Bloch, K.D.; Moskowitz, M.A.; Bevan, J.A.; Fishman, M.C. (1995). Hypertension in mice lacking the gene for endothelial nitric oxide synthase. *Nature*, (September 1995), Vol.377, No.6546, pp.239-242.

Juhaszova, M.; Zorov, D.B.; Yaniv, Y.; Nuss, H.B.; Wang ,S.; Sollott, S.J. (2009). Role of Glycogen Synthase Kinase-3β in Cardioprotection. *Circ. Res.*, (June 2009), Vol.104, No.11, pp.1240-1252

Kass, D.A.; Takimoto, E. (2010). Regulation and role of myocyte cyclic GMP-dependent protein kinase-I. *Proc. Natl. Acad. Sci. USA.* (June 2010), Vol.107, No.24, pp.E98.

Kinoshita, H.; Kuwahara, K.; Nishida, M.; Jian, Z.; Rong, X.; Kiyonaka, S.; Kuwabara, Y.; Kurose, H.; Inoue, R.; Mori, Y.; Li, Y.; Nakagawa, Y.; Usami, S.; Fujiwara, M.; Yamada, Y.; Minami, T.; Ueshima, K.; Nakao, K. (2010). Inhibition of TRPC6 channel activity contributes to the antihypertrophic effects of natriuretic peptides-guanylyl cyclase-A signaling in the heart. *Circ. Res.*, (June 2010), Vol.106, No.12, pp.1849-1860.

Klaiber, M.; Kruse, M.; Völker, K.; Schröter, J.; Feil, R.; Freichel, M.; Gerling, A.; Feil, S.; Dietrich, A.; Londoño, J.E.; Baba, H.A.; Abramowitz, J.; Birnbaumer, L.; Penninger, J.M.; Pongs, O.; Kuhn, M. (2010). Novel insights into the mechanisms mediating the local antihypertrophic effects of cardiac atrial natriuretic peptide: role of cGMP-dependent protein kinase and RGS2. *Basic Res. Cardiol.*, (September 2010), Vol.105, No.5, pp.583–595.

Kuo, J.F.; Greengard, P. (1970). Cyclic nucleotide-dependent protein kinases. VI. Isolation and partial purification of a protein kinase activated by guanosine 3',5'-monophosphate. *J. Biol. Chem.*, (May 1970), Vol.245, No.10, pp.:2493-248.

Layland, J.; Li, J.M.; Shah, A.M. (2002). Role of cyclic GMP-dependent protein kinase in the contractile respose to exogenous nitric oxide in rat cardiac myocytes. *J. Physiol.*, (April 2002), Vol.540(Pt. 2), pp.457-467.

Layland, J.; Solaro, R.J.; Shah, A.M. (2005). Regulation of cardiac contractile function by troponin I phosphorylation. *Cardiovasc. Res.*, (April 2005), Vol.66, No.1, pp.12–21.

Lee, E.; Hayes, D.B.; Langsetmo, K.; Sundberg, E.J; Tao, T.C. (2007). Interactions between the leucine-zipper motif of cGMP-dependent protein kinase and the C-terminal region of the targeting subunit of myosin light chain phosphatase. *J. Mol. Biol.*, (November 2007), Vol.373, No.5, pp.1198-1212.

Li, Z.; Xi, X.; Gu, M.; Feil, R.; Ye, R.D.; Eigenthaler, M.; Hofmann, F.; Du, X. (2003). A stimulatory role for cGMP-dependent protein kinase in platelet activation. *Cell*, (January 2003), Vol.112, No.1, pp.77-86.

Lin, M.T.; Longo, L.D.; Pearce, W.J.; Hessinger, D.A. (2005). Ca^{2+}-activated K^+ channel-associated phosphatase and kinase activities during development *Am. J. Physiol. Heart Circ. Physiol.*, (July 2005), Vol.289, No.1, H414-H425.

Lincoln ,T.M.; Dey, N.; Sellak, H. (2001). cGMP-dependent protein kinase signaling mechanisms in smooth muscle: from the regulation of tone to gene expression. *J. Appl. Physiol.*, (September 2001), Vol.91, No.3, pp.1421-1430.

Liu, S.; Ma, X.; Gong, M.; Shi, L.; Lincoln, T.; Wang, S. (2007). Glucose down-regulation of cGMP-dependent protein kinase I expression in vascular smooth muscle cells involves NAD(P)H oxidase-derived reactive oxygen species. *Free Radic. Biol. Med.*, (March 2007), Vol.42, No.6, pp.852-863.

Lohmann, S.M.; Walter, U. (2005).Tracking functions of cGMP-dependent protein kinases (cGK). *Front. Biosci.*, (May 2005), Vol.10, pp.1313-1328.

Loirand, G.; Guilluy, C.; Pacaud, P. (2006). Regulation of Rho proteins by phosphorylation in the cardiovascular system. *Trends. Cardiovasc. Med.*, (August 2006), Vol.16, No.6, pp.199-204.

Lukowski, R.; Rybalkin, S.D.; Loga, F.; Leiss, V.; Beavo, J.A.; Hofmann, F. (2010). Cardiac hypertrophy is not amplified by deletion of cGMP-dependent protein kinase I in cardiomyocytes. *Proc. Natl. Acad. Sci. USA*, (March 2010), Vol.107, No.12, pp.5646-5651.

Luo, C.; Yi, B.; Chen, Z.; Tang, W.; Chen, Y.; Hu, R.; Liu, Z.; Feng, H.; Zhang, J.H. (2011). PKGIα inhibits the proliferation of cerebral arterial smooth muscle cell induced by oxyhemoglobin after subarachnoid hemorrhage. *Acta. Neurochir. Suppl.*, Vol.110(Pt 1), pp.167-171.

MacMillan-Crow, L.A.; Murphy-Ullrich, J.E.; Lincoln, T.M. (1994). Identification and possible localization of cGMP-dependent protein kinase in bovine aortic endothelial cells. *Biochem. Biophys. Res. Commun.*, (June 1994), Vol.201, No.2, pp.531-537.

Massberg, S.; Gruner, S.; Konrad, I.; Garcia Arguinzonis, M.I.; Eigenthaler, M.; Hemler, K.; Kersting, J.; Schulz, C.; Muller, I.; Besta, F.; Nieswandt, B.; Heinzmann, U.; Walter, U.; Gawaz, M. (2004). Enhanced in vivo platelet adhesion in vasodilator-stimulated phosphoprotein (VASP)-deficient mice. *Blood,* (January 2004), Vol.103, No.1, pp.136–142.

Massberg, S.; Sausbier, M.; Klatt, P.; Bauer, M.; Pfeifer, A.; Siess, W.; Fassler, R.; Ruth, P.; Krombach, F.; Hofmann, F. (1999). Increased adhesion and aggregation of platelets lacking cyclic guanosine 3': 5'- monophosphate kinase I. *J. Exp. Med.*, (April 1999), Vol.189, No.8, pp.1255–1264.

Mazzetti, L.; Ruocco, C.; Giovannelli, L.; Ciuffi, M.; Franchi-Micheli, S.; Marra, F.; Zilletti, L.; Failli, P. (2001). Guanosine 3': 5'-cyclic monophosphate-dependent pathway alterations in ventricular cardiomyocytes of spontaneously hypertensive rats. *Br. J. Pharmacol.*, (October 2001), Vol.134, No.3, pp.596-602.

Melichar, V.O.; Behr-Roussel, D.; Zabel, U.; Uttenthal, L.O.; Rodrigo, J.; Rupin, A.; Verbeuren ,T.J.; Kumar, H.S.A.; Schmidt, H.H. (2004). Reduced cGMP signaling associated with neointimal proliferation and vascular dysfunction in late-stage atherosclerosis. *Proc. Natl. Acad. Sci. USA,* (November 2004), Vol.101, No.47, pp.16671-16676.

Mery, P.F.; Lohmann, S.M.; Walter, U.; Fischmeister, R. (1991). Ca^{2+} current is regulated by cyclic GMP-dependent protein kinase in mammalian cardiac myocytes. *Proc. Natl. Acad. Sci. USA,* (February 1991), Vol.88, No.4, pp.1197–1201.

Michael, S.K.; Surks, H.K.; Wang, Y.; Zhu, Y.; Blanton, R.; Jamnongjit, M.; Aronovitz, M.; Baur, W.; Ohtani, K.; Wilkerson, M.K.; Bonev, A.D.; Nelson, M.T.; Karas, R.H.; Mendelsohn, M.E. (2008). High blood pressure arising from a defect in vascular function. *Proc Natl Acad Sci USA,* (May 2008), Vol.105, No.18, 6702-6707.

Moldobaeva, A.; Welsh-Servinsky, L.E.; Shimoda, L.A.; Stephens, R.S.; Verin, A.D.; Tuder, R.M.; Pearse, D.B. (2006). Role of protein kinase G in barrier-protective effects of cGMP in human pulmonary artery endothelial cells. *Am. J. Physiol. Lung Cell Mol. Physiol.,* (May 2006), Vol.290, No.5, pp.L919–L930.

Mülsch, A.; Oelze, M.; Klöss, S.; Mollnau, H.; Töpfer, A.; Smolenski, A.; Walter, U.; Stasch, J.P.; Warnholtz, A.; Hink, U.; Meinertz, T.; Münzel, T. (2001). Effects of in vivo nitroglycerin treatment on activity and expression of the guanylyl cyclase and cGMP-dependent protein kinase and their downstream target vasodilator-stimulated phosphoprotein in aorta. *Circulation,* (May 2001), Vol.103, No.17, pp.2188-2194.

Münzel, T.; Daiber, A.; Mülsch, A. (2005). Explaining the phenomenon of nitrate tolerance. *Circ. Res.*, (September 2005), Vol.97, No.7, pp.618-628.

Negash, S.; Gao, Y.; Zhou, W.; Liu, J.; Chinta, S.; Raj. J.U. (2007). Regulation of cGMP-dependent protein kinase-mediated vasodilation by hypoxia-induced reactive species in ovine fetal pulmonary veins. *Am. J. Physiol. Lung Cell Mol. Physiol.,* (October 2007), Vol.293, No.4, pp.L1012-L1020.

Payne, M.C.; Zhang, H.Y.; Prosdocimo, T.; Joyce, K.M.; Koga, Y.; Ikebe, M.; Fisher, S.A. (2006). Myosin phosphatase isoform switching in vascular smooth muscle development. *J. Mol. Cell Cardiol.*, (February 2006), Vol.40, No.2, pp.274-282.

Pfeifer, A.; Klatt, P.; Massberg, S.; Ny, L.; Sausbier, M.; Hirneiss, C.; Wang, G.X.; Korth, M.; Aszodi, A.; Andersson, K.E.; Krombach, F.; Mayerhofer, A.; Ruth, P.; Fassler, R.; Hofmann, F. (1998). Defective smooth muscle regulation in cGMP kinase I-deficient mice. *Embo. J.*, (June 1998), Vol.17, No.11, pp.3045-P3051.

Qi, H.; Zheng, X.; Qin, X.; Dou, D.; Xu, H.; Raj, J.U.; Gao, Y. (2007). PKG regulates the basal tension and plays a major role in nitrovasodilator-induced relaxation of porcine coronary veins. *Br. J. Pharmacol.*, (December 2007), Vol.152, No.7, pp.1060-1069.

Qin, X.; Zheng, X.; Qi, H.; Dou, D.; Raj, J.U.; Gao, Y. (2007). cGMP-dependent protein kinase in regulation of basal tone and in nitroglycerin and nitric oxide induced relaxation in porcine coronary artery. *Pflügers Archiv.*, (September 2007), Vol.454, No.6, pp.913-923.

Rentsendorj, O.; Mirzapoiazova, T.; Adyshev, D.; Servinsky, L.E.; Renné, T.; Verin, A.D.; Pearse, D.B. (2008). Role of vasodilator-stimulated phosphoprotein in cGMP-mediated protection of human pulmonary artery endothelial barrier function. *Am. J. Physiol. Lung Cell Mol. Physiol.*, (April 2008), Vol.294, No.4, pp.L686-L697.

Resnik, E.; Herron, J.; Keck, M.; Sukovich, D.; Linden, B.; Cornfield. D.N. (2006). Chronic intrauterine pulmonary hypertension selectively modifies pulmonary artery smooth muscle cell gene expression. *Am. J. Physiol. Lung Cell Mol. Physiol.*, (March 2006), Vol.290, No.3, pp.L426-L433.

Robertson, B.E.; Schubert, R.; Hescheler, J.; Nelson, M.T. (1993). cGMP-dependent protein kinase activates Ca-activated K channels in cerebral artery smooth muscle cells. *Am. J. Physiol.*, (July 1993), Vol.265(1 Pt 1), C299- C303.

Satoh, K.; Fukumoto, Y.; Shimokawa, H. (2011). Rho-kinase: important new therapeutic target in cardiovascular diseases. *Am. J. Physiol. Heart Circ. Physiol.*, (August 2011), Vol.301, No.2, pp.H287-H296.

Schinner, E.; Salb, K.; Schlossmann, J. (2011). Signaling via IRAG is essential for NO/cGMP-dependent inhibition of platelet activation. *Platelets*, Vol.22, No.3, pp.217-227.

Schlossmann, J.; Desch, M. (2009). cGK substrates. *Handb. Exp. Pharmacol.*, Vol.191, pp.163-193.

Schroder, F.; Klein, G.; Fiedler, B.; Bastein, M.; Schnasse, N.; Hillmer, A.; Ames, S.; Gambaryan, S.; Drexler, H.; Walter, U.; Lohmann, S.M.; Wollert, K.C. (2003). Single L-type Ca^{2+} channel regulation by cGMP-dependent protein kinase type I in adult cardiomyocytes from PKG I transgenic mice. *Cardiovasc. Res.*, (November 2003), Vol.60, No.2, pp.268-277.

Schulz, E.; Tsilimingas, N.; Rinze, R.; Reiter, B.; Wendt, M.; Oelze, M.; Woelken-Weckmüller. S.; Walter, U.; Reichenspurner, H.; Meinertz, T.; Münzel, T. (2002). Functional and biochemical analysis of endothelial (dys)function and NO/cGMP signaling in human blood vessels with and without nitroglycerin pretreatment. *Circulation*, (March 2002), Vol.105, No.10, pp.1170-1175.

Schwappacher, R.; Weiske, J.; Heining, E.; Ezerski, V.; Marom, B.; Henis, Y.I.; Huber, O.; Knaus, P. (2009). Novel crosstalk to BMP signalling: cGMP-dependent kinase I modulates BMP receptor and Smad activity. *EMBO. J.* (June 2009), Vol.28, No.11, pp.1537-1550

Somlyo, A.P.; Somlyo, A.V. (2003). Ca^{2+} sensitivity of smooth muscle and nonmuscle myosin II: modulated by G proteins, kinases, and myosin phosphatase. *Physiol. Rev.*, (October 2003), Vol.83, No.4, pp.1325-1358.

Takimoto, E.; Champion, H.C.; Li, M.; Belardi, D.; Ren, S.; Rodriguez, E.R.; Bedja, D.; Gabrielson, K.L.; Wang, Y.; Kass, D.A. (2005). Chronic inhibition of cyclic GMP phosphodiesterase 5A prevents and reverses cardiac hypertrophy. *Nat. Med.*, (February 2005), Vol.11, No.2, pp:214–222.

Thomson, J.R.; Machado, R.D.; Pauciulo, M.W.; Morgan, N.V.; Humbert, M.; Elliott, G.C.; Ward, K.; Yacoub, M.; Mikhail, G.; Rogers, P.; Newman, J.; Wheeler, L.; Higenbottam, T.; Gibbs, J.S.; Egan, J.; Crozier, A.; Peacock, A.; Allcock, R.; Corris, P.; Loyd, J.E.; Trembath, R.C.; Nichols, W.C. (2000). Sporadic primary pulmonary hypertension is associated with germline mutations of the gene encoding BMPR-II, a receptor member of the TGF-beta family. *J. Med. Genet.*, (October 2000), Vol.37, No.10, pp.741-745.

Tokudome, T.; Kishimoto, I.; Horio, T.; Arai, Y.; Schwenke, D.O.; Hino, J.; Okano, I.; Kawano, Y.; Kohno, M.; Miyazato, M.; Nakao, K.; Kangawa, K. (2008). Regulator of G-protein signaling subtype 4 mediates antihypertrophic effect of locally secreted natriuretic peptides in the heart. *Circulation,* (May 2008), Vol.117, No.18, pp.2329-2339

Toshner, M.; Tajsic, T.; Morrell, N.W. (2010).Pulmonary hypertension: advances in pathogenesis and treatment. *Br. Med. Bull.*, Vol.94, pp.21–32.

Vanhoutte, P.M.; Shimokawa, H.; Tang, E.H.; Feletou, M. (2009). Endothelial dysfunction and vascular disease. Acta. Physiol. (Oxf.). (Jun 2009), Vol196, No.2, pp.193-222.

Wagner, C.; Pfeifer, A.; Ruth, P.; Hofmann, F.; Kurtz, A. (1998). Role of cGMP-kinase II in the control of renin secretion and renin expression. *J. Clin. Invest.*, (October 1998),Vol.102, No.8, pp.1576–1582.

Walter U, Gambaryan S. (2009). cGMP and cGMP-dependent protein kinase in platelets and blood cells. *Handb. Exp. Pharmacol.*, Vol.191, pp.533-548.

Wang, S.; Li, Y. (2009). Expression of constitutively active cGMP-dependent protein kinase inhibits glucose-induced vascular smooth muscle cell proliferation. *Am. J. Physiol. Heart Circ. Physiol.*, (December 2009), Vol.297, No.6, pp.H2075-H2083.

Wang, S.; Lincoln, T.M.; Murphy-Ullrich, J.E. (2010). Glucose downregulation of PKG-I protein mediates increased thrombospondin1-dependent TGF-β activity in vascular smooth muscle cells. *Am. J. Physiol. Cell Physiol.*, (May 2010), Vol.298, No.5, pp.C1188-C1197.

Warnholtz, A.; Mollnau, H.; Heitzer, T.; Kontush, A.; Möller-Bertram, T.; Lavall, D.; Giaid, A.; Beisiegel, U.; Marklund, S.L.; Walter, U.; Meinertz, T.; Munzel, T. (2002). Adverse effects of nitroglycerin treatment on endothelial function, vascular nitrotyrosine levels and cGMP-dependent protein kinase activity in hyperlipidemic Watanabe rabbits. *J. Am. Coll. Cardiol.*, (October 2002), Vol.40, No.7, pp.1356-1363.

Weber, S.; Bernhard, D.; Lukowski, R.; Weinmeister, P.; Worner, R.; Wegener, J.W.; Valtcheva. N.; Feil, S.; Schlossmann, J.; Hofmann, F.; Feil, R. (2007). Rescue of cGMP kinase I knockout mice by smooth muscle specific expression of either isozyme. *Circ. Res.*, (November 2007), Vol.101, No.11, pp.1096–1103.

Wegener, J.W.; Nawrath, H.; Wolfsgruber, W.; Kuhbandner, S.; Werner, C.; Hofmann, F.; Feil, R. (2002). cGMP-dependent protein kinase I mediates the negative inotropic effect of cGMP in the murine myocardium. *Circ. Res.*, (January 2002), Vol.90, No.1, pp18–20.

White, R.E.; Kryman, J.P.; El-Mowafy, A.M.; Han, G.; Carrier, G.O. (2000). cAMP-Dependent Vasodilators Cross-Activate the cGMP-Dependent Protein Kinase to Stimulate BKCa Channel Activity in Coronary Artery Smooth Muscle Cells *Circ. Res.*, (April 2000), Vol.86, No.8, pp. 897-905.

Wolfsgruber, W.; Feil, S.; Brummer, S.; Kuppinger, O.; Hofmann, F.; Feil, R. (2003). A proatherogenic role for cGMP-dependent protein kinase in vascular smooth muscle cells. *Proc. Natl. Acad. Sci. USA*, (November 2003), Vol.100, No. 23, pp.13519-13524.

Wollert, K.C.; Yurukova, S.; Kilic, A.; Begrow, F.; Fiedler, B.; Gambaryan, S.; Walter, U.; Lohmann, S.M.; Kuhn, M. (2003). Increased effects of C-type natriuretic peptide on contractility and calcium regulation in murine hearts overexpressing cyclic GMP-dependent protein kinase I. *Br. J. Pharmacol.*, (December 2003) Vol.140, No.7, pp.1227–1236.

Wong, J.C.; Fiscus, R.R. (2010). Protein kinase G activity prevents pathological-level nitric oxide-induced apoptosis and promotes DNA synthesis/cell proliferation in vascular smooth muscle cells. *Cardiovasc. Pathol.*, (November-December 2010), Vol.19, No.6, pp.e221-e231.

Wooldridge, A.A.; MacDonald, J.A.; Erdodi, F.; Ma, C.; Borman, M.A.; Hartshorne, D.J.; Haystead, T.A.J. (2004). Smooth muscle phosphatase is regulated in vivo by exclusion of phosphorylation of threonine 696 of MYPT1 by phosphorylation of serine 695 in response to cyclic nucleotides. *J. Biol. Chem.*, (August 2004), Vol.279, No.33, pp.34496–34504.

Xi, J.; Tian, W.; Zhang, L.; Jin, Y.; Xu, Z. (2010). Morphine prevents the mitochondrial permeability transition pore opening through NO/cGMP/PKG/Zn^{2+}/GSK-3beta signal pathway in cardiomyocytes. *Am. J. Physiol. Heart Circ. Physiol.*, (February 2010), Vol.298, No.2, pp.H601-H607.

Yang ,L.; Liu, G.; Zakharov, S.I.; Bellinger, A.M.; Mongillo, M.; Marx, S.O. (2007). Protein kinase G phosphorylates Ca_v 1.2 α_{1c} and β_2 subunits. *Circ. Res.*, (August 2007), Vol.101, No.5, pp.465-474.

Ying, L.; Xu, X.; Liu, J.; Dou, D.; Yu, X.; Ye L.; He, Q.; Gao, Y. (2011). Heterogeneity in relaxation of different sized porcine coronary arteries to nitrovasodilators: Role of PKG and MYPT1. *Pflügers Archiv. – Eur. J. Physiol.*, (February 2012), Vol. 463, No. 2, pp.257-268.

Yuasa, K.; Michibata, H.; Omori, K.; Yanaka, N. (1999).A novel interaction of cGMP-dependent protein kinase I with troponin T. *J. Biol. Chem.*, (December 1999), Vol.274, No.52, pp.37429-37434.

Zhang, G.; Xiang, B.; Dong, A.; Skoda, R.C.; Daugustherty, A.; Smyth, S.S.; Du, X.; Li, Z. (2011). Biphasic roles for soluble guanylyl cyclase in platelet activation. *Blood*, (September 2011), Vol. 118, No.13, pp.3670-3690.

Zhang, M.; Takimoto, E.; Hsu, S.; Lee, D.I.; Nagayama, T.; Danner, T.; Koitabashi, N.; Barth, A.S.; Bedja, D.; Gabrielson, K.L.; Wang, Y.; Kass, D.A. (2010). Myocardial remodeling is controlled by myocyte-targeted gene regulation of phosphodiesterase type 5. *J. Am. Coll. Cardiol.*, (December 2010), Vol.56, No.24, pp.2021-2030.

Zhang, Q.; Scholz, P.M.; He, Y.; Tse, J.; Weiss, H.R. (2005). Cyclic GMP signaling and regulation of SERCA activity during cardiac myocyte contraction. *Cell Calcium*, (March 2005), Vol.37, No.3, pp.259-266.

Zhang, T.; Zhuang, S.; Casteel, D.E.; Looney, D.J.; Boss, G.R.; Pilz, R.B. (2007). A Cysteine-rich LIM-only Protein Mediates Regulation of Smooth Muscle-specific Gene Expression by cGMP-dependent Protein Kinase. *J. Biol. Chem.,* (November 2007), Vol.282, No.46, pp.33367-33380.

Zhao, Y.Y.; Zhao, Y.D.; Mirza, M.K.; Huang, J.H.; Potula, H.H.; Vogel, S.M.; Brovkovych, V.; Yuan, J.X.; Wharton, J.; Malik, A.B. (2009). Persistent eNOS activation secondary to caveolin-1 deficiency induces pulmonary hypertension in mice and humans through PKG nitration. J Clin Invest., (July 2009),Vol.119, No. 7, pp.2009-2018.

Zhou, W.; Dasgupta, C.; Negash, S.; Raj, J.U. (2007). Modulation of pulmonary vascular smooth muscle cell phenotype in hypoxia: role of cGMP-dependent protein kinase. *Am. J. Physiol. Lung Cell Mol. Physiol.,* (June 2007), Vol.292, No.6, pp.L1459-L1466.

Zhou, W.; Negash, S.; Liu, J.; Raj, J.U. (2009). Modulation of pulmonary vascular smooth muscle cell phenotype in hypoxia: role of cGMP-dependent protein kinase and myocardin. *Am. J. Physiol. Lung Cell Mol. Physiol.,* (May 2009), Vol.296, No.5, pp.L780-L789.

Zhou, X.B.; Ruth, P.; Schlossmann, J.; Hofmann, F.; Korth, M. (1996). Protein phosphatase 2A is essential for the activation of Ca2+-activated K^+ currents by cGMP-dependent protein kinase in tracheal smooth muscle and Chinese hamster ovary cells. J. Biol. Chem., (August 1996), Vol.271, No.33, pp.19760-19767.

16

The Crucial Role of c-Src Tyrosine Kinase in Bone Metabolism

Barbara Peruzzi[1], Nadia Rucci[2] and Anna Teti[2]
[1]*Regenerative Medicine Unit, Ospedale Pediatrico Bambino Gesù. Rome*
[2]*Department of Experimental Medicine, University of L'Aquila*
Italy

1. Introduction

c-Src belongs to the SRC Family of non receptor tyrosine kinases (SFKs), which includes at least ten members (Lyn, Fyn, Lck, Hck, Fgr, Blk, Yrk, Gfr, Yes and c-Src) sharing high homology in their domain structure (Brown & Cooper, 1996).

Due to its proto-oncogene nature, c-Src is the SFK most frequently associated with malignancy (Yeatman, 2004). Over 100 years ago, Peyton Rous observed that injection of cell-free extracts from tumours grown in chickens caused the development of the same type of tumour in host animals. This observation prompted the hypothesis that a filterable agent was the cause of the tumour (Rous, 1911a, 1911b). In support of this notion, in 1955 Rubin showed that the Rous'filterable agent was a virus, called Rous Sarcoma Virus (RSV), which was found to play a direct role in inducing cell malignancy (Rubin, 1955). In the '60s and '70s, the tools of modern molecular biology provided the genetic definition of v-Src, a viral oncogene included within the RSV genome. v-Src was observed not to be required for virus replication but to be the causative agent of cancer (Martin, 1970; Duesberg & Vogt, 1970). Shortly thereafter, it was shown that v-Src had a counterpart in eukaryotic cells, named c-Src (Takeda and Hanafusa, 1983). c-Src is involved in many physiological functions of the cells. It carries a regulatory domain lacking in v-Src (Fig. 1), therefore, its activity is under tight molecular control. c-Src was the first of several proto-oncogenes discovered in the vertebrate genome and, in 1989, this discovery earned Bishop and Varmus the Nobel Prize in Physiology or Medicine, for the description of "the cellular origin of retroviral oncogenes".

In 1977, Brugge and Erikson immunoprecipitated a 60-kDa phosphoprotein from RSV-transformed fibroblasts. The protein was called $pp60^{v\text{-}src}$, but it is now usually referred as v-Src. Both v- and c-Src have tyrosine kinase activity (Collett et al, 1978; Oppermann et al, 1979). c-Src also autophosphorylates itself at tyrosine residues (Hunter & Sefton, 1980), and is the prototype of a large family of kinases that we now know are involved in the regulation of cell growth and differentiation.

v-Src lacks the C-terminal domain that in c-Src has a negative regulatory role on its tyrosine kinase activity. Consequently, v-Src shows constitutive activity and transforming ability (Jove & Hanafusa, 1987). In addition, v-Src contains point mutations throughout its coding region that probably contribute to the high intrinsic activity and transforming potential of the protein.

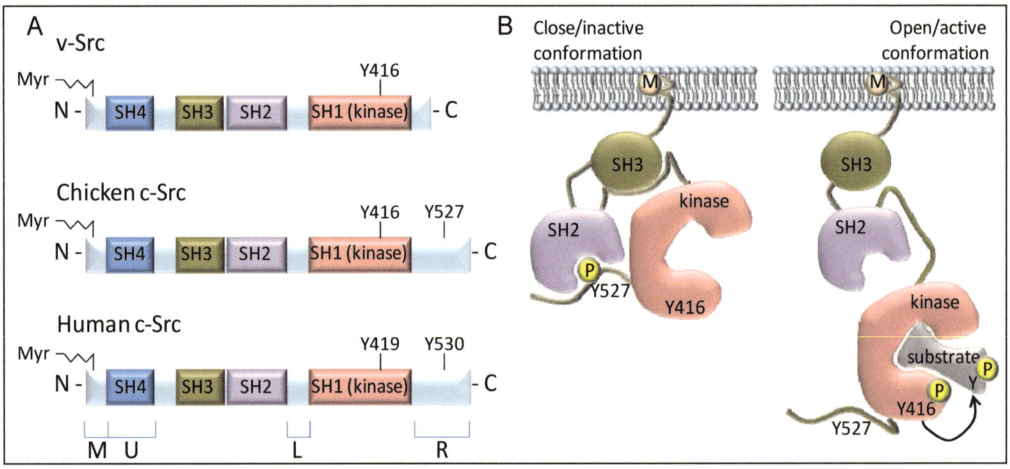

Fig. 1. Structure and activation of Src proteins
A) Comparison of protein structure of viral (v-)Src, chicken and human cellular (c-)Src, with indication of Src homology (SH) and membrane-binding (M), unique (U), linker (L) and regulatory (R) domains. (B) Representation of inactive (left) and active (right) conformation of chicken c-Src.

Aberrant activation of c-Src results in a wide variety of cellular phenotypic changes, including morphological transformation and acquisition of anchorage and growth-factor independence, that are implicated in the development, maintenance, progression, and metastatic spread of several human cancers, such as prostate, lung, breast, and colorectal carcinomas (Irby & Yeatman, 2000). Indeed, a number of human malignancies display increased c-Src expression and activation, confirming its involvement in oncogenesis (Alvarez et al, 2006).

c-Src is ubiquitous and physiologically expressed at high levels in a variety of cell types, including neurons and platelets (Brown & Cooper, 1996). Nonetheless, the very first mouse model of c-Src deficiency (Soriano, 1991) showed an unexpected prominent bone phenotype characterized by increased bone mass and lack of bone resorption, unveiling a previously unrecognized role of c-Src in bone cells (Boyce et al, 1992; Marzia et al, 2000).

In this review, we will describe the structure and function of c-Src and will highlight its crucial physiological and pathogenetic role in bone metabolism.

2. c-Src structure, activation and function

2.1 c-Src structure

c-Src shares with the other SFK members a conserved domain structure consisting of four consecutive Src Homology (SH) domains (Fig. 1A). The N-terminal segment includes the SH4 domain, as well as an "unique" domain of 50-70 residues that display the greatest divergence among the family members (Koegl et al., 1994 ; Resh, 1999). The SH3, SH2 and SH1 (catalytic) domains follow in order in the polypeptide chain. There is also a short, C-terminal "tail" which includes a hallmark of Src kinases, that is an autoinhibitory phosphorylation site [Tyrosine (Tyr) 527 in chicken, Tyr 530 in human] (Cooper & King, 1986). This tail is not present in the v-Src isoform (Fig.1A).

The SH4 domain is a 15-amino acid sequence whose myristoylation allows binding of SFK members to the inner surface of the plasma membrane. The unique domain has been proposed to be important for mediating interactions with receptors or proteins that are specific for each family member. Serine and threonine phosphorylation sites have also been identified in the unique domains of c-Src and Lck (Winkler et al, 1993).

SH3 and SH2 are protein-binding domains widely present in other molecules, such as lipid kinases, protein and lipid phosphatases, cytoskeletal proteins, adaptor molecules and transcription factors (Mayer & Baltimore, 1993). The SH3 domain consists of small, β-barrel modules and is important for intra- as well as inter-molecular interactions, regulating c-Src catalytic activity, localization and recruitment of substrates. Proline-rich sequences in target molecules mediate the interactions with SH3 (Ren et al, 1993). The other domain regulating c-Src interaction with proteins is SH2, which preferentially binds to polypeptide segments containing a phosphotyrosine (Mayer et al, 1991; Pawson, 1995).

The catalytic domain (SH1) is the most conserved domain in all tyrosine kinases. It contains an ATP-binding pocket and the tyrosine-specific protein kinase activity. As it will be described in the next paragraph, the first step of c-Src activation is the autophosphorylation of Tyr416 (in chicken, Tyr419 in human), while phosphorylation of Tyr527 by c-Src kinase (CSK) and CSK homologous kinase (CHK) results in its inhibition (Kmiecik & Shalloway, 1987; Cartwright et al, 1987; Piwnica-Worms et al, 1987; Okada & Nakagawa, 1989).

2.2 c-Src activation

c-Src is normally maintained in an inactive or "closed" conformation, where the SH2 domain is engaged with the phosphorylated Tyr527, the SH3 domain binds the SH2-kinase linker sequence and the Tyr416 is dephosphorylated. Dephosphorylation of Tyr527 disrupts its intramolecular interaction with the SH2 domain and this open conformational state allows autophosphorylation of Tyr416, resulting in c-Src activation (Fig. 1B) (Yamaguchi & Hendrickson, 1996).

Phosphorylation of Tyr527 can be removed by several protein phosphatases that function as activators of c-Src, such as protein tyrosine phosphatase-α (PTPα) (Zheng et al, 1992), PTP1, SH2-containing phosphatase 1 (SHP1) and SHP2 (Jung & Kim, 2002). The most direct evidence for a role of c-Src activation in cancer among these phosphatases is for PTP1B, which is present at high levels in breast cancer cell lines (Jung & Kim, 2002). In addition, the direct binding of focal-adhesion kinase (FAK) (Schaller et al, 1994) or its molecular partner CRK-associated substrate (CAS, also known as p130CAS) to the SH2 and the SH3 domains of c-Src also results in the open, active configuration of c-Src, since the intramolecular interactions that maintain the closed configuration are displaced (Thomas et al, 1998).

2.3 c-Src functions

c-Src plays a key role in regulating the assembly and disassembly of cell-cell (adherens junctions) and cell-matrix (focal adhesions) adhesion (Yeatman, 2004) (Fig.2). Adherens junctions are maintained by homotypic interactions between E-cadherin molecules present on neighboring cells, and loss of E-cadherin is a key event in the epithelial-to-mesenchymal transition of cancer cells. It has been shown that increased c-Src signalling correlates with decreased E-cadherin expression and decreased cell-cell adhesion (Irby and Yeatman, 2002; Nam et al, 2002). Moreover, constitutively active c-Src can phosphorylate the cadherins, resulting in loss of the cadherin-catenin complex function, thereby promoting cell invasiveness (Irby & Yeatman, 2002; Behrens et al, 1989) (Fig.2).

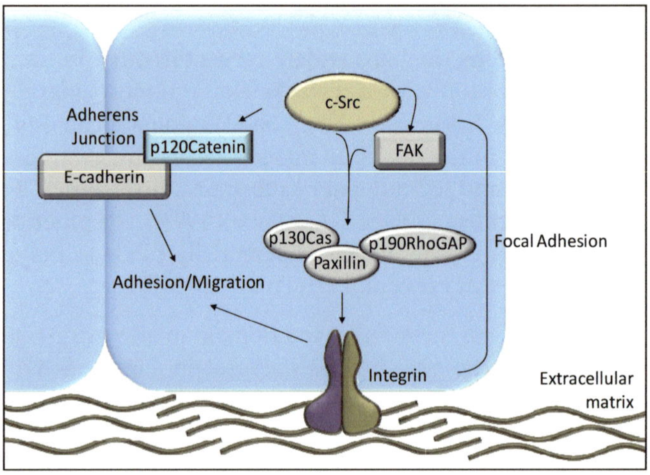

Fig. 2. c-Src involvement in cell adhesion signals.
Molecular interactions among c-Src and the components of both adherens junction and focal adhesion structure.

At the cell periphery, activated c-Src forms complexes with FAK, which in turn interacts with a multitude of substrates, including CAS, paxillin, and p190RhoGAP, that play critical roles in promoting actin remodelling and cell migration (Fig. 2) (Guarino, 2010; Playford & Schaller, 2004). In cancer, deregulated focal adhesion signalling has been implicated in increased invasion and metastasis, and decreased patient survival (McLean et al, 2005). c-Src can also be activated downstream of tyrosine kinase growth factor receptors, such as epidermal growth factor (EGF) (Tice et al, 1999), platelet-derived growth factor (PDGF) (De Mali et al, 1999; Bowman et al, 2001), insulin-like growth factor (IGF)-1 (Arbet-Engels et al, 1999), fibroblast growth factor (FGF) (Landgren et al, 1995), colony-stimulating factor (CSF)-1 (Courtneidge et al, 1993) and hepatocyte growth factor (HGF) receptors (Mao et al, 1997) (Fig. 3). Ligand binding to receptor tyrosine kinases leads to receptor dimerization, kinase activation, and autophosphorylation of tyrosine residues. These phosphorylated tyrosines then serve as docking sites for the SH2 domains of several signalling molecules, including c-Src (van der Geer et al, 1994). For instance, the EGF receptor can bind to c-Src and phosphorylate tyrosine sites on its C-terminal loop. Conversely, c-Src can directly bind to the EGF receptor and phosphorylate the Y845 residue, resulting in increased Ras/ERK/MAPK activity and enhanced cell mitogenesis and transformation (Biscardi et al, 2000).

c-Src has also been implicated in signalling activated by integrins and G-protein coupled receptors (GPCRs). Indeed, clustering of integrins can lead to downstream signalling pathways inducing activation of c-Src, FAK, Abl, and Syk (Miyamoto et al, 1995, Schlaepfer & Hunter, 1998). There is an increasing body of evidence for synergy between receptor tyrosine kinases and integrins, demonstrated by an increase in MAPK activation in response to various growth factors if integrins are preclustered (Miyamoto et al, 1996). The crosstalk between these pathways could be mediated by a common signalling molecule, including c-Src. A FAK-independent signalling pathway from integrins has also been described, in which caveolins act as adaptors, linking integrins and c-Src family kinases. Indeed, Wei et al. (1999) showed that caveolin is important for the association between β1 integrin and c-Src, and disruption of this interaction affected focal adhesions. On the other hand, c-Src can suppress the integrins attached to the extracellular matrix via phosphorylation of integrin subunits (Sakai et al, 2001; Datta et al, 2002). c-Src can also interrupt Rho-A function, which has an important role in actin filament assembly and stabilization of focal adhesions (Arthur & Burridge, 2001). c-Src activates FAK, Ras and phosphatidylinositol phosphate kinase, which indirectly affect integrin–actin cytoskeleton assembly (Brunton et al, 2004).

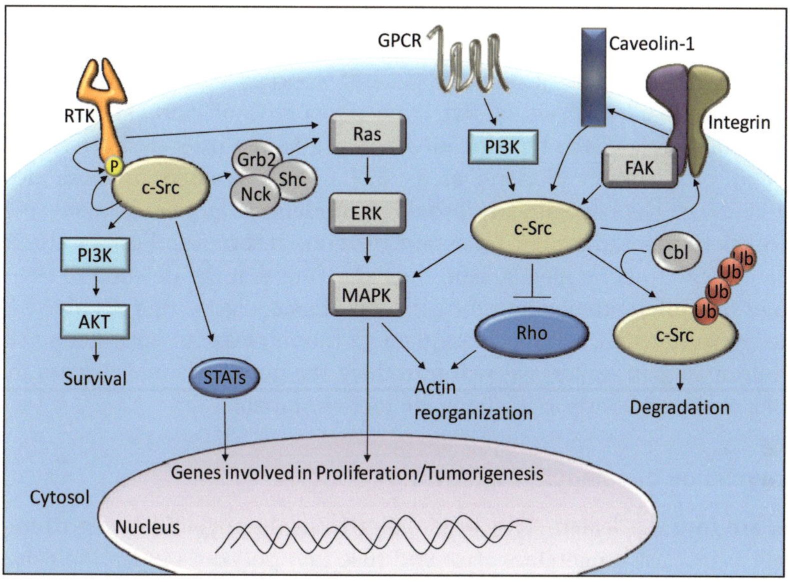

Fig. 3. c-Src-activating signals.
Extracellular signals involving Receptor Tyrosine Kinases (RTK), or G-Protein Coupled Receptors (GPCR) or integrins are shown to activate c-Src and its downstream pathways.

The localization of c-Src at the membrane–cytoskeletal interface in focal adhesions, lamellipodia and filopodia seems to be regulated by the small G-proteins RhoA, Rac1 and CdC42 (Timpson et al, 2001). The Cbl ubiquitin ligase has been shown to be important in suppressing v-Src transformation through ubiquitin-dependent protein degradation (Kim et al, 2004). Recent evidence indicates that the ubiquitin–proteasome pathway is deregulated in cancer cells, which might allow c-Src activation (Kamei et al, 2000).

Finally, there is also evidence that c-Src is activated through nitric-oxide signalling (Akhand et al, 1999) in mechanisms mainly implicated in cellular adhesion and motility (Gianni et al, 2010; Giannoni et al, 2005).

3. c-Src in the bone metabolism

Understanding the physiological role of SFKs has been aided by the advent of gene targeting and embryonic stem cell technology in the mouse. Targeted disruptions of all known mammalian SFK genes have been obtained in mice, with their phenotypes ranging from no overt defects to very distinct abnormalities in specific cell types and tissues. As stated above, the first SFK member to be disrupted was c-Src (Soriano et al, 1991). Although c-Src is ubiquitously expressed, mice lacking the *src* gene presented with overt alterations only in the bone tissue, suggesting a crucial role of this tyrosine kinase in the bone microenvironment that cannot be replaced by other SFK members, as probably occurs in the rest of the body.

The main phenotype associated with c-Src deletion is osteopetrosis, a bone remodelling disease in which excess of bone accumulates as a result of defective osteoclast bone resorption (Soriano et al, 1991). This mutation manifests itself by the failure of incisors to erupt, and the mutants have a much reduced survival rate after weaning. However, animals maintained on a soft food diet have been found to survive for at least a year and, on rare occasions, can breed, although some alterations in reproduction have been documented (Roby et al, 2005). In contrast with a general concept of c-Src involvement in cell proliferation, a detailed analysis of the bone phenotype of c-Src knock-out (KO) mice revealed the crucial role of the tyrosine kinase in regulating osteoclast activity, rather than formation and proliferation (Soriano et al, 1991). As discussed in more detail below, substantial evidence has been already provided in identifying c-Src as a key player in the correct cytoskeletal rearrangement necessary for bone resorption. Further studies pointed out the role of c-Src in bone metabolism, thus showing that the deletion of Src expression also enhances the differentiation and the function of osteoblasts, the cells of the bone tissue having osteogenic function, with a consequent further increase of bone mass (Marzia et al, 2000). Therefore, in this section we will introduce the bone cells and discuss in detail the multiple roles that c-Src exerts in the bone microenvironment.

3.1 c-Src regulation of osteoclast behaviour

Osteoclasts are multinucleated cells, originating from the myeloid tissue from which the mononuclear osteoclast progenitors arise and fuse into polykaria when their maturation is completed (Baron & Horne, 2005). They are terminally differentiated cells that resorb the mineralized matrix during physiological and pathological bone turnover by a peculiar extracellular mechanism involving specific domains of the plasma membrane. Indeed, during the bone resorbing process, the osteoclast is markedly polarized, in order to create three morphologically distinct areas of the plasma membrane: the basolateral membrane, which is not in contact with the bone; the tight sealing zone, which is closely apposed to the bone surface; and the ruffled border, a highly convoluted membrane that faces the resorbing surface (Baron & Horne, 2005).

The characterization of the phenotype of the c-Src KO mouse revealed that the most critical role of c-Src is related to osteoclast function rather than differentiation, since the number of osteoclasts in bones of c-Src KO mice is more than twice that in normal mice (Boyce et al, 1992). While osteoclasts also express other c-Src family kinases (Lowell et al, 1996), the deletion of any one of the genes encoding Fyn, Yes, Hck, and Fgr fails to produce osteopetrosis (Stein et al, 1994). Moreover, re-expression of c-Src in c-Src KO osteoclasts restores *in vitro* the bone-resorbing activity (Miyazaki et al, 2004), implying that c-Src performs some specific functions in osteoclasts that cannot be compensated by these other SFKs. A possible exception is Hck, since its expression is upregulated in c-Src KO osteoclasts and c-Src KO/Hck KO double-mutant mice are significantly more osteopetrotic than the c-Src KO animals (Lowell et al, 1996). At the cellular level, c-Src KO osteoclasts present with a critical feature, that is the absence of the ruffled border (Boyce et al, 1992), suggesting a c-Src contribution to the regulation of exocytic and/or endocytic vesicle trafficking, as well as to the attachment and motility mediated by the adhesion structures.

Osteoclastic bone resorption involves a series of regulatory phases: migration of osteoclasts to the resorption site, their attachment to the calcified tissue and development of the ruffled border and the clear zone, followed by the secretion of acids and lysosomal enzymes into the space beneath the ruffled border (reviewed in Peruzzi & Teti, 2011). The formation of the sealing zone is essential for the osteoclastic bone resorption, since it forms a diffusion barrier and permits the directional secretion of lysosomal enzymes into the space beneath the ruffled border. In the ruffled border membrane, the vacuolar-type proton ATPase mediates the transport of protons into the resorption lacunae. Lysosomal enzymes of osteoclasts, such as cathepsin K, and metalloproteinase-9 are also secreted through this membrane and degrade the organic matrix of bone. To organize these highly polarized cellular structures, osteoclasts must adhere to the bone surface as the initial and essential phase for their activity (Coxon & Taylor, 2008), which involves the interaction of integrins with the extracellular matrix proteins within the bone. Among several integrins, osteoclasts express very high levels of $\alpha V\beta 3$ integrin, and lower levels of the collagen/laminin receptor $\alpha 2\beta 1$ and the vitronectin/fibronectin receptor $\alpha V\beta 1$ (Nakamura et al, 2007; Horton, 1997; Horton & Rodan, 1996).

Like all members of the αV integrin family, the $\alpha V\beta 3$ receptor recognizes the RGD (Arg-Gly-Asp) adhesion motif present in several matrix proteins such as vitronectin, bone sialoprotein II and osteopontin (Rupp & Little, 2001; Wilder, 2002; Horton, 1997). This interaction induces an integrin conformational change leading to the so-called outside-in signalling, which in turn triggers a number of intracellular events, including changes in cytosolic calcium, protein tyrosine phosphorylation and cytoskeletal remodelling (Duong & Rodan, 2000; Teitelbaum, 2007; Faccio et al, 2003). The engagement of the matrix by the $\alpha V\beta 3$ integrin in osteoclasts and osteoclast precursors activates the non-receptor tyrosine kinase Pyk2, a member of the FAK family, by a mechanism that involves an increase in cytosolic Ca^{2+} and the binding of Pyk2 to the cytoplasmic domain of the β subunit (Fig.4) (Faccio et al, 2003; Duong & Rodan, 2000).

Both the capacity of c-Src to bind the $\alpha V\beta 3$ integrin and the subsequent activation of the kinase are mediated by Pyk2, which mobilizes c-Src to the integrin. $\alpha V\beta 3$ integrin occupancy induces phosphorylation of Pyk2, which then binds the SH2 domain of c-Src. The proposed association between phosphorylated Pyk2 and c-Src would prevent c-Src-Y527 inactivating phosphorylation, thus relieving auto-inhibition of kinase function.

The signalling downstream of c-Src activation involves tyrosine phosphorylation of a distinct set of proteins, including Pyk2 itself, Cbl, PI3K, paxillin, cortactin, vinculin, talin, tensin, and p130Cas, which are present in the osteoclast adhesion structures, called podosomes (Thomas & Brugge, 1997; Linder & Aepfelbacher, 2003; Buccione et al, 2004).

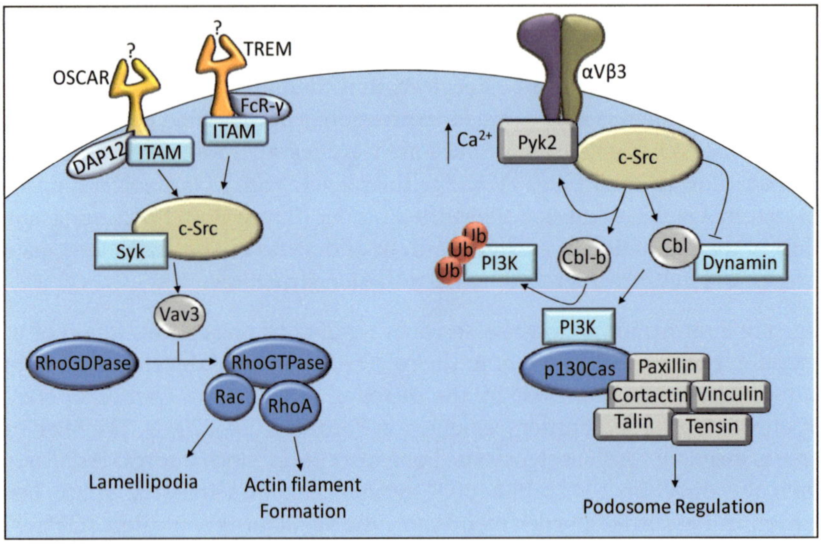

Fig. 4. c-Src involvement in osteoclast function.
Among several pathways regulating osteoclast activity, the cartoon shows c-Src activation and downstream effect depending on receptor signals.

Podosomes, which serve as attachment structures in osteoclasts and other highly motile cells, are more transient and dynamic than focal adhesion plaques (Destaing et al, 2003). As originally described (Marchisio et al, 1984; Marchisio et al, 1987), podosomes are small punctate structures with an F-actin-rich core surrounded by a ring of integrins and certain focal adhesion-associated proteins (e.g., paxillin, talin and vinculin). Cortactin, gelsolin, the actin-regulatory proteins Neuronal Wiskott–Aldrich Syndrome Protein (N-WASP), and Arp2/3 have also been identified in the podosome core (Linder & Aepfelbacher, 2003; Buccione et al, 2004). In osteoclasts, Src, Cbl, Pyk2, and various actin-associated proteins, including dynamin and filamin, are also associated with podosomes, whose rapid turnover (within minutes) is probably essential for the high mobility of the cells in which they occur (Fig. 4) (Destaing et al, 2003).

Among the substrates of c-Src activity involved in the mechanisms of bone resorption, Cbl plays a key role in promoting turnover or disassembly of podosomes (Sanjay et al, 2001). Indeed, c-Src, in association with Pyk2, recruits Cbl through its SH3 domain and promotes its activation by phosphorylation. Once activated, Cbl recruits PI3K and dynamin to the adhesion complex.

Since Cbl is an ubiquitin E3 ligase, it has been described to drive the negative feedback that has the potential to promote proteasomal degradation of the integrin-associated Pyk2/c-Src/Cbl complex (Fig. 4) (Yokouchi et al, 2001). Thus, Cbl is crucial in the integrin-mediated "inside-out" signalling, playing a key role in podosome detachment and subsequent disassembly. In this way, the c-Src/Pyk2/Cbl complex forms the basis for the cyclic attachment–detachment of single adhesion sites at the leading edge of lamellipodia in motile cells, and thereby participates in the assembly–disassembly of individual podosomes, ensuring cell adhesion while still allowing cell motility (Sanjay et al, 2001).

In the integrin-mediated outside-in signalling, also the non-receptor tyrosine kinase Syk plays a pivotal role in osteoclast activity, since it associates with the β3 integrin subunit domain in a region close to the c-Src binding site and is activated by c-Src itself, a key event in organizing the cytoskeleton (Zou W et al., 2007). Recently, it has been shown that c-Src and Syk are also involved in the signal downstream the immunoreceptor tyrosine-based activation motif (ITAM)-bearing co-receptors, DAP12 and FcRγ (Fig.4). DAP12 and FcRγ are associated with the immunoreceptors OSCAR/PIR-A and TREM2/SIRPβ1, respectively, recently identified on the osteoclast surface (Mócsai et al, 2004). The c-Src-mediated phosphorylation of Syk kinase leads to activation of a number of cytoskeleton-regulating proteins, including the Vav family of guanine nucleotide exchange factor (GEFs). These proteins convert Rho GTPases from their inactive GDP to their active GTP conformation. Among these proteins, Vav3 is expressed in osteoclasts (Faccio et al, 2005), where it is triggered upon matrix-induced Syk activation and regulates RhoGTPase-dependent effect on actin cytoskeleton (Zou et al, 2007). In this context, the small GTPases Rac and Rho exert distinctive effects on osteoclasts. Indeed, Rac stimulation in osteoclast precursors prompts the appearance of lamellipodia, thus forming the migratory front of the cell, while RhoA stimulates actin filament formation which, in osteoclasts, allow organization of the sealing zone (Fig.4) (Fukuda et al, 2005).

The integrin-activated c-Src signalling also functions in other processes necessary for normal osteoclast function, among which adenosine triphosphate (ATP)-dependent events, especially those involved in cell motility, proton secretion, and the maintenance of electrochemical homeostasis (Baron, 1993). Indeed, c-Src promotes the maintenance of energy stores in osteoclasts by phosphorylating cytochrome c oxidase within the mitochondria (Miyazaki et al, 2003), which are very abundant in osteoclasts, consistent with the energy requirements of their activity (Miyazaki et al, 2006).

3.2 c-Src regulation of osteoblast differentiation

Osteoblasts are mononucleated cells of mesenchymal origin that synthesize and mineralize the bone matrix during bone accrual and remodelling events. Bone formation involves osteoblast maturation that requires a spectrum of signalling proteins including morphogens, hormones, growth factors, cytokines, matrix proteins, transcription factors, and their co-regulatory proteins. They act coordinately to support the temporal expression (i.e., sequential activation, suppression, and modulation) of other genes that represent the phenotypic and functional properties of osteoblasts during the differentiation process from osteoblast precursors (Jiang et al, 2002). Pre-osteoblasts are also responsible of the production of cytokines regulating osteoclastogenesis, that is receptor activator of nuclear factor kappa B ligand (RANKL), osteoprotegerin (OPG) and CSF-1, thereby coupling osteoblast and osteoclast function. Given the osteoblast-mediated regulation of osteoclast differentiation and bone resorption (Rodan & Martin, 1981; Suda et al, 1997) and the bone phenotype resulted by c-Src distruption (Soriano et al, 1991), several studies aimed at investigating the involvement of osteoblasts in c-Src KO phenotype have been performed. In 1993, Lowe et al. demonstrated that osteoblasts derived from these mice successfully contributed to normal osteoclast differentiation and showed unremarkable morphological features relative to wild-type (WT) mice, suggesting that the inherited defect is independent of the bone marrow microenvironment (Lowe et al, 1993). The first evidence of an osteoblast involvement in c-Src KO mouse bone phenotype derived from Marzia and coworkers, who performed a detailed molecular analysis of the c-Src null osteoblasts (Marzia et al, 2000). This study clearly demonstrated that a decreased c-Src activity is responsible of enhanced osteoblast differentiation and *in vivo* bone formation, thereby highlighting the role of c-Src in maintaining osteoblasts in a poorly differentiated status.

Bone formation requires transcriptional mechanisms for sequential induction and repression of genes that support progressive osteoblast phenotype development. The Runx transcription factors and their co-regulators control cell differentiation and lineage commitment (Westendorf & Hiebert, 1999) by influencing the functional architecture of target gene promoters (Stein et al, 2000). Runx proteins are directed to subnuclear domains through the C-terminal nuclear matrix-targeting signal (NMTS) and interact with the DNA through the N-terminal runt homology domain (Zaidi et al, 2001). The Runx2 family member is essential for osteoblast maturation *in vivo* and its alteration is associated with the cleidocranial dysplasia (Komori et al, 1997). Runx2 is a target of several extracellular signals that regulate skeletal formation and homeostasis. The C-terminus of Runx2, which includes the NMTS, interacts with proteins involved in the transforming growth factor β/bone morphogenetic proteins (TGFβ/BMPs) (i.e., Smads), the transducin-like enhancer (TLE)/groucho and the c-Src/Yes tyrosine kinase (e.g., the Yes-associated protein, YAP) signalling pathways (Hanai et al, 1999; Yagi et al, 1999). Indeed, in response to c-Src/Yes signalling, YAP is phosphorylated and recruited by Runx2 to subnuclear sites of Runx2 target genes, resulting in their repression (Fig.5). Thus, c-Src controls osteoblast differentiation by regulation of Runx2–YAP interaction.

Another mechanism by which c-Src regulates osteoblast differentiation involves estrogens, which are known to control a variety of tissues, including the bone (Hall et al, 2001). Indeed, estrogen deficiency leads to accelerated bone loss which is the primary cause of postmenopausal osteoporosis (Manolagas et al, 2002). The estrogen receptors (ERs) belong to the nuclear receptor superfamily, acting as ligand-inducible transcription factors (Hall et al, 2001).

Indeed, ER expression is regulated by a c-Src/PKC-dependent mechanism involving osteoblast differentiation, with an increased responsiveness to estrogens in mature osteoblasts (Fig.5) (Longo et al., 2004). Estrogens are also responsible for an anti-apoptotic effect on osteoblasts, which is due to a rapid and sequential phosphorylation of the c-Src, Shc and ERK1/2 kinases. The c-Src/Shc/ERK signalling cascade rapidly phosphorylates the transcription factors Elk1, CREB and C/EBPβ with a mechanism that is retained when the receptor is localized outside the nucleus (Kousteni et al, 2003).

Beside the estrogen-mediated effect, other extracellular stimuli, such as mitogens and changes in the mechanical stress, are responsible of c-Src activation and of the downstream cascade involving the MAPK signalling. In this circumstance, the transcriptional target is the AP-1 complex, a heterodimer composed of members of the c-Fos, c-Jun, and activating transcription factor (ATF) families (Hess et al, 2004), which is an important regulator of bone development and homeostasis (Wagner & Eferl, 2005).

Of special interest in the regulation of c-Src activation and activity is the role of caveolae. They are small bulb-shaped invaginations located close to the cell surface representing specialized domains of the plasma membrane (Severs, 1988). Caveolin, a 21–24-kDa integral membrane protein, is a major structural and regulatory component of caveolae membranes (Rothberg et al., 1992). Several data suggest that caveolin may act as a scaffolding protein

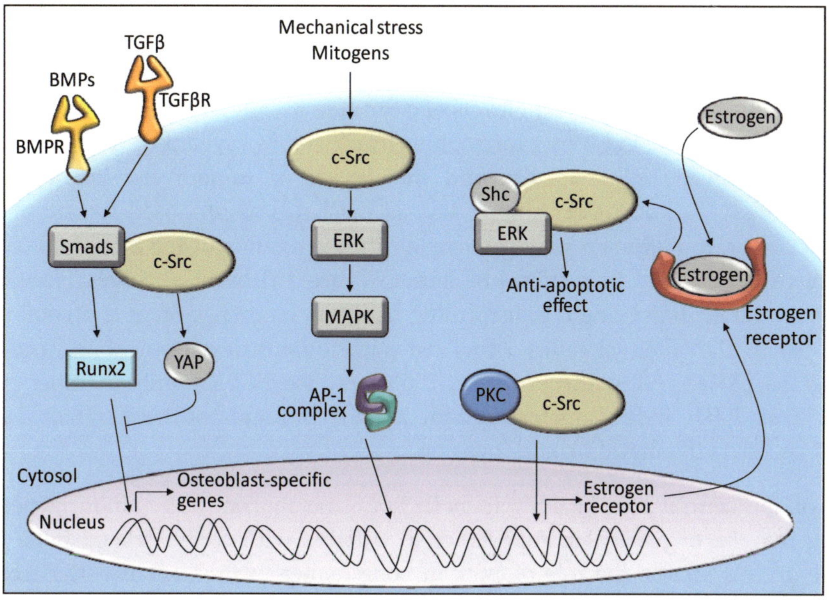

Fig. 5. c-Src involvement in osteoblast differentiation.
TGFβ/BMP signals, mechanical stress, mitogens and estrogen modulate c-Src activity and osteoblast differentiation.

within caveolae membranes, since both the N-terminal and C-terminal domains of caveolin face the cytoplasm (Dupree et al., 1993). Caveolin interacts with cytoplasmic signalling molecules including trimeric G proteins, Src family tyrosine kinases, and Ras-related GTPases. Thus, caveolin may serve as an oligomeric docking site for organizing and concentrating inactive signalling molecules within the caveolae membranes (Sargiacomo et al, 1995). Modification and/or inactivation of caveolin may be a common feature of the transformed phenotype. Caveolin can be phosphorylated by c-Src at Tyr14, an event that induces caveolar internalization by reorganizing the actin cytoskeleton (mediated by dynamin and PKC) (Mayor & Pagano, 2007).

On the other hand, caveolin has a role in regulating c-Src activation, since its interaction with c-Src, as well as with the other components of the SFKs, leads to the inhibition of auto-phosphorylation of these kinases, thus holding these molecules in the inactive conformation (Li et al, 1996).

Recently, we demonstrated that in osteoblasts c-Src regulates interleukin (IL)-6 and insulin-like growth factor binding protein (IGFBP)-5 expression (Peruzzi et al, 2012). More in details, c-Src controls IL-6 expression acting on STAT3, which is a downstream component of the IL-6 pathway and a transcription factor for IL-6 itself. At the same time, IL-6 stimulates the expression of IGFBP5 which, in turn, acts in an autocrine manner on osteoblasts inducing the c-Src activating-phosphorylation and inhibiting further osteoblast differentiation. On the other hand, in mature osteoblasts c-Src is barely expressed and therefore this loop is inactive, although IGFBP5 is still expressed under the control of Runx2. In this context, IGFBP5 has been observed to enhance osteoclast formation and bone resorption, thus unveiling its new role in the coupling between osteoblast and osteoclast activities (Peruzzi et al, 2012).

4. c-Src in cancer-related bone diseases

When deregulated, c-Src has the potential to participate in cancer pathogenesis and progression (Rucci et al, 2008). High levels of c-Src activity are found in numerous epithelial cancers, including colon and breast carcinomas (reviewed in Tsygankov & Shore, 2004), and a correlation between c-Src activity and the degree of tumour malignancy has been described (Aligayer et al, 2002). Among several mechanisms proposed for increased c-Src activity in cancer, mutations resulting in constitutive activation of the kinase domain of the protein have been found in a subset of human cancers (Irby et al, 1999). However, it is thought that rather than being a transforming agent on its own, c-Src is a pivotal regulator of a number of signalling cascades associated with tumour development and progression, including the JAK/STAT pathway (Silva, 2004) and the EGFR family member pathways (Biscardi et al, 2000). Indeed, increased c-Src activity upregulates these signals, leading to increased cell growth, migration and invasion.

Because of the central role of c-Src in both bone metabolism and tumourigenesis, c-Src signalling may be of particular importance in patients with cancer-related bone diseases. Therefore, in this section the role of c-Src in bone malignancies such as osteosarcoma and bone metastases will be elucidated.

Osteosarcoma is the most common and most often fatal primary malignant bone tumour, especially affecting children, adolescents and young adults (Raymond et al, 2002). It is a highly aggressive tumour that metastasizes primarily to the lung, with a consequent very poor prognosis (Shanley & Mulligan, 1991). Osteosarcoma arises from primitive transformed cells that exhibit osteoblastic features and produce malignant osteoid (Kansara & Thomas, 2007). Among the factors involved in the progression of this pathology, c-Src has been described as a mediator of tumour cell invasive features. Caveolin-1 downregulation has been proposed to function as a permissive mechanism by which c-Src signalling is activated, enabling osteosarcoma cells to become metastatic and invade the neighboring tissues (Cantiani et al, 2007).

Bone represents the principal site of relapse of many cancers, especially prostate and breast carcinomas. In the latter, skeletal metastases are osteolytic in nature, with a resulting bone destruction mediated by an exaggerated osteoclast activity. Tumour cells colonizing the bone produce factors that stimulate osteoclast formation, with a consequent increase of bone resorption. This leads to the release of growth factors normally stored in the bone matrix, which in turn stimulate tumour cell proliferation and survival. This mutual enhancement of tumour cell and osteoclast activities is termed "vicious cycle", which progressively increases both bone destruction and tumour burder (Weilbaecher et al, 2011). c-Src plays a role in the context of bone metastases. In fact, MDA-MB-231 breast cancer cells overexpressing a kinase dead dominant negative c-Src were found to form far less osteolytic metastases in immunocompromised mice than their wild-type counterpart (Myuoi et al, 2003; Rucci et al, 2006). Recent data demonstrate that c-Src is crucial for breast cancer cell growth in the bone marrow since it mediates AKT regulation and tumour cell survival in response to chemokine (C-X-C motif) ligand 12 CXCL12, thus confering resistance to the pro-apoptotic member of the tumour necrosis factor family, TRAIL (Zhang et al, 2009).

5. c-Src as pharmacological target in bone metastases

Several compounds have been tested in pre-clinical studies *in vitro* and *in vivo* to inhibit c-Src activity and reduce the incidence of metastases. In an model of experimentally induced breast cancer metastases, the pyrrolo-[2,3-d]pyrimidine c-Src inhibitor, CGP76030, proved effective at inhibiting the incidence of relapse in bone and visceral organs, thus improving mice survival. CGP76030 was observed to have potent anti-osteoclastic and anti-tumoral effects (Recchia et al, 2004; Rucci et al, 2006), explaining its *in vivo* effects on both fronts.

The efficacy of c-Src inhibitors in the treatment of breast cancer-induced bone metastases was strengthened by recent data showing that c-Src is a key signalling molecule for the growth of antiestrogen-resistant tumour cells. Indeed, it has been demonstrated that the expression of a constitutively active c-Src attenuated the sensitivity of MCF7 breast cancer cells to tamoxifen (Morgan et al, 2009). In the same cell line the synergistic interaction between ER, EGF receptor and c-Src enhanced tumour cell responsiveness to mitogenic stimuli, allowing their survival also in the presence of tamoxifen (Yue et al, 2007).

Another compound employed in preclinical studies is Dasatinib, a potent inhibitor of SFKs, which proved effective at inhibiting breast cancer cell growth, with a synergic effect when used in combination with chemotherapeutics (Nautiyal et al, 2009), while in mice inoculated with a triple-negative (for ER, progesterone receptor and Her2/neu expression) breast cancer cell line it prevented the formation of osteolytic metastases (Zhang et al, 2009).

Based on this body of evidence, clinical studies have recently been started using sarcatinib, which is a dual c-Src/Abl inhibitor. In a phase I study performed on patients with solid tumours, sarcatinib decreased the levels of bone resorption markers (Baselga et al, 2010). Sarcatinib is currently being tested in phase II clinical studies in patients with breast cancer bone metastases. Another dual c-Src/Abl inhibitor in clinical development with similar properties is Bosutinib.

6. Conclusions

c-Src is a non receptor tyrosine kinase ubiquitously expressed and involved in the regulation of many cellular functions, such as adhesion, growth, migration and survival. Indeed, its activity increases in response to several signals, especially downstream of growth factor and cytokine receptors, integrin receptors and G-coupled receptors. The proto-oncogene nature of c-Src is also well known, so that its aberrant activation is involved in the development and progression of many human cancers.

c-Src plays an unique role in bone metabolism, regulating the activity of both the bone-resorbing osteoclasts and the bone-forming osteoblasts. The interest on c-Src within the bone metabolism derived from the pioneer work of Soriano et al. (1991), who demonstrated that, despite the ubiquitous expression of c-Src, knock-out mice showed only a bone phenotype. This molecule is currently thought to be a suitable pharmacological target for bone metastases. Clinical studies are currently in progress to establish the efficacy of c-Src inhibitors as therapeutic agents in humans.

7. Acknowledgements

The original work was supported by a grant from the "Associazione Italiana per la Ricerca sul Cancro" (AIRC) to AT, and a grant from the European Calcified Tissue Society to BP. BP is recipient of the "Helga Baden" Fellowship provided by the "Fondazione Italiana per la Ricerca sul Cancro".

8. References

Akhand, AA., Pu, M., Senga, T., Kato, M., Suzuki, H., Miyata, T., Hamaguchi, M. & Nakashima, I. (1999). Nitric oxide controls src kinase activity through a sulfhydryl group modification-mediated Tyr-527- independent and Tyr-416-linked mechanism, *Journal of Biological Chemistry,* Vol.274, No.36, (September 1999), pp. 25821–25826, ISSN 0021-9258

Aligayer, H., Boyd, DD., Heiss, MM., Abdalla, EK., Curley, SA. & Gallick, GE. (2002). Activation of Src kinase in primary colorectal carcinoma: an indicator of poor clinical prognosis. *Cancer,* Vol.94, No.2, (January 2002), pp. 344–51, ISSN 0008-543X

Alvarez RH, Kantarjian HM, and Cortes JE. The role of Src in solid and hematologic malignancies: development of new-generation Src inhibitors. *Cancer* 107, 1918–1929. (2006) ISSN 0008-543X.

Arbet-Engels, C., Tartare-Deckert, S. & Eckhart, W. (1999). C-terminal Src kinase associates with ligand-stimulated insulin-like growth factor-I receptor. *Journal of Biological Chemistry,* Vol.274, No. 9, (February 1999), pp. 5422-5428, ISSN 0021-9258

Arthur, W T. & Burridge, K. (2001). RhoA inactivation by p190RhoGAP regulates cell spreading and migration by promoting membrane protrusion and polarity. *Molecular biology of the cell,* Vol.12, No. 9, (September 2001), pp. 2711–2720, ISSN 1059-1524

Baron, R. (1993). Cellular and molecular biology of the osteoclast, In: *Cellular and Molecular Biology of Bone,* Noda, M., Academic Press, ISBN 0-12-520225-3, San Diego

Baron, R. & Horne, WC. (2005). Regulation of osteoclast activity, In: *Topics in Bone Biology,* Vol. 2, Bronner, F., Farach-Carson, MC. & Rubin, J., Springer-Verlag, ISBN 9812562095, London

Baselga, J., Cervantes, A., Martinelli, E., Chirivella, I., Hoekman, K., Hurwitz, HI., Jodnell, DI., Hamberg, P., Casado, E., Elvin, P., Swazland, A., Iacone, R. & Tabernero, J. (2010). Phase Isafety, pharmacokinetics and inhibition of SRC activity study of sarcatinib in patients with solid tumors. *Clinical cancer research,* Vol.16, No.19, (October 2010), pp. 4876-4883, ISSN 1078-0432

Behrens, J., Mareel, MM., Van Roy, FM. & Birchmeier, W. (1989). Dissecting tumor cell invasion: epithelial cells acquire invasive properties after the loss of uvomorulin-mediated cell-cell adhesion. *Journal of Cell Biology,* Vol.108, No.6, (June 1989), pp. 2435–2447, ISSN 0021-9525

Biscardi, JS., Ishizawar, RC., Silva, CM. & Parsons, SJ. (2000). Tyrosine kinase signalling in breast cancer: epidermal growth factor receptor and c-Src interactions in breast cancer. *Breast Cancer Research,* Vol.2, No.3, (March 2000), pp. 203–210, ISSN 1465-5411

Bowman, T., Broome, MA., Sinibaldi, D., Wharton, W., Pledger, WJ., Sedivy, JM., Irby, R., Yeatman, T., Courtneidge, SA. & Jove, R. (2001). Stat3-mediated Myc expression is required for Src transformation and PDGF-induced mitogenesis. *Proceedings of the National Academy of Sciences of the United States of America,* Vol.98, No.13, (June 2001), pp. 7319-7324, ISSN 0027-8424

Boyce, BF., Yoneda, T., Lowe, C., Soriano, P. & Mundy, GR. (1992). Requirement of pp60c-src expression for osteoclasts to form ruffled borders and resorb bone in mice. *Journal of clinical investigation,* Vol.90, No.4, (October 1992), pp. 1622-1627, ISSN 0021-9738

Brown, MT. & Cooper, JA. (1996). Regulation, substrates and functions of src. *Biochimica et biophysica acta,* Vol.1287, No.2-3, (June 1996), pp. 121-149, ISSN 0006-3002

Brugge, JS. & Erikson, RL. (1977). Identification of a transformation-specific antigen induced by an avian sarcoma virus. *Nature,* Vol.269, No.5626, (September 1977), pp. 346-348, ISSN 0028-0836

Brunton, VG., MacPherson, IR. & Frame, MC. (2004). Cell adhesion receptors, tyrosine kinases and actin modulators: a complex three-way circuitry. *Biochimica et biophysica acta,* Vol. 1692, No.2-3, (July 2004), pp. 121-144, ISSN 0006-3002

Buccione, R., Orth, JD. & McNiven, MA. (2004). Foot and mouth: podosomes, invadopodia and circular dorsal ruffles. *Nature reviews. Molecular cell biology,* Vol.5, No.8, (August 2004), pp. 647-657, ISSN 1471-0072

Cantiani, L., Manara, MC., Zucchini, C., De Sanctis, P., Zuntini, M., Valvassori, L., Serra, M., Olivero, M., Di Renzo, MF., Colombo, MP., Picci, P. & Scotlandi, K. (2007). Caveolin-1 reduces osteosarcoma metastases by inhibiting c-Src activity and met signaling. *Cancer Research,* Vol.67, No.16, (August 2007), pp. 7675-7685, ISSN 0008-5472

Cartwright, CA., Eckhart, W., Simon, S. & Kaplan, PL. (1987). Cell transformation by pp60c-src mutated in the carboxyterminal regulatory domain. *Cell* Vol.49, No.1, (April 1987), pp. 83-91, ISSN 0092-8674

Collett, MS., Brugge, JS. & Erikson, RL. (1978). Characterization of a normal avian cell protein related to the avian sarcoma virus transforming gene product. *Cell.* Vol.15, No.4, (December 1978), pp. 1363-1369, ISSN 0092-8674

Cooper, JA. & King, CS. (1986). Dephosphorylation or antibody binding to the carboxy terminus stimulates pp60c-src. *Molecular and cellular biology,* Vol.6, No.12, (December 1986), pp. 4467-4477, ISSN 0270-7306

Courtneidge, SA., Dhand, R., Pilat, D., Twamley, GM., Waterfield, MD. & Roussel, MF. (1993). Activation of Src family kinases by colony stimulating factor-1, and their association with its receptor. *EMBO Journal,* Vol.12, No.3, (March 1993), pp. 943-950, ISSN 0261-4189

Coxon, FP. & Taylor, A. (2008). Vesicular trafficking in osteoclasts. *Seminars in cell & developmental biology,* Vol.19, No.5, (October 2008), pp. 424-433, ISSN 1084-9521

Datta, A., Huber, F. & Boettiger, D. (2002). Phosphorylation of beta3 integrin controls ligand binding strength. *The Journal of biological chemistry.* Vol.277, No.6, (February 2002), pp. 3943-3949, ISSN 0021-9258

De Mali, KA., Godwin, SL., Soltoff, SP. & Kazlauskas, A. (1999). Multiple roles for Src in a PDGF-stimulated cell. *Experimental cell research,* Vol.253, No.1, (November 1999), pp. 271-279, ISSN 0014-4827

Destaing, O., Saltel, F., Geminard, JC. & Jurdic, BF. (2003). Podosomes display actin turnover and dynamic self-organization in osteoclasts expressing actin-green fluorescent protein. *Molecular biology of the cell,* Vol.14, No.2, (February 2003), pp. 407–416, ISSN 1059-1524

Duesberg, PH. & Vogt, PK. (1970). Differences between the ribonucleic acids of transforming and nontransforming avian tumor viruses. *Proceedings of the National Academy of Sciences of the United States of America,* Vol.67, No.4, (December 1970), pp. 1673–1680, ISSN 0027-8424

Duong, LT. & Rodan, GA. (2000). PYK2 is an adhesion kinase in macrophages, localized in podosomes and activated by beta(2)-integrin ligation. *Cell motility and the cytoskeleton,* Vol.47, No.3, (November 2000), pp.174-88, ISSN 0886-1544

Dupree, P., Parton, RG., Raposo, G., Kurzchalia, TV. & Simons K. (1993). Caveolae and sorting in the trans-Golgi network of epithelial cells. *The EMBO journal,* Vol.12, No.4, (April 1993), pp. 1597-1605, ISSN 0261-4189

Faccio, R., Novack, DV., Zallone, A., Ross, FP. & Teitelbaum, SL. (2003). Dynamic changes in the osteoclast cytoskeleton in response to growth factors and cell attachment are controlled by beta3 integrin. *The Journal of cell biology,* Vol.162, No.3, (August 2003), pp. 499-509, ISSN 0021-9525

Faccio, R., Teitelbaum, SL., Fujikawa, K., Chappel, J., Zallone, A., Tybulewicz, VL., Ross, FP. & Swat, W. (2005). Vav3 regulates osteoclast function and bone mass. *Nature medicine,* Vol.11, No.3, (March 2005), pp. 284-290, ISSN 1078-8956

Fukuda, A., Hikita, A., Wakeyama, H., Akiyama, T., Oda, H., Nakamura, K. & Tanaka, S. (2005). Regulation of osteoclast apoptosis and motility by small GTPase binding protein Rac1. *Journal of bone and mineral research,* Vol.20, No.12, (December 2005), pp. 2245-2253, ISSN 0884-0431

Gianni, D., Taulet, N., DerMardirossian, C. & Bokoch, GM. (2010). c-Src-mediated phosphorylation of NoxA1 and Tks4 induces the reactive oxygen species (ROS)-dependent formation of functional invadopodia in human colon cancer cells. *Molecular biology of the cell,* Vol.21, No.23, (December 2010), pp. 4287-4298, ISSN 1059-1524

Giannoni, E., Buricchi, F., Raugei, G., Ramponi, G. & Chiarugi, P. (2005). Intracellular reactive oxygen species activate Src tyrosine kinase during cell adhesion and anchorage-dependent cell growth. *Molecular biology of the cell* Vol.25, No.15, (August 2005), pp. 6391-6403, ISSN 1059-1524

Guarino, M. (2010). Src signaling in cancer invasion. *Journal of cellular physiology,* Vol.223, No.1, (April 2010), pp. 14–26, ISSN 0021-9541

Hall, JM., Couse, JF. & Korach, KS. (2001). The multifaceted mechanisms of estradiol and estrogen receptor signaling. *Journal of Biological Chemistry,* Vol.276, No.40, (October 2001), pp. 36869–36872, ISSN 0021-9258

Hanai, J., Chen, LF., Kanno, T., Ohtani-Fujita, N., Kim, WY., Guo, WH., Imamura, T., Ishidou, Y., Fukuchi, M., Shi, MJ., Stavnezer, J., Kawabata, M., Miyazono, K. & Ito, Y. (1999). Interaction and functional cooperation of PEBP2/CBF with Smads. Synergistic induction of the immunoglobulin germline Calpha promoter. *Journal of Biological Chemistry,* Vol.274, No.44, (October 1999), pp. 31577–31582, ISSN 0021-9258

Hess, J., Angel, P. & Schorpp-Kistner, M. (2004). AP-1 subunits: quarrel and harmony among siblings. *Journal of cell science,* Vol.117, Pt.25, (December 2004), pp. 5965–5973, ISSN 0021-9533

Horton, MA. (1997). The $\alpha V\beta 3$ integrins 'vitronectin receptor'. *The international journal of biochemistry & cell biology*, Vol.29, No.5, (May 1997), pp. 721-725, ISSN 1357-2725

Horton, MA. & Rodan, GA. (1996). Integrins as therapeutic targets in bone disease. In: *Adhesion Receptors as Therapeutic Targets*, Horton, MA., CRC Press Inc., ISBN 9780849376559, Boca Raton

Hunter, T. & Sefton, BM. (1980) Transforming gene product of Rous sarcoma virus phosphorylates tyrosine. *Proceedings of the National Academy of Sciences of the United States of America*, Vol.77, No.3, (March 1980), pp. 1311-1315, ISSN 0027-8424

Irby, RB. & Yeatman, TJ. (2002). Increased Src activity disrupts cadherin/catenin-mediated homotypic adhesion in human colon cancer and transformed rodent cells. *Cancer Research*, Vol.62, No.9, (May 2002), pp. 2669-2674, ISSN 0008-5472

Irby, RB., Mao, W., Coppola, D., Kang, J., Loubeau, JM., Trudeau, W., Karl, R., Fujita, DJ., Jove, R. & Yeatman, TJ. (1999). Activating SRC mutation in a subset of advanced human colon cancers. *Nature Genetics*, Vol.21, No.2, (February 1999), pp. 187-190, ISSN 1061-4036

Irby, RB. & Yeatman, TJ. (2000) Role of Src expression and activation in human cancer. *Oncogene*, Vol.19, No.49, (November 2000), pp. 5636-5642, ISSN 0950-9232

Jiang, Y., Jiang, Y., Jahagirdar, BN., Reinhardt, RL., Schwartz, RE., Keene, CD., Ortiz-Gonzalez, XR., Reyes, M., Lenvik, T., Lund, T., Blackstad, M., Du, J., Aldrich, S., Lisberg, A., Low, WC., Largaespada, DA. & Verfaillie, CM. (2002). Pluripotency of mesenchymal stem cells derived from adult marrow. *Nature*, Vol.418, No.6893, (July 2002), pp. 41-49, ISSN 0028-0836

Jove, R. & Hanafusa, H. (1987). Cell transformation by the viral *src* oncogene. *Annual review of cell biology*, Vol.3, (November 1987), pp. 31-56, ISSN 0743-4634

Jung, EJ. & Kim, CW. (2002). Interaction between chicken protein tyrosine phosphatase 1 (CPTP1)-like rat protein phosphatase 1 (PTP1) and p60(v-src) in v-src transformed Rat-1 fibroblasts. *Experimental & molecular medicine*, Vol.34, No.6, (December 2002), pp. 476-480, ISSN 1226-3613

Kamei, T. Machida K, Nimura Y, Senga T, Yamada I, Yoshii S, Matsuda S, Hamaguchi M. (2000). C-Cbl protein in human cancer tissues is frequently tyrosine phosphorylated in a tumor-specific manner. *International journal of oncology*, Vol.17, No.2, (August 2000), pp. 335-339, ISSN 1019-6439

Kansara, M. & Thomas, DM. (2007). Molecular pathogenesis of osteosarcoma. *DNA and cell biology*, Vol.26, No.1, (January 2007), pp. 1-18, ISSN 1044-5498

Kim, M., Tezuka, T., Tanaka, K. & Yamamoto, T. (2004). Cbl-c suppresses v-Src-induced transformation through ubiquitin-dependent protein degradation. *Oncogene*, Vol.23, No.9, (March 2004), pp. 1645-1655, ISSN 0950-9232

Kmiecik, TE. & Shalloway, D. (1987). Activation and suppression of pp60c-src transforming ability by mutation of its primary sites of tyrosine phosphorylation. *Cell*, Vol.49, No.1, (April 1987), pp. 65-73, ISSN 0092-8674

Koegl, M., Zlatkine, P., Ley, SC., Courtneidge, SA. & Magee, AI. (1994). Palmitoylation of multiple Src-family kinases at a homologous N-terminal motif. *The Biochemical journal*, Vol.303, Pt.3, (November 1994), pp. 749-753, ISSN 0264-6021

Komori, T., Yagi, H., Nomura, S., Yamaguchi, A., Sasaki, K., Deguchi, K., Shimizu, Y., Bronson, RT., Gao, YH., Inada, M., Sato, M., Okamoto, R., Kitamura, Y., Yoshiki, S. & Kishimoto, T. (1997). Targeted disruption of Cbfa1 results in a complete lack of bone formation owing to maturational arrest of osteoblasts. *Cell*, Vol.89, No.5, (May 1997), pp. 755–764, ISSN 0092-8674

Kousteni, S., Han, L., Chen, JR., Almeida, M., Plotkin, LI., Bellido, T. & Manolagas, SC. (2003). Kinase-mediated regulation of common transcription factors accounts for the bone-protective effects of sex steroids. *The Journal of clinical investigation*, Vol.111, No.11, (June 2003), pp. 1651-1664, ISSN 0021-9738

Landgren, E., Blume-Jensen, P., Courtneidge, SA. & Claesson-Welsh, L. (1995). Fibroblast growth factor receptor-1 regulation of Src family kinases. *Oncogene*, Vol.10, No.10, (May 1995), pp. 2027–2035, ISSN 0950-9232

Li, S., Couet, J. & Lisanti, MP. (1996). Src tyrosine kinases, Galpha subunits, and H-Ras share a common membrane-anchored scaffolding protein, caveolin. Caveolin binding negatively regulates the auto-activation of Src tyrosine kinases. *The Journal of biological chemistry*, Vol.271, No.46, (November 1996), pp. 29182-29190, ISSN 0021-9258

Linder, S. & Aepfelbacher, M. (2003). Podosomes: adhesion hot-spots of invasive cells. *Trends in cell biology*, Vol.13, No.7, (July 2003), pp. 376–385, ISSN 0962-8924

Longo, M., Brama, M., Marino, M., Bernardini, S., Korach, KS., Wetsel, WC., Scandurra, R., Faraggiana, T., Spera, G., Baron, R., Teti, A. & Migliaccio, S. (2004). Interaction of estrogen receptor alpha with protein kinase C alpha and c-Src in osteoblasts during differentiation. *Bone*, Vol.34, No.1, (January 2004), pp. 100-111, ISSN 8756-3282

Lowe, C., Yoneda, T., Boyce, BF., Chen, H., Mundy, GR. & Soriano, P. (1993). Osteopetrosis in Src deficient mice is due to an autonomous defect of osteoclasts. *Proceedings of the National Academy of Sciences of the United States of America*, Vol.90, No.10, (May 1993), pp. 4485–4489, ISSN 0027-8424

Lowell, CA., Niwa, M., Soriano, P. & Varmus, HE. (1996). Deficiency of the Hck and Src tyrosine kinases results in extreme levels of extramedullary hematopoiesis. *Blood*, Vol.87, No.5, (May 1996), pp. 1780–1792, ISSN 0006-4971

Manolagas, SC., Kousteni, S. & Jilka, RL. (2002). Sex steroids and bone. *Recent Progress in Hormone Research*, Vol.57, (2002), pp. 385–409, ISSN 0079-9963

Mao, W., Irby, R., Coppola, D., Fu, L., Wloch, M., Turner, J., Yu, H., Garcia, R., Jove, R. & Yeatman, TJ. (1997). Activation of c-Src by receptor tyrosine kinases in human colon cancer cells with high metastatic potential. *Oncogene*, Vol.15, No.25, (December 1997), pp. 3083-3090, ISSN 0950-9232

Marchisio, PC., Cirillo, D., Naldini, L., Primavera, MV., Teti, A. & Zambonin-Zallone, A. (1984). Cell–substratum interaction of cultured avian osteoclasts is mediated by specific adhesion structures. *The Journal of cell biology*, Vol.99, No.5, (November 1984), pp. 1696–1705, ISSN 0021-9525

Marchisio, PC., Cirillo, D., Teti, A., Zambonin-Zallone, A. & Tarone, G. (1987). Rous sarcoma virus transformed fibroblasts and cells of monocytic origin display a peculiar dot-like organization of cytoskeletal proteins involved in microfilament–membrane interactions. *Experimental cell research*, Vol.169, No.1, (March 1987), pp. 202–214, ISSN 0014-4827

Martin, GS. (1970). Rous sarcoma virus: a function required for the maintenance of the transformed state. *Nature*, Vol.227, No.5262, (September 1970), pp. 1021–1023, ISSN 0028-0836

Marzia, M., Sims, NA., Voit, S., Migliaccio, S., Taranta, A., Bernardini, S., Faraggiana, T., Yoneda, T., Mundy, GR., Boyce, BF., Baron, R. & Teti, A. (2000). Decreased c-Src expression enhances osteoblast differentiation and bone formation. *The Journal of cell biology*, Vol. 151, No.2, (October 2000), pp. 311-320, ISSN 0021-9525

Mayer, BJ., Jackson, PK. & Baltimore, D. (1991). The noncatalytic *src* homology region 2 segment of *abl* tyrosine kinase binds to tyrosine-phosphorylated cellular proteins with high affinity. *Proceedings of the National Academy of Sciences of the United States of America*, Vol.88, No.2, (January 1991), pp. 627–631, ISSN 0027-8424

Mayer, BJ. & Baltimore, D. (1993). Signalling through SH2 and SH3 domains. *Trends in cell biology*, Vol.3, No.1, (January 1993), pp. 8-13, ISSN 0962-8924

Mayor, S. & Pagano, RE. (2007). Pathways of clathrin independent endocytosis. *Nature reviews. Molecular cell biology*, Vol.8, No.8, (August 2007), pp. 603-612, ISSN 1471-0072

McLean, GW., Carragher, NO., Avizienyte, E., Evans, J., Brunton, VG. & Frame, MC. (2005). The role of focal-adhesion kinase in cancer—a new therapeutic opportunity. *Nature reviews. Cancer*, Vol.5, No.7, (July 2005), pp. 505–515, ISSN 1474-175X

Miyamoto, S., Akiyama, SK. & Yamada, KM. (1995). Synergistic roles for receptor occupancy and aggregation in integrin transmembrane function. *Science*, Vol.267, No.5199, (February 1995), pp. 883–885, ISSN 0036-8075

Miyamoto, S., Teramoto, H., Gutkind, JS. & Yamada, KM. (1996). Integrins can collaborate with growth factors for phosphorylation of receptor tyrosine kinases and MAP kinase activation: Roles of integrin aggregation and occupancy of receptors. *The Journal of cell biology*, Vol.135, No.6 Pt.1, (December 1996), pp. 1633–1642, ISSN 0021-9525

Miyazaki, T., Neff, L., Tanaka, S., Horne, WC. & Baron, R. (2003). Regulation of cytochrome c oxidase activity by c-Src in osteoclasts. *The Journal of cell biology*, Vol.160, No.5, (March 2003), pp. 709–718, ISSN 0021-9525

Miyazaki, T., Sanjay, A., Neff, L., Tanaka, S., Horne, WC. & Baron, R. (2004). Src kinase activity is essential for osteoclast function. *The Journal of biological chemistry*, Vol.279, No.17, (April 2004), pp. 17660–17666, ISSN 0021-9258

Miyazaki, T., Tanaka, S., Sanjay, A. & Baron, R. (2006). The role of c-Src kinase in the regulation of osteoclast function. *Modern rheumatology*, Vol.16, No.2, (2006), pp. 68-74, ISSN 1439-7595

Mocsai, A., Humphrey, MB., Van Ziffle, JA., Hu, Y., Burghardt, A., Spusta, SC., Majumdar, S., Lanier, LL., Lowell, CA., Nakamura, MC. (2004). The immunomodulatory adapter proteins DAP12 and Fc receptor gamma-chain (FcR gamma) regulate development of functional osteoclasts through the Syk tyrosine kinase. *Proceedings of the national academy of science USA*, Vol. 101, No.16, (April 2004) pp.6158-6163

Morgan, L., Gee, J., Pumford, S., Farrow, L., Finlay, P., Robertson, J., Ellis, I., Kawakatsu, H., Nicholson, R. & Hiscox, S. (2009). Elevated Src kinase activity attenuates Tamoxifen response in vitro and is associated with poor prognosis clinically. *Cancer biology & therapy*, Vol.8, No.16, (August 2009), pp. 1550-1558, ISSN 1538-4047

Myuoi, A., Nishimura, P., Williams, PJ., Tamura, D., Michigami, T., Mundy, GR. & Yoneda, T. (2003). c-Src tyrosine kinase activity is associated with tumor colonization in bone and lung in an animal model of human breast cancer bone metastasis. *Cancer research*, Vol.63, No.16, (August 2003), pp. 5028-5033, ISSN 0008-5472

Nakamura, I., Duong, le T., Rodan, SB. & Rodan, GA. (2007). Involvement of alpha(v)beta3 integrins in osteoclast function. *Journal of bone and mineral metabolism*, Vol.25, No.6, (October 2007), pp. 337-344, ISSN 0914-8779

Nam, JS., Ino, Y., Sakamoto, M. & Hirohashi, S. (2002). Src family kinase inhibitor PP2 restores the E-cadherin/catenin cell adhesion system in human cancer cells and reduces cancer metastasis. *Clinical cancer research*, Vol.8, No.7, (July 2002), pp. 2430–2436, ISSN 1078-0432

Nautiyal, J., Majumder, P., Patel, BB., Lee, FY. & Majumdar, AP. (2009). Src inhibitor dasatinib inhibits growth of breast cancer cells by modulating EGFR signaling. *Cancer letters*, Vol283, No.2, (October 2009), pp. 143-151, ISSN 0304-3835

Okada, M. & Nakagawa, H. (1989). A protein tyrosine kinase involved in regulation of pp60c-src function. *The Journal of biological chemistry*, Vol.264, No.35, (December 1989), pp. 20886–20893, ISSN 0021-9258

Oppermann, H., Levinson, AD., Varmus, HE., Levintow, L. & Bishop, JM. (1979). Uninfected vertebrate cells contain a protein that is closely related to the product of the avian sarcoma virus transforming gene (src). *Proceedings of the National Academy of Sciences of the United States of America*, Vol.76, No.4, (April 1979), pp. 1804–1808, ISSN 0027-8424

Pawson, T. (1995). Protein modules and signalling networks. *Nature*, Vol.373, No.6515, (February 1995), pp. 573-580, ISSN 0028-0836

Peruzzi B, Cappariello A, Del Fattore A, Rucci N, De Benedetti F, Teti A. (2012) c-Src and IL-6 inhibit osteoblast differentiation and integrate IGFBP5 signalling. *Nature Communications*, Vol. 3, (January 2012) article number 630. ISSN 2041-1723, doi: 10.1038/ncomms1651.

Peruzzi, B. & Teti, A. (2011). The Physiology and Pathophysiology of the Osteoclast, *Clinical Reviews in Bone and Mineral Metabolism*, ISSN 1534-8644, DOI: 10.1007/s12018-011-9086-6

Piwnica-Worms, H., Saunders, KB., Roberts, TM., Smith, AE. & Cheng, SH. (1987). Tyrosine phosphorylation regulates the biochemical and biological properties of pp60c-src. *Cell*, Vol.49, No.1, (April 1987), pp. 75–82, ISSN 0092-8674

Playford, MP. & Schaller, MD. (2004). The interplay between Src and integrins in normal and tumor biology. *Oncogene*, Vol.23, No.48, (October 2004), pp. 7928–7946, ISSN 0950-9232

Raymond, AK., Ayala, AG. & Knuutila, S. (2002). Conventional osteosarcoma. In: *World Health Organization classification of tumors*, Fletcher, CDM., Unni, KK. & Mertens, F., IARC Press, ISBN 92 832 2413 2, Lyon:, 2002. p. 264–70.

Recchia, I., Rucci, N., Funari, A., Migliaccio, S., Taranta, A., Longo, M., Kneissel, M., Susa, M., Fabbro, D. & Teti, A. (2004). Reduction of c-Src activity by substituted 5,7-diphenyl-pyrrolo[2,3-d]-pyrimidines induces osteoclast apoptosis in vivo and in vitro. Involvement of ERK1/2 pathway. *Bone*, Vol.34, No.1, (January 2004), pp. 65-79, ISSN 8756-3282

Ren, R., Mayer, BJ., Cicchetti, P. & Baltimore, D. (1993). Identification of a ten amino acid proline-rich SH3 binding site. *Science*, Vol.259, No.5098, (February 1993), pp. 1157-1161, ISSN 0036-8075

Resh, MD. (1999). Fatty acylation of proteins: new insights into membrane targeting of myristoylated and palmitoylated proteins. *Biochimica et biophysica acta*, Vol.1451, No.1, (August 1999), pp. 1–16, ISSN 0006-3002

Roby, KF., Son, DS., Taylor, CC., Montgomery-Rice, V., Kirchoff, J., Tang, S. & Terranova, PF. (2005). Alterations in reproductive function in SRC tyrosine kinase knockout mice. *Endocrine*, Vol. 26, No.2, (March 2005), pp. 169-176, ISSN 1355-008X

Rodan, GA. & Martin, TJ. (1981). Role of osteoblasts in hormonal control of bone resorption--a hypothesis. *Calcified tissue International*, Vol.33, No.4, (1981), pp. 349-351, ISSN 0171-967X

Rothberg, KG., Heuser, JE., Donzell, WC., Ying, YS., Glenney, JR. & Anderson, RG. (1992). Caveolin, a protein component of caveolae membrane coats. *Cell*, Vol68, No. 4, (February 1992), pp. 673-682, ISSN 0092-8674

Rous, PA. (1911a). Transmission of a malignant new growth by means of a cell-free filtrate. *Journal of the American Medical Association*, Vol.56, (1911), pp. 98, ISSN 0098-7484

Rous, PA. (1911b). A sarcoma of the fowl transmissible by an agent separable from the tumor cells. *The Journal of experimental medicine*, Vol.13, No.4, (April 1911), pp. 397-411, ISSN 0022-1007

Rubin, H. (1995). Quantitative relations between causative virus and cell in the Rous No. 1 chicken sarcoma. *Virology*, Vol.1, No.5, (December 1955), pp. 445–473, ISSN 0042-6822

Rucci, N., Recchia, I., Angelucci, A., Alamanou, M., Del Fattore, A., Fortunati, D., Susa, M., Fabbro, D., Bologna, M. & Teti, A. (2006). Inhibition of protein kinase c-Src reduces th incidence of breast cancer metastases and increases survival in mice: implication for therapy. *The Journal of pharmacology and experimental therapeutics*, Vol. 318, No.1, (July 2006), pp. 161-172, ISSN 0022-3565

Rucci, N., Susa, M. & Teti, A. (2008). Inhibition of protein kinase c-Src as a therapeutic approach for cancer and bone metastases. *Anti-cancer agents in medicinal chemistry*, Vol.8, No.3, (April 2008), pp. 342–349, ISSN 1871-5206

Rupp, PA. & Little, CD. (2001). Integrins in vascular development. *Circulation research*, Vol.89, No.7, (September 2001), pp. 566–572, ISSN 0009-7330

Sakai, T., Jove, R., Fassler, R. & Mosher, DF. (2001). Role of the cytoplasmic tyrosines of beta 1A integrins in transformation by v-src. *Proceedings of the National Academy of Sciences of the United States of America*, Vol.98, No.7, (March 2001), pp. 3808–3813, ISSN 0027-8424

Sanjay, A., Houghton, A., Neff, L., DiDomenico, E., Bardelay, C., Antoine, E., Levy, J., Gailit, J., Bowtell, D., Horne, WC. & Baron, R. (2001). Cbl associates with Pyk2 and Src to regulate Src kinase activity, alpha(v)beta(3) integrin-mediated signaling, cell adhesion, and osteoclast motility. *The Journal of cell biology*, Vol.152, No. 1, (January 2001), pp. 181-195, ISSN 0021-9525

Sargiacomo, M., Scherer, PE., Tang, Z., Kübler, E., Song, KS., Sanders, MC. & Lisanti, MP. (1995). Oligomeric structure of caveolin: implications for caveolae membrane organization. *Proceedings of the National Academy of Sciences of the United States of America*, Vol.92, No.20, (September 1995), pp. 9407-9411, ISSN ISSN 0027-8424

Schaller, MD., Hildebrand, JD., Shannon, JD., Fox, JW., Vines, RR. & Parsons, JT. (1994). Autophosphorylation of the focal adhesion kinase, pp125FAK, directs SH2-

dependent binding of pp60src. *Molecular and cellular biology*, Vol.14, No.3, (March 1994), pp. 1680-1688, ISSN 0270-7306

Schlaepfer, DD. & Hunter, T. (1998). Integrin signaling and tyrosine phosphorylation: Just the FAKs? *Trends in cell biology*, Vol.8, No.4, (April 1998), pp. 151–157, ISSN 0962-8924

Severs, NJ. (1988). Caveolae: static inpocketings of the plasma membrane, dynamic vesicles or plain artifact? *Journal of cell science*, Vol.90, Pt.3, (July 1988), pp. 341-348, ISSN 0021-9533

Shanley, DJ. & Mulligan, ME. (1991). Osteosarcoma with isolated metastases to the pleura. *Pediatric radiology*, Vol.21, No.3, (1991), pp.226, ISSN 0301-0449

Silva, CM. (2004). Role of STATs as downstream signal transducers in Src family kinase-mediated tumorigenesis *Oncogene*, Vol.23, No.48, (October 2004), pp. 8017–8023, ISSN 0950-9232

Soriano, P., Montgomery C., Geske R. & Bradley A.. (1991). Targeted disruption of the c-src proto-oncogene leads to osteopetrosis in mice. *Cell*, Vol.64, No.4, (February 1991), pp. 693–702, ISSN 0092-8674

Stein, GS., van Wijnen, AJ., Stein, JL., Lian, JB., Montecino, M., Choi, JY., Zaidi, K. & Javed, A. (2000). Intranuclear trafficking of transcription factors: implications for biological control. *Journal of cell science*, Vol.113, Pt.14, (July 2000), pp. 2527–2533, ISSN 0021-9533

Stein, PL., Vogel, H. & Soriano, P. (1994). Combined deficiencies of Src, Fyn, and Yes tyrosine kinases in mutant mice. *Genes & development*, Vol.8, No.17, (September 1994), pp. 1999–2007, ISSN 0890-9369

Suda, T., Nakamura, I., Jimi, E. & Takahashi, N. (1997). Regulation of osteoclast function. *Journal of bone and mineral research*, Vol.12, No.6, (June 1997), pp. 869-879, ISSN 0884-0431

Takeya, T. & Hanafusa, H. (1983). Structure and sequence of the cellular gene homologous to the RSV *src* gene and the mechanism for generating the transforming virus. *Cell*, Vol.32, No.3, (March 1983), pp. 881–890, ISSN 0092-8674

Teitelbaum SL. (2007). Osteoclasts: what do they do and how do they do it? *The American journal of pathology*, Vol.170, No.2, (February 2007), pp. 427-435, ISSN 0002-9440

Thomas, SM. & Brugge, JS. (1997). Cellular functions regulated by Src family kinases. *Annual Review of Cell and Developmental Biology*, Vol.13, (November 1997), pp. 513–609, ISSN 1081-0706

Thomas, JW., Ellis, B., Boerner, RJ., Knight, WB., White, GC.II & Schaller, MD. (1998). SH2- and SH3-mediated interactions between focal adhesion kinase and Src. The *Journal of biological chemistry*, Vol.273, No.1, (January 1998), pp. 577–583, ISSN 0021-9258

Tice, DA., Biscardi, JS., Nickles, AL. & Parsons, SJ. (1999). Mechanism of biological synergy between cellular Src and epidermal growth factor receptor. *Proceedings of the National Academy of Sciences of the United States of America*, Vol.96, No.4, (February 1999), pp. 1415–1420, ISSN ISSN 0027-8424

Timpson, P., Jones, GE., Frame, MC. & Brunton, VG. (2001). Coordination of cell polarization and migration by the Rho family GTPases requires Src tyrosine kinase activity. *Current biology*, Vol.11, No.23, (November 2001), pp. 1836–1846, ISSN 0960-9822

Tsygankov, AY. & Shore, SK. Src: regulation, role in human carcinogenesis and pharmacological inhibitors. *Current pharmaceutical design*, Vol.10, No.15, (2004), pp. 1745–1756, ISSN 1381-6128

van der Geer, P., Hunter, T., & Lindberg, RA. (1994). Receptor protein-tyrosine kinases and their signal transduction pathways. *Annual review of cell biology*, Vol.10, (1994), pp. 251–337, ISSN 0743-4634

Wagner, EF. & Eferl R. (2005). Fos/AP-1 proteins in bone and the immune system. *Immunological reviews*, Vol.208, (December 2005), pp. 126–140, ISSN 0105-2896

Wei, Y., Yang, X., Liu, Q., Wilkins, JA., & Chapman, HA. (1999). A role for caveolin and the urokinase receptor in integrin-mediated adhesion and signaling. *The Journal of cell biology*, Vol.144, No.6, (March 1999), pp. 1285–1294, ISSN 0021-9525

Weilbaecher, KN., Guise, TA. & McCauley, LK. (2011). Cancer and bone: a fatal attraction. *Nature reviews. Cancer*, Vol.11, No.6, (June 2011), pp. 411-425, ISSN 1474-175X

Westendorf, JJ. & Hiebert, SW. (1999). Mammalian runt-domain proteins and their roles in hematopoiesis, osteogenesis, and leukemia. *Journal of cellular biochemistry*, Suppl. 32-33, (1999), pp. 51–58, ISSN 0730-2312

Wilder, RL. (2002). Integrin alpha V beta 3 as a target for treatment of rheumatoid arthritis and related rheumatic diseases. *Annals of the Rheumatic* Diseases, Vol.61, Suppl.2, (November 2002), pp. 96-99, ISSN 0003-4967

Winkler, DG., Park, I., Kim, T., Payne, NS., Walsh, CT., Strominger, JL. & Shin, J. (1993). Phosphorylation of Ser-42 and Ser-59 in the N-terminal region of the tyrosine kinase p56lck. *Proceedings of the National Academy of Sciences of the United States of America*, Vol.90, No.11, (June 1993), pp. 5176-5180, ISSN ISSN 0027-8424

Yagi, R., Chen, LF., Shigesada, K., Murakami, Y. & Ito, Y. (1999). A WW domain-containing yes-associated protein (YAP) is a novel transcriptional co-activator. *The EMBO Journal*, Vol.18, No.9, (May 1999), pp. 2551–2562, ISSN 0261-4189

Yamaguchi, H. & Hendrickson, WA. (1996). Structural basis for activation of human lymphocyte kinase Lck upon tyrosine phosphorylation. *Nature*, Vol.384, No.6608, (December 1996), pp. 484–489, ISSN 0028-0836

Yeatman, TJ. (2004). A renaissance for SRC. *Nature reviews. Cancer*, Vol.4, No.6, (June 2004), pp. 470–480, ISSN 1474-175X

Yokouchi, M., Kondo, T., Sanjay, A., Houghton, A., Yoshimura, A., Komiya, S., Zhang, H. & Baron, R. (2001). Src-catalyzed phosphorylation of c-Cbl leads to the interdependent ubiquitination of both proteins. *The Journal of biological chemistry*, Vol.276, No.37, (September 2001), pp. 35185-35193, ISSN 0021-9258

Yue, W., Fan, P., Wang, J., Li, Y. & Santen, RJ. (2007). Mechanisms of acquired resistance to endocrine therapy in hormone-dependent breast cancer cells. *The Journal of steroid biochemistry and molecular biology*, Vol.106, No.1-5, (August-September 2007), pp. 102-110,, ISSN 0960-0760

Zaidi, SK., Javed, A., Choi, JY., van Wijnen, AJ., Stein, JL., Lian, JB. & Stein, GS. (2001). A specific targeting signal directs Runx2/Cbfa1 to subnuclear domains and contributes to transactivation of the osteocalcin gene. *Journal of cell science*, Vol.114, Pt.17, (September 2001), pp. 3093–3102, ISSN 0021-9533

Zhang, XH., Wang, Q., Gerald, W., Hodis, CA., Norton, L., Smid, M., Foekens, JA. & Massagué, J. (2009). Latent bone metastasis in breast cancer tied to Src-dependent survival signals. *Cancer Cell*, Vol.16, No.1, (July 2009), pp. 67-78, ISSN 1535-6108

Zheng, XM., Wang, Y. & Pallen, CJ. (1992). Cell transformation and activation of pp60c-src by overexpression of a protein tyrosine phosphatase. Nature, Vol.359, No.6393, (September 1992), pp. 336-339, ISSN 0028-0836

Zou, W., Kitaura, H., Reeve, J., Long, F., Tybulewicz, VL., Shattil, SJ., Ginsberg, MH., Ross, FP. & Teitelbaum, SL. (2007). Syk, c-Src, the alphavbeta3 integrin, and ITAM immunoreceptors, in concert, regulate osteoclastic bone resorption. *The Journal of cell biology*, Vol.176, No.6, (March 2007), pp. 877-888, ISSN 0021-9525

17

The ERK MAPK Pathway in Bone and Cartilage Formation

Takehiko Matsushita and Shunichi Murakami
Department of Orthopaedics, Case Western Reserve University, USA

1. Introduction

The Extracellular signal-related kinase (ERK) Mitogen-activated protein kinase (MAPK) pathway is a signaling cascade that is activated by various extracellular stimuli including fibroblast growth factors. The ERK MAPK pathway has recently been shown to play critical roles in skeletal development. A number of human skeletal syndromes have been shown to result from mutations in this pathway. These include Noonan, Costello, and cardio-facio-cutaneous syndromes (Aoki et al., 2005; Pandit et al., 2007; Rodriguez-Viciana et al., 2006). In addition, activating mutations in FGFR2 cause craniosynostosis syndromes such as Apert and Crouzon syndromes, while activating mutations in fibroblast growth factor receptor 3 (FGFR3) are responsible for the most common forms of human dwarfism, achondroplasia, thanatophoric dysplasia, and hypochondroplasia (Bellus et al., 1995; Jabs et al., 1994; Rousseau et al., 1994, 1995; Rutland et al., 1995; Shiang et al., 1994; Wilcox et al., 1998; Wilkie et al., 1995).

Although a number of in vitro experiments indicated profound effects of the ERK MAPK pathway on chondrocyte and osteoblast phenotype, sometimes conflicting results were reported presumably due to variable culture conditions (Bobick & Kulyk, 2008; Schindeler & Little, 2006), and the roles of the ERK MAPK pathway in vivo remained elusive. Therefore, to examine the role of ERK MAPK in skeletal development, we used both gain-of-function and loss-of-function approaches to activate or inactivate the ERK MAPK pathway in skeletal tissues of genetically engineered mice. We used the Cre-*loxP* system to inactivate *ERK1* and *ERK2* in skeletal tissues (Logan et al., 2002; Matsushita et al., 2009a). By using the *Prx1-Cre* transgene, mice lacking *ERK1* and *ERK2* in the limb and head mesenchyme were created. We also generated a loss of function model in chondrocytes by using the *Col2a1-Cre* transgene (Matsushita et al., 2009a; Ovchinnikov et al., 2000). To induce postnatal inactivation of *ERK1* and *ERK2* in chondrocytes, we used the *Col2a1-CreER* transgene to express a tamoxifen-inducible form of Cre recombinase (Nakamura et al., 2006; Sebastian et al., 2011). For gain-of-function experiments, we generated *Prx1-MEK1* transgenic mice that express a constitutively active mutant of MEK1 in undifferentiated mesenchymal cells under the control of a *Prx1* promoter (Matsushita et al., 2009a). We also generated *Col2a1-MEK1* transgenic mice that express a constitutively active mutant of MEK1 in chondrocytes under the control of the regulatory sequences of *Col2a1* (Murakami et al., 2004). In this review, we will summarize the roles of the ERK MAPK pathway in skeletal development based on our recent studies using genetically engineered mouse models.

2. Skeletal development and ossification processes

Bone formation takes place in two major ossification processes, endochondral ossification and intramembranous ossification (Colnot, 2005; Hunziker, 1994; Opperman, 2000; Shapiro et al., 2005). Both chondrocytes and osteoblasts arise from common undifferentiated mesenchymal progenitor cells. In endochondral ossification, the skeletal element is formed as a cartilaginous template that is subsequently replaced by bone. Undifferentiated mesenchymal cells first aggregate to form mesenchymal condensation and differentiate into chondrocytes. Chondrocytes proliferate in columnar stacks to form the growth plate, then exit the cell cycle, and differentiate into hypertrophic chondrocytes. The cartilaginous matrix of hypertrophic chondrocytes is calcified and subsequently invaded by blood vessels. Hypertrophic chondrocytes are removed by apoptotic cell death, and the cartilaginous matrix is resorbed by chondroclasts/osteoclasts and replaced by trabecular bone. Chondroclast/osteoclast formation is supported by receptor activator of nuclear factor-kappa B ligand (RANKL) secreted from osteoblasts and bone marrow stromal cells (Kim et al., 2000; Yasuda et al., 1998). In intramembranous ossification, mesenchymal cells directly differentiate into bone-forming osteoblasts; cortical bone is formed by osteoblasts that arise from the osteochondro progenitor cells in the perichondrium. The entire process of endochondral ossification and intramembranous ossification is under the control of various hormones and growth factors. These include systemic factors such as growth hormone, estrogen, and glucocorticoids; and local factors such as Indian hedgehog (Ihh), parathyroid hormone-related peptide (PTHrP), fibroblast growth factors (FGF), transforming growth factor-β (TGF-β), and bone morphogenetic proteins (BMP) (DeLise et al., 2000; van der Eerden et al., 2003).

3. The ERK MAPK pathway and human syndromes

3.1 The ERK MAPK pathway

The ERK MAPK pathway (Fig. 1), which is activated by various stimuli in eukaryotic cells, transduces extracellular signals into cells and coordinates cellular responses. The MAPK pathways are generally organized into three kinase modules: MAPKK kinase (MAPKKK), MAPK kinase (MAPKK), and MAPK. MAPKKK phosphorylates and activates MAPKK, which in turn phosphorylates and activates MAPK. A diverse array of growth factors and cytokines transduce their signals through the activation of the small G protein Ras, which leads to the activation of the Raf members of MAPKKK, and then to the activation of MAPKK, MEK1 and MEK2. MEK1 and MEK2 then phosphorylate and activate MAPK, ERK1 and ERK2. ERK1 and ERK2 then phosphorylate various cytoplasmic and nuclear target proteins, ranging from cytoplasmic adaptor proteins and transcription factors to kinases including RSK (Cargnello & Roux, 2011; Roux & Blenis, 2004). The ERK MAPK pathway has been shown to mediate the intracellular signaling induced by a variety of growth factors such as FGFs, BMPs, and TGFs (Jun et al., 2010; Mu et al., 2011; Murakami et al., 2000; Osyczka & Leboy, 2005; Qureshi et al., 2005; Tuli et al., 2003).

3.2 Human syndromes caused by mutations in the MAPK pathway

Recently, a number of human mutations have been identified in the molecules in the MAPK cascade (Fig. 1). Missense activating mutations in KRAS, BRAF, MEK1, and MEK2 have been identified in Cardio-facio-cutaneous syndrome (Rodriguez-Viciana et al., 2006). KRAS mutations have been also identified in Noonan syndrome (Schubbert et al., 2006). HRAS mutations cause Costello syndrome (Aoki et al., 2005), and loss-of-function mutations in RSK2 cause Coffin-Lowry syndrome (Trivier et al., 1996). In addition, haploinsufficient expression of ERK2 has been associated with DiGeorge syndrome (Newbern et al., 2008). All of these syndromes present with various skeletal manifestations, including short stature and craniofacial and limb abnormalities, underscoring the importance of the MAPK pathway in human skeletal development (Hanauer & Young, 2002; Hennekam, 2003; Noonan, 2006; Reynolds et al., 1986; van der Burgt, 2007).

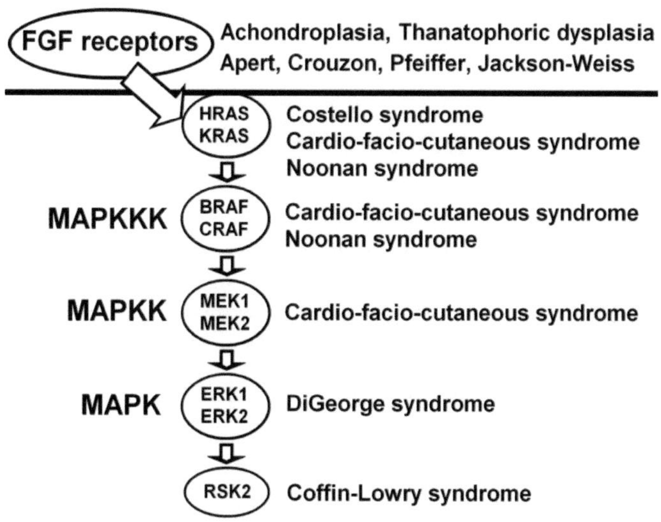

Fig. 1. The ERK MAPK pathway and human skeletal syndromes

4. Genetic manipulation of molecules in the ERK MAPK pathway in mice

To examine the role of the MAPK pathway, various mutant mice have been generated. Inactivation of Rsk2, a downstream kinase of the MAPK pathway, caused a widening of cranial sutures at birth, similar to the delayed closure of fontanelles in patients with Coffin-Lowry syndrome (David et al., 2005; Yang et al., 2004). These observations indicate that Rsk2 plays a critical role in osteoblast differentiation. In contrast to *Rsk2*-null mice, the inactivation of ERK, MEK, and Raf family members has provided little information regarding skeletal development. *ERK1*-null mice are viable and fertile and develop normally without obvious skeletal abnormalities, suggesting that ERK1 is dispensable for skeletal development (Pages et al., 1999; Selcher et al., 2001). In remarkable contrast, *ERK2*-null mice show early embryonic lethality at E6.5, precluding the analysis of skeletal development (Saba-El-Leil et al., 2003; Yao et al., 2003). *Mek1*-null embryos die at E10.5 due to placental defects, while *Mek2*-null mice develop normally without any obvious abnormalities (Belanger et al., 2003; Giroux et al., 1999). *Araf*-null mice show neurological and gastrointestinal defects, but do not show an obvious skeletal phenotype (Pritchard et al., 1996). *Braf*-deficient embryos die at midgestation due to vascular defects, precluding the analysis of skeletal development (Wojnowski et al., 1997). *Craf*-null mice show placental defects and die at around E10.5-12.5 on the C57BL/6 and 129 backgrounds. On the outbred

CD1 background, two-thirds of embryos reach term and die soon after birth. These surviving embryos show a mild delay in ossification; however, it is not clear whether the observed skeletal phenotype is primarily caused by *Craf* deficiency in the skeletal tissues (Wojnowski et al., 1998). Furthermore, mice with chondrocytes deficient in both *Araf* and *Braf* showed normal endochondral bone development (Provot et al., 2008). These observations suggest that members of the ERK, MEK, and Raf family are functionally redundant, while some of the tissue-specific functions are not fully compensated by other family members. To circumvent early embryonic lethality caused by the systemic inactivation of the target gene, tissue-specific inactivation would be essential. Furthermore, the inactivation of multiple family members may be necessary to uncover the roles of ERK, MEK, and Raf family members in skeletal development.

5. Inactivation of ERK1 and ERK2 in undifferentiated mesenchymal cells disrupts bone formation and induces ectopic cartilage formation

Since early embryonic lethality hampered researchers from analyzing the role of the ERK MAPK pathway in skeletal development in vivo as described above, we used the Cre-*loxP* system to inactivate *ERK1* and *ERK2* in skeletal tissues. We used *Prx1-Cre* transgenic mice that express Cre recombinase under the control of the 2.4 kb *Prx1* promoter (Logan et al., 2002) to inactivate *ERK1* and *ERK2* in the limb and head mesenchyme (Matsushita et al., 2009a). The *Prx1* promoter has been shown to direct transgene expression in undifferentiated mesenchyme in the developing limb buds and head mesenchyme. The transgene expression is detectable as early as E10.5, and the transgene expression is confined to the periosteum of the long bones and tendons of the limbs at E15.5 (Logan et al., 2002). We analyzed *ERK1*-null mice and *ERK2*$^{flox/flox}$; *Prx1-Cre* mice, and these mice did not show obvious skeletal abnormalities. Therefore, we further inactivated *ERK2* in the *ERK1*-null background to totally inactivate *ERK1* and *ERK2* in the head and limb mesenchyme of mouse embryos using the *Prx1-Cre* transgene (Matsushita et al., 2009a).

5.1 Inactivation of ERK1 and ERK2 in mesenchymal cells disrupts osteoblast differentiation

Skeletal preparation of *ERK1* $^{-/-}$; *ERK2* $^{flox/flox}$; *Prx1-Cre* mutants revealed severe limb deformities as well as calvaria defects characterized by delayed closure of the cranial sutures (Fig. 2). Histological analysis of the long bones showed disruption of bone formation. These findings indicate that ERK1 and ERK2 play an essential role in bone formation. In situ hybridization analysis indicated that master osteogenic transcription factors *Runx2, Osterix,* and *Atf4* were expressed at normal levels, while expression of *Osteocalcin*, a marker of mature osteoblasts, was strongly decreased in *ERK1* $^{-/-}$; *ERK2* $^{flox/flox}$; *Prx1-Cre* mice. These observations suggest that osteoblast differentiation was blocked after *Runx2, Osterix,* and *Atf4* expression and before *Osteocalcin* expression. The impaired bone formation was associated with decreased beta-catenin protein levels in the periosteum, suggesting decreased Wnt signaling. We also found that other transcriptional regulators such as *Krox20, Fra1, Fra2, cFos,* and *Cbfb* were downregulated in *ERK1* $^{-/-}$; *ERK2* $^{flox/flox}$; *Prx1-Cre* embryos. While the regulatory mechanisms of osteoblast differentiation require further investigation, ERK1 and ERK2 are likely to control skeletal development and osteoblast differentiation through multiple downstream molecules. Consistent with our observation, Ge et al. reported that transgenic mice that express dominant negative MEK1 under an *Osteocalcin* promoter showed delayed bone formation and reduced mineralization of calvaria, while mice that express a constitutively active MEK1 under the *Osteocalcin* promoter showed accelerated bone formation (Ge et al., 2007). These observations also indicated a critical role of ERK1 and ERK2 in osteoblast differentiation. In vitro studies have also suggested that the ERK MAPK pathway regulates osteoblast differentiation through phosphorylation and acetylation of Runx2 (Ge et al., 2007; Park et al., 2010; Xiao et al., 2000, 2002). The importance of the ERK MAPK pathway in osteoblast differentiation in head mesenchyme was also demonstrated by Shukla et al., who showed that craniosynostosis in a mouse model of Apert syndrome that carries a mutation in *Fgfr2* was prevented by the treatment of MEK1/2 inhibitor (Shukla et al., 2007). These observations further link the activation of the ERK MAPK pathway to the pathogenesis of craniosynostosis syndromes caused by activating mutations in FGFR2.

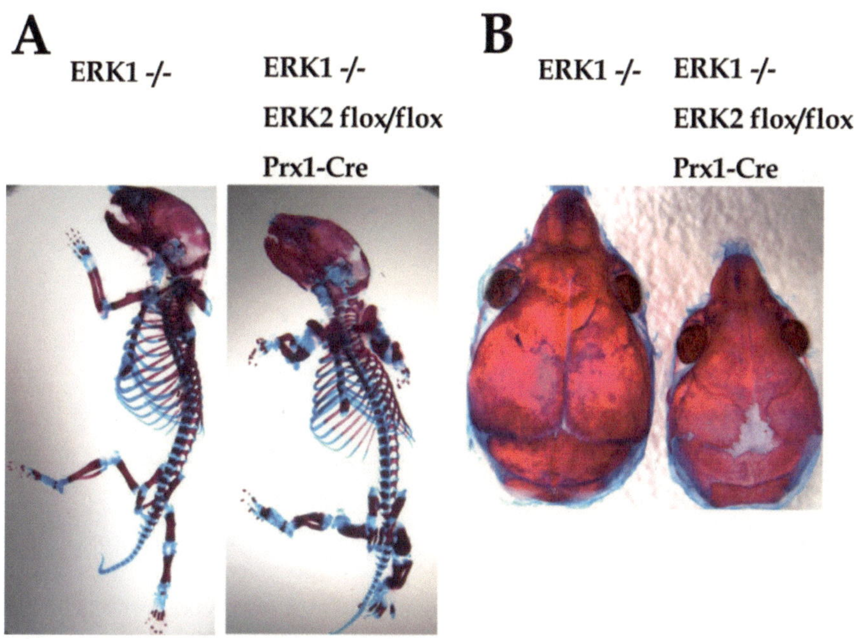

Fig. 2. (A) Skeletal preparation after alizarin red and alcian blue staining at postnatal day 1. (B) Skeletal preparation after alizarin red and alcian blue staining at postnatal day 5.

5.2 Inactivation of ERK1 and ERK2 disrupts osteocyte differentiation

Osteoblasts undergo sequential steps of differentiation and subsequently become embedded in bone matrix as osteocytes. Osteocytes function as a mechanosensor in the bone and secrete dentin matrix protein 1 (Dmp1) and FGF23 to regulate phosphate homeostasis (Dallas & Bonewald, 2010; Feng et al., 2006; Tatsumi et al., 2007; Xiao and Quarles, 2010). Although *ERK1*$^{-/-}$; *ERK2*$^{flox/flox}$; *Prx1-Cre* mice showed a remarkable impairment of bone formation, bone-like architecture was observed in the diaphyses of long bones, and osteocyte-like cells were found within the bone-like matrix. Our real-time PCR and immunohistochemical analysis indicated a strong decrease in Dmp1 expression in the skeletal elements of *ERK1*$^{-/-}$; *ERK2* $^{flox/flox}$; *Prx1-Cre* mice. Furthermore, scanning electron microscopic analysis revealed that osteocytes in *ERK1*$^{-/-}$; *ERK2*$^{flox/flox}$; *Prx1-Cre* mice lack dendritic processes, indicating that *ERK1* and *ERK2* inactivation disrupts the formation of osteocyte-lacunar-canalicular system (Kyono et al., 2011). These observations indicate that ERK signaling is essential for *Dmp1* expression and osteocyte differentiation.

5.3 Inactivation of ERK1 and ERK2 in mesenchyme causes ectopic cartilage formation

While inactivation of *ERK1* and *ERK2* inhibited osteoblast differentiation and bone formation, we found ectopic cartilage formation in the perichondrium of *ERK1* $^{-/-}$; *ERK2* $^{flox/flox}$; *Prx1-Cre* mice (Fig. 3). The ectopic cartilage expressed *Sox9*, a transcription factor for chondrocyte differentiation, and *Col2a1*, the gene for type II collagen (Matsushita et al., 2009a). These findings suggest that inactivation of *ERK1* and *ERK2* in mesenchyme inhibited osteoblast differentiation and promoted chondrocyte differentiation. Ectopic cartilage formation has been also reported in the perichondrium of *Osterix*-null mice (Nakashima et al., 2002) and in the perichondrium of mice in which beta-catenin was disrupted in mesenchymal cells (Day et al., 2005; Hill et al., 2005). While normal *Osterix* expression was observed in *ERK1* $^{-/-}$; *ERK2* $^{flox/flox}$; *Prx1-Cre* mice, beta-catenin protein levels were decreased in the perichondrium of *ERK1* $^{-/-}$; *ERK2* $^{flox/flox}$; *Prx1-Cre* mice. These observations suggest a role for decreased beta-catenin in the ectopic cartilage formation in the perichondrium.

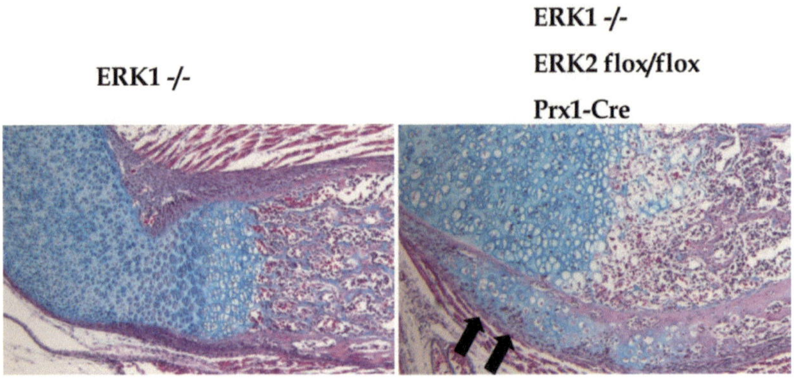

Fig. 3. Hematoxylin, eosin and alcian blue staining of femur of newborn mice. Arrows indicate ectopic cartilage formation in the perichondrium of *ERK1*$^{-/-}$;*ERK2*$^{flox/flox}$;*Prx1-Cre* mice.

6. Constitutive activation of MEK1 in undifferentiated mesenchymal cells leads to increased bone formation and inhibition of cartilage formation

As a complementary experiment, we generated *Prx1-MEK1* transgenic mice that express a constitutively active mutant of MEK1 in undifferentiated limb and cranial mesenchyme (Matsushita et al., 2009a). *Prx1-MEK1* mice showed a marked increase in cortical bone formation, fusion of long bones as well as carpal and tarsal bones, and an accelerated closure of cranial sutures, mimicking the phenotype of human craniosynostosis syndromes caused by activating mutations in FGFR2. The increase in bone formation was associated with increased expression of osteoblast markers, such as *Runx2, Osterix, Bsp*, and *Osteocalcin*. In contrast, cartilage formation was inhibited in *Prx1-MEK1* transgenic mice. There was a clear delay in the formation of cartilage anlagen as well as a decrease in anlagen size. In

Effects on the ERK MAPK pathway	Mouse models	Phenotypes
	Mesenchymal cells	
Loss of function	ERK1 $^{-/-}$; ERK2 $^{flox/flox}$; Prx1-Cre (Matsushita 2009a)	Limb deformity Delayed closure of the cranial suture Calvaria defect Inhibition of bone formation Ectopic cartilage formation in perichondrium Increase in terminally differentiated hypertrophic chondrocytes Absence of Osteocalcin expression Decreased Dmp1 expression Absence of osteocyte-lacunar-canalicur system
Gain of function	CA MEK1/Prx1 (Matsushita 2009a)	Increase in cortical bone formation Fusion of long bones, carpal and tarsal bones Accelerated closure of cranial suture Delayed and decreased formation of cartilage anlagen Increased Runx2, Osterix, Bsp, Osteocalcin expression
	Chondrocytes	
Loss of function	ERK1 $^{-/-}$; ERK2 $^{flox/flox}$; Col2a1-Cre (Matsushita 2009a)	Die immediately after birth Deformed rib cage Kyphotic deformity of the spine Disorganization of epiphyseal cartilage Widening of zone of hypertrophic chondrocytes
	ERK1 $^{-/-}$; ERK2 $^{flox/flox}$; Col2a1-CreER (Sebastian 2011)	Delayed synchondrosis closure of the vertebrae Increased vertebral foramen cross-sectional area
Gain of function	CA MEK1/Col2a1 (Murakami 2004)	Dwarfism Premature synchondrosis closure Narrower zone of hypertrophic chondrocytes Smaller hypertrophic chondrocytes Reduced rate of chondrocyte hypertrophy
	Osteoblasts	
Loss of function	DN MEK1/Osteocalcin (Ge 2007)	Delayed calvarial mineralization Delayed formation of primary ossification centers in the long bones
Gain of function	CA MEK1/Osteocalcin (Ge 2007)	Increased calvarial mineralization Accelerated trabecular bone formation

Table 1. Genetically engineered mouse models with increased or decreased ERK MAPK signaling in skeletal cells. CA MEK1/Prx1, Transgenic mice that express a constitutively active mutant of MEK1 under the control of a *Prx1* promoter. CA MEK1/Col2a1, Transgenic mice that express a constitutively active mutant of MEK1 under the control of a *Col2a1* promoter. DN MEK1/Osteocalcin, Transgenic mice that express a dominant-negative mutant of MEK1 under the control of an *Osteocalcin* promoter. CA MEK1/Osteocalcin, Transgenic mice that express a constitutively active mutant of MEK1 under the control of an *Osteocalcin* promoter.

addition, expression of *Col2a1* was reduced in the cartilage primordia. These observations are consistent with the phenotypes of *ERK1* $^{-/-}$; *ERK2* $^{flox/flox}$; *Prx1-Cre* mice, in which the lack of *ERK1* and *ERK2* disrupts bone formation and induces ectopic cartilage formation. These observations are also consistent with the in vitro studies showing inhibitory effects of the ERK MAPK pathway on chondrogenesis. The ERK MAPK inhibitors U0126 and PD98059 increased the expression of *Col2a1* and *aggrecan* in embryonic limb mesenchyme, and transfection of limb mesenchyme with constitutively active mutant of MEK decreased the activity of a Sox9-responsive *Col2a1* enhancer reporter gene (Bobick & Kulyk, 2004). Collectively, these observations indicate that ERK MAPK signaling plays an important role in the lineage specification of mesenchymal cells.

7. Inactivation of ERK1 and ERK2 in chondrocytes causes severe chondrodysplasia and enhances bone growth

Type II collagen is the most abundant collagen in cartilage. *Col2a1*, the gene encoding the proalpha1(II) collagen chain, is a principal marker of chondrocyte differentiation. We created a loss of function model of ERK1 and ERK2 in chondrocytes by using the *Col2a1-Cre* transgenic mice that express Cre recombinase under the regulatory sequences of *Col2a1* (Matsushita et al., 2009a; Ovchinnikov et al., 2000). *ERK1* $^{-/-}$; *ERK2* $^{flox/flox}$; *Col2a1-Cre* mutant mice died immediately after birth, likely secondary to respiratory insufficiency caused by rib cage deformity (Matsushita et al., 2009a). To circumvent the perinatal lethality, we also used the *Col2a1-CreER* transgene to express a tamoxifen-inducible form of Cre recombinase and examined the role of *ERK1* and *ERK2* in chondrocytes during postnatal growth (Nakamura et al., 2006; Sebastian et al., 2011). For the gain-of-function experiments, we generated *Col2a1-MEK1* transgenic mice that express a constitutively active mutant of MEK1 in chondrocytes under the control of the regulatory sequences of *Col2a1* (Murakami et al., 2004).

7.1 ERK1 and ERK2 are essential for proper organization of epiphyseal cartilage

A strong skeletal phenotype was observed in *ERK1* $^{-/-}$; *ERK2* $^{flox/flox}$; *Col2a1-Cre* embryos. We observed severe kyphotic deformities of the spine. Histological analysis of *ERK1* $^{-/-}$; *ERK2* $^{flox/flox}$; *Col2a1-Cre* embryos at embryonic day 16.5 showed an absence of primary ossification centers in the axial skeleton and a widening of the zone of hypertrophic chondrocytes in the long bones (Fig. 4). In addition, disorganization of the epiphyseal cartilage with lack of columnar growth structures was observed in *ERK1* $^{-/-}$; *ERK2* $^{flox/flox}$; *Col2a1-Cre* embryos at embryonic day 18.5. These findings indicate that ERK1 and ERK2 are essential for the proper organization of the epiphyseal cartilage.

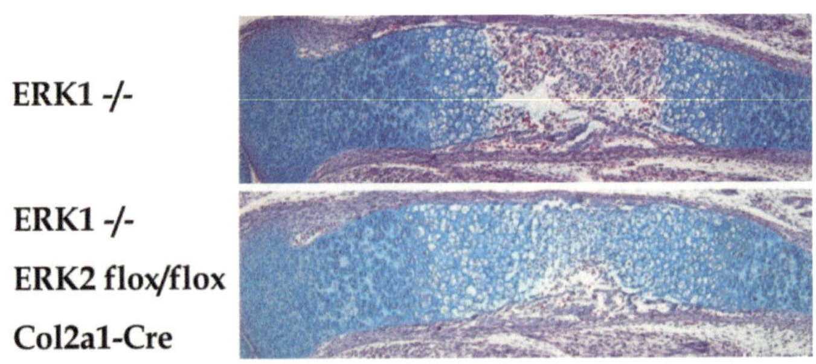

Fig. 4. Hematoxylin, eosin and alcian blue staining of the tibia showed delayed formation of primary ossification center in *ERK1* $^{-/-}$; *ERK2* $^{flox/flox}$; *Col2a1-Cre* embryos at embryonic day 16.5.

7.2 ERK1 and ERK2 inhibit hypertrophic chondrocyte differentiation

Both *ERK1* $^{-/-}$; *ERK2* $^{flox/flox}$; *Prx1-Cre* and *ERK1* $^{-/-}$; *ERK2* $^{flox/flox}$; *Col2a1-Cre* mice showed a remarkable expansion of the zone of hypertrophic chondrocytes. In both animal models, chondrocytes that were closer to the articular surface expressed hypertrophic chondrocyte marker *Col10a1,* suggesting premature chondrocyte hypertrophy. These observations are consistent with the growth plate phenotype of *Col2a1-MEK1* transgenic mice that express a constitutively active MEK1 mutant in chondrocytes (Murakami et al., 2004). The growth plate of *Col2a1-MEK1* transgenic mice was characterized by smaller than normal hypertrophic chondrocytes and narrower zone of hypertrophic chondrocytes. BrdU labeling of proliferating chondrocytes and subsequent identification of BrdU-labeled hypertrophic chondrocytes indicated reduced rate of chondrocyte hypertrophy in *Col2a1-MEK1* mice. Collectively, these observations indicate that ERK MAPK signaling inhibits hypertrophic chondrocyte differentiation.

The pronounced expansion of the zone of hypertrophic chondrocytes in *ERK1* $^{-/-}$; *ERK2* $^{flox/flox}$; *Prx1-Cre* mice was also characterized by an increase in terminally differentiated hypertrophic chondrocytes expressing *Vegf, Mmp13,* and *Osteopontin,* suggesting impaired removal of terminally differentiated hypertrophic chondrocytes. We also observed a decrease in TRAP-positive osteoclasts in association with reduced expression of *receptor activator of nuclear factor-kappa B ligand (RANKL).* Therefore, decreased osteoclastogenesis may also account for the expansion of the zone of hypertrophic chondrocytes.

7.3 ERK1 and ERK2 inhibit growth of cartilaginous skeletal element

A number of genetic studies have indicated the growth inhibitory role of FGFR3 signaling. Mice with activating mutations in Fgfr3 show a dwarf phenotype similar to the human syndromes of achondroplasia and thanatophoric dysplasia (Chen et al., 1999; Iwata et al., 2000, 2001; Li et al., 1999; Naski et al., 1998; Wang et al., 1999). In contrast, *Fgfr3*-null mice show a skeletal overgrowth (Colvin et al., 1996; Deng et al., 1996). Our observations in genetically engineered mouse models have provided evidence indicating that the ERK MAPK pathway is a critical downstream effector of Fgfr3 signaling. We have shown that *Col2a1-MEK1* transgenic mice that express a constitutively active MEK1 mutant in chondrocytes show an achondroplasia-like dwarf phenotype (Murakami et al., 2004). We have also shown that *ERK1* and *ERK2* inactivation in chondrocytes promotes bone growth. We found increased length of the proximal long bones, specifically the humerus and femur, of *ERK1* $^{-/-}$; *ERK2* $^{flox/flox}$; *Col2a1-Cre* embryos (Sebastian et al., 2011). These embryos also showed an increase in the width of epiphyses in the humerus and femur. Histological analysis of the vertebrae also showed an overgrowth of cartilage in the vertebral body. These observations indicate that ERK1 and ERK2 negatively regulate the growth of cartilaginous skeletal elements.

8. Postnatal ERK1 and ERK2 inactivation delays synchondrosis closure and enlarges the spinal canal

Our studies have indicated that FGFR3 and the MAPK pathway are important regulators of synchondrosis closure. In bones such as the vertebrae, sternum, pelvis, and bones in the cranial base, a synchondrosis—a growth plate-like cartilaginous structure—connects the ossification centers and contributes to the bone growth. During postnatal skeletal development, the width of synchondroses reduces with age. Ossification centers eventually unite when synchondroses close. Histologically, a synchondrosis consists of two opposed growth plates with a common zone of resting chondrocytes (Fig. 5A). We have found premature synchondrosis closure in the vertebrae and cranial base of human samples of achondroplasia and thanatophoric dysplasia (Matsushita et al., 2009b). In addition, we have also observed premature synchondrosis closure in a mouse model of achondroplasia and *Col2a1-MEK1* transgenic mice that express a constitutively active MEK1 mutant in chondrocytes (Fig. 5B). Because growth at the synchondrosis determines the final dimension and shape of the endochondral skeletons, premature synchondrosis closure should play a critical role in the development of spinal canal stenosis that is frequently seen in patients with achondroplasia.

Since increased Fgfr3 and MEK1 signaling accelerates synchondrosis closure, we hypothesized that *ERK1* and *ERK2* inactivation delays synchondrosis closure and enlarges the spinal canal. To test this hypothesis, we inactivated *ERK2* in chondrocytes of *ERK1*-null

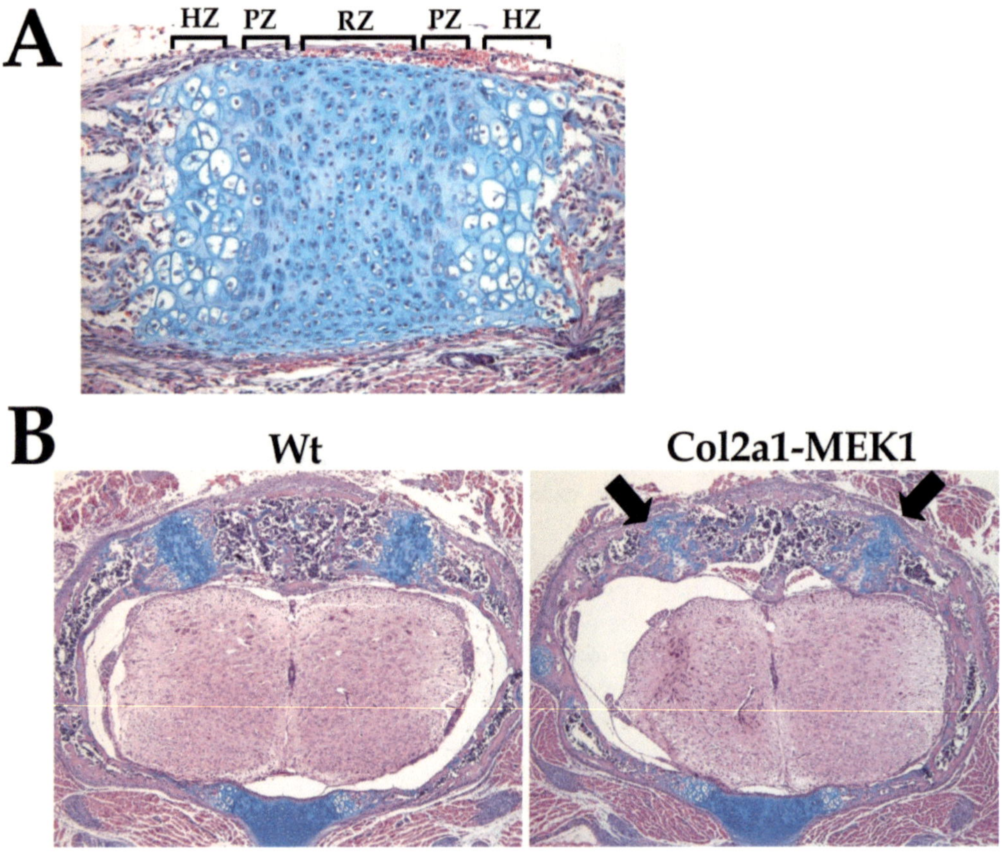

Fig. 5. (A) Spheno-occipital synchondrosis of a 4-day-old wild type mouse. HZ; hypertrophic zone. PZ; proliferation zone. RZ; resting zone. (B) Thoracic spine of wild type and *Col2a1-MEK1* transgenic mice at postnatal day 4. Arrows indicate prematurely closing synchondroses. Wt; wild type.

mice using the *Col2a1-CreER* transgene (Sebastian et al., 2011). Tamoxifen injection into *ERK1* $^{-/-}$; *ERK2* $^{flox/flox}$; *Col2a1-CreER* mice resulted in 60% inhibition of *ERK2* expression in the epiphyseal cartilage. Although these mice did not show an increased growth of the long bones presumably due to incomplete *ERK2* inactivation, we observed a significant delay in synchondrosis closure of the vertebrae and an increase in the cross-sectional area of vertebral foramen. The delayed synchondrosis closure was associated with a decreased expression of the endothelial marker CD31 surrounding the synchondroses, suggesting that *ERK1* and *ERK2* inactivation in chondrocytes causes reduced vascular invasion. These observations indicate the potential of ERK1 and ERK2 as therapeutic targets for spinal canal stenosis in achondroplasia.

9. Conclusion

By creating and analyzing gain-of-function and loss-of-function mouse models, we have identified multiple roles of the ERK MAPK pathway at successive steps of skeletal development. In undifferentiated mesenchymal cells, *ERK1* and *ERK2* inactivation causes a block in osteoblast differentiation and induces ectopic cartilage formation. In contrast, increased MEK1 signaling promotes bone formation and inhibits cartilage formation. These observations indicate that ERK MAPK signaling plays a critical role in the lineage specification of mesenchymal cells (Fig. 6). ERK MAPK signaling also inhibits hypertrophic chondrocyte differentiation and bone growth. Furthermore, ERK MAPK signaling regulates the timing of growth plate and synchondrosis closure, the very last step of endochondral ossification. A better understanding of the roles of the ERK MAPK pathway in skeletal tissues will lead to new insights in skeletal development and the treatment of various skeletal disorders.

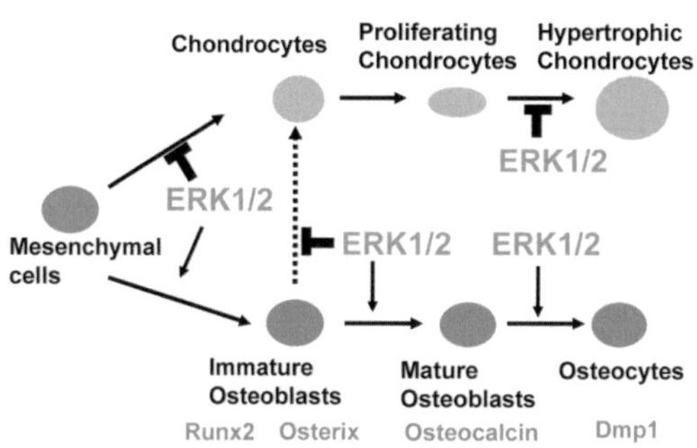

Fig. 6. Proposed model of the role of ERK1 and ERK2 in the regulation of osteoblast and chondrocyte differentiation.

10. Acknowledgments

We thank Ms. Valerie Schmedlen for editorial assistance. Work in the laboratory of the authors was supported by Arthritis Investigator Award of the Arthritis Foundation, March of Dimes Birth Defects Foundation, and NIH grants R21DE017406, R01AR055556, R03DE019814 to S.M.

11. References

Aoki, Y., Niihori, T., Kawame, H., Kurosawa, K., Ohashi, H., Tanaka, Y., Filocamo, M., Kato, K., Suzuki, Y., Kure, S., & Matsubara, Y. (2005). Germline mutations in HRAS proto-oncogene cause Costello syndrome. *Nat Genet*, Vol. 37, No. 10, pp. 1038-40, ISSN 1061-4036

Belanger, L.F., Roy, S., Tremblay, M., Brott, B., Steff, A.M., Mourad, W., Hugo, P., Erikson, R., & Charron, J. (2003). Mek2 is dispensable for mouse growth and development. *Mol Cell Biol*, Vol. 23, No. 14, pp. 4778-87. ISSN 0270-7306

Bellus, G.A., McIntosh, I., Smith, E.A., Aylsworth, A.S., Kaitila, I., Horton, W.A., Greenhaw, G.A., Hecht, J.T., & Francomano, C.A. (1995). A recurrent mutation in the tyrosine kinase domain of fibroblast growth factor receptor 3 causes hypochondroplasia. *Nat Genet,* Vol. 10, No. 3, pp. 357-9, ISSN 1061-4036

Bobick, B.E., & Kulyk, W.M. (2004). The MEK-ERK signaling pathway is a negative regulator of cartilage-specific gene expression in embryonic limb mesenchyme. *J Biol Chem*, Vol. 279, No. 6, pp. 4588-95, ISSN 0021-9258

Bobick, B.E., & Kulyk, W.M. (2008). Regulation of cartilage formation and maturation by mitogen-activated protein kinase signaling. *Birth Defects Res C Embryo Today,* Vol. 84, No. 2, pp. 131-54, ISSN 1542-975X

Cargnello, M., & Roux, P.P. (2011). Activation and function of the MAPKs and their substrates, the MAPK-activated protein kinases. *Microbiol Mol Biol Rev*, Vol. 75, No. 1, pp. 50-83, ISSN 1092-2171

Chen, L., Adar, R., Yang, X., Monsonego, E.O., Li, C., Hauschka, P.V., Yayon, A., & Deng, C.X. (1999). Gly369Cys mutation in mouse FGFR3 causes achondroplasia by affecting both chondrogenesis and osteogenesis. *J Clin Invest*, Vol. 104, No. 11, pp. 1517-25, ISSN 0021-9738

Colnot, C. (2005). Cellular and molecular interactions regulating skeletogenesis. *J Cell Biochem*, Vol.95, No. 4, pp. 688–97, ISSN 0730-2313

Colvin, J.S., Bohne, B.A., Harding, G.W., McEwen, D.G., & Ornitz, D.M. (1996). Skeletal overgrowth and deafness in mice lacking fibroblast growth factor receptor 3. *Nat Genet*, Vol. 12, No. 4, pp. 390-7, ISSN 1061-4036

Dallas, S.L., & Bonewald, L.F. (2010). Dynamics of the transition from osteoblast to osteocyte. *Ann N Y Acad Sci*, Vol. 1192, pp. 437-43, ISSN 0077-8923

David, J.P., Mehic, D., Bakiri, L., Schilling, A.F., Mandic, V., Priemel, M., Idarraga, M.H., Reschke, M.O., Hoffmann, O., Amling, M., & Wagner, E.F. (2005). Essential role of RSK2 in c-Fos-dependent osteosarcoma development. *J Clin Invest*, Vol. 115, No. 3, pp. 664-72, ISSN 0021-9738

Day, T.F., Guo, X., Garrett-Beal, L., & Yang, Y. (2005). Wnt/beta-catenin signaling in mesenchymal progenitors controls osteoblast and chondrocyte differentiation during vertebrate skeletogenesis. *Dev Cell*, Vol. 8, No. 5, pp. 739–50, ISSN 1534-5807

DeLise, A.M., Fischer, L., & Tuan, R.S. (2000). Cellular interactions and signaling in cartilage development. *Osteoarthritis Cartilage,* Vol. 8, No. 5, pp. 309-34, ISSN1063-4584

Deng, C., Wynshaw-Boris, A., Zhou, F., Kuo, A., & Leder, P. (1996). Fibroblast growth factor receptor 3 is a negative regulator of bone growth. *Cell,* Vol. 84, No. 6, pp. 911-21, ISSN 0092-8674

Feng, J.Q., Ward, L.M., Liu, S., Lu, Y., Xie, Y., Yuan, B., Yu, X., Rauch, F., Davis, S.I., Zhang, S., Rios, H., Drezner, M.K., Quarles, L.D., Bonewald, L.F., & White, K.E. (2006). Loss of DMP1 causes rickets and osteomalacia and identifies a role for osteocytes in mineral metabolism. *Nat Genet*, Vol. 38, No. 11, pp. 1310-5, ISSN 1061-4036

Ge, C., Xiao, G., Jiang, D., & Franceschi, R.T. (2007). Critical role of the extracellular signal-regulated kinase-MAPK pathway in osteoblast differentiation and skeletal development. *J Cell Biol*, Vol. 176, No. 5, pp. 709-18, ISSN 0021-9525

Giroux, S., Tremblay, M., Bernard, D., Cardin-Girard, J.F., Aubry, S., Larouche, L., Rousseau, S., Huot, J., Landry, J., Jeannotte, L., & Charron, J. (1999). Embryonic death of Mek1-deficient mice reveals a role for this kinase in angiogenesis in the labyrinthine region of the placenta. *Curr Biol*, Vol. 9, No. 7, pp. 369-72, ISSN 0960-9822

Hanauer, A., & Young, I.D. (2002). Coffin-Lowry syndrome: clinical and molecular features. *J Med Genet*, Vol. 39, No. 10, pp. 705-13, Review, ISSN 0022-2593

Hennekam, R.C. (2003). Costello syndrome: an overview. *Am J Med Genet C Semin Med Genet*, Vol. 117C, No. 1, pp. 42-8, Review, ISSN 1552-4868

Hill, T.P., Später, D., Taketo, M.M., Birchmeier, W., & Hartmann, C. (2005). Canonical Wnt/beta-catenin signaling prevents osteoblasts from differentiating into chondrocytes. *Dev Cell*, Vol. 8, No. 5, pp. 727–38, ISSN 1534-5807

Hunziker, E.B. (1994). Mechanism of longitudinal bone growth and its regulation by growth plate chondrocytes. *Microsc Res Tech*, Vol. 28, No. 6, pp. 505–19, ISSN 1059-910X

Iwata, T., Chen, L., Li, C., Ovchinnikov, D.A., Behringer, R.R., Francomano, C.A., & Deng, C.X. (2000). A neonatal lethal mutation in FGFR3 uncouples proliferation and differentiation of growth plate chondrocytes in embryos. *Hum Mol Genet*, Vol. 9, No. 11, pp. 1603-13, ISSN 0964-6906

Iwata, T., Li, C.L., Deng, C.X., & Francomano, C.A. (2001). Highly activated Fgfr3 with the K644M mutation causes prolonged survival in severe dwarf mice. *Hum Mol Genet*, Vol. 10, No. 12, pp. 1255-64. ISSN 0964-6906

Jabs, E.W., Li, X., Scott, A.F., Meyers, G., Chen, W., Eccles, M., Mao, J.I., Charnas, L.R., Jackson, C.E., & Jaye, M. (1994). Jackson-Weiss and Crouzon syndromes are allelic with mutations in fibroblast growth factor receptor 2. *Nat Genet*, Vol. 8, No. 3, pp. 275-9, ISSN 1061-4036

Jun, J.H., Yoon, W.J., Seo, S.B., Woo, K.M., Kim, G.S., Ryoo, H.M., & Baek, J.H. (2010). BMP2-activated Erk/MAP kinase stabilizes Runx2 by increasing p300 levels and histone acetyltransferase activity. *J Biol Chem*, Vol. 285, No. 47, pp. 36410-9, ISSN 0021-9258

Kim, N., Odgren, P.R., Kim, D.K., Marks, S.C.Jr., & Choi, Y. (2000). Diverse roles of the tumor necrosis factor family member TRANCE in skeletal physiology revealed by TRANCE deficiency and partial rescue by a lymphocyte-expressed TRANCE transgene. *Proc Natl Acad Sci U S A*, Vol. 97, No. 20, pp. 10905-10, ISSN 0027-8424

Kyono, A., Avishai, N., Ouyang, Z., Landreth, G.E., & Murakami, S. (2011). FGF and ERK signaling coordinately regulate mineralization-related genes and play essential roles in osteocyte differentiation. *J Bone Miner Metab*, (Jun 17). ISSN 0914-8779

Li, C., Chen, L., Iwata, T., Kitagawa, M., Fu, X.Y., & Deng, C.X. (1999). A Lys644Glu substitution in fibroblast growth factor receptor 3 (FGFR3) causes dwarfism in mice by activation of STATs and ink4 cell cycle inhibitors. *Hum Mol Genet*, Vol. 8, No. 1, pp. 35-44, ISSN 0964-6906

Logan, M., Martin, J.F., Nagy, A., Lobe, C., Olson, E.N., & Tabin, C.J. (2002). Expression of Cre Recombinase in the developing mouse limb bud driven by a Prxl enhancer. *Genesis*, Vol.33, No. 2, pp. 77-80, ISSN 1526-954X

Matsushita, T., Chan, Y., Kawanami, A., Balmes, G., Landreth, G.E., & Murakami, S. (2009a). Extracellular Signal-Regulated Kinase 1 (ERK1) and ERK2 play essential roles in osteoblast differentiation and in supporting osteoclastogenesis. *Mol Cell Biol,* Vol. 29, No. 21, pp. 5843-57, ISSN 0270-7306

Matsushita, T., Wilcox, W.R., Chan, Y.Y., Kawanami, A., Bükülmez, H., Balmes, G., Krejci, P., Mekikian, P.B., Otani, K., Yamaura, I., Warman, M.L., Givol, D., & Murakami, S. (2009b). FGFR3 promotes synchondrosis closure and fusion of ossification centers through the MAPK pathway. *Hum Mol Genet,* Vol. 18, No. 2, pp. 227-40, ISSN 0964-6906

Mu, Y., Gudey, S.K., & Landström, M. (2011). Non-Smad signaling pathways. *Cell Tissue Res,* (Jun 24).

Murakami, S., Kan, M., McKeehan, W.L., & de Crombrugghe, B. (2000). Up-regulation of the chondrogenic Sox9 gene by fibroblast growth factors is mediated by the mitogen-activated protein kinase pathway. *Proc Natl Acad Sci U S A*, Vol. 97, No. 3, pp. 1113-8, ISSN 0027-8424

Murakami, S., Balmes, G., McKinney, S., Zhang, Z., Givol, D., & de Crombrugghe, B. (2004). Constitutive activation of MEK1 in chondrocytes causes Stat1-independent achondroplasia-like dwarfism and rescues the Fgfr3-deficient mouse phenotype. *Genes Dev,* Vol. 18, No. 3, pp. 290-305, ISSN 0860-9369

Nakamura, E., Nguyen, M.T., & Mackem, S. (2006). Kinetics of tamoxifen-regulated Cre activity in mice using a cartilage-specific CreER(T) to assay temporal activity windows along the proximodistal limb skeleton. *Dev Dyn*, Vol. 235, No. 9, pp. 2603-12, ISSN 1058-8388

Nakashima, K., Zhou, X., Kunkel, G., Zhang, Z., Deng, J.M., Behringer, R.R., & de Crombrugghe, B. (2002). The novel zinc finger-containing transcription factor osterix is required for osteoblast differentiation and bone formation. *Cell*, Vol. 108, No. 1, pp. 17-29, ISSN 0092-8674

Naski, M.C., Colvin, J.S., Coffin, J.D., & Ornitz, D.M. (1998). Repression of hedgehog signaling and BMP4 expression in growth plate cartilage by fibroblast growth factor receptor 3. *Development*, Vol. 125, No. 24, pp. 4977-88, ISSN 1011-6370

Newbern, J., Zhong, J., Wickramasinghe, R.S., Li, X., Wu, Y., Samuels, I., Cherosky, N., Karlo, J.C., O'Loughlin, B., Wikenheiser, J., Gargesha, M., Doughman, Y.Q., Charron, J., Ginty, D.D., Watanabe, M., Saitta, S.C., Snider, W.D., & Landreth, G.E. (2008). Mouse and human phenotypes indicate a critical conserved role for ERK2 signaling in neural crest development. *Proc Natl Acad Sci U S A*, Vol. 105, No. 44, pp. 17115-20, ISSN 0027-8424

Noonan, J.A. (2006). Noonan syndrome and related disorders: Alterations in growth and puberty. *Rev Endocr Metab Disord*, Vol. 7, No. 4, pp. 251-5, ISSN 1389-9155

Opperman, L.A. (2000). Cranial sutures as intramembranous bone growth sites. *Dev Dyn*, Vol. 219, No. 4, pp. 472–85, ISSN 1058-8388

Osyczka, A.M., & Leboy, P.S. (2005). Bone morphogenetic protein regulation of early osteoblast genes in human marrow stromal cells is mediated by extracellular signal-regulated kinase and phosphatidylinositol 3-kinase signaling. *Endocrinology,* Vol. 146, No. 8, pp. 3428-37, ISSN 0013-7227

Ovchinnikov, D.A., Deng, J.M., Ogunrinu, G., & Behringer, R.R. (2000). Col2a1-directed expression of Cre recombinase in differentiating chondrocytes in transgenic mice. *Genesis,* Vol. 26, No. 2, pp. 145-6, ISSN 1526-954X

Pages, G., Guerin, S., Grall, D., Bonino, F., Smith, A., Anjuere, F., Auberger, P., & Pouyssegur, J. (1999). Defective thymocyte maturation in p44 MAP kinase (Erk 1) knockout mice. *Science,* Vol. 286, No. 5443, pp. 1374-7, ISSN 0036-8075

Pandit, B., Sarkozy, A., Pennacchio, L.A., Carta, C., Oishi, K., Martinelli, S., Pogna, E.A., Schackwitz, W., Ustaszewska, A., Landstrom, A., Bos, J.M., Ommen, S.R., Esposito, G., Lepri, F., Faul, C., Mundel, P., López Siguero, J.P., Tenconi, R., Selicorni, A., Rossi, C., Mazzanti, L., Torrente, I., Marino, B., Digilio, M.C., Zampino, G., Ackerman, M.J., Dallapiccola, B., Tartaglia, M., & Gelb, B.D. (2007). Gain-of-function RAF1 mutations cause Noonan and LEOPARD syndromes with hypertrophic cardiomyopathy. *Nat Genet,* Vol. 39, No. 8, pp. 1007-12, ISSN 1061-4036

Park, O.J., Kim, H.J., Woo, K.M., Baek, J.H., & Ryoo, H.M. (2010). FGF2-activated ERK mitogen-activated protein kinase enhances Runx2 acetylation and stabilization. *J Biol Chem,* Vol. 285, No. 6, pp. 3568-74, ISSN 0021-9258

Pritchard, C.A., Bolin, L., Slattery, R., Murray, R., & McMahon, M. (1996). Post-natal lethality and neurological and gastrointestinal defects in mice with targeted disruption of the A-Raf protein kinase gene. *Curr Biol,* Vol. 6, No. 5, pp. 614-7, ISSN 0960-9822

Provot, S., Nachtrab, G., Paruch, J., Chen, A.P., Silva, A., & Kronenberg, H.M. (2008). A-raf and B-raf are dispensable for normal endochondral bone development, and parathyroid hormone-related peptide suppresses extracellular signal-regulated kinase activation in hypertrophic chondrocytes. *Mol Cell Biol,* Vol. 28, No. 1, pp. 344-57, ISSN 0270-7306

Qureshi, H.Y., Sylvester, J., El Mabrouk, M., & Zafarullah, M. (2005). TGF-beta-induced expression of tissue inhibitor of metalloproteinases-3 gene in chondrocytes is mediated by extracellular signal-regulated kinase pathway and Sp1 transcription factor. *J Cell Physiol,* Vol. 203, No. 2, pp. 345-52, ISSN 0021-9541

Reynolds, J.F., Neri, G., Herrmann, J.P., Blumberg, B., Coldwell, J.G., Miles, P.V., & Opitz, J.M. (1986). New multiple congenital anomalies/mental retardation syndrome with cardio-facio-cutaneous involvement--the CFC syndrome. *Am J Med Genet,* Vol. 25, No. 3, pp. 413-27, ISSN 1552-4833

Rodriguez-Viciana, P., Tetsu, O., Tidyman, W.E., Estep, A.L., Conger, B.A., Santa Cruz, M., McCormick, F., & Rauen, K.A. (2006). Germline Mutations in Genes within the MAPK Pathway Cause Cardio-facio-cutaneous Syndrome. *Science,* Vol. 311, No. 5765, pp. 1287-90, ISSN 0036-8075

Rousseau, F., Bonaventure, J., Legeai-Mallet, L., Pelet, A., Rozet, J.M., Maroteaux, P., Le Merrer, M., & Munnich, A. (1994). Mutations in the gene encoding fibroblast growth factor receptor-3 in achondroplasia. *Nature,* Vol. 371, No. 6494, pp.252-4, ISSN 0028-0836

Rousseau, F., Saugier, P., Le Merrer, M., Munnich, A., Delezoide, A.L., Maroteaux, P., Bonaventure, J., Narcy, F., & Sanak, M. (1995). Stop codon FGFR3 mutations in thanatophoric dwarfism type 1. *Nat Genet,* Vol. 10, No. 1, pp. 11-2, ISSN 1061-4036

Roux, P.P., & Blenis, J. (2004). ERK and p38 MAPK-activated protein kinases: a family of protein kinases with diverse biological functions. *Microbiol Mol Biol Rev,* Vol. 68, No. 2, pp. 320-44, ISSN 1092-2171

Rutland, P., Pulleyn, L.J., Reardon, W., Baraitser, M., Hayward, R., Jones, B., Malcolm, S., Winter, R.M., Oldridge, M., Slaney, S.F., Poole, M.D., & Wilkie A.O.M. (1995). Identical mutations in the FGFR2 gene cause both Pfeiffer and Crouzon syndrome phenotypes. *Nat Genet,* Vol. 9, No. 2, 173-6, ISSN 1061-4036

Saba-El-Leil, M.K., Vella, F.D., Vernay, B., Voisin, L., Chen, L., Labrecque, N., Ang, S.L., & Meloche, S. (2003). An essential function of the mitogen-activated protein kinase Erk2 in mouse trophoblast development. *EMBO Rep,* Vol. 4, No. 2, pp. 964-8, ISSN 1469-221X

Schindeler, A., & Little, D.G. (2006). Ras-MAPK signaling in osteogenic differentiation: friend or foe? *J Bone Miner Res,* Vol. 21, No. 9, pp. 1331-8, ISSN 0914-8779

Schubbert, S., Zenker, M., Rowe, S.L., Boll, S., Klein, C., Bollag, G., van der Burgt, I., Musante, L., Kalscheuer, V., Wehner, L.E., Nguyen, H., West, B., Zhang, K.Y., Sistermans, E., Rauch, A., Niemeyer, C.M., Shannon, K., & Kratz, C.P. (2006). Germline KRAS mutations cause Noonan syndrome. *Nat Genet,* Vol. 38, No. 3, pp. 331-6, ISSN 1061-4036

Sebastian, A., Matsushita, T., Kawanami, A., Mackem, S., Landreth, G., & Murakami S. (2011). Genetic inactivation of ERK1 and ERK2 in chondrocytes promotes bone growth and enlarges the spinal canal. *J Orthop Res,* Vol. 29, No. 3, pp. 375-9, ISSN 0736-0266

Selcher, J.C., Nekrasova, T., Paylor, R., Landreth, G.E., & Sweatt, J.D. (2001). Mice lacking the ERK1 isoform of MAP kinase are unimpaired in emotional learning. *Learn Mem,* Vol. 8, No. 1, pp. 11-9, ISSN 1072-0502

Shapiro, I.M., Adams, C.S., Freeman, T. & Srinivas, V. (2005). Fate of the hypertrophic chondrocyte: microenvironmental perspectives on apoptosis and survival in the epiphyseal growth plate. *Birth Defects Res. C,* Vol. 75, No. 4, pp. 330–9, ISSN 1542-975X

Shiang, R., Thompson, L.M., Zhu, Y.Z., Church, D.M., Fielder, T.J., Bocian, M., Winokur, S.T., & Wasmuth, J.J. (1994). Mutations in the transmembrane domain of FGFR3 cause the most common genetic form of dwarfism, achondroplasia. *Cell,* Vol. 78, No. 2, pp. 335-42, ISSN 0092-8674

Shukla, V., Coumoul, X., Wang, R.H., Kim, H.S., & Deng, C.X. (2007). RNA interference and inhibition of MEK-ERK signaling prevent abnormal skeletal phenotypes in a mouse model of craniosynostosis. *Nat Genet,* Vol. 39, No. 9, pp. 1145-50, ISSN 1061-4036

Tatsumi, S., Ishii, K., Amizuka, N., Li, M., Kobayashi, T., Kohno, K., Ito, M., Takeshita, S., & Ikeda, K. (2007). Targeted ablation of osteocytes induces osteoporosis with defective mechanotransduction. *Cell Metab*, Vol. 5, No. 6, pp. 464-75, ISSN 1550-4131

Trivier, E., De Cesare, D., Jacquot, S., Pannetier, S., Zackai, E., Young, I., Mandel, J.L., Sassone-Corsi, P., & Hanauer, A. (1996). Mutations in the kinase Rsk-2 associated with Coffin-Lowry syndrome. *Nature*, Vol. 384, No. 1, pp. 567-70, ISSN 0028-0836

Tuli, R., Tuli, S., Nandi, S., Huang, X., Manner, P.A., Hozack, W.J., Danielson, K.G., Hall, D.J., & Tuan, R.S. (2003). Transforming growth factor-beta-mediated chondrogenesis of human mesenchymal progenitor cells involves N-cadherin and mitogen-activated protein kinase and Wnt signaling cross-talk. *J Biol Chem*, Vol. 278, No. 42, pp. 41227-36, ISSN 0021-9258

van der Burgt, I. (2007). Noonan syndrome. *Orphanet J Rare Dis*, Vol. 2, pp. 4, ISSN 1750-1172

van der Eerden, B.C., Karperien, M., & Wit, J.M. (2003). Systemic and local regulation of the growth plate. *Endocr Rev*, Vol. 24, No. 6, pp. 782-801, ISSN 0163-769X

Wang, Y., Spatz, M.K., Kannan, K., Hayk, H., Avivi, A., Gorivodsky, M., Pines, M., Yayon, A., Lonai, P., & Givol, D. (1999). A mouse model for achondroplasia produced by targeting fibroblast growth factor receptor 3. *Proc Natl Acad Sci U S A*, Vol. 96, No. 8, pp. 4455-60. ISSN 0027-8424

Wilcox, W.R., Tavormina, P.L., Krakow, D., Kitoh, H., Lachman, R.S., Wasmuth, J.J., Thompson, L.M., & Rimoin, D.L. (1998). Molecular, radiologic, and histopathologic correlations in thanatophoric dysplasia. *Am J Med Genet*, Vol. 78, No. 3, pp. 274-81, ISSN 1552-4833

Wilkie, A.O., Slaney, S.F., Oldridge, M., Poole, M.D., Ashworth, G.J., Hockley, A.D., Hayward, R.D., David, D.J., Pulleyn, L.J., Rutland, P., Malcolm, S., Winter, R.M., & Reardon, W. (1995). Apert syndrome results from localized mutations of FGFR2 and is allelic with Crouzon syndrome. *Nat Genet*, Vol. 9, No. 2, pp. 165-72, ISSN 1061-4036

Wojnowski, L., Zimmer, A.M., Beck, T.W., Hahn, H., Bernal, R., Rapp, U.R., & Zimmer, A. (1997). Endothelial apoptosis in Braf-deficient mice. *Nat Genet*, Vol. 16, No. 3, pp. 293-7, ISSN 1061-4036

Wojnowski, L., Stancato, L.F., Zimmer, A.M., Hahn, H., Beck, T.W., Larner, A.C., Rapp, U.R., & Zimmer, A. (1998). Craf-1 protein kinase is essential for mouse development. *Mech Dev*, Vol. 76, No. 1-2, pp. 141-9, ISSN 0925-4773

Xiao, G., Jiang, D., Gopalakrishnan, R., & Franceschi, R.T. (2002). Fibroblast growth factor 2 induction of the osteocalcin gene requires MAPK activity and phosphorylation of the osteoblast transcription factor, Cbfal/Runx2. *J Biol Chem*, Vol. 277, No. 39, pp. 36181-7, ISSN 0021-9258

Xiao, G., Jiang, D., Thomas, P., Benson, M.D., Guan, K., Karsenty, G., & Franceschi, R.T. (2000). MAPK pathways activate and phosphorylate the osteoblast-specific transcription factor, Cbfal. *J Biol Chem*, Vol. 275, No. 6, pp. 4453-9, ISSN 0021-9258

Xiao, Z.S., & Quarles, L.D. (2010). Role of the polycytin-primary cilia complex in bone development and mechanosensing. *Ann N Y Acad Sci*, Vol. 1192, pp. 410-21, ISSN 0077-8923

Yang, X., Matsuda, K., Bialek, P., Jacquot, S., Masuoka, H.C., Schinke, T., Li, L., Brancorsini, S., Sassone-Corsi, P., Townes, T.M., Hanauer, A., & Karsenty, G. (2004). ATF4 is a substrate of RSK2 and an essential regulator of osteoblast biology; implication for Coffin-Lowry Syndrome. *Cell*, Vol. 117, No. 3, pp. 387-98, ISSN 0092-8674

Yao, Y., Li, W., Wu, J., Germann, U.A., Su, M.S., Kuida, K., & Boucher, D.M. (2003). Extracellular signal-regulated kinase 2 is necessary for mesoderm differentiation. *Proc Natl Acad Sci U S A*, Vol. 100, No. 22, pp. 12759-64, ISSN 0027-8424

Yasuda, H., Shima, N., Nakagawa, N., Yamaguchi, K., Kinosaki, M., Mochizuki, S., Tomoyasu, A., Yano, K., Goto, M., Murakami, A., Tsuda, E., Morinaga, T., Higashio, K., Udagawa, N., Takahashi, N., & Suda, T. (1998). Osteoclast differentiation factor is a ligand for osteoprotegerin/osteoclastogenesis-inhibitory factor and is identical to TRANCE/RANKL. *Proc Natl Acad Sci U S A*, Vol. 95, No. 7, pp. 3597-602, ISSN 0027-8424

18

Protein Kinases and Pancreatic Islet Function

Gabriela Da Silva Xavier
Imperial College London, Section of Cell Biology,
Division of Diabetes, Endocrinology and Metabolism, London
UK

1. Introduction

Intense research on pancreatic islet function is fuelled by its link to the disease diabetes mellitus. Diabetes is a chronic disease that is characterised by inappropriate regulation of blood glucose levels. This dysregulation of blood glucose homeostasis occurs when the pancreas produces insufficient amounts of the hormone, insulin, or when insulin sensitive organs lose sensitivity to insulin (insulin resistance). Increasingly there is evidence to support the idea that abnormal release of another pancreatic hormone, glucagon, may be involved in the dysregulation of blood glucose homeostasis.

There are two types of diabetes. Type 1 diabetes is characterised by the failure of the pancreas to produce insulin. In most cases, type 1 diabetes is caused by autoimmune destruction of the pancreatic β cells that produce insulin and release the hormone in response to changes in blood glucose levels. Type 2 diabetes is characterised by relative insulin insufficiency and insulin resistance. This is the more common form of diabetes, comprising of 90 % of people with diabetes worldwide. The prominence of this form of diabetes is associated with lifestyle choices: the obesity epidemic has led to an increase in the incidence of diseases associated with metabolic imbalance such as diabetes. Energy homeostasis and obesity are intimately linked. Between 45-80 % of our energy intake is in the form of carbohydrates (1;2), which are converted to glucose and transported in the blood stream (3). Thus, there has been much interest in the mechanisms that regulate glucose homeostasis, particularly those involving the endocrine pancreas. The pancreatic hormones glucagon and insulin, produced and released in the pancreatice α and β cells, repectively, are involved in maintaining blood glucose homeostasis.

Uncontrolled diabetes leads to hyperglycaemia and can lead to a number of diabetes related complications over time, such as increased risk of cardiovascular disease, diabetic retinopathy, kidney failure, and diabetic neuropathy. Current figures published by the World Health Organisation (WHO) indicate that 346 million people worldwide have diabetes. It is estimated that 5 % of all deaths worldwide each year are as a consequence of diabetes and its associated complications, with more than 80 % of these deaths occuring in low- and middle-income countries (WHO figures). It is projected that the number of diabetes related deaths will double between 2005 and 2030 (WHO figures), making diabetes a major burden on society and the health system. There is currently no cure for diabetes.

2. The pancreatic endocrine compartment

Blood glucose homeostasis is regulated by the pancreatic hormones, glucagon and insulin, which are secreted from the pancreatic islets of Langerhans. There are approximately one million islets in the adult human pancreas; this equates to about 2 % of total pancreatic mass. Pancreatic islets of Langerhans consist of four different cell types: glucagon producing α cells, insulin producing β cells, somatostatin producing δ cells and pancreatic polypeptide producing PP cells. The hormones insulin and glucagon are the principal islet hormones regulating blood glucose levels. Insulin is characteristic of the fed state and is released in response to hyperglycaemia. Glucagon is characteristic of the fasted state and is released in response to hypoglycaemia. Multiple other factors, including neural factors, regulate hormone release. The regulation of hormone release occurs at the level of the single hormone producing cell, the islet of Langerhans, the pancreas and the whole organism (Fig. 1).

Each islet of Langerhans is composed of about 2000 cells; typically, 60 % of these cells are insulin producing β cells. Insulin ensures that glucose is taken up and stored by peripheral tissues (4). Insulin secretion from islet β cells is regulated by nutrient availability,

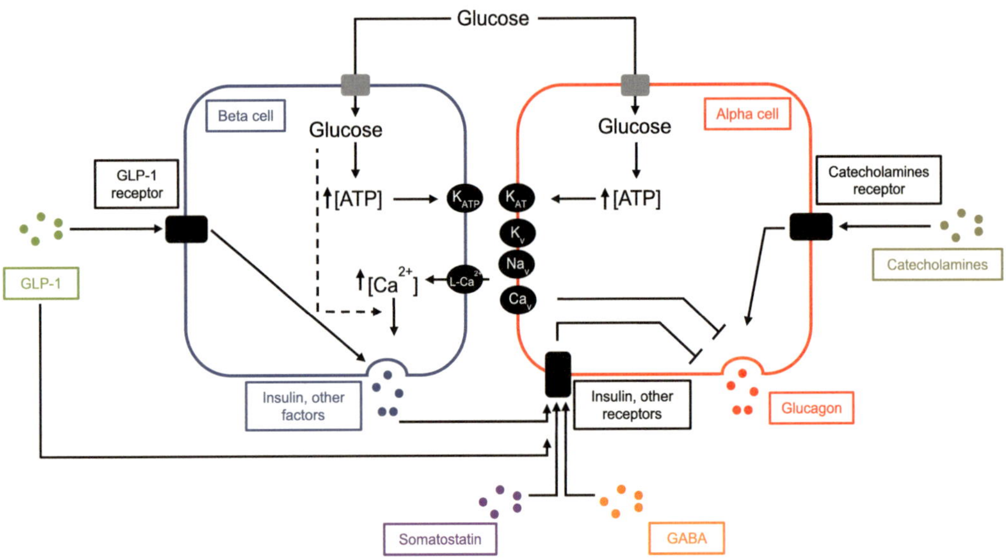

Fig. 1. **Insulin and glucagon release are regulated by nutrient availability, and a range of other factors.** Glucose entry in pancreatic α and β cells leads to increased ATP production, which in turn leads to closure of K-ATP (K_{ATP}) channels and membrane depolarisation. In the β cell this leads to opening of L-type voltage-gated calcium channels ($L\text{-}Ca^{2+}$) and subsequent release of insulin via the triggering pathway. Glucose metabolism augment the calcium signal to enhance insulin secretion through the amplification pathways (dashed line). The incretin, glucagon-like peptide 1 (GLP-1), which is secreted from L-cells in the gut, can enhance insulin secretion from pancreatic β cells. In the α cells the presence of voltage gated sodium (Na_v) and potassium (K_v) channels keep voltage-gated calcium (Ca_v) channels closed, and leads to inhibition of glucagon release when extracellular glucose concentrations are high. Insulin and other factors secreted by the pancreatic β cell, e.g. zinc ions and gamma aminobutyric acid (GABA), somatostatin (secreted by pancreatic δ cells), and catecholamines all play a part in the regulation of glucagon release (42;220-224).

neurotransmitters, and hormones. β cells are electrically excitable and transduce variations in circulating glucose concentrations into secretory signals through changes in their own metabolic state. β cells are equipped with a high capacity glucose transporter (5-7) and the high K_m type IV hexokinase (or glucokinase) (8;9). Thus, glucose phosphorylation is a rate limiting step in glucose metabolism in the β cell and glucokinase is generally known as the β cell glucose sensor (10). However, other steps in the lower glycolytic pathway are also important in regulating glucose metabolism (11-13; reviewed in 14). Ultimately, the resultant increase in cellular ATP leads to changes in the activity of the K_{ATP} and L-type voltage gated Ca^{2+} channels, influx of calcium into the cell and, subsequently, insulin secretion (15-19; Fig. 1).

2.1 Insulin release from the islet β cell

Glucose stimulated insulin secretion from islets of Langerhans is biphasic (20), with a rapid first phase and more sustained second phase. The first phase is activated by the triggering pathway- K_{ATP} channel closure following increased glucose influx into pancreatic β cells, as described above. The second phase involves the activation of amplification pathways, also called K_{ATP} channel independent mechanisms (18;21), whereby the increase in intracellular calcium concentrations following K_{ATP} channel closure leads to changes in the sensitivity of the secretory machinery (18;22). Protein kinase A and C (21), AMP activated protein kinase (AMPK; (23-27), and insulin sensitive protein kinases such as protein kinase B and p70S6 kinase (28-33), are involved in regulating the amplifying pathways. The phosphoinositide 3 kinases (PI3Ks) are also thought to be important in the regulation of insulin secretion and synthesis (34-40). The roles of some of these protein kinases will be discussed in section 3.

2.2 Glucagon release from the islet α cell

Glucagon is released from pancreatic α cells in response to low blood glucose concentration, amongst other stimuli, to maintain blood glucose levels in the fasted state (Fig. 1). Elevated glucose concentrations (> 3.5 mM) normally suppress the release of glucagon from pancreatic α cells and dysregulation of this process is a feature of both types 1 and 2 diabetes (41;42). The lack of a counter regulatory response leads to potential danger from episodes of hypoglycaemia and is a limiting factor for good glycaemic management in diabetes (43).

Many of the proteins that are involved in glucose sensing in the β cell, such as glucokinase (44), are also present in the α cell and glucose is able to raise intracellular ATP concentrations in the α cells (45). Both intrinsic (45;46) and extrinsic (42;47;48) mechanisms for the regulation of glucagon secretion have been proposed. Recent evidence suggest that two fuel sensitive protein kinase, PAS-domain containing protein kinase (PASK) and AMPK, may be involved in the regulation of glucagon release and may have a role in the pathophysiology of type 2 diabetes (49; see section 3.1 and 3.2).

3. Glucose sensing and hormone production/release

A number of protein kinases are known to play crucial roles in the regulation of islet α and β cell function. These protein kinases represent potential drug targets for the treatment of type 2 diabetes, as glucose sensitivity and secretory capacity of the pancreatic islets may be improved (or restored) through the manipulation of the action of these proteins.

3.1 AMP activated protein kinase (AMPK)

AMPK is an evolutionarily conserved fuel-sensitive protein kinase that plays a role in glucose homeostasis (23-25;50-53). It is a target of the glucose-lowering drugs, metformin and thiazolidinediones, which act to improve insulin sensitivity in insulin-sensitive tissues such as muscle and liver (54). However, the long term effects of these drugs on β cell survival and function are less clear (53).

AMPK is a heterotrimeric protein consisting of an α catalytic subunit (two isoforms, α1 and 2), a β (two isoforms, β1 and 2) scaffold subunit, and a gamma (three isoforms, gamma 1, 2, 3) regulatory subunit (55;56). It is activated by increased intracellular AMP concentrations, i.e. at times of fuel deprivation. Salt and colleagues (23) showed that both AMPK catalytic subunits are present in the clonal rat pancreatic β cell line, INS-1 (57), and that AMPK activity is regulated by extracellular glucose concentrations in this cell type (23). Moreover, they demonstrated that insulin secretion and AMPK activity were inversely related (23). AICA riboside (5-aminoimidazole-4-carboxamide riboside; AICAR) (23;25), an activator of AMPK, and metformin (58), inhibited glucose stimulated insulin secretion in both clonal β cell lines and primary rat islets. Similarly, overexpression of a constitutively active form of AMPK in clonal β cells (25) and primary islets (59) led to impaired β cell function, due in part to the inhibition of secretory granule transport to the cell surface (27). In contrast, overexpression of a dominant negative form of AMPK led to increased insulin secretion at non-premissive glucose concentrations (25) without affecting release at elevated glucose concentrations (60).

In support of the above findings, it was recently demonstrated that transgenic mice over-expressing constitutively active AMPK specifically in pancreatic β cells were glucose intolerant and displayed defective insulin secretion (26). Mice in which expression of both AMPK catalytic subunits was ablated selectively in pancreatic β cells displayed defective glucose homeostasis due to defective insulin secretion in response to hyperglycaemia *in vivo* (26). Ablation of the expression of AMPK catalytic subunits in the β cell was protective against the deleterious effects of exposure to a high fat diet on insulin secretion (26).

Aside from insulin secretion, data from work in clonal cell lines indicate that AMPK may also be involved in the regulation of insulin gene expression in response to changing glucose concentrations (24;25). Thus, inhibition of the AMPK α2 catalytic subunit (24), which displays substantial nuclear localisation (24;57), led to increased insulin gene expression, while inhibition of the AMPK α1 catalytic subunit, which is predominantly cytosolic (57), had no effect on insulin gene expression (24).

Recently, Leclerc and colleagues showed that AMPK activity is modulated by glucose in a mouse clonal α cell line, αTC1-9 cells (61). Overexpression of constitutively active AMPK in αTC1-9 cells and specifically in pancreatic α cells in primary mouse islets of Langerhans led to activation of glucagon release at inhibitory glucose concentrations (61). Activation of AMPK with metformin, phenformin (62) and a selective activator of AMPK, A-769662 (63-65), in αTC1-9 cells led to increased AMPK activity and stimulated glucagon secretion at both permissive and non-permissive glucose concentrations (61). In contrast, overexpression of dominant negative AMPK and/or the AMPK inhibitor, compound C (66), inhibited glucagon release (61).

Thus, AMPK is involved in the regulation of both the release and biosynthesis of insulin in pancreatic β cells, and glucagon release from pancreatic α cells, in response to glucose challenge. Drugs that specifically inhibit AMPK activity in the pancreatic islet are likely to be useful in the treatment of diabetes by performing the dual role of increasing insulin secretion and inhibiting glucagon secretion.

3.2 PAS domain containing protein kinase (PASK)

Whilst homologous genes are common in prokaryotes, there is only one known mammalian PASK (67;68). The enzyme has one well-defined ligand-binding domain with potential for drug-targeting (67). We (49;69) and others (70;71) have shown that PASK is important for energy-sensing and maintenance of normal cellular energy balance in mammalian systems. Thus, *PASK/Pask* is expressed in human and rodent pancreatic islets of Langerhans and its expression is regulated by glucose (49;69;71). *PASK* is expressed in both α and β cells in human pancreatic islets (49). Importantly, *PASK* expression is lower in the β cells of patients with type 2 diabetes in comparison to β cells of non-diabetic individuals (49). Very recently, an activating mutation in *PASK* was identified which is associated with non-autoimmune early-onset diabetes in humans (72). However, this mutant did not fully co-segregate with the disease and appears to serve as a modifier for a separate disease-causing mutation in another gene. Thus, the mutated PASK (G1117E) has ~25 % higher activity than wild-type PASK and overexpression of this kinase variant led to a left-shift in the glucose response in mouse pancreatic islets. As a result, glucose-stimulated insulin secretion and insulin gene expression were increased at normally non-permissive (3 mM) glucose concentrations (72).

Pask activity is regulated by glucose and the enzyme is involved in the regulation of glucose-induced preproinsulin and pancreatic duodenum *homeo*box-1 (PDX-1) gene expression in the mouse pancreatic β cell line, MIN6 (69;71). Recently, PASK was implicated in the regulation of lipogenic gene expression (70) and might, therefore, influence glucose signaling through lipid intermediates as proposed for glucose-induced insulin secretion (73). *Pask* null (*Pask$^{-/-}$*) mice (74) have normal glucose tolerance (49;70;74) and lower plasma insulin content than control littermates (49;70), but increased insulin sensitivity in peripheral tissues (70).

Glucose-stimulated insulin secretion from *Pask$^{-/-}$* islets was variously shown to be not different (74) or lower (70) *vs* control islets. However, the lack of corresponding total islet insulin measurements in these studies made it difficult to evaluate islet secretory capacity (70). In our hands, *Pask$^{-/-}$* islets were able to release insulin in response to glucose (49) but islet insulin content (49) and, thus, the amount of insulin released, was lower, in agreement with (70). Our data also indicate that the total pancreatic insulin content in *Pask$^{-/-}$* is lower than control mice (49).

Pask$^{-/-}$ mice, when maintained on a high fat diet (HFD), develop glucose intolerance (70). Furthermore, inhibition of insulin gene expression by palmitate was reversed by *PASK* over-expression in MIN6 β cells (71), suggesting a role for PASK in protection against lipotoxicity. Thus, PASK appears to exert a protective effect in mature β cells and aberrant PASK expression and function may play a role in the development of diabetes.

Our recent data (49) indicate that PASK may regulate glucagon secretion in rodent and human pancreatic α cells. Eight-week old *Pask$^{-/-}$* mice displayed higher plasma glucose after

16 h of fasting than wild-type littermate controls, but normal glucose tolerance after intraperitoneal glucose injection. After fasting, plasma glucagon was higher in *Pask-/-* mice than littermate controls. This increased glucagon concentration may account for the increased levels of blood glucose after the 16 h fast. This is physiologically relevant as long-term elevated plasma glucagon and glucose is a feature of type 2 diabetes (75).

The regulation of glucagon secretion by glucose was also impaired in *Pask-/-* islets (49). Interestingly, we observed a slight inhibitory effect of glucose on glucagon secretion, suggesting either a cell autonomous role for PASK in the α cell and/or reflecting altered secretion of regulatory factors from neighbouring β cells, e.g. insulin (49). Forced changes in PASK content affected the regulation of glucagon secretion by glucose in αTC1-9 cells (a mouse clonal α cell line) and human islets (49). RNAi-mediated silencing of *Pask* expression in αTC1-9 cells led to constitutive release of glucagon (49), while over-expression of PASK in αTC1-9 cells or in human islets led to inhibition of glucagon secretion, suggesting that PASK may be involved in glucose-sensing in pancreatic α cells (49). Inhibition of glucagon secretion by insulin was not affected in *Pask* silenced αTC1-9 cells indicating that the insulin signalling pathway was intact (49). Interestingly, there was still an apparent effect of glucose on glucagon secretion in human islets over-expressing PASK, possibly due to insulin release from β cells. Thus, the dysregulation of glucagon secretion in *Pask-/-* islets may be due, in part, to the decrease in insulin secretion.

To identify the mechanism(s) through which PASK may regulate glucagon secretion, we measured the expression of a number of potential target genes in *Pask-/-* islets (49) and αTC1-9 cells (49). The expression of both the insulin and *Pdx-1* genes was impaired in *Pask-/-* islets (49;69;71). Thus, the decrease in insulin release may be due to decreased insulin synthesis and/or islet number in *Pask-/-* pancreata (49). Preproglucagon gene expression was increased, although total glucagon protein content was unaltered (49), consistent with the relatively slow turnover of mature glucagon. AMPKα2, but not AMPKα1, gene expression was increased by loss of *Pask* expression (49). This is an interesting observation since AMPK activity is glucose-responsive in pancreatic α and β cells (24;61) and regulates insulin (25;27) and glucagon (61) release.

Interestingly, we also observed that E13.5 rat pancreatic epithelial explants (76) in which PASK gene expression was silenced also display similar gene expression changes following culture to allow the development of endocrine cells (49) indicating that loss of PASK gene expression may have effects on pancreatic development that lead to the dysregulation of glucagon release. These data, and those from pancreatic β cells (69;71;72), suggest that changes in PASK activity may be important for appropriate glucose signalling in both α and β cells.

3.3 Protein kinase A (PKA)

Glucose administered via the gastointestinal tract leads to a greater induction of glucose-stimulated insulin secretion than a comparable intravenous adminstration of glucose (77;78) due to stimulation of the release of the incretin hormones, glucagon-like peptide 1 (GLP-1) and gastric inhibitory peptide (GIP), from intestinal L- and K-cells respectively (79;80). Potentiation of glucose stimulated insulin secretion only occurs at permissive glucose concentrations (81), making the use of incretins attractive in the treatment of diabetes (80). Both hormones potentiate glucose stimulated insulin secretion (82) via binding to their cognate G-protein coupled receptors on pancreatic β cells, activating adenylyl cyclase and thereby increasing cytosolic cyclic AMP (cAMP) levels (83-85). cAMP regulates insulin secretion, in part, by inducing the phosphorylation of proteins involved in the secretory process by PKA (86-88). For example, there is evidence that PKA activation is involved in controlling vesicle exocytosis by regulating Munc 13-1 function (89;90).

3.4 Protein Kinase C (PKC)

There is evidence both for and against the involvement of PKC in stimulus secretion coupling in the β cell (reviewed in (91;92). Early studies using pharmacological modulation of PKC activity in β cells gave contradictory results (91;92). Conventional PKCs are recruited to the plasma membrane by calcium-dependent binding of the C2 domain to the phospholipids, which is potentiated by the binding of diacyglycerol to the C1 domains (93). Conventional, novel and atypical PKCs are found in pancreatic islet cells (94-97), and PKC activity is present in primary pancreatic islets (98) and clonal β cell lines (99;100). It was proposed that the increase in cytosolic calcium concentrations, following exposure of pancreatic β cells to high glucose concentrations, leads to increased diacyglycerol (DAG) production which leads to the activation of PKC, and the translocation of PKC to the plasma membrane leading to potentiation of insulin release (92;101). DAG is produced by activated phospholipase C (PLC) from phosphatidyl-4,5-bisphosphate, and PLC activity in pancreatic β cells has been shown to occur in a dose-dependent manner in parallel to physiologically relevant increases in glucose that lead to insulin secretion (102-104).

The use of PKC isoform and green fluorescent protein chimeras, coupled with the use of total internal reflection fluorescence (TIRF) microscopy, demonstrated that elevated glucose concentrations led to complex oscillatory translocation of the conventional PKC, PKCβ2, to the plasma membrane, in primary β cells and clonal β cell lines, in response to the formation of calcium microdomains following transient depolarisation of the plasma membrane (101). Moreover, PKCβ2 was shown to migrate to the surface of secretory vesicles, suggesting that the process of vesicle fusion may be regulated locally (101) by this kinase.

3.5 Phosphoinositide 3 Kinases (PI3Ks)

The autocrine feedback action of insulin on pancreatic β cell function is a subject of much debate. β cells secrete insulin at basal glucose concentrations and increase the level of secretion in response to a glucose challenge. Thus, much of the discussion has been on whether the insulin signalling pathway in β cells is desensitised as these cells are constantly exposed to the hormone. Recent studies, using approaches that circumvent some of the confounding factors in earlier studies, have provided evidence that insulin acts as a positive regulator of its own secretion and synthesis in pancreatic β cells, and of β cell mass and survival (30;32;34-36;38;39;105-130). Activation of the insulin and insulin-like growth factor 1 (IGF-1) receptor (IGF-1R), through insulin and/or IGF-1 binding to these receptors, leads to activation of downstream signalling cascades including, those involving the PI3Ks (35;36;110;111;116;126;130), resulting in the moderation of β cell function. There are three classes of mammalian PI3Ks, class I-III (131). Class Ia and II PI3Ks have been reported to be activated by insulin and will be reviewed in the following subsections. The class I PI3Ks generate phosphatidyl 3,4,5-trisphosphate (132), while the class II PI3Ks generate phosphatidyl 3-phosphate (PI3P) *in vivo* (133;134), both of which interact with distinct molecules. Therefore, the activation of class I and II PI3Ks result in the activation of distinct signaling cascades.

3.5.1 PI3Ks and insulin gene expression

There is evidence in the literature suggesting that insulin regulates the expression of its own gene via activation of Class Ia PI3K (PI3K-1a) (35;36). Of a number of transcription factors that regulate the expression of the insulin gene, the action of the transcription factors pancreatic duodenal homeobox-1 (PDX-1) (110;135-141) and FoxO-1 (142) are thought to be regulated by insulin in a PI3K-1a dependent manner.

Insulin gene expression is upregulated in response to increasing glucose concentrations in a PI3K-1a-dependent manner (30;35;36;110). The increase in insulin gene expression in response to elevated glucose concentrations is due, at least in part, to the the activation of PI3K-1a by secreted insulin (35;36;107;143;144). Thus, PDX-1 translocates from the cytosol to the nucleus in response to an increase in circulating glucose concentrations (110;144;145) and binds to a region upstream of the insulin gene called the A3 box (139). This binding is upregulated as glucose concentrations are increased in the near physiological range (140;141), and in response to insulin (146), to activate insulin gene expression. The expression of the PDX-1 gene itself is regulated by insulin (147), further implicating insulin and PI3K-1a in the feed-forward regulation of insulin gene expression.

Activation of PI3K-1a by insulin also leads to enhanced expressionof glucose/insulin-responsive genes through the removal of transcriptional repressors. For example, the transcription factor, FoxO1, is phosphorylated at Ser-256 in response to activation of PI3Ks and PKB by insulin (142;148), translocates from the nucleus to the cytosol (149;150) and is degraded (151). The degradation of FoxO1 leads to enhanced expression of glucose/insulin-responsive genes (152) such as the L-type pyruvate kinase (36), insulin 2 (*Ins2*) and *Pdx-1* genes in pancreatic β cells (142).

3.5.2 PI3Ks and insulin secretion

Data is available in the published literature suggesting that insulin has an inhibitory, activatory or no role in the regulation of its own secretion (153-160). Whilst most studies have focussed on the action of the class I PI3Ks on insulin secretion, recent data implicated the class II PI3K, PI3K-C2α, in the regulation of insulin secretion (126;130).

PI3K-C2α gene expression was shown to be down-regulated in islets of Langerhans of patients with type 2 diabetes (130). Leibiger and colleagues showed that PI3K-C2α is able to enhance glucose-stimulated insulin release in a feed-forward (126) mechanism, in response to stimuli from insulin itself, via the activation of PKB by PI3P. Subsequently, it was shown that PI3K-C2α may have a role in the late stages of insulin granule release through its action on the synaptosomal-associated protein of 25 kDA (SNAP25), which is independent of nutrient control (130). Insulin granules dock at the plasma mebrane via the interaction of SNAP25 within a protein complex (161-163), with granule fusion and insulin release occuring post proteolysis of SNAP25 (164). PI3K-C2α was shown to regulate the degradation rate of SNAP2, thereby controlling insulin granule fusion with the plasma membrane (130).

3.6 Homeodomain interacting protein kinase 2 (HIPK2)

Homeodomain interacting protein kinase 2 (HIPK2) belongs to the family of homeodomain interacting protein kinases, which was originally identified as binding partners of the homeodomain protein neurokinin-3 (NK-3) (165). Studies have shown that members of the HIPK family interact with, phosphorylate and modulate the function of other homeodomain containing proteins and transcription factors (166-170), indicating that the HIPKs may have an important role in the control of transcription. Recently it was shown that HIPK2 is expressed in the developing pancreatic epithelium from E12 to E15 and that its expression is confined preferentially to pancreatic endocrine cells later in development (171). Phosphorylation of the transcription factor, PDX-1, in the C-terminus by HIPK2, possibly at Ser-214 (172), was reported to increase the stability and transcriptional activity of PDX-1 (171). Our own data indicate that HIPK2 posphorylates PDX-1 at Ser-269 in the C-terminal portion of PDX-1 in pancreatic β cells *in vivo* and that phosphorylation at this site leads to nuclear exclusion of PDX-1 and a decrease in PDX-1 target gene expression (172). As PDX-1 has been shown to directly bind to and regulate the promoter activity of various β cell genes, e.g. insulin (110;135;140;144-146), glucose transporter 2 (GLUT2;173), glucokinase (GCK)(174;175), and islet amyloid polypeptide (176;177), HIPK2 represents an important regulator of β cell gene expression during development and in adult β cells.

4. β cell survival, growth and proliferation

Maintenance of an adequate functional β cell mass is a potential therapeutic target for diabetes. In this section we will look at the evidence that indicate protein kinases such as AMPK, Serine/threonine protein kinase 11 (STK11/LKB1), mammalian target of rapamycin (mTOR), and protein kinases in the Wnt signalling pathway may be involved in the regulation of β cell proliferation.

4.1 AMPK and LKB1

In section 3.1, we discussed the role of AMPK in the regulation of pancreatic β cell function. Recent data indicate that AMPK may also have a role in the maintenance of adequate β cell mass. Activation of AMPK has been shown to lead to decreased β cell viability (178;179), potentially through its action on the cell cycle regulator, p53 (180). Mice in which the expression of both AMPK catalytic subunits was ablated selectively in pancreatic β cells have normal β cell mass but smaller β cells and pancreatic islets (26). Islets of Langerhans in which the expression of both AMPK catalytic subunits was ablated did not display apparent differences in the ratio of α to β cells or in islet architecture. β cell proliferation was enhanced in islets from mice in which the expression of both AMPK catalytic subunits was ablated selectively in pancreatic β cells (26). In contrast, islets of Langerhans from mice in which the expression of an AMPK upstream kinase liver kinase B1 (LKB1) was ablated, had increased islet and β cell size (181) and altered islet architecture (181-184). Thus, the number of large islets, which may account for as much as 50 % of the total pancreatic β cell mass in normal pancreata (185), was increased in pancreata from mice in which LKB1 expression was selectively ablated in the β cell (181). These data indicate that AMPK and LKB1 play distinct roles in the control of islet development and cell proliferation and may impact on the development of pharmacological reagents targetting these pathways for the treatment of diabetes. In particular, inhibition of LKB1, or its downstream targets, may be a means by which β cell mass may be increased for the treatment of diabetes.

4.2 mTOR

Activation of the PI3K-PKB pathway by growth factors such as insulin and IGF-1, and subsequent activation of mammalian target of rapamycin (mTOR), is involved in β cell compensation in animals with genetic or high-fat diet induced insulin resistance (122;186). mTOR is an important nutrient sensor that plays a central role in the regulation of cellular metabolism, growth, proliferation and apoptosis (187-190). Signalling by (mTOR) to eukaryotic initiation factor 4-binding protein-1 (4E-BP1) and ribosomal S6 kinase (S6K) was enhanced in islets of Langerhans from mice in which LKB1 expression was specifically ablated in the β cell (181). Likewise, there is evidence for crosstalk between mTOR and AMPK as a regulatory pathway that couples cellular fuel availability to β cell apoptosis (58;59;191). S6K1 knockout mice are glucose intolerant, despite increased insulin sensitivity, which is associated with depletion of pancreatic insulin content, hypoinsulinaemia and reduced β cell mass (192;193), indicating that S6K1 is required for β cell growth and function. In addition, crosstalk between the mTOR and JNK pathways (194) is thought to regulate β cell survival through action on FoxO1 (195;196), whereby activation of FoxO1 in β cells protects the cells from oxidative stress by reducing cellular metabolic activity and energy-consuming processes, e.g. proliferation, and cell-specific function, e.g. insulin secretion (195;196).

4.3 Wnt signalling

The Wnt proteins are a family of cysteine-rich glycoproteins involved in intracellular signalling during vertebrate development. Activation of Wnt signalling leads to the expression of genes that are involved in promoting stem cell fate and inhibiting cell differentiation (197). One of the prominent biological phenomena controlled by Wnt signaling is the expansion of cells with predefined fates (198;199). It has previously been shown to be involved in the regulation of pancreatic development at all stages during organogenesis from specification to maintenance of normal function (200-206;206-211). Thus, tight temporal regulation of the Wnt signalling pathway is required for normal development.

Activation of the Wnt signalling pathway was shown to upregulate β cell proliferation in mouse islets (208;212;213) with upregulation of cell cycle genes that have been shown to regulate β cell proliferation (214) such as cyclin D1 and D2, and cyclin-dependent kinase 4 (CDK4) (208;213). The Wnt signalling pathway was also shown to be involved in the neogenesis of human β cells *in vitro* (215).

It was recently demonstrated that the incretin, GLP-1, induced β cell proliferation via activation of the Wnt signaling pathway (216). GLP-1 has previously been shown to increase β cell proliferation and survival (217-219). Liu and colleagues showed that Wnt signaling can be activated by downstream events from GLP-1 receptor activation in a Protein kinase B (PKB) and PKA-dependent manner (216).

5. Prospects for the development of treatment

Type 2 diabetes is fast becoming a major global problem and understanding the signalling molecules that regulate the maintenance of glucose homeostasis, particularly those that regulate pancreatic islet function, could lead to better therapeutic intervention for the disease. Increasing functional β cell mass is a particularly promising strategy: although islet transplantation is an effective means to restore glucose homeostasis, the lack of transplantable material makes this a non-viable treatment module for the masses. Thus, strategies that can lead to the generation and proliferation of β cells *in vivo* and/or *in vitro* may be important for the treatment of the disease.

6. Acknowledgements

Work in the author's laboratory is funded by the European Foundation for the Study of Diabetes, Diabetes U.K. and the Juvenile Diabetes Research Foundation.

7. References

[1] Henderson L, Gregory J, Irving K, and Swan G. The national diet and nutrition survey: adults aged 19 to 64 years. Volume 2. Energy, protein, carbohydrate , Fat and alcohol intake. HMSO: Norwich, 2003.

[2] Wright JD, Wang CY, Kennedy-Stephenson J, Ervin RB (2003) Dietary intake of ten key nutrients for public health, United States: 1999-2000. Adv.Data 1-4

[3] Burelle Y, Lamoureux MC, Peronnet F, Massicotte D, Lavoie C (2006) Comparison of exogenous glucose, fructose and galactose oxidation during exercise using 13C-labelling. Br.J.Nutr. 96: 56-61

[4] Ashcroft FM, Ashcroft SJM (1992) Insulin, Molecular Biology to Pathology. Oxford University Press, Oxford,

[5] Bell GI, Kayano T, Buse JB, et al (1990) Molecular biology of mammalian glucose transporters. Diabetes Care 13: 198-208

[6] Thorens B, Sarkar HK, Kaback HR, Lodish HF (1988) Cloning and functional expression in bacteria of a novel glucose transporter present in liver, intestine, kidney, and beta-pancreatic islet cells. Cell 55: 281-290

[7] Schuit FC (1997) Is GLUT2 required for glucose sensing? Diabetologia 40: 104-111

[8] Matschinsky FM, Meglasson M, Ghosh A, et al (1986) Biochemical design features of the pancreatic islet cell glucose-sensory system. Adv.Exp.Med.Biol. 211: 459-469

[9] Iynedjian PB (1993) Mammalian glucokinase and its gene. Biochem.J. 293 (Pt 1): 1-13

[10] Meglasson MD, Matschinsky FM (1986) Pancreatic islet glucose metabolism and regulation of insulin secretion. Diabetes Metab Rev. 2: 163-214

[11] Sekine N, Cirulli V, Regazzi R, et al (1994) Low lactate dehydrogenase and high mitochondrial glycerol phosphate dehydrogenase in pancreatic beta-cells. Potential role in nutrient sensing. J.Biol.Chem. 269: 4895-4902

[12] Schuit F, De VA, Farfari S, et al (1997) Metabolic fate of glucose in purified islet cells. Glucose-regulated anaplerosis in beta cells. J.Biol.Chem. 272: 18572-18579

[13] Berman HK, Newgard CB (1998) Fundamental metabolic differences between hepatocytes and islet beta-cells revealed by glucokinase overexpression. Biochemistry 37: 4543-4552

[14] Rutter GA (2001) Nutrient-secretion coupling in the pancreatic islet beta-cell: recent advances. Mol.Aspects Med. 22: 247-284

[15] Rorsman P (1997) The pancreatic beta-cell as a fuel sensor: an electrophysiologist's viewpoint. Diabetologia 40: 487-495

[16] Aguilar-Bryan L, Bryan J (1999) Molecular biology of adenosine triphosphate-sensitive potassium channels. Endocr.Rev. 20: 101-135

[17] Seino S, Iwanaga T, Nagashima K, Miki T (2000) Diverse roles of K(ATP) channels learned from Kir6.2 genetically engineered mice. Diabetes 49: 311-318

[18] Henquin JC (2000) Triggering and amplifying pathways of regulation of insulin secretion by glucose. Diabetes 49: 1751-1760

[19] Ashcroft FM (2007) The Walter B. Cannon Physiology in Perspective Lecture, 2007. ATP-sensitive K+ channels and disease: from molecule to malady. Am.J.Physiol Endocrinol.Metab 293: E880-E889

[20] Curry DL, Bennett LL, Grodsky GM (1968) Dynamics of insulin secretion by the perfused rat pancreas. Endocrinology 83: 572-584

[21] Aizawa T, Komatsu M, Asanuma N, Sato Y, Sharp GW (1998) Glucose action 'beyond ionic events' in the pancreatic beta cell. Trends Pharmacol.Sci. 19: 496-499

[22] Jonas JC, Gilon P, Henquin JC (1998) Temporal and quantitative correlations between insulin secretion and stably elevated or oscillatory cytoplasmic Ca2+ in mouse pancreatic beta-cells. Diabetes 47: 1266-1273

[23] Salt IP, Johnson G, Ashcroft SJ, Hardie DG (1998) AMP-activated protein kinase is activated by low glucose in cell lines derived from pancreatic beta cells, and may regulate insulin release. Biochem.J. 335 (Pt 3): 533-539

[24] da Silva Xavier G, Leclerc I, Salt IP, et al (2000) Role of AMP-activated protein kinase in the regulation by glucose of islet beta cell gene expression. Proc.Natl.Acad.Sci.U.S.A 97: 4023-4028

[25] da Silva Xavier G, Leclerc I, Varadi A, Tsuboi T, Moule SK, Rutter GA (2003) Role for AMP-activated protein kinase in glucose-stimulated insulin secretion and preproinsulin gene expression. Biochem.J. 371: 761-774

[26] Sun G, Tarasov AI, McGinty J, et al (2010) Ablation of AMP-activated protein kinase alpha1 and alpha2 from mouse pancreatic beta cells and RIP2.Cre neurons suppresses insulin release in vivo. Diabetologia 53: 924-936

[27] Tsuboi T, da Silva Xavier G, Leclerc I, Rutter GA (2003) 5'-AMP-activated protein kinase controls insulin-containing secretory vesicle dynamics. J.Biol.Chem. 278: 52042-52051

[28] Rutter GA (1999) Insulin secretion: feed-forward control of insulin biosynthesis? Curr.Biol. 9: R443-R445

[29] Leibiger B, Leibiger IB, Moede T, et al (2001) Selective insulin signaling through A and B insulin receptors regulates transcription of insulin and glucokinase genes in pancreatic beta cells. Mol.Cell 7: 559-570

[30] Leibiger B, Wahlander K, Berggren PO, Leibiger IB (2000) Glucose-stimulated insulin biosynthesis depends on insulin-stimulated insulin gene transcription. J.Biol.Chem. 275: 30153-30156

[31] Khan FA, Goforth PB, Zhang M, Satin LS (2001) Insulin activates ATP-sensitive K(+) channels in pancreatic beta-cells through a phosphatidylinositol 3-kinase-dependent pathway. Diabetes 50: 2192-2198

[32] Aspinwall CA, Lakey JR, Kennedy RT (1999) Insulin-stimulated insulin secretion in single pancreatic beta cells. J.Biol.Chem. 274: 6360-6365

[33] Kulkarni RN, Bruning JC, Winnay JN, Postic C, Magnuson MA, Kahn CR (1999) Tissue-specific knockout of the insulin receptor in pancreatic beta cells creates an insulin secretory defect similar to that in type 2 diabetes. Cell 96: 329-339

[34] Kulkarni RN, Bruning JC, Winnay JN, Postic C, Magnuson MA, Kahn CR (1999) Tissue-specific knockout of the insulin receptor in pancreatic beta cells creates an insulin secretory defect similar to that in type 2 diabetes. Cell 96: 329-339

[35] Leibiger IB, Leibiger B, Moede T, Berggren PO (1998) Exocytosis of insulin promotes insulin gene transcription via the insulin receptor/PI-3 kinase/p70 s6 kinase and CaM kinase pathways. Mol.Cell 1: 933-938

[36] da Silva Xavier G, Varadi A, Ainscow EK, Rutter GA (2000) Regulation of gene expression by glucose in pancreatic beta -cells (MIN6) via insulin secretion and activation of phosphatidylinositol 3'-kinase. J.Biol.Chem. 275: 36269-36277

[37] Andreolas C, da Silva Xavier G, Diraison F, et al (2002) Stimulation of acetyl-CoA carboxylase gene expression by glucose requires insulin release and sterol regulatory element binding protein 1c in pancreatic MIN6 beta-cells. Diabetes 51: 2536-2545

[38] da Silva Xavier G, Qian Q, Cullen PJ, Rutter GA (2004) Distinct roles for insulin and insulin-like growth factor-1 receptors in pancreatic beta-cell glucose sensing revealed by RNA silencing. Biochem.J. 377: 149-158

[39] Kulkarni RN (2002) Receptors for insulin and insulin-like growth factor-1 and insulin receptor substrate-1 mediate pathways that regulate islet function. Biochem.Soc.Trans. 30: 317-322

[40] Burks DJ, White MF (2001) IRS proteins and beta-cell function. Diabetes 50 Suppl 1: S140-S145

[41] Unger RH (1985) Glucagon physiology and pathophysiology in the light of new advances. Diabetologia 28: 574-578

[42] Gromada J, Franklin I, Wollheim CB (2007) Alpha-cells of the endocrine pancreas: 35 years of research but the enigma remains. Endocr.Rev. 28: 84-116

[43] Cryer PE (2002) Hypoglycaemia: the limiting factor in the glycaemic management of Type I and Type II diabetes. Diabetologia 45: 937-948

[44] Heimberg H, De VA, Moens K, et al (1996) The glucose sensor protein glucokinase is expressed in glucagon-producing alpha-cells. Proc.Natl.Acad.Sci.U.S.A 93: 7036-7041

[45] Ravier MA, Rutter GA (2005) Glucose or insulin, but not zinc ions, inhibit glucagon secretion from mouse pancreatic alpha-cells. Diabetes 54: 1789-1797

[46] Rorsman P, Salehi SA, Abdulkader F, Braun M, MacDonald PE (2008) K(ATP)-channels and glucose-regulated glucagon secretion. Trends Endocrinol.Metab 19: 277-284

[47] Ishihara H, Maechler P, Gjinovci A, Herrera PL, Wollheim CB (2003) Islet beta-cell secretion determines glucagon release from neighbouring alpha-cells. Nat.Cell Biol. 5: 330-335

[48] Zhou H, Zhang T, Harmon JS, Bryan J, Robertson RP (2007) Zinc, not insulin, regulates the rat alpha-cell response to hypoglycemia in vivo. Diabetes 56: 1107-1112

[49] da Silva Xavier G, Farhan H, Kim H, et al (2011) Per-arnt-sim (PAS) domain-containing protein kinase is downregulated in human islets in type 2 diabetes and regulates glucagon secretion. Diabetologia 54: 819-827

[50] Shaw RJ, Lamia KA, Vasquez D, et al (2005) The kinase LKB1 mediates glucose homeostasis in liver and therapeutic effects of metformin. Science 310: 1642-1646
[51] Towler MC, Hardie DG (2007) AMP-activated protein kinase in metabolic control and insulin signaling. Circ.Res. 100: 328-341
[52] Kurth-Kraczek EJ, Hirshman MF, Goodyear LJ, Winder WW (1999) 5' AMP-activated protein kinase activation causes GLUT4 translocation in skeletal muscle. Diabetes 48: 1667-1671
[53] Rutter GA, Leclerc I (2009) The AMP-regulated kinase family: enigmatic targets for diabetes therapy. Mol.Cell Endocrinol. 297: 41-49
[54] Long YC, Zierath JR (2006) AMP-activated protein kinase signaling in metabolic regulation. J.Clin.Invest 116: 1776-1783
[55] Stapleton D, Woollatt E, Mitchelhill KI, et al (1997) AMP-activated protein kinase isoenzyme family: subunit structure and chromosomal location. FEBS Lett. 409: 452-456
[56] Hardie DG, Carling D (1997) The AMP-activated protein kinase--fuel gauge of the mammalian cell? Eur.J.Biochem. 246: 259-273
[57] Salt I, Celler JW, Hawley SA, et al (1998) AMP-activated protein kinase: greater AMP dependence, and preferential nuclear localization, of complexes containing the alpha2 isoform. Biochem.J. 334 (Pt 1): 177-187
[58] Leclerc I, Woltersdorf WW, da Silva Xavier G, et al (2004) Metformin, but not leptin, regulates AMP-activated protein kinase in pancreatic islets: impact on glucose-stimulated insulin secretion. Am.J.Physiol Endocrinol.Metab 286: E1023-E1031
[59] Richards SK, Parton LE, Leclerc I, Rutter GA, Smith RM (2005) Over-expression of AMP-activated protein kinase impairs pancreatic {beta}-cell function in vivo. J.Endocrinol. 187: 225-235
[60] Gleason CE, Lu D, Witters LA, Newgard CB, Birnbaum MJ (2007) The role of AMPK and mTOR in nutrient sensing in pancreatic beta-cells. J.Biol.Chem. 282: 10341-10351
[61] Leclerc I, Sun G, Morris C, Fernandez-Millan E, Nyirenda M, Rutter GA (2011) AMP-activated protein kinase regulates glucagon secretion from mouse pancreatic alpha cells. Diabetologia 54: 125-134
[62] Hardie DG (2006) Neither LKB1 nor AMPK are the direct targets of metformin. Gastroenterology 131: 973-975
[63] Hawley SA, Boudeau J, Reid JL, et al (2003) Complexes between the LKB1 tumor suppressor, STRAD alpha/beta and MO25 alpha/beta are upstream kinases in the AMP-activated protein kinase cascade. J.Biol. 2: 28
[64] Scott JW, van Denderen BJ, Jorgensen SB, et al (2008) Thienopyridone drugs are selective activators of AMP-activated protein kinase beta1-containing complexes. Chem.Biol. 15: 1220-1230
[65] Cool B, Zinker B, Chiou W, et al (2006) Identification and characterization of a small molecule AMPK activator that treats key components of type 2 diabetes and the metabolic syndrome. Cell Metab 3: 403-416
[66] Zhou G, Myers R, Li Y, et al (2001) Role of AMP-activated protein kinase in mechanism of metformin action. J.Clin.Invest 108: 1167-1174

[67] Rutter J, Michnoff CH, Harper SM, Gardner KH, McKnight SL (2001) PAS kinase: an evolutionarily conserved PAS domain-regulated serine/threonine kinase. Proc.Natl.Acad.Sci.U.S.A 98: 8991-8996

[68] Hofer T, Spielmann P, Stengel P, et al (2001) Mammalian PASKIN, a PAS-serine/threonine kinase related to bacterial oxygen sensors. Biochem.Biophys.Res.Commun. 288: 757-764

[69] da Silva Xavier G, Rutter J, Rutter GA (2004) Involvement of Per-Arnt-Sim (PAS) kinase in the stimulation of preproinsulin and pancreatic duodenum homeobox 1 gene expression by glucose. Proc.Natl.Acad.Sci.U.S.A 101: 8319-8324

[70] Hao HX, Cardon CM, Swiatek W, et al (2007) PAS kinase is required for normal cellular energy balance. Proc.Natl.Acad.Sci.U.S.A 104: 15466-15471

[71] Fontes G, Semache M, Hagman DK, et al (2009) Involvement of Per-Arnt-Sim Kinase and extracellular-regulated kinases-1/2 in palmitate inhibition of insulin gene expression in pancreatic beta-cells. Diabetes 58: 2048-2058

[72] Semplici F, Vaxillaire M, Fogarty S, et al (2011) A human mutation within the per-ARNT-sim (PAS) domain-containing protein kinase (PASK) causes basal insulin hypersecretion. J.Biol.Chem.

[73] Corkey BE, Deeney JT, Yaney GC, Tornheim K, Prentki M (2000) The role of long-chain fatty acyl-CoA esters in beta-cell signal transduction. J.Nutr. 130: 299S-304S

[74] Borter E, Niessen M, Zuellig R, et al (2007) Glucose-stimulated insulin production in mice deficient for the PAS kinase PASKIN. Diabetes 56: 113-117

[75] Li XC, Liao TD, Zhuo JL (2008) Long-term hyperglucagonaemia induces early metabolic and renal phenotypes of Type 2 diabetes in mice. Clin.Sci.(Lond) 114: 591-601

[76] Miralles F, Serup P, Cluzeaud F, Vandewalle A, Czernichow P, Scharfmann R (1999) Characterization of beta cells developed in vitro from rat embryonic pancreatic epithelium. Dev.Dyn. 214: 116-126

[77] Elrick H, Stimmler L, Hlad CJ, Rai Y (1964) Plasma insulin response to oral and intravenous glucose administration. J.Clin.Endocrinol.Metab 24: 1076-1082

[78] Perley MJ, Kipnis DM (1967) Plasma insulin responses to oral and intravenous glucose: studies in normal and diabetic sujbjects. J.Clin.Invest 46: 1954-1962

[79] Holst JJ, Gromada J (2004) Role of incretin hormones in the regulation of insulin secretion in diabetic and nondiabetic humans. Am.J.Physiol Endocrinol.Metab 287: E199-E206

[80] Baggio LL, Drucker DJ (2007) Biology of incretins: GLP-1 and GIP. Gastroenterology 132: 2131-2157

[81] Weir GC, Mojsov S, Hendrick GK, Habener JF (1989) Glucagonlike peptide I (7-37) actions on endocrine pancreas. Diabetes 38: 338-342

[82] Dupre J, Ross SA, Watson D, Brown JC (1973) Stimulation of insulin secretion by gastric inhibitory polypeptide in man. J.Clin.Endocrinol.Metab 37: 826-828

[83] Thorens B (1992) Expression cloning of the pancreatic beta cell receptor for the gluco-incretin hormone glucagon-like peptide 1. Proc.Natl.Acad.Sci.U.S.A 89: 8641-8645

[84] Usdin TB, Mezey E, Button DC, Brownstein MJ, Bonner TI (1993) Gastric inhibitory polypeptide receptor, a member of the secretin-vasoactive intestinal peptide receptor family, is widely distributed in peripheral organs and the brain. Endocrinology 133: 2861-2870

[85] Yasuda K, Inagaki N, Yamada Y, Kubota A, Seino S, Seino Y (1994) Hamster gastric inhibitory polypeptide receptor expressed in pancreatic islets and clonal insulin-secreting cells: its structure and functional properties. Biochem.Biophys.Res.Commun. 205: 1556-1562

[86] Seino S, Shibasaki T (2005) PKA-dependent and PKA-independent pathways for cAMP-regulated exocytosis. Physiol Rev. 85: 1303-1342

[87] Malaisse WJ, Malaisse-Lagae F (1984) The role of cyclic AMP in insulin release. Experientia 40: 1068-1074

[88] Prentki M, Matschinsky FM (1987) Ca2+, cAMP, and phospholipid-derived messengers in coupling mechanisms of insulin secretion. Physiol Rev. 67: 1185-1248

[89] Kwan EP, Gao X, Leung YM, Gaisano HY (2007) Activation of exchange protein directly activated by cyclic adenosine monophosphate and protein kinase A regulate common and distinct steps in promoting plasma membrane exocytic and granule-to-granule fusions in rat islet beta cells. Pancreas 35: e45-e54

[90] Kwan EP, Xie L, Sheu L, Ohtsuka T, Gaisano HY (2007) Interaction between Munc13-1 and RIM is critical for glucagon-like peptide-1 mediated rescue of exocytotic defects in Munc13-1 deficient pancreatic beta-cells. Diabetes 56: 2579-2588

[91] Jones PM, Persaud SJ, Howell SL (1991) Protein kinase C and the regulation of insulin secretion from pancreatic B cells. J.Mol.Endocrinol. 6: 121-127

[92] Wollheim CB, Regazzi R (1990) Protein kinase C in insulin releasing cells. Putative role in stimulus secretion coupling. FEBS Lett. 268: 376-380

[93] Nishizuka Y (1988) The molecular heterogeneity of protein kinase C and its implications for cellular regulation. Nature 334: 661-665

[94] Tian YM, Urquidi V, Ashcroft SJ (1996) Protein kinase C in beta-cells: expression of multiple isoforms and involvement in cholinergic stimulation of insulin secretion. Mol.Cell Endocrinol. 119: 185-193

[95] Onoda K, Hagiwara M, Hachiya T, Usuda N, Nagata T, Hidaka H (1990) Different expression of protein kinase C isozymes in pancreatic islet cells. Endocrinology 126: 1235-1240

[96] Arkhammar P, Juntti-Berggren L, Larsson O, et al (1994) Protein kinase C modulates the insulin secretory process by maintaining a proper function of the beta-cell voltage-activated Ca2+ channels. J.Biol.Chem. 269: 2743-2749

[97] Kaneto H, Suzuma K, Sharma A, Bonner-Weir S, King GL, Weir GC (2002) Involvement of protein kinase C beta 2 in c-myc induction by high glucose in pancreatic beta-cells. J.Biol.Chem. 277: 3680-3685

[98] Tanigawa K, Kuzuya H, Imura H, et al (1982) Calcium-activated, phospholipid-dependent protein kinase in rat pancreas islets of langerhans. Its possible role in glucose-induced insulin release. FEBS Lett. 138: 183-186

[99] Lord JM, Ashcroft SJ (1984) Identification and characterization of Ca2+-phospholipid-dependent protein kinase in rat islets and hamster beta-cells. Biochem.J. 219: 547-551

[100] Zawalich W, Brown C, Rasmussen H (1983) Insulin secretion: combined effects of phorbol ester and A23187. Biochem.Biophys.Res.Commun. 117: 448-455

[101] Pinton P, Tsuboi T, Ainscow EK, Pozzan T, Rizzuto R, Rutter GA (2002) Dynamics of glucose-induced membrane recruitment of protein kinase C beta II in living pancreatic islet beta-cells. J.Biol.Chem. 277: 37702-37710

[102] Best L, Malaisse WJ (1984) Nutrient and hormone-neurotransmitter stimuli induce hydrolysis of polyphosphoinositides in rat pancreatic islets. Endocrinology 115: 1814-1820

[103] Zawalich W (1990) Multiple effects of increases in phosphoinositide hydrolysis on islets and their relationship to changing patterns of insulin secretion. Diabetes Res. 13: 101-111

[104] Zawalich WS, Zawalich KC (1996) Regulation of insulin secretion by phospholipase C. Am.J.Physiol 271: E409-E416

[105] Rothenberg PL, Willison LD, Simon J, Wolf BA (1995) Glucose-induced insulin receptor tyrosine phosphorylation in insulin-secreting beta-cells. Diabetes 44: 802-809

[106] Harbeck MC, Louie DC, Howland J, Wolf BA, Rothenberg PL (1996) Expression of insulin receptor mRNA and insulin receptor substrate 1 in pancreatic islet beta-cells. Diabetes 45: 711-717

[107] Xu GG, Rothenberg PL (1998) Insulin receptor signaling in the beta-cell influences insulin gene expression and insulin content: evidence for autocrine beta-cell regulation. Diabetes 47: 1243-1252

[108] Porzio O, Federici M, Hribal ML, et al (1999) The Gly972-->Arg amino acid polymorphism in IRS-1 impairs insulin secretion in pancreatic beta cells. J.Clin.Invest 104: 357-364

[109] Xu GG, Gao ZY, Borge PD, Jr., Jegier PA, Young RA, Wolf BA (2000) Insulin regulation of beta-cell function involves a feedback loop on SERCA gene expression, Ca(2+) homeostasis, and insulin expression and secretion. Biochemistry 39: 14912-14919

[110] Rafiq I, da Silva Xavier G, Hooper S, Rutter GA (2000) Glucose-stimulated preproinsulin gene expression and nuclear trans-location of pancreatic duodenum homeobox-1 require activation of phosphatidylinositol 3-kinase but not p38 MAPK/SAPK2. J.Biol.Chem. 275: 15977-15984

[111] Aspinwall CA, Qian WJ, Roper MG, Kulkarni RN, Kahn CR, Kennedy RT (2000) Roles of insulin receptor substrate-1, phosphatidylinositol 3-kinase, and release of intracellular Ca2+ stores in insulin-stimulated insulin secretion in beta -cells. J.Biol.Chem. 275: 22331-22338

[112] Assmann A, Ueki K, Winnay JN, Kadowaki T, Kulkarni RN (2009) Glucose effects on beta-cell growth and survival require activation of insulin receptors and insulin receptor substrate 2. Mol.Cell Biol. 29: 3219-3228

[113] Hennige AM, Ozcan U, Okada T, et al (2005) Alterations in growth and apoptosis of insulin receptor substrate-1-deficient beta-cells. Am.J.Physiol Endocrinol.Metab 289: E337-E346

[114] Hennige AM, Burks DJ, Ozcan U, et al (2003) Upregulation of insulin receptor substrate-2 in pancreatic beta cells prevents diabetes. J.Clin.Invest 112: 1521-1532

[115] Hisanaga E, Nagasawa M, Ueki K, Kulkarni RN, Mori M, Kojima I (2009) Regulation of calcium-permeable TRPV2 channel by insulin in pancreatic beta-cells. Diabetes 58: 174-184

[116] Kaneko K, Ueki K, Takahashi N, et al (2010) Class IA phosphatidylinositol 3-kinase in pancreatic beta cells controls insulin secretion by multiple mechanisms. Cell Metab 12: 619-632

[117] Kulkarni RN, Almind K, Goren HJ, et al (2003) Impact of genetic background on development of hyperinsulinemia and diabetes in insulin receptor/insulin receptor substrate-1 double heterozygous mice. Diabetes 52: 1528-1534

[118] Kulkarni RN, Holzenberger M, Shih DQ, et al (2002) beta-cell-specific deletion of the Igf1 receptor leads to hyperinsulinemia and glucose intolerance but does not alter beta-cell mass. Nat.Genet. 31: 111-115

[119] Kulkarni RN, Winnay JN, Daniels M, et al (1999) Altered function of insulin receptor substrate-1-deficient mouse islets and cultured beta-cell lines. J.Clin.Invest 104: R69-R75

[120] Leibiger B, Leibiger IB, Moede T, et al (2001) Selective insulin signaling through A and B insulin receptors regulates transcription of insulin and glucokinase genes in pancreatic beta cells. Mol.Cell 7: 559-570

[121] Liu S, Okada T, Assmann A, et al (2009) Insulin signaling regulates mitochondrial function in pancreatic beta-cells. PLoS.One. 4: e7983

[122] Okada T, Liew CW, Hu J, et al (2007) Insulin receptors in beta-cells are critical for islet compensatory growth response to insulin resistance. Proc.Natl.Acad.Sci.U.S.A 104: 8977-8982

[123] Otani K, Kulkarni RN, Baldwin AC, et al (2004) Reduced beta-cell mass and altered glucose sensing impair insulin-secretory function in betaIRKO mice. Am.J.Physiol Endocrinol.Metab 286: E41-E49

[124] Roper MG, Qian WJ, Zhang BB, Kulkarni RN, Kahn CR, Kennedy RT (2002) Effect of the insulin mimetic L-783,281 on intracellular Ca2+ and insulin secretion from pancreatic beta-cells. Diabetes 51 Suppl 1: S43-S49

[125] Ueki K, Okada T, Hu J, et al (2006) Total insulin and IGF-I resistance in pancreatic beta cells causes overt diabetes. Nat.Genet. 38: 583-588

[126] Leibiger B, Moede T, Uhles S, et al (2010) Insulin-feedback via PI3K-C2alpha activated PKBalpha/Akt1 is required for glucose-stimulated insulin secretion. FASEB J. 24: 1824-1837

[127] Leibiger B, Moede T, Uhles S, Berggren PO, Leibiger IB (2002) Short-term regulation of insulin gene transcription. Biochem.Soc.Trans. 30: 312-317

[128] Leibiger IB, Leibiger B, Berggren PO (2008) Insulin signaling in the pancreatic beta-cell. Annu.Rev.Nutr. 28: 233-251

[129] Leibiger IB, Leibiger B, Berggren PO (2002) Insulin feedback action on pancreatic beta-cell function. FEBS Lett. 532: 1-6

[130] Dominguez V, Raimondi C, Somanath S, et al (2011) Class II phosphoinositide 3-kinase regulates exocytosis of insulin granules in pancreatic beta cells. J.Biol.Chem. 286: 4216-4225

[131] Fruman DA, Meyers RE, Cantley LC (1998) Phosphoinositide kinases. Annu. Rev. Biochem. 67: 481-507

[132] Cantley LC (2002) The phosphoinositide 3-kinase pathway. Science 296: 1655-1657

[133] Maffucci T, Cooke FT, Foster FM, Traer CJ, Fry MJ, Falasca M (2005) Class II phosphoinositide 3-kinase defines a novel signaling pathway in cell migration. J.Cell Biol. 169: 789-799

[134] Falasca M, Hughes WE, Dominguez V, et al (2007) The role of phosphoinositide 3-kinase C2alpha in insulin signaling. J.Biol.Chem. 282: 28226-28236

[135] Ohlsson H, Karlsson K, Edlund T (1993) IPF1, a homeodomain-containing transactivator of the insulin gene. EMBO J. 12: 4251-4259

[136] Leonard J, Peers B, Johnson T, Ferreri K, Lee S, Montminy MR (1993) Characterization of somatostatin transactivating factor-1, a novel homeobox factor that stimulates somatostatin expression in pancreatic islet cells. Mol.Endocrinol. 7: 1275-1283

[137] Miller CP, McGehee RE, Jr., Habener JF (1994) IDX-1: a new homeodomain transcription factor expressed in rat pancreatic islets and duodenum that transactivates the somatostatin gene. EMBO J. 13: 1145-1156

[138] Boam DS, Clark AR, Docherty K (1990) Positive and negative regulation of the human insulin gene by multiple trans-acting factors. J.Biol.Chem. 265: 8285-8296

[139] Petersen HV, Serup P, Leonard J, Michelsen BK, Madsen OD (1994) Transcriptional regulation of the human insulin gene is dependent on the homeodomain protein STF1/IPF1 acting through the CT boxes. Proc.Natl.Acad.Sci.U.S.A 91: 10465-10469

[140] MacFarlane WM, Read ML, Gilligan M, Bujalska I, Docherty K (1994) Glucose modulates the binding activity of the beta-cell transcription factor IUF1 in a phosphorylation-dependent manner. Biochem.J. 303 (Pt 2): 625-631

[141] Marshak S, Totary H, Cerasi E, Melloul D (1996) Purification of the beta-cell glucose-sensitive factor that transactivates the insulin gene differentially in normal and transformed islet cells. Proc.Natl.Acad.Sci.U.S.A 93: 15057-15062

[142] Meur G, Qian Q, da Silva Xavier G, et al (2011) Nucleo-cytosolic shuttling of FoxO1 directly regulates mouse Ins2 but not Ins1 gene expression in pancreatic beta cells (MIN6). J.Biol.Chem. 286: 13647-13656

[143] Rutter GA (1999) Insulin secretion: feed-forward control of insulin biosynthesis? Curr.Biol. 9: R443-R445

[144] MacFarlane WM, McKinnon CM, Felton-Edkins ZA, Cragg H, James RF, Docherty K (1999) Glucose stimulates translocation of the homeodomain transcription factor PDX1 from the cytoplasm to the nucleus in pancreatic beta-cells. J.Biol.Chem. 274: 1011-1016

[145] Rafiq I, Kennedy HJ, Rutter GA (1998) Glucose-dependent translocation of insulin promoter factor-1 (IPF-1) between the nuclear periphery and the nucleoplasm of single MIN6 beta-cells. J.Biol.Chem. 273: 23241-23247

[146] Wu H, MacFarlane WM, Tadayyon M, Arch JR, James RF, Docherty K (1999) Insulin stimulates pancreatic-duodenal homoeobox factor-1 (PDX1) DNA-binding activity and insulin promoter activity in pancreatic beta cells. Biochem.J. 344 Pt 3: 813-818

[147] Campbell SC, MacFarlane WM (2002) Regulation of the pdx1 gene promoter in pancreatic beta-cells. Biochem.Biophys.Res.Commun. 299: 277-284

[148] Cahill CM, Tzivion G, Nasrin N, et al (2001) Phosphatidylinositol 3-kinase signaling inhibits DAF-16 DNA binding and function via 14-3-3-dependent and 14-3-3-independent pathways. J.Biol.Chem. 276: 13402-13410

[149] Trumper A, Trumper K, Trusheim H, Arnold R, Goke B, Horsch D (2001) Glucose-dependent insulinotropic polypeptide is a growth factor for beta (INS-1) cells by pleiotropic signaling. Mol.Endocrinol. 15: 1559-1570

[150] Rena G, Guo S, Cichy SC, Unterman TG, Cohen P (1999) Phosphorylation of the transcription factor forkhead family member FKHR by protein kinase B. J.Biol.Chem. 274: 17179-17183

[151] Matsuzaki H, Daitoku H, Hatta M, Tanaka K, Fukamizu A (2003) Insulin-induced phosphorylation of FKHR (Foxo1) targets to proteasomal degradation. Proc.Natl.Acad.Sci.U.S.A 100: 11285-11290

[152] Brunet A, Bonni A, Zigmond MJ, et al (1999) Akt promotes cell survival by phosphorylating and inhibiting a Forkhead transcription factor. Cell 96: 857-868

[153] Zawalich WS, Zawalich KC (2000) Glucose-induced insulin secretion from islets of fasted rats: modulation by alternate fuel and neurohumoral agonists. J.Endocrinol. 166: 111-120

[154] Zawalich WS, Tesz GJ, Zawalich KC (2002) Inhibitors of phosphatidylinositol 3-kinase amplify insulin release from islets of lean but not obese mice. J.Endocrinol. 174: 247-258

[155] Hagiwara S, Sakurai T, Tashiro F, et al (1995) An inhibitory role for phosphatidylinositol 3-kinase in insulin secretion from pancreatic B cell line MIN6. Biochem.Biophys.Res.Commun. 214: 51-59

[156] Turner MD, Arvan P (2000) Protein traffic from the secretory pathway to the endosomal system in pancreatic beta-cells. J.Biol.Chem. 275: 14025-14030

[157] Li LX, MacDonald PE, Ahn DS, Oudit GY, Backx PH, Brubaker PL (2006) Role of phosphatidylinositol 3-kinasegamma in the beta-cell: interactions with glucagon-like peptide-1. Endocrinology 147: 3318-3325

[158] MacDonald PE, Joseph JW, Yau D, et al (2004) Impaired glucose-stimulated insulin secretion, enhanced intraperitoneal insulin tolerance, and increased beta-cell mass in mice lacking the p110gamma isoform of phosphoinositide 3-kinase. Endocrinology 145: 4078-4083

[159] Pigeau GM, Kolic J, Ball BJ, et al (2009) Insulin granule recruitment and exocytosis is dependent on p110gamma in insulinoma and human beta-cells. Diabetes 58: 2084-2092

[160] Aoyagi K, Ohara-Imaizumi M, Nishiwaki C, Nakamichi Y, Nagamatsu S (2010) Insulin/phosphoinositide 3-kinase pathway accelerates the glucose-induced first-phase insulin secretion through TrpV2 recruitment in pancreatic beta-cells. Biochem.J. 432: 375-386

[161] Rorsman P, Renstrom E (2003) Insulin granule dynamics in pancreatic beta cells. Diabetologia 46: 1029-1045

[162] Iezzi M, Kouri G, Fukuda M, Wollheim CB (2004) Synaptotagmin V and IX isoforms control Ca2+ -dependent insulin exocytosis. J.Cell Sci. 117: 3119-3127

[163] Gauthier BR, Duhamel DL, Iezzi M, et al (2008) Synaptotagmin VII splice variants alpha, beta, and delta are expressed in pancreatic beta-cells and regulate insulin exocytosis. FASEB J. 22: 194-206

[164] Marshall C, Hitman GA, Partridge CJ, et al (2005) Evidence that an isoform of calpain-10 is a regulator of exocytosis in pancreatic beta-cells. Mol.Endocrinol. 19: 213-224

[165] Kim YH, Choi CY, Lee SJ, Conti MA, Kim Y (1998) Homeodomain-interacting protein kinases, a novel family of co-repressors for homeodomain transcription factors. J.Biol.Chem. 273: 25875-25879

[166] Kim EA, Noh YT, Ryu MJ, et al (2006) Phosphorylation and transactivation of Pax6 by homeodomain-interacting protein kinase 2. J.Biol.Chem. 281: 7489-7497

[167] D'Orazi G, Cecchinelli B, Bruno T, et al (2002) Homeodomain-interacting protein kinase-2 phosphorylates p53 at Ser 46 and mediates apoptosis. Nat.Cell Biol. 4: 11-19

[168] Hofmann TG, Moller A, Sirma H, et al (2002) Regulation of p53 activity by its interaction with homeodomain-interacting protein kinase-2. Nat.Cell Biol. 4: 1-10

[169] Zhang Q, Yoshimatsu Y, Hildebrand J, Frisch SM, Goodman RH (2003) Homeodomain interacting protein kinase 2 promotes apoptosis by downregulating the transcriptional corepressor CtBP. Cell 115: 177-186

[170] Kanei-Ishii C, Ninomiya-Tsuji J, Tanikawa J, et al (2004) Wnt-1 signal induces phosphorylation and degradation of c-Myb protein via TAK1, HIPK2, and NLK. Genes Dev. 18: 816-829

[171] Boucher MJ, Simoneau M, Edlund H (2009) The homeodomain-interacting protein kinase 2 regulates insulin promoter factor-1/pancreatic duodenal homeobox-1 transcriptional activity. Endocrinology 150: 87-97

[172] An R, da Silva Xavier G, Semplici F, et al (2010) Pancreatic and duodenal homeobox 1 (PDX1) phosphorylation at serine-269 is HIPK2-dependent and affects PDX1 subnuclear localization. Biochem.Biophys.Res.Commun. 399: 155-161

[173] Waeber G, Thompson N, Nicod P, Bonny C (1996) Transcriptional activation of the GLUT2 gene by the IPF-1/STF-1/IDX-1 homeobox factor. Mol.Endocrinol. 10: 1327-1334

[174] Watada H, Kajimoto Y, Miyagawa J, et al (1996) PDX-1 induces insulin and glucokinase gene expressions in alphaTC1 clone 6 cells in the presence of betacellulin. Diabetes 45: 1826-1831

[175] Watada H, Kajimoto Y, Umayahara Y, et al (1996) The human glucokinase gene beta-cell-type promoter: an essential role of insulin promoter factor 1/PDX-1 in its activation in HIT-T15 cells. Diabetes 45: 1478-1488

[176] Bretherton-Watt D, Gore N, Boam DS (1996) Insulin upstream factor 1 and a novel ubiquitous factor bind to the human islet amyloid polypeptide/amylin gene promoter. Biochem.J. 313 (Pt 2): 495-502

[177] Carty MD, Lillquist JS, Peshavaria M, Stein R, Soeller WC (1997) Identification of cis- and trans-active factors regulating human islet amyloid polypeptide gene expression in pancreatic beta-cells. J.Biol.Chem. 272: 11986-11993

[178] Kefas BA, Heimberg H, Vaulont S, et al (2003) AICA-riboside induces apoptosis of pancreatic beta cells through stimulation of AMP-activated protein kinase. Diabetologia 46: 250-254

[179] Riboulet-Chavey A, Diraison F, Siew LK, Wong FS, Rutter GA (2008) Inhibition of AMP-activated protein kinase protects pancreatic beta-cells from cytokine-mediated apoptosis and CD8+ T-cell-induced cytotoxicity. Diabetes 57: 415-423

[180] Jones RG, Plas DR, Kubek S, et al (2005) AMP-activated protein kinase induces a p53-dependent metabolic checkpoint. Mol.Cell 18: 283-293

[181] Sun G, Tarasov AI, McGinty JA, et al (2010) LKB1 deletion with the RIP2.Cre transgene modifies pancreatic beta-cell morphology and enhances insulin secretion in vivo. Am.J.Physiol Endocrinol.Metab 298: E1261-E1273

[182] Williams T, Brenman JE (2008) LKB1 and AMPK in cell polarity and division. Trends Cell Biol. 18: 193-198

[183] Fu A, Ng AC, Depatie C, et al (2009) Loss of Lkb1 in adult beta cells increases beta cell mass and enhances glucose tolerance in mice. Cell Metab 10: 285-295

[184] Granot Z, Swisa A, Magenheim J, et al (2009) LKB1 regulates pancreatic beta cell size, polarity, and function. Cell Metab 10: 296-308

[185] Alanentalo T, Asayesh A, Morrison H, et al (2007) Tomographic molecular imaging and 3D quantification within adult mouse organs. Nat.Methods 4: 31-33

[186] Jetton TL, Lausier J, LaRock K, et al (2005) Mechanisms of compensatory beta-cell growth in insulin-resistant rats: roles of Akt kinase. Diabetes 54: 2294-2304

[187] Montagne J, Stewart MJ, Stocker H, Hafen E, Kozma SC, Thomas G (1999) Drosophila S6 kinase: a regulator of cell size. Science 285: 2126-2129

[188] Shima H, Pende M, Chen Y, Fumagalli S, Thomas G, Kozma SC (1998) Disruption of the p70(s6k)/p85(s6k) gene reveals a small mouse phenotype and a new functional S6 kinase. EMBO J. 17: 6649-6659

[189] Kwon G, Marshall CA, Liu H, Pappan KL, Remedi MS, McDaniel ML (2006) Glucose-stimulated DNA synthesis through mammalian target of rapamycin (mTOR) is regulated by KATP channels: effects on cell cycle progression in rodent islets. J.Biol.Chem. 281: 3261-3267

[190] Rajasekhar VK, Viale A, Socci ND, Wiedmann M, Hu X, Holland EC (2003) Oncogenic Ras and Akt signaling contribute to glioblastoma formation by differential recruitment of existing mRNAs to polysomes. Mol.Cell 12: 889-901

[191] Eto K, Yamashita T, Matsui J, Terauchi Y, Noda M, Kadowaki T (2002) Genetic manipulations of fatty acid metabolism in beta-cells are associated with dysregulated insulin secretion. Diabetes 51 Suppl 3: S414-S420

[192] Ruvinsky I, Sharon N, Lerer T, et al (2005) Ribosomal protein S6 phosphorylation is a determinant of cell size and glucose homeostasis. Genes Dev. 19: 2199-2211

[193] Pende M, Kozma SC, Jaquet M, et al (2000) Hypoinsulinaemia, glucose intolerance and diminished beta-cell size in S6K1-deficient mice. Nature 408: 994-997

[194] Fraenkel M, Ketzinel-Gilad M, Ariav Y, et al (2008) mTOR inhibition by rapamycin prevents beta-cell adaptation to hyperglycemia and exacerbates the metabolic state in type 2 diabetes. Diabetes 57: 945-957

[195] Kawamori D, Kaneto H, Nakatani Y, et al (2006) The forkhead transcription factor Foxo1 bridges the JNK pathway and the transcription factor PDX-1 through its intracellular translocation. J.Biol.Chem. 281: 1091-1098

[196] Buteau J, Shlien A, Foisy S, Accili D (2007) Metabolic diapause in pancreatic beta-cells expressing a gain-of-function mutant of the forkhead protein Foxo1. J.Biol.Chem. 282: 287-293

[197] De BJ, Wang HJ, Van BC (2004) Effects of Wnt signaling on proliferation and differentiation of human mesenchymal stem cells. Tissue Eng 10: 393-401

[198] van de WM, Sancho E, Verweij C, et al (2002) The beta-catenin/TCF-4 complex imposes a crypt progenitor phenotype on colorectal cancer cells. Cell 111: 241-250

[199] Korinek V, Barker N, Moerer P, et al (1998) Depletion of epithelial stem-cell compartments in the small intestine of mice lacking Tcf-4. Nat.Genet. 19: 379-383

[200] Heller RS, Dichmann DS, Jensen J, et al (2002) Expression patterns of Wnts, Frizzleds, sFRPs, and misexpression in transgenic mice suggesting a role for Wnts in pancreas and foregut pattern formation. Dev.Dyn. 225: 260-270

[201] Papadopoulou S, Edlund H (2005) Attenuated Wnt signaling perturbs pancreatic growth but not pancreatic function. Diabetes 54: 2844-2851

[202] Heiser PW, Lau J, Taketo MM, Herrera PL, Hebrok M (2006) Stabilization of beta-catenin impacts pancreas growth. Development 133: 2023-2032

[203] McLin VA, Rankin SA, Zorn AM (2007) Repression of Wnt/beta-catenin signaling in the anterior endoderm is essential for liver and pancreas development. Development 134: 2207-2217

[204] Dessimoz J, Bonnard C, Huelsken J, Grapin-Botton A (2005) Pancreas-specific deletion of beta-catenin reveals Wnt-dependent and Wnt-independent functions during development. Curr.Biol. 15: 1677-1683

[205] Wells JM, Esni F, Boivin GP, et al (2007) Wnt/beta-catenin signaling is required for development of the exocrine pancreas. BMC.Dev.Biol. 7: 4

[206] Pedersen AH, Heller RS (2005) A possible role for the canonical Wnt pathway in endocrine cell development in chicks. Biochem.Biophys.Res.Commun. 333: 961-968

[207] Fujino T, Asaba H, Kang MJ, et al (2003) Low-density lipoprotein receptor-related protein 5 (LRP5) is essential for normal cholesterol metabolism and glucose-induced insulin secretion. Proc.Natl.Acad.Sci.U.S.A 100: 229-234

[208] Rulifson IC, Karnik SK, Heiser PW, et al (2007) Wnt signaling regulates pancreatic beta cell proliferation. Proc.Natl.Acad.Sci.U.S.A 104: 6247-6252

[209] Grant SF, Thorleifsson G, Reynisdottir I, et al (2006) Variant of transcription factor 7-like 2 (TCF7L2) gene confers risk of type 2 diabetes. Nat.Genet. 38: 320-323

[210] Florez JC (2007) The new type 2 diabetes gene TCF7L2. Curr.Opin.Clin.Nutr.Metab Care 10: 391-396

[211] Murtaugh LC, Law AC, Dor Y, Melton DA (2005) Beta-catenin is essential for pancreatic acinar but not islet development. Development 132: 4663-4674

[212] Welters HJ, Kulkarni RN (2008) Wnt signaling: relevance to beta-cell biology and diabetes. Trends Endocrinol.Metab 19: 349-355

[213] Schinner S, Ulgen F, Papewalis C, et al (2008) Regulation of insulin secretion, glucokinase gene transcription and beta cell proliferation by adipocyte-derived Wnt signalling molecules. Diabetologia 51: 147-154

[214] Cozar-Castellano I, Fiaschi-Taesch N, Bigatel TA, et al (2006) Molecular control of cell cycle progression in the pancreatic beta-cell. Endocr.Rev. 27: 356-370

[215] Dodge R, Loomans C, Sharma A, Bonner-Weir S (2009) Developmental pathways during in vitro progression of human islet neogenesis. Differentiation 77: 135-147

[216] Liu Z, Habener JF (2008) Glucagon-like peptide-1 activation of TCF7L2-dependent Wnt signaling enhances pancreatic beta cell proliferation. J.Biol.Chem. 283: 8723-8735

[217] Egan JM, Bulotta A, Hui H, Perfetti R (2003) GLP-1 receptor agonists are growth and differentiation factors for pancreatic islet beta cells. Diabetes Metab Res.Rev. 19: 115-123

[218] Xu G, Stoffers DA, Habener JF, Bonner-Weir S (1999) Exendin-4 stimulates both beta-cell replication and neogenesis, resulting in increased beta-cell mass and improved glucose tolerance in diabetic rats. Diabetes 48: 2270-2276
[219] List JF, Habener JF (2004) Glucagon-like peptide 1 agonists and the development and growth of pancreatic beta-cells. Am.J.Physiol Endocrinol.Metab 286: E875-E881
[220] Ashcroft FM (2006) K(ATP) channels and insulin secretion: a key role in health and disease. Biochem.Soc.Trans.34: 243-246.
[221] Braun M, Rorsman P (2006) The glucagon-producing alpha cell: an electrophysiologically exceptional cell. Diabetologia 53:1827-1830.
[222] Kawamori D, Kurpad AJ, Hu J, Liew CW, Shih JL, Ford EL, Herrera PL, Polonsky KS, McGuinness OP, Kulkarni RN (2009) Cell Metab 9:350-361.
[223] Henquin J.C, Nenquin M, Ravier M.A, Szollosi A (2009) Shortcomings of current models of glucose-induced insulin secretion. Diabetes Obes.Metab. 11 (Suppl 4): 168-179.

Raf Serine/Threonine Protein Kinases: Immunohistochemical Localization in the Mammalian Nervous System

András Mihály
University of Szeged, Faculty of Medicine, Department of Anatomy, Histology and Embryology, Hungary

1. Introduction

The Raf protein kinases are members of the mitogen-activated protein kinase (MAPK)/extracellular signal-regulated kinase (ERK) signaling pathway, which links the activation of cell surface receptors to intracellular cytoplasmic and nuclear molecular targets. Raf kinases relay signals from growth factor receptor-coupled GTPases of the Ras family to the MAPK/ERK kinase (MEK). This pathway is very important in developmental processes, cell survival and proliferation (Zebisch & Troppmair, 2006). Three different Raf isoenzymes exist in mammals: C-Raf, B-Raf and A-Raf. They originate from 3 independent genes (for review, see Matallanas et al., 2011). The Raf genes are protooncogenes; they are the cellular counterparts of the v-Raf oncogene, originally discovered in murine sarcoma cell lines (Rapp et al. 1988). The Raf genes are ubiquitously present in different tissues, showing a relatively segregated expression pattern: B-Raf gene is expressed mainly in the brain and testes, C-Raf gene is ubiquitous, although the degree of expression is uneven, A-Raf gene is expressed in gonads, kidney, spleen and bone mainly, as detected in mice (Storm et al., 1990). Accordingly, the Raf proteins are present in three isoforms: the A-Raf, B-Raf and C-Raf are protein kinase isozymes, showing similar molecular structure (Zebisch & Troppmair, 2006). The Raf proteins consist of three conserved regions (CR1, CR2, CR3), which display different functions: CR1 contains the Ras-binding region, which is responsible for the interaction with Ras and membrane phopholipids; CR2 contains phosphorylation site and CR3 is the kinase region of the molecule (Heidecker et al., 1991; Zebisch and Troppmair, 2006). The inactive Raf molecule is in a closed conformation: the N-terminal and C-terminal regions are above each other (Matallanas et al., 2011). When this molecule opens, activation and dimerization can happen (Heidecker et al., 1991; Matallanas et al., 2011; Zebisch & Troppmair, 2006). Activation-inactivation are regulated by phosphorylation-dephosphorylation of the molecule (Matallanas et al., 2011). Activation also needs the recruitment of Raf to the cell membrane (Leevers et al., 1994). Many details of the Raf kinase activation steps have been described in the last twenty years (for review, see Matallanas et al., 2011). The regulation of Raf signaling by intracellular proteins is complicated: (1) because of the membrane recruitment of Raf during activation, scaffolding proteins exert regulatory role on the Ras-Raf-MEK pathway (Matallanas et al., 2011). (2) Recently discovered Raf kinase inhibitor protein (RKIP) negatively modulates Raf kinase, being important mainly in cancer cells (Klysik et al., 2008; Matallanas et al., 2011). Over the years, B-Raf has been increasingly associated with cancer development (Niault & Baccarini, 2010). Mutations of B-Raf proved to be important in malignant melanoma, thyroid carcinomas and colorectal tumors (Niault & Baccarini, 2010). This underlines the utmost importance of B-Raf in mitogenic signal transduction.

Experimental data suggest, that Raf kinases are activated by growth factor- and cytokine receptors. Platelet-derived growth factor (PDGF; Morrison et al., 1989), epidermal growth factor (EGF; App et al., 1991), nerve growth factor (NGF; Oshima et al., 1991), insulin (Blackshear et al., 1990), interleukins (Carrol et al., 1990), neuronal angiotensin receptor (Yang et al., 1997) and vascular endothelial growth factor (VEGF; Lu et al., 2011) act through the Raf-MEK signaling pathway. Raf mediated signals are able to move in the cytoplasm; the Golgi-apparatus, the mitochondria and the cell nucleus are targets of Raf-mediated phosphorylation signals (Matallanas et al., 2011; Zebisch & Troppmair, 2006). The question arises, if these organelles contain Raf kinase, or Raf kinase translocates to them during the signalling process (Mor & Philips, 2006).

Immunohistochemical studies have revealed that B-raf and C-Raf are widely distributed in central nervous system (CNS) areas including the hippocampus, neocortex and spinal cord (Mihály et al. 1991, 1993; Mihály & Rapp 1993). They are present not only in neurons but also in astrocytes (Mihály & Rapp 1994). Although most CNS neurons express Raf immunoreactivity, the possibility remains that Raf expression varies according to the type of growth factor receptor that a neuron expresses (Mihály & Endrész, 2000). Our ultrastructural immunohistochemical studies proved that Raf protein-like immunoreactivity was localized primarily in postsynaptic densities, dendritic spines, dendrites and soma of the neurons (Mihály et al., 1991). On the basis of these observations the possibility arose, that Raf kinases participated in some receptor-mediated postsynaptic phenomena which might initiate long-term changes in postsynaptic neurons (Mihály et al., 1990). The present review collects the relevant localization data of Raf protein kinases in CNS structures. The localization of the Raf proteins will be discussed in light of some plasticity experiments (Mihály et al., 1990; 1991; 1996).

2. Detection of Raf protein kinases in the brain with Western blotting

The first literature data about the presence of Raf protein kinase in the mammalian brain are those of Mihály et al. (1991). In these studies we used the polyclonal v-Raf antiserum which was produced in the laboratory of Dr. Ulf Rapp. The serum was raised in rabbit, against a large (30 kDa) C-terminal Raf protein, expressed in *E. coli*, by inserting the v-Raf oncogene into the bacterial DNA (Kolch et al., 1988). Adult Wistar rats were deeply anesthetized with halothane, decapitated and the brains were homogenized, lysed and immunprecipitated. The precipitates were subjected to gel electrophoresis, then proteins were transferred to nitrocellulose membranes and incubated with the polyclonal v-Raf antiserum (Mihály et al., 1991). One conspicuous 95 kDa protein band was detected with this method (Mihály et al., 1991). The same antibody was used on isolated subcellular fractions of the rat brain, where the cytosolic- and microsome fractions displayed positive signals, again at 95 kDa (Mihály et al., 1996). Efforts aiming the detection of C-Raf and B-Raf separately, were also successful (Morice et al., 1999): Western blotting revealed the presence of C-Raf and B-Raf proteins in different areas of the rat brain (Morice et al., 1999). Immunoblotting of guinea pig brain homogenates also resulted in positive signals at 95 kDa with the aforementioned polyclonal anti-v-Raf serum (Mihály & Endrész, 2000). Several years later, using specific, polyclonal B-raf antibodies (Santa Cruz Biotechnology, Santa Cruz, CA, USA), we detected the 95 kDa protein again in rat brain homogenates (Mihály et al., 2007). Therefore, we conclude, that brain B-Raf kinase is ubiquitous and detectable through Western blotting in the rodent brain, as a 95 kDa protein (Fig. 1).

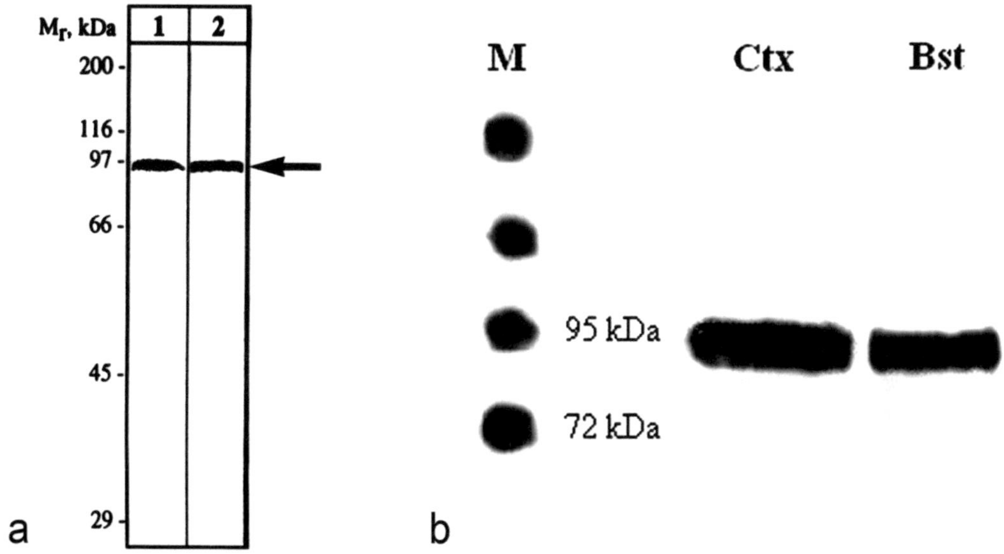

Fig. 1. Rat brain Western blots made with v-Raf-specific (a) and B-raf-specific (b) polyclonal antibodies. Molecular weight markers are on the left side (Mr, kDa; M). Fig. 1a: lane 1 represents the cytosolic fraction, lane 2 the microsome fraction (Mihály et al., 1996). Fig. 1b: „Ctx" represents the signal obtained from brain cortex homogenate; „Bst" represents the signal obtained from brain stem homogentes (Mihály et al., 2007). The two antibodies detect one 95 kDa protein, indicating the presence of B-raf kinase in the rat brain.

3. Light microscopic localization of Raf proteins in the central nervous system (CNS)

Light microscopic immunohistochemistry of brain areas from laboratory rats, laboratory guinea pigs and domestic cats have been performed with help of the polyclonal v-Raf antibodies (Mihály et al., 1991), polyclonal B-raf antibodies (Santa Cruz Biotechnology, Santa Cruz, CA, USA), monoclonal B-Raf antibodies (produced against the last 12 amino acids of the C-terminal region of B-Raf) and polyclonal C-Raf antibodies (anti-SP 63 serum). The polyclonal anti-v-Raf-, anti-SP 63- and monoclonal anti-B-Raf antibodies were prepared in the laboratories of Dr. Ulf Rapp (National Cancer Institute, Frederick Cancer Research and Development Center, Frederick, MD, USA; see descriptions of the antibodies in Mihály et al., 1993; 1996). Two human hippocampi were investigated, too. The human brain samples were obtained from autopsy in the Department of Pathology, Szeged University. The autopsy was performed 16 h after death. The hippocampal tissue was kept in fixative solution (4% paraformaldehyde in 0.1 M phosphate buffer) for 12 days, then transverse plane frozen sections were made and stained with the monoclonal B-Raf antibody. Rats, guinea pigs and cats were deeply anesthetized and perfused through the heart with fixative solution (4% paraformaldehyde in 0.1 M phosphate buffer, pH 7.4). The brains were sectioned with freezing microtome, cryostat and vibratome. The tissue sections were immunostained as glass-mounted- or free-floating sections. Immunofluorescence was performed on cryostat sections, mounted on glass slides. The slides were incubated with primary antibodies (anti-Raf serum), then with secondary antibodies conjugated with fluorescein isothiocyanate (FITC). In peroxidase-based techniques, free-floating sections were used, incubated with the primary antibody, then treated according to the avidin-biotin method, using the Vectastain ABC kits (Vector, Burlingame, CA, USA). Independent from the antibody and method used, the staining pattern of the different brain areas was consistent, showing neuronal and glial localization of Raf kinases. The differences between B-Raf and C-Raf localizations are indicated in the text. No differences were detected in the staining pattern comparing the different species (rat, cat, guinea pig, human). The descriptions below refer to the results obtained with polyclonal v-Raf antibodies, if not indicated otherwise. Staining obtained with v-Raf antibodies will be referred to as B-Raf-like staining, staining with anti-SP 63 will be described as C-Raf-like immunoreactivity. Controls of the immunohistochemical procedure included absorption controls performed with the recombinant v-Raf protein and the SP 63 peptide. One mg of the protein and 10 mg of the peptide were reconstituted in 1 ml of deionized water, mixed with the undiluted antibody (0.01 ml), and kept at room temperature overnight with slow agitation. The mixture was diluted to working concentration (1:500), centrifuged at 70,000xg for 20 min, and used for the immunohistochemical procedure (Mihály et al., 1993). No immunostaining was seen after the procedure (Mihály et al., 1993). Other controls included incubations without the primary anti-Raf sera: in these experiments the primary antibody was omitted, and the sections were incubated in normal horse serum, diluted to 1:100, and secondary antibodies as usual. In these sections, no specific immunostaining was detected (Fig. 6e).

3.1 Localization of B-Raf in the spinal cord and brain stem

Neuronal and glial localization of Raf-protein-like immunoreactivity (RPI) was encountered in the spinal cord of the laboratory rat and guinea pig (Mihály & Rapp, 1993; 1994). Neurons in laminae I, V, VI and IX displayed strong cytoplasmic staining. Glial cells in the white matter and grey substance were stained, too (Fig. 7). Axonal staining (e.g.: in the dorsal horn, where afferent axons terminate in large numbers) was not observed (Mihály et al., 1996). Different levels of the brain stem displayed similar staining pattern: cytoplasmic staining in large- and medium-sized nerve cells. The motor neurons of the cranial nerve nuclei were strongly stained. Scattered RPI was detected in sensory nuclei, such as the superior and medial vestibular nuclei and the ventral cochlear nucleus. The raphe nuclei of the pons and medulla also contained RPI (Mihály & Endrész, 2000; Mihály et al., 2007). Large neurons of the reticular formation and the inferior olive were containing strong RPI (Fig. 3). The neurons of the mesencephalic trigeminal nucleus were strongly stained (Mihály et al., 2007). On the other hand, the spinal trigeminal nucleus displayed only faint staining (except lamina I neurons, which were strongly stained; Mihály et al., 2007). The red nucleus displayed strong staining, whilst the substantia nigra cells contained less RPI (Fig. 3). The immunostaining was always localized in cell bodies, no axonal staining was observed in any of the brain stem areas. Detailed examination of the brain stem of the rat and the guinea pig did not reveal any species differences (Mihály & Endrész, 2000; Mihály et al., 2007).

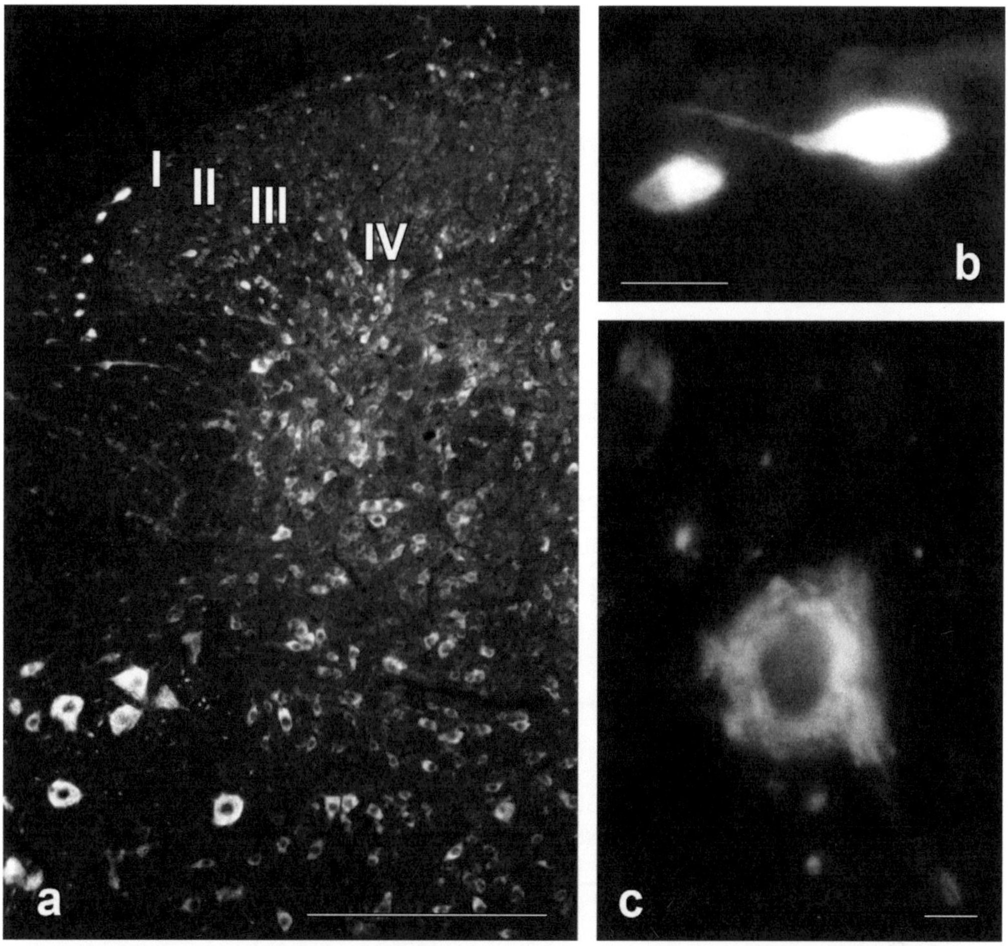

Fig. 2. B-Raf immunoreactivity in the spinal cord of the rat. The RPI is localized in cell bodies and proximal dendrites. No axonal staining is visible. Bar: 0.5 mm. Rexed laminae are indicated with roman numerals (a). Strong staing of lamina I neurons in the dorsal horn is conspicuous. Bar: 10 µm (b). Large motor neurons of lamina IX are labelled strongly, too. Bar: 10 µm (c). Immunolabeling was obtained with polyclonal anti-v-Raf serum.

3.2 Localization of B-Raf and C-Raf proteins in the cerebellum

The C-Raf and B-Raf proteins were detected in the vermis of the guinea pig cerebellum, using polyclonal anti-SP 63 (anti-C-Raf), and polyclonal anti-v-Raf (Mihály et al., 1993) antibodies. The B-Raf-like staining was very strong in neuronal cell bodies and glia-like cells. The staining was localized in neuronal cell bodies and dendrites. Dendritic staining was conspicuous in Purkinje cells, mainly in primary and secondary, large dendrites. Cell bodies of Purkinje cells were stained, too. Strong RPI was seen in the granular layer, where cell bodies of the granule cells and Golgi cells were immunoreactive. Small groups of glia cells stained in the white matter, and strong RPI was detected in the neurons of the fastigial nucleus (Fig. 4).

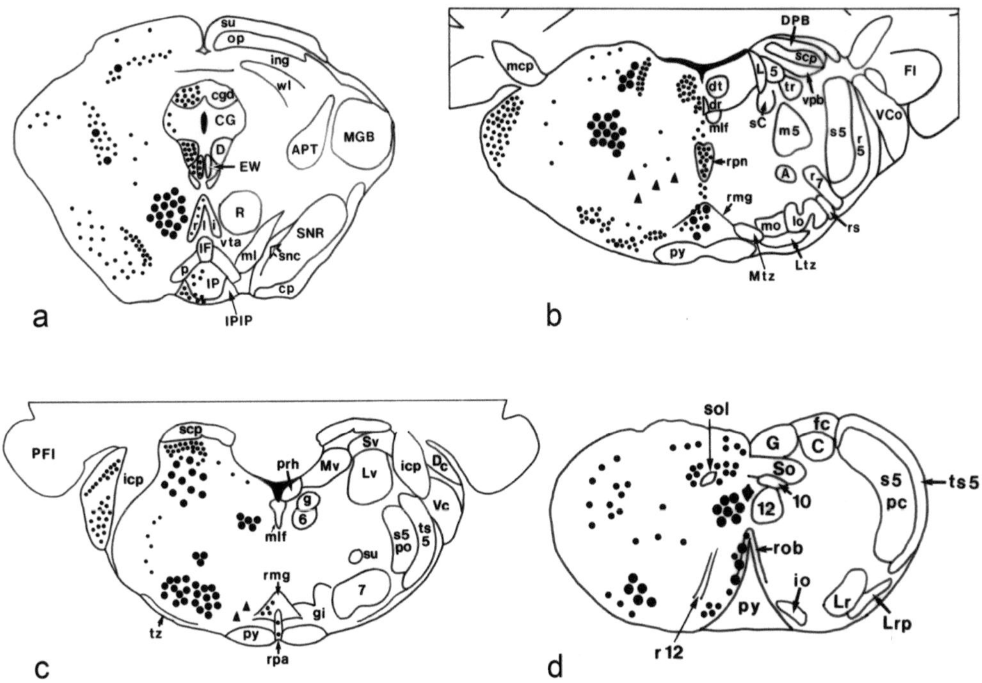

Fig. 3. Representative cross sections of the rat brain stem (Swanson, 1992), showing the cellular distrubution of Raf-protein-like immunoreactivity. Large dots represent strongly stained neurons, small dots represent medium or faint staining. Triangles represent large nerve cells with outstanding RPI. The upper mesencephalon (a), the upper pons (b), the lower pons (c) and the lower medulla (d) are represented (Mihály et al., 2007). Abbreviations: 5: mesencephalic nucleus of trigeminal nerve; m5: motor nucleus of trigeminal nerve; s5: principal sensory nucleus of trigeminal nerve; r5: radix of trigeminal nerve; s5po: spinal trigeminal nucleus pars oralis; s5pc: spinal trigeminal nucelus pars caudalis; ts5: spinal trigeminal tract; 6: abducens nucleus; 7: facial motor nucleus; g: genu of facial nerve; r7: radix of facial nerve; 10: dorsal nucleus of vagus nerve; 12: hypoglossal nucleus; r12: radix of hypoglossal nerve; D: Darkschewitsch nucleus; CG, cgd: periaqueductal gray; su, op, ing, wl: layers of the superior colliculus; R: red nucleus; SNR, snc: substantia nigra; EW: Edinger-Westphal nucleus; APT: anterior pretectal nucleus; MGB: medial geniculate body; IP, IPIP: interpeduncular nucleus; vta: ventral tegmental area; IF: interfascicular nucleus; p: paranigral nucleus; cp: cerebral peduncle; rli: raphe rostral linear nucleus; rmg: raphe magnus nucleus; rpn: raphe pontis nucleus; rpa: raphe pallidus nucleus; rob: raphe obscurus nucleus; dt: dorsal tegmental nucleus; mlf: medial longitudinal fasciculus; Lr, Lrp: lateral reticular nucleus; py: pyramidal tract; mo, lo: medial and lateral superior olivary nucleus; Mtz, Ltz: medial and lateral nucleus of the trapezoid body; Vco, Vc: ventral cochlear nucleus; Dc: dorsal cochlear nucleus; DPB, vpb: dorsal and ventral parabrachial nucleus; scp: superior cerebellar peduncle; mcp: middle peduncle; icp: inferior cerebellar peduncle; rs: rubrospinal tract; Mv, Lv, Sv: medial, lateral, superior vestibular nucleus; prh: prepositus hypoglossal nucleus; gi: gigantocellularis reticular nucleus; su: superior salivatory nucleus; sol: solitary tract; So: nucleus of the solitary tract; io: inferior olivary nucleus; Lr, Lrp: lateral reticular nucleus; G: gracile nucleus; C: cuneate nucleus; fc: fasciculus cuneatus; Fl: flocculus; PFl: paraflocculus.

The C-Raf staining was pale, displaying Purkinje cells and their large (primary and secondary) dendrites. Other neuronal elements (e.g.: granule cells, basket cells, Golgi cells) did not stain. C-Raf-like staining was seen in the nerve cells of the fastigial nucleus, and faint labeling was detected in some glia-like cells in white matter (Fig. 5).

3.3 Localization of B-Raf and C-Raf proteins in the cerebral cortex

RPI has been detected with polyclonal anti-v-Raf antibody and polyclonal anti-SP 63 antiserum. The latter is specific for C-Raf (Schultz et al., 1985; 1988; Morice et al., 1999). The RPI was detected in the neocortex (Fig. 6), cingular, pyriform, perirhinal, entorhinal areas and in the hippocampus (Mihály et al., 1993). RPI was localized in neurons and glia-like cells. No attempt was made to identify the cell types with double immunostainings. However, on the basis of the shape of the RPI-containing cells, we state that pyramidal and non-pyramidal cells of the Ammon's horn, granule cells of the dentate fascia and multipolar cells of the hilum of the dentate fascia were strongly stained (Mihály et al., 1993). Similar localizations were detected with the two antibodies, although C-Raf-like staining was faint compared to the B-Raf-like immunoreactivity. The intensity differences of the staining obtained with serial dilutions of the antibodies indicated that the cytoplasmic concentration of B-Raf kinase was much higher than that of the C-Raf in every region of the cerebral cortex. Detailed analysis of C-Raf-like staining and localization was not performed (Mihály et al., 1993). The brains of three domestic cats were also studied, using vibratome sections of the motor neocortex (sigmoid gyrus). Similarly to rats and guinea pigs, RPI has been detected in neuronal and glial cells bodies. Layer V pyramidal cells were stained in their cell bodies and proximal, large dendrites (Fig. 6). Scattered, small, glia-like cells were labeled, too. The human hippocampus displayed faint immunostaining with the B-Raf antiserum. Neuronal cell bodies were stained, which were similar to pyramidal cells, located in the stratum pyramidale of CA1 (Fig. 6).

3.4 Localization of Raf proteins in glia cells

B-Raf immunostaining was observed in small, glia-like cells, measuring 8-12 μm, in the cerebral cortex (gray and white matter), cerebellum (mentioned above), brain stem and spinal cord. Glial cells in the grey substance and in the whitte matter were regularly seen (Fig. 4; Fig. 7). Marginal astrocytes contained RPI in the spinal cord. Radial glia processes were stained in the cerebellar cortex (Fig. 4). No double immunolabelling was attempted in order to identify the cells. RPI was localized in the cytoplasm of the cell body and the processes (Fig. 7). No systematic study was made for the exploration of RPI in other glial cell types (oligodendroglia, microglia).

4. Electron microscopic localization of RPI in neurons

Ultrastructural immunohistochemistry was performed on vibratome sections of rat and cat cerebral cortex, using peroxidase-based preembedding methods (Mihály et al., 1991; Mihály & Rapp, 1994). These reports were the first in the literature, describing the electron microscopic localization of Raf kinases in the brain (Mihály et al., 1991; Mihály & Rapp, 1994). Comparison of the Raf kinase localization in a rodent and a carnivore, did not reveal differences: the localization pattern of RPI was consistent and similar in the two species.

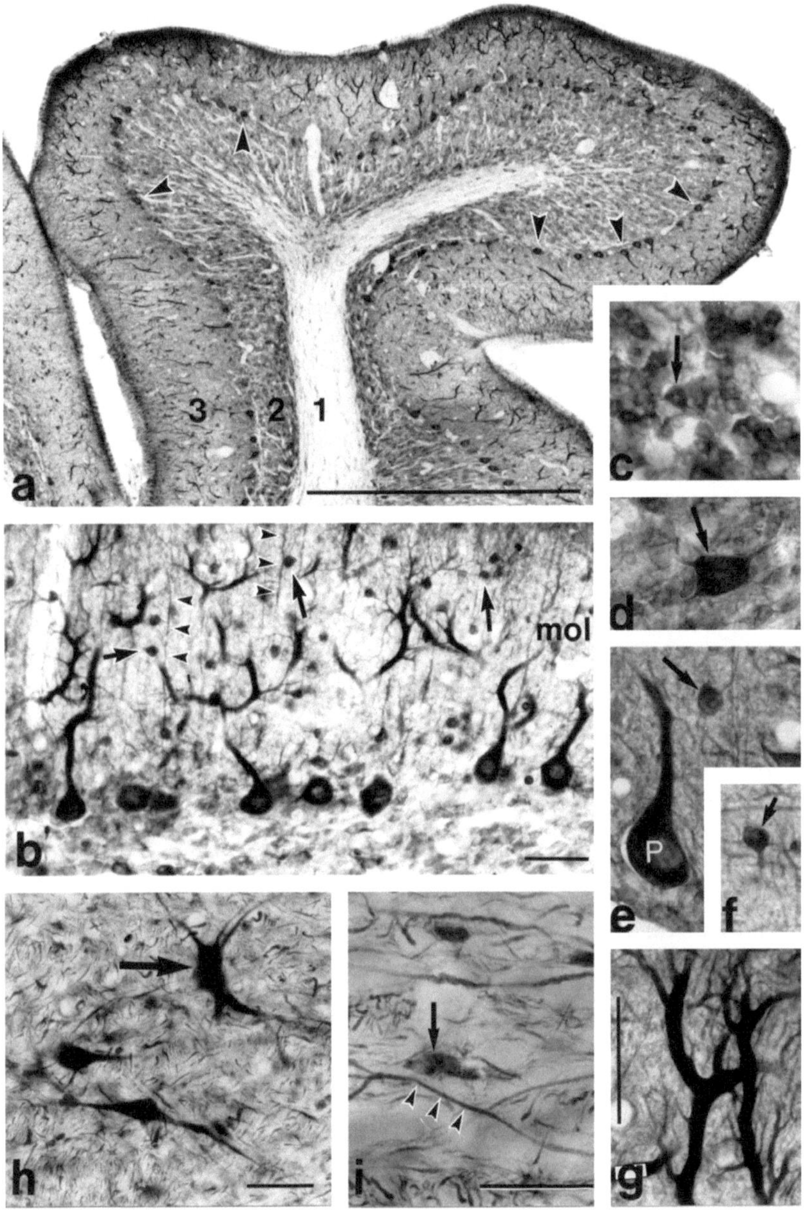

Fig. 4. Cerebellar immunostaining with polyclonal v-Raf serum in the guinea pig. Fig. 4a: Immunostaining depicts the layers of the cerebellar cortex: (1) white matter; (2) granular layer; (3) molecular layer; arrowheads: Purkinje cell layer; bar: 0.5 mm. Fig. 4b: the Purkinje cell RPI was detected not only in cell bodies, but also in dendrites (Fig. 4g). Arrowheads: radial glia processes; arrows: small neurons in the molecular layer (mol). Bar: 50 µm in Fig. 4b; 5 µm in Fig. 4g. The RPI in granule cells (Fig. 4c), Golgi cells (Fig. 4d), Purkinje cells (Fig. 4e) and outer stellate cells (Fig. 4f) are pointed by arrows. Bar as in Fig. 4g. Fig. 4h: the B-Raf staining of the neurons of the fastigial nucleus (arrow). Bar: 10 µm. Fig. 4i: white matter RPI apparently localized in glia cells (arrow). Arrowheads point to white matter axons containing RPI. Bar: 50 µm.

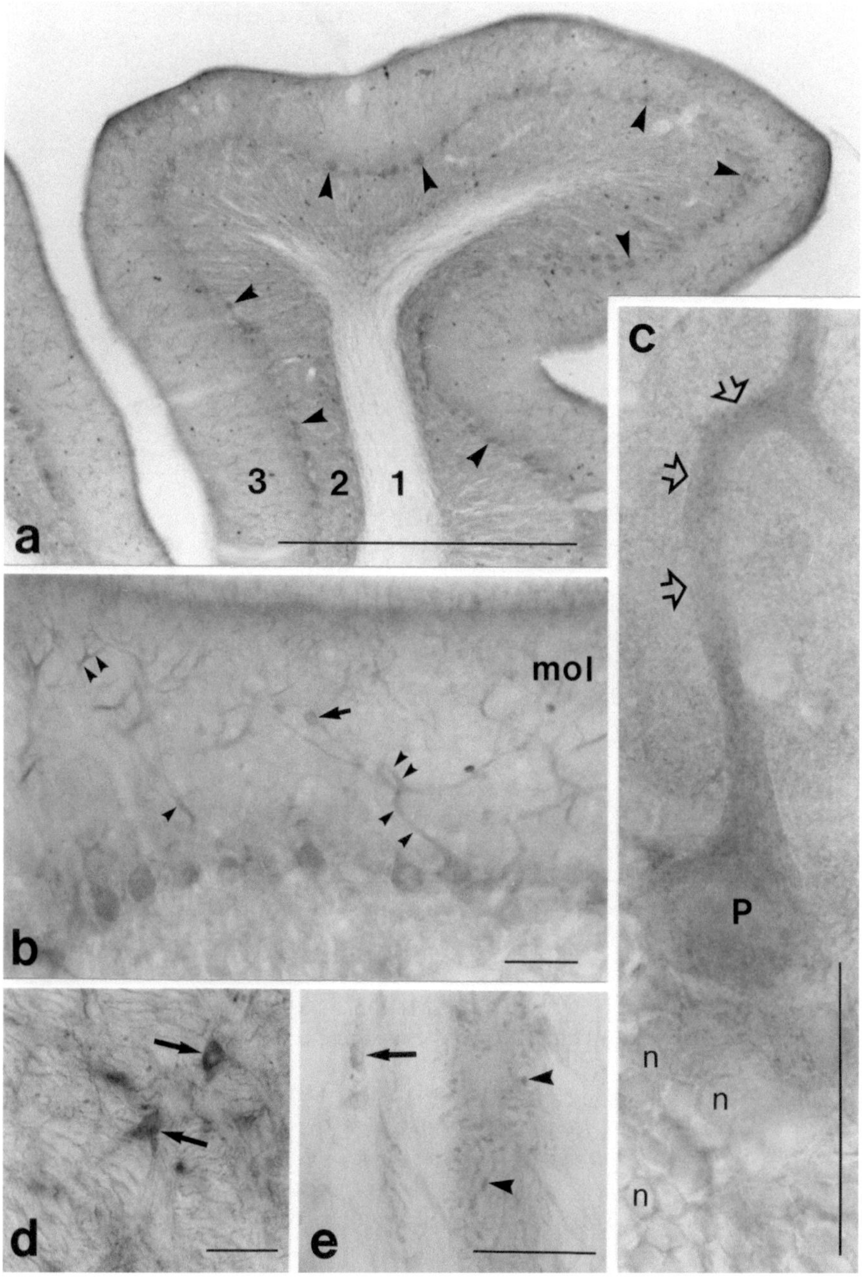

Fig. 5. Cerebellar immunostaining with anti-C-Raf serum in the guinea pig. Fig. 5a shows the overall weak staining (arrowheads: Purkinje cell layer; 1,2,3: as in Fig. 4a; bar: 0.5 mm). Fig. 5b: Purkinje cells contain C-Raf, which is localized in cell bodies and dendrites (arrowheads). Arrow points to outer stellate cell, which displays weak staining. Bar: 50 µm. Fig. 5c: Purkinje cell (P) staining with high magnification. Arrows point to dendrites (n: unstained granule cells; bar: 50 µm). Fig. 5d: neurons of the fastigial nucleus (arrows) containing C-Raf. Fig. 5e: white matter containing weakly stained glia cell (arrow) and weak axonal staining (arrowheads). Bars: 50 µm.

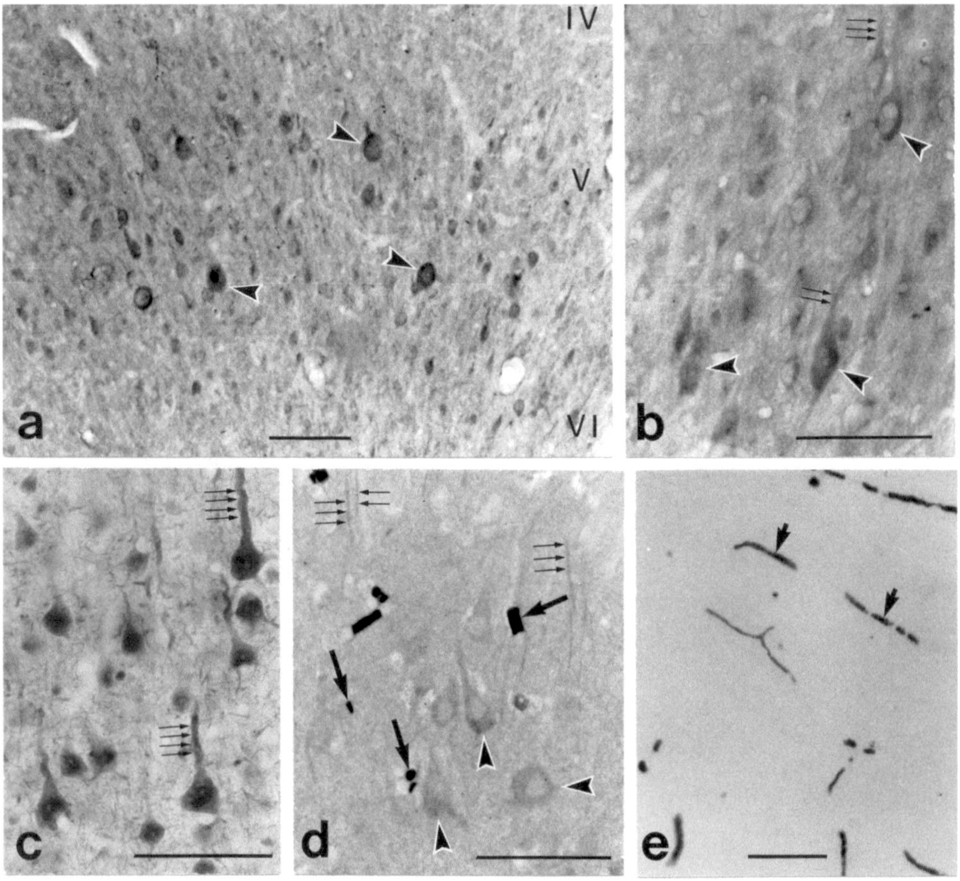

Fig. 6. B-Raf localization in the cat neocortex (a, b), rat neocortex (c) and human hippocampus (d, e). Bars: 50 µm. Roman numerals show neocortex layers. Arrowheads point to large nerve cells with immunostaining. Thin arrows in (b), (c) and (d) point to immunostained apical dendrites. Thick arrows in (d) and (e) point to red blood cells remaining in capillaries. Fig. 6e: control section was incubated in normal horse serum, the anti-B-Raf serum was omitted.

Neurons contain strong Raf kinase immunoreactivity in the cytoplasm of the cell body and dendrites (Figs 8-11). Not every neuron did stain: in the granule cell layer of the dentate fascia, cca. 20% of the granule cells remained unstained. This feature was visible already under the light microsocope, and was applicable to other cortical areas, as well (Mihály et al., 1993). As to the staining of the dendrites, very strong RPI was seen in dendritic spines (Figs 8-9; Fig. 11). The Raf-kinase-like staining was strong in postsynaptic densities, and in spine apparatuses (Fig. 9; Fig. 11; Mihály et al., 1991). Not only the spine apparatuses, but also dendritic subsynaptic cisternae contained RPI (Fig. 10d). In these membrane cisternal organelles the immunprecipitate was localized between the membrane cisternae; the cavity of the cisternae did not contain RPI (Fig. 11). We did not observe immunostained axon terminals in the tissues studied so far. None of the synapsing axons terminating on dendrites, dendritic spines or neuron somas did contain Raf-kinase-like immunostaining (see Figs 8-11). There are no comprehensive electron microscopic data about the ultrastructural localization of C-Raf, so this issue requires and deserves further, extensive experimentation.

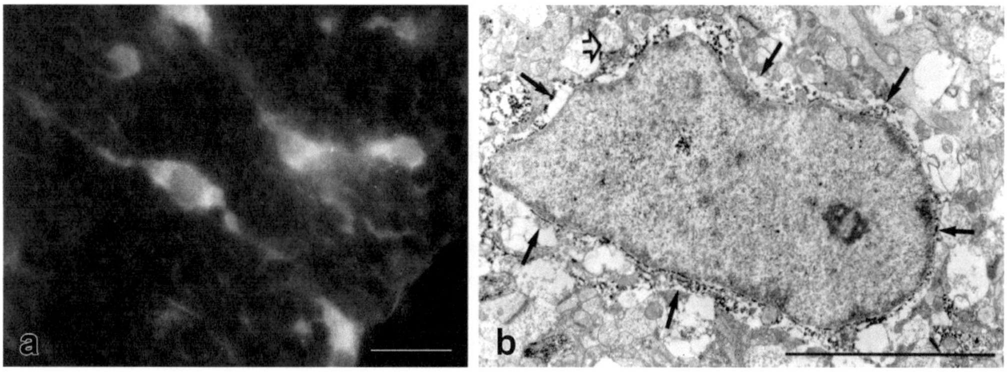

Fig. 7. Glial B-Raf kinase localization in rat spinal cord (a) and rat hippocampus (b). Fig. 7a displays RPI in astroglia-like cells of the white matter (bar: 10 μm). Fig. 7b shows the ultrastructural localization of B-Raf immunoreactivity in the cytoplasm of astrocyte-like cell of the rat hippocampus (solid arrows point to the contours of the astrocyte; empty arrow point to a synapse in the vicinity of the astrocyte; bar: 5 μm).

4.1 Electron microscopic localization of RPI in astrocytes

We observed several immunostained astrocytes in the cerebral cortex of rats (Fig. 7) and cats (Mihály & Rapp, 1994). The cytoplasmic staining of the cell body was conspicuous: very often in the vicinity of bundles of intermediate GFAP filaments (Mihály & Rapp, 1994). RPI was also observed in perivascular glial processes, and glial processes in the neuropil, around synapses (Mihály & Rapp, 1994). The RPI containing astrocytes were detected in the white matter, too - similarly to light microscopy observations. No RPI containing oligodendroglia cells were observed - in some sections we have seen RPI containing astrocytes together with unstained oligodendroglial cells (Mihály & Rapp, 1994). However, no systematic studies were performed to prove or disprove the lack of Raf proteins in oligodendroglia cells. The endothelial cells and pericytes were not labelled with any of the antibodies.

5. Comparison of B-Raf and C-Raf brain immunolocalization patterns

The localization of the two isozymes did not show substantial differences. The main observation in this respect was, that C-Raf kinase was expressed only in a few neurons and glia cells - compared to the expression of B-Raf. The staining was also weaker in most of the experiments. The C-Raf staining was specific, because the absorption controls did not show immunostaining (Mihály et al., 1993). The C-Raf was detected in the cerebellum and the cerebral cortex: these structures are compared in Table 1. B-Raf antibodies stained several neurons, irrespective of the size and type. However, staining was stronger in large cells (Figs 2-4). The B-Raf was also localized in astrocytes, as strong cytoplasmic staining. Under the electron microscope, the neuronal immunostaining was cytoplasmic, and exceptionally strong in dendritic spines, spine apparatuses and postsynaptic densities. Axon terminals did not stain. The C-Raf was localized in large neurons and their proximal dendritic branches. A few glia-like cells also displayed faint C-Raf-like staining (Fig. 5). We did not perform systematic electron microscopic studies for the localization of C-Raf kinase: only one series of experiments was done, in which we investigated the ultrastructural localization of C-Raf in guinea pig hippocampal slices (Mihály et al., 1991). In this electron microscopic study we observed neuronal and dendritic localization of C-Raf, thus corroborating the light microscopy findings (Mihály et al., 1991).

Isozymes	Neuronal localization	Glial localization	Other (e.g.: endothelium)
B-Raf	++++	++++	None
C-Raf	+	+	None

Table 1. Comparison of B-Raf and C-Raf localization and expression in the cerebellum and cerebral cortex of the rat and the guinea pig. The staining intensities are characterized by (+) symbols; (++++) means strong immunostaining; (+) means weak, but consistent staining.

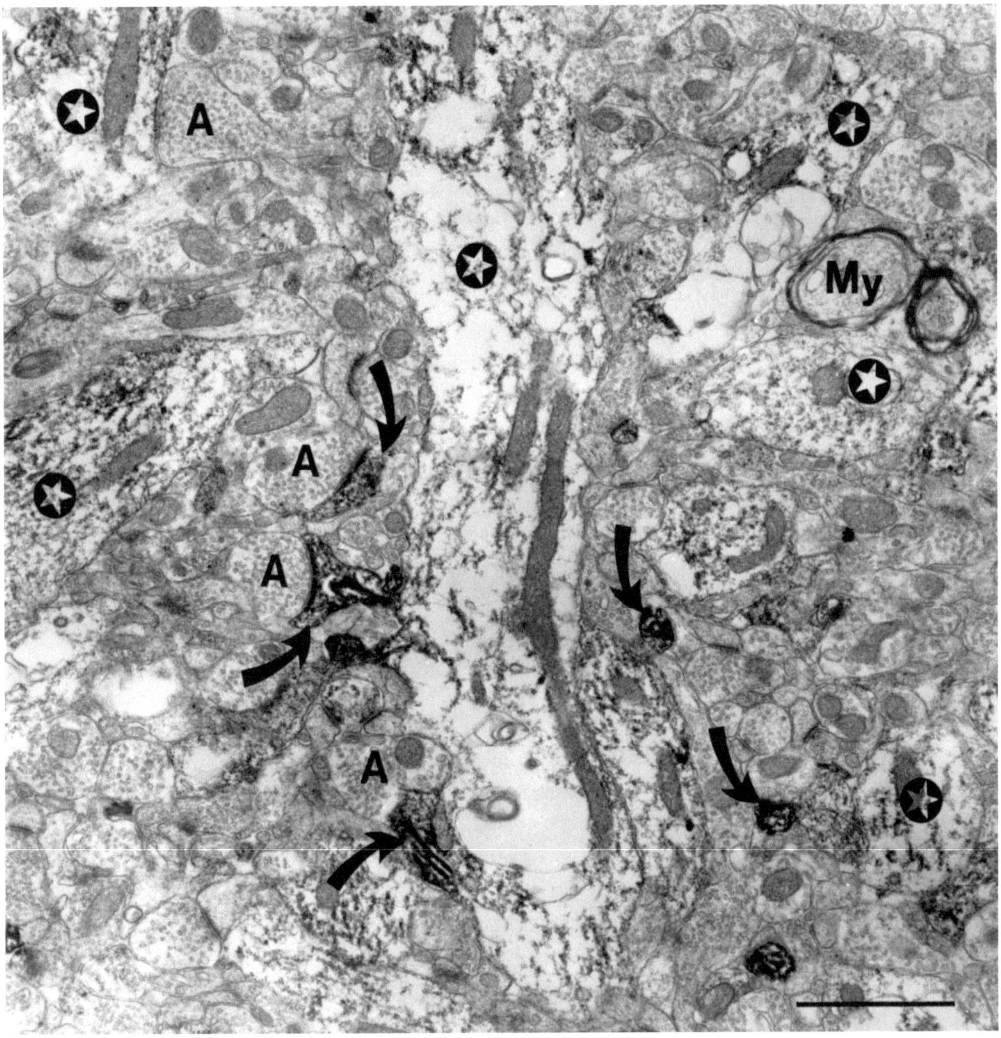

Fig. 8. B-Raf immunostaining in the rat cerebral cortex. Dendrites (stars) are labeled, axons (A) do not stain. The immunostaining of the dendritic spines (arrows) is extremely strong. My: myelinated axon. Bar: 1 μm.

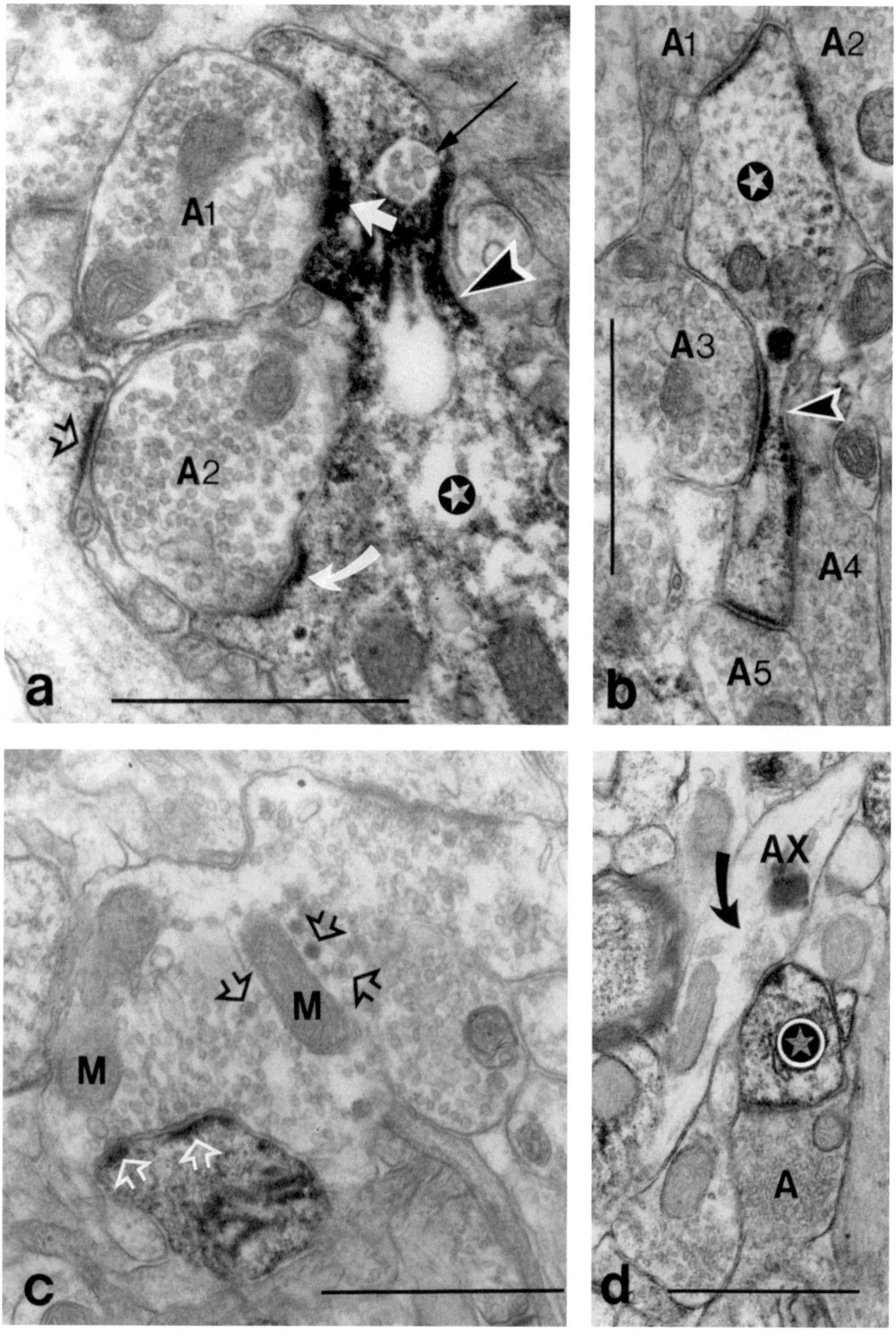

Fig. 9. a-d. Immunohistochemical localization of B-raf kinase in the hippocampus of the rat. In every case, postsynaptic dendrites and dendritic spines are immunoreactive. In (a) thick

arrows point to postsynaptic densities (thin arrow shows multivesicular body); arrowhead points to neck of spine, asterisk is in dendritic stem (A1, A2 are axon terminals); bar: 1 µm. In (b) dendrite (asterisk) is seen from which spine originates (arrowhead); A1-A5 are presynaptic axons; bar: 1 µm. In (c) mossy fiber terminal is unstained (M: mitochondria, black arrows point to dense core vesicles), postsynaptic dendrite (spine) is strongly immunoreactive (white arrows point to postsynaptic densities); bar: 1 µm. In (d) unstained axons (AX, A) synapsing with strongly stained small dendrite (asterisk); bar: 1 µm.

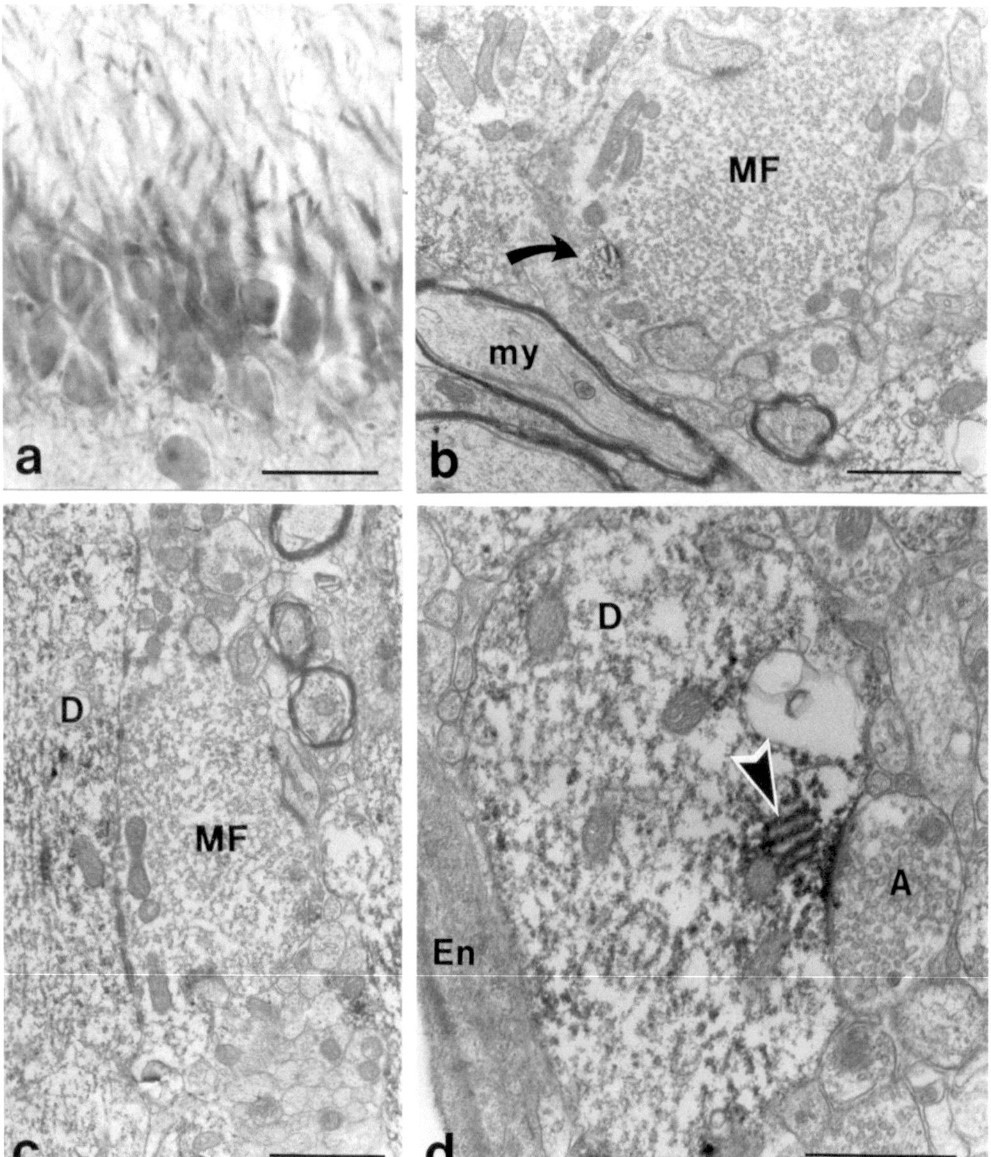

Fig. 10. a-d. Ultrastructural localization of RPI in the CA3 sector of the rat hippocampus. Fig. 9a displays the light microscopic appearance of the stratum pyramidale and stratum

lucidum after staining with polyclonal v-Raf antibodies. Cell bodies and apical dendrites are immunostained; mossy fibers do not stain (bar: 100 µm). Fig. 9b and c show two unstained mossy fiber ending (MF) synapsing with RPI containing dendrites (arrow and D). My: myelinated axon (unstained). Bars: 1 µm. Fig. 9d: large dendrite (D) containing RPI and immunostained subsynaptic cisternae (arrowhead). Synapsing axon (A) is unstained. En: unstained endothelial cell, bar: 1 µm.

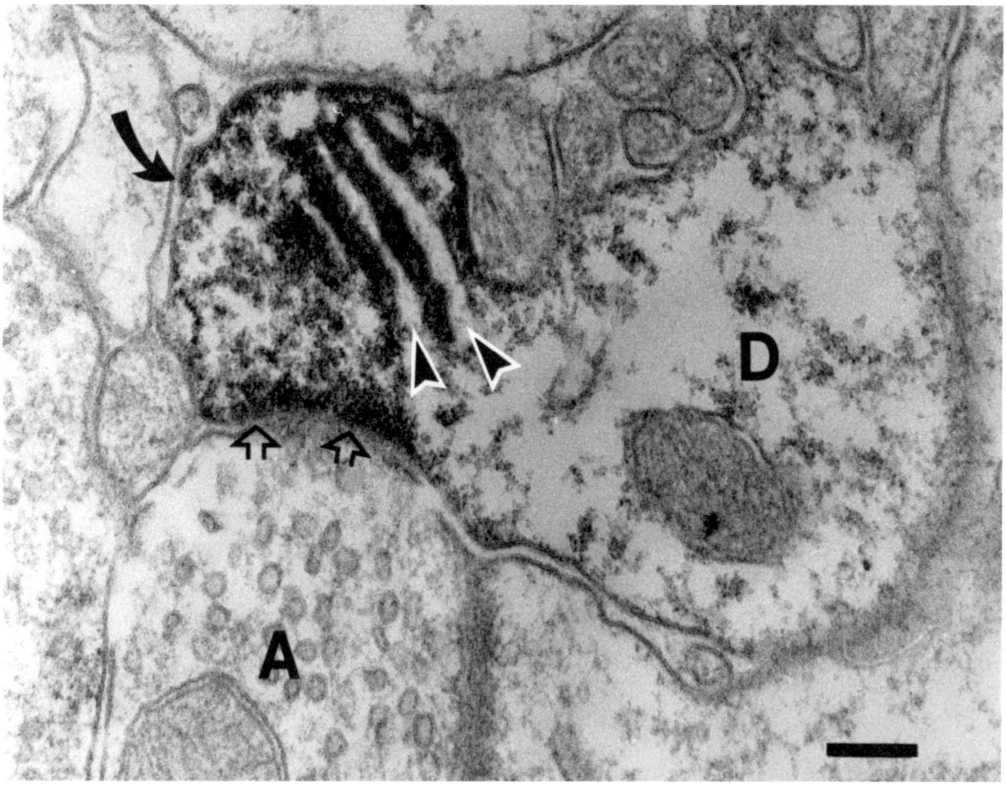

Fig. 11. Typical spine apparatus localization of B-Raf kinase in the rat cerebral cortex. The dendritic spine is pointed by the large solid arrow. The membranes of the spine apparatus are pointed by arrowheads. The postsynaptic density (empty arrows) is containing RPI, too. A: axon terminal; D: dendrite, containg RPI. Bar: 0.1 µm.

6. Discussion of the immunohistochemical findings

6.1 Neuronal B-Raf kinases

The immunolocalization of Raf kinases shows consistent patterns in different mammalian species (rat, guinea pig, cat and human). The main Raf molecule of the brain is certainly the B-Raf isozyme. The 95 kDa protein was detected in rat and guinea pig (Mihály et al., 1991; 1996; 2007; Mihály & Endrész, 2000). This observation conforms the literature data which give the same molecular weight in other experiments (Stephens et al., 1992; Dwivedi et al., 2006). Based on the specificity of the polyclonal v-Raf serum we refer to the immunostaining detected as to the B-Raf kinase immunostaining. This statement is strongly supported by the repeated Western blot experiments, which gave very consistent results: the polyclonal v-Raf serum and the B-Raf antibodies resulted in the same 95 kDa band in brain homogenates (Mihály et al., 1991; 2007; Mihály & Endrész, 2000). We did not perform immunoblotting with the C-Raf antibody because the polyclonal anti-SP 63 serum was thoroughly investigated (Schultz et al., 1985; 1988; Morice et al., 1999). The synthetic SP 63 peptide is the C-terminal part of the C-Raf protein (Schultz et al., 1985; 1988). This 12 amino acids peptide was used for immunization in rabbits (Schultz et al., 1985; 1988). The polyclonal serum was affinity purified, tested with immunoprecipitation and was found to be specific for the murine and human C-Raf terminal sequences in cell cultures and in tissue sections (Schultz et al., 1985; 1988; Mihály et al., 1993). This immunostaining is therefore referred to as the immunolocalization of C-Raf kinase. Neuronal B-Raf and C-Raf localization at the light microscopic level was found to be very similar in other experiments, years later (Morice et al., 1999). We think therefore, that our descriptions concerning the brain localization of Raf kinases are the first in the literature, and give a precise picture of neuronal B-Raf kinases (Mihály et al., 1991; 1993).

The main features of neuronal B-Raf kinase localization are as follows:

1. Cytoplasmic localization in the cell body was a general feature. This reflects the existence of a standard pool of the B-Raf kinase in the neuron. In case of growth factor signals, this cytoplasmic kinase pool can be mobilized quickly, in form of recruitment to the cell membranes or to the nucleus, if necessary.
2. Localization in dendrites was another characteristic feature of neuronal B-Raf. Dendritic staining was extremely strong in dendritic spines and in postsynaptic densities, meaning that a certain amount of B-Raf was already attached to the membrane, in proximity of the synaptic receptors. This localization suggested a very effective signal transduction process, mediated by the B-Raf.

6.2 Neuronal signal transduction with B-Raf: experimental proofs

The first observations pointing to the possible significance of Raf kinases in learning and memory were those of Mihály et al., (1990). In these experiments long term potentiation (LTP) was induced in rats, then the animals were subjected to immunohistochemical detection of Raf kinase with the polyclonal v-Raf antibodies (Mihály et al., 1990). The RPI was investigated in the granule cell layer of the dentate gyrus, by counting the immunoreactive neuronal cell bodies (Mihály et al., 1990). The number of RPI-containing cell bodies increased significantly in successfully potentiated animals (Mihály et al., 1990). This observation has been discussed either as the post-LTP increase of Raf kinase expression in dentate granule cells; or as the translocation of the Raf kinase from the distal dendrites to the cell body, following LTP (Mihály et al., 1990). We were not able to decide between these two possibilities in these in vivo experiments (Mihály et al., 1990). Years later, with help of in situ hybridization, it turned out that 24 h following LTP, the B-Raf expression in the dentate gyrus increased significantly (Thomas et al., 1994). Recently, experiments on B-Raf knockout mice proved, that the B-Raf is absolutely necessary for learning and memory consolidation (Chen et al., 2006; Valluet et al., 2010). However, the detailed mechanisms of the participation of B-Raf in LTP are unknown.

6.3 The possible intracellular translocation of B-Raf in neurons

Since the early immunohistochemical experiments on LTP (Mihály et al., 1990), the possibility of intracellular Raf translocation was an important issue, because the translocation could be one intracellular mechanism for signal transduction. The translocation was observed in vitro (Leevers et al., 1994) and in vivo (Oláh et al., 1991) in different experimental conditions. The recruitment to the cell membrane is considered to be a necessary step of Raf activation (Matallanas et al., 2011). However, the translocation to the cell nucleus is difficult to explain, although the experimental facts seem to be firm (Oláh et al., 1991). Another, translocation-like phenomenon was observed in sensory ganglion cells (Mihály et al., 1996). Primary sensory neurons contain B-Raf in their cell body, in form of homogeneous immunoreactivity (Mihály et al., 1996). Two-eight days after the transection of the peripheral nerve, this cytoplasmic staining pattern displays translocation towards the periphery, and at 8 days, most of the RPI was localized beneath the cell membrane of the sensory ganglion cells (Mihály et al., 1996). The significance of this phenomenon is not known, possibly it is connected to the process of chromatolysis (Mihály et al., 1996). The phenomenon of the translocation has been proved in cell cultures (see Mor & Philips, 2006), but not in vivo, in plasticity or pathological conditions. Further experiments are needed for the understanding of the possible in vivo translocation processes.

6.4 Possible functions of C-Raf in the neuron

Although C-Raf was found to be ubiquitous (Storm et al., 1990; Matallanas et al., 2011), not every neuron displayed immunoreactivity in our studies. These findings are in conform with other immunohistochemical studies (Morice et al., 1999). On the other hand, it seems, that C-Raf expression can be induced by some (yet unknown) conditions. Hippocampal slices maintained in vitro do not express C-Raf at the beginning of the in vitro incubation. Incubation of the slice for 2-4 h in oxygenized environment induced the appearance of the C-Raf staining in the nerve cells (Mihály et al., 1991). Upregulation of C-Raf was observed in Alzheimer-brains, too (Mei et al., 2006). In vitro grown neurons were protected by C-Raf inhibitors against the toxicity of amyloid peptides (Echeverria et al., 2008). It seems therefore, that the upregulation of C-Raf in the cell causes cell damage and death: it can happen through the activation of apoptosis signals (Echeverria et al., 2008). No data exist about the causes of C-Raf upregulation: excess excitatory amino acids (excitotoxic effects), hypoxic conditions, and ageing may operate through unknown signaling pathways in brain cells.

7. Conclusions

Raf protein kinases are widespread in the mammalian brain and spinal cord. Raf protein kinases are activated by several cytokines and growth factors through their membrane receptors. We investigated the immunohistochemical localization of B-Raf kinase and C-Raf kinase in neurons and glial cells. The B-Raf kinase is localized in the cell body, in dendrites, dendritic spines, spine apparatuses, subsynaptic cisternae and postsynaptic densities. The localization does not depend on neuronal type, but the intensity of the immunostaining is greater in large neurons. Neuronal B-Raf was not found in presynaptic axons, suggesting, that in these cells B-Raf kinase is coupled to postsynaptic signaling pathways. The similarities of the presence and localization of B-Raf in three mammalian orders (rodents, carnivores and hominids) suggest that this protein kinase has an important, and phylogenetically conserved function in normal, adult neurons. Data are available that learning and memory consolidation are processes in which B-Raf participates importantly. The C-Raf kinase has a similar CNS localization, but the amount of C-Raf in the cells is generally lower, than that of B-Raf. On the other hand, results from different experiments suggest, that the neuron adapts and alters the C-Raf expression, and thus regulates the amount of C-Raf present in the cytoplasm and in the membrane compartments. Therefore, we can state, that the regulation of the two isozymes (B-Raf and C-Raf) are different (although the downstream cellular events are similar):

1. CNS cells have a constant cytoplasmic pool of B-Raf, which translocates to different neighbouring organelles, in order to participate in signaling. The short-term regulation of B-Raf depends on membrane receptor stimulation.
2. CNS cells have another signaling molecule, the C-Raf, which displays low cytoplasmic concentration in normal neurons. However, the amount of cytoplasmic C-Raf is probably regulated by gene expression, the mechanisms of which are largely unknown. Therefore, C-Raf can be regulated twice: first by gene expression (upstream), second by membrane receptor activation (downstream). The significance and effects of this double regulation in the brain is not known: the effects probably manifest on the system level (e.g.: as alterations of the viability of neuronal networks).

8. Acknowledgement

This review was supported by the project „TÁMOP-4.2.1/B-09/1/KONV-2010-0005 - Creating the Center of Excellence at the University of Szeged". Financial support is from the European Union and the European Regional Fund. The author is thankful to Mrs. Katica Lakatos (Department of Anatomy) for technical help. We are thankful to Professor Dr. Ulf R. Rapp and Dr. Jakob Troppmair, Frederick Cancer Research and Development Center, Frederick, MD, USA, for providing the Raf antibodies for these experiments (see Acknowledgements in our referred articles).

9. References

App, H., Hazan, R., Zilberstein, A., Ullrich, A., Schlessinger, J., Rapp, U.R. (1991). Epidermal growth factor (EGF) stimulates association and kinase activity of Raf-1 with the EGF receptor. *Molecular Cell Biology*, Vol.11, pp. 913-919

Blackshear, P.J., Haupt, D.M.N., App, H., Rapp, U.R. (1990). Insulin activates the Raf-1 protein kinase. *Journal of Biological Chemistry*, Vol.265, pp. 12131-12134

Carrol, M.P., Clark-Lewis, I., Rapp, U.R., May, W.S. (1990). Interleukin-3 and granulocyte-macrophage colony-stimulating factor mediate rapid phosphorylation and activation of cytosolic c-Raf. *Journal of Biological Chemistry*, Vol.265, pp. 19812-19817

Chen, A.P., Ohno, M., Giese, K.P., Kühn, R., Chen, R.L., Silva, A.J. (2006). Forebrain-specific knockout of B-raf kinase leads to deficits in hippocampal long-term potentiation, learning and memory. *Journal of Neuroscience Research*, Vol.83, pp. 28-38

Dwivedi, Y., Rizavi, H.S., Conley, R.R., Pandey, G.N. (2006). ERK MAP kinase signaling in post-mortem brain of suicide subjects: differential regulation of upstream Raf kinases Raf-1 and B-Raf. *Molecular Psychiatry*, Vol.11, pp. 86-98

Echeverria, V., Burgess, S., Gamble-George, J., Arendash, G.W., Citron, B.A. (2008). Raf inhibition protects cortical cells against β-amyloid toxicity. *Neuroscience Letters*, Vol.444, pp. 92-96

Heidecker, G., Kolch, W., Morrison, D.K., Rapp, U.R. (1992). The role of Raf-1 phosphorylation in signal transduction. *Advances in Cancer Research*, Vol.58, pp. 53-73

Klysik, J., Theroux, S.J., Sedivy, J.M., Moffit, J.S., Boekelheide, K. (2008). Signaling crossroads: the function of Raf kinase inhibitory protein in cancer, the central nervous system and reproduction. *Cell Signal*, Vol.20, pp. 1-9

Kolch, W., Schultz, A.M., Oppermann, H., Rapp, U.R. (1988). Preparation of Raf-oncogene-specific antiserum with Raf protein produced in E. coli. *Biochimica Biophysica Acta*, Vol.949, pp. 233-239

Leevers, S.J., Paterson, H.F., Marshall, C.J. (1994). Requirement for *ras* in *raf* activation is overcome by targeting *raf* to the plasma membrane. *Nature*, Vol.369, pp. 411-414

Matallanas, D., Birtwistle, M., Romano, D., Zebisch, A., Rauch, J., von Kriegsheim, A., Kolch, W. (2011). Raf family kinases: old dogs have learned new tricks. *Genes & Cancer*, Vol.2, No.3, pp. 232-260

Mei, M., Su, B., Harrison, K., Chao, M., Siedlak, S.L., Previll, L.A., Jackson, L., Cai, D.X., Zhu, X. (2006). Distribution, levels and phosphorylation of Raf-1 in Alzheimer's disease. *Journal of Neurochemistry*, Vol.99, pp. 1377-1388

Mihály, A., Oláh, Z., Krug, M., Matthies, H., Rapp, U.R., Joó, F. (1990) Transient increase of raf protein kinase-like immunoreactivity in the rat dentate gyrus during long-term potentiation. *Neuroscience Letters*, Vol.116, pp. 45-50

Mihály, A., Kuhnt, U., Oláh, Z., Rapp, U.R. (1991) Induction of raf-1 protein immunoreactivity in guinea pig hippocampal slices during the *in vitro* maintenance. *Histochemistry* Vol.96, pp. 99-105

Mihály, A., Oravecz, T., Oláh, Z., Rapp, U.R. (1991) Immunohistochemical localization of raf protein kinase in dendritic spines and spine apparatuses of the rat cerebral cortex. *Brain Research*, Vol.547, pp. 309-314

Mihály, A., Endrész, V., Oravecz, T., Rapp, U.R., Kuhnt, U. (1993) Immunohistochemical detection of raf protein kinase in cerebral cortical areas of adult guinea pigs and rats. *Brain Research*, Vol.627, pp. 225-238

Mihály, A., Rapp, U.R. (1994) Expression of the raf protooncogene in glial cells of the adult rat cerebral cortex, brain stem and spinal cord. *Acta histochemica*, Vol.96, pp. 155-164

Mihály, A., Priestley, J.V., Molnár, E. (1996) Expression of raf serine/threonine protein kinases in the cell bodies of primary sensory neurons of the adult rat. *Cell & Tissue Research*, Vol.285, pp. 261-271

Mihály, A., Endrész, V. (2000) Neuronal expression of Raf protooncogene in the brain stem of adult guinea pig. *Acta histochemica*, Vol.102, pp. 203-217

Mihály, A., Karcsú-Kiss, G., Bakos, M., Bálint, E., (2007) Cellular distribution of B-raf protein kinase in the brainstem of the adult rat. A fluorescent immunohistochemical study. *Acta Biologica Szegediensis*, Vol.51, pp. 7-15

Mor, A., Philips, M.R. (2006). Compartmentalized Ras/MAPK signaling. *Annual Review of Immunology*, Vol.24, pp. 771-800

Morice, C., Nothias, F., König, S., Vernier, P., Baccarini, M., Vincent, J-D., Barnier, J.V. (1999). Raf-1 and B-Raf proteins have similar regional distributions but differential subcellular localization in adult rat brain. *European Journal of Neuroscience*, Vol.11, pp. 1995-2006

Morrison, D.K., Kaplan, D.R., Escobedo, J.A., Rapp, U.R., Roberts, T.M., Williams, L.T. (1989). Direct activation of the serine/threonine kinase activity of Raf-1 through tyrosine phosphorylation by the PDGF β-receptor. *Cell*, Vol.58, pp. 649-657

Niault, T.S., Baccarini, M. (2010). Targets of Raf in tumorigenesis. *Carcinogenesis*, Vol.31, No.7, pp. 1165-1174

Oláh, Z., Komoly, S., Nagashima, N., Joó, F., Rapp, U.R., Anderson, W.B. (1991). Cerebral ischemia induces transient intracellular redistribution and intranuclear translocation of the raf protoncogene product in hippocampal pyramidal cells. *Experimental Brain Research*, Vol.84, pp. 403-410

Oshima, M., Shitanandam, G., Rapp, U.R., Guroff, G. (1991). The phosphorylation and activation of B-Raf in PC 12 cells stimulated by nerve growth factor. *Journal of Biological Chemistry*, Vol.266, pp. 23753-23760

Rapp, U.R., Cleveland, J.L., Bonner, T.I., Storm, S.M. (1988). The raf oncogenes, In: *The Oncogene Handbook*, E.P. Reddy, A.M. Skalka, T. Curran, (Eds), pp. 213-251, Elsevier, Amsterdam, The Netherlands

Schultz, A.M., Copeland, T.D., Mark, G.E., Rapp, U.R., Oroszlan, S. (1985). Detection of the myristylated gag-raf transforming protein with raf-specific antipeptide sera. *Virology*, Vol.146, pp. 78-89

Schultz, A.M., Copeland, T.D., Oroszlan, S., Rapp, U.R. (1988). Identification and characterization of c-raf phosphoproteins in transformed murine cells. *Oncogene*, Vol.2, pp. 187-193

Stephens, R.M., Sithanandam, G., Copeland, T.D., Kaplan, D.R., Rapp, U.R., Morrison, D.K. (1992). 95-kilodalton B-Raf serine/threonine kinase: identification of the protein and its major autophosphorylation site. *Molecular Cell Biology*, Vol.12, pp. 3733-3742

Storm, S.M., Cleveland, J.L., Rapp, U.R. (1990). Expression of Raf family proto-oncogenes in normal mouse tissues. *Oncogene*, Vol.5, pp. 345-351

Swanson, L.W. (1992). *Brain Maps: Structure of the Rat Brain*, Elsevier, New York, USA

Thomas, K.L., Laroche, S., Errington, M.L., Bliss, T.V., Hunt, S.P. (1994). Spatial and temporal changes in signal transduction pathways during LTP. *Neuron*, Vol.13, pp. 737-745

Yang, H., Lu, D., Raizada, M.K. (1997). Angiotensin II-induced phosphorylation of the AT1 receptor from rat brain neurons. *Hypertension*, Vol.30, pp. 351-357

Valluet, A., Hmitou, I., Davis, S., Druillennec, S., Larcher, M., Laroche, S., Eychéne, A. (2010). B-raf alternative splicing is dispensable for development but required for learning and memory associated with the hippocampus int he adult mouse. *PLoS ONE*, Vol.5, pp. 1-9

Zebisch, A., Troppmair, J. (2006). Back to the roots: the remarkable RAF oncogene story. *Cellular and Molecular Life Sciences*, Vol.63, pp. 1314-1330

20

Protein Kinases in Spinal Plasticity: A Role for Metabotropic Glutamate Receptors

J. Russell Huie and Adam R. Ferguson
Brain and Spinal Injury Center (BASIC), University of California, San Francisco, USA

1. Introduction

Neural plasticity is characterized by the lasting modulation of synaptic strength which alters the central nervous system's (CNS) capacity to encode and store information. Changes in synaptic plasticity have implications for brain-dependent learning and memory as well as a number of other forms of CNS information processing, including alterations in spinal cord function (reviewed in Patterson & Grau, 2001). As protein kinases have been shown to greatly affect neurotransmitter receptor dynamics, their role in synaptic plasticity is essential. A large body of work has highlighted the link between group I metabotropic glutamate receptor (mGluR) activation, subsequent protein kinase activation, and synaptic plasticity (Gereau & Heinemann, 1998; Sheng et al., 2002; Gallagher et al., 2004). Upon glutamate binding to mGluRs, a G-protein coupled to the receptor sets off an intracellular cascade, activating phospholipase C (PLC), diacyl glycerol (DAG), and ultimately protein kinases. Once activated, protein kinases can then exert a modulatory effect on both excitatory and inhibitory receptors that ultimately affects synaptic strength. In this way, mGluRs and protein kinases both play critical roles in a number of forms of neural plasticity, including those that are thought to underlie learning and memory (Bortolotto & Collingridge, 1993; Wang, et al. 2004). What many in neuroscience have overlooked is the fact that, just like in the brain, the *spinal cord* also exhibits the capacity for an amazing amount of plasticity, including simple forms of learning and memory that are relevant to pain, motor learning, and recovery of function after spinal cord damage (Woolf & Salter, 2000; Grau et al., 2006; Raineteau & Schwab, 2001). And, just as in the brain, spinal plasticity has been shown across a number of preparations to be mediated by mGluR/protein kinase signaling (Giles et al., 2007; Ferguson et al., 2008a).

This chapter will detail the specific role of protein kinase C (PKC) as an intermediary between initial mGluR activity and long-term changes in synaptic strength, and how this critical interaction affects a number of spinal processes. We will highlight how PKC and its isoforms provide a critical link between initial glutamatergic input and the alterations in receptor phosphorylation and trafficking that lead to spinal plasticity. We will first look at the role of mGluR/PKC signaling in the dorsal horn of the spinal cord, and the implications of this process on pain. We will then explore how the mGluR/PKC pathway also exerts modulatory control over the changes in plasticity (or *metaplasticity)* in the spinal cord, as evidenced in a spinal cord learning preparation. In addition, we will consider how the reorganization and reclamation of appropriate function after spinal cord injury is affected in large part by various forms of mGluR-mediated, protein kinase-dependent plasticity. Finally, we will investigate the neurobiological consequences and possible therapeutic potential to be found in altering protein kinase activity.

2. Nociceptive plasticity: mGluRs, protein kinase C, and pain

mGluR-mediated activation of protein kinases can modulate synaptic strength, by altering presynaptic and postsynaptic signaling. While these forms of synaptic plasticity have been suggested to underlie learning and memory in the hippocampus, similar mechanisms at work in the dorsal horn of the spinal cord produce a decidedly different form of learning that contributes to neuropathic pain. As the dorsal horn is the locus for the integration of incoming sensory information, sensitization of these neurons and their synaptic targets can have a dramatic behavioral effect (Mendell, 1966; Woolf, 1983). If a strong nociceptive signal is relayed from the periphery to the superficial dorsal horn, neurons in this area become sensitized, in a fashion that is mechanistically very similar to long-term potentiation (LTP), an electrophysiological mechanism believed to underlie learning and memory in the hippocampus (Sandkuhler & Liu, 1998; Bliss & Collingridge, 1993). This effect has been termed *central sensitization* (Woolf, 1983). As in any form of potentiation, subsequent input following sensitization can elicit a response even if it is much weaker than the initial input. That is, even those stimuli that would not normally be considered painful may now have the capacity to elicit a nociceptive response (*allodynia*). Likewise, normally painful stimuli can now induce a much more robust nociceptive response (*hyperalgesia*). These phenomena are conserved across species, and this makes sense from an evolutionary standpoint. To be rapidly sensitized to a painful stimulus is indeed adaptive, in that an organism will 'learn' to avoid this stimulus. Although this type of plasticity is essential for self-preservation, it can be problematic if dysregulated. Neural insult, whether it be an injury to the peripheral nervous system (e.g., nerve damage) or central nervous system (e.g., spinal cord injury), can produce an unregulated barrage of nociceptive input that may have no external initiator. This can produce a persistent pain response, and lasting nociceptive plasticity, that is generated wholly within the organism (neuropathic pain) (Kim & Chung, 1992; Christensen & Hulsebosch; Willis & Coggeshall, 1991; Willis & Westlund, 1997; Lindsey et al., 2000). A number of studies have outlined an important role for spinal protein kinase C (PKC) in mediating persistent pain states, including hyperalgesia, allodynia, and neuropathic pain (Coderre, 1992; Sun et al., 2004; Hua et al. 1999). As a major activator of PKC, group I mGluRs have been implicated as a driver of nociceptive plasticity (Yashpal et al., 2001; Fisher & Coderre, 1996; Adwanikar et al., 2004).

mGluRs activate PKC in two ways: both directly, through the G-coupled protein-PLC-DAG pathway, as well as indirectly by freeing intracellular Ca^{++} via activation of phosphoinositide-3 kinase (PI3K) pathway. *In vitro* work has shed light on how PKC works to sensitize spinal neurons and elicit persistent nociceptive behavior. PKC appears to have the capacity to induce sensitization through both presynaptic and postsynaptic effects. Presynaptically, PKC has been shown to phosphorylate voltage-gated calcium channels, thus increasing intracellular calcium and promoting neurotransmitter release (Yang et al., 2005). Postsynaptically, PKC can facilitate excitatory tone through actions on ionotropic glutamate receptors (Lu et al., 1999; Li et al., 1999). Following post-synaptic binding of glutamate to mGluRs, and subsequent PKC activation, PKC then phosphorylates NMDA receptors, increasing their open probability (Liao et al., 2001). Further, PKC has been shown to induce the rapid exocytosis of AMPA receptors (Li et al., 1999). Taken together, these excitatory effects have been suggested to play a critical role in sensitizing dorsal horn neurons (Ji et al., 2003).

In the brain, however, mGluR activation can have contradictory effects. For example, in hippocampal culture, mGluR activation has been shown to cause *internalization* of AMPA receptors leading to long-term *depression* (LTD) of postsynaptic potentials rather than sensitization (Huber et al., 2000; Oliet et al., 1997; Moult et al., 2002). This effect has been shown to depend on activation of the immediate early gene Arc/Arg 3.1 (Waung et al., 2008) as well as PKC (Camodeca et al., 1999; Oliet et al., 1997). On the other hand, mGluR-induced activation of PKC has also been implicated in LTP (Jia et al., 1998; Balschun et al., 1999; Aiba, et al., 1994; Conquet et al., 1994; Bortolotto, et al., 1994; Anwyl et al., 2009). This confusion about the bi-directional role of mGluRs and downstream PKC in hippocampal-dependent synaptic plasticity has been reviewed elsewhere (Bortolotto et al., 1999; Malenka & Bear, 2004) and the role of mGluRs in LTP and LTD remains a topic of ongoing research in the hippocampal plasticity literature (Mockett et al., 2011). It is clear that much more work is required to reconcile the hippocampal literature with the spinal plasticity literature regarding mGluR-PKC activation.

Tests of the necessity of mGluRs and PKC *in vivo* have confirmed the critical role of this pathway in spinally-mediated nociceptive plasticity. In order to study persistent inflammation, researchers will often give a subcutaneous injection of a noxious substance to the periphery, and then assess the effect this treatment has on the activity of dorsal horn neurons within nociceptive fields (Harris & Ryall, 1988; LaMotte et al., 1992). To investigate the specific role of mGluRs on this type of nociceptive plasticity, Young et al. (1995) gave an injection of the noxious substance algogen (mustard oil) to the hindpaws of rats, a substance known to evoke sustained activity in neurons of the dorsal horn. This treatment was followed by microinjections of the mGluR antagonist CHPG directly into the dorsal horn. They found that blocking mGluR activity with CHPG strongly inhibited mustard oil-evoked activity in these nociceptive areas. Beyond the necessity for mGluRs in sustained nociceptive activity, they also demonstrated the sufficiency for an mGluR agonist (ACPD) to *produce* sustained activity in these neurons (Young et al., 1995). Interestingly, Munro et al. found that treatment with a PKC inhibitor (chelerythrine or GF109203X) was able to inhibit both mustard oil- and ACPD-evoked activity (Munro et al. 1994). Together, these findings provide a strong case for mGluR/PKC signaling in mediating at least one form of long-term nociceptive plasticity. In 2001, Yashpal and colleagues demonstrated a further role for mGluR/PKC interaction in chronic neuropathic pain. Following a chronic constriction of the sciatic nerve (known to produce long-term nociceptive activity in the dorsal horn), they found that membrane localization of PKC was increased. Behaviorally, such an injury manifests as a chronic and robust hypersensitivity to both touch (mechanical allodynia) and temperature (thermal hyperalgesia). But, if an mGluR inhibitor was given prior to and after the injury, they found a decrease in the membrane-bound PKC expression, as well as a decrease in the subjects' injury-induced mechanical and thermal hypersensitivity.

Based on the mechanistic similarities, many researchers have drawn the distinct parallels between learning/memory and pain (Sandkuhler, 2000; Ji et al., 2003; Latremoliere and

Woolf, 2009). It is perhaps not surprising then that recent breakthroughs in our understanding of long-term memory storage in the brain continue to shed light on persistent pain syndromes that are mediated by the spinal cord. The protein kinase PKMζ (an isoform of PKC) is unique in that, unlike most enzymes, it can remain active for extremely long periods of time. This persistence is possibly due to the fact that activated PKMζ inhibits the protein PIN1, an inhibitor of PKMζ mRNA translation. Thus, by blocking the action of its regulator, PKMζ effectively creates a positive feedback loop, perpetuating PKMζ activity (Sacktor, 2011). Once PKMζ becomes active, it can promote trafficking of the major fast excitatory ionotropic glutamate receptor (AMPAR) to post-synaptic membranes, which in turn aids in potentiating the synapse (Sacktor, 2008; Migues et al., 2010; Malinow & Malenka, 2002). Further, PKMζ has been shown to also *inhibit* the internalization of these trafficked AMPARs (Yao et al., 2008). In this way, synaptic strength is not only induced, but also *maintained.* This finding has led many to investigate PKMζ has a critical component in long-term memory. Researchers have found that inhibiting PKMζ produces profound behavioral effects. A number of recent studies have shown that treatment with a PKMζ inhibitor effectively eliminates long-term memories from a number of learning paradigms (Pastalkova et al., 2006; Shema et al. 2007; Parsons & Davis, 2011; Madronal et al., 2010).

While these findings provide compelling evidence for the necessity for PKMζ in long-term memory, erasing long-term memories may not be a likely sought-after therapy. But consider the problem of chronic pain. As we have discussed, long-term neuropathic pain bears a striking mechanistic resemblance to memory, yet it is often regarded to be biologically dysfunctional. Therefore, inhibiting PKMζ may prove to be a very attractive therapeutic tool in overcoming persistent neuropathic pain. Asiedu and colleagues have recently shown this idea to be entirely possible. They initially primed rat subjects with a peripheral intraplantar injection of the inflammatory cytokine IL-6 or vehicle to the hindpaw. This treatment has previously been shown to induce allodynia for up to 3 days after injection (Asiedu et al., 2011). They then injected the mGluR agonist DHPG intrathecally to the spinal cord 6 days after initial peripheral injection, and found that DHPG injection produced a markedly enhanced nociceptive response in those subjects that had been primed days earlier with IL-6. This finding suggested that the maintenance of peripheral nociceptive sensitization was mediated centrally in the spinal cord and led them to investigate the possibility that PKMζ might mediate the storage of this nociceptive 'memory'. They found that intrathecal administration of the PKMζ inhibitor ZIP attenuated the capacity for DHPG to evoke the expression of nociceptive behavior. This lends support for the argument that, as in learning and memory in the brain, this persistent nociceptive sensitization reflects an LTP-like mechanism. Given that PKMζ has been shown to maintain LTP by inhibiting the internalization of AMPARs, and this effect can be disrupted by the peptide pep2m (Yao et al. 2008), Asiedu and colleagues hypothesized that this same mechanism is involved in the maintenance of nociceptive hypersensitivity. To test this, they gave an intrathecal injection of pep2m, and found that this treatment also blocked the DHPG evoked expression of sensitization (Asiedu et al., 2011). Together these findings suggest that the maintenance of nociception involves a PKMζ-dependent process within the spinal cord, and lends confirmatory evidence that the maintenance of LTP, memory, and nociception may be mediated by a common mechanism. In the future, PKMζ inhibition could hold very promising therapeutic potential for those suffering from chronic pain.

3. Learning in the spinal cord: Metaplasticity is PKC-dependent

The spinal cord supports a number of other forms of plasticity beyond just nociception. Throughout development, ventral motor neurons undergo a great amount of plasticity, as complex motor skills and locomotion are honed. Although the spinal cord was once believed to be fairly hard-wired after development, we now understand that the capacity for ongoing plasticity in spinal motor neurons persists throughout life (Edgerton et al. 2001; Courtine et al. 2008; De Leon et al. 2001; Wolpaw, 2007; Grau et al., 2006). This is evident in the spinal cord injury literature from the previous decade, where researchers have demonstrated the ability for spinal cord injured-subjects to regain locomotor function through the use of behavioral training, often in combination with pharmacological agents that facilitate plasticity (Wernig et al., 2000; Rossignol, 2007; Edgerton & Harkema, 2011). Promoting this kind of adaptive, use-dependent spinal plasticity is essential in order to realize functional recovery after injury.

Despite advances in our awareness of the spinal cord's capacity for plasticity, the underlying mechanisms dictating use-dependent spinal cord plasticity still require investigation. In order to better understand the unique role of the spinal cord in neural plasticity, outside of any supraspinal input, researchers have developed an *in vivo*, behavioral method for measuring plasticity in the isolated spinal cord. Building upon earlier work from Chopin and Buerger, Grau and colleagues demonstrated that following a complete spinal transection, spinal neurons below the lesion were able to support a simple form of instrumental (response-outcome) learning (Buerger and Fennesey, 1970; Chopin & Buerger, 1976; Grau et al., 1998). In this preparation, transected rats receive an electrical shock to the tibilias anterior muscle of their hindlimb whenever that limb is in an unflexed position (see Figure 1A). This stimulation causes a flexion of the hindlimb, at which point the stimulation is terminated. When the limb again falls to a resting, unflexed position, another shock is delivered.

Without input from the brain, spinalized subjects will learn to keep the hindlimb flexed in order to reduce exposure to the stimulation (Fig 1C). This form of spinal learning can also be inhibited: if subjects are given electrical stimulation that is not contingent upon limb position (*intermittent stimulation*), they will later fail to learn to keep their hindlimb flexed when tested with response-contingent stimulation (*controllable stimulation;* Figure 1B; Crown et al., 2002). Although these subjects are not learning the target response, they are still exhibiting a form of plasticity. Essentially, they have *learned* from the exposure to intermittent stimulation, that their limb position is not related to stimulation exposure, and thus fail even when later tested with controllable stimulation. This phenomenon has been considered analogous to the phenomenon of *learned helplessness* (Grau et al., 1998; Seligman & Maier, 1967). In contrast to the maladaptive effects of intermittent stimulation, training with controllable stimulation can *enhance* future learning (Grau et al., 1998). Subjects that have previously learned this instrumental task can be tested in the future with a more difficult response criterion (one that untrained subjects would not be able to exhibit), and this prior training *facilitates* learning. While both the learning deficit and the facilitation of learning are forms of plasticity, they are something more. Both of these phenomena affect lasting *change* in plasticity: a lasting alteration in the threshold at which learning occurs, either shifting the threshold up (in the case of intermittent stimulation inducing a future learning deficit) or down (in the instance of instrumental training facilitating future learning). In essence, these experience-dependent spinal changes represent a *plasticity* of plasticity. Abraham and Bear (1996) first described this type of plasticity of plasticity, characterizing it as "a higher-order form of synaptic plasticity" that they termed *metaplasticity* (Abraham and Bear, 1996). Uncontrollable stimulation induces a lasting alteration that undermines spinal learning, that can be described as a metaplastic inhibition of adaptive plasticity. Importantly, the same stimulation parameters that induce spinal learning deficits also undermine long-term recovery of locomotor function following a spinal contusion injury (Grau et al., 2004). Thus, a better understanding of the neurobiology underlying metaplastic inhibition of adaptive plasticity in the spinal cord will aid in the development of strategies to aid in functional recovery after spinal cord injury.

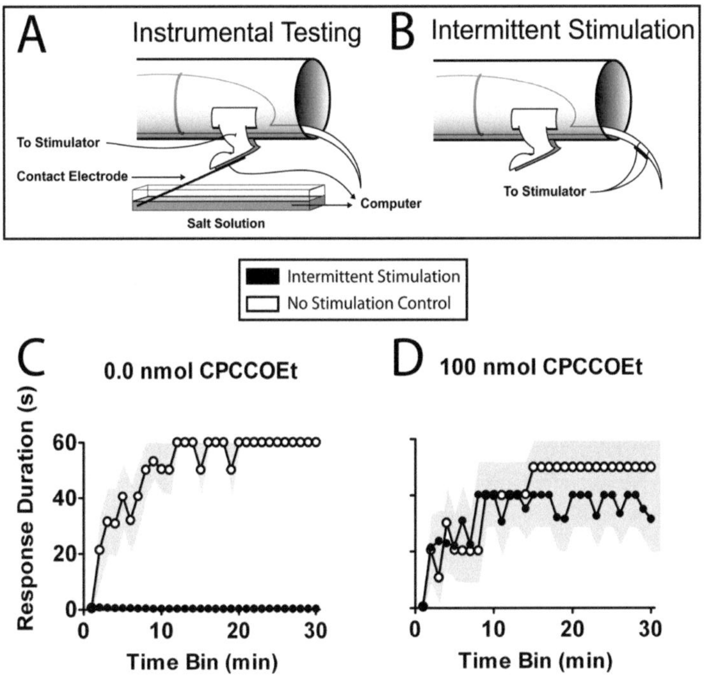

Fig. 1. Instrumental learning model of spinal plasticity, and the role of mGluRs in this phenomenon. A) Spinalized rat subjects are given an electrical shock each time their hindlimb is in an unflexed position (*controllable stimulation*). Over time, they learn to increase their response durations in order to reduce exposure to the stimulation, thus encoding an instrumental (response-outcome) relationship. B) If uncontrollable, intermittent stimulation is administered prior to instrumental testing, the subjects are unable to learn the relationship. C) In subjects that are given vehicle treatment (0.0 nmol of mGluR antagonist CPCCOEt), intermittent stimulation produces a significant spinal learning deficit. D) Treatment with 100 nmol CPCCOEt blocks the deficit induced by intermittent stimulation, suggesting a necessary role for mGluR activity in this effect. Adapted from Ferguson et al., 2008a.

Previous work has shown that metaplasticity in the hippocampus involves ionotropic glutamate receptor trafficking (Hellier et al., 2007). Given that mGluRs modulate ionotropic receptor function and trafficking through a PKC-mediated mechanism, we recently investigated the role of this mechanism in the metaplastic inhibition of spinal instrumental learning (Ferguson et al., 2008a). We first tested whether group I mGluRs are necessary for the metaplastic inhibition of spinal learning. We found that intermittent stimulation had no effect on future spinal learning if given after an intrathecal injection of an mGluR antagonist (CPCCOEt or MPEP, Fig. 1D). We next considered the contribution of PKC to this learning deficit. We found that in response to intermittent stimulation (which produces a lasting metaplastic inhibition of spinal learning) PKC activity in the spinal cord was significantly increased (Figure 2A). Similarly, if PKC inhibitors (bisindolylmaleimide or chelerythrine) were delivered intrathecally prior to intermittent stimulation, subjects exhibited no metaplastic inhibition of learning when tested 24 hours later. These data provided strong evidence that both mGluR and PKC activity are necessary in producing this form of spinal metaplasticity. To further examine the role of mGluR/PKC in this phenomenon, we tested whether pharmacological activation of mGluRs was sufficient to produce metaplastic inhibition of spinal learning. We found that a single bolus of the mGluR agonist DHPG was able to produce a spinal learning deficit that lasted at least 24 hours. We also found that PKC activity blockade (with chelerythrine or bisindolylmaleimide) prior to administration of DHPG, prevented metaplastic inhibition of spinal learning. These findings suggest an essential role for the mGluR/PKC pathway in mediating metaplasticity in the spinal cord.

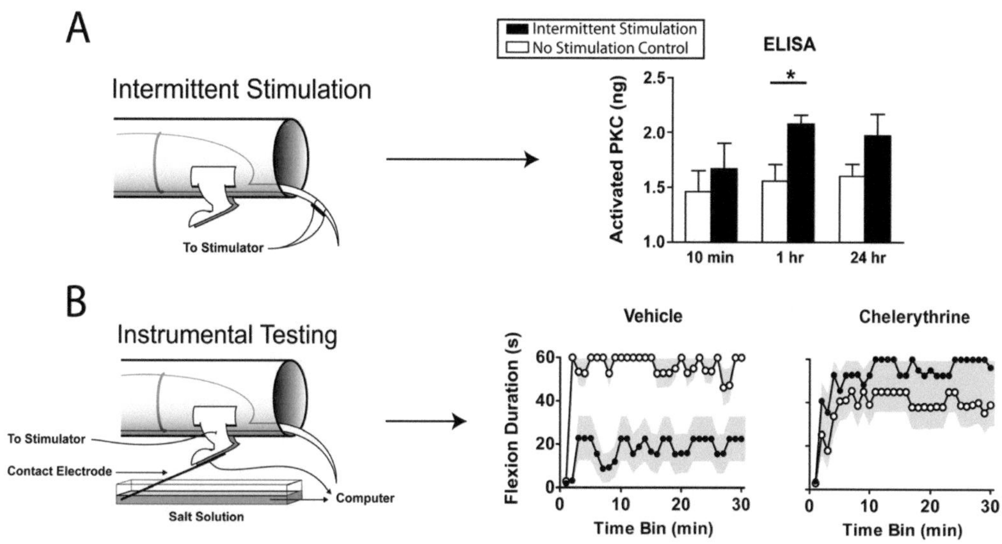

Fig. 2. Role for PKC in the metaplastic inhibition of spinal learning. A) Intermittent stimulation (which induces a lasting metaplastic inhibition of spinal learning) produces an increase in the expression of activated PKC that is significantly greater than unstimulated controls at 1 hour. B) When tested for spinal instrumental learning, vehicle-treated subjects that had received intermittent stimulation fail to learn; intrathecal injection of the PKC inhibitor chelerythrine blocks intermittent stimulation from producing a learning deficit. Adapted from Ferguson et al., 2008a.

As discussed above, PKC is known to alter the open-channel probability of NMDARs. Spinal learning, like many other forms of plasticity, has been characterized by its dependence on subtle, precise alterations in NMDAR function. Thus, it is likely that increased potentiation of NMDARs by the mGluR/PKC pathway upsets a delicate balance, pushing NMDARs (and subsequent intracellular calcium levels) beyond the range in which spinal learning can occur. Other work has also implicated other protein kinase activity in both the adaptive and maladaptive metaplastic changes in spinal learning. Inhibition of calcium/calmodulin-dependent protein kinase II (CaMKII) has been shown to block the development of the long-term inhibition of spinal learning if given after uncontrollable shock, and also blocks the facilitation effect of instrumental training if given prior to training (Baumbauer et al., 2007; Gomez-Pinilla et al., 2007). These findings suggest that protein kinase activity may engage a common mechanism in different forms of spinal metaplasticity. Future work will be necessary to elucidate how the various protein kinases interact and integrate to produce these lasting behavioral changes. Further, if these mechanisms are at work in the injured spinal cord, we can begin to develop therapeutic strategies that can reduce maladaptive metaplasticity, and promote the adaptive plasticity necessary for successful rehabilitation.

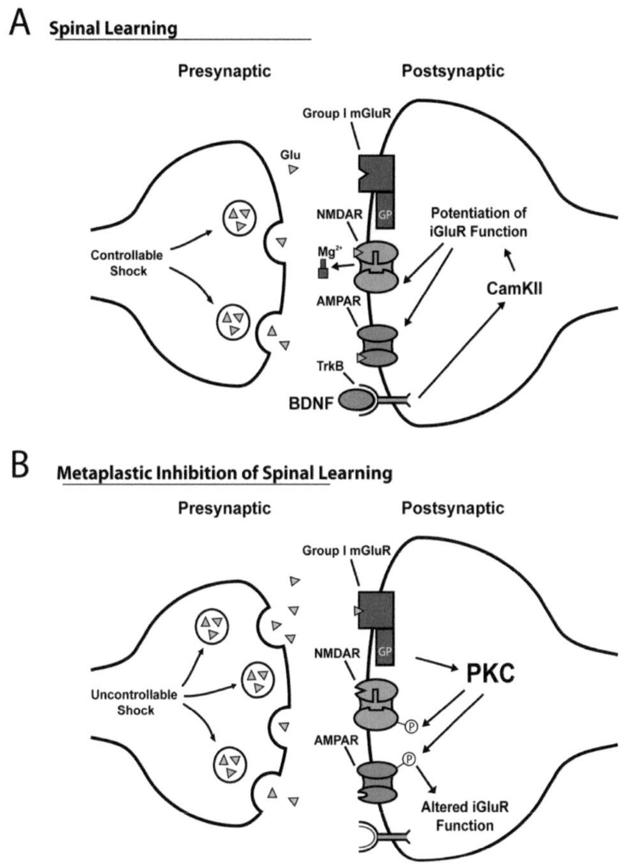

Fig. 3. Possible cellular mechanism for A) spinal learning and B) metaplastic inhibition of spinal learning. Both forms of spinal plasticity have been shown to involve protein kinase signaling. In contrast to controllable stimulation, uncontrollable stimulation is believed to engage group I metabotropic glutamate receptors, which leads to the downstream activation of PKC. PKC in turn alters ionotropic glutamate receptor function, which is believed to induce a lasting saturation of the synapse, inhibiting future learning. Adapted from Ferguson et al., 2008a.

4. Spinal cord injury: Protein kinase modulation as therapy

The previous sections have focused on spinal plasticity in isolated systems: how the mGluR/PKC pathway modulates nociceptive processing in the dorsal horn, and how it mediates the metaplastic inhibition of a ventral motor learning task. These models yield insight into the mechanisms by which plasticity in the spinal cord occur, and demonstrate that protein kinase activity is an essential step in mediating long-term neural modifications. In this final section, we will consider the therapeutic potential in altering protein kinase activity for spinal cord injury and related CNS disorders.

Within the wave of secondary processes following spinal cord injury, high levels of glutamate release can have a devastating effect on cell survival (Crowe et al., 1997; McAdoo et al., 1999; Ferguson et al., 2008b). As mGluR activation leads to PLC-mediated release of intracellular calcium stores, PKC activity increases, and in turn ionotropic glutamate receptors can be further potentiated. While this pathway can induce long-term spinal plasticity, the neural microenvironment around the spinal lesion is more vulnerable to excitotoxicity, and this cascade can ultimately lead to cellular degradation and excitotoxic cell death (Choi, 1992; Gereau & Heinemann, 1998; Mills et al., 2001). This has lead researchers to investigate the effect of PKC inhibition on cell survival after injury. Hara et al. showed that following ischemic injury, treatment with the PKC inhibitor staurosporine produced a neuroprotective effect (Hara et al. 1990). This group later showed that the broad protein kinase inhibitor fasudil was also effective in improving locomotor function and tissue sparing following a spinal cord injury (Hara et al., 2000).

As secondary injury processes develop, a glial scar formed by chondroitin sulfate proteoglycans (CSPGs) is created to protect the damaged tissue (Fawcett & Asher, 1999). This scar formation, along with myelin-associated proteins, exerts inhibitory effects on axonal regeneration (McKerracher et al. 1994; Chen et al. 2000). Interestingly, PKC has been shown to be a key signaling mediator that is activated by these inhibitory agents (Sivanskaran et al., 2004). Sivasankaran and colleagues used immobilized substrates coated in either inhibitory myelin proteins or CSPGs to assay neurite outgrowth. They tested a range of PKC inhibitors, and found that inhibiting PKC activity stimulated neurite outgrowth on both the inhibitory myelin protein and CSPG substrates. Further, they were able to confirm these *in vitro* findings in an *in vivo* model of spinal cord injury. Rat subjects were given a dorsal hemisection, followed by an osmotic infusion of the PKC inhibitor Go6976 over the next 14 days. Results showed axonal regeneration in the dorsal column, with fibers crossing the lesion gap and extending as far as 6 mm (Sivasankaran et al., 2004). While this treatment appears promising, they showed very little axonal regeneration of the the corticospinal tract (CST), which is thought to be necessary in order to maximize functional recovery of descending motor control in primates, including humans (Blesch & Tuszynski, 2009). Further work will be needed to determine whether the regeneration promoted by PKC inhibition results in functional connectivity and improved behavioral outcomes.

Recently, many spinal cord researchers have begun focusing on the signaling pathways that are activated by the inhibitory myelin proteins. Interestingly, many of these inhibitory proteins act through their receptors to activate a small GTPase called Rho (Niederost et al. 2002). Rho is known to be important for regulating cytoskeletal structure and guiding axons in the developing CNS, and thus has been become a target of interest for those that seek to promote the regeneration of axons across spinal cord lesions after injury (Hall, 1998; Dubreuil et al., 2003). Many researchers have focused on inhibiting or altering Rho function after spinal cord injury, and have shown this treatment to be effective in blocking the growth inhibitory factors that are rampant after spinal cord injury (McKerracher et al. 2006; Fehlings et al., 2011). Others have looked further downstream, to the protein kinase that is activated by Rho. Rho-associated protein kinase (ROCK) has been shown to mediate the retraction of neurites *in vitro,* and experimental activation of ROCK is known to regulate myelin phosphotase, an essential component of axonal sprouting (Hirose et al., 1998; Kimura et al. 1996). Thus, ROCK has become an attractive therapeutic target, as specific ROCK inhibition is believed to mitigate the inhibitory effects of those myelin-derived proteins that undermine axonal regeneration. In 2000, Bito and colleagues used cultured, immature cerebellar granule neurons to directly study the effects of ROCK inhibition on neurite outgrowth. By co-transfecting these cells with green fluorescent protein (GFP) and an active form of Rho (V14Rho), they observed a marked retardation of axonal growth. When they then introduced the ROCK inhibitor Y-27632, they found that inhibiting ROCK attenuated the stunted growth, and produced significant axon genesis.

Building on these findings, Dergham and colleagues (2002) tested this same ROCK inhibitor on a variety of substrates coated with myelin inhibitory proteins or CSPG. They too found that *in vitro* administration of Y-27632 promoted the growth of primary neurons across these substrates. Further, they extended these findings to an *in vivo* spinal cord injury model. They gave mouse subjects dorsal hemisections, followed by spinal injection of Y-27632. They found that this ROCK inhibitor not only attenuated axonal dieback, but promoted the regeneration of neurons within the corticospinal tract, generating sprouting that stretched 2-3 mms across the lesion site. The ROCK-inhibited subjects also exhibited a long-term improvement in locomotor function. In 2003, Fournier published a study that extended these findings to rats, showing similar results (Fournier et al., 2003). They found that ROCK inhibition with Y-27632 was not only sufficient to promote neurite outgrowth *in vitro,* but as with the Dergham study, they showed that ROCK inhibition after hemisection could promote CST axons to regenerate across the lesion, as well as produce significant behavioral improvement in comparision to vehicle-treated subjects.

Taken together as a group, these data indicate a strong role for PKC as well as Rho kinases in morphological regeneration of the CST that is thought to be necessary for recovery of function after spinal cord injury (Blesch & Tuszynski, 2009; Nielson et al., 2010). However, recent findings have revealed that, even in the absence of CST regeneration through a spinal cord lesion, there may be substantial sprouting of surviving CST fibers below the lesion site in primates, which has been linked to improved recovery of function in forelimb control (Rosenzweig et al., 2010). This suggests a role for local spinal plasticity in restoring recovery of function in CST-dependent function. The role of mGluRs and PKC flux in these effects is largely unknown, and may represent a fruitful area for further study.

5. Conclusions

This chapter has reviewed the role of mGluRs and down-stream PKC activity as a major factor in a number of forms of plasticity throughout the spinal cord. Its ubiquity indicates a potential common mechanism for a host of complex processes, in both the intact and injured spinal cord. As a critical link between mGluR activation and ionotropic GluR trafficking and phosphorylation, PKC activity can mediate either the potentiation or depression of synaptic strength, the promotion of neural regeneration or the exacerbation of excitotoxicity. In the future, targeting PKC activity within the appropriate circumstances, and at the right time, will be essential to tailoring effective treatments for both central and peripheral injuries, as well as in the promotion of use-dependent spinal plasticity.

6. Acknowledgements

The authors wish to thank Aiwen Liu for assistance in preparing this chapter. Supported by NIH Grants NS069537 and NS067092 to A.R.F. Correspondence should be addressed to J. Russell Huie (huier@neurosurg.ucsf.edu) or Adam Ferguson (adam.ferguson@ucsf.edu).

7. References

Abraham, W. C., & Bear, M. F. (1996). Metaplasticity: the plasticity of synaptic plasticity. *Trends Neurosci, 19*(4), 126-130.

Adwanikar, H., Karim, F., & Gereau, R. W. t. (2004). Inflammation persistently enhances nocifensive behaviors mediated by spinal group I mGluRs through sustained ERK activation. *Pain, 111*(1-2), 125-135.

Aiba, A., Chen, C., Herrup, K., Rosenmund, C., Stevens, C. F., & Tonegawa, S. (1994). Reduced hippocampal long-term potentiation and context-specific deficit in associative learning in mGluR1 mutant mice. *Cell, 79*(2), 365-375.

Anwyl, R. (2009). Metabotropic glutamate receptor-dependent long-term potentiation. *Neuropharmacology, 56*(4), 735-740.

Asiedu, M. N., Tillu, D. V., Melemedjian, O. K., Shy, A., Sanoja, R., Bodell, B., et al. Spinal protein kinase M zeta underlies the maintenance mechanism of persistent nociceptive sensitization. *J Neurosci, 31*(18), 6646-6653.

Balschun, D., Manahan-Vaughan, D., Wagner, T., Behnisch, T., Reymann, K. G., & Wetzel, W. (1999). A specific role for group I mGluRs in hippocampal LTP and hippocampus-dependent spatial learning. *Learn Mem, 6*(2), 138-152.

Baumbauer, K. M., Young, E. E., Hoy, K. C., Jr., Abood, A., & Joynes, R. L. (2007). Administration of a Ca-super(2+)/calmodulin-dependent protein kinase II (CaMKII) inhibitor prevents the learning deficit observed in spinal rats after noncontingent shock administration. *Behav Neurosci, 121*(3), 570-578.

Bito, H., Furuyashiki, T., Ishihara, H., Shibasaki, Y., Ohashi, K., Mizuno, K., et al. (2000). A critical role for a Rho-associated kinase, p160ROCK, in determining axon outgrowth in mammalian CNS neurons. *Neuron, 26*(2), 431-441.

Blesch, A., & Tuszynski, M. H. (2009). Spinal cord injury: plasticity, regeneration and the challenge of translational drug development. *Trends Neurosci, 32*(1), 41-47.

Bliss, T. V., & Collingridge, G. L. (1993). A synaptic model of memory: long-term potentiation in the hippocampus. *Nature, 361*(6407), 31-39.

Bortolotto, Z. A., Bashir, Z. I., Davies, C. H., & Collingridge, G. L. (1994). A molecular switch activated by metabotropic glutamate receptors regulates induction of long-term potentiation. *Nature, 368*(6473), 740-743.

Bortolotto, Z. A., & Collingridge, G. L. (1993). Characterisation of LTP induced by the activation of glutamate metabotropic receptors in area CA1 of the hippocampus. *Neuropharmacology, 32*(1), 1-9.

Bortolotto, Z. A., Fitzjohn, S. M., & Collingridge, G. L. (1999). Roles of metabotropic glutamate receptors in LTP and LTD in the hippocampus. *Curr Opin Neurobiol, 9*(3), 299-304.

Buerger, A. A., & Fennessy, A. (1970). Learning of leg position in chronic spinal rats. *Nature, 225*(5234), 751-752.

Camodeca, N., Breakwell, N. A., Rowan, M. J., & Anwyl, R. (1999). Induction of LTD by activation of group I mGluR in the dentate gyrus in vitro. *Neuropharmacology, 38*(10), 1597-1606.

Chen, M. S., Huber, A. B., van der Haar, M. E., Frank, M., Schnell, L., Spillmann, A. A., et al. (2000). Nogo-A is a myelin-associated neurite outgrowth inhibitor and an antigen for monoclonal antibody IN-1. *Nature, 403*(6768), 434-439.

Choi, D. W. (1992). Excitotoxic cell death. *J Neurobiol, 23*(9), 1261-1276.

Chopin, S. F., & Buerger, A. A. (1976). Instrumental avoidance conditioning in the spinal rat. *Brain Res Bull, 1*(2), 177-183.

Christensen, M. D., & Hulsebosch, C. E. (1997). Chronic central pain after spinal cord injury. *J Neurotrauma, 14*(8), 517-537.

Coderre, T. J. (1992). Contribution of protein kinase C to central sensitization and persistent pain following tissue injury. *Neurosci Lett, 140*(2), 181-184.

Conquet, F., Bashir, Z. I., Davies, C. H., Daniel, H., Ferraguti, F., Bordi, F., et al. (1994). Motor deficit and impairment of synaptic plasticity in mice lacking mGluR1. *Nature, 372*(6503), 237-243.

Courtine, G., Song, B., Roy, R. R., Zhong, H., Herrmann, J. E., Ao, Y., et al. (2008). Recovery of supraspinal control of stepping via indirect propriospinal relay connections after spinal cord injury. *Nat Med, 14*(1), 69-74.

Crowe, M. J., Bresnahan, J. C., Shuman, S. L., Masters, J. N., & Beattie, M. S. (1997). Apoptosis and delayed degeneration after spinal cord injury in rats and monkeys. *Nat Med, 3*(1), 73-76.

Crown, E. D., Ferguson, A. R., Joynes, R. L., & Grau, J. W. (2002). Instrumental learning within the spinal cord: IV. Induction and retention of the behavioral deficit observed after noncontingent shock. *Behav Neurosci, 116*(6), 1032-1051.

de Leon, R. D., Roy, R. R., & Edgerton, V. R. (2001). Is the recovery of stepping following spinal cord injury mediated by modifying existing neural pathways or by generating new pathways? A perspective. *Phys Ther, 81*(12), 1904-1911.

Dergham, P., Ellezam, B., Essagian, C., Avedissian, H., Lubell, W. D., & McKerracher, L. (2002). Rho signaling pathway targeted to promote spinal cord repair. *J Neurosci, 22*(15), 6570-6577.

Dubreuil, C. I., Winton, M. J., & McKerracher, L. (2003). Rho activation patterns after spinal cord injury and the role of activated Rho in apoptosis in the central nervous system. *J Cell Biol, 162*(2), 233-243.

Edgerton, V. R., & Harkema, S. (2011). Epidural stimulation of the spinal cord in spinal cord injury: current status and future challenges. *Expert Rev Neurother, 11*(10), 1351-1353.

Edgerton, V. R., Leon, R. D., Harkema, S. J., Hodgson, J. A., London, N., Reinkensmeyer, D. J., et al. (2001). Retraining the injured spinal cord. *J Physiol, 533*(Pt 1), 15-22.

Fawcett, J. W., & Asher, R. A. (1999). The glial scar and central nervous system repair. *Brain Res Bull, 49*(6), 377-391.

Fehlings, M. G., Theodore, N., Harrop, J., Maurais, G., Kuntz, C., Shaffrey, C. I., et al. A phase I/IIa clinical trial of a recombinant Rho protein antagonist in acute spinal cord injury. *J Neurotrauma, 28*(5), 787-796.

Ferguson, A. R., Bolding, K. A., Huie, J. R., Hook, M. A., Santillano, D. R., Miranda, R. C., et al. (2008). Group I metabotropic glutamate receptors control metaplasticity of spinal cord learning through a protein kinase C-dependent mechanism. *J Neurosci, 28*(46), 11939-11949.

Ferguson, A. R., Christensen, R. N., Gensel, J. C., Miller, B. A., Sun, F., Beattie, E. C., et al. (2008). Cell death after spinal cord injury is exacerbated by rapid TNF alpha-induced trafficking of GluR2-lacking AMPARs to the plasma membrane. *J Neurosci, 28*(44), 11391-11400.

Fisher, K., & Coderre, T. J. (1996). The contribution of metabotropic glutamate receptors (mGluRs) to formalin-induced nociception. *Pain, 68*(2-3), 255-263.

Fournier, A. E., Takizawa, B. T., & Strittmatter, S. M. (2003). Rho kinase inhibition enhances axonal regeneration in the injured CNS. *J Neurosci, 23*(4), 1416-1423.

Gallagher, S. M., Daly, C. A., Bear, M. F., & Huber, K. M. (2004). Extracellular signal-regulated protein kinase activation is required for metabotropic glutamate receptor-dependent long-term depression in hippocampal area CA1. *J Neurosci, 24*(20), 4859-4864.

Gereau, R. W. t., & Heinemann, S. F. (1998). Role of protein kinase C phosphorylation in rapid desensitization of metabotropic glutamate receptor 5. *Neuron, 20*(1), 143-151.

Giles, P. A., Trezise, D. J., & King, A. E. (2007). Differential activation of protein kinases in the dorsal horn in vitro of normal and inflamed rats by group I metabotropic glutamate receptor subtypes. *Neuropharmacology, 53*(1), 58-70.

Gomez-Pinilla, F., Huie, J. R., Ying, Z., Ferguson, A. R., Crown, E. D., Baumbauer, K. M., et al. (2007). BDNF and learning: Evidence that instrumental training promotes learning within the spinal cord by up-regulating BDNF expression. *Neuroscience, 148*(4), 893-906.

Grau, J. W., Barstow, D. G., & Joynes, R. L. (1998). Instrumental learning within the spinal cord: I. Behavioral properties. *Behav Neurosci, 112*(6), 1366-1386.

Grau, J. W., Crown, E. D., Ferguson, A. R., Washburn, S. N., Hook, M. A., & Miranda, R. C. (2006). Instrumental learning within the spinal cord: underlying mechanisms and implications for recovery after injury. *Behav Cogn Neurosci Rev, 5*(4), 191-239.

Grau, J. W., Washburn, S. N., Hook, M. A., Ferguson, A. R., Crown, E. D., Garcia, G., et al. (2004). Uncontrollable stimulation undermines recovery after spinal cord injury. *J Neurotrauma, 21*(12), 1795-1817.

Hall, A. (1998). Rho GTPases and the actin cytoskeleton. *Science, 279*(5350), 509-514.

Hara, H., Onodera, H., & Kogure, K. (1990). Protein kinase C activity in the gerbil hippocampus after transient forebrain ischemia: morphological and autoradiographic analysis using [3H]phorbol 12,13-dibutyrate. *Neurosci Lett, 120*(1), 120-123.

Hara, M., Takayasu, M., Watanabe, K., Noda, A., Takagi, T., Suzuki, Y., et al. (2000). Protein kinase inhibition by fasudil hydrochloride promotes neurological recovery after spinal cord injury in rats. *J Neurosurg, 93*(1 Suppl), 94-101.

Harris, N. C., & Ryall, R. W. (1988). Mustard oil excites but does not inhibit nociceptive dorsal horn neurones in the rat: a presumed effect on A-delta fibres. *Br J Pharmacol, 94*(1), 180-184.

Hellier, J. L., Grosshans, D. R., Coultrap, S. J., Jones, J. P., Dobelis, P., Browning, M. D., et al. (2007). NMDA receptor trafficking at recurrent synapses stabilizes the state of the CA3 network. *J Neurophysiol, 98*(5), 2818-2826.

Hirose, M., Ishizaki, T., Watanabe, N., Uehata, M., Kranenburg, O., Moolenaar, W. H., et al. (1998). Molecular dissection of the Rho-associated protein kinase (p160ROCK)-

regulated neurite remodeling in neuroblastoma N1E-115 cells. *J Cell Biol, 141*(7), 1625-1636.

Hua, X. Y., Chen, P., & Yaksh, T. L. (1999). Inhibition of spinal protein kinase C reduces nerve injury-induced tactile allodynia in neuropathic rats. *Neurosci Lett, 276*(2), 99-102.

Huber, K. M., Kayser, M. S., & Bear, M. F. (2000). Role for rapid dendritic protein synthesis in hippocampal mGluR-dependent long-term depression. *Science, 288*(5469), 1254-1257.

Ji, R. R., Kohno, T., Moore, K. A., & Woolf, C. J. (2003). Central sensitization and LTP: do pain and memory share similar mechanisms? *Trends Neurosci, 26*(12), 696-705.

Jia, Z., Lu, Y., Henderson, J., Taverna, F., Romano, C., Abramow-Newerly, W., et al. (1998). Selective abolition of the NMDA component of long-term potentiation in mice lacking mGluR5. *Learn Mem, 5*(4-5), 331-343.

Kim, S. H., & Chung, J. M. (1992). An experimental model for peripheral neuropathy produced by segmental spinal nerve ligation in the rat. *Pain, 50*(3), 355-363.

Kimura, K., Ito, M., Amano, M., Chihara, K., Fukata, Y., Nakafuku, M., et al. (1996). Regulation of myosin phosphatase by Rho and Rho-associated kinase (Rho-kinase). *Science, 273*(5272), 245-248.

LaMotte, R. H., Lundberg, L. E., & Torebjork, H. E. (1992). Pain, hyperalgesia and activity in nociceptive C units in humans after intradermal injection of capsaicin. *J Physiol, 448*, 749-764.

Latremoliere, A., & Woolf, C. J. (2009). Central sensitization: a generator of pain hypersensitivity by central neural plasticity. *J Pain, 10*(9), 895-926.

Li, P., Kerchner, G. A., Sala, C., Wei, F., Huettner, J. E., Sheng, M., et al. (1999). AMPA receptor-PDZ interactions in facilitation of spinal sensory synapses. *Nat Neurosci, 2*(11), 972-977.

Liao, G. Y., Wagner, D. A., Hsu, M. H., & Leonard, J. P. (2001). Evidence for direct protein kinase-C mediated modulation of N-methyl-D-aspartate receptor current. *Mol Pharmacol, 59*(5), 960-964.

Lindsey, A. E., LoVerso, R. L., Tovar, C. A., Hill, C. E., Beattie, M. S., & Bresnahan, J. C. (2000). An analysis of changes in sensory thresholds to mild tactile and cold stimuli after experimental spinal cord injury in the rat. *Neurorehabil Neural Repair, 14*(4), 287-300.

Lu, W. Y., Xiong, Z. G., Lei, S., Orser, B. A., Dudek, E., Browning, M. D., et al. (1999). G-protein-coupled receptors act via protein kinase C and Src to regulate NMDA receptors. *Nat Neurosci, 2*(4), 331-338.

Madronal, N., Gruart, A., Sacktor, T. C., & Delgado-Garcia, J. M. (2011). PKMzeta inhibition reverses learning-induced increases in hippocampal synaptic strength and memory during trace eyeblink conditioning. *PLoS One, 5*(4), e10400.

Malenka, R. C., & Bear, M. F. (2004). LTP and LTD: an embarrassment of riches. *Neuron, 44*(1), 5-21.

Malinow, R., & Malenka, R. C. (2002). AMPA receptor trafficking and synaptic plasticity. *Annu Rev Neurosci, 25*, 103-126.

McAdoo, D. J., Xu, G. Y., Robak, G., & Hughes, M. G. (1999). Changes in amino acid concentrations over time and space around an impact injury and their diffusion through the rat spinal cord. *Exp Neurol, 159*(2), 538-544.

McKerracher, L., David, S., Jackson, D. L., Kottis, V., Dunn, R. J., & Braun, P. E. (1994). Identification of myelin-associated glycoprotein as a major myelin-derived inhibitor of neurite growth. *Neuron, 13*(4), 805-811.

McKerracher, L., & Higuchi, H. (2006). Targeting Rho to stimulate repair after spinal cord injury. *J Neurotrauma, 23*(3-4), 309-317.

Mendell, L. M. (1966). Physiological properties of unmyelinated fiber projection to the spinal cord. *Exp Neurol, 16*(3), 316-332.

Migues, P. V., Hardt, O., Wu, D. C., Gamache, K., Sacktor, T. C., Wang, Y. T., et al. (2011). PKMzeta maintains memories by regulating GluR2-dependent AMPA receptor trafficking. *Nat Neurosci, 13*(5), 630-634.

Mills, C. D., Xu, G. Y., McAdoo, D. J., & Hulsebosch, C. E. (2001). Involvement of metabotropic glutamate receptors in excitatory amino acid and GABA release following spinal cord injury in rat. *J Neurochem, 79*(4), 835-848.

Mockett, B. G., Guevremont, D., Wutte, M., Hulme, S. R., Williams, J. M., & Abraham, W. C. (2011). Calcium/calmodulin-dependent protein kinase II mediates group I metabotropic glutamate receptor-dependent protein synthesis and long-term depression in rat hippocampus. *J Neurosci, 31*(20), 7380-7391.

Moult, P. R., Schnabel, R., Kilpatrick, I. C., Bashir, Z. I., & Collingridge, G. L. (2002). Tyrosine dephosphorylation underlies DHPG-induced LTD. *Neuropharmacology, 43*(2), 175-180.

Munro, F. E., Fleetwood-Walker, S. M., & Mitchell, R. (1994). Evidence for a role of protein kinase C in the sustained activation of rat dorsal horn neurons evoked by cutaneous mustard oil application. *Neurosci Lett, 170*(2), 199-202.

Niederost, B., Oertle, T., Fritsche, J., McKinney, R. A., & Bandtlow, C. E. (2002). Nogo-A and myelin-associated glycoprotein mediate neurite growth inhibition by antagonistic regulation of RhoA and Rac1. *J Neurosci, 22*(23), 10368-10376.

Nielson, J. L., Sears-Kraxberger, I., Strong, M. K., Wong, J. K., Willenberg, R., & Steward, O. Unexpected survival of neurons of origin of the pyramidal tract after spinal cord injury. *J Neurosci, 30*(34), 11516-11528.

Oliet, S. H., Malenka, R. C., & Nicoll, R. A. (1997). Two distinct forms of long-term depression coexist in CA1 hippocampal pyramidal cells. *Neuron, 18*(6), 969-982.

Parsons, R. G., & Davis, M. (2011). Temporary disruption of fear-potentiated startle following PKMzeta inhibition in the amygdala. *Nat Neurosci, 14*(3), 295-296.

Pastalkova, E., Serrano, P., Pinkhasova, D., Wallace, E., Fenton, A. A., & Sacktor, T. C. (2006). Storage of spatial information by the maintenance mechanism of LTP. *Science, 313*(5790), 1141-1144.

Raineteau, O., & Schwab, M. E. (2001). Plasticity of motor systems after incomplete spinal cord injury. *Nat Rev Neurosci, 2*(4), 263-273.

Rosenzweig, E. S., Courtine, G., Jindrich, D. L., Brock, J. H., Ferguson, A. R., Strand, S. C., et al. Extensive spontaneous plasticity of corticospinal projections after primate spinal cord injury. *Nat Neurosci, 13*(12), 1505-1510.

Rossignol, S., Schwab, M., Schwartz, M., & Fehlings, M. G. (2007). Spinal cord injury: time to move? *J Neurosci, 27*(44), 11782-11792.

Sacktor, T. C. (2011). How does PKMzeta maintain long-term memory? *Nat Rev Neurosci, 12*(1), 9-15.

Sacktor, T. C. (2008). PKMzeta, LTP maintenance, and the dynamic molecular biology of memory storage. *Prog Brain Res, 169*, 27-40.

Sandkuhler, J. (2000). Learning and memory in pain pathways. *Pain, 88*(2), 113-118.

Sandkuhler, J., & Liu, X. (1998). Induction of long-term potentiation at spinal synapses by noxious stimulation or nerve injury. *Eur J Neurosci, 10*(7), 2476-2480.

Seligman, M. E., & Maier, S. F. (1967). Failure to escape traumatic shock. *J Exp Psychol, 74*(1), 1-9.

Shema, R., Sacktor, T. C., & Dudai, Y. (2007). Rapid erasure of long-term memory associations in the cortex by an inhibitor of PKM zeta. *Science, 317*(5840), 951-953.

Sheng, M., & Kim, M. J. (2002). Postsynaptic signaling and plasticity mechanisms. *Science, 298*(5594), 776-780.

Sivasankaran, R., Pei, J., Wang, K. C., Zhang, Y. P., Shields, C. B., Xu, X. M., et al. (2004). PKC mediates inhibitory effects of myelin and chondroitin sulfate proteoglycans on axonal regeneration. *Nat Neurosci, 7*(3), 261-268.

Sun, R. Q., Tu, Y. J., Lawand, N. B., Yan, J. Y., Lin, Q., & Willis, W. D. (2004). Calcitonin gene-related peptide receptor activation produces PKA- and PKC-dependent mechanical hyperalgesia and central sensitization. *J Neurophysiol, 92*(5), 2859-2866.

Wang, Q., Walsh, D. M., Rowan, M. J., Selkoe, D. J., & Anwyl, R. (2004). Block of long-term potentiation by naturally secreted and synthetic amyloid beta-peptide in hippocampal slices is mediated via activation of the kinases c-Jun N-terminal kinase, cyclin-dependent kinase 5, and p38 mitogen-activated protein kinase as well as metabotropic glutamate receptor type 5. *J Neurosci, 24*(13), 3370-3378.

Waung, M. W., Pfeiffer, B. E., Nosyreva, E. D., Ronesi, J. A., & Huber, K. M. (2008). Rapid translation of Arc/Arg3.1 selectively mediates mGluR-dependent LTD through persistent increases in AMPAR endocytosis rate. *Neuron, 59*(1), 84-97.

Wernig, A., Nanassy, A., & Muller, S. (2000). Laufband (LB) therapy in spinal cord lesioned persons. *Prog Brain Res, 128*, 89-97.

Willis, W. D., & Westlund, K. N. (1997). Neuroanatomy of the pain system and of the pathways that modulate pain. *J Clin Neurophysiol, 14*(1), 2-31.

Wolpaw, J. R. (2007). Spinal cord plasticity in acquisition and maintenance of motor skills. *Acta Physiol (Oxf), 189*(2), 155-169.

Woolf, C. J. (1983). Evidence for a central component of post-injury pain hypersensitivity. *Nature, 306*(5944), 686-688.

Woolf, C. J., & Salter, M. W. (2000). Neuronal plasticity: increasing the gain in pain. *Science, 288*(5472), 1765-1769.

Xiao, M. Y., Zhou, Q., & Nicoll, R. A. (2001). Metabotropic glutamate receptor activation causes a rapid redistribution of AMPA receptors. *Neuropharmacology, 41*(6), 664-671.

Yang, L., Liu, G., Zakharov, S. I., Morrow, J. P., Rybin, V. O., Steinberg, S. F., et al. (2005). Ser1928 is a common site for Cav1.2 phosphorylation by protein kinase C isoforms. *J Biol Chem, 280*(1), 207-214.

Yao, Y., Kelly, M. T., Sajikumar, S., Serrano, P., Tian, D., Bergold, P. J., et al. (2008). PKM zeta maintains late long-term potentiation by N-ethylmaleimide-sensitive factor/GluR2-dependent trafficking of postsynaptic AMPA receptors. *J Neurosci, 28*(31), 7820-7827.

Yashpal, K., Fisher, K., Chabot, J. G., & Coderre, T. J. (2001). Differential effects of NMDA and group I mGluR antagonists on both nociception and spinal cord protein kinase C translocation in the formalin test and a model of neuropathic pain in rats. *Pain, 94*(1), 17-29.

Young, M. R., Fleetwood-Walker, S. M., Mitchell, R., & Dickinson, T. (1995). The involvement of metabotropic glutamate receptors and their intracellular signalling pathways in sustained nociceptive transmission in rat dorsal horn neurons. *Neuropharmacology, 34*(8), 1033-1041.

21

Protein Kinases and Pain

Mani Indiana Funez, Fabiane Hiratsuka Veiga de Souza,
José Eduardo Pandossio and Paulo Gustavo Barboni Dantas Nascimento
School of Ceilandia, Brasilia University,
Brazil

1. Introduction

There is abundant evidence that protein kinases are involved in the physiopathology of acute and chronic pain. In the first section, we discuss the role of protein kinases in pain and the signalling pathways involved in both the acute and chronic states. The second section will present evidence supporting the contribution of protein kinase inhibition to pain control by different classes of drugs. Both well-known drugs and new molecules can control pain in the peripheral and central nervous systems. The third section highlights the progress in pharmaceutical development and protein kinase research for new pain control drugs in the first decade of the 21st century.

2. Role of protein kinases in acute and chronic pain

In this section, we will discuss the differential activation of protein kinases by pain mediators and the modulation of the acute and chronic pain processes by several kinases.

2.1 PKC

Protein Kinase C (PKC) is a family of phospholipid-dependent serine/threonine phosphotransferases; it can be divided into the following groups of isoforms: a) conventional or classical (α, βI, βII, γ), b) novel (δ, ε, η, θ), and c) atypical (ζ, λ (mouse)/ι (human)) isoforms (Nishizuka, 1992). Five subspecies of PKC, PKC-βI, PKC-βII, PKC-δ, PKC-ε, and PKC-ζ, are expressed in the dorsal root ganglion (DRG) of rats (Cesare et al. 1999). The PKC isoforms that are expressed in the DRG of mice include PKC-α, PKC-βI, PKC-βII, PKC-δ, PKC-ε, PKC-η, PKC-θ, PKC-ζ, and PKC-λ (Khasar et al., 1999a). Thus, there are some differences in the expression of DRG PKC isoforms between species.

Signal transduction through the PKC pathway has been strongly linked to pain. Inflammatory stimuli and mediators can activate PKC to induce pain. Nociceptive response caused by formalin injection into the mouse paw is characterised by two phases; the neurogenic response, which is due to direct nociceptor activation, and the inflammatory response, which is caused by inflammatory mediators (Hunskaar and Hole, 1987). In this model, PKC blockade by local treatment with chelerythrine inhibited the second phase of nociceptive response (Souza et al., 2002), which is driven largely by tissue inflammation, indicating a relationship between PKC activation and the inflammatory process. In the same way, mechanical sensitisation induced by the inflammatory mediator bradykinin in rats is inhibited by a PKC inhibitor (Souza et al., 2002). *In vitro* experiments conducted in DRG neurons strongly suggest that bradykinin-induced heat sensitisation is dependent on PKC activation because it can be reversed by pharmacological inhibition with staurosporine or phosphatase inhibitors (Burguess et al., 1989; Cesare and McNaughton, 1996). *In vitro* experiments have shown noticeable PKC-activity in rat DRGs after 3 hours of prostaglandin E_2 (PGE_2) paw administration. This activity was accompanied by paw sensitisation to mechanical stimuli, as measured by behavioural experiments (Sachs et al., 2009).

Carrageenan injection in rat or mouse paws is another tool used to study inflammatory sensitisation. In the same way, pharmacological inhibition of PKCε reduces carrageenan-induced mechanical sensitisation in mice (Khasar et al., 1999a). Phosphorylation of PKCε in DRG neurons is increased after carrageenan-induced acute sensitisation (Zhou et al. 2003), and a PKCε agonist sensitises nociceptors to mechanical stimuli (Aley and Levine, 2003). Inflammatory sensitisation has a "sympathetic" component that involves the release of amines such as epinephrine and dopamine (Coderre et al. 1984; Nakamura and Ferreira 1987). Evidence suggests that mechanical sensitisation induced either by epinephrine in rats (Khasar et al. 1999b) or dopamine in mice (Villarreal et al., 2009b) is blocked by PKCε-selective inhibition. Cesare et al. (1999) found that bradykinin exposure induces PKCε translocation from the cytosol to a membrane-associated position in cultured DRG neurons, thus contributing to heat sensitisation.

The mechanical sensitisation induced by PGE_2 involves the peripheral activation of PKCε in rats and mice, as shown by specific pharmacological inhibition (Sachs et al., 2009; Villareal et al., 2009b). In PKCε-mutant mice, the nociceptive threshold is preserved, whereas the nociceptive response was significantly impaired, as evaluated in a model of visceral pain using peritoneal administration of acetic acid (Khasar et al. 1999a). Kassuya et al. (2007), found a noticeable increase in membrane-bound PKCα expression of mouse paw tissue after PGE_2 administration (Kassuya et al., 2007). Thus, the PGE_2–induced pain-related effects during inflammation may be mediated by PKCε and PKCα.

Multiple voltage-gated sodium channel (VGSC) isoforms are expressed in DRG neurons. For example, isoforms Na_v 1.8 and Na_v 1.9 are responsible for tetrodotoxin-resistant (TTX-resistant) currents due to Na^+ channel blocker insensitivity. These sodium currents can be modulated by PKC phosphorylation, which is induced by inflammatory mediators (Gold et al., 1998; Khasar et al., 1999a). Using whole-cell voltage-clamp recordings from DRG neurons, Gold et al. (1998) found that PKC inhibitors decreased the density of tetrodotoxin-resistant sodium current, whereas the PKC activator PMA produces changes that are opposite, suggesting that PKC modulates it. In addition, it was show a relationship between inflammatory mediators-induced changes in TTX-resistant sodium currents and PKC activity (Gold et al., 1998; Khasar et al., 1999a).

PKC peripheral activation contributes to central pain processing. During serotonin-induced rat paw sensitisation, another pain-sensitising mediator associated with inflammation, the response of animals to thermal stimulation and c-fos activation in the dorsal horn is attenuated by intraplantar application of the PKC inhibitor chelerythrine (Chen et al., 2006). During inflammation or in naïve animals, activation of glutamate receptors mGluRs in the spinal dorsal horn modulates acute nociception. These receptors are coupled to Gq/II protein phospholipase C (PLC)-phosphoinositide (PI) hydrolysis and PKC pre- and postsynaptic activation (Neugebauer, 2002; Giles et al., 2007), suggesting that PKC modulates the synaptic transmission at the spinal level.

PKC activation is associated with chronic pain conditions. Mao et al. (1992) found an increase in membrane-bound PKC in the spinal cord of rats in a model of post-injury neuropathic pain. The role of PKC was confirmed using an intracellular inhibitor of PKC translocation/activation and analysing membrane-bound PKC translocation and pain behaviour. The data suggest a role for PKC in neuropathic pain states. Ahlgren and Levine (1994) found a reduction in streptozotocin-induced diabetic rat pain sensitisation after treatment with PKC inhibitors.

Using the partial sciatic nerve section model, Malmberg et al. (1997) verified that mice lacking PKCγ completely fail to develop neuropathic-associated sensitisation even though they respond normally to acute pain stimuli. In addition, PKCγ expression is restricted to a subset of dorsal horn neurons. Malmberg and co-workers suggest that targeting PKCγ is a promising tool for treating chronic pain. This isoform inhibition also attenuates opioid tolerance in the spinal cord (section 3).

The physiopathology of alcoholic neuropathy in rats seems to depend on PKCε activation and up-regulation in DRG neurons, as shown by selective pharmacological inhibition and western blot analysis performed after 70 days of ethanol administration (Dina et al., 2000). In addition, the role of PKCε in pain sensitisation is associated with neuropathy induced by the antineoplastic agent paclitaxel in rats (Dina et al., 2001).

The role of PKCε is well demonstrated during chronic inflammatory pain conditions. Aley et al. (2000) developed a model to study chronic inflammatory sensitisation that can be induced by a single episode of acute inflammation; after the induction, in a time-lapse of 5 days there is inflammatory-mediator prolonged-response. During this state, PKCε seems to be responsible for the maintenance of this "primed state" and the prolonged response to inflammatory mediators (Aley et al. 2000). Accordingly, the phosphorylation of PKCε in DRG neurons correlated with pain-associated prolonged inflammation after 3 days of the administration of Complete Freund's Adjuvant (CFA) to rat paws (Zhou et al., 2003).

Mechanical persistent inflammatory sensitisation can also be induced by intraplantar administration of inflammatory mediators like prostaglandins and sympathetic amines in rats and mice (Ferreira et al., 1990; Villarreal et al., 2009b). Studies suggest that PKC activity in the DRG is up-regulated by and is at least partially responsible for the persistent condition, as shown by analyses of PKC activity in rat DRGs (Villarreal et al. 2009a). Moreover, the local administration of a selective PKCε inhibitor abolished the persistent state induced by PGE_2 in rats and mice (Villarreal et al. 2009a; Villarreal et al., 2009b). Evaluation of the mechanisms downstream of PKCε activation found that $Na_v1.8$ mRNA levels in the DRG from rats was up-regulated and inhibition of PKCε activity reduced these levels (Villarreal et al., 2009a).

2.2 PKA

Cyclic adenosine-monophosphate (cAMP)-dependent protein kinase (PKA) is a serine/threonine phosphotransferase; in its inactive form, it is a tetrameric holoenzyme composed of two regulatory and two catalytic subunits (Taylor et al., 1990). When the second messenger cAMP is generated the PKA-regulatory subunits bind cAMP, and the holoenzyme separates into the regulatory subunits and the catalytic subunits (Taylor et al., 1990). The catalytic subunits can phosphorylate their biological targets and regulate many cellular functions. There are different regulatory (RIα, RIβ, RIIα, RIIβ) and catalytic (Cα, Cβ) subunits; α subunits are expressed in non-neuronal and neuronal tissue, whereas β subunits are expressed predominantly in neuronal cells (Cadd and McKnight, 1989).

cAMP/PKA signalling is involved in nociceptor sensitisation by inflammatory mediators. Ferreira and Nakamura (1979) provided evidence that sensitisation of rat hind-paws by prostaglandins is dependent on cAMP generation. Since this original study, many subsequent studies have shown that cAMP generation is induced by a plethora of inflammatory stimuli. The mechanical nociceptor sensitisation that occurs during inflammation or induced by either inflammatory mediators (PGE_2, dopamine, serotonin) is blocked by treatment with PKA inhibitors in rats and mice (Taiwo and Levine 1991; Taiwo et al., 1992; Aley and Levine, 1999; Aley et al., 2000; Sachs et al., 2009; Villarreal et al., 2009b).

Adenylyl cyclase (AC)/cAMP/PKA activation may be necessary to induce and maintain mechanical nociceptor sensitisation (Aley and Levine, 1999). Moreover, PGE_2-induced inflammatory sensitisation increased PKA activity in mouse paws (Kassuya et al., 2007); and in rat DRGs (Sachs et al., 2009), which correlates with the behavioural data. Accordingly, the intraplantar administration of the catalytic subunit of PKA (PKACS) induces mechanical nociceptor sensitisation (Aley and Levine, 1999; Aley and Levine, 2003). Supporting the animal model data, *in vitro* studies using sensory neurons that were cultured and bathed in classic inflammatory mediators showed that prostaglandins can sensitise these cells to bradykinin and that this effect is dependent on PKA activation (Cui and Nicol, 1995; Smith et al., 2000). The role of PKA in formalin-induced nociceptive pain and inflammatory sensitisation was demonstrated in experiments in mice with a null mutation in the type I regulatory subunit (RIβ) of PKA. This mutation dampens the response during nociceptive pain and thermal stimulation (Malmberg et al., 1997).

Once activated, the PKA substrate in the nociceptive pathways can be voltage-gated sodium channels. In fact, *in vitro* studies have shown that TTX-resistant sodium current is modulated via PKA activation during inflammation (England et al., 1996; Gold et al., 1998). Additionally, during inflammation, PKA enhances the gating of transient receptor potential vanilloid channel-1 (TRPV-1) via direct phosphorylation (Lopshire and Nicol, 1998; Rathee et al., 2002). Therefore, PKA can directly phosphorylate ion channels, thus increasing the excitability of sensory neurons and contributing to some pain conditions. Studies using a model of persistent inflammatory sensitisation in rats and mice show that PKA could exert a role in the maintenance of the chronic state. The persistent sensitisation is abolished by injection of PKA inhibitors, and PKA expression and activity were up-regulated in DRG (Villarreal et al., 2009a, 2009b). The contribution of PKA to sensitisation maintenance seems to be due to the regulation of the $Na_V1.8$ sodium channel expression (Villarreal et al., 2009a).

In the neuropathic pain model of sciatic nerve ligature, PKARIβ-null animals present nociceptive responses that are similar to control animals (Malmberg et al., 1997). However, in a model of paclitaxel-induced pain neuropathy, pharmacological inhibition of PKA attenuates the response to thermal stimulation (Dina et al., 2001). Other subunits of PKA, different from PKARIβ, may be activated during neuropathy because PKA inhibitors do not present selectivity.

2.3 MAPKs

Mitogen-Activated Protein Kinases (MAPKs) are protein-serine/threonine kinases. There are many subfamily isoforms known, and are currently 14 mammalian members. They are important for pain regulation and control and are divided in extracellular-signal-regulated kinases (ERKs, 7 isoforms), stress-activated protein kinases or c-Jun N-terminal kinases (JNKs, 3 isoforms) and p38 mitogen-activated protein kinases (p38 MAPK, 4 isoforms). These enzymes are activated by direct phosphorylation of two sites in the kinase activation loop, at a tyrosine and a threonine residue; separated by a single, variable residue (Pearson et al. 2001).

Classically, upon receptor-dependent tyrosine kinase activation on the cellular surface, a cascade of biochemical reactions culminates in small GTPase (Ras) activation. This molecular event initiates a series of catalytic phosphorylation-based signalling, involving kinases such as the proto-oncogene serine/threonine-protein kinase (C-Raf), mitogen-activated protein kinase kinase 1 (MEK1), MEK2 and MAPKs. The dual phosphorylation of these proteins leads to conformational changes, allowing their respective catalytic domains to be accessible to their substrates, which are mainly transcription factors that regulate diverse genes, and others proteins that are regulated by phosphorylation. In addition, the MAPKs interact with inactivating phosphatases, which finely tunes their cellular activity. The same hierarchical cascade exists for JNK and p38 MAPK activation, consisting of three consecutive steps of phosphorylation and activation of different kinases (MAPKKK → MAPKK → MAPK).

Extracellular mitogens such as growth factors (cytokines and hormones) and phorbol esters (e.g., 12-O-tetradecanoylphorbol-13-acetate, TPA) activate ERK1 and ERK2, which regulates cell proliferation and promotes effects such as induction or inhibition of differentiation, stimulation of secretory responses in a variety of cell types such as neutrophils, modulating membrane activity, and generating active oxygen species (Blumberg, 1988). The stress-activated protein kinases, or JNKs, and p38 MAPK signalling pathways are responsive to stress stimuli, such as cytokines, ultraviolet irradiation, heat shock, osmotic shock and cellular redox state, and are involved in cell differentiation and apoptosis. There are 10 isoforms of the three JNKs due to alternative splicing of JNK-1, JNK-2, and JNK-3, and there are four p38 MAPK isoforms.

These MAPKs are involved in processing cellular pain. Dai et al. (2002) demonstrated that ERK is activated in DRG neurons by electrical, thermal and chemical stimuli using electrophysiological recordings and western blot analysis. The peripheral stimulation of ERK1/2 and p38 MAPKs is involved in the nociceptor sensitisation produced by epinephrine, nerve grow factor (NGF) and capsaicin (Aley & Levine 2003, Zhu & Oxford, 2007). Activation of nuclear factor-kappaB (NF-κB), a transcription factor linked to inflammation, and p38 MAPK leads to the formation of various pro-inflammatory cytokines, such as TNF-α, IL-1β, and IL-6 (Doyle et al. 2011). TNF-α may induce acute peripheral mechanical sensitisation by acting directly on its receptor TNFR1, which is localised in primary afferent neurons, resulting in the p38-dependent modulation of TTX-resistant Na^+ channel currents (Jin & Gereau 2006; Zhang et al. 2011).

In neurons, synaptic activity-induced increases in the intracellular Ca^{2+} concentration activate MAPKs. Ca^{2+}/calmodulin-activated protein kinase (CaMKII) is essential for synaptic plasticity because it regulates transcriptional and translational modifications in gene expression and regulation. MAPKs are downstream effectors of multiple kinases, including CaMKII. Membrane depolarisation and calcium influx activate MAPK/ERK kinases. ERK and p38 MAPKs are up-regulated both in primary afferent nerves and the spinal cord in response to noxious stimulation, nerve injury and tissue injury. Inhibition of ERK or p38 MAPK phosphorylation or activity induces an antinociceptive effect in many of the animal pain models described throughout this section. Thus, in addition to the PKA and PKC signalling pathways, some cross-talk may exist with MAPK cascades upon inflammation or injury.

As an example, IL-6 exerts an important role in the development and maintenance of muscular sensitisation to nociception. The IL-6-mediated muscular pain response involves resident cell activation, polymorphonuclear cell infiltration, cytokine production, prostanoids and sympathomimetic amines release (Manjavachi et al. 2010). This response to IL-6 triggers the activation of intracellular pathways, especially MAPKs. Upon IL-6 stimulation, ERK, p38 MAPK and JNK phosphorylation is measurable by flux cytometry, and selective inhibitors of ERK and p38 MAPK partially reduced mechanical nociceptive behaviour (Manjavachi et al. 2010). Inflamed tissues release NGF that act upon nociceptors, activating the p38 MAPK cascade and leading to an increase of TRPV-1 translation and transport to nerve terminals, which contributes to the maintenance of nociceptive behaviour in animal models (Ji et al. 2002). Additionally, two separate p38 MAPK pharmacological inhibitors were effective at inhibiting the development of burn-induced sensitisation when administered as intrathecal pre-treatments (Sorkin et al. 2009).

A screen of MAPK activation in the dorsal horn in both phases of the formalin test demonstrated that p38 MAPK is activated in spinal microglia. Thus, a reduction in the level of spinal p38β, but not p38α, prevented the development of sensitisation following peripheral inflammation (Li et al., 2010). Any study of MAPK signalling must also consider the effect of nervous system cells other than neurons in the pain process.

The same kind of consideration is needed for chronic pain. Synaptic and nerve plasticity is a key element in pain chronification. Changes in structure and function as a result of input from the environment, lesions and pathologies may lead to neuropathic pain. These changes depend upon transcriptional and translational modifications in cell function that are mediated by MAPK signalling. Thus, MAPK modulation became a natural choice for research and the development of new drugs and pharmacological tools.

Pfizer Global Research and Development published a research paper in 2003 showing that the development of neuropathic pain is associated with an increase in the activity of the MAPK/ERK-kinase cascade within the spinal cord. They explored the chronic constriction injury model and the streptozocin-induced diabetic model to mimic neuropathic pain states. Global changes in gene expression and the effect of MAPK/ERK-kinase (MEK) inhibitor were analysed (Ciruela et al., 2003). These efforts lead to the selection of these kinases as targets of drug design for pain, with a focus on neuropathic pain.

The MAPK intracellular signalling cascades are also associated with synaptic long-term potentiation and memory and are associated with nociceptive behaviour in spinal cord injury (Crown et al., 2006). ERK 1/2 and p38 MAPK phosphorylation levels are upregulated in rat-spinal cords during mechanical sensitisation after spinal cord injury. Neurons are not the only cells involved in this process; microglial but not astrocytic p38α contributes to the maintenance of neuronal hyperexcitability in caudal areas after spinal cord injury (Gwak et al. 2009).

The IL-6/p38 MAPK/CX3C Receptor 1 signalling cascade is involved in neural-glial communication and plays an important role in triggering spinal glial activation and facilitating pain processing following peripheral nerve injury. Up-regulation of CX3CR1 expression by IL-6-p38 MAPK signalling enhances the responsiveness of microglia to chemokine CXCL1, or fractalkine, after nerve injury (Lee et al., 2010). TNF-α is important during the development of neuropathic pain by spinal nerve ligature (SNL) (Schäfers et al., 2003). The inhibition of spinal p38 MAPK activation prevents this event. However, the activation of ERK but not p38 MAPK is critically involved in the TNFα-induced increase in TRPV1 expression in cultured DRG neurons (Hensellek et al., 2007).

Injury to peripheral nerves may result in the formation of neuromas. Elevated levels of phosphorylated ERK1/2 can be identified in individual neuroma axons that also possess the voltage-gated sodium channel $Na_v1.7$. Painful human neuromas show accumulation of this sodium channel, and its function is modulated by ERK1/2 phosphorylation (Persson et al., 2011).

MAPK expression analysed in the spinal cord after SNL showed differential activation in injured and uninjured DRG neurons. Uninjured neurons had only p38 MAPK detectable induction. In contrast ERK, p38 MAPK and JNK were activated in several populations of injured DRG neurons (Obata et al., 2004). Differential activation of MAPK in lesioned and sound primary nerve afferents may be linked to the pathogenesis of neuropathic pain after partial nerve injury (Svensson et al., 2003).

2.4 Interplay between pathways

The specificity of activation for each signalling pathway may be determined by the stimuli (Juntilla et al., 2008), and the crosstalk between them could be induced during pathological states (Noselli, 2000). Pimienta and Pascual (2007) described MAPK intracellular signalling as "different signalling cascades crosstalk with each other in a way that their functional compensation makes possible the simultaneous integration of multiple inputs".

Considering only one inflammatory mediator, PGE_2, in three models performed in the same species (mice) with analyses of not the same tissue, differences between the signalling pathways involved can be detected:

a. Acute nociception induced by high-dose PGE_2 administration is dependent on ERK signalling mechanisms because its overexpression was detected in hind paw by western immunoblotting analyses (Kassuya et al., 2007). This effect was reversed by EP receptor antagonists (Kassuya et al., 2007).
b. In the same way, PGE_2 is a final mediator of nociceptor sensitisation that acts on the peripheral nerve endings through the prostanoid receptors, leading to sensitisation of sensory nerves. PGE_2-induced acute mechanical sensitisation, which is also associated with kinase activation, was completely prevented by PKA and PKCε, but not by ERK, pharmacological inhibition (Villarreal et al. 2009b). The persistent pain state induced by

chronic PGE_2 administration is completely abolished by PKA or PKCε inhibitors, but not by ERK inhibitors (Villarreal et al. 2009b).

Thus, we conclude that PKA, PKC and ERK are involved in the effects of PGE_2 (including nociceptor sensitisation and nociception). The inflammatory processes include several others mediators in addition to PGE_2. In the same work, Villarreal et al. (2009b) showed that dopamine-induced acute sensitisation involves PKA, PKC and ERK activation, whereas the dopamine-induced persistent sensitisation state is abolished by ERK inhibition and temporarily inhibited by PKA or PKCε inhibitors, suggesting that ERK plays the major role. So, what is the real meaning of these results?

The study of MAPK and other kinases must keep its momentum. The interplay between different signalling pathways is challenging to understand. The available experimental models allow individual probing of each mediator and the kinase transduction of its signalling. Biological systems and pathological states have multiple variables in a complex regulated environment that hinder our understanding of each molecule and their combined role. Nevertheless, the continuous efforts have already achieved interesting findings. In the next sections, additional mechanisms and protein kinases will be described in discussing the pharmacological mechanisms of different drug classes in pain control.

3. Role of protein kinases in pain control

Both acute and chronic pain are usually controlled by administration of pharmacological agents (analgesics and adjuvants) that attempt to tackle pain in both the central and peripheral divisions of the nociceptive pathway. Although a classical mechanism of action is well described for most drugs, additional mechanisms link their analgesic effects with some kinases-dependent pathways, mainly pathways related to PKA, PKC, MAPKs and cyclic guanosine monophosphate (cGMP)-dependent protein kinase (PKG). In this section, the involvement of protein kinases in the mechanisms of action of drugs used for pain control, such as opioids, dipyrone, general and local anaesthetics and antidepressants will be analysed.

3.1 Opioids

Among opioids, morphine is widely used as a classical opioid analgesic for the clinical management of acute and chronic pain. Despite its wide use, tolerance to the analgesic actions of morphine is an important side effect of prolonged exposure. Individuals who are tolerant to the effects of morphine require larger doses to elicit the same amount of analgesia. Thus, antinociceptive tolerance and the high doses required to achieve effects have limited the use of morphine.

Many factors have been related to morphine tolerance, such as a change of the descending pain modulatory pathway, receptor desensitisation, down-regulation of opioid functional receptors, release of excitatory neurotransmission and other adaptive changes in cell signalling pathways. Interestingly, PKC, especially PKCγ, plays a major role in the changes associated with morphine tolerance. Song et al. (2010) demonstrated that an isoform-specific inhibitor could successfully down-regulate PKCγ in the spinal cord and reverse the development of morphine tolerance in rats. This result not only implicates this PKC isoform in the opioid tolerance mechanism but also has potential applications in pain management.

Beyond the involvement of PKC in opioid tolerance, PKC is involved in inflammatory and neuropathic pain. The capacity of opioids to alleviate inflammatory pain is negatively regulated by the glutamate-binding N-methyl-D-aspartate receptor (NMDAR). And increased activity of this receptor complicates the clinical use of opioids for treating neuropathic pain. Rodríguez-Muñoz et al. (2011) indicated that morphine disrupts the glutamate-binding NMDAR complex by PKC-mediated phosphorylation and potentiates the NMDAR-CaMKII pathway, which is implicated in morphine tolerance. Inhibition of PKC restored the antinociceptive effect of morphine on the μ-opioid receptor (MOR). Thus, the opposing activities of the MOR and NMDAR in pain control affect their relation within neurons of structures such as the periaqueductal grey (PAG), a region that is implicated in the opioid control of nociception. This finding could be exploited in developing bifunctional drugs that would act exclusively on NMDARs associated with MORs.

MORs are not the only opioid receptors that influence PKC. Berg et al. (2011), who were investigating the regulation of the κ-opioid receptor (KOR) in rat primary sensory neurons *in vitro* and in a rat model of thermal sensitisation, showed that the application of a KOR agonist (U50488) did not inhibit AC activity or release of calcitonin gene-related peptide (CGRP) *in vitro* and did not inhibit thermal sensitisation *in vivo*. It is important to note that AC activity, CGRP release, and thermal sensitisation process are related to PKC activation (see section 2). However, after a 15-min pretreatment with bradykinin, the agonist became capable of inhibiting AC activity, CGRP release, and thermal sensitisation. The *in vitro* effects of bradykinin on the KOR agonist were abolished by a PKC inhibitor; thus, Berg and co-workers suggest that PKC activation mediates BK-induced regulation of the KOR system. More studies are necessary to understand the mechanisms by which peripheral KOR agonist efficacy is regulated and the relationship of the KOR agonist effects with PKC activation.

In this regard, formalin-induced inflammatory nociception may inhibit morphine tolerance in mice. In this model, conventional PKC (cPKC) is up-regulated and treatment with an antisense oligonucleotide (AS-ODN) directed against cPKC abolished the development of morphine tolerance, suggesting that cPKC is involved in morphine tolerance development (Fujita-Hamabe et al., 2010). Additionally, formalin-induced inflammatory nociception inhibit morphine tolerance by a mechanism involving KOR activation, down-regulation of cPKC, and up-regulation of MOR activity (Fujita-Hamabe et al., 2010). The data suggest a key role to cPKC in opioid-induced tolerance and that nociception-activated mechanisms may modulate opioid-response, improving it. In addition, studying the effects of chronic ethanol-induced neuropathy in rats, Narita et al. (2007) showed that chronic ethanol exposure dysregulated MOR but not DOR and KOR, and was related to PKC up-regulation in the spinal cord, which may explain the reduced sensitivity to the morphine antinociceptive effect. Taken together, these findings suggest the PKC activation disrupts MOR function, which could be counteracted by the KOR system. How the DOR participates remains unclear.

Like PKC, PKA may also play a role in morphine antinociceptive tolerance. Previous studies have shown that chronic exposure to morphine results in intracellular adaptations within neurons that cause an increase in PKA activity. Unexpectedly, sustained morphine treatment produces paradoxical pain sensitisation (opioid-induced hyperalgesia) and causes an increase in spinal pain-related neurotransmitter concentrations, such as CGRP, in experimental animals. Studies have also shown that PKA plays a major role in the regulation of presynaptic neurotransmitter (such as CGRP and substance P) synthesis and release. Tumati et al. (2011) previously showed that in cultured DRG neurons, sustained *in vitro* opioid agonist treatment up-regulates cAMP levels (AC superactivation) and augments CGRP release in a PKA-dependent manner. The authors also showed that selective knock-down of spinal PKA activity by intrathecal pretreatment of rats with a PKA-selective small interference RNA (siRNA) mixture significantly attenuates sustained morphine-mediated augmentation of spinal CGRP immunoreactivity, thermal and mechanical sensitisation and antinociceptive tolerance. These findings indicate that sustained morphine-mediated activation of spinal cAMP/PKA-dependent signalling may play an important role in opioid-induced pain sensitisation. More specifically, morphine acts acutely on MORs, which couple with G-proteins to inhibit AC and reduce PKA activity. However, during tolerance, MORs become uncoupled from G-proteins, AC inhibition is reduced, and PKA activity is increased. These findings also provide potential molecular targets for pharmacological intervention to prevent the development of such paradoxical pain sensitisation.

The majority of studies that have demonstrated an increase in PKA activity during opioid tolerance have been conducted in rats using brain regions associated with the reinforcing properties of opioids, such as the *locus coeruleus* and *nucleus accumbens*. Studying the expression of morphine antinociceptive tolerance at the behavioural level (tail-flick test) and the alterations in PKA activity at the cellular level in mouse brain (PAG, medulla, thalamus) and lumbar spinal cord, Dalton et al. (2005) support the hypothesis that an increase in PKA activity contributes to the tolerance to morphine-induced antinociception. However, the effect of chronic morphine treatment for 15 days on PKA activity was region-specific because increases in cytosolic PKA activity were observed in the lumbar spinal cord. In contrast, PKA activity/kinetics was not altered in the PAG, medulla or thalamus. These results demonstrate that spinal and supraspinal PKA activity are differentially altered during morphine tolerance. Thus, the neurons in mouse brain and lumbar spinal cord that make up the pain pathway from the brainstem to the spinal cord respond differently to chronic morphine treatment. To confirm these findings, future studies need to elucidate the differential responses to chronic morphine treatment using *in vivo* models of morphine antinociceptive tolerance concerning the PKA involvement.

Using a behavioural paw pressure test in rats, Yamdeu et al. (2011) demonstrated that up-regulation of NGF, through activation of the p38 MAPK pathway, lead to adaptive changes in sensory neuron opioid receptors that enhance susceptibility to local opioids. After intraplantar NGF treatment, this effect occurs in three consecutive steps: MOR expression is increased in DRG at 24 h, increased axonal MOR transport at 48 h, and increased MOR density at 96 h. Consequently, the dose-dependent peripheral antinociceptive effects of locally applied full opioid agonists such as fentanyl are potentiated, and the effects of partial opioid agonists such as buprenorphine are more efficacious, which is reversed by the intrathecal administration of p38 MAPK inhibitor SB203580. Thus, in rats, peripheral inflammation increases MOR expression in nociceptors by NGF activation of p38 MAPK. This mechanism may act as a counter-regulatory response to painful p38 MAPK-induced conditions, such as inflammatory pain, to facilitate exogenously or endogenously mediated opioid antinociception.

Recently, the roles of several MAPKs, including p38 MAPK and ERK, have been investigated in animal models of morphine tolerance and postoperative nociceptive sensitisation. It is unknown, however, whether prior morphine-induced MAPK activation affects the resolution of postoperative nociceptive sensitisation. Horvath et al. (2010) investigated the effect of morphine-induced antinociceptive tolerance on the resolution of postoperative nociceptive sensitisation. They hypothesised that prior chronic morphine administration would inhibit or delay the resolution of postoperative nociceptive sensitisation via enhanced spinal glial proteins expression and MAPK signalling. Chronic morphine treatment attenuated the resolution of postoperative nociceptive sensitisation, as determined by thermal and mechanical behavioural tests, and enhanced microglial p38 MAPK and ERK phosphorylation. To better understand these results, prior chronic morphine exposure could prime microglia, causing exacerbated MAPK signalling pathway activation following subsequent paw incision injury. This would cause more robust microglial responses in rats with a history of morphine tolerance versus naïve rats, and this response is manifested by further neuronal sensitisation, behavioural hypersensitivity and inhibition of the resolution of the postoperative-associated nociceptive condition. The Horvath and co-workers study indicates that microglial MAPKs play a role in the mechanisms by which morphine attenuates the resolution of postoperative pain and suggests that patients who abuse opioids or are on chronic opioid therapy may be more susceptible to developing chronic pain syndromes following acute injury.

In conclusion, protein kinases (PKs) exert a crucial role in pain control responses mediated by opioids, mainly in tolerance-induced mechanisms. Thus, PKs could be the key to better understanding opioid pharmacodynamics.

3.2 General and local anaesthetics

3.2.1 General anaesthetics

Ketamine is an NMDAR antagonist that is available for clinical use as a general anaesthetic. Ketamine presents analgesic effect in acute and chronic pain models in both animals and humans (Mathisen et al. 1995, Rabben et al., 1999; Visser & Schug, 2006; Pascual et al., 2010).

The involvement of kinases in the analgesic effect of ketamine has been investigated. Using a model of neuropathic pain induced by SNL in rats, Mei et al. (2011) showed that SNL induced ipsilateral JNK phosphorylation up-regulation in astrocytes, but not microglia or neurons, within the spinal dorsal horn. Intrathecal ketamine relieved SNL-induced mechanical sensitisation and produced a dose-dependent effect on the suppression of SNL-induced spinal astrocytic JNK phosphorylation but had no effect on JNK protein expression, suggesting that the inhibition of spinal JNK activation may be involved in the analgesic effects of ketamine in this model.

The inhibition of MAPK phosphorylation by ketamine has also been related to a reduction in cytokine gene expressions in lipopolysaccharide (LPS)-activated macrophages (Wu et al., 2008). A therapeutic concentration of ketamine can decrease LPS-induced JNK phosphorylation, thus inhibiting TNF-α and IL-6 gene expression, which leads to the suppression of LPS-induced macrophage activation (Wu et al., 2008). In addition, ketamine reduced IL-1β biosynthesis in LPS-stimulated macrophages through the suppression of Ras, Raf, MEK1/2, and ERK1/2/IKK phosphorylation and the subsequent translocation and transactivation of the transcription factor NF-κB (Chen et al., 2009). The involvement of TNF-α, IL-6 and IL-1β in inflammatory nociceptive sensitisation is well known. Thus, the inhibition of cytokine production by ketamine in different cells may be an additional mechanism that contributes toward its analgesic effect.

Beyond its own specific effects, ketamine also has analgesic effects when given in combination with opioids. As mentioned earlier, several studies have demonstrated that ERK1/2 is involved in nociception. However, activation of MOR by opioids leads to ERK1/2 phosphorylation (Fukuda et al. 1996; Gutstein et al. 1997; Gupta et al., 2011), and this can be potentiated by ketamine. Gupta et al. (2011) investigated whether the ability of ketamine to increase the duration of opioid-induced effects could be related to the modulation of opioid-induced signalling. The authors found that, in a cell culture model, ketamine increases the effectiveness of opioid-induced signalling by enhancing the level of opioid-induced ERK1/2 phosphorylation. Ketamine also delays the desensitisation and improves the resensitisation of ERK1/2 signalling. These effects were observed in heterologous cells expressing MOR, suggesting a non-NMDA receptor-mediated action of ketamine (Gupta et al., 2011). The authors concluded that the overall effect of ketamine appears to be keeping opioid-induced ERK1/2 signalling active for a longer time period, and this could account for the observed effects of ketamine on the duration of opioid-induced analgesia. However, these data were obtained from *in vitro* experiments, and the link with analgesia is not clearly understood. Data provided from *in vivo* studies could contribute to improve the understanding of opioid-induced analgesia and its potentiation by ketamine.

3.2.2 Local anaesthetics

Systemic or topical administration of lidocaine and other local anaesthetics reduce hypersensitivity states induced by both acute inflammation and peripheral nerve injury in animals and brings significant relief in some patients with neuropathic pain syndromes (Mao & Chen, 2000; Ma et al., 2003; Gu et al., 2008; Fleming & O'Connor, 2009; Suter et al., 2009; Buchanan& MacIvor, 2010; Suzuki et al., 2011).

The analgesic effect of lidocaine in neuropathic pain can be partially explained by its ability to attenuate MAPK activation. Intrathecal injection of lidocaine in rats with chronic constriction injury suppressed the phosphorylation of p38 MAPK in the activated microglia in the spinal cord (Gu et al., 2008). In ATP-activated cultured rat microglia, lidocaine inhibited p38 MAPK activation and attenuated the production of proinflammatory cytokines, including TNF-α, IL-1β and IL-6 (Su et al., 2010). Furthermore, lidocaine significantly inhibited LPS-induced Toll-like receptor 4, NF-κB, ERK and p38 MAPK activation, but not JNK activation in LPS-stimulated murine macrophages (Lee et al., 2008).

Spared nerve injury (SNI) induces mechanical sensitisation and p38 MAPK activation in spinal microglia. Bupivacaine microspheres induced a complete sensory and motor blockade and significantly inhibited p38 MAPK activation and microglial proliferation in the spinal cords of rats (Suter et al., 2009). Carrageenan-induced hind paw inflammation and sensitisation triggers phosphorylation of spinal p38 MAPK and enhances TNF and IL-1 production in the bilateral DRGs and spinal cord. Although bupivacaine inhibits oedema, hyperalgesia and the carrageenan-induced production of systemic cytokines (Beloeil et al., 2006a; Combettes et al., 2010), the inhibitory effects of bupivacaine on the expression of cytokines or phosphorylated p38 MAPK in spinal cord or DRGs have not been verified (Beloeil et al., 2006b).

ERK activation as a potential target for bupivacaine antinociception was also investigated. The activation of both ionotropic (AMPA, NMDA, TRPV1) and metabotropic (NK-1, bradykinin 2 receptor, mGluR) receptors results in ERK phosphorylation in superficial dorsal horn neurons in rats. Bupivacaine blocked ionotropic but not metabotropic, receptor-induced ERK activation by apparently blocking Ca^{2+} influx through the plasma membrane in the spinal cord (Yanagidate & Strichartz, 2006).

Taken together, the inhibition of MAPK activation by general and local anaesthetics seems to represent a common and important pathway to at least partially explain the mechanism of analgesic action exerted by these drugs through ion influx inhibition.

3.3 Antidepressants

Selected antidepressants suppress pain through diverse mechanisms and are now considered as an essential component of the therapeutic strategy for treatment of many types of persistent pain. Their main mechanism of action involves reinforcement of the descending inhibitory pathways by increasing the amount of norepinephrine and serotonin in the synaptic cleft at both the supraspinal and spinal levels. Based on this, tricyclic antidepressants (TCAs) are widely used for treating chronic pain, such as neuropathic and inflammatory pain. Intrathecal (i.t.) co-infusion of amitriptyline with morphine not only attenuates the development of morphine tolerance but also preserves its antinociceptive efficacy (Tai et al., 2006). Tai et al. (2007) showed that amitriptyline pretreatment reverses the spinal cord PKA and PKC upregulation and preserves morphine's antinociceptive effect in morphine-tolerant rats submitted to thermal behaviour test; this reversal may occur via preventing the up-regulation of PKA and PKC protein expression. It results in the trafficking of glutamate transporters from the cytosol to the plasma membrane of glial cells, thus reducing the excitatory amino acid (EAA) concentration in the cerebrospinal fluid (CSF) spinal cord by the morphine challenge. This study suggested that amitriptyline is a useful analgesic adjuvant in the treatment of patients who need long-term opioid administration for pain relief.

In addition to the traditionally used TCAs, such as amitriptyline, selective serotonin reuptake inhibitors (SSRIs) and mixed monoamine uptake inhibitors are also used as a first-line treatment for managing pain syndromes. As mentioned above, voltage-gated sodium channels (VGSCs) are subject to modulation by G protein-coupled receptor signalling cascades involving PKA- and PKC-mediated phosphorylation. Depending on the neuron type and its anatomical location, phosphorylation of the VGSCs by PKC may facilitate slow inactivation (Cantrell and Catterall, 2001). Activating the 5-HT2C subtype of serotonin receptors in prefrontal cortex neurons results in a negative shift in the voltage-dependence of fast inactivation accompanied with a reduction of the peak current due to a PKC-mediated phosphorylation process (Carr et al., 2002). Concurrent phosphorylation by PKA seems necessary for the maximal current reduction (Cantrell et al., 2002). These mechanisms can be activated by various neurotransmitters including serotonin (Cantrell and Catterall, 2001). Because SSRIs increase the extracellular concentration of serotonin it is logical that they would indirectly modulate sodium channels in the central nervous system. This action mediated by increased serotonin and, PKA and PKC activity, could account for the analgesic effect of SSRIs.

Thán et al. (2007) studied the pharmacological interaction between SSRIs and sodium channel blocking agents such as lamotrigine. They examined the interaction of VGSCs blockers and SSRIs at the level of spinal segmental neurotransmission in the rat hemisected spinal cord model. The reflex inhibitory action of VGSCs blocker was markedly enhanced when SSRI compounds were co-applied; and it was found serotonin receptors and PKC involvement in the modulation of sodium channel function (Thán et al., 2007).

In conclusion, it seems that antidepressants exert analgesic effects by a mechanism involving serotonin, PKA and PKC activation, and modulation of VGSCs. Understanding the PK dynamics in these processes would be key to improve pain management.

3.4 PKG signalling and pain control

As described in section 1, the activation of signalling pathways that are dependent on PKA, PKC and MAPKs is important for the sensitisation of nociceptors and pain processing. The PKG pathway, in turn, is related to the nitric oxide (NO)/cGMP/PKG/ATP-sensitive K^+ channel pathway, which plays an important role in peripheral antinociception (Rodrigues & Duarte, 2000; Sachs et al., 2004).

The relationship between the NO/cGMP pathway and peripheral antinociception was first demonstrated by Ferreira and co-workers (Durate et al., 1990; Ferreira et al., 1991). They showed that the antinociceptive effect of acetylcholine and morphine was blocked by a guanylyl cyclase inhibitor and an NO synthase inhibitor, and was potentiated by a specific cGMP phosphodiesterase inhibitor. Moreover, the antinociception achieved with these drugs was mimicked by NO donors such as sodium nitroprusside. The involvement of this pathway has also been demonstrated for other analgesics, such as dipyrone (Duarte et al., 1992), diclofenac (Tonussi et al., 1994), and some antinociceptive agents, such as *Crotalus durissus terrificus* snake-venom (Picolo et al., 2000), the potent κ-opioid receptor agonist bremazocine (Amarante & Duarte, 2002), xylazine (Romero & Duarte, 2009), the cannabinoid receptor agonist anandamida (Reis et al., 2009) and ketamine (Romero et al., 2011). In agreement with *in vivo* studies, data from electrophysiological experiments studying inflammatory sensitisation showed that capsaicin-induced elevations in intracellular Ca^{2+} levels of rat sensory neurons lead to an enhanced production of cGMP via the NO pathway. The elevated cGMP levels and the subsequent activation of PKG appear to inactivate the sensitisation, confirming the important regulatory role of this kinase in reversing the neuronal sensitisation (Lopshire & Nicol, 1997).

In addition to studies on the mechanism of antinociceptive action of analgesics, Duarte and co-workers showed that the ability of morphine and dipyrone to induce peripheral antinociception is dependent on the activation of ATP-sensitive K^+ channels (Rodrigues & Duarte, 2000). cGMP can directly or indirectly (via PKG stimulation) modulate the activity of ion channels. PKG is a protein kinase that is stimulated selectively but not exclusively by cGMP. Once stimulated, PKG inhibits phospholipase C activity, stimulates Ca^{2+}-ATPase activity, inhibits inositol 1,4,5-triphosphate, inhibits Ca^{2+} channels, and/or stimulates K^+ channels activity (Cury et al., 2011). Furthermore, Sachs et al. (2004) demonstrated that the antinociceptive effect of dipyrone on persistent inflammatory sensitisation is dependent on the PKG activation and its modulation of ATP-sensitive K^+ channels.

Taken together, these findings suggest the relevant role of PKG as an intermediate between cGMP generation and the opening of ATP-sensitive K^+ channels. The activation of this modulatory pathway may be an interesting target for new drug development.

4. Conclusions: A perspective of promising drug targets

In this section, protein kinases will be viewed as targets for pain control drug development. Several pre-clinical and clinical trials will be reviewed, focusing on the effectiveness and adverse effects of such drugs.

The genomic analysis of the eukaryotic protein kinase superfamily together with drug design approaches such as the bioisosteric replacement of pharmacophoric groups of lead compounds and 3D-quantitative structure-activity relationship analysis provide several new chemical entities to be tested and developed as drug candidates.

The continuous progress in protein structure determination and improved resolution allows the identification of pharmacological targets. The experimental results from genetically modified animals support new hypotheses and help to validate new concepts to better understand the pathological genesis and natural processes of our body.

Such progress in medicinal chemistry, biochemistry and pharmacology paradoxically leads to poor results in terms of new pharmaceutical entities and therapeutics. The pharmaceutical innovation decrease in recent decades is due to many aspects that are beyond the scope of this chapter. As targets of pharmaceutics, protein kinases play an important role in this history, providing several new therapeutic cancer targets. Drug discovery companies have targeted protein kinase inhibitors, which have led to billion dollar merges and a new branch of research and development that spread beyond the boundaries of cancer therapeutics (Garber, 2003).

At the beginning of the second decade of the 21st century, there are synthetic and medicinal chemistry service companies with strong backgrounds in kinase targets and kinase inhibitor drug discovery; these companies can develop new compounds on demand. There are sixteen pharmaceuticals actually licensed as protein kinase inhibitors, mainly to treat different cancers. The first drug, Trastuzumab, was licensed in 1998; this drug is a monoclonal antibody targeting membrane receptors that activates the MAPK pathway as well as the PI3 Kinase/AKT pathways. After this initial drug, many small molecules followed, targeting kinases as mechanism of action, mainly as ATP competitors.

The International Federation of Pharmaceutical Manufacturers & Associations has listed in its Clinical Trials Portal three entries for clinical trials focusing on pain and protein kinases. Two of these trials involve p38 MAPK inhibitors from a large pharmaceutical company and are testing for neuropathic pain following nerve trauma and from lumbosacral radiculopathy. The third trial involves tyrosine kinase (TrkA) receptor expression in children with retrosternal pain.

Experimental evidence suggests that p38 MAPK is activated in spinal microglia after nerve injury and contributes to neuropathic pain development and maintenance (Ji & Suter, 2007). p38 MAPK phosphorylates targets that transduce cellular signals to molecules and transcription factors that are involved in regulating the biosynthesis of inflammatory cytokines such as IL-1 and TNF-α. The inhibitor dilmapimod was associated with a significant reduction in pain intensity in patients with neuropathic pain following nerve injury (Anand et al., 2011). The clinical efficacy of p38 MAPK inhibitors in acute pain was also demonstrated in an assay of acute postsurgical dental pain; these inhibitors increased the time to rescue medication and decreased pain intensity when compared with the placebo group (Tong et al. 2011).

Despite these clinical assays that are directly associated with pain, many other clinical and pre-clinical studies have some degree of relevance when pain management is the goal. The action of different protein kinases inhibitors in cancer, rheumatoid arthritis, postsurgical conditions, diabetes and so forth has significant impact on decreasing pain in subjects suffering from these pathologies.

In addition to these efforts, one long-sought goal is the development of inhibitors of PKC isoforms because this family of protein kinases is involved in the cellular signalling of nociception, anxiety and cognition (Van Kolen et al., 2008). Non-isoform-specific PKC inhibitors have proven to be too toxic for *in vivo* use. PKCε is the primary target for drug design. This isoform is activated during nerve sensitisation and phosphorylates ion channels in the peripheral nervous system such as TRPV-1, and N-type voltage-dependent calcium channels (VDCCs) in isolectin B4-positive nociceptors; in addition, it mediates interplay between other kinases that are important to nociceptor function, such as PKA and MAPK (Hucho et al., 2005). There are no specific ATP-binding competitors for PKCε. There are other compounds that target alternative domains, such as the pseudosubstrate sequence, which is responsible for keeping the kinase in an inactivate state. The lipid-binding, cellular localisation and actin-binding domains are also valid targets. The main goal is to develop isoform-specific inhibitors among the ten known isozymes and to provide tissue specificity because cardiac-specific PKCε inhibition blocks norepinephrine-mediated regulation of heart contraction (Johnson et al., 1996).

The pharmaceutical paradigm of "new targets for old drugs", where known medications are employed in new pathologies as an innovation strategy to keep new products flowing to the market also applies to protein kinases and pain control because many drugs utilised for chronic and neuropathic pain management, such as antidepressants and anaesthetics, depend on protein kinases for their mechanisms of action.

New lead drugs are also being proposed; these drugs utilise molecular hybridisation and bioisosteric replacement of pharmacophoric groups, where different bioactive molecular moieties of mechanistically diverse drugs are fused, giving birth to new chemical entities with dual activity profiles (Brando Lima et al., 2011) that incorporate protein kinase inhibition with another type of biological activity. The challenge of developing new molecular approaches to create drugs that manage pain is great, and as seen throughout this chapter, protein kinases are an important aspect of this problem.

5. References

Ahlgren, S.C. & Levine, J.D. (1994). Protein Kinase C Inhibitors Decrease Hyperalgesia and C-Fiber Hyperexcitability in the Streptozotocin-Diabetic Rat. *J Neurophysiol*, Vol.72, No.2, (August 1994), pp.684-692, ISSN 0022-3077.

Aley, K.O. & Levine, J.D. (1999). Role of Protein Kinase a in the Maintenance of Inflammatory Pain. *J Neurosci*, Vol.19, No.6, (March 1999), pp.2181-2186, ISSN 0270-6474.

Aley, K.O., Messing, R.O., Mochly-Rosen, D. & Levine, J.D. (2000). Chronic Hypersensitivity for Inflammatory Nociceptor Sensitization Mediated by the Epsilon Isozyme of Protein Kinase C. *J Neurosci*, Vol.20, No.12, (June 2000), pp.4680-4685, ISSN 0270-6474.

Aley, O. & Levine, J.D. (2003). Contribution of 5- and 12-Lipoxygenase Products to Mechanical Hyperalgesia Induced by Prostaglandin E(2) and Epinephrine in the Rat. *Exp Brain Res*, Vol.148, No.4, (February 2003), pp.482-487, ISSN 0014-4819.

Amarante, L.H. & Duarte, I.D. (2002). The Kappa-Opioid Agonist (+/-)-Bremazocine Elicits Peripheral Antinociception by Activation of the L-Arginine/Nitric Oxide/Cyclic Gmp Pathway. *Eur J Pharmacol*, Vol.454, No.1, (November 2002), pp.19-23, ISSN 0014-2999.

Anand, P., Shenoy, R., Palmer, J.E., Baines, A.J., Lai, R.Y., Robertson, J., Bird, N., Ostenfeld, T. & Chizh, B.A. (2011). Clinical Trial of the P38 Map Kinase Inhibitor Dilmapimod in Neuropathic Pain Following Nerve Injury. *Eur J Pain*, (May 2011), ISSN 1532-2149.

Beloeil, H., Ababneh, Z., Chung, R., Zurakowski, D., Mulkern, R.V. & Berde, C.B. (2006a). Effects of Bupivacaine and Tetrodotoxin on Carrageenan-Induced Hind Paw Inflammation in Rats (Part 1): Hyperalgesia, Edema, and Systemic Cytokines. *Anesthesiology*, Vol.105, No.1, (July 2006), pp.128-138, ISSN 0003-3022.

Beloeil, H., Ji, R.R. & Berde, C.B. (2006b). Effects of Bupivacaine and Tetrodotoxin on Carrageenan-Induced Hind Paw Inflammation in Rats (Part 2): Cytokines and P38 Mitogen-Activated Protein Kinases in Dorsal Root Ganglia and Spinal Cord. *Anesthesiology*, Vol.105, No.1, (July 2006), pp.139-145, ISSN 0003-3022.

Berg, K.A., Rowan, M.P., Sanchez, T.A., Silva, M., Patwardhan, A.M., Milam, S.B., Hargreaves, K.M. & Clarke, W.P. (2011). Regulation of Kappa-Opioid Receptor Signaling in Peripheral Sensory Neurons in Vitro and in Vivo. *J Pharmacol Exp Ther*, Vol.338, No.1, (July 2011), pp.92-99, ISSN 1521-0103.

Blumberg, P.M. (1988). Protein Kinase C as the Receptor for the Phorbol Ester Tumor Promoters: Sixth Rhoads Memorial Award Lecture. *Cancer Res*, Vol.48, No.1, (January 1988), pp.1-8, ISSN 0008-5472.

Brando Lima, A.C., Machado, A.L., Simon, P., Cavalcante, M.M., Rezende, D.C., Sperandio da Silva, G.M., Nascimento, P.G., Quintas, L.E., Cunha, F.Q., Barreiro, E.J., Lima, L.M. & Koatz, V.L. (2011). Anti-Inflammatory Effects of Lassbio-998, a New Drug Candidate Designed to Be a P38 Mapk Inhibitor, in an Experimental Model of Acute Lung Inflammation. *Pharmacol Rep*, Vol.63, No.4, (July 2011), pp.1029-1039, ISSN 1734-1140.

Buchanan, D.D. & F, J.M. (2010). A Role for Intravenous Lidocaine in Severe Cancer-Related Neuropathic Pain at the End-of-Life. *Support Care Cancer*, Vol.18, No.7, (July 2010), pp.899-901, ISSN 1433-7339.

Burgess, G.M., Mullaney, I., McNeill, M., Dunn, P.M. & Rang, H.P. (1989). Second Messengers Involved in the Mechanism of Action of Bradykinin in Sensory Neurons in Culture. *J Neurosci*, Vol.9, No.9, (September 1989), pp.3314-3325, ISSN 0270-6474.

Cadd, G. & McKnight, G.S. (1989). Distinct Patterns of Camp-Dependent Protein Kinase Gene Expression in Mouse Brain. *Neuron*, Vol.3, No.1, (July 1989), pp.71-79, ISSN 0896-6273.

Cantrell, A.R. & Catterall, W.A. (2001). Neuromodulation of Na+ Channels: An Unexpected Form of Cellular Plasticity. *Nat Rev Neurosci*, Vol.2, No.6, (June 2001), pp.397-407, ISSN 1471-003X.

Cantrell, A.R., Tibbs, V.C., Yu, F.H., Murphy, B.J., Sharp, E.M., Qu, Y., Catterall, W.A. & Scheuer, T. (2002). Molecular Mechanism of Convergent Regulation of Brain Na(+) Channels by Protein Kinase C and Protein Kinase a Anchored to Akap-15. *Mol Cell Neurosci*, Vol.21, No.1, (September 2002), pp.63-80, ISSN 1044-7431.

Cesare, P., Dekker, L.V., Sardini, A., Parker, P.J. & McNaughton, P.A. (1999). Specific Involvement of Pkc-Epsilon in Sensitization of the Neuronal Response to Painful Heat. *Neuron*, Vol.23, No.3, (July 1999), pp.617-624, ISSN 0896-6273.

Cesare, P. & McNaughton, P. (1996). A Novel Heat-Activated Current in Nociceptive Neurons and Its Sensitization by Bradykinin. *Proc Natl Acad Sci U S A*, Vol.93, No.26, (December 1996), pp.15435-15439, ISSN 0027-8424.

Chen, T.L., Chang, C.C., Lin, Y.L., Ueng, Y.F. & Chen, R.M. (2009). Signal-Transducing Mechanisms of Ketamine-Caused Inhibition of Interleukin-1 Beta Gene Expression in Lipopolysaccharide-Stimulated Murine Macrophage-Like Raw 264.7 Cells. *Toxicol Appl Pharmacol*, Vol.240, No.1, (October 2009), pp.15-25, ISSN 1096-0333.

Chen, X., Bing, F., Dai, P. & Hong, Y. (2006). Involvement of Protein Kinase C in 5-Ht-Evoked Thermal Hyperalgesia and Spinal Fos Protein Expression in the Rat. *Pharmacol Biochem Behav*, Vol.84, No.1, (May 2006), pp.8-16, ISSN 0091-3057.

Ciruela, A., Dixon, A.K., Bramwell, S., Gonzalez, M.I., Pinnock, R.D. & Lee, K. (2003). Identification of Mek1 as a Novel Target for the Treatment of Neuropathic Pain. *Br J Pharmacol*, Vol.138, No.5, (March 2003), pp.751-756, ISSN 0007-1188.

Coderre, T.J., Abbott, F.V. & Melzack, R. (1984). Effects of Peripheral Antisympathetic Treatments in the Tail-Flick, Formalin and Autotomy Tests. *Pain*, Vol.18, No.1, (Jananuary 1984), pp.13-23, ISSN 0304-3959.

Combettes, E., Benhamou, D., Mazoit, J.X. & Beloeil, H. (2010). Comparison of a Bupivacaine Peripheral Nerve Block and Systemic Ketoprofen on Peripheral Inflammation and Hyperalgesia in Rats. *Eur J Anaesthesiol*, Vol.27, No.7, (July 2010), pp.642-647, ISSN 1365-2346.

Crown, E.D., Ye, Z., Johnson, K.M., Xu, G.Y., McAdoo, D.J. & Hulsebosch, C.E. (2006). Increases in the Activated Forms of Erk 1/2, P38 Mapk, and Creb Are Correlated with the Expression of at-Level Mechanical Allodynia Following Spinal Cord Injury. *Exp Neurol*, Vol.199, No.2, (June 2006), pp.397-407, ISSN 0014-4886.

Cui, M. & Nicol, G.D. (1995). Cyclic Amp Mediates the Prostaglandin E2-Induced Potentiation of Bradykinin Excitation in Rat Sensory Neurons. *Neuroscience*, Vol.66, No.2, (May 1995), pp.459-466, ISSN 0306-4522.

Cury, Y., Picolo, G., Gutierrez, V.P. & Ferreira, S.H. (2011). Pain and Analgesia: The Dual Effect of Nitric Oxide in the Nociceptive System. *Nitric Oxide*, Vol.25, No.3, (October 2011), pp.243-254, ISSN 1089-8611.

Dai, Y., Iwata, K., Fukuoka, T., Kondo, E., Tokunaga, A., Yamanaka, H., Tachibana, T., Liu, Y. & Noguchi, K. (2002). Phosphorylation of Extracellular Signal-Regulated Kinase in Primary Afferent Neurons by Noxious Stimuli and Its Involvement in Peripheral

Sensitization. *J Neurosci*, Vol.22, No.17, (September 2002), pp.7737-7745, ISSN 1529-2401.

Dalton, G.D., Smith, F.L., Smith, P.A. & Dewey, W.L. (2005). Protein Kinase a Activity Is Increased in Mouse Lumbar Spinal Cord but Not Brain Following Morphine Antinociceptive Tolerance for 15 Days. *Pharmacol Res*, Vol.52, No.3, (September 2005), pp.204-210, ISSN 1043-6618.

Dina, O.A., Barletta, J., Chen, X., Mutero, A., Martin, A., Messing, R.O. & Levine, J.D. (2000). Key Role for the Epsilon Isoform of Protein Kinase C in Painful Alcoholic Neuropathy in the Rat. *J Neurosci*, Vol.20, No.22, (November 2000), pp.8614-8619, ISSN 1529-2401.

Dina, O.A., Chen, X., Reichling, D. & Levine, J.D. (2001). Role of Protein Kinase Cepsilon and Protein Kinase a in a Model of Paclitaxel-Induced Painful Peripheral Neuropathy in the Rat. *Neuroscience*, Vol.108, No.3, (December 2001), pp.507-515, ISSN 0306-4522.

Doyle, T., Chen, Z., Muscoli, C., Obeid, L.M. & Salvemini, D. (2011). Intraplantar-Injected Ceramide in Rats Induces Hyperalgesia through an Nf-Kappab- and P38 Kinase-Dependent Cyclooxygenase 2/Prostaglandin E2 Pathway. *FASEB J*, Vol.25, No.8, (Aug), pp.2782-2791, ISSN 1530-6860.

Duarte, I.D., dos Santos, I.R., Lorenzetti, B.B. & Ferreira, S.H. (1992). Analgesia by Direct Antagonism of Nociceptor Sensitization Involves the Arginine-Nitric Oxide-Cgmp Pathway. *Eur J Pharmacol*, Vol.217, No.2-3, (July 1992), pp.225-227, ISSN 0014-2999.

Durate, I.D., Lorenzetti, B.B. & Ferreira, S.H. (1990). Peripheral Analgesia and Activation of the Nitric Oxide-Cyclic Gmp Pathway. *Eur J Pharmacol*, Vol.186, No.2-3, (September 1990), pp.289-293, ISSN 0014-2999.

England, S., Bevan, S. & Docherty, R.J. (1996). Pge2 Modulates the Tetrodotoxin-Resistant Sodium Current in Neonatal Rat Dorsal Root Ganglion Neurones Via the Cyclic Amp-Protein Kinase a Cascade. *J Physiol*, Vol.495 (Pt 2), (September 1996), pp.429-440, ISSN 0022-3751.

Ferreira, S.H., Duarte, I.D. & Lorenzetti, B.B. (1991). Molecular Base of Acetylcholine and Morphine Analgesia. *Agents Actions Suppl*, Vol.32, (January 1991), pp.101-106, ISSN 0379-0363.

Ferreira, S.H., Lorenzetti, B.B. & De Campos, D.I. (1990). Induction, Blockade and Restoration of a Persistent Hypersensitive State. *Pain*, Vol.42, No.3, (September 1990), pp.365-371, ISSN 0304-3959.

Ferreira, S.H. & Nakamura, M. (1979). I - Prostaglandin Hyperalgesia, a Camp/Ca2+ Dependent Process. *Prostaglandins*, Vol.18, No.2, (August 1979), pp.179-190, ISSN 0090-6980.

Fleming, J.A. & O'Connor, B.D. (2009). Use of Lidocaine Patches for Neuropathic Pain in a Comprehensive Cancer Centre. *Pain Res Manag*, Vol.14, No.5, (September 2009), pp.381-388, ISSN 1203-6765.

Fujita-Hamabe, W., Nagae, R., Nawa, A., Harada, S., Nakamoto, K. & Tokuyama, S. (2010). Involvement of Kappa Opioid Receptors in the Formalin-Induced Inhibition of Analgesic Tolerance to Morphine Via Suppression of Conventional Protein Kinase C Activation. *J Pharm Pharmacol*, Vol.62, No.8, (August 2010), pp.995-1002, ISSN 2042-7158.

Fukuda, K., Kato, S., Morikawa, H., Shoda, T. & Mori, K. (1996). Functional Coupling of the Delta-, Mu-, and Kappa-Opioid Receptors to Mitogen-Activated Protein Kinase and

Arachidonate Release in Chinese Hamster Ovary Cells. *J Neurochem*, Vol.67, No.3, (September 1996), pp.1309-1316, ISSN 0022-3042.

Garber, K. (2003). Research Retreat: Pfizer Eliminates Sugen, Shrinks Cancer Infrastructure. *J Natl Cancer Inst*, Vol.95, No.14, (July 2003), pp.1036-1038, ISSN 1460-2105.

Giles, P.A., Trezise, D.J. & King, A.E. (2007). Differential Activation of Protein Kinases in the Dorsal Horn in Vitro of Normal and Inflamed Rats by Group I Metabotropic Glutamate Receptor Subtypes. *Neuropharmacology*, Vol.53, No.1, (July 2007), pp.58-70, ISSN 0028-3908

Gold, M.S., Levine, J.D. & Correa, A.M. (1998). Modulation of Ttx-R Ina by Pkc and Pka and Their Role in Pge2-Induced Sensitization of Rat Sensory Neurons in Vitro. *J Neurosci*, Vol.18, No.24, (December 1998), pp.10345-10355, ISSN 0270-6474.

Gu, Y.W., Su, D.S., Tian, J. & Wang, X.R. (2008). Attenuating Phosphorylation of P38 Mapk in the Activated Microglia: A New Mechanism for Intrathecal Lidocaine Reversing Tactile Allodynia Following Chronic Constriction Injury in Rats. *Neurosci Lett*, Vol.431, No.2, (January 2008), pp.129-134, ISSN 0304-3940.

Gupta, A., Devi, L.A. & Gomes, I. (2011). Potentiation of Mu-Opioid Receptor-Mediated Signaling by Ketamine. *J Neurochem*, Vol.119, No.2, (October 2011), pp.294-302, ISSN 1471-4159.

Gutstein, H.B., Rubie, E.A., Mansour, A., Akil, H. & Woodgett, J.R. (1997). Opioid Effects on Mitogen-Activated Protein Kinase Signaling Cascades. *Anesthesiology*, Vol.87, No.5, (November 1997), pp.1118-1126, ISSN 0003-3022.

Gwak, Y.S., Unabia, G.C. & Hulsebosch, C.E. (2009). Activation of P-38alpha Mapk Contributes to Neuronal Hyperexcitability in Caudal Regions Remote from Spinal Cord Injury. *Exp Neurol*, Vol.220, No.1, (November 2009), pp.154-161, ISSN 1090-2430.

Hensellek, S., Brell, P., Schaible, H.G., Brauer, R. & Segond von Banchet, G. (2007). The Cytokine Tnfalpha Increases the Proportion of Drg Neurones Expressing the Trpv1 Receptor Via the Tnfr1 Receptor and Erk Activation. *Mol Cell Neurosci*, Vol.36, No.3, (November 2007), pp.381-391, ISSN 1044-7431.

Horvath, R.J., Landry, R.P., Romero-Sandoval, E.A. & DeLeo, J.A. (2010). Morphine Tolerance Attenuates the Resolution of Postoperative Pain and Enhances Spinal Microglial P38 and Extracellular Receptor Kinase Phosphorylation. *Neuroscience*, Vol.169, No.2, (August 2010), pp.843-854, ISSN 1873-7544.

Hucho, T.B., Dina, O.A. & Levine, J.D. (2005). Epac Mediates a Camp-to-Pkc Signaling in Inflammatory Pain: An Isolectin B4(+) Neuron-Specific Mechanism. *J Neurosci*, Vol.25, No.26, (June 2005), pp.6119-6126, ISSN 1529-2401.

Hunskaar, S. & Hole, K. (1987). The Formalin Test in Mice: Dissociation between Inflammatory and Non-Inflammatory Pain. *Pain*, Vol.30, No.1, (July 1987), pp.103-114, ISSN 0304-3959.

Ji, R.R., Samad, T.A., Jin, S.X., Schmoll, R. & Woolf, C.J. (2002). P38 Mapk Activation by Ngf in Primary Sensory Neurons after Inflammation Increases Trpv1 Levels and Maintains Heat Hyperalgesia. *Neuron*, Vol.36, No.1, (September 2002), pp.57-68, ISSN 0896-6273.

Ji, R.R. & Suter, M.R. (2007). P38 Mapk, Microglial Signaling, and Neuropathic Pain. *Mol Pain*, Vol.3, (November 2007), pp.33, ISSN 1744-8069.

Jin, X. & Gereau, R.W.t. (2006). Acute P38-Mediated Modulation of Tetrodotoxin-Resistant Sodium Channels in Mouse Sensory Neurons by Tumor Necrosis Factor-Alpha. *J Neurosci*, Vol.26, No.1, (January 2006), pp.246-255, ISSN 1529-2401.

Johnson, J.A., Gray, M.O., Chen, C.H. & Mochly-Rosen, D. (1996). A Protein Kinase C Translocation Inhibitor as an Isozyme-Selective Antagonist of Cardiac Function. *J Biol Chem*, Vol.271, No.40, (October 1996), pp.24962-24966, ISSN 0021-9258.

Junttila, M.R., Li, S.P. & Westermarck, J. (2008). Phosphatase-Mediated Crosstalk between Mapk Signaling Pathways in the Regulation of Cell Survival. *FASEB J*, Vol.22, No.4, (April 2007), pp.954-965, ISSN 1530-6860.

Kassuya, C.A., Ferreira, J., Claudino, R.F. & Calixto, J.B. (2007). Intraplantar Pge2 Causes Nociceptive Behaviour and Mechanical Allodynia: The Role of Prostanoid E Receptors and Protein Kinases. *Br J Pharmacol*, Vol.150, No.6, (March 2007), pp.727-737, ISSN 0007-1188.

Khasar, S.G., Lin, Y.H., Martin, A., Dadgar, J., McMahon, T., Wang, D., Hundle, B., Aley, K.O., Isenberg, W., McCarter, G., Green, P.G., Hodge, C.W., Levine, J.D. & Messing, R.O. (1999a). A Novel Nociceptor Signaling Pathway Revealed in Protein Kinase C Epsilon Mutant Mice. *Neuron*, Vol.24, No.1, (September 2000), pp.253-260, ISSN 0896-6273.

Khasar, S.G., McCarter, G. & Levine, J.D. (1999b). Epinephrine Produces a Beta-Adrenergic Receptor-Mediated Mechanical Hyperalgesia and in Vitro Sensitization of Rat Nociceptors. *J Neurophysiol*, Vol.81, No.3, (March, 1999), pp.1104-1112, ISSN 0022-3077.

Lee, K.M., Jeon, S.M. & Cho, H.J. (2010). Interleukin-6 Induces Microglial Cx3cr1 Expression in the Spinal Cord after Peripheral Nerve Injury through the Activation of P38 Mapk. *Eur J Pain*, Vol.14, No.7, (August 2009), pp.682 e681-612, ISSN 1532-2149.

Lee, P.Y., Tsai, P.S., Huang, Y.H. & Huang, C.J. (2008). Inhibition of Toll-Like Receptor-4, Nuclear Factor-Kappab and Mitogen-Activated Protein Kinase by Lignocaine May Involve Voltage-Sensitive Sodium Channels. *Clin Exp Pharmacol Physiol*, Vol.35, No.9, (September 2008), pp.1052-1058, ISSN 1440-1681.

Lopshire, J.C. & Nicol, G.D. (1997). Activation and Recovery of the Pge2-Mediated Sensitization of the Capsaicin Response in Rat Sensory Neurons. *J Neurophysiol*, Vol.78, No.6, (December 1998), pp.3154-3164, ISSN 0022-3077.

Lopshire, J.C. & Nicol, G.D. (1998). The Camp Transduction Cascade Mediates the Prostaglandin E2 Enhancement of the Capsaicin-Elicited Current in Rat Sensory Neurons: Whole-Cell and Single-Channel Studies. *J Neurosci*, Vol.18, No.16, (August 1998), pp.6081-6092, ISSN 0270-6474.

Ma, W., Du, W. & Eisenach, J.C. (2003). Intrathecal Lidocaine Reverses Tactile Allodynia Caused by Nerve Injuries and Potentiates the Antiallodynic Effect of the Cox Inhibitor Ketorolac. *Anesthesiology*, Vol.98, No.1, (January 2002), pp.203-208, ISSN 0003-3022.

Malmberg, A.B., Chen, C., Tonegawa, S. & Basbaum, A.I. (1997). Preserved Acute Pain and Reduced Neuropathic Pain in Mice Lacking Pkcgamma. *Science*, Vol.278, No.5336, (October 1997), pp.279-283, ISSN 0036-8075.

Manjavachi, M.N., Motta, E.M., Marotta, D.M., Leite, D.F. & Calixto, J.B. (2010). Mechanisms Involved in Il-6-Induced Muscular Mechanical Hyperalgesia in Mice. *Pain*, Vol.151, No.2, (November 2010), pp.345-355, ISSN 1872-6623.

Mao, J. & Chen, L.L. (2000). Systemic Lidocaine for Neuropathic Pain Relief. *Pain*, Vol.87, No.1, (July 2000), pp.7-17, ISSN 0304-3959.

Mao, J., Price, D.D., Mayer, D.J. & Hayes, R.L. (1992). Pain-Related Increases in Spinal Cord Membrane-Bound Protein Kinase C Following Peripheral Nerve Injury. *Brain Res*, Vol.588, No.1, (August 1992), pp.144-149, ISSN 0006-8993.

Mathisen, L.C., Skjelbred, P., Skoglund, L.A. & Oye, I. (1995). Effect of Ketamine, an Nmda Receptor Inhibitor, in Acute and Chronic Orofacial Pain. *Pain*, Vol.61, No.2, (May 1995), pp.215-220, ISSN 0304-3959.

Mei, X.P., Zhang, H., Wang, W., Wei, Y.Y., Zhai, M.Z., Xu, L.X. & Li, Y.Q. (2011). Inhibition of Spinal Astrocytic C-Jun N-Terminal Kinase (Jnk) Activation Correlates with the Analgesic Effects of Ketamine in Neuropathic Pain. *J Neuroinflammation*, Vol.8, No.1, (January 2011), pp.6, ISSN 1742-2094.

Nakamura, M. & Ferreira, S.H. (1987). A Peripheral Sympathetic Component in Inflammatory Hyperalgesia. *Eur J Pharmacol*, Vol.135, No.2, (March 1987), pp.145-153, ISSN 0014-2999.

Narita, M., Miyoshi, K. & Suzuki, T. (2007). Functional Reduction in Mu-Opioidergic System in the Spinal Cord under a Neuropathic Pain-Like State Following Chronic Ethanol Consumption in the Rat. *Neuroscience*, Vol.144, No.3, (February 2006), pp.777-782, ISSN 0306-4522.

Neugebauer, V. (2002). Metabotropic Glutamate Receptors--Important Modulators of Nociception and Pain Behavior. *Pain*, Vol.98, No.1-2, (July 2002), pp.1-8, ISSN 0304-3959.

Nishizuka, Y. (1992). Intracellular Signaling by Hydrolysis of Phospholipids and Activation of Protein Kinase C. *Science*, Vol.258, No.5082, (October 1992), pp.607-614, ISSN 0036-8075.

Noselli, S. (2000). Signal Transduction: Are There Close Encounters between Signaling Pathways? *Science*, Vol.290, No.5489, (October 2000), pp.68-69, ISSN 00368075.

Obata, K., Yamanaka, H., Dai, Y., Mizushima, T., Fukuoka, T., Tokunaga, A. & Noguchi, K. (2004). Differential Activation of Mapk in Injured and Uninjured Drg Neurons Following Chronic Constriction Injury of the Sciatic Nerve in Rats. *Eur J Neurosci*, Vol.20, No.11, (December 2004), pp.2881-2895, ISSN 0953-816X.

Pascual, D., Goicoechea, C., Burgos, E. & Martin, M.I. (2010). Antinociceptive Effect of Three Common Analgesic Drugs on Peripheral Neuropathy Induced by Paclitaxel in Rats. *Pharmacol Biochem Behav*, Vol.95, No.3, (May 2010), pp.331-337, ISSN 1873-5177.

Pearson, G., Robinson, F., Beers Gibson, T., Xu, B.E., Karandikar, M., Berman, K. & Cobb, M.H. (2001). Mitogen-Activated Protein (Map) Kinase Pathways: Regulation and Physiological Functions. *Endocr Rev*, Vol.22, No.2, (April 2001), pp.153-183, ISSN 0163-769X.

Persson, A.K., Gasser, A., Black, J.A. & Waxman, S.G. (2011). Nav1.7 Accumulates and Co-Localizes with Phosphorylated Erk1/2 within Transected Axons in Early Experimental Neuromas. *Exp Neurol*, Vol.230, No.2, (August 2011), pp.273-279, ISSN 1090-2430.

Picolo, G., Giorgi, R. & Cury, Y. (2000). Delta-Opioid Receptors and Nitric Oxide Mediate the Analgesic Effect of Crotalus Durissus Terrificus Snake Venom. *Eur J Pharmacol*, Vol.391, No.1-2, (March 2000), pp.55-62, ISSN 0014-2999.

Pimienta, G. & Pascual, J. (2007). Canonical and Alternative Mapk Signaling. *Cell Cycle*, Vol.6, No.21, (November 2007), pp.2628-2632, ISSN 1551-4005.

Rabben, T., Skjelbred, P. & Oye, I. (1999). Prolonged Analgesic Effect of Ketamine, an N-Methyl-D-Aspartate Receptor Inhibitor, in Patients with Chronic Pain. *J Pharmacol Exp Ther*, Vol.289, No.2, (May 1999), pp.1060-1066, ISSN 0022-3565.

Rathee, P.K., Distler, C., Obreja, O., Neuhuber, W., Wang, G.K., Wang, S.Y., Nau, C. & Kress, M. (2002). Pka/Akap/Vr-1 Module: A Common Link of Gs-Mediated Signaling to Thermal Hyperalgesia. *J Neurosci*, Vol.22, No.11, (June 2002), pp.4740-4745, ISSN 1529-2401.

Reis, G.M., Pacheco, D., Perez, A.C., Klein, A., Ramos, M.A. & Duarte, I.D. (2009). Opioid Receptor and No/Cgmp Pathway as a Mechanism of Peripheral Antinociceptive Action of the Cannabinoid Receptor Agonist Anandamide. *Life Sci*, Vol.85, No.9-10, (August 2009), pp.351-356, ISSN 1879-0631.

Rodrigues, A.R. & Duarte, I.D. (2000). The Peripheral Antinociceptive Effect Induced by Morphine Is Associated with Atp-Sensitive K(+) Channels. *Br J Pharmacol*, Vol.129, No.1, (January 2000), pp.110-114, ISSN 0007-1188.

Rodriguez-Munoz, M., Sanchez-Blazquez, P., Vicente-Sanchez, A., Berrocoso, E. & Garzon, J. (2011). The Mu-Opioid Receptor and the Nmda Receptor Associate in Pag Neurons: Implications in Pain Control. *Neuropsychopharmacology*, (August 2011), ISSN 1740-634X.

Romero, T.R. & Duarte, I.D. (2009). Alpha(2)-Adrenoceptor Agonist Xylazine Induces Peripheral Antinociceptive Effect by Activation of the L-Arginine/Nitric Oxide/Cyclic Gmp Pathway in Rat. *Eur J Pharmacol*, Vol.613, No.1-3, (June 2009), pp.64-67, ISSN 1879-0712.

Romero, T.R., Galdino, G.S., Silva, G.C., Resende, L.C., Perez, A.C., Cortes, S.F. & Duarte, I.D. (2011). Ketamine Activates the L-Arginine/Nitric Oxide/Cyclic Guanosine Monophosphate Pathway to Induce Peripheral Antinociception in Rats. *Anesth Analg*, Vol.113, No.5, (November 2011), pp.1254-1259, ISSN 1526-7598.

Sachs, D., Cunha, F.Q. & Ferreira, S.H. (2004). Peripheral Analgesic Blockade of Hypernociception: Activation of Arginine/No/Cgmp/Protein Kinase G/Atp-Sensitive K+ Channel Pathway. *Proc Natl Acad Sci U S A*, Vol.101, No.10, (March 2004), pp.3680-3685, ISSN 0027-8424.

Sachs, D., Villarreal, C., Cunha, F., Parada, C. & Ferreira, S. (2009). The Role of Pka and Pkcepsilon Pathways in Prostaglandin E2-Mediated Hypernociception. *Br J Pharmacol*, Vol.156, No.5, (March 2009), pp.826-834, ISSN 1476-5381.

Smith, J.A., Davis, C.L. & Burgess, G.M. (2000). Prostaglandin E2-Induced Sensitization of Bradykinin-Evoked Responses in Rat Dorsal Root Ganglion Neurons Is Mediated by Camp-Dependent Protein Kinase A. *Eur J Neurosci*, Vol.12, No.9, (September 2000), pp.3250-3258, ISSN 0953-816X.

Song, Z., Zou, W., Liu, C. & Guo, Q. (2010). Gene Knockdown with Lentiviral Vector-Mediated Intrathecal Rna Interference of Protein Kinase C Gamma Reverses Chronic Morphine Tolerance in Rats. *J Gene Med*, Vol.12, No.11, (November 2010), pp.873-880, ISSN 1521-2254.

Sorkin, L., Svensson, C.I., Jones-Cordero, T.L., Hefferan, M.P. & Campana, W.M. (2009). Spinal P38 Mitogen-Activated Protein Kinase Mediates Allodynia Induced by First-

Degree Burn in the Rat. *J Neurosci Res*, Vol.87, No.4, (March 2008), pp.948-955, ISSN 1097-4547.

Souza, A.L., Moreira, F.A., Almeida, K.R., Bertollo, C.M., Costa, K.A. & Coelho, M.M. (2002). In Vivo Evidence for a Role of Protein Kinase C in Peripheral Nociceptive Processing. *Br J Pharmacol*, Vol.135, No.1, (January 2002), pp.239-247, ISSN 0007-1188.

Su, D., Gu, Y., Wang, Z. & Wang, X. (2010). Lidocaine Attenuates Proinflammatory Cytokine Production Induced by Extracellular Adenosine Triphosphate in Cultured Rat Microglia. *Anesth Analg*, Vol.111, No.3, (September 2010), pp.768-774, ISSN 1526-7598.

Suter, M.R., Berta, T., Gao, Y.J., Decosterd, I. & Ji, R.R. (2009). Large a-Fiber Activity Is Required for Microglial Proliferation and P38 Mapk Activation in the Spinal Cord: Different Effects of Resiniferatoxin and Bupivacaine on Spinal Microglial Changes after Spared Nerve Injury. *Mol Pain*, Vol.5, (September 2009), pp.53, ISSN 1744-8069.

Suzuki, N., Hasegawa-Moriyama, M., Takahashi, Y., Kamikubo, Y., Sakurai, T. & Inada, E. (2011). Lidocaine Attenuates the Development of Diabetic-Induced Tactile Allodynia by Inhibiting Microglial Activation. *Anesth Analg*, Vol.113, No.4, (October 2011), pp.941-946, ISSN 1526-7598.

Svensson, C.I., Marsala, M., Westerlund, A., Calcutt, N.A., Campana, W.M., Freshwater, J.D., Catalano, R., Feng, Y., Protter, A.A., Scott, B. & Yaksh, T.L. (2003). Activation of P38 Mitogen-Activated Protein Kinase in Spinal Microglia Is a Critical Link in Inflammation-Induced Spinal Pain Processing. *J Neurochem*, Vol.86, No.6, (September 2003), pp.1534-1544, ISSN 0022-3042.

Tai, Y.H., Wang, Y.H., Tsai, R.Y., Wang, J.J., Tao, P.L., Liu, T.M., Wang, Y.C. & Wong, C.S. (2007). Amitriptyline Preserves Morphine's Antinociceptive Effect by Regulating the Glutamate Transporter Glast and Glt-1 Trafficking and Excitatory Amino Acids Concentration in Morphine-Tolerant Rats. *Pain*, Vol.129, No.3, (June 2007), pp.343-354, ISSN 1872-6623.

Tai, Y.H., Wang, Y.H., Wang, J.J., Tao, P.L., Tung, C.S. & Wong, C.S. (2006). Amitriptyline Suppresses Neuroinflammation and up-Regulates Glutamate Transporters in Morphine-Tolerant Rats. *Pain*, Vol.124, No.1-2, (September 2006), pp.77-86, ISSN 1872-6623.

Taiwo, Y.O., Heller, P.H. & Levine, J.D. (1992). Mediation of Serotonin Hyperalgesia by the Camp Second Messenger System. *Neuroscience*, Vol.48, No.2, (January 1992), pp.479-483, ISSN 0306-4522.

Taiwo, Y.O. & Levine, J.D. (1991). Further Confirmation of the Role of Adenyl Cyclase and of Camp-Dependent Protein Kinase in Primary Afferent Hyperalgesia. *Neuroscience*, Vol.44, No.1, (January 1991), pp.131-135, ISSN 0306-4522.

Taylor, S.S., Buechler, J.A. & Yonemoto, W. (1990). Camp-Dependent Protein Kinase: Framework for a Diverse Family of Regulatory Enzymes. *Annu Rev Biochem*, Vol.59, (January 1990), pp.971-1005, ISSN 0066-4154.

Than, M., Kocsis, P., Tihanyi, K., Fodor, L., Farkas, B., Kovacs, G., Kis-Varga, A., Szombathelyi, Z. & Tarnawa, I. (2007). Concerted Action of Antiepileptic and Antidepressant Agents to Depress Spinal Neurotransmission: Possible Use in the

Therapy of Spasticity and Chronic Pain. *Neurochem Int*, Vol.50, No.4, (March 2007), pp.642-652, ISSN 0197-0186.

Tong, S.E., Daniels, S.E., Black, P., Chang, S., Protter, A. & Desjardins, P.J. (2011). Novel P38{Alpha} Mitogen-Activated Protein Kinase Inhibitor Shows Analgesic Efficacy in Acute Postsurgical Dental Pain. *J Clin Pharmacol*, (June 2011), ISSN 1552-4604.

Tonussi, C.R. & Ferreira, S.H. (1994). Mechanism of Diclofenac Analgesia: Direct Blockade of Inflammatory Sensitization. *Eur J Pharmacol*, Vol.251, No.2-3, (January 1994), pp.173-179, ISSN 0014-2999.

Tumati, S., Roeske, W.R., Largent-Milnes, T.M., Vanderah, T.W. & Varga, E.V. (2011). Intrathecal Pka-Selective Sirna Treatment Blocks Sustained Morphine-Mediated Pain Sensitization and Antinociceptive Tolerance in Rats. *J Neurosci Methods*, Vol.199, No.1, (July 2011), pp.62-68, ISSN 1872-678X.

Van Kolen, K., Pullan, S., Neefs, J.M. & Dautzenberg, F.M. (2008). Nociceptive and Behavioural Sensitisation by Protein Kinase Cepsilon Signalling in the Cns. *J Neurochem*, Vol.104, No.1, (January 2007), pp.1-13, ISSN 1471-4159.

Villarreal, C.F., Sachs, D., Funez, M.I., Parada, C.A., de Queiroz Cunha, F. & Ferreira, S.H. (2009a). The Peripheral Pro-Nociceptive State Induced by Repetitive Inflammatory Stimuli Involves Continuous Activation of Protein Kinase a and Protein Kinase C Epsilon and Its Na(V)1.8 Sodium Channel Functional Regulation in the Primary Sensory Neuron. *Biochem Pharmacol*, Vol.77, No.5, (March 2008), pp.867-877, ISSN 1873-2968.

Villarreal, C.F., Funez, M.I., Figueiredo, F., Cunha, F.Q., Parada, C.A. & Ferreira, S.H. (2009b). Acute and Persistent Nociceptive Paw Sensitisation in Mice: The Involvement of Distinct Signalling Pathways. *Life Sci*, Vol.85, No.23-26, (December 2009), pp.822-829, ISSN 1879-0631.

Visser, E. & Schug, S.A. (2006). The Role of Ketamine in Pain Management. *Biomed Pharmacother*, Vol.60, No.7, (August 2006), pp.341-348, ISSN 0753-3322.

Wu, G.J., Chen, T.L., Ueng, Y.F. & Chen, R.M. (2008). Ketamine Inhibits Tumor Necrosis Factor-Alpha and Interleukin-6 Gene Expressions in Lipopolysaccharide-Stimulated Macrophages through Suppression of Toll-Like Receptor 4-Mediated C-Jun N-Terminal Kinase Phosphorylation and Activator Protein-1 Activation. *Toxicol Appl Pharmacol*, Vol.228, No.1, (April 2008), pp.105-113, ISSN 0041-008X.

Yamdeu, R.S., Shaqura, M., Mousa, S.A., Schafer, M. & Droese, J. (2011). P38 Mitogen-Activated Protein Kinase Activation by Nerve Growth Factor in Primary Sensory Neurons Upregulates Mu-Opioid Receptors to Enhance Opioid Responsiveness toward Better Pain Control. *Anesthesiology*, Vol.114, No.1, (January 2010), pp.150-161, ISSN 1528-1175.

Yanagidate, F. & Strichartz, G.R. (2006). Bupivacaine Inhibits Activation of Neuronal Spinal Extracellular Receptor-Activated Kinase through Selective Effects on Ionotropic Receptors. *Anesthesiology*, Vol.104, No.4, (April 20006), pp.805-814, ISSN 0003-3022.

Zhang, X.C., Kainz, V., Burstein, R. & Levy, D. (2011). Tumor Necrosis Factor-Alpha Induces Sensitization of Meningeal Nociceptors Mediated Via Local Cox and P38 Map Kinase Actions. *Pain*, Vol.152, No.1, (January 2010), pp.140-149, ISSN 1872-6623.

Zhou, Y., Li, G.D. & Zhao, Z.Q. (2003). State-Dependent Phosphorylation of Epsilon-Isozyme of Protein Kinase C in Adult Rat Dorsal Root Ganglia after Inflammation and Nerve Injury. *J Neurochem*, Vol.85, No.3, (May 2003), pp.571-580, ISSN 0022-3042.

Zhu, W. & Oxford, G.S. (2007). Phosphoinositide-3-Kinase and Mitogen Activated Protein Kinase Signaling Pathways Mediate Acute Ngf Sensitization of Trpv1. *Mol Cell Neurosci*, Vol.34, No.4, (April 2007), pp.689-700, ISSN 1044-7431.